X-ray binaries are some of the most varied and perplexing systems known to astronomers. The compact object, which accretes mass from its companion star, may be a white dwarf, a neutron star, or a black hole, whereas the donor star can be a 'normal' star or a white dwarf. The various combinations differ widely in their behaviour. This timely volume provides a unique and up-to-date reference of our knowledge of all of them.

Fifteen specially written chapters by a team of the world's foremost researchers in the field explore all aspects of X-ray binaries. They cover the X-ray, ultraviolet, optical and radio properties of these violent systems and address key issues such as: how were these systems formed, and what will their fate be; how can we understand X-ray bursts and quasi-periodic oscillations; what is the connection between millisecond radio pulsars and low-mass X-ray binaries; and how does the magnetic field of a neutron star decay?

This long awaited review provides graduate students and researchers with the standard reference on X-ray binaries for many years to come.

X-ray binaries

CAMBRIDGE ASTROPHYSICS SERIES

Series editors

Andrew King, Douglas Lin, Stephen Maran, Jim Pringle and Martin Ward

Titles available in this series

5. The Solar Granulation
   by R. J. Bray, R. E. Loughhead and C. J. Durrant

7. Spectroscopy of Astrophysical Plasmas
   by A. Dalgarno and D. Layzer

10. Quasar Astronomy
    by D. W. Weedman

13. High Speed Astronomical Photometry
    by B. Warner

14. The Physics of Solar Flares
    by E. Tandberg-Hanssen and A. G. Emslie

15. X-ray Detectors in Astronomy
    by G. W. Fraser

16. Pulsar Astronomy
    by A. Lyne and F. Graham-Smith

17. Molecular Collisions in the Interstellar Medium
    by D. Flower

18. Plasma Loops in the Solar Corona
    by R. J. Bray, L. E. Cram, C. J. Durrant and R. E. Loughhead

19. Beams and Jets in Astrophysics
    edited by P. A. Hughes

20. The Observation and Analysis of Stellar Photospheres
    by David F. Gray

21. Accretion Power in Astrophysics 2nd Edition
    by J. Frank, A. R. King and D. J. Raine

22. Gamma-ray Astronomy 2nd Edition
    by P. V. Ramana Murthy and A. W. Wolfendale

23. The Solar Transition Region
    by J. T. Mariska

24. Solar and Stellar Activity Cycles
    by Peter R. Wilson

25. 3K: The Cosmic Microwave Background Radiation
    by R. B. Partridge

26. X-ray Binaries
    by Walter H. G. Lewin, Jan van Paradijs and Edward P. J. van den Heuvel

27. RR Lyrae Stars
    by Horace A. Smith

28. Cataclysmic Variable Stars
    by Brian Warner

X-RAY BINARIES

Edited by
WALTER H. G. LEWIN
Department of Physics, Massachusetts Institute of Technology

JAN VAN PARADIJS
Astronomical Institute 'Anton Pannekoek', University of Amsterdam,
Center for High-Energy Astrophysics, Amsterdam, and
Physics Department, University of Alabama in Huntsville

and
EDWARD P. J. VAN DEN HEUVEL
Astronomical Institute 'Anton Pannekoek', University of Amsterdam,
Center for High-Energy Astrophysics, Amsterdam

CAMBRIDGE
UNIVERSITY PRESS

PUBLISHED BY THE PRESS SYNDICATE OF THE UNIVERSITY OF CAMBRIDGE
The Pitt Building, Trumpington Street, Cambridge CB2 1RP, United Kingdom

CAMBRIDGE UNIVERSITY PRESS
The Edinburgh Building, Cambridge CB2 2RU, United Kingdom
40 West 20th Street, New York, NY 10011–4211, USA
10 Stamford Road, Oakleigh, Melbourne 3166, Australia

www.cambridge.org
Information on this title: www.cambridge.org/9780521416849

First published 1995
First paperback edition 1997

A catalogue record for this book is available from the British Library

Library of Congress Cataloguing in Publication data

Lewin, Walter H. G.
X-ray binaries/Walter H. G. Lewin, Jan van Paradijs, and Edward P. J. van den Heuvel.
  p. cm.
ISBN 0 521 41684 1
1. X-ray binaries. I. Paradijs, J. van II. Heuvel, Edward Peter Jacobus van den, 1940–
III. Title.
QB830.L49 1995
523.8′41–dc20 95–6370 CIP

ISBN-13 978-0-521-41684-9 hardback
ISBN-10 0-521-41684-1 hardback

ISBN-13 978-0-521-59934-4 paperback
ISBN-10 0-521-59934-2 paperback

Transferred to digital printing 2005

Contents

Preface   page xi

1 The properties of X-ray binaries
   N. E. White, F. Nagase and A. N. Parmar   1
1.1 Introduction   1
1.2 Orbital periods   3
1.3 Pulse periods   18
1.4 Third periods and other modulations   25
1.5 The emission region   27
1.6 Environmental radiative processes   38
    References   49

2 Optical and ultraviolet observations of X-ray binaries
   J. van Paradijs and J. E. McClintock   58
2.1 Introduction   58
2.2 High-mass X-ray binaries   59
2.3 Low-mass X-ray binaries   73
2.4 Neutron star and black-hole masses   107
2.5 The Magellanic Cloud sources   113
2.6 Triple-star systems   117
    References   121

3 Black-hole binaries
   Y. Tanaka and W. H. G. Lewin   126
3.1 Some historic notes   126
3.2 Introduction   127
3.3 Description of the individual black-hole candidates   134
3.4 X-ray spectra – interpretations   159
3.5 BHC diagnostics   164
3.6 How many black-hole binaries are there?   167
    References   168

4 X-ray bursts
   Walter H. G. Lewin, Jan Van Paradijs and Ronald E. Taam   175
4.1 Introduction   175

4.2 Characteristics of burst sources   177
4.3 Type I X-ray bursts   178
4.4 Mass–radius relation of neutron stars   199
4.5 Theory of type I X-ray bursts   201
4.6 The Rapid Burster (1730–335)   211
4.7 Models for the Rapid Burster   222
     References   228

5 **Millisecond pulsars**
   D. Bhattacharya   233
5.1 Overview   233
5.2 The general characteristics of millisecond pulsars   234
5.3 Spin-up of neutron stars and the origin of millisecond pulsars   236
5.4 Evolution of LMXBs: the standard model   238
5.5 Statistics   240
5.6 Evolution of LMXBs: beyond the standard model   243
5.7 Pulsars in globular clusters   245
5.8 Future prospects   248
     References   249

6 **Rapid aperiodic variability in X-ray binaries**
   M. van der Klis   252
6.1 Introduction   252
6.2 Black-hole candidates   259
6.3 Z and atoll sources   274
6.4 Other sources   289
6.5 Overview and outlook   295
     References   300

7 **Radio properties of X-ray binaries**
   R. M. Hjellming and X. Han   308
7.1 Introduction   308
7.2 Transient radio emission   313
7.3 Coupled radio–optical–UV–X-ray state changes   315
7.4 Radio jets and extended radio emission   320
7.5 Other models of compact radio emission from X-ray binaries   324
7.6 Acceleration of relativistic electrons   326
7.7 Conclusions   328
     References   329

8 **Cataclysmic variable stars**
   France Anne-Dominic Córdova   331
8.1 Introduction   331
8.2 What multiwavelength observations reveal   337
8.3 The variability of CVs and its origins   360
8.4 The importance of magnetic fields in CVs   373

Contents ix

8.5 CV evolution 378
8.6 Endpiece: areas for future studies 381
    References 384

9 Normal galaxies and their X-ray binary populations
    G. Fabbiano 390
9.1 Introduction 390
9.2 X-ray sources in Local Group galaxies 391
9.3 Other spiral galaxies: detecting the upper end of the luminosity
    distribution of X-ray sources 398
9.4 The X-ray emission of galaxies: a very brief summary 404
9.5 Recent spectral results 405
9.6 The future 410
    References 416

10 Accretion in close binaries
    Andrew King 419
10.1 Introduction 419
10.2 Summary of accretion disc theory 419
10.3 Disc structure in LMXBs 423
10.4 Disc instabilities 427
10.5 Tides, resonances and superhumps 430
10.6 Spiral shocks 437
10.7 Coronae and winds in discs 438
10.8 Boundary layers 438
10.9 Accretion from a wind 441
10.10 Accretion on to a magnetic star 444
    References 453

11 Formation and evolution of neutron stars and black holes in binaries
    F. Verbunt and E. P. J. van den Heuvel 457
11.1 Introduction and brief observational overview 457
11.2 Origin and evolution of high-mass X-ray binaries 464
11.3 Origin and evolution of low-mass X-ray binaries 476
11.4 X-ray sources in globular clusters 486
11.5 Statistical considerations 488
    References 492

12 The magnetic fields of neutron stars and their evolution
    D. Bhattacharya and G. Srinivasan 495
12.1 Introduction 495
12.2 Estimation of the magnetic field 496
12.3 The origin and structure of the magnetic field 499
12.4 Evolution of the magnetic field 503
12.5 Possible mechanisms for field decay 513
12.6 Some consequences of field evolution 518

Appendix. Interaction between fluxoids and vortices   520
References   521

13 Cosmic gamma-ray bursts
K. Hurley   523
13.1 Foreword   523
13.2 Introduction   523
13.3 Time histories   524
13.4 X- and gamma-ray energy spectra   526
13.5 Counterparts and lack thereof   528
13.6 Statistical properties of gamma-ray bursters   529
13.7 The great debate redux   532
References   533

14 A catalogue of X-ray binaries
Jan van Paradijs   536

15 A compilation of cataclysmic binaries with known or suspected orbital periods
Hans Ritter and Ulrich Kolb   578
References   579

Index   639

Preface

In the decade since the publication of *Accretion Driven Stellar X-ray Sources* – the predecessor of the present book – the study of X-ray binaries has received an enormous impetus due to observations over a very large range of photon energies. Particularly in the X-ray range, a host of new observing facilities has become available: for example, EXOSAT, Ginga, ROSAT, the MIR station, Granat and GRO. These observatories, and also a large variety of observatories working in the radio, optical and ultraviolet parts of the electromagnetic spectrum, have produced many new results that have transformed our picture of these intriguing binary systems. Of major importance has been the understanding of the change in the evolutionary framework. These binary systems are now considered to be the parent population from which recycled millisecond radio pulsars, both single and those in binaries, are generated. Globular clusters seem to be ideal birth places for these millisecond pulsars; 12 have been found in 47 Tuc alone! The importance of the millisecond pulsars was recently highlighted when Russell Hulse and Joseph Taylor were awarded the Nobel Prize for Physics for their discovery in 1974 of the binary millisecond pulsar PSR 1913+16; this system provided the first observational evidence for the existence of gravitational radiation, which manifests itself in the form of a decreasing orbital period. A crucial change has also occurred in our ideas on the decay of the magnetic field of neutron stars. It is no longer believed that magnetic fields decay spontaneously on a time scale of a few million years. This had been the party line for almost two decades. Instead, the present consensus is that non-accreting neutron stars may have *no* field decay at all, and that the field decay of a neutron star in an accreting binary depends on the accretion or its spin-period history. The discovery of quasi-periodic oscillations has led to some welcome order in the confusion that before had plagued our understanding of the irregular variability of low-mass X-ray binaries. On the basis of their correlated fast-variability and spectral properties, two types of low-mass X-ray binaries have been recognized: the *Z* and *atoll* sources. Several of the components found in the power spectra of these low-mass X-ray binaries appear also to be present in the power spectra of black-hole candidates and X-ray pulsars. This may lead to a unification of the variability properties of all types of accreting compact stars. A decade ago, the orbital periods of only a few low-mass X-ray binaries were known. This number has since increased enormously, thanks to dedicated searches by many researchers for periodic variations of the optical and X-ray brightness of these objects. In particular EXOSAT – due to its special Earth orbit, which allowed for long uninterrupted observations – has been very productive in finding orbital periods of low-mass X-ray binaries, particularly through the detection of the so-called *dipping* sources. In addition, great progress has been made in our ability to distinguish accreting neutron stars from accreting black holes on the basis of their X-ray spectra alone. To date, more than two decades after the discovery of γ-ray bursts, their origin is still a mystery. In comparison, the origin of X-ray bursts (all the ins and outs of which are thoroughly covered in this book) was well understood within only three years of their discovery. During the past few years (after Dr Hurley had been invited to write a review for this book), the theory of γ-ray burst sources has come a long way. Most scientists now believe that they are at cosmological distances as opposed to very nearby (only a few hundred parsec) within our Galaxy. This book is by far the most comprehensive survey ever written on this subject. It covers in great detail the most recent developments in multi-wavelength observations, as well as the theory of X-ray binaries.

Walter H. G. Lewin, Jan van Paradijs,
and Edward P. J. van den Heuvel

I

The properties of X-ray binaries

N. E. White
Laboratory for High Energy Astrophysics, Goddard Space Flight Center,
Greenbelt, Maryland 20771, USA

F. Nagase
Institute of Space and Astronautical Science, 3-1-1 Yoshinodai,
Sagamihara, Kanagawa 229, Japan

A. N. Parmar
Astrophysics Division, Space Science Department of ESA, ESTEC,
2200 AG Noordwijk, The Netherlands

1.1   Introduction

An X-ray binary contains either a neutron star or a black hole accreting material from a companion star. X-ray binaries constitute the brightest class of X-ray sources in the sky, and were the main focus of the first 15 years of X-ray astronomy, until the advent of X-ray imaging instruments in the late 1970s allowed fainter classes to be studied. Sco X-1, the first non-solar X-ray point source discovered (Giacconi *et al.* 1962), was subsequently classified as an X-ray binary (Gursky *et al.* 1966; Sandage *et al.* 1966; Gottlieb *et al.* 1975). Approximately 175 X-ray binaries have now been identified from various X-ray surveys and optical identification programs (see Ch. 14). An optical identification is crucial to establish the nature of the mass-donating companion star, the overall geometry of the accretion flow and the mass of the X-ray source (see Ch. 2 and 3, and references therein).

The primary factors that determine the emission properties of an accreting compact object are (1) whether the central object is a black hole or a neutron star, (2) if it is a neutron star, the strength and geometry of its magnetic field, and (3) the geometry of the accretion flow from the companion (disk vs. spherical accretion). These determine whether the emission region is the small magnetic polar cap of a neutron star, a hot accretion disk surrounding a black hole, a shock heated region in a spherical inflow, or the boundary layer between an accretion disk and a neutron star. Two more factors are the mass of the central object, and the mass accretion rate; these influence the overall luminosity, spectral shape and time variability of the emission.

A neutron star with a strong magnetic field ($\sim 10^{12}$ G) will disrupt the accretion flow at several hundred neutron star radii and funnel material onto the magnetic poles (Pringle and Rees 1972; Davidson and Ostriker 1973; Lamb, Pethick and Pines 1973). If the magnetic and rotation axes are misaligned, X-ray pulsations will be observed if the beamed emission from the magnetic poles rotates through the line of sight (e.g. Mészáros, Nagel and Ventura 1980; Nagel 1981a,b; Wang and Welter 1981). When the magnetic field of the neutron star is relatively weak ($< 10^{10}$ G), the disk may touch or come close to the neutron star surface. The energy released from the inner accretion disk and the boundary layer between the disk and the neutron star will dominate the emission (e.g. Mitsuda *et al.* 1984). If the central object is a black hole, the X-rays come from the inner disk and are the results of viscous heating (Shakura and Sunyaev 1973).

Instabilities in the emission region, or its influence on the nearby accretion flow, can give rise to rapid fluctuations, or quasi-periodic oscillations (see Ch. 6). The material, as it accumulates on the neutron star, may reach a critical mass and undergo a thermonuclear flash, resulting in an X-ray burst (see Ch. 4). Instabilities in the accretion flow can also give rise to X-ray bursts, or flares (Taam and Fryxell 1988).

The spectral type of the companion determines the mode of mass transfer to the compact object and the overall environment in the vicinity of the compact object. In the low-mass X-ray binaries, LMXBs, the companion is later than type A, and can, in some very evolved systems, even be a white dwarf. A late type or degenerate star does not have a natural wind strong enough to power the observed X-ray source. Significant mass transfer will occur only if the companion fills its critical gravitational potential lobe, the Roche lobe. X-ray heating of the accretion disk and the companion star dominates the optical light, and LMXBs appear as faint blue stars (see Bradt and McClintock 1983 and references therein).

In high-mass X-ray binaries, HMXBs, the companion is an O or B star whose optical/UV luminosity may be comparable to, or greater than, that of the X-ray source (Conti 1978; Petterson 1978). X-ray heating is minimal, with the optical properties dominated by the companion star. The OB star companion has a substantial stellar wind, removing between 10^{-6} and 10^{-10} $M_\odot$ yr^{-1} with a terminal velocity up to 2000 km s^{-1}. A neutron star or black hole in a relatively close orbit will capture a significant fraction of the wind, sufficient to power the X-ray source. The X-rays must propagate through the wind to the observer, which causes photoelectric absorption in the X-ray spectrum. Roche lobe overflow can also be a supplement to the mass transfer rate in HMXBs. However, if the mass ratio of the compact object to its companion is greater than unity, then mass transfer via Roche lobe will become unstable $\sim 10^5$ yr after it starts (see Savonije 1983 and references therein). Quasi-Roche lobe overflow may occur as the supergiant approaches its Roche lobe, where the reduced gravity can cause a focusing of the wind towards the compact object (Friend and Castor 1982).

Many X-ray binaries are transient sources that appear on a timescale of a few days, and then decay over many tens or hundreds of days (White, Kaluzienski and Swank 1984; Van Paradijs and Verbunt 1984). These transient sources can, for a few weeks, be amongst the brightest in the sky, before they fade away. They are particularly important in the study of X-ray binaries since they cover an enormous dynamic range in luminosity (typically 10^4–10^5). This allows models for the emission region and the accretion process to be tested over a large range of mass accretion rate. The transient episodes may result from an instability in the accretion disk, or a mass ejection episode from the companion. Many transients are seen to recur on a timescale that ranges from days to tens of years. Some transients recur periodically, others do so randomly.

The flow geometry is determined by the angular momentum per specific mass of the accretion flow (see Ch. 10). If the companion star fills its critical Roche lobe, then a stream of material will be driven through the inner Lagrangian point. This stream will orbit the compact object at a radius determined by its specific angular momentum (Lubow and Shu 1975). Viscous interactions and angular momentum conservation cause the ring to expand into a disk. The disk's outer radius is limited by tidal forces, which will transfer angular momentum back to the binary orbit.

The specific angular momentum captured from a stellar wind is determined by gradients and asymmetries in the wind across the accretion cylinder (e.g. Shapiro and Lightman 1976). The magnitude of the captured specific angular momentum is much less in this case (Davies and Pringle 1980), and any resulting accretion disk may, depending on the circumstances, be very tenuous.

The emission from the vicinity of the compact object propagates to the observer through the surrounding environment, which modifies the spectrum by absorption and scattering. Important environmental zones are the magnetosphere of the neutron star, the accretion disk, the accretion disk corona/wind, the wind and/or atmosphere of the companion star and last, but not least, the interstellar medium. The prime result of this is that the X-ray spectrum undergoes substantial absorption at low energies, caused by the increasing absorption cross-section of the medium-Z elements such as iron, oxygen and carbon (Brown and Gould 1970). This results in K and L absorption edges in the spectra, and emission lines from fluoresence and recombination. The relative edge to line strength can be used to infer the geometry of the surrounding material. The line and edge energies are used to constrain the ionization state of the material, which in turn provides insight into its density and location. Lastly, the intervening interstellar medium causes low-energy absorption and dust scattering halos.

In this review, we describe the properties of X-ray binaries containing neutron stars and how they can be used to classify and understand the nature of the compact object, the companion star and the mass transfer process. We will concentrate on the X-ray properties measured using a number of X-ray astronomy satellites over the past two decades. The review by Tanaka and Lewin (Ch. 3) gives an overview of the X-ray properties of black hole systems. Sect. 1.2–1.4 outline the orbital, pulse and other periods found in X-ray binaries; Sect. 1.5 describes the properties of the underlying emission region; and in Sect. 1.6 the influence of the environment, i.e. the material flowing in and around the compact object, is described.

1.2   Orbital periods

1.2.1   Overview

The orbital periods of X-ray binaries have been determined from the observation of one or more of the following:

- eclipses,
- a smooth periodic modulation,
- periodically recurring X-ray absorption dips,
- periodically recurring transient X-ray outbursts,
- pulsar arrival time variations,
- radial-velocity variations, and/or
- a pulsar-orbital beat period.

The classification as low- or high-mass X-ray binary is based on the spectral type of the companion obtained from an optical identification, and/or on the mass function from X-ray pulse arrival time measurements. If neither is available, a classification may be inferred based on the similarity of the X-ray properties to other identified systems. An unidentified system is classified as an LMXB containing a neutron star if one or more of the following properties are observed:

- type I X-ray bursts (which to date have only been seen from LMXBs),
- the 1–10 keV spectrum is *soft* with a characteristic temperature of 5–10 keV, and/or
- the orbital period is less than about 12 hr.

The last criterion is adopted because a *normal* O or B star will not fit into such a small orbit, although at the extremes of the evolutionary scale this may not always be the case (cf. Van Kerkwijk *et al.* 1992). An unidentified system is classified as an HMXB containing a neutron star if the X-ray source shows one or more of the following features:

- strong flaring and absorption variability on a timescale of minutes,
- transient outbursts,
- pulsations,
- and/or has a hard 1–10 keV spectrum with a power-law energy index of order 0–1.

1.2.2 LMXBs

1.2.2.1 The period distribution

Table 1.1 lists the 32 LMXBs with well established orbital periods. The orbital periods range from 0.19 hr to 398 hr. The table also lists the bands in which the modulation is detected, the band where it was first found and key source properties. Note that about twice as many periods are seen from optical studies than from X-ray studies. This is because an X-ray modulation is only seen when the system is viewed close to the orbital plane, so that eclipses by the companion star and absorption by material in the accretion flow occur. An optical modulation can be caused by much more subtle effects, such as viewing the rotating face of an X-ray heated or tidally distorted companion, and are detectable over a greater range of inclination angle.

Figure 1.1 compares the distributions of the orbital periods of LMXB and cataclysmic variable (CV) systems, where the compact object is a white dwarf (see Ch. 8). There are no X-ray binaries with orbital periods corresponding to the gap in the CV period distribution between 2 and 3 hr. In the case of the LMXBs the period gap may extend down to $\lesssim 1$ hr (White 1985; White and Mason 1985); in the CV systems the 1–2 hr period range is populated by the SU Uma and AM Her type systems. This difference appears to be significant, since there is no selection effect against detecting LMXBs with periods in this range.

The lower X-ray luminosity systems ($\sim 10^{36}$–10^{37} erg s^{-1}), which are predominantly X-ray burst sources, typically have orbital periods of $\lesssim 15$ hr. The orbital period distribution of the high ($\sim 10^{38}$ erg s^{-1}) luminosity systems (many located in the optically obscured galactic bulge region) is still largely unknown, but in two cases (Sco X-1 and Cyg X-2) orbital periods of 19 and 235 hr have been found from optical studies.

Table 1.1. *The orbital periods of LMXBs*

Source	Alternative name	Period (hr)	Modulation Optical[a]	X-ray[a]	Discovery	Source properties[b]	Reference
X1820−303		0.19	-	M	X	B,G,Q	1,2,3
X1627−673		0.70	B	-	O	P,Q?	4
X1916−053		0.83	M	D	X	B	5,6,7
X1323−619		2.93	-	D	X	B	8
X1636−536		3.80	M	-	O	B	9,10
X0748−676		3.82	M,E	D,E	X	B,T	11,12
X1254−690		3.93	M	D	O/X	B	13,14
X1728−169	GX 9+9	4.19	M	M	X	-	15,16
X1755−338		4.46	M	D	X	-	17,18
X1735−444		4.65	M	-	O	B	19
X2129+470		5.2	M,E	M,PE	O	B,T	20,21
X1822−371		5.6	M,E,S	M,PE	O	-	22,23,24
X1746−370		5.7	-	D	X	B,G	25,26
X2023+338	V404 Cyg	5.7/155	M,S	-	O	T,BHC	27, 56
X1658−298		7.1	-	D,E	X	B,T	28
X0620−003	N'Mon 75	7.7	M,S	-	O	T,BHC	29
X2000+251	N'Vul 88	8.3	M	-	O	T,BHC	30,31
X1556−605		9.1	M	-	O	-	32
X1957+115		9.3	M	-	O	-	33
X1124−684	N'Mus 91	10.4	M	-	O	T,BHC,Q	34,35
X0547−711	CAL 87	10.6	M	-	O	SS	36,37
X1659−487	GX 339−4	14.8	M	-	O	T,BHC,Q	38
X1455−314	Cen X−4	15.1	M,S	-	O	B,T	39,40
X2127+119	AC211	17.1	M,S,PE?	-	O	B,G	41
X1617−155	Sco X−1	18.9	M,S	-	O	Q	42,43
X1908+005	Aql X−1	19.0	M	-	-	B,T	44
X1624−490		21.0	-	D	X	B	45
X0543−682	CAL 83	25.0	M	-	O	SS	46
X1656+354	Her X−1	40.8	M,S,E	E	X	P	47,48,49
X0921−630		216	M,S	PE	O	-	50,51,52
X2142+380	Cyg X−2	236	M,S	D	O	B?,Q	53,54
X1516−569	Cir X−1	398	-	O	X	B,Q	55

[a] 'E' - total eclipse, 'PE' - partial eclipse, 'D'- periodic dips, 'M' - other modulation, 'B' - beat period, 'O' - periodic outbursts.
[b] The source properties are indicated by 'B' - burster, 'G' - globular cluster, 'P' - pulsar, 'T' - transient, 'SS' - super-soft, 'BHC' - black-hole candidate, 'Q' - QPO.

References: [1]Stella *et al.* 1987; [2]Sansom *et al.* 1989; [3]Tan *et al.* 1991; [4]Middleditch *et al.* 1981; [5]Walter *et al.* 1982; [6]Grindlay *et al.* 1988; [7]White & Swank 1982; [8]Parmar *et al.* 1989a; [9]Pedersen *et al.* 1981; [10]Smale & Mukai 1988; [11]Parmar *et al.* 1986; [12]Parmar *et al.* 1991; [13]Courvoisier *et al.* 1986; [14]Motch *et al.* 1987; [15]Hertz & Wood 1988; [16]Schaefer 1987; [17]White *et al.* 1984; [18]Mason *et al.* 1985; [19]Corbet *et al.* 1986; [20]Thorstensen *et al.* 1979; [21]Ulmer *et al.* 1980; [22]White *et al.* 1981; [23]Mason *et al.* 1980; [24]Hellier *et al.* 1990; [25]Parmar *et al.* 1989b; [26]Sansom *et al.* 1993; [27]Casares & Charles 1992; [28]Cominsky & Wood 1984; [29]McClintock & Remillard 1986; [30]Chevalier & Ilovaisky 1990; [31]Charles *et al.* 1991; [32]Smale 1991; [33]Ilovaisky *et al.* 1987; [34]Bailyn 1991; [35]McClintock *et al.* 1992; [36]Callanan *et al.* 1989; [37]Cowley *et al.* 1990; [38]Callanan *et al.* 1992; [39]Chevalier *et al.* 1989b; [40]Cowley *et al.* 1988; [41]Ilovaisky *et al.* 1993; [42]Gottlieb *et al.* 1975; [43]Cowley & Crampton 1975; [44]Chevalier & Ilovaisky 1991; [45]Jones & Watson 1989; [46]Smale *et al.* 1988a; [47]Tananbaum *et al.* 1972; [48]Bahcall *et al.* 1974; [49]Voges *et al.* 1985; [50]Mason *et al.* 1987; [51]Branduardi-Raymont *et al.* 1983; [52]Chevalier & Ilovaisky 1982; [53]Cowley *et al.* 1979; [54]Vrtilek *et al.* 1986b; [55]Kaluzienski *et al.* 1976; [56]Casares *et al.* 1992.

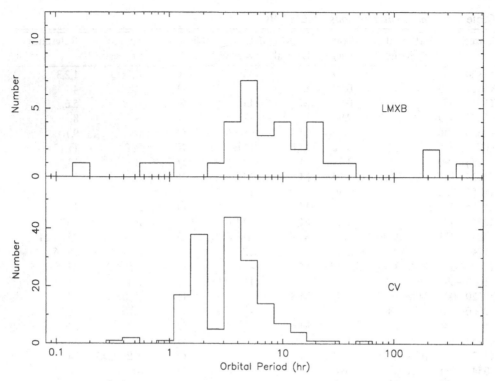

Fig. 1.1. The orbital-period distributions of LMXB and cataclysmic variable (CV) systems. The CV periods are taken from Ritter (1990).

1.2.2.2 *Globular cluster systems*

There are 12 luminous ($> 10^{35}$ erg s^{-1}) X-ray sources located in globular clusters, which is two orders of magnitude more than expected from the total mass in globular clusters relative to that in the Galaxy (Katz 1975). The tidal capture of neutron stars in close encounters with main-sequence or giant stars in the cluster core may favor the production of X-ray binaries (Fabian, Pringle and Rees 1975). Three orbital periods are known for X-ray sources in globular clusters: an 11 min X-ray modulation from X1820−303 (located in NGC 6624; Stella, Priedhorsky and White 1987); a 5.7 hr X-ray modulation from X1747−371 (in NGC 6441; Parmar, Stella and Giommi 1989b; Sansom *et al.* 1993); and 17.1 hr from the optical modulation of X2127+119 (located in M15; Ilovaisky *et al.* 1987; Naylor *et al.* 1988, Ilovaisky *et al.* 1993).

1.2.2.3 *X-ray orbital light curves*

LMXBs exhibit fewer X-ray eclipses than might be expected if the systems simply consist of a dwarf companion overflowing its Roche lobe and transferring material to a compact object via a *thin* accretion disk (Joss and Rappaport 1979). Milgrom (1978) suggested that this discrepancy could be resolved if LMXBs contain *thick* accretion disks which block the X-ray source in systems that are viewed close to the orbital plane. The discovery of a partial eclipse with *HEAO 1* from the LMXB

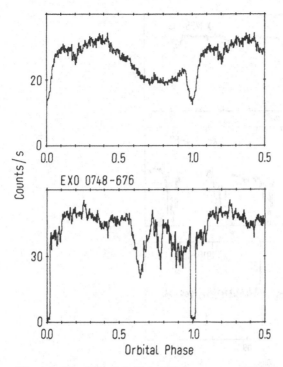

Fig. 1.2. The folded 1–10 keV light curves of X1822–371 (top) and X0748–676 (taken from Parmar *et al.* 1986). One and a half orbital cycles are shown.

X1822–371, proved Milgrom's thick-disk model to be essentially correct (White *et al.* 1981). In X1822–371, the system is viewed almost edge on, and the compact X-ray source is hidden by the disk. X-rays are still seen because they are scattered in a photo-ionized corona above the disk. This makes the source appear extended, and results in the eclipse being partial (Figure 1.2). The orbital light curve also shows a sinusoidal modulation, with a minimum preceding the partial eclipse. This can be ascribed to the partial occultation of the accretion disk corona (ADC), by a bulge at the rim of the disk caused by its interaction with the incoming gas stream (White and Holt 1982).

The LMXB X1916–053 was discovered with the *Einstein* Observatory to show irregular dips that recur periodically every 50 min (Walter *et al.* 1982; White and Swank 1982). These dips are ascribed to material which is projected up above the disk plane by a splash point, where the gas stream from the companion hits the accretion disk. A total of ten dipping sources are now known, most of them having been discovered with *EXOSAT* (Table 1.1). The long, 90 hr, orbital period of the *EXOSAT* satellite (compared with 100 min for most other X-ray observatories) allowed unprecedented continuous coverage, which was ideal for discovering the irregular, but periodic, dipping behavior. The X-ray light curves of three of these *dippers* are shown in Figure 1.3.

The strongest confirmation of the thick accretion disk model was the discovery of

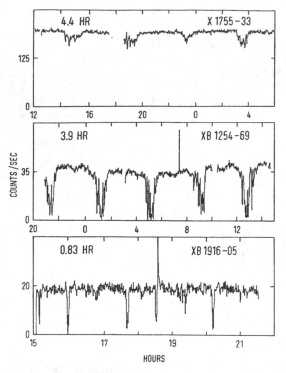

Fig. 1.3. The 1–10 keV light curves of X1755–338, XB 1254–690 and XB 1916–053 recorded by the *EXOSAT* observatory.

two dipping sources which also show an eclipse by the companion star: X0748–676 (Parmar *et al.* 1986) and X1658–298 (Cominsky and Wood 1984). In both, the eclipse follows an interval of dipping activity, consistent with the dips being due to the splash from the accretion stream passing through the line of sight. These observations show that the most likely reason that the accretion disk is *thick* is that the incoming gas stream creates turbulence at the outer edge of the disk.

Figure 1.2 compares the folded light curves of the ADC source X1822–371 and those of the dip source X0748–676. The eclipse from X0748–676 has a sharp ingress and egress (with a transition time of 6 s). The dips cause highly irregular structure prior to the eclipse. In contrast, the X1822–371 eclipse has a gradual ingress and egress, and is partial. A smooth modulation occurs over approximately the same range of orbital phase as the dips seen from X0748–676. These differences are consistent with the view that the observed X-ray source is point-like in the case of X0748–676 and extended in the case of X1822–371. The typical L_x/L_{opt} ratio of an LMXB is typically 100–1000, except for the ADC sources, which have L_x/L_{opt} ratios of ~ 20. This lower ratio in the ADC sources results from the fact that the overall X-ray luminosity is reduced because the central X-ray source is hidden.

To summarize, the observed properties of an LMXB depend on the viewing angle. At a low inclination ($< 70°$), no X-ray dips or eclipses are seen, but an optical modulation from the X-ray heated companion may still betray the orbital period.

At an intermediate inclination, periodic dipping behavior is seen which is caused by structure at the edge of the accretion disk; in a few cases, a very brief eclipse by the companion may be seen. In high inclination systems (> 80°), the central X-ray source is hidden behind the disk rim, but X-rays scattered via an ADC are still seen giving rise to a partial eclipse.

1.2.2.4 Individual cases

The orbital modulation of each LMXB has its own peculiarities, and in this section we discuss some of the typical and exceptional examples. We will not discuss the black-hole candidates, X2023+338, X0620−003, X2000+251, X1124−684 and GX339−4, which are described in Ch. 3.

X1820−303, P_o=0.19 hr. The 11 min orbital period of the X-ray burst source X1820−303, located in the globular cluster NGC 6624, is the shortest known of any binary system. The orbital period is detected from a low amplitude (3% peak-to-peak) energy independent modulation of the X-ray flux (Stella, Priedhorsky and White 1987). The modulation may be caused by obscuration of the central neutron star by material in the disk, in a similar way to the irregular X-ray dips seen from other LMXBs, or from the obscuration of the outer regions of an ADC that scatters a small fraction of the observed X-ray luminosity. The extremely short, 11 min, orbital period suggests that the companion star is a degenerate dwarf. A discussion of the likely formation mechanisms of such a system by Verbunt (1987) (see also Bailyn and Grindlay 1987 and Rappaport *et al.* 1987) supports the globular cluster tidal-capture model. This system may be the result of the subsequent spiral-in of the neutron star in the atmosphere of a red-giant companion.

X1626−673, P_o=0.70 hr. The orbital period of this 7.7 s X-ray pulsar was determined from photometric observations of the optical counterpart by Middleditch *et al.* (1981). These revealed not only the 7.7 s X-ray period, but also a low-level sideband period. This sideband can be explained as the beat between the orbital period and the pulse period. The optical sideband pulsation is probably the result of reprocessing of the X-ray pulse on a bulge in the accretion disk and/or on the face of the companion star (see Sect. 2.3.5.2).

X1916−053, P_o=0.83 hr. The orbital period was discovered from the periodic dipping behavior, which shows great variation from cycle to cycle (Figure 1.3). Typically, the dips are narrow (with a duration of < 10 min) and recur every 55 min, but occasionally anomalous/secondary dips are seen 180° out of phase with the main dip interval (Walter *et al.* 1982; White and Swank 1982). The depth and duty cycle of the dips can change dramatically from cycle to cycle and observation to observation (see, *e.g.,* Smale *et al.* 1988b, 1992).

The optical counterpart to X1916−053 is an m_v=21 blue object (Grindlay *et al.* 1988), which exhibits a photometric period that is 1% longer than the X-ray period Grindlay *et al.* (1988), suggest that the system is part of a hierarchical triple and that the third body modulates the mass transfer (see Sect. 2.6.3). This could cause the X-ray dips to have a slightly shorter period. White (1989) and Smale *et al.* (1992)

discuss alternative models based on similar double periodicities separated by a small percentage found in SU UMa dwarf novae. Numerical simulations by Whitehurst (1988) suggest that precession of an elliptical accretion disk may be responsible for the brightness maxima or 'superhumps' seen in CVs with orbital periods below the period gap. This precession only occurs when the mass ratio of the companion to the compact object is < 0.15. If this condition is satisfied, the disk becomes tidally unstable, causing it to become asymmetrical in shape, with the axis of the asymmetry rotating slightly faster than the orbital period.

X0748−676, P_o=3.82 hr.   This transient X-ray burst source, discovered serendipitously with *EXOSAT* in 1985, exhibits both dips and eclipses (Parmar *et al.* 1986). The center of the dipping activity precedes the eclipses by the companion, confirming that the dips arise from the impact of the gas stream with the disk. Depending on the assumptions made about the mass–radius relation of the companion star, the eclipse duration implies an orbital inclination of 73–83°. Thus, the thickened region responsible for the dips must subtend an angle of at least 7° above the plane of the orbit in order to obscure our line of sight. Since dips are sometimes seen for about half the orbital cycle, this region must also extend halfway around the disk at times. During eclipse, 4% of the flux remains which can be attributed to residual emission scattered by an optically thin ADC (Parmar *et al.* 1986).

This source was not seen before 1985, but has been active since then at about the same level, being detected with *Ginga* and, most recently, with *ASCA* in March 1993. This is typical of this class of LMXBs where the source appears for a number of years and then turns off for a similar number of years. During the X-ray outburst, the optical counterpart is a blue m_v=17 object (Pedersen and Mayer 1985), which is modulated at the orbital period with an optical minimum centered on the time of X-ray eclipse (Pedersen *et al.* 1985; Crampton *et al.* 1986; Schmidtke and Cowley 1987). When the X-ray source is quiescent, the optical counterpart is fainter than m_v=23 (Wade *et al.* 1985). This indicates that the X-ray outburst is caused by an accretion episode, as opposed to a thickening of the disk causing the central X-ray source to be occulted.

X1254−690, P_o=3.88 hr.   Courvoisier *et al.* (1986) discovered that the X-ray light curve of X1254−690 shows irregular dips that repeat with a period of 3.88 hr (Figure 1.3). This recurrence interval is consistent with the optical period of this system independently discovered by Motch *et al.* (1987). A 0.4 magnitude quasi-sinusoidal modulation of the m_v=19.1 counterpart is observed, with the minimum occurring 0.15 cycles after the X-ray dips. This phase difference is expected if the material responsible for the dips is located close to where the gas stream from the companion impacts the accretion disk. The depth and duration of the dips show great variety, typical of this class of X-ray source (Courvoisier *et al.* 1986).

X1755−338, P_o=4.4 hr.   X1755−338 is a very unusual X-ray dip source. *EXOSAT* observations (see Figure 1.3) revealed shallow energy independent dips in X-ray intensity every 4.4 hr (White *et al.* 1984; Mason, Parmar and White 1985). The duration and irregular variability of the X1755−338 dips are similar to those seen from other dip sources, but the energy independent nature of the dips is unusual. It is identified with an $m_v=19$ counterpart, and simultaneous optical/X-ray observations show that the optical light curve is modulated with the 4.4 hr period, with the optical minimum 0.15 cycles after the center of the X-ray dips (as for X1254−690; see Figure 2.20).

X2129+470, $P_o=5.2\,hr$.   The 5.2 hr binary period of X2129+470 was discovered from photometric optical observations by Thorstensen *et al.* (1979). Subsequently, Ulmer *et al.* (1980) used *HEAO 1* all-sky survey data to show that the X-ray light curve is modulated at the optical period. A partial X-ray eclipse, the signature of scattering in an ADC, was discovered by McClintock *et al.* (1982) and White and Holt (1982). The duration of the partial X-ray eclipse gives a scattering region size of $\sim 0.5 R_\odot$. Garcia and Grindlay (1987) discovered a type I X-ray burst from X2129+470 in *Einstein* data. The rise time of the burst is longer, and the peak burst luminosity is ~ 500 times less, than is typical for such bursts (see Ch. 4). This is consistent with the X-ray burst emission being indirectly observed via scattering in an ADC.

The optical light curve showed a strong color dependent modulation, which was interpreted as being due to viewing different aspects of an X-ray heated companion (Thorstensen *et al.* 1979, McClintock, Remillard and Margon 1981). The optical modulation of X2129+470 had been regarded as anomalous because the cool side of the K star companion would not be expected to contribute a significant fraction to the optical light (Mason and Córdova 1982). An *EXOSAT* observation in 1983 revealed that X2129+470 had turned off (Pietsch *et al.* 1986), and subsequent studies of the optical counterpart by Thorstensen *et al.* (1988) and Chevalier *et al.* (1989a) show a late F star with no orbital modulation. This was unexpected because an F star is bigger than the entire system. Chevalier *et al.* (1989a) show that, if the light of the F star is subtracted from the high state modulation, then the amplitude increases and the color dependence becomes negligible. Recently, this source has been detected with *ROSAT* at a very low level, consistent with the upper limits found with *EXOSAT* (Garcia and Grindlay 1992). It is unclear if this residual emission is coronal emission from companion stars, or if it results from low level accretion.

Both Thorstensen *et al.* (1988) and Chevalier *et al.* (1989a) conclude that a third star is required to reconcile their low state data with the earlier high state optical and X-ray observations (see Sect. 2.6.1). If it is a field star, then it must be located within 0.26 arc sec, which has a 10^{-3} probability of being a chance superposition (Thorstensen *et al.* 1988). The other possibility is that this is a hierarchical triple system, which, if correct, would have important implications for the evolution of X-ray binaries in general. It is currently a high priority to determine the orbital parameters of the F star. Aside from confirming the triple hypothesis, it could provide the first opportunity of measuring the total mass of an X-ray binary.

X1822−371, $P_o=5.57\,hr$.   A 5.57 hr modulation of the optical light curve of X1822−371 (see Figure 2.19) was discovered by Mason *et al.* (1980). The X-ray light curve of X1822−371 is modulated with the same period, and consists of a partial eclipse plus a broad energy independent modulation with a minimum ~ 0.2 cycles before the eclipse (Figure 1.2; White *et al.* 1981; White and Holt 1982). The partial X-ray eclipses occur simultaneously with the optical minima, and indicate the eclipse of an X-ray scattering ADC in this system. The broader X-ray modulation originates from the obscuration of the ADC by structure at the rim of the accretion disk; the minimum of this modulation is at the phase when the impact point of the gas stream passes through the line of sight (Mason and Córdova 1982; White and Holt 1982).

If the companion is assumed to fill its Roche lobe, and if $M_x/M_c < 0.8$ (where M_x is the mass of the compact object and M_c is the mass of the companion), then combining Kepler's third law with the mass–radius relation of Paczyński (1971) and the observed eclipse depth and width gives a radius for the ADC of $0.3R_\odot$ (White and Holt 1982; Hellier and Mason 1989). The optical eclipse lasts twice as long as the X-ray eclipse, which requires that the accretion disk has a size about a factor of two larger than that of the ADC responsible for scattering the X-rays. Modeling of the X-ray and optical light curves shows that the obscuring material must be located at the rim of the accretion disk with a maximum vertical extent of $0.15R_\odot$ (Mason and Córdova 1982; White and Holt 1982; Hellier and Mason 1989).

X1746−371, P_o=5.7 hr.   This system is located in the globular cluster NGC 6441. The X-ray light curve shows shallow dips separated by 5.74 ± 0.15 hr (Parmar, Stella and Giommi 1989b; Sansom *et al.* 1993). A notable feature is that the dips appear to be energy independent, similar to those seen from X1755−338.

X1658−298, P_o=7.1 hr.   X1658−298 is similar to X0748−676, showing dipping activity for $\sim 25\%$ of each 7.1 hr cycle, followed by an eclipse that lasts for ~ 15 min (Cominsky and Wood 1984). This source has been in a prolonged off state throughout the 1980s, and was not studied in detail with *Ginga* or *EXOSAT*.

CAL87, P_o=10.6 hr.   The source CAL87 was discovered in an X-ray survey of the LMC made using the *Einstein* imaging proportional counter by the Columbia Astrophysics Laboratory (Long, Helfand and Grabelsky 1981). It was identified with a faint blue star by Pakull *et al.* (1988), and has properties that are similar to those of an LMXB. Cowley *et al.* (1990) discovered a 10.6 hr optical photometric sinusoidal modulation. CAL87 has a low L_x/L_{opt}, which suggests this may be an ADC source. The X-ray spectrum of CAL87 is very soft, with a temperature of ~ 10 eV, much softer than that of a typical X-ray binary (Greiner, Hasinger and Kahabka 1991). It is an order of magnitude softer than the spectra of the *ultra-soft* black-hole candidates (see Ch. 3). CAL87 (along with CAL83 described later), constitute a new class of *super-soft* X-ray sources discovered with *ROSAT* (Greiner, Hasinger and Kahabka 1991). The nature of the compact object in these systems is unknown. It has been suggested that the X-ray emission comes from continuous nuclear burning on a white dwarf (Van den Heuvel *et al.* 1992). These may be a new class of CVs, not LMXBs.

Cen X-4, P_o=15.1 hr.   The transient Cen X-4 was discovered during an outburst in 1969 by Conner, Evans and Belian (1969). During a second outburst in 1979 (Kaluzienski, Holt and Swank 1980), it was identified with a star that had brightened from 19th to 12th magnitude (Canizares, McClintock and Grindlay 1980). A type I X-ray burst was discovered by Matsuoka *et al.* (1980), indicating that the compact object is a neutron star. Chevalier *et al.* (1989b) discovered the orbital period to be 15.1 hr from V-band photometric measurements made over 4 years. The folded light curve shows a double-humped modulation with a full amplitude of ~ 0.15 mag. The absorption line velocity of the K star, which is detected between outbursts, was found by Cowley *et al.* (1988) to be modulated at the 15.1 hr period, with a mass function of $0.2M_\odot$. For the K star secondary to fill its Roche lobe, it must have a peculiarly low mass of $0.1M_\odot$ (Cowley *et al.* 1988; Chevalier *et al.* 1989b; McClintock and Remillard 1990). Kaluzienski, Holt and Swank (1980) reported an ~ 8.2 hr modulation of the X-ray flux during part of the 1979 outburst, which is approximately one-half the orbital period found later.

X2127+119 (AC211 in M15), P_o=17.1 hr. The star AC211 in the globular cluster M15 is the optical counterpart to the X-ray source X2127+119 (Aurière, Le Fevre and Terzan 1984; Aurière *et al.* 1986; Charles, Jones and Naylor 1986). An 8.5 hr periodicity in AC211 was discovered photometrically by Ilovaisky *et al.* (1987) and spectroscopically by Naylor *et al.* (1988). Hertz (1987) showed that an X-ray modulation with a comparable period was present in *HEAO 1* sky-survey data with a full amplitude of ~ 50%. The L_x/L_{opt} ratio is of order 50, which is comparable to that of X1822−371 and X2129+470, suggesting X2127+119 is also an ADC source (Charles, Jones and Naylor 1986; Hertz 1987).

Detailed X-ray observations by *EXOSAT* did not reveal the expected partial eclipse of the ADC (Callanan *et al.* 1987). This discrepancy may have been resolved by Ilovaisky *et al.* (1993), who report that the orbital period is 17.1 hr, twice the value suggested earlier. They found that the optical photometry shows times when no modulation is observed, and these intervals are always on the same odd-cycle. A reanalysis of the *EXOSAT* data by Ilovaisky *et al.* (1993) shows that all these observations were taken mostly on the odd-cycle, away from when the primary eclipse would be expected. Folded light curves from *EXOSAT*, *Ginga* and *HEAO 1* show evidence for a broad secondary minimum at phase 0.5, as well as for a possible partial eclipse at the new phase 0.0. These latest results confirm the hypothesis that this is an ADC source. A strong burst was detected by Dotani *et al.* (1990). This burst may be scattered in the ADC, like the one detected from X2129+470. However, unlike the burst seen from X2129+470, the parameters of the burst are normal, which remains puzzling.

Sco X-1, P_o=21 hr. The orbital period of Sco X-1 was determined by Gottlieb, Wright and Liller (1975) to be 0.78 d based on archival photometric data taken over 85 years. The periodicity manifests itself as a low amplitude (0.3 mag on a 12.5 magnitude star) sinusoidal modulation (see Figure 2.23). This period was detected spectroscopically by Cowley and Crampton (1975). Holt *et al.* (1976) searched *Ariel V* all-sky monitor data for evidence of this periodicity and found none. The low amplitude optical modulation and the lack of an X-ray modulation indicate Sco X-1 is a low inclination system.

X1624−490, P_o=21 hr.   This source is often referred to as the *big dipper* because it exhibits dipping intervals lasting up to 6–8 hr that recur every 21 hr (Watson *et al.* 1985; Jones and Watson 1989). During the dips, the X-ray intensity reaches a lower level that is ∼ 25% of the quiescent value. The dips occur rapidly (in a few seconds) and are associated with an increase in photo-electric absorption. The constant lower level component probably represents the contribution from scattering in an ADC, which is only partially obscured during dipping intervals.

CAL83, P_o=25.0 hr.   CAL83 (Long, Helfand and Grabelsky 1981) is identified with a faint blue star that has properties similar to those of an LMXB (Cowley *et al.* 1984). A 25 hr photometric sinusoidal modulation was discovered in the optical by Smale *et al.* (1988a). CAL83 is very similar to CAL87, which was described earlier in this section. The X-ray spectrum indicates it to be one of the *super-soft* class of X-ray sources (Greiner, Hasinger and Kahabka 1991), but a low L_x/L_{opt} suggests it may be an ADC source.

Her X-1, P_o=40.8 hr.   Her X-1 exhibits 1.24 s pulsations, eclipses every 40.8 hr (1.7 d) and a 35 d high–low intensity cycle (Tananbaum *et al.* 1972). Irregular absorption dips lasting ∼ 7 hr are seen (Giacconi *et al.* 1973; Voges *et al.* 1985; Vrtilek and Halpern 1985) that are similar to those from the X-ray dippers. The dips occur just before eclipse (pre-eclipse dips) and occasionally at Φ ∼ 0.3–0.6 (anomalous dips). The recurrence interval of the dips is not the same as the orbital period of the system, by an amount that may be related to the 35 d cycle (Boynton, Crosa and Deeter 1980; Crosa and Boynton 1980).

Her X-1 underwent an extended low-state during part of 1983. During this interval, the 35 d high–low cycle appeared to cease, and the source intensity remained similar to that of the low-state (Jones, Forman and Liller 1973; Parmar *et al.* 1985a). Since the effects of X-ray heating on the companion were still present (Delgado, Schmidt and Thomas 1983), it is likely that during this interval the accretion disk became thicker and obscured the direct line of sight to the neutron star (Parmar *et al.* 1985a). An eclipse observed during the extended low-state was partial and gradual, which, if interpreted in terms of an extended emission region, implies a size of $0.7R_\odot$, similar to that of the accretion disk. This suggests scattering in an ADC; apparently, the X-ray source was still active, but was hidden behind the accretion disk.

X0921−630, P_o=216 hr.   Optical observations show eclipses and an orbital period of 9.0 d (Chevalier and Ilovaisky 1982; Cowley, Crampton and Hutchings 1982; Branduardi-Raymont *et al.* 1983). A partial X-ray eclipse lasts for ∼ 1 d centered on the time of the optical eclipse (Mason *et al.* 1987). A low L_x/L_{opt} ∼ 1 confirms that this is an ADC source.

Cyg X-2, P_o=235 hr.   Radial-velocity measurements have shown that the orbital period of Cyg X-2 is 9.8 d (Cowley, Crampton and Hutchings 1979). *Einstein* X-ray data reported by Vrtilek *et al.* (1986b) revealed shallow X-ray dips prior to the time of inferior conjunction of the companion star with respect to the X-ray source, similar to the behavior seen from the other dip sources.

Table 1.2. *Changes in orbital periods of LMXBs*

Source	P_{orb} (hr)	$-P_{orb}/\dot{P}_{orb}$ (yr)	Reference
X1820−303	0.18	-1×10^7	Tan *et al.* (1991)
			Sansom *et al.* (1989)
			Van der Klis *et al.* (1993)
X0748−676	3.8	-5×10^6	Parmar *et al.* (1991)
		$+1 \times 10^7$	Asai *et al.* (1993)
X1822−371	5.6	$+3 \times 10^6$	Hellier *et al.* (1990)
Her X−1	40.8	-8×10^7	Deeter *et al.* (1991)

Cir X-1, $P_o=398\,hr$. Cir X-1 (X1516−569) is an unusual X-ray binary that shows recurrent X-ray transitions at a period of 16.6 d, which represents the orbital period (Kaluzienski *et al.* 1976). Each X-ray transition is accompanied by a radio flare, which typically begins when a sharp dip in the X-ray flux is seen (Whelan *et al.* 1977). Cir X-1 shows extreme ranges of X-ray variability, with variations in excess of 1000 up to super-Eddington luminosities within a few seconds. The classification of Cir X-1 as either an HMXB or an LMXB is uncertain. Some of the X-ray properties are typical of an LMXB: type I X-ray bursts (Tennant, Fabian and Shafer 1986) and quasi-periodic oscillations (QPOs) (Tennant 1987). Whelan *et al.* (1977) identified the primary as an OB supergiant, and Murdin *et al.* (1980) suggested that it is an eccentric HMXB, with the outbursts caused by enhanced accretion at periastron passage. Doubt was cast on this classification by Nicolson, Feast and Glass (1980), who showed the infrared emission to vary by more than 2 magnitudes, inconsistent with an OB star. The original optical counterpart has now been resolved into three stars within 1.5 arc sec by Moneti (1992), with one of the stars showing variations with the 16.6 d cycle. High-resolution radio studies by Stewart *et al.* (1993) show evidence for radio jets that curve back towards a nearby supernova remnant. Cir X-1 may be a runaway binary from the supernova explosion.

1.2.2.5 Orbital period changes
The orbital periods of four LMXBs have been measured with sufficient precision to detect orbital period variations. Two of the systems have *decreasing* orbital periods, one has an *increasing* orbital period and one a more complex behavior (Table 1.2). In all cases where an orbital timescale, $P_o/\dot{P}_o$, measurement has been made, the results have been contrary to evolutionary expectations.

The orbital period changes of X1822−371 and X0748−676 have been measured using eclipse timing, which provides an accurate fiducial marker. For Her X-1, pulse arrival time measurements provide a more accurate fiducial marker than does the eclipse. In X1820−303, the orbital modulation is not marked by a sharp eclipse, but rather by a smooth modulation, which nonetheless seems to be stable enough to provide an epoch measurement.

The 5.6 hr orbital period of X1822−371 is *increasing* on a timescale of 2.9×10^6 yr (Hellier *et al.* 1990). This is opposite to expectations, since to drive the mass transfer requires the orbit to shrink, which should result in a period decrease. The 40.8 hr orbital period of Her X-1 is *decreasing* on a timescale of -8×10^7 yr (Deeter *et al.* 1991), which is larger than can be explained by the currently observed mass transfer rate. Deeter *et al.* (1991) suggest additional angular momentum loss either through the L2 point, or by a radiation driven wind from the companion. The 11 min orbital period of X1820$-$303 is *decreasing* on a timescale of -1×10^7 yr (Sansom *et al.* 1989; Tan *et al.* 1991; Van der Klis *et al.* 1993). This is contrary to expectations for a degenerate companion, which will expand as it loses mass, resulting in an increase in the orbital period. The period change may indicate that the binary is part of a hierarchical triple, or is being accelerated by the cluster potential. Another possibility is that random or systematic changes in the position and shape of an occulting bulge on the disk rim give rise to apparent period changes (Van der Klis *et al.* 1993).

For X0748$-$676, the situation is complex. By combining *EXOSAT* observations with the first *Ginga* observation, Parmar *et al.* (1991) determined the orbital period to be *decreasing* with a characteristic timescale of -5×10^6 yr. However, when timing data from the three subsequent *Ginga* observations are included, this trend is not continued (Asai *et al.* 1993). Instead, a quadratic fit to the eclipse times gives $+1 \times 10^7$ yr, but with a poor fit. A sinusoidal variation with a period of about 12 yr gives a better fit, perhaps suggesting a triple system, although this requires further observations to confirm the trend (Asai *et al.* 1993). Another possibility is that the period changes in X0748$-$676 provide clues about the transient nature of this source (see Sect. 1.2.2.4) because, if the period continues to lengthen, the companion will eventually go out of contact, and the mass transfer will cease.

1.2.3   HMXBs

1.2.3.1   The period distribution

The orbital periods of the confirmed and suspected HMXB systems are listed in Table 1.3. The periods range from 4.8 hr to 187 d. All but one of the supergiant systems have orbital periods less than 15 d, and dominate this part of the period distribution, whereas most Be star systems have longer periods of several tens or hundreds of days. Table 1.3 also lists the type of the HMXB. The periods of the HMXBs are mostly determined from pulse arrival time analysis and/or the detection of eclipses.

The supergiant systems typically are eclipsing and show extreme intensity and absorption variability on all timescales. The shorter orbital period systems have circular orbits, whereas the longer period systems show some eccentricity. The Be star X-ray binaries are very often bright pulsating transient sources (Maraschi *et al.* 1976). Doppler variations in the pulse period give orbital periods of a few tens of days, with a moderate eccentricity ($e \sim 0.3$). Such transients include X0115+634 (Rappaport *et al.* 1978), X1553$-$542 (Kelley *et al.* 1983b), X0331+530 (Stella *et al.* 1985) and X2030+375 (Parmar *et al.* 1989c). These outbursts are caused by mass ejection episodes of the Be star, and are probably related to the fact that Be stars rotate close to break up. Eclipses are rare in these systems.

There are also Be X-ray binaries that are persistent sources. The luminosities are generally low, 10^{33}–10^{35} erg s^{-1}. The prototype in this class is the star X Per. The

Table 1.3. *The orbital periods of HMXBs*

Source	Alternative name	Orbital period (d)	Properties[a]	Reference
X2030+407	Cyg X−3	0.2	WR	1,2,3
X0532−664	LMC X−4	1.4	SG, P	4,5,6
X0538−641	LMC X−3	1.7	Be, BHC	7
X1119−603	Cen X−3	2.1	SG, P	8
X1700−377	HD153919	3.41	SG	9
X1538−522	QV Nor	3.73	SG, P	10,11
X0115−737	SMC X−1	3.89	SG, P	12
X0540−697	LMC X−1	4.22	SG, BHC	13
X1956+350	Cyg X−1	5.6	SG, BHC	14
X1907+097		8.38	B, P	15
X0900−403	Vela X−1	8.96	SG, P	16
X1657−415		10.4	SG?, P	17
X0114+650	V662 Cas	11.6	SG	18
X1909+048	SS433	13.1	SG, J	19
X0535−668	A0538−66	16.7	Be, T, P	20
X0115+634	V635 Cas	24.3	Be, T, P	21
X0236+610	LS I +61 303	26.45	Be	22
X1553−542		30.6	Be?, T, P	23
X0331+530	BQ Cam	34.25	Be, T, P	24
X1223−624	GX301−2	41.5	SG, P	25,26,27
X2030+375		45−47	Be, T, P	28
X0535+262	HD245770	111	Be, T, P	29
X1258−613	GX304−1	133?	Be, P	30
X1145−619	Hen 715	187.5	Be, P	31

[a]The source properties are indicated by 'SG' - supergiant, 'Be' - Be star, 'P' - pulsar, 'BHC' - black-hole candidate, 'T' - transient, 'WR' - Wolf–Rayet, 'J' - Jets.
References : [1]Parsignault *et al.* 1972; [2]Sanford & Hawkins 1972; [3]van Kerkwijk *et al.* 1992; [4]Li *et al.* 1978; [5]White 1978; [6]Chevalier & Ilovaisky 1977; [7]Cowley *et al.* 1983; [8]Schreier *et al.* 1972b; [9]Jones, Forman and Liller 1973; [10]Becker *et al.* 1977; [11]Davison, Watson and Pye 1977; [12]Schreier *et al.* 1972b; [13]Hutchings *et al.* 1983; [14]Webster & Murdin 1972; [15]Marshall & Ricketts 1980; [16]Ulmer *et al.* 1972; [17]Chakrabarty *et al.* 1993; [18]Crampton *et al.* 1985; [19]Crampton *et al.* 1980; [20]Johnston, *et al.* 1980; [21]Rappaport *et al.* 1978; [22]Taylor & Gregory 1982; [23]Kelley *et al.* 1983b; [24]Stella *et al.* 1985; [25]Watson *et al.* 1982; [26]Kelley *et al.* 1980; [27]White *et al.* 1978; [28]Parmar *et al.* 1989c,d; [29]Priedhorsky & Terrell 1983a; [30]Priedhorsky & Terrell 1983b; [31]Watson *et al.* 1981.

orbital periods are not so well known, but the lack of eclipses and pulse timing measurements indicate periods of several hundred days. This is borne out by the persistent Be star source X1145−61, which shows a periodic outburst every 200 d (Watson, Warwick and Ricketts 1981).

The most unusual HMXB is the 4.8 hr system Cyg X-3. This system will be discussed at the end of this chapter.

1.2.3.2 Orbital period changes
Cen X-3 and SMC X-1 are the only HMXBs for which a finite rate of change of orbital period has been measured. Kelley *et al.* (1983c) used pulse timing

measurements to measure for Cen X-3 a timescale $P_o/\dot{P}_o = -5.62 \times 10^5$ yr using eclipse epoch data obtained between 1971 and 1981. This trend has been confirmed by Nagase *et al.* (1992) using *Ginga, Tenma* and *Hakucho* measurements. Kelley *et al.* (1983c) interpret this decrease in the orbital period with time as the result of tidal torque between the distorted supergiant and the neutron star. A decay in the orbit of SMC X-1 has been measured by Levine *et al.* (1993), who combined *Ginga* observations with earlier measurements made with *Ariel V* and *SAS 3* to give a timescale of -3.0×10^5 yr.

There are several other HMXBs where marginal evidence for a change in period has been reported, including LMC X-4, Vela X-1 and X0115+634 (see Nagase 1992 and references therein). However, the baseline of the measurements must be extended before they become definitive.

1.3   Pulse periods

1.3.1   *The period distribution*

Table 1.4 lists the 32 X-ray pulsars discovered to date, their pulse periods, any associated orbital periods and whether the system is an LMXB or an HMXB. The pulse periods are distributed between 0.069 s and 835 s, with no evidence for a clustering at any particular period. In Figure 1.4, the pulse periods are shown versus orbital period for those systems where both are known. The figure is divided into three parts based on whether the underlying system is an LMXB, a Be star system or a supergiant system. For the Be star systems, there is a strong correlation between orbital period, P_o, and pulse period, P_p, with $P_p \propto P_o$ (Corbet 1984). The supergiant systems show no obvious dependence. For the LMXBs there are only three points, which prevents the drawing of any firm conclusions about a trend (see also Sect. 10.9.2).

1.3.2   *Period changes*

Long term monitoring of the pulse periods of X-ray pulsars has revealed three types of behavior: (1) the pulse period shows a linear decrease with time (spin-up) with erratic variations around the trend; (2) no long term trend is seen, only a random walk in the period; and (3) a steady increase in pulse period (spin-down). These types of behavior are illustrated in Figure 1.5.

1.3.2.1   *Spin-up*

A neutron star will spin-up, i.e. its period will decrease, because of the angular momentum gained from the accretion flow (Pringle and Rees 1972; Davidson and Ostriker 1973; Lamb *et al.* 1973; see Ch. 10). The angular momentum captured by the neutron star is determined by the magnetospheric radius, r_m, where the accretion flow is threaded onto the magnetic field lines. The simplest method of calculating r_m is to equate the ram pressure of a spherically symmetric inflow to the magnetic pressure, which gives

$$r_m = 3 \times 10^8 \mu_{30}^{4/7} M_x^{1/7} R_6^{-2/7} L_{37}^{-2/7} \, \text{cm}.$$

$\mu_{30} = \mu/10^{30}$ is the magnetic moment of the neutron star in G cm^3, M_x is the mass of the X-ray source in solar units, R_6 is the radius of the neutron star in 10^6 cm,

Table 1.4. *Pulse periods from X-ray binaries*

Source	Alternative name	Pulse period (s)	Orbital period (d)	Type	Reference
X0535−668	A0538−66	0.069	16.7	HMXB	1
X0115−737	SMC X−1	0.71	3.89	HMXB	2
X1656+354	Her X−1	1.24	1.7	LMXB	3
X0115+634	V635 Cas	3.6	24.3	HMXB	4
X0332+530	BQ Cam	4.4	34.25	HMXB	5
X1119−603	Cen X−3	4.8	2.1	HMXB	6
X1048−594		6.4		?	7
X2259+587		7.0		LMXB	8
X1627−673		7.7	0.029	LMXB	9
X1553−542		9.3	30.6	HMXB	10
X0834−430	GR0834−430	12.2	-	?	11
X0532−664	LMC X−4	13.5	1.4	HMXB	12
X1417−624		17.6		HMXB	13
X1843+009		29.5		?	14
X1657−415		38	10.4	HMXB	15
X2030+375		42	45.6	HMXB	16
X2138+568	Cep X−4	66		?	17
X1836−045		81		?	14
X1843−024		95		?	14,34
X0535+262		104	111	HMXB	18
X1833−076	Sct X−1	111		?	19
X1728−247	GX1+4	114	304?	LMXB	20,21,22
X0900−403	Vela X−1	283	8.96	HMXB	23
X1258−613	GX 304−1	272	133?	HMXB	24,25
X1145−614		298		HMXB	26,27
X1145−619		292	187.5	HMXB	26,27
X1118−615	A1118−61	405		HMXB	28
X1722−363		413		?	29
X1907+097		438	8.38	HMXB	30
X1538−522	QV Nor	529	3.73	HMXB	31
X1223−624	GX301−2	696	41.5	HMXB	32
X0352−309	X Per	835		HMXB	33

References: [1]Skinner *et al.* 1982; [2]Lucke *et al.* 1976; [3]Tananbaum *et al.* 1972; [4]Cominsky *et al.* 1978; [5]Stella *et al.* 1985; [6]Giacconi *et al.* 1971; [7]Corbet & Day 1990; [8]Gregory & Fahlman 1980; [9]Rappaport *et al.* 1977; [10]Kelley *et al.* 1983b; [11]Grebenev & Sunyaev 1991; [12]Kelley *et al.* 1983a; [13]Kelley *et al.* 1981; [14]Koyama *et al.* 1990a; [15]White & Pravdo 1979; [16]Parmar *et al.* 1989d; [17]Koyama *et al.* 1991a; [18]Rosenberg *et al.* 1975; [19]Koyama *et al.* 1991b; [20]Lewin *et al.* 1971; [21]White *et al.* 1976a; [22]Strickman *et al.* 1980; [23]McClintock *et al.* 1976; [24]Huckle *et al.* 1977; [25]McClintock *et al.* 1977; [26]White *et al.* 1978b; [27]Lamb *et al.* 1980; [28]Ives *et al.* 1975; [29]Tawara *et al.* 1989; [30]Makishima *et al.* 1984; [31]Davison *et al.* 1977; [32]White *et al.* 1976a; [33]White *et al.* 1976b; [34]Koyama *et al.* 1990b.

and L_{37} is the X-ray source luminosity in units of 10^{37} erg s^{-1}. If the flow is via an accretion disk, the magnetospheric radius is approximately one-half the spherical value (Ghosh and Lamb 1978, 1979a,b). For a given neutron star spin period, this equation for r_m and angular momentum conservation give an empirical relationship

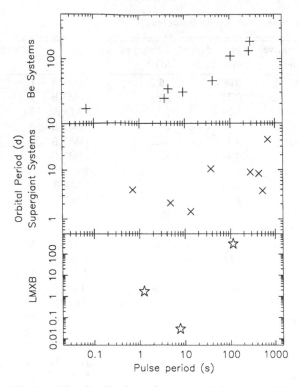

Fig. 1.4. The distribution of pulse period versus orbital period for the LMXB (bottom), the supergiant (middle) and the Be systems (top).

between the spin-up, $\dot{P}_p$, and the observed luminosity L. This gives $\dot{P}_p \propto (P_p L^{3/7})^2$, with the normalization determined by the magentic dipole moment and the mass of the neutron star. The measured values of these parameters for the pulsars that exhibit spin-up are in good agreement with this simple model (Mason 1977; Rappaport and Joss 1977). The observed spin-up timescale ranges from 100 to 100 000 years. Figure 1.6 compares the theoretical relationship determined by Ghosh and Lamb (1979a,b) with the observed values for pulsars showing strong spin-up. The agreement is good. In the X-ray transient system X2030+375, this relationship has been confirmed by directly measuring the $\dot{P}_p$ over a wide range of luminosity (Parmar *et al.* 1989d).

1.3.2.2 Spin-down

An equilibrium is reached when the neutron star magnetosphere corotates with the inner edge of the accretion disk, i.e. when r_m equals the corotation radius, r_c. Elsner and Lamb (1976) divided pulsars up into two types: *slow rotators*, where $r_c > r_m$, and *fast rotators*, where $r_c \simeq r_m$. The boundary layer between the unperturbed disk and the magnetosphere is complex, and there will be a transition region where the magnetic field begins to thread the disk. As the neutron star reaches corotation, the spin-up torque diminishes. Elsner and Lamb (1977) introduced a dimensionless *fastness parameter,* which is the ratio of the angular velocity of the neutron star and the Keplerian angular velocity at the magnetospheric radius. Close to corotation,

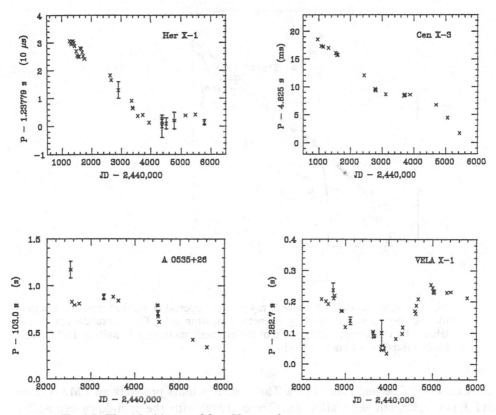

Fig. 1.5. The spin history of four X-ray pulsars.

when this parameter approaches unity, the field lines in the transition region are swept backwards, and a negative torque is exerted on the neutron star. The neutron star will be spun down, even though accretion continues. This causes a turnover in the predicted $P_p/\dot{P}_p$ at low luminosities (Figure 1.6).

In March 1983, the 114 s X-ray pulsar GX1+4 entered a low state where the intensity was a factor of 20 below its typical high-state value (Hall and Davelaar 1983). Subsequent observations with *Ginga* in 1987 and 1988 by Makishima *et al.* (1988) detected the pulsations in the low state. The measured pulse period showed that in the low state the pulsar is spinning down (Figure 1.5), even though it is still accreting material. Similar erratic behavior has also been observed in Her X-1 and Cen X-3 (Figure 1.5), and also seems to be associated with a low state of the source, presumably because of a reduction in accretion rate.

The pulsar X2259+587, which has a 7 s period, lies in the young supernova remnant G109.1−1.0 (Gregory and Fahlman 1980), and is spinning down on a timescale of 3.7×10^5 yr (Davies *et al.* 1990; Iwasawa, Koyama and Halpern 1992). This rate is similar to that of the pulsar in the Crab supernova remnant. However, the period of 7 s and the observed X-ray luminosity of 10^{35} erg s^{-1} are inconsistent with the expected rotational energy loss. No optical counterpart has been identified (Davies

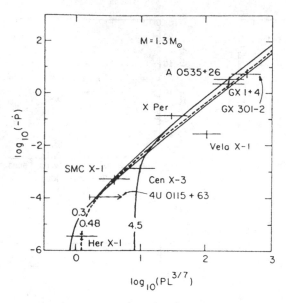

Fig. 1.6. $\dot{P}_p$ versus $P_p L^{3/7}$ for nine X-ray pulsars, superposed on the theoretical relation for the spin-up of an accreting magnetized neutron star. The theoretical curves are labeled by the corresponding value of the magnetic moment μ in units of 10^{30} G cm^3. From Ghosh and Lamb (1979a,b).

and Coe 1991; Coe and Jones 1992). This system is most probably an LMXB (Coe and Jones 1992, but see Carlini and Treves 1989) with the spin-down caused by the pulsar being near its equilibrium state with the accretion torque, slightly smaller than the deceleration torque (Iwasawa, Koyama and Halpern 1992). The observed luminosity requires a relatively weak magnetic field of 5×10^{11} G. The X-ray spectrum is much softer than for most X-ray pulsars, and shows a hint of cyclotron resonance absorption lines at 5 and 10 keV, consistent with this magnetic field strength (Iwasawa, Koyama and Halpern 1992).

The unidentified 6.4 s pulsar X1048−594 lies close to η Carina (Seward, Charles and Smale 1986). This is spinning down on a timescale of 1×10^4 yr (Corbet and Day 1990). Seward, Charles and Smale (1986) suggest an $m_v \sim 19$ optical counterpart that might be a Be star, but this has not been confirmed (Mereghetti, Caraveo and Bignami 1992). The spectrum, like that of X2259+586, is softer than that of most X-ray pulsars (Corbet and Day 1990). As Corbet and Day point out, this pulsar seems in most respects similar to X2259+586, except it is not located in or near a known supernova remnant.

1.3.2.3   Period fluctuations

In the lower luminosity ($< 10^{37}$ erg s^{-1}) X-ray binary pulsars identified with OB systems, no overall trend in the pulse period is seen. Large fluctuations in pulse period are evident in both the disk and wind driven systems (see Figure 1.5) that persist down to timescales at least as short as a few days. Power spectrum analysis of the pulse timing noise from Vela X-1 shows that its frequency fluctuations can be described as white noise in the angular acceleration of the neutron star (Boynton *et al.* 1984; Deeter *et al.* 1987). This suggests that the pulse period fluctuations reflect inhomogeneities in the accretion flow, rather than a torque internal to the neutron star caused by the coupling of the super-fluid interior and the solid outer crust (which would give red noise). Baykal and Ögelman (1993) have applied an autoregressive time series model to the pulse period variations of a large sample of X-ray pulsars, and they find no difference between the pulse period fluctuations in the disk and wind driven systems. They also conclude that the pulse period fluctuations are caused by episodic mass transfer with positive and negative torques.

The amount of angular momentum captured by a neutron star accreting from a stellar wind determines whether or not an accretion disk will form around the neutron star and consequently the spin history of the neutron star. Illarionov and Sunyaev (1975) and Shapiro and Lightman (1976) show that sufficient angular momentum to form a disk can be captured from the wind if a density or velocity gradient exists across the accretion radius. Shapiro and Lightman (1976) also suggest the accretion flow could reverse around a black hole if there are fluctuations in the density across the accretion radius. But Davies and Pringle (1980) point out that, for accretion to take place at all, the transverse component of the momentum must be lost, and the only way to do this in an ideal situation is for the matter passing either side to collide. Wang (1981) disputes the *ideal model*, and suggests instead that turbulence in the accretion process allows the bulk of the angular momentum to be captured.

Hydrodynamical simulations of accretion from a stellar wind have been made to better understand the accretion process. Early results by, e.g., Livio *et al.* (1986) and Soker *et al.* (1986) showed a steady-state situation, similar to that envisioned by Davies and Pringle. However, other calculations using a much finer grid reveal a non-steady situation (Matsuda, Inoue and Sawada 1987; Taam and Fryxell 1988, 1989). These models show that, in the presence of inhomogeneities in the wind, there is a tendency for instabilities in the accretion flow associated with oscillatory motion in the accretion wake. In these models the accretion flow shows flip-flop behavior where an accretion disk briefly forms, disperses, then forms again but rotating in the opposite sense. These models seem to be in qualitative agreement with the observed period behavior from the *wind driven* pulsars. These simulations also can reproduce the unusual periodic flaring behavior seen from the Be transient X2030+375 (Parmar *et al.* 1989d; Taam, Fryxell and Brown 1988).

1.3.3   *Period evolution*

The rotation period evolution of a neutron star throughout its lifetime is determined by its period at birth, the efficiency of spin-down (assuming initial rapid rotation) and the history of the mass loss from the primary. The spin-down efficiency is not well understood, nor is the period distribution at birth, but based on the observations of so many long period X-ray pulsars either spin-down is a relatively efficient process, or pulsars are born slow rotators (Lea 1976).

The correlation between pulse period and orbital period for the Be star X-ray binaries (Corbet 1984; see Sect. 1.3.1 and Figure 1.5) suggests these pulsars have reached some sort of equilibrium spin period that depends on the orbital separation. The only obvious equilibrium is the condition $r_c = r_m$ (see Sect. 1.3.2.2). Combining this with Kepler's law, and assuming a wind that is radially expanding at a constant velocity, gives $P_p \propto P_o^{4/7}$ (Corbet 1984; Stella, White and Rosner 1986; Van den Heuvel and Rappaport 1987; Waters and Van Kerkwijk 1989). This is not consistent with the observed relation for Be stars where $P_p \propto P_o$ (Corbet 1984).

The observed luminosity of many Be X-ray binaries is one to two orders of magnitude higher than that predicted for stellar-wind accretion (White *et al.* 1982). Be stars are rapidly rotating, close to break-up and exhibit an infrared excess that comes from a dense equatorial ring. The material in this ring flows outwards with a much more gradual velocity law than does the radially expanding wind from the polar regions. The observed X-ray luminosities are consistent with accretion from an enhanced equatorial flow (Waters *et al.* 1988). For a wind density distribution $\rho(r) = \rho_o(r/R_*)^{-n}$, $P_p \propto P_o^{2/7(4n-6)}$ (Waters and Van Kerkwijk 1989). The observed P_p, P_o distribution gives $n \sim 3.5$, similar to the value found independently from the infrared measurements.

If the neutron star orbital axis is offset from that of the Be star, then a double peaked orbital modulation may be seen as the neutron star slices through the dense Be star disk twice every orbit. The double peaked 8.4 d X-ray modulation seen from X1907+097 may arise from such an effect (Marshall and Ricketts 1980).

The P_p, P_o distribution found for the longer period pulsars in supergiant systems is similar to that predicted for a radially expanding wind. However, the normalization requires a mass accretion rate two orders of magnitude lower than observed (Stella, White and Rosner 1986; Van den Heuvel and Rappaport 1987). Assuming the condition $r_c = r_m$ was satisfied at some point in the spin history of the pulsar, then the supergiant distribution on the P_p, P_o diagram may reflect an earlier evolutionary phase when the companion was on the main sequence and its wind much weaker (Waters and Van Kerkwijk 1989).

King (1991) points out that in a Be system the accreted angular momentum will be insufficient to form a disk and that the current P_p, P_o distribution must also reflect an earlier evolutionary phase for the Be stars (see Sect. 10.9). But all of the shorter pulse period Be X-ray binaries (which are driving the observed correlation) are transients, and during each outburst strong spin-up is directly observed (e.g. Parmar *et al.* 1989d). The observed spin-up directly testifies to the presence of a disk, and demonstrates that during each outburst sufficient angular momentum is captured. An effective $r_c \sim r_m$ equilibrium is reached where the interoutburst spin-down and outburst spin-up balance.

1.3.4   *A centrifugal barrier to accretion*

The magnetosphere radius increases as the accretion rate decreases, and eventually $r_c > r_m$ may occur during the outburst decay. If this happens, centrifugal forces will prevent material from entering the magnetosphere, and accretion onto the neutron star will cease. The X-ray source will turn off at a minimum luminosity given by

$$L_{min} = 2.5 \times 10^{37} \, R_6^{-1} \, M_x^{-2/3} \, \mu_{30}^2 \, P_s^{-7/3} \text{ erg s}^{-1},$$

where P_s is the pulse period in seconds (Stella, White and Rosner 1986). If the luminosity can be measured just before the transient turns off, it gives an estimate of the magnetic dipole moment. An extreme Be star system is the 69 ms pulsar X0535−668 (A0538−66), which is located in the LMC (White and Carpenter 1978; Skinner *et al.* 1982). This is a highly eccentric system with an orbital period of 16.7 d (Johnston, Griffiths and Ward 1980; Pakull and Parmar 1981; Charles *et al.* 1983). It is a transient system that is sometimes dormant for long periods, but when active shows super-Eddington luminosities. A 69 ms rotation period requires a magnetic field of $\sim 10^{11}$ G, a factor of ten less than typical, simply to overcome the centrifugal barrier.

As a pulsar orbits the companion, the change in accretion rate expected around the eccentric orbit can, for a particular mass loss rate from the Be star, cause the pulsar to cross the centrifugal limit. This will result in a large orbital modulation, much larger than expected simply from the expected variation in stellar-wind accretion. This effect was first observed from the Be transient X0332+530 (Stella, White and Rosner 1986).

The centrifugal barrier means that an X-ray outburst may be much shorter than any associated optical phenomenon. This has been seen for X0115+634, where an outburst in December 1980 was preceded some months before by an optical brightening of the Be star (Kriss *et al.* 1983). The delay between the X-ray and optical outbursts is naturally explained by the centrifugal barrier preventing accretion until late in the mass ejection episode. Long term monitoring of the optical counterpart to X0115+634 between 1985 and 1990 by Mendelson and Mazeh (1991) revealed three optical outbursts; only two of them were accompanied by X-ray outbursts (see Tamura *et al.* 1992). The relatively short 1988 optical outburst did not appear in X-rays, probably because the accretion rate was insufficient to overcome the centrifugal barrier.

1.4 Third periods and other modulations

In many X-ray binaries, a third period, P_3, has been detected that is longer than the orbital period. This periodicity usually manifests itself as a long term modulation of the overall X-ray or optical flux. In some cases, the third periodicity is detected at a high level of significance, and is a well established feature of the overall source properties. But in other sources the modulation is very weak and/or long compared with the observation duration, so its stability and significance are not well established. Table 1.5 lists X-ray binaries where periods have been reported that are due neither to the rotation of a neutron star nor to the binary orbit. Those periods which are not well established are indicated by a '?'. The well established third periods in X-ray binaries are the 35 d cycle of Her X-1, the 30.5 d period of LMC X-4 and the 164 d period of SS433. The other reported detections of third periods are less certain because they have lower amplitudes and are less significant. The ratio of P_3/P_o is also given in Table 1.5, but shows a wide dispersion in value.

The 35 d cycle of Her X-1 manifests itself as an 11 d *main-high* followed by a 5 d *short-high*, separated by intervals of relatively low flux (Jones and Forman 1976). The 35 d cycle is also seen in variations in the optical light curve (Gerend and Boynton 1976; see Figure 2.21) and the shape of the 1.24 s pulse profile (Trümper *et al.* 1986). The 30.5 d modulation of LMC X-4 (see Figure 2.8) is similar to that of Her X-1. It shows a factor of >5 amplitude, with a low state lasting 40% of the cycle. Modeling

Table 1.5. *Third periods from X-ray binaries*

Source	P_3 (d)	P_o (d)	Ratio P_3/P_o	Type	Reference
LMC X-4	30	1.4	21	HMXB	1
Her X-1	35	1.7	21	LMXB	2
Cyg X-2	77?	9.8	8	LMXB	3
SS433	164	13.1	13	HMXB	4
X1820−303	175	0.008	22100	LMXB	3,5
X1907+097	42?	8.4	5	HMXB	6
X1916−053	199?	0.035	5750	LMXB	3,6
LMC X-3	198 or 99?	1.7	116	HMXB	7
Cyg X-1	294	5.6	53	HMXB	8

References: [1]Lang *et al.* 1981; [2]Tananbaum *et al.* 1972; [3]Smale & Lochner 1992; [4]Margon *et al.* 1979; [5]Priedhorsky & Terrell 1984b; [6]Priedhorsky & Terrell 1984a; [7]Cowley *et al.* 1991; [8]Priedhorsky *et al.* 1983.

of the light curve of the optical counterpart to LMC X-4 requires shadowing of the X-ray heated companion by the disk, similar to the effects seen in Her X-1 (Heemskerk and Van Paradijs 1989).

SS433 is an unusual X-ray binary, and not typical of the class. It is famous for *moving lines* in the optical spectrum, which indicate the ejection of jets of material at a velocity of 0.26 times the speed of light. The movement of the optical emission lines from the jets gives a period of 162 d, which is thought to represent a precession period either of the disk or the underlying compact object (Margon 1984). The binary period is 13.1 d, and the companion is a massive early-type star. The nature of the compact object is unknown. *EXOSAT* observations of SS433 revealed a 6.7 keV iron K emission line whose energy varies with time, in concert with the optical lines (Watson *et al.* 1986). The optical lines from both the receding and advancing jets are always seen, but for most of the precession cycle only the X-ray line from the advancing jet is observed. This means that the iron emission comes from very close to the compact object, such that the X-ray line from the receding jet is usually hidden behind the disk. *Ginga* observations by Kawai *et al.* (1989) show a partial eclipse of the X-ray emission by the companion star and a spectral softening during the eclipse. This is further evidence for a temperature gradient along the jet.

The 35 d cycle of Her X-1 has been attributed to either the retrograde precession of a tilted accretion disk (Katz 1973; Roberts 1974; Petterson 1975, 1977, 1978; Gerend and Boynton 1976; Boynton, Crosa and Deeter 1980; Crosa and Boynton 1980; Petterson, Rothschild and Gruber 1991) or the free precession of a neutron star whose spin axis is offset from the orbital axis (Trümper *et al.* 1986). Bisnovatyi-Kogan, Mersov and Sheffer (1990) find that the neutron star precession model is unlikely to be correct. Precessing disk models are also invoked for SS433 and LMC X-4. Schwarzenberg-Czerny (1992) has compared the third periods from X-ray binaries with similar periods from CVs, and suggests that only disk precession can systematically account for the third periods seen in all the different types of system. However, even if this is correct, it is not clear how or why the accretion disk is tilted in these systems, and what drives the precession.

Smale and Lochner (1992) used *Vela 5B* data to survey the long term variability from LMXBs, and to search for longer periods. Out of 17 LMXBs, only three show evidence for periodicities (Cygnus X-2, X1820−303 and X1916−053). This demonstrates that such long periods are rare in LMXBs.

1.5 The emission region

1.5.1 *Strongly magnetized neutron stars*

1.5.1.1 *Pulse profiles*

There are dramatic differences in pulse shape and amplitude between one pulsar and another. In some cases, the pulse profile shows a strong dependence on energy and overall source luminosity. Figure 1.7 shows representative X-ray pulse profiles in three energy bands for six pulsars (e.g. White, Swank and Holt 1983; Nagase 1989). The pulse periods and the log of the luminosity are given. Working from high to low luminosity, Cen X-3 shows a relatively simple single-peak profile with a weak interpulse and no strong energy dependence. The high energy pulse of Her X-1 is similar to that of Cen X-3, but at low energies it becomes sinusoidal, with a minimum at the phase of the high energy pulse peak. A pulse reversal at low energies is evident from 4U1626−67 (X1627−673), with an additional reversal in the highest energy band. The pulse profile of the intermediate luminosity system Vela X-1 is a simple double sinusoid at high energies, but becomes complex at low energies. By comparison, the pulse profiles of X Per and 4U1145−61 (X1145−619), two low luminosity Be star systems, are much simpler, being sinusoidal-like and of lower amplitude at all energies.

The pulse profile of an X-ray pulsar can also depend on X-ray luminosity. Observations of the transient 42 s X-ray pulsar X2030+375 made through the decay phase of an outburst showed that as the outburst decayed the relative strength of the main pulse and interpulse changed (Figure 1.8; Parmar, White and Stella 1989c). At high luminosity, what were the main pulse and interpulse became, at low luminosity, the reverse. This can only occur if the beam pattern has changed phase by 180°.

Other pulsars show similar luminosity dependencies. When the luminosity of Cen X-3 is a factor of two lower than that shown in Figure 1.7, the low-energy pulse shows a double-peaked structure (Nagase *et al.* 1992). The LMXB pulsar GX1+4 showed a simple broad-peak structure when it was in a high luminosity state (White, Swank and Holt 1983). When it was in a lower luminosity state, this changed to a sharp dip structure superposed on the broad sinusoidal modulation (Dotani *et al.* 1989). The profile of LMC X-4 is a simple sinusoid during outbursts, but during quiescence it shows a complex series of notches (Levine *et al.* 1991).

1.5.1.2 *Geometric pulse profile models*

The basic accretion model for X-ray pulsars was first proposed by Pringle and Rees (1972), Davidson and Ostriker (1973) and Lamb, Pethick and Pines (1973) after the discovery of pulsations from Her X-1 and Cen X-3. The inflowing material from

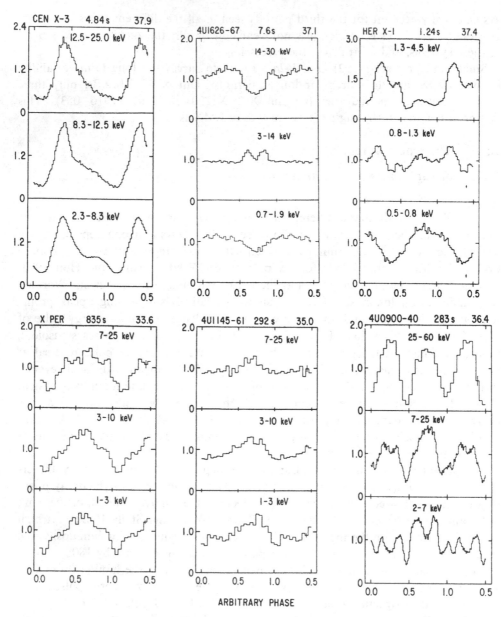

Fig. 1.7. Folded pulse profiles of six X-ray pulsars in three energy bands; 1.5 cycles are shown, with increasing energy from bottom to top. The pulse period in seconds and the log of the luminosity in erg s^{-1} are indicated above each panel. Adapted from White, Swank and Holt (1983).

the companion is threaded onto the neutron star magnetic field lines and channeled onto the magnetic poles. The accretion energy is released as X-rays in a shock near or at the surface.

The pulse profile of many pulsars can be reproduced simply by assuming a beam

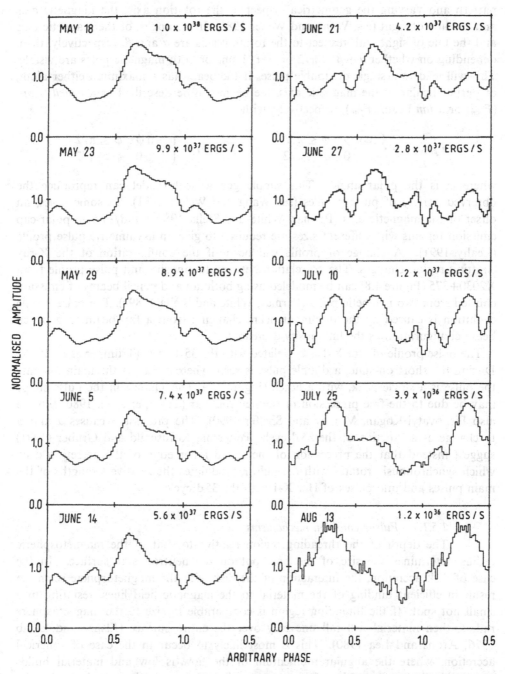

Fig. 1.8. Folded 1–10 keV *EXOSAT* pulse profiles of X2030+375 (EXO2030+375) showing the evolution of the pulse profiles during the decay of an outburst; 1.5 cycles are shown. (From Parmar *et al.* 1989c.)

pattern and varying the geometrical aspect of the rotation axis, the magnetic axis and the line of sight (e.g. Wang and Welter 1981). If the angles of the magnetic axis and the line of sight with respect to the rotation axis are α and β, respectively, then, depending on whether $\alpha + \beta < \pi/2$ or $> \pi/2$, one or both magnetic poles are visible. This will lead to a single or double pulse. If the beam has a maximum either along or perpendicular to the magnetic axis, the beam can be described as a *pencil* beam (F_{pen}) or a *fan* beam (F_{fan}), respectively, with

$$F_{pen}(\phi) = \begin{cases} \cos\phi & \phi \le \pi/2 \\ 0 & \phi > \pi/2 \end{cases} \qquad F_{fan}(\phi) = \begin{cases} \sin\phi & \phi \le \pi/2 \\ 0 & \phi > \pi/2 \end{cases}$$

where ϕ is the polar angle. This simple geometrical model can reproduce the observed variety of pulse profiles (see Wang and Welter 1981). In some cases, an offset of the magnetic axis (Parmar, White and Stella 1989c) and/or two polar-cap emission regions with different sizes are required to give an asymmetric pulse profile (Leahy 1991). A change of profile will occur if the configuration of the X-ray emitting region changes. The correlation between luminosity and pulse profile from X2030+375 (Figure 1.8) can be modeled using both fan and pencil beams of emission emitted from two magnetic poles (Parmar, White and Stella 1989c). The pulse profile evolution is caused by the dominant beam changing from a fan beam to a pencil beam configuration as the luminosity decreased.

The pulse profile of Her X-1 is correlated with the 35 d cycle (Trümper *et al.* 1986). During the short on-state, a double pulse is seen, whereas during the main on-state the pulse is a single peak, with a notch (Figure 1.7). This change in the pulse shape may be due to the free precession of the neutron star (Trümper *et al.* 1986; but see also Bisnovatyi-Kogan, Mersov and Sheffer 1990). The precession causes α and β to change as a function of the 35 d cycle. Petterson, Rothschild and Gruber (1991) suggest instead that the precession of the raised inner edge of the accretion disk, which synchronously rotates with the pulsar, modulates the relative strengths of the main pulses and interpulses of Her X-1 with the 35 d cycle.

1.5.1.3 *Pulsar emission mechanisms*

The depth of the threading region relative to that of the magnetospheric radius determines the size of the hot spot on the neutron star surface. In the case of disk accretion, the interaction of the disk with the magnetosphere seems to result in efficient binding of the material to the magnetic field lines, resulting in a small hot spot. If the threading region is comparable in size to the magnetosphere radius, then material will fall unevenly over the entire surface (Elsner and Lamb 1976; Arons and Lea 1980). This is most likely to occur in the case of spherical accretion, where the angular momentum of the flow is low and material builds up outside the magnetosphere before penetrating. Consequently, spherical accretors should have much lower amplitude and less structured pulsations. This seems to be the case for many widely separated Be star stellar wind accretors (e.g. X Per in Figure 1.7).

When the luminosity is in excess of $\sim 10^{36}$ erg s^{-1}, a radiative shock will form at the neutron star surface (Basko and Sunyaev 1976; Wang and Frank 1981). At lower luminosities the inflowing material is thermalized either by Coulomb interactions at the surface (Mészáros, Nagel and Ventura 1980) or by a collisionless shock (Langer and Rappaport 1982). The feasibility of a collisionless shock in a 10^{12} G magnetic field is unknown, so most calculations have concentrated on Coulomb interactions. In this case, beaming along the magnetic field lines is responsible for the observed deep modulation. The beaming comes about because of the anisotropy in the photon-scattering cross-sections in a strongly magnetized plasma (Canuto, Lodenquai and Ruderman 1971; Basko and Sunyaev 1975; Basko 1976; Ventura 1979). This leads to a reduced scattering cross-section along the magnetic field lines, and results in a pencil beam (Bonazzola *et al.* 1979; Hamada 1980; Kanno 1980; Mészáros *et al.* 1980; Yahel 1980; Mészáros and Bonazzola 1981; Nagel 1981a; Harding *et al.* 1984; Kii *et al.* 1986).

The effect of the magnetic field is negligible for photons with an energy larger than the cyclotron energy, and the scattering is independent of photon propagation direction. Thus, the pulse profiles at higher energies should be relatively simple and determined by the geometrical configuration of the emission region, as seems to be the case (Figure 1.7). The scattering cross-section of photons with an energy less than the cyclotron energy E_B depends on the direction of the photon propagation with respect to the direction of the magnetic field (Ventura 1979; Yahel 1979; Nagel 1981a,b; Kaminker *et al.* 1982; Daugherty and Harding 1986). This results in energy dependent changes in the pulse profile, similar to those shown in Figure 1.7 (Nagel 1981a). If the luminosity is high and a radiative shock forms, then the emission is predominantly a fan beam perpendicular to the magnetic field lines. Nagel (1981a) finds that it is possible at some luminosities to have a fan beam and a pencil beam simultaneously, but with the different beams dominant at different energies, as is seen from X1627−673 (4U1626−67 in Figure 1.7).

Gravitational bending of the X-rays plays a significant role in shaping the beam pattern emitted from the polar caps on the neutron star surface in the case of a fan beam (Brainerd and Mészáros 1991). For a typical neutron star, a photon emitted perpendicular to the magnetic axis is bent $\sim 30°$ from the original direction. Brainerd and Mészáros (1991) demonstrate that the shadowing of surface radiation by the accretion column and gravitational focusing of column radiation determine the broad range of angular behavior seen at different energies.

The beamed emission will be subject to scattering and absorption by the accretion column, the magnetosphere, the accretion disk and material in the stellar wind, which may modify the pulse profile. If the accretion column is optically thick to Thomson scattering, then radiation propagating vertically through the accretion column may be scattered. The sharp dip or *notch* observed in the main pulse of several pulsars may be caused by such an effect (White, Swank and Holt 1983; Dotani *et al.* 1989). Occultation of the pulsar beam by the precessing inner disk is the model invoked by Petterson, Rothschild and Gruber (1992) to explain the 35 d dependence of the Her X-1 pulse profile. X-ray emission from X-ray heated material in the magnetosphere or inner disk may be responsible for a strong pulsed component at low energies seen from Her X-1 (McCray *et al.* 1982). A smearing of the pulse observed from Cen X-3 during a pre-eclipse dip by Nagase *et al.* (1992) may be caused by scattering of the pulse in the stellar wind.

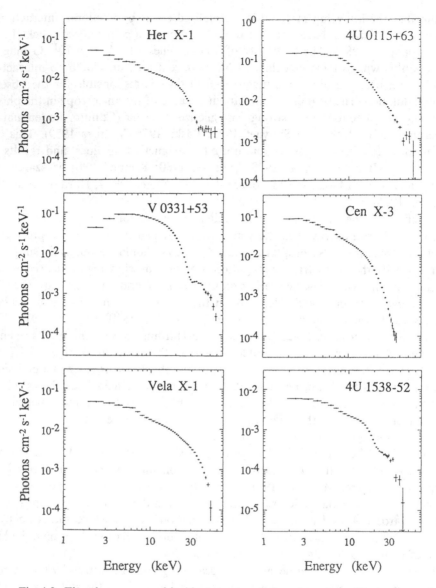

Fig. 1.9. The phase-averaged incident spectra of six pulsars from *Ginga* observations. Cyclotron line features are visible in the spectra of Her X-1, X0115+634, X0332+530 (V0331+530) and X1538−522.

1.5.1.4   Pulsar continuum spectra and cyclotron lines

Figure 1.9 shows a representative selection of X-ray pulsar spectra. The energy spectrum of an X-ray pulsar is characterized by a flat power-law with an energy index, Γ, of 0.0–1.0 up to a high-energy cutoff at 10–20 keV, above which the spectrum decays more steeply (e.g. White, Swank and Holt 1983). The spectrum above the high-energy cutoff, E_c, can be approximated by the function $\exp[(E_c - E)/E_f]$, where the e-folding energy is given by E_f (White, Swank and Holt 1983). Typical values for E_c and E_f both lie in the range 10–20 keV. A soft excess is seen below 1 keV from several pulsars, which may represent thermal emission from hot material in the magnetosphere (e.g. Her X-1; McCray *et al.* 1982).

A few, mostly lower luminosity ($< 10^{35}$ erg s^{-1}), pulsars have softer, 1–20 keV, spectra (White, Swank and Holt 1983). X Per (White *et al.* 1982) and Sct X-1 (Koyama *et al.* 1991b) show a significantly steeper spectrum ($\Gamma \simeq 2$). A soft spectrum is also observed from the pulsar X2259+587 (Iwasawa, Koyama and Halpern 1992), the pulsar located in the supernova remnant G109.1−1.0. This is one of several unusual pulsars that are spinning down. One explanation for the steep spectrum is that the cutoff energy is below 1 keV. Another case of a soft spectrum is the 6.4 s pulsar X1048−594, which is also spinning down and may be related to X2259+587.

Improved spectral measurements with *Ginga* over a broader energy band (2–30 keV) and with better statistical precision have shown that a Lorentzian function multiplied by E^2 gives a better fit to the high-energy turnover, which may reflect the presence of a cyclotron line (Makishima *et al.* 1990a). In this model, the overall continuum can be represented by a function of the form

$$f(E) = AE^{-\Gamma} \exp[-H(E)], \quad \text{where} \quad H(E) = \frac{\tau W_o^2 (E/E_o)^2}{(E - E_o)^2 + W_o^2}.$$

E_o is the observed cyclotron energy, W_o is the line width and τ is the optical depth of the absorption line (Makishima *et al.* 1990a). This function can simultaneously describe a trough at $\sim E_o$ and a high-energy turnover starting at $\sim E_o/2$. It provides an approximate measurement of the cyclotron absorption, and resembles the cross-section of cyclotron resonant scattering (Herold 1979; Daugherty and Harding 1986; Alexander and Mészáros 1991). The cyclotron energy, E_B, is defined as $E_B \sim 11.6 B_{12}$ keV, where B_{12} is the magnetic field strength in units of 10^{12} G. The cyclotron line energy will be redshifted by the neutron star's gravitational field, such that $E_o = E_B/(1 + z)$, where the redshift, z, is ~ 0.3.

Cyclotron scattering resonance features, CSRFs, have now been detected from nine pulsars at energies between 7 keV and 40 keV, corresponding to a range of surface magnetic field strengths of $(0.8–4.4) \times 10^{12}$ G. The first detection was made by Trümper *et al.* (1978) from Her X-1 at 35–45 keV. This feature has been extensively studied (Gruber *et al.* 1980; Tueller *et al.* 1984; Soong *et al.* 1990) and discussed in terms of either an emission line at 45 keV or an absorption line at 35 keV. The absorption line interpretation is more likely, based on modeling the variation of the feature with pulse phase (Voges *et al.* 1982; Mihara *et al.* 1990; Soong *et al.* 1990).

CSRFs at $E_o \sim 11.5$ keV and ~ 23 keV were detected in the spectrum of the transient Be star system X0115+634 from *HEAO 1* observations (Wheaton *et al.* 1979; White, Swank and Holt 1983). Both line features appear first in absorption during the main pulse and then in emission during a broad interpulse. *Ginga* observations of X0115+634 by Nagase *et al.* (1991) showed a different behavior, with the same two lines in absorption during most pulse phases. The *Ginga* satellite had increased sensitivity above 10 keV compared with that of earlier X-ray missions, and seven more pulsars were discovered to have CSRFs: X1538−522 (Clark *et al.* 1990), X0332+530 (Makishima *et al.* 1990b), Cep X-4 (Mihara *et al.* 1991a), X2259+586 (Iwasawa,

Table 1.6. *Cyclotron features detected in X-ray pulsar spectra*

Name	Alternative name	P_p (s)	E_c (keV)	E_B (keV)	$B(1+z)$ 10^{12}G	Reference
Her X-1	HZ Her	1.24	17–21	35	3.0	1
X0115+634		3.6	7–9	12, 23	1.0	2,3
X0332+530	BQ Cam	4.38	14–17	28.5	2.5	4
X2259+587		6.9	≤ 4	$\sim 7?$	0.6	5
X2138+568	Cep X-4	66.3	15–17	32	2.8	6
X0900−403	Vela X-1	283	15–20	27	2.3	7
X1907+097		438	14–16	21	1.8	7
X1538−522		530	14–16	21	1.8	8
X1223−624	GX301−2	690	19–21	40	3.4	7

References: [1]Trümper *et al.* (1978); [2]Wheaton *et al.* (1979); [3]White, Swank & Holt (1983); [4]Makishima *et al.* (1990b); [5]Iwasawa *et al.* (1992); [6]Mihara *et al.* (1991a); [7]Makishima and Mihara (1992); [8]Clark *et al.* (1990).

Koyama and Halpern 1992), Vela X-1, X1907+097 and GX301−2 (Makishima and Mihara 1992).

An overview of the X-ray pulsar CSRF line energies and corresponding magnetic field strengths is given in Table 1.6. For some (e.g. Vela X-1), the CSRF is detected only for part of the pulse phase. The absorption line features are broad ($\sim 25\%$) and shallow (e.g. Tamura *et al.* 1992). The energy of the fundamental (and also the second harmonic) resonance energies vary by up to 30% with pulse phase (Voges *et al.* 1982; Clark *et al.* 1990; Soong *et al.* 1990; Nagase *et al.* 1991; Tamura *et al.* 1992). There is evidence of a second harmonic in the spectra of X1538−522 (Clark *et al.* 1990), Her X-1 (Mihara *et al.* 1990), X0332+530 (Makishima *et al.* 1990b), Vela X-1 and X1907+097 (Makishima and Mihara 1992), but its detection is less significant than that in X0115+634 because it is close to the upper end of the observable energy range.

The observed cyclotron resonance energy, E_o, and the high-energy cutoff, E_c (in the old model for describing the high-energy spectrum), are correlated with $E_o \simeq 2 \times E_c$ (Makishima and Mihara 1992). This correlation suggests that the cutoff energy is related to the cyclotron resonant scattering. If this is the case, then E_c provides a measure of the magnetic field strength for those pulsars where a CRSF has not been detected. The high-energy cutoff energies are all typically between 10 and 20 keV, giving a narrow scatter in magnetic field strength of $B = (1\text{–}4) \times 10^{12}$ G. The luminosities of the nine pulsars for which CSRFs have been detected were all less than 4×10^{37} erg s^{-1}. A CSRF feature has not been detected from the more luminous pulsars such as Cen X-3. The power-law plus Lorentzian gives a good fit, but with the line energy at 30 keV, i.e. at the top end of the observable energy range (Makishima and Mihara 1992).

Since the discovery of the cyclotron line feature in Her X-1, many papers have been published that model both the line feature and the overall continuum (e.g. Yahel 1979, 1980; Mészáros, Nagel and Ventura 1980; Nagel 1981b; Kaminker *et al.* 1983;

Mészáros and Nagel 1985). The most recent is a phase-resolved model constructed by Bulik *et al.* (1992), which utilizes a magnetized radiation transfer code. This model was fitted to the 16 phase resolved spectra of X1538−522 obtained by Clark *et al.* (1990). The result gives a distribution of magnetic field that is broader than that expected from a dipole field at constant radius. It also requires a significant difference between the two polar caps, both in terms of opening angle and temperature. This model can account for the phase dependence, the energy, the width and depth of the cyclotron resonance.

1.5.2   The LMXB spectral emission

Early observations of Sco X-1 showed that the entire spectrum from the infra-red to the hard X-rays could be modeled as thermal bremsstrahlung emission from a plasma cloud with an optical depth of a few and a temperature of 5–10 keV (Neugebauer *et al.* 1969). Higher-quality spectra obtained in the late 1970s showed the X-ray spectra of several LMXBs (including Sco X-1) require, in addition to the bremsstrahlung emission, a highly variable blackbody component (Swank and Serlemitsos 1985). The spectra of other LMXBs, in particular the lower luminosity systems, can be described by very simple power-law spectral models.

The overall spectral properties of the non-pulsing LMXBs can be broadly classified either according to the overall luminosity of the source, with the split occurring at around 10^{37} erg s^{-1} (White, Stella and Parmar 1988), or according to their timing and spectral properties (Hasinger and Van der Klis 1989).

1.5.2.1   High luminosity/Z-source systems

Mitsuda *et al.* (1984) showed that the spectra of the high luminosity LMXBs can be well represented by a two-component model. One component is the sum of blackbodies from an accretion disk with a surface temperature that depends on radius, r, as $r^{-3/4}$. The other is a variable isothermal blackbody component from a boundary layer between the disk and the neutron star. Variability in the blackbody component is responsible for flaring activity seen from many luminous LMXBs. This flaring behavior was first seen from Sco X-1 and became known as *Sco X-1 like behavior* (Mason *et al.* 1976a). It is characterized by an increase in spectral hardness correlated with source intensity. The blackbody was detected simply by subtracting the quiescent spectra from the flare spectra (Mitsuda *et al.* 1984).

White, Peacock and Taylor (1985) and White *et al.* (1986) also used a two-component disk plus isothermal blackbody model, but assumed that the disk emission was dominated by Comptonization processes, where low energy photons are upscattered on high-energy electrons (analogous to the Comptonization model invoked to explain the hard power-law spectrum from Cyg X-1; Sunyaev and Titarchuk 1980). There have been other related models that assume the observed spectra to be the result of a single component that has been modified by Comptonization in a surrounding optically thick cloud (Fabian, Guilbert and Ross 1982; Ponman, Foster and Ross 1990; Raymond 1993; Vrtilek, Soker and Raymond 1993). These models invoke a central source of emission that is reprocessed in a surrounding cloud. The spectral change associated with the flaring behavior is due to changes in the optical depth of the surrounding plasma. These models do not address the formation of the central emission spectrum, but rather assume an arbitrary input power-law spectral model.

The Sco X-1 like behavior was subsequently resolved into two branches called the *flaring branch* and the *normal branch* (Hasinger 1987). On the flaring branch, the blackbody component ranges from a small percentage, up to 40% of the total luminosity, with a radius of order a few kilometers. An exception is Cyg X-2, where the flaring branch shows a reduction in intensity, associated with the spectral hardening (Hasinger *et al.* 1990). In a color–color diagram the flaring branch of Cyg X-2 appears to be similar to that of other luminous LMXBs (Schulz, Hasinger and Trümper 1989; Hasinger *et al.* 1990).

On the normal branch, a large increase in hardness ratio is seen for a small increase in count rate (Hasinger 1987; Hasinger and Van der Klis 1989; Schulz, Hasinger, and Trümper 1989). Many high luminosity LMXBs also show a *horizontal branch* (Branduardi *et al.* 1980; Shibazaki and Mitsuda 1984), where the hardness ratio is relatively constant for large intensity variations. The normal branch appears to be the most common branch found for these objects (Schulz, Hasinger and Trümper 1989) and joins the horizontal branch to the flaring branch. Each branch has been identified with a different QPO and variability behavior (see Ch. 6 for a full discussion). Sources that exhibit these branches are often referred to as the *Z sources*, because their color–color behavior describes a Z pattern.

The ratio of the blackbody to the Comptonized-disk component increases as the source moves through the horizontal–normal–flaring branches (HB–NB–FB; Van der Klis *et al.* 1987; Hasinger *et al.* 1990; Schulz, Hasinger and Trümper 1989). There are also other subtle changes in the temperature and luminosity of the Comptonized component, but these are minor compared with the variations in the blackbody component. The opposite is the case with the blackbody-disk model, where the isothermal blackbody luminosity decreases across the HB–NB–FB for Cyg X-2 (Hasinger *et al.* 1990; Hoshi and Mitsuda 1991). However, as noted above, the flaring branch of Cyg X-2 is peculiar in that the overall flux decreases, as opposed to the 'flares' seen from other luminous LMXBs. In Sco X-1, the flaring branch in the blackbody-disk model seems to correspond to an increase in the luminosity of the isothermal blackbody component (Mitsuda *et al.* 1984). The interpretation of the physical origin, and even the reality, of the two-component spectra remains controversial (e.g. Vacca *et al.* 1987).

The isothermal blackbody component seen in both proposed models may come from a boundary layer between the disk and the neutron star (Mitsuda *et al.* 1984; White *et al.* 1986). The expected emission from the boundary layer for a neutron star rotating far away from its break-up period should be 70% of the total (Sunyaev and Shakura 1986), so the detected blackbody is under-luminous by a factor of two or more (White, Stella and Parmar 1988). Why this should be is not clear, but it could be that the boundary layer is not a pure blackbody, or that the neutron star rotates close to break-up.

1.5.2.2 *Low luminosity/atoll systems*

The continuum spectra of the less luminous ($< 10^{37}$ erg s^{-1}) LMXBs, mostly the ones that show X-ray bursts, are much simpler (White, Stella, and Parmar 1988).

These have harder spectra than the higher luminosity systems and are well modeled as simple power-law spectra with energy index ~ 1 and an exponential high-energy cutoff with a temperature of 5 to 20 keV. *Atoll sources* are LMXB classified based on their behavior in the color–color diagrams and their timing properties (Hasinger and Van der Klis 1989; see also Ch. 6). Unlike the Z sources, the atoll sources do not exhibit QPOs, and do not show the characteristic Z color–color pattern. Many of these atoll sources are the lower luminosity sources that include the X-ray burst sources. These sources show no strong correlation between hardness ratio and intensity. However, those that show large factors of five to ten excursions in intensity show a trend to increasing hardness ratio at low intensities and high intensities, with a minimum in between (e.g. Mitsuda *et al.* 1989; Schulz, Hasinger and Trümper 1989).

A few atoll sources have comparable luminosities to the Z sources (Van der Klis 1992). At high luminosities, the spectra of the atoll sources still require two components, whereas at lower luminosities the spectral shape is much simpler, and can be well represented by a power-law, or a power-law with an exponential cutoff (White *et al.* 1986; Mitsuda *et al.* 1989). The atoll source X1608−522 was tracked by Mitsuda *et al.* (1989) over a luminosity variation of a factor of ten and fit to both spectral models. In the blackbody-disk model, the major change that occurs is that the blackbody component becomes heavily distorted by Comptonization. In the Comptonized-disk model, the spectral change is modeled by a reduction in the isothermal blackbody component, plus a change in the parameters of the Comptonized-disk component.

1.5.2.3 Accretion disk models

A major uncertainty in applying accretion disk models to the spectra of LMXBs is the nature of the viscosity in the inner radiation pressure dominated disk, where the X-rays come from. In the standard Shakura and Sunyaev (1973) prescription, the viscosity uncertainties are hidden by scaling the viscous stress with the disk pressure. It is unknown how the viscosity scales in the radiation pressure, dominated region, which is unfortunate because this critically determines the spectral properties and the stability of the disk. If the viscosity scales with the radiation pressure the disk is unstable to thermal and viscous instabilities, which may drive the disk into a puffed-up state. If the viscosity scales with gas pressure (Stella and Rosner 1984), then the disk is stable, but the validity of this disk model is unclear. Another complication in these LMXBs is the presence of additional emission from the neutron star. Czerny, Czerny and Grindlay (1986) suggest that photons from the boundary layer may stabilize a disk where the viscous stress is proportional to the radiation pressure; again, this is far from certain.

White, Parmar and Stella (1988) fitted a variety of accretion disk models to the spectra of a selection of bright LMXBs including the Shakura and Sunayev (1973) α-disk (where α represents the viscosity scaling with the total gas plus radiation pressure), a model where the viscosity only scales with gas pressure (Stella and Rosner 1984) and the simple blackbody-disk model proposed by Mitsuda *et al.* (1984). All the disk models give an acceptable fit to the data; however, they all require mass accretion rates of a factor of two or more in excess of the Eddington limit.

The Mitsuda *et al.* (1984) model does not include the fact that the opacity of the disk in its inner region is dominated by electron scattering. This will cause the spectrum from any radius to be a modified blackbody, rather than a blackbody. Since the emissivity is reduced, this will make the disk hotter at a given radius. Including a constant color correction to the blackbody temperature overcomes this problem and brings the required mass accretion rates to a plausible value (Mitsuda *et al.* 1989; Hoshi and Mitsuda 1991). However, this is an arbitrary correction and not a full self-consistent solution of the accretion disk radiative transfer. There is also a need for a more accurate treatment of the electron scattering in the Shakura and Sunayev (1973) model (L. G. Titarchuk 1993, private communication).

The low state spectrum of the black-hole candidate Cyg X-1 cannot be well described by an optically thick accretion disk model, and it seems likely that the emission is dominated by the Compton cooling of electrons by an unspecified source of UV, or longer-wavelength, photons (Shapiro, Lightman and Eardley 1976). A Comptonization model for the LMXB spectra also gives a good fit (White, Stella and Parmar 1988). The optical depth and temperature of the scattering plasma are typically 13 and 4 keV, respectively, with a Comptonization y parameter of order 3. While the optical depth and temperature of the scattering medium are quite different to those of Cyg X-1 (where they are 5 and 27 keV, respectively), the y parameter is very similar. The problem with this model as it is currently used is that it conveys no information on the geometry of the emission region, nor on the source of the high energy electrons or low energy photons. However, future models will keep track of the number and energy of the injected photons, and will allow a disk area to be derived (L. G. Titarchuk 1993, private communication).

Further progress in the modeling of the spectra of LMXBs requires both improved accretion disk models and observations that cover a greater energy range. The current measurements typically do not cover the entire 0.5–200 keV band over which the bright LMXBs could be observed. All the models discussed have relatively similar spectral shapes in the 1–20 keV band, but begin to diverge at higher and lower energies. Such a large dynamic range has been hard to achieve, but will be possible with future missions.

1.6   Environmental radiative processes

1.6.1   The accretion disk

1.6.1.1   The dip energy spectra

The X-ray dips seen from the LMXBs (Figure 1.3) are caused by material from the splash where the accretion stream strikes the accretion disk passing through the line of sight. However, a simple cold-absorber model fails in most cases to give an adequate representation of the spectra. Spectra taken during dips show a low-energy excess below ~ 4 keV in, e.g., X0748−676 (Parmar *et al.* 1986). In other cases (X1755−338 and X1747−371), the dips are energy independent (White *et al.* 1984; Sansom *et al.* 1993). Many different explanations have been proposed for these discrepancies from a simple absorption model, including:

- ionization of the absorbing material by the central X-ray source (Frank *et al.* 1987),

- rapid variations in absorption, on a timescale faster than the accumulation time for the spectra (Parmar *et al.* 1986),
- scattered emission in an ADC (Sztajno and Frank 1984), and
- reduced abundances in the absorbing material (White and Swank 1982).

The dip spectra can be modeled by a two-component absorption model, where one component is allowed to have a variable absorption column density and the other is kept fixed at the minimum absorption seen from the system. The ratio of flux in the two components translates to an equivalent ratio of scattered to directly absorbed flux. The absorption is dominated by photoelectric absorption from the medium-Z elements (e.g. O, Fe, Si, S). Electron scattering also plays an important role, especially at high column densities. The flux ratio in the two components translates to an equivalent measure of the metallicity of the absorbing medium. The resulting equivalent abundances are listed in Table 1.7. Five of the seven dip sources studied have equivalent abundances comparable (within a factor of two) to the solar value, whereas the dips observed from X1755−338 and X1746−371 are energy independent, resulting in abundances at least factors of 600 and 150 less than solar, respectively (White *et al.* 1984; Mason, Parmar and White 1985; Parmar, Stella and Giommi 1989b).

One explanantion for the energy independence of the dips from X1755−338 and X1746−371 is simply that the abundance of the absorbing material is indeed two orders of magnitude less than cosmic. While extreme, reduced abundances are a characteristic of Population II objects (see Verbunt, Van Paradijs and Elson 1984). Day, Fabian and Ross (1992), by fitting the spectrum of the tail of a burst from X1636−536, find an abundance of 0.3, again suggesting abundance deficiencies in the LMXB. This is, however, not a unique explanation for the energy independent component in the dip spectra.

If the spectra were accumulated over timescales longer than the intrinsic timescale for variability during the dips, then the spectra will be the sum of a range of column densities. This summation can mimic a low-energy excess (Parmar *et al.* 1986; Smale *et al.* 1992). Alternatively, the absorbing material may be in an intermediate ionization

Table 1.7. *X-ray dip sources*

Source	Period (hr)	Abundance deficiency	Reference
X1916−053	0.83	1.6−2.5	Smale *et al.* (1992)
X1323−619	2.96	15−0.5	Parmar *et al.* (1989a)
X0748−676	3.82	7−2	Parmar *et al.* (1986)
X1254−690	3.88	2−0.25	Courvoisier *et al.* (1986)
X1755−338	4.40	> 600	White *et al.* (1984)
X1746−371	5.74	> 150	Parmar *et al.* (1989b)
X1658−298	7.1	...	Cominsky and Wood (1989)
X1624−490	21	1.3−0.5	Jones and Watson (1989)
Her X-1	40.8	...	Tananbaum *et al.* (1972)
Cyg X-2	235	...	Vrtilek *et al.* (1986b)

state where the medium-Z elements are fully ionized. This will reduce the absorption cross-sections at low energies and allow photons to leak through. In X1755−338, the maximum radius for ionization is $0.15R_\odot$ (Mason, Parmar and White 1985), which is significantly smaller than the expected size of the accretion disk of $\sim 0.5R_\odot$. Frank, King and Lasota (1987) consider different locations for the absorbing material. Lubow and Shu (1975) suggest that a significant fraction of the gas stream flows over the accretion disk in a ballistic trajectory before circularizing at a distance of $\sim 0.15R_\odot$ from the central X-ray source. This is significantly smaller than the size of the disk, and results in a ring-like thickening at this radius. Material in the gas stream strikes this ring – causing a bulge that extends about half-way around the disk between orbital phase 0.3 and 0.8. The low-energy excess discussed earlier could be the result of scattering off these hot clouds (Frank, King and Lasota 1987).

Sztajno and Frank (1984) suggested that the energy independence results from partial covering of an extended ADC. Church and Balucinska-Church (1993) reanalyzed the *EXOSAT* spectra of X1755−338 and found out that the invariant X-ray spectrum observed during dips could be explained as the result of viewing a two-component source. They showed that it is possible to arrange the relative fluxes and shapes of two spectral components such that the overall spectrum changes only slightly during shallow dips. One component is extended, the other compact, with the dips caused by only the compact component being absorbed. A similar two-component spectrum was seen much more clearly in X1624−490 by Watson *et al.* (1985), although the dips were clearly seen to be energy dependent in that case.

1.6.1.2   Line emission from LMXBs

Iron K line emission at 6.7 keV was reported from Sco X-1 and other related LMXBs using proportional counters on *OSO 8* (Swank and Serlemitsos 1985). The line emission was confirmed using *EXOSAT* and *Tenma* gas scintillation proportional counter, GSPC, measurements with energy resolution better by a factor of two (Suzuki *et al.* 1984; White, Peacock and Taylor 1985; White *et al.* 1986; Hirano *et al.* 1987). The line emission in most cases was relatively weak, with equivalent widths of order 50–100 eV. There are a couple of notable exceptions, such as GX9+1 (White *et al.* 1986) and X1755−338 (White *et al.* 1984), where upper limits of ~ 10 eV were measured.

The *EXOSAT* GSPC detector resolved the emission to be broad with a full width half maximum, FWHM, of order 0.5–1 keV, or about 10% of the line energy. The *Tenma* results by Hirano *et al.* (1987) and Suzuki *et al.* (1984) confirmed the width of the line for some LMXBs. Makishima (1986) points out that complexities in the underlying continuum might lead to an apparent broadening. Recent results from the *BBXRT* mission, where a solid-state spectrometer measured the iron K line from Cyg X-2 with a resolution a factor of four better than that from a GSPC, confirm the line width (Smale *et al.* 1993a). The ADC source X1822−371 shows a stronger line, with an equivalent width of 270 eV and a FWHM of 1 keV (Hellier and Mason 1989). During the eclipse, the line flux remains constant, causing an increase in the line equivalent width to 360 eV, and suggesting that the bulk of the emission is confined to a limited part of the ADC.

The spectra of many LMXBs also show line emission at lower energies between

0.6 and 1.1 keV from the iron L shell and the K shell of oxygen, nitrogen and other medium-Z elements (Kahn, Seward and Chlebowski 1984; Vrtilek *et al.* 1986a,b, 1991). The low energy line emission was first detected in *Einstein* objective grating spectrometer, OGS, observations of Sco X-1 by Kahn, Seward and Chlebowski (1984). These and other OGS observations of a variety of LMXBs have resolved many narrow emission lines between 0.7 and 1.5 keV (Vrtilek *et al.* 1986a,b, 1991). These lines have also been detected with the *Einstein* solid-state spectrometer, SSS (Vrtilek, Swank and Kallman 1988; Christian 1993; Christian, White and Swank 1994), and the *EXOSAT* transmission grating spectrometer, TGS (Brinkman *et al.* 1985; Van der Woerd, White and Kahn 1989; Barr and Van der Woerd 1990).

1.6.1.3 Accretion disk coronae

The X-ray line emission from LMXBs most likely arises in a photo-ionized ADC (Kahn, Seward and Chlebowski 1984; Kallman and White 1989; Melia, Zylstra and Fryxell 1991; Liedahl *et al.* 1992; Vrtilek, Soker and Raymond 1993). In these systems, only emission scattered in the ADC is seen. The fact that the line equivalent width in ADC systems is higher than that from other LMXB, (Hellier and Mason 1989) where the central source is seen, confirms that the ADC is an important source of line emission. Fabian, Guilbert and Ross (1982) simulated the spectrum expected from a photo-ionized accretion disk corona and found qualitative agreement with the X1822−371 observations. Liedahl *et al.* (1992) have modeled the L-shell emission from a photo-ionized plasma, and found that the conditions in an ADC are ideal to produce these lines.

The structure of an ADC is very uncertain, and the line emission can be used to constrain the various models (London 1982; Begelman, McKee and Shields 1983; Kallman and White 1989). Most ADC models assume hydrostatic equilibrium in the inner region. In the outer region, where the thermal energy exceeds the local gravity, a wind is driven from the disk. The optical depths in the inner region could build up to be greater than unity, but this is probably a self-limiting process since high optical depths would prevent photons reaching the disk to maintain the ADC. There are several uncertain issues involved in modeling the ADC, including:

- a self-consistent treatment of the optically thick region of the corona and its shadowing effect on the outer part of the disk,
- the effects of heating on the disk surface by scattered X-rays, and
- cooling of the corona by inverse Compton scattering.

In spite of these uncertainties, the observed line equivalent widths and 6.7 keV energies are consistent with what might be expected from recombination of He-like ions of iron in an accretion disk corona of moderate optical depth (Hirano *et al.* 1987). The origin of the observed line width is less clear, but could be caused by Keplerian rotation of the accretion disk, Comptonization in the ADC or the blending of several individual line components. Either the iron K lines come from very close to the compact object, and rotational broadening is dominant, or the optical depth of the corona is higher than expected for an X-ray heated ADC (Kallman and White 1989). Melia, Zylstra and Fryxell (1991) have made radiative hydrodynamical simulations of an accretion disk corona that indicate a volatile corona, where bulk motions of order a small percentage of the speed of light are present. Their simulations predict, in the innermost region, oscillations on the dynamical timescale of 10 Hz, and, in the outer regions, considerable mass loss from a wind.

Ponman, Foster and Ross (1990) considered an idealized model where the central X-ray source is embedded in a spherically symmetric optically thick cloud. Their model requires scattering depths of 2–10, which correlate with the overall source luminosity. At such high optical depths Compton scattering in the cloud causes substantial modification of the original continuum spectrum (see Sect. 1.5.2.1). The model predicts a significant iron absorption edge, which is not observed. Vrtilek, Soker and Raymond (1993) consider an ADC geometry with ionization, thermal and hydrostatic balance, along with the full radiative transfer. The predicted spectrum is strongly dependent on inclination angle. As with the Ponman *et al.* (1990) model, an iron absorption edge is predicted at 7.1 keV, but only for the high inclination, edge on, systems. The predicted equivalent widths seem consistent with the observations. They favor the interpretation of Makishima (1986) that the line widths are caused by incorrect modeling of the continuum.

1.6.2   Stellar winds

The O and B stars in HMXBs have substantial stellar winds that will modify the properties of the X-ray source. If the system is wind driven, then any inhomogeneities in the wind will directly translate to X-ray source variability. The X-ray emission from the compact object must propagate through the wind and will undergo absorption. Again, if the wind is inhomogeneous, then variations in absorption may occur. The HMXBs provide a useful laboratory to test models for radiatively driven stellar winds. In LMXBs the natural stellar wind from the companion is less important, although a radiation driven wind caused by the X-ray illumination of the companion may be present in some systems (see Tavani and London 1993 and references therein). It has been suggested that in some HMXBs X-ray illumination of the primary may also lead to a radiation driven wind (Day and Stevens 1993).

When the OB star does not come close to filling its Roche lobe, the X-ray properties are determined by the wind of the star and its interaction with the compact object. These tend to be lower luminosity systems, typically less than 10^{36} erg s^{-1}. In systems where the star is close to filling its critical Roche lobe, which tend to be the more luminous systems, the gas stream dominates the accretion flow and a disk forms around the neutron star. If the companion is out of synchronous rotation with the orbit, the gas stream will trail behind the compact object and escape the system (see Savonije 1983).

1.6.2.1   Time variability

The time variability of an HMXB depends on the average X-ray luminosity. The lower luminosity systems ($< 10^{37}$ erg s^{-1}) show erratic flaring activity, with, in some cases, luminosity variations up to a factor of 100 on a timescale of tens of minutes. In these lower luminosity systems, the X-ray source is driven by material captured from the wind (Conti 1978), and the observed flares reflect inhomogeneities in the wind (White, Kallman and Swank 1983). These inhomogeneities may be a

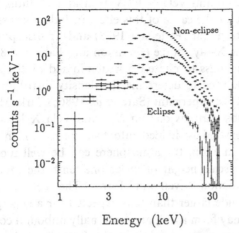

Fig. 1.10. *Ginga* pulse-averaged spectra from Vela X-1 from a variety of absorption states and orbital phases.

common feature of all radiation driven stellar winds, or be found only in HMXBs because of the disrupting effect of the X-ray source (Haberl, White and Kallman 1989).

The more luminous systems (10^{37}–10^{38} erg s^{-1}) are much less variable. The luminosities of these systems are one or two orders of magnitude higher than can be accounted for by stellar wind capture (Conti 1978). This suggests the presence of some form of enhanced flow to the neutron star. This flow will carry sufficient angular momentum to cause a stable accretion disk to form which mediates the flow and smooths out any inhomogeneities.

1.6.2.2 Absorption properties

The absorption of soft X-rays by circumstellar matter is most prominent in the spectra of wind-fed X-ray pulsars, such as Vela X-1 and GX301−2. The amount of the absorption column density is variable and can range up to $N_H \sim 10^{23}$–10^{24} cm^{-2}. Figure 1.10 illustrates a variety of spectra from Vela X-1 taken at different orbital phases and absorption states. The unabsorbed spectra are featureless compared with those taken at other times, when a strong iron K absorption edge and emission line can be seen around 6–7 keV. During eclipse, the continuum shape is similar to the uneclipsed spectrum, but with a strong emission line.

OB star winds are driven by the transfer of momentum from the supergiant's radiation field to the wind by scattering of radiation in UV spectral lines (Lucy and Solomon 1970). The absorption of the bright X-ray source provides a probe into the outer atmosphere of the OB star and its vicinity. The observations of several systems show a smooth decrease in absorption after eclipse egress, to a minimum around phase 0.5. Around phase 0.6 there is a sharp increase which persists through to eclipse ingress, where the absorption is twice that at egress. This asymmetry in absorption either side of phase 0.5 has been known for over ten years, from X1700−377 (Mason, Branduardi and Sanford 1976a; Haberl, White and Kallman

1989), Cen X-3 (e.g. Pounds *et al.* 1976) and Vela X-1 (Watson and Griffiths 1977; Haberl and White 1990). Suggestions for the cause of this effect include a gas stream from the supergiant trailing the neutron star (Petterson 1978) and/or disruption of the radiatively accelerated wind by the X-ray source (Fransson and Fabian 1980).

The X-ray absorption from eclipse egress to phase 0.5 is dominated by the atmosphere of the companion, and can be used to determine the density distribution of the wind and the outer atmosphere of the supergiant (Sato *et al.* 1986; Clark, Minato and Mi 1988; Lewis *et al.* 1992). Observations of Cen X-3 and Vela X-1 indicate that the radial density profile of the wind is divided into two zones. In an inner zone with a radius less than 1.5 stellar radii, the atmosphere can be well modeled as an exponential atmosphere with a scale height of order one-tenth the radius of the underlying star. Above this a radiation driven wind forms. The scale height of the atmosphere is an order of magnitude larger than that expected for a supergiant. X-ray irradiation of the atmosphere may form a hot, gravitationally unbound coronal region, giving rise to a thermally driven stellar wind from the X-ray heated face of the companion (Day and Stevens 1993).

For a purely wind driven system, the observed and predicted X-ray luminosity should be consistent with the observed and predicted absorption. A simultaneous fit to the flare averaged luminosity and the X-ray absorption of Vela X-1 does indeed give good agreement (Haberl, White and Kallman 1989; Haberl and White 1990). An additional absorption component is required to account for the sudden absorption increase at orbital phase 0.6. This can be modeled as due to a gas stream from the supergiant trailing behind the X-ray source (Haberl, White and Kallman 1989; Haberl and White 1990). The presence of a gas stream is also suggested from earlier optical absorption line observations by Fahlman and Walker (1980). Two-dimensional gas-dynamical simulations by Blondin, Stevens and Kallman (1991) show that a tidal stream will be drawn off the primary if it is close to filling its Roche lobe. This tidal stream will evolve into full-blown Roche lobe overflow when the surface of the primary approaches its critical radius. The tidal stream will be deflected behind the X-ray source, and will cause the observed sharp increase in absorption at orbital phase 0.6.

The X-ray source in the wind will ionize a cavity, within which the radiative acceleration process is inhibited because the UV line transitions which drive the wind acceleration are destroyed. This introduces complications since the magnitude of the disruption of the acceleration process must be included in any modeling. For wind driven systems this is a relatively small effect, although it must be included in the modeling (Haberl, White and Kallman 1989). The luminosity of Cen X-3 in the high state is such that the wind not eclipsed by the supergiant will be photo-ionized. The low-state spectra reported by Schreier *et al.* (1976) indicate higher absorption and much lower-amplitude pulsations. The high–low state transitions of Cen X-3 were explained as being due to the X-ray source becoming immersed in a strong stellar wind that strongly absorbs (snuffs out) the X-ray source (Schreier *et al.* 1976; Hatchett and McCray 1977). Observations of the low state by Day (1988), however, show little absorption, in contrast to the high absorption that might be expected. The lack of any pulsations rules out that the low state is caused by a reduced accretion rate. The spectrum does show an iron line with an equivalent width an order of magnitude larger (550 eV) than that seen in the high state. This suggests that the central X-ray source is hidden from view, probably by a thick accretion disk, with X-rays scattered to the observer either via the wind or by a corona above the disk.

1.6.2.3 Fluorescent iron lines

An iron K absorption edge at 7.1 keV was first resolved from the X-ray pulsar GX301−2 by Swank *et al.* (1976) using *OSO 8* data. An iron K emission line was reported by Pravdo *et al.* (1977, 1978) in the spectrum of the pulsar Her X-1 (an LMXB). Thereafter, iron K emission line features were detected with *OSO 8* and *HEAO 1* from many other X-ray pulsars, including Vela X-1, X0115−737, X1538−522, Cen X-3, GX301−2 and X1626−673 (Becker *et al.* 1978; Pravdo 1979; Rose *et al.* 1979; White and Pravdo 1979; White *et al.* 1980; White, Swank and Holt 1983; White and Swank 1984).

The GSPC on *Tenma* measured the line and edge energies with sufficient precision to identify the ionization state of the emitting material. Data from Vela X-1 give a line energy of 6.42 ± 0.02 and an absorption edge at 7.24 ± 0.03 keV (Ohashi *et al.* 1984; Nagase *et al.* 1986; Sato *et al.* 1986). A study of GX301−2 gives a similarly precise line energy of 6.46 ± 0.05 keV and edge at 7.36 ± 0.05 keV (Leahy *et al.* 1989a,b). These line energies constrain the iron ionization state to be I–XIX, and the absorption edge energy gives an ionization state of V–X (Nagase 1989). The widths of the emission lines are less than the resolution of a GSPC (i.e. 0.5 keV FWHM at 6.4 keV). Observations of other pulsars confirm this picture (see, e.g., Nagase 1989), although as usual there is the occasional exception. For example, *EXOSAT* observations of the transient X-ray pulsar EXO2030+375 show a broad iron emission line centered at 6.7 keV (Reynolds, Parmar and White 1993).

The observed iron line equivalent width varies significantly from source to source and from observation to observation. The N_H and EW measured for GX301−2, Her X-1 and Vela X-1 are shown in Figure 1.11 (taken from Makishima 1986). There is a strong correlation between the two when $N_H > 10^{23}$ cm^{-2}; below this N_H, the equivalent width stays at a minimum value. This correlation suggests that the line is the result of fluorescence in the absorbing medium (Ohashi *et al.* 1984; White and Swank 1984). The observed relationship depends critically on the geometry of the fluorescing material. In Figure 1.11, four different geometries are shown. Type I applies when the X-ray source is in front, type II when the X-ray source is surrounded, type III when the source is surrounded but is also eclipsed by the companion, and type IV when a blob of material passes in front of the X-ray source. The intensity of the fluorescence iron line, I_{Fe}, is for a type II spherical geometry

$$
\begin{aligned}
I_{Fe} &= \int_{E_K}^{\infty} \eta_K \frac{\sigma_{Fe}(E) N_{Fe}}{\tau_o} f(E)[1 - \exp(-\tau_o)] dE \\
&\simeq \eta_K N_{Fe} \int_{E_K}^{\infty} \sigma_{Fe} f(E) dE \quad \text{for } \tau_o \ll 1,
\end{aligned}
$$

where E_K is the K-edge energy of iron, η_K is the fluorescence yield of iron ($\eta_K = 0.34$), τ_o is the optical depth, $f(E)$ is the input continuum spectrum and σ_{Fe} is the photoelectric absorption cross-section of iron. The equivalent width, EW, of the

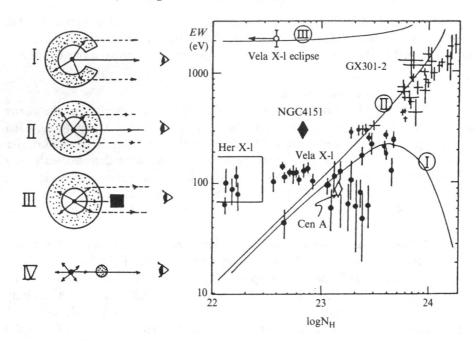

Fig. 1.11. The observed iron K line equivalent width as a function of the absorption column density for GX301−2, Vela X-1 and Her X-1. The solid lines indicate the predicted fluorescent equivalent width for the various geometries described in the text. Taken from Makishima (1986).

fluorescence iron line is given by

$$EW = \frac{I_{Fe}}{f(E_l)\exp[-\sigma(E_l)N_H] + f'(E_l)},$$

where $f'(E_l)$ is the continuum scattered by the surrounding matter (Inoue 1985) and E_l is the line energy of 6.4 keV. In the N_H range $\simeq 10^{22}$–10^{24} cm^{-2}, one finds $EW \simeq 100(N_H/10^{23})$ eV (Inoue 1985). Figure 1.11 shows that this relationship is in good agreement with the observations for $N_H > 10^{23}$ cm^{-2}. During eclipse, only scattered X-rays and fluorescent emissions from the stellar wind are visible (type III). The equivalent width is calculated to be $EW = I_{Fe}/f'(E_l)$, which gives $EW = 1$–2 keV, consistent with the observations. At low N_H, the EW is constant with varying N_H, which suggests type I behavior, where the absorbing material subtends a small solid angle to the X-ray source, i.e. a gas stream, accretion disk or blobs in the wind (see Inoue 1985; Makishima 1986; Nagase 1989; Haberl and White 1990).

There is a dependence of the iron line intensity on orbital phase in Vela X-1 (Sato *et al.* 1986). Part of this can be explained as being due to expected variations in the intensity of the scattered/fluorescent emission with orbital phase. But some of the line emission appears to be eclipsed, which indicates an additional component of fluorescent emission in the vicinity of the pulsar. This may be from stagnant matter in the magnetosphere shell (Inoue 1985). The absence of a pulsed modulation of the iron line intensity gives another clue to the distribution of the reprocessing material

(Ohashi *et al.* 1984; Leahy and Matsuoka 1990). This requires that the reprocessing material covers a fairly large solid angle so the pulsed modulation is smeared out, or that the reprocessing matter corotates with the pulsar. Day *et al.* (1993) found a weak pulsation of the iron line intensity in Cen X-3 from *Ginga* data. As the amplitude of the iron line pulse is relatively small ($\sim$ 30 % at peak-to-minimum amplitude) compared with the amplitude of the continuum pulse, this implies the coexistence of a local reprocessing site that produces a modulation and a spherically (or axially symmetrically) extended reprocessing site that smears the iron line pulsations.

Recent observations with *Ginga* during low states, eclipse phases or dip phases of X-ray pulsars show the iron lines then become more complex. There is evidence for the coexistence of a 6.4 keV and a 6.7 keV line in the dip and eclipse spectra of Cen X-3 (Nagase *et al.* 1992). The orbital phase dependence of the two line intensities indicates that the 6.4 keV line originates from a local region smaller than the companion radius and that the 6.7 keV line comes from a large region which extends over the companion radius. The eclipse spectrum of Cen X-3 shows evidence for a line at 8.4 keV, which may be from the fluorescent Kβ line of iron and/or the Kα line of nickel. Evidence for the coexistence of a 6.7 keV line with a 6.4 keV line is found in the low-state spectrum of Her X-1 (Mihara *et al.* 1991b). Emission from an accretion disk corona may be responsible for this 6.7 keV line emission.

1.6.2.4 Soft excess

A soft excess below 5 keV can be seen in the Vela X-1 spectra shown in Figure 1.10 during high-absorption phases such as X-ray dips, eclipse transitions and low-intensity states (Nagase *et al.* 1986; Haberl and White 1990). A soft excess has also been observed under similar circumstances from X1700$-$377 (Haberl, White and Kallman 1989), GX301$-$2 (Haberl 1991) and Cen X-3 (Nagase *et al.* 1992). This low-energy excess is not pulsed (Haberl and White 1990). There are two causes of the soft excess. During later (> 0.5) orbital phases and during absorption dips, when the absorption is very high, the soft excess originates in scattering of X-rays by the ambient wind around an obscuring gas stream. At earlier phases it is caused by partial ionization of the absorbing wind.

The non-detection of pulsations in the soft excess when the absorption is high led Haberl and White (1990) to the conclusion that it is caused by scattering in the wind, as opposed to partial covering or partial ionization of the absorbing medium. Lewis *et al.* (1992) used a Monte Carlo scattering code to model the spectra at different orbital phases, and confirmed this. Their simulations also demonstrated that scattering and fluorescence in the stellar wind can account for the eclipse spectrum.

At orbital phases before phase 0.5, the soft excess is a much weaker effect. In this case there is a correlation between the soft excess and the overall source luminosity in spectra of X1700$-$377, suggesting the increase of partial ionization with luminosity (Haberl, White and Kallman 1989). A photo-ionized zone will form around the X-ray source (e.g. Hatchett and McCray 1977; Friend and Castor 1982; McCray *et al.* 1984), which reduces the X-ray opacity at low energies (Krolik and Kallman 1984). A similar effect is also evident from Vela X-1 (Haberl and White 1990). The spectra during these intervals can be modeled by reduced opacities caused by partial ionization of the absorbing medium (Haberl, White and Kallman 1989; Haberl and White 1990).

If the X-ray source is located behind a high interstellar absorption column, scattering by interstellar dust grains will produce a soft X-ray halo. Evidence for dust scattering was detected from the *EXOSAT* observations of Cen X-3 in the X-ray light curve during the eclipse (Day 1988). The spectrum of X1538−522 during the eclipse shows that dust scattering can contribute significantly to the soft excess at very low energies during these times (Clark, Woo and Nagase 1994). But the contribution of the dust scattering low-energy excess is relatively minor compared with the low-energy excess caused by scattering in the wind of the companion.

1.6.3   Cyg X-3: one of a few abnormal cases

We end this chapter with a discussion of the puzzling X-ray binary Cyg X-3. This source is one of a few X-ray binaries that do not fit into the well established classes, and appears to be undergoing a cataclysmic accretion process. These exceptional X-ray sources perhaps present the most interesting challenges. Two other exceptional binary X-ray sources are Cir X-1 (Sect. 1.2.2.4) and SS433 (Sect. 1.4). These three objects may all be examples of super-Eddington accretion, and they represent an extreme of stellar evolution.

Cyg X-3 is a 4.8 hr period system that shows an asymmetric sinusoidal like modulation. It is in a reddened region of the galaxy, and no optical counterpart has been identified. Up until recently, Cyg X-3 had been assumed to be an LMXB, because its orbital period is similar to that of other LMXBs. The 4.8 hr modulation is also present in the infrared (Becklin *et al.* 1972; Mason, Cordova and White 1986) with a similar light curve to the X-ray band. Infrared spectroscopy has recently shown that the companion is probably a Wolf–Rayet star (Van Kerkwijk *et al.* 1992). Cyg X-3 shows giant radio outbursts, with evidence for jet-like emission expanding at one-third the speed of light (Gregory *et al.* 1972; Geldzahler *et al.* 1983).

The 4.8 hr orbital period of Cyg X-3 is *increasing* on a relatively short timescale of 4.5×10^5 yr (Kitamoto *et al.* 1987; Molnar 1988; Van der Klis and Bonnet-Bidaud 1989). This is the opposite of what might be expected if the mass transfer is driven by Roche lobe overflow (unless the companion is degenerate; see Molnar 1988). The timescale is consistent with a wind that has a mass loss rate of $\sim 10^{-6}$ yr^{-1} and carries away orbital angular momentum (Van Kerkwijk *et al.* 1992).

Cyg X-3 was the first X-ray binary to reveal strong iron K line emission (Serlemitsos *et al.* 1975). An absorption edge at around 9 keV was reported from *EXOSAT* observations by Willingale, King and Pounds (1985), and was confirmed by *Tenma* and *Ginga* observations (Kitamoto *et al.* 1987; Nakamura *et al.* 1993). The spectrum is complex, and, as the sensitivity of X-ray detectors has improved, the number of spectral components required to give a good fit has increased to include: (i) an absorbed power-law, with an exponential cut-off at high energies; (ii) an absorbed blackbody component; (iii) a broad iron emission line; (iv) a low-energy excess caused by dust scattering in the interstellar medium; and (v) an absorption edge at 9 keV. The X-ray source undergoes high and low states, where the relative strength of the various components changes, with no obvious correlations. A recent result from *BBXRT* by Smale *et al.* (1993b) using a solid state detector failed to detect any iron K line emission, with upper limits well below those reported using proportional counters.

Many models have been proposed to explain the X-ray light curve, with varying degrees of success. The two most discussed are a stellar wind scattering model (Davidson and Ostriker 1974; Pringle 1974), and an accretion disk corona model (White and Holt 1982). In the wind scattering model, the X-ray source is surrounded by an optically thick scattering cloud, centered on the mass-donating star. As the X-ray source is offset from the center of the cloud, the resulting asymmetry causes the modulation with the binary period. The asymmetry in the light curve may be caused by an accretion wake (Willingale, King and Pounds 1985). The wind model predicts an increase in absorption at the minimum of the light curve, which is not observed. In the ADC model, the modulation is caused by the partial eclipse of the ADC by the edge of the disk, with the asymmetries at the edge of the disk causing the modulation. This is an adaptation of the model proposed to explain the light curve of the ADC source X1822-371 (White and Holt 1982). The recent identification of a Wolf–Rayet star in this system favors the stellar wind model. However, there may well also be an accretion disk present that has a strong influence. Cyg X-3 seems to be an extreme case where the environment has almost completely enshrouded the central X-ray source.

Acknowledgements
We thank all our colleagues for their direct and indirect contributions to this chapter. In particular, we thank Lorella Angelini, Charles Day and Alan Smale for their comments on an earlier version of the manuscript.

References
Alexander, S.G. and Mészáros, P. 1991, *Ap. J.*, **372**, 565.
Arons, J. and Lea, S.M. 1980, *Ap. J.*, **235**, 1016.
Asai, K. *et al.* 1993, *Pub. Astr. Soc. Japan*, **44**, 633.
Aurière, M., Le Fèvre, O. and Terzan, A. 1984, *Astr. Ap.*, **138**, 415.
Aurière, M., Maucherat, A., Corddoni, J.-P., Fort, B. and Picat, J.P. 1986, *Astr. Ap.*, **158**, 158.
Bahcall, J.N., Joss, P. and Avni, Y. 1974, *Ap. J.*, **191**, 211.
Bailyn, C.D. 1991, *IAU Circ. No.* 5259.
Bailyn, C.D. and Grindlay, J.E. 1987, *Ap. J. (Letters)*, **316**, L25.
Barr, P. and Van der Woerd, H. 1990, *Ap. J.*, **352**, L41.
Basko, M.M. 1976, *Astrofizika*, **12**, 273; English translation in *Astrophysics*, **12**, 169.
Basko, M.M. and Sunyaev, R.A. 1975, *Astr. Ap.*, **42**, 311.
Basko, M.M. and Sunyaev, R.A. 1976, *M.N.R.A.S.*, **175**, 395.
Baykal, A. and Ögelman, H. 1993, *Astr. Ap.*, **267**, 119.
Becker, R.H. *et al.* 1977, *Ap. J. (Letters)*, **216**, L11.
Becker, R.H. *et al.* 1978, *Ap. J.*, **221**, 912.
Becklin, E.E. *et al.* 1972, *Nature Phys. Sci.*, **239**, 134.
Begelman, M.C., McKee, C.F. and Shields, G.A. 1983, *Ap. J.*, **271**, 70.
Bisnovatyi-Kogan, G.S., Mersov, G.A. and Sheffer, E.K. 1990, *Soviet Astr.*, **34**, 44.
Blondin, J.M, Stevens, I.R. and Kallman, T.R. 1991, *Ap. J.*, **371**, 684.
Bonazzola, S., Heyvarts, J. and Puget, J.L. 1979, *Astr. Ap.*, **78**, 53.
Boynton, P.E., Crosa, L.M. and Deeter, J.E. 1980, *Ap. J.*, **237**, 169.
Boynton, P.E. *et al.* 1984, *Ap. J. (Letters)*, **283**, L53.
Bradt, H.V.D. and McClintock, J.E. 1983, *Ann. Rev. Astr. Ap.*, **21**, 13.
Brainerd, J.J. and Mészáros, P. 1991, *Ap. J.*, **369**, 179.

Branduardi, G., Kylafis, N.D., Lamb, D.Q. and Mason, K.O. 1980, *Ap. J. (Letters)*, **235**, L153.
Branduardi-Raymont, G. *et al.* 1983, *M.N.R.A.S.*, **205**, 403.
Brinkman, A.C. *et al.* 1985, *Space Sci. Rev.*, **40**, 201.
Brown, R.L. and Gould, R.J. 1970, *Phys. Rev. D.*, 1, 2252.
Bulik, T., Mészáros, P., Woo, J., Nagase, F. and Makishima, K. 1992, *Ap. J.*, **395**, 564.
Callanan, P.J., Charles, P.A., Honey, W.B. and Thorstensen, J.R. 1992, *M.N.R.A.S.*, **259**, 395.
Callanan, P.J., Machin, G., Naylor, T. and Charles, P.A. 1989, *M.N.R.A.S.*, **241**, 37P.
Callanan, P.J. *et al.* 1987, *M.N.R.A.S.*, **224**, 781.
Canizares, C.R., McClintock, J.E. and Grindlay, J.E. 1980, *Ap. J. (Letters)*, **236**, L55.
Canuto, V., Lodenquai, J. and Ruderman, M. 1971, *Phys. Rev. D*, 3, 2303.
Carlini, A. and Treves, A. 1989, *Astr. Ap.*, **215**, 283.
Casares, J. and Charles, P. A. 1992, *M.N.R.A.S.*, **255**, 7.
Casares, J., Charles, P. A. and Naylor, T. 1992, *Nature*, **355**, 614.
Chakrabarty, D. *et al.* 1993, *Ap. J. (Letters)*, **403**, L33.
Charles, P.A., Jones, D.C. and Naylor, T. 1986, *Nature*, **323**, 417.
Charles, P.A. *et al.* 1983, *M.N.R.A.S.*, **202**, 657.
Charles, P.A. *et al.* 1991, *M.N.R.A.S.*, **249**, 567.
Chevalier, C. and Ilovaisky, S.A. 1977 *Astr. Ap.*, **59**, L9.
Chevalier, C. and Ilovaisky, S.A. 1982, *Astr. Ap.*, **112**, 68.
Chevalier, C. and Ilovaisky, S.A. 1990, *Astr. Ap.*, **238**, 163.
Chevalier, C. and Ilovaisky, S.A. 1991, *Astr. Ap.*, **251**, L11.
Chevalier, C. *et al.* 1989a, *Astr. Ap.*, **217**, 108.
Chevalier, C. *et al.* 1989b, *Astr. Ap.*, **210**, 114.
Christian, D. 1993, *Ph.D. thesis, Univ. of Maryland.*
Christian, D., White, N.E. and Swank, J.H. 1994, *Ap. J.* 4, **422**, 791.
Church, M. J. and Balucinska-Church, M. 1993, *M.N.R.A.S.*, **260**, 59..
Clark, G.W., Minato, J.R. and Mi, G. 1988, *Ap. J.*, **324**, 974.
Clark, G.W., Woo, J. and Nagase, F. 1994, *Ap. J.*, **422**, 336.
Clark, G.W. *et al.* 1990, *Ap. J.*, **353**, 274.
Coe, M.J. and Jones, L.R. 1992, *M.N.R.A.S.*, **259**, 191.
Cominsky, L., Clark, G.W., Li,F., Mayer, W. and Rappaport, S. 1978, *Nature*, **273**, 367.
Cominsky, L. and Wood, K. 1984, *Ap. J.*, **283**, 765.
Cominsky, L. and Wood, K. 1989, *Ap. J.*, **337**, 485.
Connor, J.P., Evans, W.D. and Belian, R.D. 1969, *Ap. J. (Letters)*, **157**, L157.
Conti, P.S., 1978, *Astr. Ap.*, **63**, 225.
Corbet, R.H.D. 1984, *Astr. Ap.*, **141**, 91.
Corbet, R.H.D. and Day, C.S.R. 1990, *M.N.R.A.S.*, **243**, 553.
Corbet, R.H.D. *et al.* 1986, *M.N.R.A.S.*, **222**, 15P.
Courvoisier, T. J-L., Parmar, A. N., Peacock, A. and Pakull, M. 1986, *Ap. J.*, **309**, 265.
Cowley, A.P. and Crampton, D. 1975, *Ap. J. (Letters)*, **201**, L65.
Cowley, A.P., Crampton, D. and Hutchings, J.B. 1979, *Ap. J.*, **231**, 539.
Cowley, A.P., Crampton, D. and Hutchings, J.B. 1982, *Ap. J.*, **256**, 605.
Cowley, A.P., Schmidtke, P.C., Crampton, D. and Hutchings, J.B. 1990, *Ap. J.*, **350**, 288.
Cowley, A.P. *et al.* 1983, *Ap. J.*, **272**, 118.
Cowley, A.P. *et al.* 1984, *Ap. J.*, **286**, 196.
Cowley, A.P. *et al.* 1988, *Astr. J*, **95** (1231).
Cowley, A.P. *et al.* 1991, *Ap. J.*, **381**, 533.
Crampton, D., Cowley, A.P. and Hutchings, J.B. 1980, *Ap. J. (Letters)*, **235**, L131.
Crampton, D., Cowley, A. P., Stauffer, J., Ianna, P. and Hutchings, J. B. 1986, *Ap. J.*, **306**, 599.
Crampton, D. *et al.* 1985, *Ap. J.*, **299**, 839.
Crosa, L. and Boynton, P. E., 1980, *Ap. J.*, **235**, 999.
Czerny, B., Czerny, M. and Grindlay, J.E. 1986, *Ap. J.*, **311**, 241.
Daugherty, J.K. and Harding, A.K. 1986, *Ap. J.*, **309**, 362.
Davidson, K. and Ostriker, J.P. 1973, *Ap. J.*, **179**, 585.
Davidson, K. and Ostriker, J.P. 1974, *Ap. J.*, **189**, 331.
Davies, R.E. and Pringle, J.E. 1980, *M.N.R.A.S.*, **191**, 599.
Davies, S.R. and Coe, M.J. 1991, *M.N.R.A.S.*, **249**, 313.

Davies, S.R. *et al.* 1990, *M.N.R.A.S.*, **245**, 313.

Davison, P.J.N. 1977, *M.N.R.A.S.*, **179**, 35P.

Davison, P.J.N., Watson, M.G. and Pye, J. 1977, *M.N.R.A.S.*, **181**, 73P.

Day, C.S.R. 1988, Ph.D. thesis, Univ. of Cambridge.

Day, C.S.R., Fabian, A.C. and Ross, R. 1992, *M.N.R.A.S.*, **257**, 471.

Day, C.S.R., Nagase, F., Asai, K. and Takeshima, T. 1993, *Ap. J.*, **408**, L122.

Day, C.S.R. and Stevens, I.R. 1993, *Ap. J.*, **403**, 322.

Day, C.S.R., Tennant, A.F. and Fabian, A.C. 1988, *M.N.R.A.S.*, **231**, 69.

Deeter, J.E., Boynton, P.E., Lamb, F.K. and Zylstra, G. 1987, *Ap. J.*, **314**, 634.

Deeter, J.E. *et al.* 1991, *Ap. J.*, **383**, 324.

Delgado, A.J., Schmidt, H.U. and Thomas, H.-C. 1983, *Astr. Ap.*, **127**, L15.

Dotani, T. *et al.* 1989, *Pub. Astr. Soc. Japan*, **41**, 427.

Dotani, T. *et al.* 1990, *Nature*, **347**, 534.

Elsner, R.F. and Lamb, F.K. 1976, *Nature*, **262**, 356.

Elsner, R.F. and Lamb, F.K. 1977, *Ap. J.*, **215**, 897.

Fabian, A.C., Guilbert, P.W. and Ross, R.R. 1982, *M.N.R.A.S.*, **199**, 1045.

Fabian, A.C., Pringle, J.E. and Rees, M.J. 1975, *M.N.R.A.S.*, **172**, 15P.

Fahlman, G.G. and Walker, G.A.H. 1980, *Ap. J.*, **240**, 169.

Frank, J., King, A. R. and Lasota, J-P, 1987, *Astr. Ap.*, **178**, 137.

Fransson, C. and Fabian, A.C. 1980, *Astr. Ap.*, **87**, 102.

Friend, D.B. and Castor, J.I. 1982, *Ap. J.*, **261**, 293.

Garcia, M.R. and Grindlay, J.E. 1987, *Ap. J. (Letters)*, **313**, L59.

Garcia, M.R. and Grindlay, J.E. 1992, *IAU Circ. No. 5578*.

Geldzahler, B.J. *et al.* 1983, *Ap. J.*, **273**, L65.

Gerend, D. and Boynton, P. 1976, *Ap. J.*, **209**, 562.

Ghosh, P. and Lamb, F.K. 1978, *Ap. J. (Letters)*, **223**, L83.

Ghosh, P. and Lamb, F.K. 1979a, *Ap. J.*, **232**, 259.

Ghosh, P. and Lamb, F.K. 1979b, *Ap. J.*, **234**, 296.

Giacconi, R. *et al.* 1962, *Phys. Rev. Lett.*, 9, 439.

Giacconi, R. *et al.* 1971, *Ap. J. (Letters)*, **167**, L67.

Giacconi, R. *et al.* 1973, *Ap. J.*, **184**, 227.

Gottlieb, E.W., Wright, E.L. and Liller, W. 1975, *Ap. J. (Letters)*, **195**, L33.

Grebenev, S.A. and Sunyaev, R. 1991, *IAU Circ. No. 5294*.

Gregory, P.C. and Fahlman, G.G. 1980, *Nature*, **287**, 805.

Gregory, P.C. *et al.* 1972, *Nature Phys. Sci.*, **239**, 114.

Greiner, J., Hasinger, G. and Kahabka, P. 1991, *Astr. Ap.*, **246**, L17.

Grindlay, J.E. *et al.* 1988, *Ap. J. (Letters)*, **334**, L25.

Gruber, D.E. *et al.* 1980, *Ap. J. (Letters)*, **240**, l127.

Gursky, H. *et al.* 1966, *Ap. J.*, **146**, 310.

Haberl, F. 1991, *Ap. J.*, **376**, 245.

Haberl, F. and White, N. E. 1990, *Ap. J.*, **361**, 225.

Haberl, F., White, N.E. and Kallman, T.R. 1989, *Ap. J.*, **343**, 409.

Hall, R. and Davelaar, J. 1983, *IAU Circ. No. 3872*.

Hamada, T. 1980, *Pub. Astr. Soc. Japan*, **32**, 117.

Harding, A.K. *et al.* 1984, *Ap. J.*, **278**, 369.

Hasinger, G. 1987, *Astr. Ap.*, **186**, 153.

Hasinger, G. and Van der Klis, M. 1989, *Astr. Ap.*, **225**, 79.

Hasinger, G. *et al.* 1990, *Astr. Ap.*, **235**, 131.

Hatchett, S. and McCray, R. 1977, *Ap. J.*, **211**, 552.

Heemskerk, M.H.M. and Van Paradijs, J. 1989, *Astr. Ap.*, **223**, 154.

Hellier, C. and Mason, K.O. 1989, *M.N.R.A.S.*, **239**, 715.

Hellier, C., Mason, K. O., Smale, A. P. and Kilkenny, D. 1990, *M.N.R.A.S.*, **244**, 39P.

Herold, H 1979, *Phys. Rev. D*, **19**, 2868.

Hertz, P. 1987, *Ap. J. (Letters)*, **315**, L119.

Hertz, P. and Wood, K. S. 1988, *Ap. J.*, **764**, 31.

Hirano, T. *et al.* 1987, *Pub. Astr. Soc. Japan*, **39**, 619.

Holt, S.S., Boldt, E.A., Serlemitsos, P.J. and Kaluzienski, L.J. 1976, *Ap. J. (Letters)*, **205**, L27.

Hoshi, R., and Mitsuda, K. 1991, *Pub. Astr. Soc. Japan,* **43**, 485.

Huckle, H.E. *et al.* 1977, *M.N.R.A.S.,* **180**, 21P.

Hutchings, J.B., Crampton, D. and Cowley, A.P. 1983, *Ap. J. (Letters),* **275**, L43.

Illarionov, A.F. and Sunyaev, R.A. 1975, *Astr. Ap.,* **39**, 185.

Ilovaisky, S.A. *et al.* 1987, *Astr. Ap.,* **179**, L1.

Ilovaisky, S.A. *et al.* 1993, *Astr. Ap.,* **270**, 139.

Inoue, H. 1975, *Pub. Astr. Soc. Japan,* **27**, 311.

Inoue, H. 1985, *Space Sci. Rev.,* **40**, 317.

Ives, J.C., Sanford, P.W. and Bell-Burnell, S.J., 1975, *Nature,* **254**, 578.

Iwasawa, K., Koyama, K. and Halpern, J.P. 1992, *Pub. Astr. Soc. Japan,* **44**, 9.

Johnston, M.D., Griffiths, R.E. and Ward, M.J. 1980, *Nature,* **285**, 26.

Jones, C.A. and Forman, W. 1976, *Ap. J.,* **209**, L131.

Jones, C., A., Forman, W. and Liller, W. 1973, *Ap. J. (Letters),* **182**, L109.

Jones, M. H. and Watson, M. G. 1989, in *Proc. of the 23rd ESLAB Symposium. Two Topics in X-ray Astronomy* (ESA SP-296), p. 439.

Joss, P. C. and Rappaport, S. A. 1979, *Astr. Ap.,* **71**, 217.

Kahn, S.M., Seward, F.D. and Chlebowski, T. 1984, *Ap. J.,* **283**, 286.

Kallman, T. and White, N.E. 1989, *Ap. J.,* **341**, 955.

Kaluzienski, L.J., Holt, S.S., Boldt, E.A. Serlemitsos, P.J. 1976, *Ap. J. (Letters),* **208**, L71.

Kaluzienski, L.J., Holt, S.S. and Swank, J.H. 1980, *Ap. J.,* **241**, 779.

Kaminker, A.D., Pavlov, G.G. and Shibanov, Yu.A. 1983, *Astrophys. Space Sci.,* **91**, 167.

Kaminker, A.D. *et al.* 1982, *Astrophys. Space Sci.,* **86**, 249.

Kanno, S. 1980, *Pub. Astr. Soc. Japan,* **32**, 105.

Katz, J.L. 1973, *Nature Phys. Sci.,* **246**, 87.

Katz, J.L. 1975, *Nature,* **253**, 698.

Kawai, N., Matsuoka, M., Pan, H.C. and Stewart, G.C. 1989, *Pub. Astr. Soc. Japan,* **41**, 491.

Kelley, R.L., Jernigan, J.G., Levine, A., Petro, L.D. and Rappaport, S., 1983a, *Ap. J.,* **264**, 568.

Kelley, R.L., Rappaport, S. and Ayasli, S. 1983b, *Ap. J.,* **274**, 765.

Kelley, R.L., Rappaport, S., Clark, G.W. and Petro, L.D. 1983c, *Ap. J.,* **268**, 790.

Kelley, R.L., Rappaport, S.A. and Petre, R. 1980, *Ap. J.,* **238**, 699.

Kelley, R.L. *et al.* 1981, *Ap. J.,* **243**, 251.

Kii, T. *et al.* 1986, *Pub. Astr. Soc. Japan,* **38**, 751.

King, A.R. 1991, *M.N.R.A.S.,* **250**, 3P.

Kitamoto, S., Miyamoto, S., Matsui, W. and Inoue, H. 1987, *Pub. Astr. Soc. Japan,* **39**, 259.

Koyama, K., Kunieda, H., Takeuchi, Y. and Tawara, Y., 1990b, *Pub. Astr. Soc. Japan,* **42**, L59.

Koyama, K., Kunieda, H., Takeuchi, Y. and Tawara, Y. 1991b, *Ap. J. (Letters),* **370**, L77.

Koyama, K. *et al.* 1990a, *Nature,* **343**, 148.

Koyama, K. *et al.* 1991a, *Ap. J. (Letters),* **366**, L19.

Kriss, G.A. *et al.* 1983, *Ap. J.,* **266**, 806.

Krolik, J.H. and Kallman, T.R. 1984, *Ap. J.,* **286**, 366.

Lamb, F.K., Pethick, C.J. and Pines, D. 1973, *Ap. J.,* **184**, 271.

Lamb, R.C., Markert, T., Hartman, R., Thompson, D. and Bignami, G.F. 1980, *Ap. J.,* **239**, 651.

Lang, F.L. *et al.* 1981, *Ap. J. (Letters),* **246**, L21.

Langer, S.H. and Rappaport, S. 1982, *Ap. J.,* **257**, 733.

Lea, S. 1976, *Ap. J. (Letters),* **209**, L69.

Leahy, D.A. 1991, *M.N.R.A.S.,* **251**, 203.

Leahy, D. A. and Matsuoka, M. 1990, *Ap. J.,* **355**, 627.

Leahy, D. A. *et al.* 1989a, *M.N.R.A.S.,* **236**, 603.

Leahy, D. A. *et al.* 1989b, *M.N.R.A.S.,* **237**, 269.

Levine, A., Rappaport, S., Deeter, J.E., Boynton, P.E. and Nagase, F. 1993, *Ap. J.,* **410**, 328.

Levine, A., Rappaport, S., Putney, A., Corbet, R. and Nagase, F. 1991, *Ap. J.,* **381**, 101.

Lewin, W.H.G., Ricker, G. and McClintock, J.E. 1971, *Ap. J. (Letters),* **169**, L17.

Lewis, W., Rappaport, S., Levine, A. and Nagase, F. 1992, *Ap. J.,* **389**, 665.

Li, F., Rappaport, S. and Epstein, A. 1978, *Nature,* **271**, 37.

Liedahl, D.A., Kahn, S.M., Osterheld, A.L. and Goldstein, W.H. 1992, *Ap. J.,* **391**, 306.

Livio, M., Soker, N., de Kool, M. and Savonije, G.J. 1986, *M.N.R.A.S.,* **218**, 593.

London, R. 1982, in *Cataclysmic Variables and Low Mass X-ray Binaries*, eds. J. Patterson and D.Q. Lamb (Reidel, Dordrecht), p. 121.

Long, K.S., Helfand, D. and Grabelsky, D.A. 1981, *Ap. J.*, **248**, 925.

Lubow, S. H. and Shu, F. H. 1975, *Ap. J.*, **198**, 383.

Lucke, R., Yentis, D., Friedman, H., Fritz, G. and Shulman, S. 1976, *Ap. J. (Letters)*, **206**, L25.

Lucy, L.B. and Solomon, P. 1970, *Ap. J.*, **159**, 879.

McClintock, J.E., London, R. A., Bond, H. E. and Grauer, A. D. 1982, *Ap. J.*, **258**, 245.

McClintock, J.E., Rappaport, S., Nugent, J. and Li, F. 1977, *Ap. J. (Letters)*, **216**, L15.

McClintock, J.E. and Remillard, R. A. 1986, *Ap. J.*, **308**, 110.

McClintock, J.E. and Remillard, R.A. 1990, *Ap. J.*, **350**, 386.

McClintock, J.E. and Remillard, R. A. 1992, *IAU Circ. No.* 5499.

McClintock, J.E., Remillard, R. A. and Margon, B. 1981, *Ap. J.*, **243**, 900.

McClintock, J.E. *et al.* 1976, *Ap. J. (Letters)*, **206**, L99.

Makishima, K. 1986, in *The Physics of Accretion onto Compact Objects*, eds. K.O. Mason, M.G. Watson and N. E. White (Springer-Verlag, Berlin), p. 249.

Makishima, K., and Mihara, T. 1992, in *Frontiers of X-Ray Astronomy* (Proc. of the 28th Yamada Conference), eds. Y. Tanaka and K. Koyama, (Uni. Acad. Press, Tokyo), p. 23.

Makishima, K. *et al.* 1984, *Pub. Astr. Soc. Japan*, **36**, 679.

Makishima, K. *et al.* 1988, *Nature*, **333**, 746.

Makishima, K. *et al.* 1990a, *Pub. Astr. Soc. Japan*, **42**, 295.

Makishima, K. *et al.* 1990b, *Ap. J.*, **365**, L59.

Maraschi, L., Huckle, H.E., Ives, J.C., Ives, J.C. and Sanford, P.W. 1976, *Nature*, **263**, 34.

Margon, B. 1984, *Ann. Rev. Astr. Ap.*, **22**, 507.

Margon, B., Ford, H.C., Grandi, S.A. and Stone, R.P.S. 1979 *Ap. J. (Letters)*, **233**, L63.

Marshall, N. and Ricketts, M.J. 1980, *M.N.R.A.S.*, **193**, 7P.

Mason, K.O. 1977, *M.N.R.A.S.*, **178**, 81P.

Mason, K.O., Branduardi, G. and Sanford, P.W. 1976b, *Ap. J. (Letters)*, **203**, L29.

Mason, K.O., Charles, P.A., White, N.E., Culhane, J.L., Sanford, P.W. and Strong, K.T. 1976a, *M.N.R.A.S.*, **177**, 513.

Mason, K.O. and Córdova, F.A. 1982, *Ap. J.*, **262**, 253.

Mason, K.O., Córdova, F.A. and White, N.E. 1986, *Ap. J.*, **309**, 700.

Mason, K.O., Parmar, A.N. and White, N.E. 1985, *M.N.R.A.S.*, **216**, 1033.

Mason, K.O. *et al.* 1980, *Ap. J. (Letters)*, **242**, L109.

Mason, K.O. *et al.* 1987, *M.N.R.A.S.*, **226**, 423.

Matsuda, T., Inoue, M. and Sawada, K. 1987, *M.N.R.A.S.*, **226**, 785.

Matsuoka, M. *et al.* 1980, *Ap. J. (Letters)*, **240**, L137.

McCray, R.A. *et al.* 1982, *Ap. J.*, **262**, 301.

McCray, R.A. *et al.* 1984, *Ap. J.*, **282**, 245.

Melia, F., Zylstra, G.J. and Fryxell, B. 1991, *Ap. J. (Letters)*, **377**, L101.

Mendelson, H. and Mazeh, T. 1991, *M.N.R.A.S.*, **250**, 373.

Mereghetti, S., Caraveo, P. and Bignami, G.F. 1992, *Astr. Ap.*, **263**, 172.

Mészáros, P. and Bonazzola, S. 1981, *Ap. J.*, **251**, 695.

Mészáros, P. and Nagel. W. 1985, *Ap. J.*, **298**, 147.

Mészáros, P., Nagel, W. and Ventura, J. 1980, *Ap. J.*, **238**, 1066.

Middleditch, J., Mason, K. O., Nelson, J. E. and White, N. E. 1981, *Ap. J.*, **244**, 1001.

Mihara, T. *et al.* 1990, *Nature*, **346**, 250.

Mihara, T. *et al.* 1991a, *Ap. J. (Letters)*, **379**, L65.

Mihara, T. *et al.* 1991b, *Pub. Astr. Soc. Japan*, **43**, 501.

Milgrom, M. 1978, *Astr. Ap.*, **208**, 191.

Mitsuda, K., Inoue, H., Nakamura, N. and Tanaka, Y. 1989, *Pub. Astr. Soc. Japan*, **41**, 97.

Mitsuda, K. *et al.* 1984, *Pub. Astr. Soc. Japan*, **36**, 741.

Molnar, L. A. 1988, *Ap. J. (Letters)*, **331**, L25.

Moneti, A. 1992, *Astr. Ap.*, **260**, L7.

Morrison, R. and McCammon, D. 1983, *Ap. J.*, **270**, 119.

Motch, C. *et al.* 1987, *Ap. J.*, **313**, 792.

Murakami, T. *et al.* 1988, *Nature*, **335**, 234.

Murdin, P. *et al.* 1980, *Astr. Ap.*, **87**, 292.

Nagase, F. 1985, *Adv. Space Res.*, **5**, 95.

Nagase, F. 1989, *Pub. Astr. Soc. Japan*, **41**, 1.

Nagase, F. 1992, *Ginga Memorial Symposium*, eds. F. Makino and F. Nagase, p. 1.

Nagase, F. *et al.* 1986, *Pub. Astr. Soc. Japan*, **38**, 547.

Nagase, F. *et al.* 1991, *Ap. J. (Letters)*, **375**, L49.

Nagase, F. *et al.* 1992, *Ap. J.*, **396**, 147.

Nagel, W. 1981a, *Ap. J.*, **251**, 278.

Nagel, W. 1981b, *Ap. J.*, **251**, 288.

Nakamura, H. *et al.* 1993, *M.N.R.A.S.*, **261**, 353.

Naylor, T., Charles, P. A., Drew, J.E. and Hassall, B.J.M. 1988, *M.N.R.A.S.*, **233**, 285.

Neugebauer, G., Oke, J.B., Becklin, E. and Garmire, G. 1969, *Ap. J.*, **155**, 1.

Nicolson, G.D., Feast, M.W. and Glass, I.S. 1980, *M.N.R.A.S.*, **191**, 293.

Ohashi, T. *et al.* 1984, *Pub. Astr. Soc. Japan*, **36**, 699.

Paczyński, B. 1971, *Ann. Rev. Astr. Ap.*, **9**, 183.

Pakull, M.W. Beuermann, K., Van der Klis, M. and Van Paradijs, J. 1988, *Astr. Ap.*, **203**, L27.

Pakull, M. and Parmar, A.N. 1981, *Astr. Ap.*, **102**, L1.

Parmar, A.N., Gottwald, M., Van der Klis, M. and Van Paradijs, J. 1989a, *Ap. J.*, **338**, 1024.

Parmar, A.N., Smale, A.P., Verbunt, F. and Corbet, R.H.D. 1991 *Ap. J.*, **366**, 253.

Parmar, A.N., Stella, L. and Giommi, P. 1989b, *Astr. Ap.*, **222**, 96.

Parmar, A.N., White, N.E., Giommi, P. and Gottwald, M. 1986, *Ap. J.*, **308**, 199.

Parmar, A.N., White, N.E. and Stella, L. 1989c, *Ap. J.*, **338**, 373.

Parmar, A.N., White, N.E., Stella, L., Izzo, C. and Ferri, P. 1989d, *Ap. J.*, **338**, 359.

Parmar, A.N., White, N.E., Sztajno, M. and Mason, K.O. 1985b, *Space Sci. Rev.*, **40**, 213.

Parmar, A.N. *et al.* 1985a, *Nature*, **313**, 119.

Parsignault, D. R. *et al.* 1972, *Nature Phys. Sci.*, **239**, 123.

Pedersen, H., Cristiani, S., d'Ororico, S. and Thomsen, B. 1985, *IAU Circ. No.* 4047.

Pedersen, H. and Mayer, M. 1985, *IAU Circ. No.* 4039.

Pedersen, H., Van Paradijs, J. and Lewin, W. H. G. 1981, *Nature*, **294**, 725.

Petterson, J.A. 1975, *Ap. J.*, **201**, L61.

Petterson, J.A. 1977, *Ap. J.*, **218**, 783.

Petterson, J.A. 1978, *Ap. J.*, **224**, 625.

Petterson, J.A., Rothschild, R.E. and Gruber, D.E. 1991, *Ap. J.*, **378**, 696.

Pietsch, W., Steinle, H., Gottwald, M. and Graser, U. 1986, *Astr. Ap.*, **157**, 23.

Ponman, T.J., Foster, A.J. and Ross, R.R. 1990, *M.N.R.A.S.*, **246**, 287.

Pounds, K.A. *et al.* 1976, *M.N.R.A.S.*, **172**, 473.

Pravdo, S. H. 1979, in *X-ray Astronomy*, eds. W. A. Baity and L. E. Peterson (Pergamon Press, Oxford), p. 169.

Pravdo, S.H. *et al.* 1977, *Ap. J. (Letters)*, **215**, L61.

Pravdo, S.H. *et al.* 1978, *Ap. J.*, **225**, 988.

Pravdo, S.H. *et al.* 1979, *Ap. J.*, **231**, 912.

Priedhorsky, W.C. and Terrell, J. 1983a, *Nature*, **303**, 681.

Priedhorsky, W.C. and Terrell, J. 1983b, *Ap. J.*, **273**, 709.

Priedhorsky, W.C. and Terrell, J. 1984a, *Ap. J.*, **280**, 661.

Priedhorsky, W.C. and Terrell, J. 1984b, *Ap. J. (Letters)*, **284**, L17.

Priedhorsky, W.C., Terrell, J. and Holt, S.S. 1983, *Ap. J.*, **270**, 233.

Pringle, J.E. 1974, *Nature*, **247**, 21.

Pringle, J.E. and Rees, M.J. 1972, *Astr. Ap.*, **21**, 1.

Rappaport, S. and Joss, P.C. 1977, *Nature*, **266**, 683.

Rappaport, S., Nelson, L. A., Ma, C. P. and Joss, P. C. 1987, *Ap. J.*, **322**, 842.

Rappaport, S.A. *et al.* 1977, *Ap. J. (Letters)*, **217**, L29.

Rappaport, S.A. *et al.* 1978, *Ap. J. (Letters)*, **224**, L1.

Raymond, J.C. 1993, *Ap. J.*, **412**, 267.

Reynolds, A., Parmar, A.N., and White, N.E. 1993, *Ap. J.*, **414**, 302.

Ritter, H. 1990, *Astr. Ap. Suppl.*, **85**, 1189.

Roberts, J.W. 1974, *Ap. J.*, **187**, 575.

Rose, L.A. *et al.* 1979, *Ap. J.*, **231**, 919.

Rosenberg, F.D., Eyles, C., Skinner, G. and Willmore, A.P. 1975, *Nature*, **256**, 628.

Sandage, A.R. *et al.* 1966, *Ap. J.,* **146**, 315.
Sanford, P.W. and Hawkins, F.J. 1972, *Nature Phys. Sci.,* **239**, 135.
Sansom, A.E., Dotani, T., Asai, K. and Lehto, H. J. 1993, *M.N.R.A.S,* **262**, 429.
Sansom, A.E., Watson, M.G., Makishima, K. and Dotani, T. 1989, *Pub. Astr. Soc. Japan,* **41**, 595.
Sato, N. *et al.* 1986, *Pub. Astr. Soc. Japan,* **38**, 731.
Savonije, G.J. 1983, in *Accretion Driven X-ray Sources,* eds. W. H. G. Lewin and E. P. J. Van den Heuvel (Cambridge University Press), p. 343.
Schaefer, B.E. 1987, *IAU Circ. No.* 4478.
Schmidtke, P.C. and Cowley, A.P. 1987, *Astr. J,* **92** (2), 374.
Schreier, E., Swartz, K., Giaconni, R., Fabbiano, G. and Morin, J. 1976, *Ap. J.,* **204**, 539.
Schreier, E. *et al.* 1972a, *Ap. J. (Letters),* **172**, L79.
Schreier, E. *et al.* 1972b, *Ap. J. (Letters),* **172**, L79.
Schulz, N.S., Hasinger, G. and Trümper, J. 1989, *Astr. Ap.,* **225**, 48.
Schwarzenberg-Czerny, A. 1992, *Astr. Ap.,* **260**, 268.
Serlemitsos, P.J. *et al.* 1975, *Ap. J. (Letters),* **201**, L9.
Seward, F.D., Charles, P.A. and Smale, A.P. 1986, *Ap. J.,* **305**, 814.
Shakura, N.I. and Sunyaev, R.A. 1973, *Astr. Ap.,* **24**, 337.
Shapiro, S.L. and Lightman, A.P. 1976, *Ap. J.,* **204**, 555..
Shapiro, S.L., Lightman, A.P. and Eardley, D.M. 1976, *Ap. J.,* **204**, 187.
Shapiro, S.L. and Salpeter, E.E. 1975, *Ap. J.,* **198**, 671.
Shibazaki, N. and Mitsuda, K. 1984, in *High Energy Transients in Astrophysics,* ed. S.E. Woosley (AIP, New York), p. 63.
Skinner, G.K. *et al.* 1982, *Nature,* **297**, 568.
Smale, A. P. 1991, *Pub. A. S. P.,* **103**, 636.
Smale, A.P. and Lochner, J.C. 1992, *Ap. J.,* 2, **395**, 582.
Smale, A.P., Mason, K.O., White, N.E. and Gotthelf, M. 1988b, *M.N.R.A.S.,* **232**, 647.
Smale, A.P. and Mukai, K. 1988, *M.N.R.A.S.,* **231**, 663.
Smale, A.P. *et al.* 1988a, *M.N.R.A.S.,* **233**, 51.
Smale, A.P. *et al.* 1989, *Pub. Astr. Soc. Japan,* **41**, 607.
Smale, A.P. *et al.* 1992, *Ap. J.,* **400**, 330.
Smale, A.P. *et al.* 1993a, *Ap.J.,***410**, 796.
Smale, A.P. *et al.* 1993b, *Ap.J.,***894**, 243.
Soker, N., Livio, M., de Kool, M. and Savonije, G.J. 1986 *M.N.R.A.S.,* **221**, 445.
Soong, Y. *et al.* 1990, *Ap. J.,* **348**, 641.
Stella, L., Priedhorsky, W. and White, N. E. 1987, *Ap. J. (Letters),* **312**, L17.
Stella, L. and Rosner, R. 1984, *Ap. J.,* **277**, 312.
Stella, L., White, N.E. and Rosner, R. 1986, *Ap. J.,* **308**, 669.
Stella, L. *et al.* 1985, *Ap. J. (Letters),* **288**, L45.
Stewart, R.T., Caswell, J.L., Haynes, R.F. and Nelson, G.J. 1993, *M.N.R.A.S.,* **261**, 593.
Strickman, M.S., Johnson, W.N. and Kurfess, J. 1980, *Ap. J. (Letters),* **240**, L21.
Sunyaev, R.A. and Shakura, N.I. 1986, *Sov. Astron. Lett.,* **12**(2), 117.
Sunyaev, R.A. and Titarchuk, L.G. 1980, *Astr. Ap.,* **86**, 121.
Suzuki, K. *et al.* 1984, *Pub. Astr. Soc. Japan,* **36**, 761.
Swank, J.H. and Serlemitsos, P.J. 1985, in *Galactic and Extragalactic Compact X-ray Sources,* eds. Y. Tanaka and W.H.G. Lewin (ISAS, Tokyo), p. 175.
Swank, J.H. *et al.* 1976, *Ap. J. (Letters),* **209**, L57.
Sztajno, M. and Frank, J. 1984, *Astr. Ap.,* **138**, L15.
Taam, R.E. and Fryxell, B.A. 1988, *Ap. J. (Letters),* **327**, L73.
Taam, R.E. and Fryxell, B.A. 1989, *Ap. J.,* **338**, 297.
Taam, R.E., Fryxell, B.A. and Brown, D.A. 1988, *Ap. J. (Letters),* **331**, L117.
Taam, R.E., Fu, A. and Fryxell, B.A. 1988, *Ap. J.,* **371**, 696.
Tamura, K., Tsunemi, H., Kitamoto, S., Hayashida, K. and Nagase, F. 1992, *Ap. J.,* **389**, 676.
Tan, J. *et al.* 1991 *Ap. J.,* **374**, 291.
Tanaka, Y. and Lewin, W.H.G. 1993, Ch. 3, this volume.
Tanaka, Y. *et al.* 1984, *Pub. Astr. Soc. Japan,* **36**, 641.
Tananbaum, H. *et al.* 1972, *Ap. J. (Letters),* **174**, L143.
Tavani, M. and London, R. 1993. *Ap. J.,* **410**, 281.

Tawara, Y. *et al.* 1989 *Pub. Astr. Soc. Japan,* **41**, 473.

Taylor, A.R. and Gregory, P.C. 1982, *Ap. J.,* **255**, 210.

Tennant, A.F. 1987, *M.N.R.A.S.,* **226**, 971.

Tennant, A.F. , Fabian, A.C. and Shafer, R.A. 1986 *M.N.R.A.S.,* **219**, 871.

Thorstensen, J.R. *et al.* 1979, *Ap. J. (Letters),* **233**, L57.

Thorstensen, J.R. *et al.* 1988, *Ap. J.,* **334**, 430.

Trümper, J. *et al.* 1978, *Ap. J. (Letters),* **219**, L105.

Trümper, J. *et al.* 1986, *Ap. J. (Letters),* **300**, L63.

Tueller, J. *et al.* 1984, *Ap. J.,* **279**, 177.

Ulmer, M.P, Baity, W.A., Wheaton, W.A. and Peterson, L.E. 1972, *Ap. J. (Letters),* **178**, L121.

Ulmer, M.P. *et al.* 1980, *Ap. J. (Letters),* **235**, L159.

Vacca, W.D. *et al.* 1987, *Astr. Ap.,* **172**, 143.

Van den Heuvel, E.P.J. and Rappaport, S. 1987, in *Physics of Be Stars,* eds. A. Slettebak and T.D. Snow (Cambridge University Press), p. 291.

Van den Heuvel, E.P.J. *et al.* 1992, *Astr. Ap.,* **262**, 97.

Van der Klis, M. 1992, in *Frontiers of X-Ray Astronomy* (Proc. of the 28th Yamada Conference), eds. Y. Tanaka and K. Koyama (Uni. Acad. Press, Tokyo), p. 23.

Van der Klis, M. and Bonnet-Bidaud, J. M. 1989, *Astr. Ap.,* **214**, 203.

Van der Klis, M., Stella, L., White, N., Jansen, F. and Parmar, A.N. 1987, *Ap. J.,* **316**, 411.

Van der Klis, M. *et al.* 1993, *M.N.R.A.S.,* **260,**, 686.

Van der Woerd, H., White, N.E. and Kahn, S.M. 1989, *Ap. J.,* **344**, 320.

Van Kerkwijk, M.H. *et al.* 1992, *Nature,* **355**, 703.

Van Paradijs, J. 1993, Ch. 14, this volume.

Van Paradijs, J. and McClintock, J.E. 1993, Ch. 2, this volume.

Van Paradijs, J. and Verbunt, F. 1984, in *High Energy Transients in Astrophysics,* ed. S.E. Woosley (A.I.P, NY), p. 31.

Ventura, J. 1979, *Phys. Rev. D,* **19**, 1684.

Verbunt, F. 1987, *Ap. J. (Letters),* **312**, L23.

Verbunt, F., Van Paradijs, J. and Elson, R. 1984, *M.N.R.A.S.,* **210**, 899.

Voges, W. *et al.* 1982, *Ap. J.,* **263**, 803.

Voges, W. *et al.* 1985, *Space Sci. Rev.,* **40**, 339.

Vrtilek, S.D. and Halpern, J.P. 1985, *Ap. J.,* **296**, 606.

Vrtilek, S.D., Soker, N. and Raymond, J.C. 1993, *Ap. J.,* **404**, 696.

Vrtilek, S.D., Swank, J.H. and Kallman, T.R. 1988, *Ap. J.,* **326**, 186.

Vrtilek, S.D. *et al.* 1986a, *Ap. J.,* **308**, 644.

Vrtilek, S.D. *et al.* 1986b, *Ap. J.,* **307**, 698.

Vrtilek, S.D. *et al.* 1991, *Ap. J. Suppl.,* **76**, 1127.

Wade, R. A., Quintana, H., Horne, K. and Marsh, T. R. 1985, *Pub. A. S. P.,* **97**, 1092.

Walter, F.M. *et al.* 1982, *Ap. J. (Letters),* **253**, L67.

Wang, Y.-M. 1981, *Astr. Ap.,* **102**, 36.

Wang, Y.-M. and Frank, J. 1981, *Astr. Ap.,* **93**, 255.

Wang, Y.-M. and Welter, G. L. 1981, *Astr. Ap.,* **102**, 97.

Waters, L.B.F.M. and Van Kerkwijk, M.H. 1989, *Astr. Ap.,* **223**, 196.

Waters, L.B.F.M. *et al.* 1988, *Astr. Ap.,* **198**, 200.

Watson, M.G. and Griffiths, R.E. 1977, *M.N.R.A.S.,* **178**, 513.

Watson, M.G., Stewart, G.C., Brinkmann, W. and King, A.R. 1986, *M.N.R.A.S.,* **222**, 261.

Watson, M.G., Warwick, R.S. and Corbet, R. 1982, *M.N.R.A.S.,* **199**, 197.

Watson, M.G., Warwick, R.S. and Ricketts, M.J. 1981, *M.N.R.A.S.,* **195**, 197.

Watson, M.G. *et al.* 1985, *Space Sci. Rev.,* **40**, 195.

Webster, B.L. and Murdin, P. 1972, *Nature,* **235**, 37.

Wheaton, W.A. *et al.* 1979, *Nature,* **282**, 240.

Whelan, J.A.J. *et al.* 1977, *M.N.R.A.S.,* **181**, 259.

White, N.E. 1978, *Nature,* **271**, 38.

White, N.E. 1985, in *The Evolution of Galactic X-ray Binaries,* eds. J. Trümper, W. H. G. Lewin and W. Brinkmann (NATO ASI, Reidel), p. 227.

White, N.E. 1989, *Ann. Rev. Astr. Ap.,* **1**, 85.

White, N.E. and Carpenter, G.F. 1978, *M.N.R.A.S.,* **183**, 11P.

White, N.E. and Holt, S.S. 1982, *Ap. J.*, **257**, 318.
White, N.E., Kallman, T.R. and Swank, J.H. 1983, *Ap. J.*, **269**, 264.
White, N.E., Kaluzienski, J.L. and Swank, J.H. 1984, *High Energy Transients in Astrophysics*, ed. S.E. Woosley, (A.I.P, NY), p. 31.
White, N.E. and Mason, K.O. 1985, *Space Sci. Rev.*, **40**, 167.
White, N.E., Mason, K.O. and Sanford, P.W. 1978a, *M.N.R.A.S.*, **184**, 67P.
White, N.E., Mason, K.O., Sanford, P.W. and Murdin, P. 1976b, *M.N.R.A.S.*, **176**, 201.
White, N.E., Parkes, G., Sanford, P.W., Mason, K. and Murdin, P.G. 1978b, *Nature*, **274**, 664.
White, N.E., Peacock, A. and Taylor, B.G. 1985, *Ap. J.*, **296**, 475.
White, N.E. and Pravdo, S.H. 1979, *Ap. J. (Letters)*, **233**, L121.
White. N.E., Stella, L. and Parmar, A.N., 1988, *Ap. J.*, **324**, 363.
White, N.E. and Swank, J.H. 1982, *Ap. J. (Letters)*, **253**, L61.
White, N.E. and Swank, J.H. 1984, *Ap. J.*, **287**, 856.
White, N.E., Swank, J.H. and Holt, S.S. 1983, *Ap. J.*, **270**, 711.
White, N.E., Swank, J.H., Holt, S.S. and Parmar, A.N. 1982, *Ap. J.*, **263**, 277.
White, N.E. *et al.* 1976a, *Ap. J. (Letters)*, **209**, L119.
White, N.E. *et al.* 1980, *Ap. J.*, **239**, 655.
White, N.E. *et al.* 1981, *Ap. J.*, **247**, 994.
White, N.E. *et al.* 1984, *Ap. J. (Letters)*, **283**, L9.
White, N.E. *et al.* 1986, *M.N.R.A.S.*, **218**, 129.
Whitehurst, R. 1988, *M.N.R.A.S.*, **232**, 35.
Willingale, R., King, A.R. and Pounds, K.A. 1985, *M.N.R.A.S.*, **215**, 295.
Yahel, R.Z. 1979, *Astr. Ap.*, **78**, 136.
Yahel, R.Z. 1980, *Ap. J.*, **236**, 911.

2

Optical and ultraviolet observations of X-ray binaries

J. van Paradijs
Astronomical Institute 'Anton Pannekoek', University of Amsterdam, and
Center for High-Energy Astrophysics (CHEAF),
Kruislaan 403, 1098 SJ Amsterdam, The Netherlands

J. E. McClintock
Harvard-Smithsonian Center for Astrophysics,
60 Garden Street, Cambridge MA 02138, USA

2.1   Introduction

With few exceptions, X-ray binaries have companions with masses either $\gtrsim 10 M_\odot$ or $\lesssim 1 M_\odot$. These two classes of sources are commonly referred to as high-mass X-ray binaries (HMXBs) and low-mass X-ray binaries (LMXBs), respectively. HMXBs and LMXBs differ in many respects, including their optical properties (see Ch. 1).

In general, the optical counterparts of HMXBs look like normal early-type stars; i.e., their spectra can be MK classified without particular difficulty. The optical appearance of an HMXB is scarcely affected by the X-ray source because the bolometric luminosity of the mass donor generally exceeds the X-ray luminosity, often by a large margin.

The optical spectrum of an LMXB consists of emission lines superposed on a blue continuum, and is therefore totally unlike the spectrum of an ordinary star. An accretion disk around the X-ray source dominates the optical emission from an LMXB. The disk absorbs a sizeable fraction of the X-rays from the central source and reprocesses them into optical and UV photons. The contribution from the secondary is generally negligible. On occasion, however, the presence of the secondary can be discerned in the spectrum (or colors) of the LMXBs.

Because of the great differences between HMXBs and LMXBs, we discuss many of their properties separately in Sect. 2.2 and 2.3, respectively. In Sect. 2.4 we discuss the results of mass determinations (for both the HMXBs and LMXBs), and in Sect. 2.5 we compare the galactic X-ray binaries with their counterparts in the Magellanic Clouds. In Sect. 2.6 we discuss the evidence that several galactic X-ray sources may be members of triple-star systems.

2.2   High-mass X-ray binaries

2.2.1   *Supergiant and Be primaries*

2.2.1.1   *Mass transfer mechanisms*

Because the optical emission of an HMXB is dominated by a more or less normal star, it is relatively easy to understand these systems using conventional ideas. In fact, a few months after the discovery of the binary nature of Cen X-3 [412], whose companion star is massive, the basic outline of the evolution of an HMXB was established [462] (see Ch. 11).

From the data on HMXBs collected in Ch. 14, it appears that most HMXBs fall in one of the two following groups. (i) The primary has evolved away from the main sequence; it is a supergiant with spectral type earlier than B2, or an Of star. The orbital periods are generally less than ~ 10 days. The primaries fill, or almost fill, their Roche lobes, as is apparent from estimates of their radii [470] and the amplitudes of their optical light curves (see Sect. 2.2.2). (ii) The primary is an Oe or Be star, characterized by emission lines (predominantly the Balmer series) which originate from circumstellar material. OBe stars rotate rapidly [420] (but not at break-up speed [421]) and lie close to the main sequence in the Hertzsprung–Russell diagram [528]. The known orbits of Be/X-ray binaries are eccentric, with periods generally $\gtrsim 20$ days. The primaries underfill their Roche lobes. For general reviews on OBe stars, we refer to [9,422]; reviews on Be/X-ray binaries can be found in [390,463,497]. Recent work on Be stars is regularly discussed in the Be Newsletter.

These two groups are distinguished by their mass transfer mechanisms [285]. The supergiants lose mass in a roughly spherical stellar wind (terminal velocity several 10^3 km s^{-1}; mass loss rate up to $10^{-6} M_\odot$ yr^{-1} [26,118,388]), of which a small fraction is captured by the compact star. Supergiants can also lose mass by incipient Roche lobe overflow (see Ch. 10). The mass loss of Be stars is anisotropic [264,379]. Near the poles, the wind has a low density and a high outflow velocity, i.e. it is similar to a supergiant wind, but with a much reduced total mass loss rate; near the equator, the density is high, but the outflow velocity low. The mass loss rate in the equatorial wind is highly variable, and leads to irregular shedding of line-emitting equatorial envelopes. During 'inactive' periods Be stars look like normal B stars; conversely, some normal B stars are known to have temporarily turned into Be stars. The equatorial mass loss is probably related to the rapid rotation of the Be star. It may originate in the interplay of rotation and non-radial pulsation, but currently there is no general agreement on this [9].

Some HMXBs cannot be easily placed in these groups. Two (SS 433, Cyg X-3) are peculiar sources, which are encountered during short-lasting episodes in the evolution of HMXBs. Of the remaining four, three (LMC X-3, LMC X-4, and Cen X-3) exchange mass by Roche lobe overflow.

The X-ray flux of most Be/X-ray binaries is highly variable. Some show periodic outbursts (e.g., 0535+262 [162]), which reflect the varying accretion rate onto the neutron star as it moves in its eccentric orbit through regions of varying density (see Ch. 1). The amplitude of the variation may be enhanced by the following mechanism that can turn off the accretion flow: at very low wind density, the neutron star magnetosphere may become larger than the co-rotation radius, so that

Table 2.1. *Average properties of HMXBs*

Group of HMXBs	ξ^a	$\log L_X$
Supergiants	11.6 ± 1.7 (0.6)	36.7 ± 0.9 (0.3)
Be/X in outburst	14.9 ± 1.5 (0.5)	36.6 ± 0.8 (0.3)
Be/X in quiescence	7.2 ± 2.1 (0.6)	33.3 ± 1.2 (0.3)
Magellanic sources	17.3 ± 1.5 (0.6)	38.4 ± 0.5 (0.2)

^{a} The error is the standard deviation in the distribution; the mean error is given in parentheses.

accretion becomes centrifugally inhibited ([431]; see Ch. 1 and 10). Outbursts have been observed which are not related to the orbital phase, but are caused by sudden enhancement of the (equatorial) mass loss of the Be star (e.g., 0115+634 [399]). During these active periods, the X-ray luminosity may still undergo an orbital modulation (e.g., in 0535−668 [507]). In several cases, a period of X-ray activity was found to be preceded by the onset of optical line emission [162].

2.2.1.2 Ratio of X-ray to optical luminosity

The ratios of X-ray to optical luminosities of HMXBs cover a large range (a factor of 10^6), but a more informative picture emerges when we distinguish 'natural' groups of HMXBs. For this purpose, we introduce the X-ray to optical 'color index', $\xi = B_0 + 2.5 \log F_X$, where B_0 is the reddening-corrected B magnitude and F_X is the (2–10 keV) X-ray flux (in μJy; see Ch. 14). In Figure 2.1, we show the ξ distributions separately for: (i) the galactic HMXBs with evolved companions and wind accretion (we have taken the maximum X-ray flux for these objects); (ii) galactic Be/X-ray binaries in outburst, and (iii) not in outburst; (iv) the Magellanic Cloud sources. These groups individually occupy rather narrow ξ intervals (see Table 2.1).

2.2.1.3 X-ray luminosity

Together, the absolute magnitude M_V of the optical counterpart (from its spectral type and luminosity class) and the reddening-corrected apparent magnitude V_0 provide estimates of the source distance and X-ray luminosity L_X. These estimates are not good enough for detailed studies of individual sources, but they probably suffice for a statistical description of their properties. For instance, using the M_V calibration of Lesh [271], we find average distance moduli $V_0 - M_V$ for the SMC and LMC sources of 18.9 ± 0.3 (1 s.d.) and 18.5 ± 0.6, respectively, in good agreement with their 'canonical' values [139]. The apparent magnitude of a Be/X-ray binary may be contaminated by emission from its equatorial envelope. This may lead to an underestimate of the X-ray luminosity, particularly if the observations were made during outburst. Average X-ray luminosities for the four groups of HMXBs distinguished above are listed in Table 2.1; their distributions are shown in Figure 2.2. Compared to the LMXBs, very few galactic HMXBs have (maximum) X-ray luminosities at the Eddington limit. The quiescent luminosities of Be/X-ray binaries far exceed the values expected for coronal emission of early-type stars, for which L_X

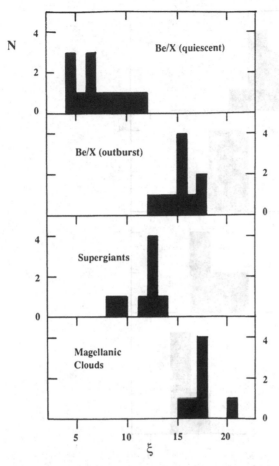

Fig. 2.1. Distribution of ξ for four groups of high-mass X-ray binaries; ξ is a measure of the ratio of observed X-ray to optical fluxes (see text).

is about 10^{-7} of the bolometric luminosity [362] (i.e. $L_X \lesssim 10^{30}$ erg s^{-1}). Some of the fainter Be/X-ray binaries may have a white-dwarf accretor [380] (see also [320]). As noticed previously [79,233], the Magellanic sources appear, on average, to be more luminous than their galactic counterparts (see Table 2.1). This is discussed in Sect. 2.5.

2.2.1.4   *Total number of high-mass X-ray binaries*

Be/X-ray binaries are the largest group of observed HMXBs, and since they are, on average, much less distant than the supergiant systems, they dominate the galactic population of HMXBs [390]. We have estimated the total number of galactic HMXBs by multiplying their surface density σ in the solar neighborhood by the area of the Galaxy (assumed to be a circular disk with 10 kpc radius). For σ we simply took the number of known Be/X-ray binaries, divided by $\pi\rho^2$, where ρ is the r.m.s. value of their distances (the individual distances were obtained from spectral type, apparent magnitude and interstellar reddening; see Sect. 2.2.1.3). Since we are also interested

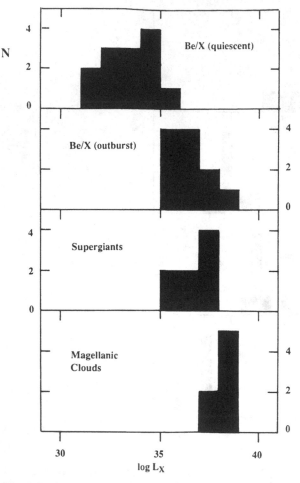

Fig. 2.2. X-ray luminosity distribution for four groups of high-mass X-ray binaries.

in the total luminosity of all quiescent Be/X-ray binaries (see below), we separated these into intervals $\Delta \log L_X = 1$, starting at $\log L_X = 31$. We find that the number of galactic HMXBs with an evolved companion is ~ 40, and the number of Be/X-ray binaries is $(2-6) \times 10^3$ (see also [319]). In view of their high X-ray luminosities, it is unlikely that we have missed nearby HMXBs with evolved companions, but this cannot be said of the Be/X-ray binaries; their number is therefore likely to be an underestimate. We find the total luminosity of the quiescent Be/X-ray binaries in the Galaxy to be $\sim 2 \times 10^{36}$ erg s^{-1}. This is only a small fraction of the (diffuse) galactic X-ray luminosity.

2.2.1.5 *Kinematic properties of HMXBs*

There is some evidence that a neutron star receives a 'kick' velocity at birth (see [123]), and, if it forms in a binary, the center of mass of the system will pick up a fraction of that velocity (in addition, the binary will receive a velocity due to the sudden loss of mass in the supernova). Since the neutron star in an HMXB was already a member of the HXMB progenitor when it formed (see Ch. 11), the HMXBs may have higher space velocities than normal OB stars, i.e. be runaway objects [434]. A statistical study of the space velocities of HMXBs suggests that HMXBs are runaway objects [469]. In particular, they do not seem to be linked with OB associations. A recent study of the positions of X-ray sources in M31 leads to the same conclusion [283]. We note that searches for (non-X-ray) binaries with compact components among runaway OB stars have, so far, been unsuccessful [161].

2.2.2 Optical light curves of high-mass X-ray binaries

Many HMXBs with evolved components show moderate ($\lesssim 10\%$) optical brightness variations, with two maxima and two minima per orbit that occur at quadratures and conjunctions, respectively. These so-called ellipsoidal light curves are caused by the rotational and tidal distortions of the primary which fills (or nearly fills) its critical lobe, and the non-uniform distribution of its surface brightness (gravity darkening, limb darkening). Often, the observed light curves cannot be described as purely ellipsoidal: X-ray heating and the presence of an accretion disk around the compact component can be important. In LMXBs, these effects completely dominate the optical properties, including the shape of the orbital light curves (see Sect. 2.3.4). Generally, in Be/X-ray binaries the irregular brightness variations (which are related to the equatorial mass loss) dominate any orbital variability.

In Sect. 2.2.2.1 we describe a simple model for ellipsoidal light curves and for the effects related to the presence of the X-ray source. Some representative light curves of HMXBs are discussed in Sect. 2.2.2.2.

2.2.2.1 Description of the model

We make the following three assumptions: (i) the gravitational forces in the binary are those of two point masses M_1 and M_2; (ii) the primary star is co-rotating with the orbit and its spin axis is perpendicular to the orbital plane; (iii) the orbit is circular. The forces acting on a fluid element of the primary can then be written as the gradient of a potential, which is a function of spatial coordinates and depends on M_1, M_2 and their mutual distance a. The shape of the primary will be that of an equipotential surface.

For a given mass ratio $q = M_2/M_1$, the shape of an equipotential surface is determined by a dimensionless potential parameter Ω. At a critical value Ω_c of Ω, the equipotential surface consists of two closed surfaces (one around each star), which have in common a single point on the line ℓ connecting the centers of the two stars, the first Lagrange point L_1. For $\Omega > \Omega_c$ the equipotential surface consists of two separate closed surfaces, one around each component. At L_1, the gradient of the potential along ℓ vanishes (Ω is saddle-shaped near L_1), and matter is free to leave the primary there and flow to the compact star (or leave the system). It is therefore assumed that the primary cannot exceed this critical surface (Roche lobe; see Figure 11.4).

The brightness distribution on the surface of the primary is determined by Von Zeipel's gravity darkening 'theorem': if a star is in hydrostatic equilibrium, and the diffusion equation for radiation transfer is valid, the local radiation flux is proportional

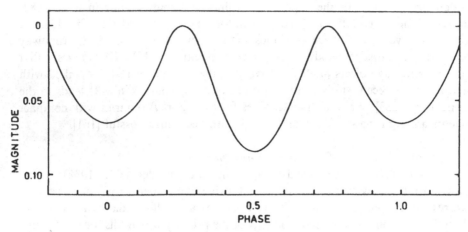

Fig. 2.3. Theoretical ellipsoidal light curve for a Roche lobe filling primary star (mass ratio $q = 0.076$, orbital inclination 90°). Adapted from [529].

to the local surface gravity, i.e. $T_{eff}(:) \, g^{0.25}$ (cf. [80,318]). Under Von Zeipel's theorem, the condition of radiative equilibrium is violated, and meridional circulation currents will develop [414]. The velocities of these currents are small enough that hydrostatic equilibrium is not affected [318,414]. When the energy transfer in the outer envelope is by convection, as is the case for low-mass companion stars, one has approximately $T_{eff}(:) \, g^{0.08}$ [278].

The double-waved shape of an ellipsoidal light curve (see Figure 2.3) reflects the pear-like shape of the equipotential surfaces: near conjunctions the (projected) stellar disk is smallest; at quadratures it is largest. At superior conjunction of the primary, L_1 is directed towards the observer; since near L_1 the surface gravity, and therefore the surface brightness, is minimal, the corresponding minimum in the light curve is the deeper of the two.

An ellipsoidal light curve is determined by scaling parameters (M_1, a and the bolometric luminosity L_{bol} of the primary), parameters that define the primary shape (q, Ω) and the inclination i.

For a discussion of the assumptions underlying this model, and of approximate ways to treat non-circularity of the orbit and non-synchronous rotation of the primary, see [448,518,529]. Numerical calculations of ellipsoidal light curves [8,34,375,435,448,515,516,519] show that their shape is virtually independent of M_1, a and L_{bol}, and depends mainly on (q, Ω, i).

X-rays incident on the primary affect the thermal structure of its outer layers and the properties of its stellar wind (see Sect. 2.2.4). To simplify matters, one assumes that a fraction of these X-rays are reflected, and that the remainder are absorbed by the primary and are reradiated as lower energy (UV and optical) photons. In this approximation, the only effect of the incident X-rays is to increase the effective temperature of the irradiated parts of the primary. This 'X-ray heating' shows up in the light curve by the filling in of the deeper minimum (near inferior conjunction

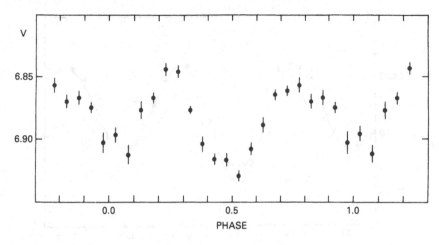

Fig. 2.4. Average V band light curve of Vela X-1 [448].

of the X-ray source). When the X-ray luminosity is sufficiently high, this minimum disappears, and the light curve has a single maximum per orbit.

An accretion disk can affect the optical light curve in several ways. (i) In the absence of mutual eclipses, the disk will decrease the amplitude of the light curve by contributing a dc component. (ii) On the other hand, if mutual eclipses of the primary and the disk do occur, then the amplitude of the light curve will increase. (iii) The disk may cast an X-ray shadow on the primary, thereby decreasing the effect of X-ray heating. For light curve calculations that include X-ray heating, mutual disk-primary eclipses and X-ray shadowing by the disk, see [154,298,448].

2.2.2.2 Observed optical light curves of HMXBs

We discuss here some representative examples of orbital light curves of HMXBs, in the light of the model described above. This section is not intended to be comprehensive. A more detailed discussion of the light curves of individual sources can be found in [471]; for references on individual sources, the reader is invited to consult Ch. 14.

Vela X-1/GP Vel. Although double-waved, the light curve of GP Vel (see Figure 2.4) is not purely ellipsoidal [448]. This is probably due to the orbital eccentricity, which invalidates several assumptions of the ellipsoidal model: in an eccentric orbit the shape of the primary may not even be a unique function of orbital phase, as is suggested by the occasional absence of expected maxima [235]. The variable gravitational disturbance by the compact object may excite pulsation modes in the primary envelope, which could result in irregular brightness variations. Independently, the supergiant primary may show irregular pulsations that are not related to the presence of the compact star, as observed in single early-type supergiants [282].

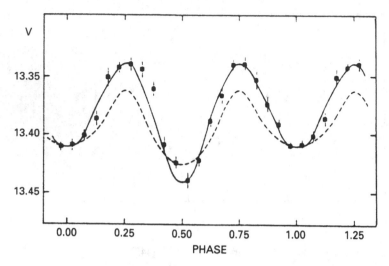

Fig. 2.5. Average V band light curve of Cen X-3. The full curve represents a theoretical light curve with the effects of X-ray heating and an accretion disk included. The dashed curve includes only ellipsoidal variations [448].

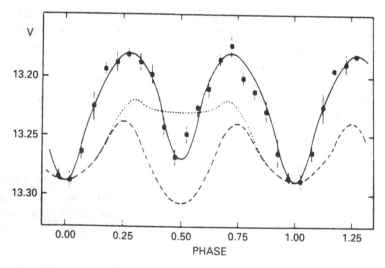

Fig. 2.6. Average V band light curve of SMC X-1. The full curve represents a theoretical light curve with the effects of X-ray heating and an accretion disk included. The dashed curve includes only ellipsoidal variations; the dotted one also includes X-ray heating [448].

Cen X-3/V779 Cen and SMC X-1.   The light curves of Cen X-3 and SMC X-1 are double-waved but not ellipsoidal (see Figures 2.5 and 2.6): X-ray heating and an accretion disk play a significant role [448] (Roche-lobe overflow contributes significantly to the mass transfer [475]). Using known system parameters, the average light curves of Cen X-3 and SMC X-1 can be satisfactorily accounted for by the model described in Sect. 2.2.2.1 [448].

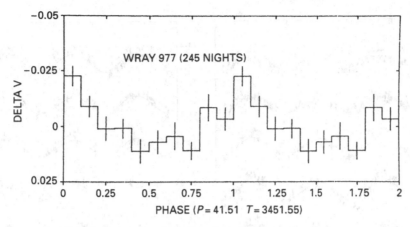

Fig. 2.7. Average V band light curve of Wray 977 [361].

GX 301-2/Wray 977. Periodic X-ray outbursts of GX 301-2 occur near periastron [384,499] of its eccentric orbit [509]. Pakull [358] found evidence for an orbital light curve with a single maximum near periastron, and an amplitude of 0.03 mag (Figure 2.7). This maximum cannot be explained by increased reprocessing of X-rays. However, if the primary fills its Roche lobe near periastron, the maximum can be explained by 'ellipsoidal' variations caused by variable distortion of the primary in its eccentric orbit [361].

LMC X-4. LMC X-4 shows a long-term periodic (30.4 day) X-ray on–off cycle [266], which is also visible as a backward-moving (in orbital phase) feature in the double-waved orbital light curve [228] (see Figure 2.8). The 30.4 day X-ray and optical period can be explained by precession of a tilted accretion disk [200,266], analogous to Her X-1 [154] and SS 433 [269,288].

A0535-668. When active, this Be/X-ray source shows outbursts at intervals of 16.6 days, which reflect the period of its eccentric orbit [419]. At times, the peak X-ray luminosity exceeds the Eddington limit by a substantial factor [63,507], and the optical brightness then increases by about an order of magnitude ([122]; see Figure 2.9). Active periods are alternated by long quiescent intervals in which no X-rays are detected; the optical star is then relatively faint and shows no orbital variation [459].

Long-term periodicities in optical light curves. Long-term periodic changes have been reported in the optical light curves of several HMXBs (see [383] for a review of these long periodicities). In the case of SS 433 [288] and the LMXB Her X-1 [154], there is independent evidence that the changes are caused by precession of a tilted accretion disk [376], and this interpretation has also been proposed in the case of LMC X-4, LMC X-3 and Cyg X 1 [98,200,247,248,266,385]. The viability of this model has been discussed in [229,254,413]. In our opinion, the long-term variations reported for Vela X-1 [249], SMC X-1 [250], Cen X-3 [68] and 1700-377 [69] need confirmation [448,472,478,479].

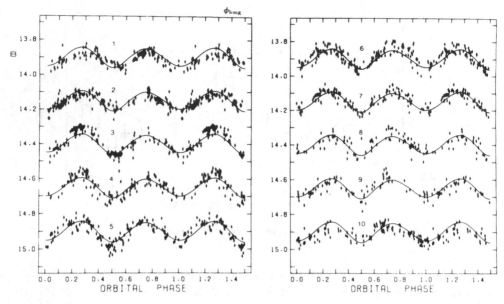

Fig. 2.8. B band light curves of LMC X-4 in ten consecutive phase bins of the 30.4 day precessional cycle [200].

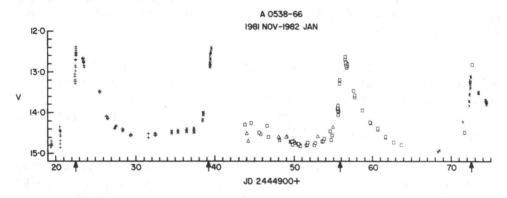

Fig. 2.9. Light curve of A0538-66, covering four consecutive 16.6 day outburst cycles [122].

2.2.3   *Spectroscopic studies of HMXB with evolved companions*

The optical spectra of HMXBs are similar to those of normal early-type stars (see Figure 2.10), which shows that the photospheres of their primaries are scarcely affected by the X-rays. On the other hand, their stellar winds are grossly affected. This is particularly clear in the UV spectra of HMXBs (see Sect. 2.2.4), but can also be seen in their optical spectra, which contain details that reveal the effects of X-ray irradiation.

The only stars that normally show He II $\lambda4686$ emission in their spectra are Of stars [82]. However, this emission is present in the spectra of most HMXB with evolved primaries (although only one HMXB, 1700–377, has an Of star primary) [72,102,218–22,338,348]. The radial-velocity variations of this emission line show a

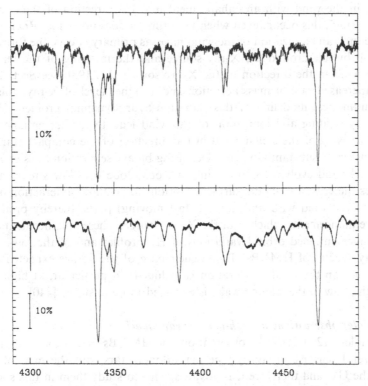

Fig. 2.10. Comparison of the optical spectra (wavelength range 4300–4500 Å) of HD 77581, the optical counterpart of the HMXB Vela X-1 (bottom panel), and of the normal star κ Ori (upper panel), which have spectral types B0.5 Ib and B0.5 Ia, respectively. Note the difference in line widths, caused by the larger rotational velocity of HD 77581. (Courtesy M. van Kerkwijk.)

large phase shift (a lead, ranging between ∼ 120° and ∼ 180°) relative to those of the (photospheric) absorption lines, which indicates that the He II emission originates within the Roche lobe of the X-ray source, from a location that generally trails the motion of the latter.

Vela X-1, Cyg X-1, and 1700−377 show relatively stable Hα emission, with a variable absorption component superposed [23,83,137,530]. The orbital variation of the strength and velocity of this component indicates that there is a density enhancement in the wind which trails the X-ray source. Further evidence for such enhancement is provided by the orbital variations of the low-energy cut off in the X-ray spectrum (see Ch. 1); these can be explained naturally as variations of the column density between the observer and the X-ray source, with maximum values generally occurring in the last third of the orbit before superior conjunction of the X-ray source [44,60,132,181,182,230,381,438,455,498].

Theoretical modeling of HMXB stellar winds has shown that the presence of the X-ray source affects the wind structure in a complicated, and interactive, way. The non-spherical shape of the primary, and of the effective gravitational potential around it (including rotation and radiation pressure), leads to a directional dependence of the mass flow rate in the wind with an enhancement into the direction of the compact star (wind 'focussing'; this occurs even when the primary does not fill its Roche lobe) [145]. X-ray heating and ionization of the wind have, as primary effects, the formation of an ionization bubble around the X-ray source and the lowering of the terminal velocity of the wind in the direction of the X-ray source [197,198] (see Sect. 2.2.4), leading to an increased rate of mass accretion and a higher level of X-ray emission. Detailed 2-D numerical modeling of this radiation-hydrodynamics problem [30,31] shows that X-ray heating and ionization of the wind lead to further enhancement and stronger focussing of the stellar wind in the direction of the compact star. The enhancement increases substantially with decreasing binary separation (it is expected that the focussed wind evolves smoothly into a Roche lobe overflow stream as the primary reaches the critical potential surface). Coriolis force allows interaction of this focussed part of the wind with undisturbed (fast moving) parts, thereby producing strong density enhancements which trail the X-ray source. These enhancements may explain the abovementioned orbital variations of Hα profiles and of the low-energy cut off in X-ray spectra of HMXBs. Time dependence of the flow is expected when radiative feedback on the wind acceleration is included; in particular, at high mass transfer rates the flow in the trailing wake is expected to be unstable [440].

2.2.4   *Ultraviolet observations of high-mass X-ray binaries*

Ultraviolet (1200–3000 Å) observations of HMXBs (see Table 2.2) are of interest for several reasons. (i) Early-type mass donor stars emit the bulk of their luminosity in the UV, and therefore it is advantageous to study them in this spectral range. This is particularly important for the Be/X-ray binaries (see Sect. 2.2.1); often in the optical their spectra are difficult to classify, and their continua are contaminated by circumstellar IR emission. (ii) The broad feature near 2200 Å provides a direct measure of interstellar reddening. (iii) The mass transfer in HMXBs occurs predominantly via a stellar wind; the UV is by far the best spectral region in which to study the properties of these winds.

As remarked in Sect. 2.2.3, due to the very high bolometric luminosity of the primary, in general the optical spectra of HMXBs are very seldom affected significantly by the X-ray source. By and large the same is true for their UV spectra (see Figure 2.11).

The continuum of the UV spectrum of an HMXB (corrected for interstellar reddening) determines the effective temperature of the mass donor star. Superposed on this continuum are photospheric absorption lines and some strong resonance lines formed in the stellar wind. The latter have so-called P Cygni profiles, consisting of an emission component (centered at the rest wavelength of the transition) and a blue-shifted absorption component formed in the outflowing column of material in the wind that is seen projected against the stellar disk (see [56,57,263] for a discussion of P Cyg profiles).

The shapes of the P Cyg profiles reflect the radial variation of the stellar-wind properties; in particular, the maximum blue-shift in the absorption component is determined by the terminal velocity of the wind (of order three times the escape velocity of the star [143,144,388]), which is reached within ∼ 5 stellar radii from the surface. From a comparison of observed and calculated P Cyg profiles, the mass

Table 2.2. *UV observations of X-ray binaries*

X-ray source	Type	Optical counterpart	References
0053+604	Be	γ Cas	[35,124,186,201,294]
0115−737/SMC X−1	H	Sk 160	[37,188,466]
0239+610	Be	LS I +61°301	[205,215,217,287]
0352+309	Be	X Per	[21,186]
J0422+32	L		[418]
0543−682/CAL 83	L	star V	[104]
0532−664/LMC X−4	H	Sk−Ph	[37,466]
0535+262	Be	HDE 245770	[121,366,524]
0538−641/LMC X−3	H (bhc)	star 1	[450,451]
0535−668	Be	star Q	[63,208,391]
0540−697/LMC X−1	H (bhc)	star 32	[25,224]
0620−000	L (bhc)	V616 Mon	[523,524]
0900−403/Vela X−1	H	HD 77581	[130,241,366,402]
1118−615	Be	He 3−640	[81]
1124−684/N Mus '91	L (bhc)		[164]
1145−619	Be	HD 102567	[24,186]
1455−314/Cen X−4	L	V822 Cen	[28]
1617−155/Sco X−1	L	V818 Sco	[242,490,517,522]
1656+354/Her X−1	L	HZ Her	[129,180,206,207]
1700−377	H	HD 153919	[129,185,189,199,209,243]
1704+240	L	HD 154791	[149]
1735−444	L	V926 Sco	[187]
1822−371	L	V691 CrA	[297,298]
1956+350/Cyg X−1	H (bhc)	HDE 226868	[115,129,449,523a]
2023+338	L (bhc)	V404 Cyg	[494]
2127+119	L	AC211	[345,345a]
2142+380/Cyg X−2	L	V1341 Cyg	[77,286,316,488]

H: high-mass X-ray binary with evolved primary; Be: Be/X-ray binary; L: LMXB; bhc: black-hole candidate.

loss rate in the wind can be estimated ([153,256]). For systems with supergiant companions, the undisturbed P Cyg profiles (observed during superior conjunction of the X-ray source) are similar to those of normal supergiants of the same spectral type.

In several Be/X-ray systems, the mass loss rate inferred from the ultraviolet P Cyg profiles is too low to account for the observed X-ray luminosity (see, e.g., [497]). This supports the idea (see, e.g., [265,379]) that in these systems the wind is anisotropic, with a normal (i.e. high-velocity low-density) polar component and a dense low-velocity equatorial outflow.

The X-rays change the ionization balance of the stellar wind in a region around the compact star [197,198], and some ionic species that are abundant in the undisturbed wind may be ionized away in this region. As the stellar wind is accelerated outward, in the vicinity of the X-ray source the velocity distribution of these ions may then

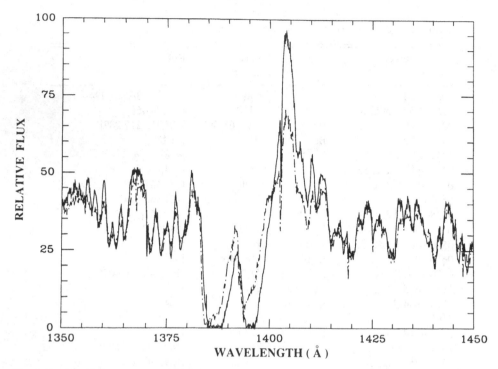

Fig. 2.11. The UV spectrum of HD 153919 (the optical counterpart of the HMXB 1700–377; spectral type O6.5 Iaf+) in the vicinity of the Si IV resonance doublet at 1394, 1403 Å is indicated by the full line. This spectrum was obtained during eclipse of the X-ray source. For comparison, the same part of the UV spectrum of the normal star λ Cep (spectral type O6 I(n)fp) is indicated by the dash-dotted line. Note the similarity between the photospheric spectra; the differences between the Si IV P Cygni profiles are related to the difference in luminosity class. (Courtesy L. Kaper.)

be truncated at the high side. As a result, the maximum blue-shift of the absorption component of a P Cyg profile is expected to vary with the orbital cycle: it will be high when the X-ray source is near superior conjunction, and low half an orbital cycle later. This predicted [196,197] orbital variation of the P Cyg profiles has been amply demonstrated with IUE observations of HMXBs (see Table 2.2), as variations of the terminal velocity (see Figure 2.12) and of the equivalent widths.

The stellar wind is accelerated by radiation pressure in spectral lines [240,280], and the additional ionization by X-rays will therefore also influence the velocity profile of the wind in the vicinity of the X-ray source. Since, in turn, the X-ray luminosity strongly depends on the wind velocity [113], it is clear that the interaction of the X-ray source with the stellar wind poses a complex problem [31].

X-rays can affect the UV spectrum of an HMXB also through Raman scattering: the X-ray spectrum in the vicinity of a very strong stellar line of a particular ion is mapped onto a spectral interval around a longer-wavelength line of the same ion. Kaper *et al.* [243] showed that Raman scattering of soft X-rays around the He II Lyβ line can explain the presence of a multitude of unidentified lines in an interval ~200 Å wide around the He II λ1640 line in the spectrum of HD 153919 (4U 1700–37) whose

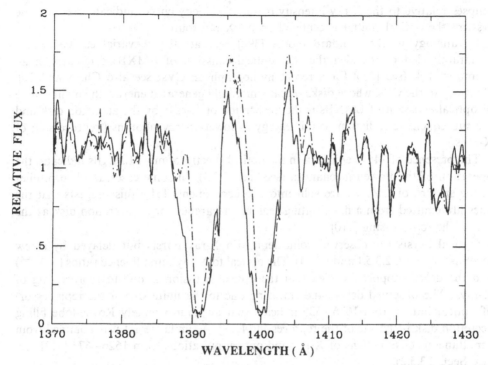

Fig. 2.12. Comparison of the UV spectra of HD 77581, the optical counterpart of the HMXB Vela X-1, as observed during X-ray eclipse (full line) and half an orbital cycle later (dash-dotted line). Note the difference in blueward extent of the absorption component in the P Cyg profile of the Si IV resonance line, which reflects a difference of the terminal velocity of the stellar wind induced by the X-ray source. The other (photospheric) lines do not differ much. (Courtesy L. Kaper.)

strengths vary regularly with orbital phase. Such lines may be used to reconstruct the very soft X-ray spectral region, which is difficult to access otherwise.

2.3 Low-mass X-ray binaries

2.3.1 *The origin of the optical emission*

The coincidence of soft X-ray transients (SXTs) with optical novae, with typical amplitudes $\gtrsim 5$ mag (see Sect. 2.3.6) shows that almost all optical emission (except for a small percentge) of SXTs near maximum is related to the X-rays. The great similarity of SXTs near maximum and the persistent LMXBs indicates that this holds for the latter as well. This explains why normal stellar features are generally absent in optical spectra of LMXBs (see Sect. 2.3.3).

Most LMXBs do not show large-amplitude periodic optical brightness variations as observed, e.g. in Her X-1 (see Sect. 2.3.4.2); this shows that, in general, the optical emission is not dominated by heating of one side of a companion star by X-rays.

Several LMXBs show partial eclipses of the optical emitter (e.g. 0748−676, 0921−631, 1822−371; see Ch. 14 for references). The phase relations of these eclipses relative to the X-ray intensity or radial-velocity curves indicate that in these systems the optical emitter is centered on the X-ray source.

By analogy to the standard model [398] for cataclysmic variables (CVs), one is naturally led to the idea that the optical emission of LMXBs originates in an accretion disk (see [260] for a recent monograph on CVs; see also Ch. 8 and 10). Contrary to the CVs, whose disks radiate internally generated energy, the main source of optical emission of LMXBs is the absorption of X-rays by the accretion disk and the subsequent re-radiation of this energy as lower-energy photons ('reprocessing of X-rays').

The presence of $\lambda 4640$ emission in the optical spectra of most LMXBs indicates the operation of the Bowen mechanism (see Sect. 2.3.3). In some cases, the line profiles in the spectra of LMXBs are split into two components [51]; this suggests that the lines are emitted from a flat rotating region, and argues for an accretion disk as the site of the reprocessing [210].

Optical bursts are observed coincident with X-ray bursts but delayed by a few seconds (see Sect. 2.3.5.1 and Ch. 4). The optical to X-ray burst fluence ratios ($\sim 10^{-4}$) and the delay support the idea that the optical emission is due to reprocessing of X-rays. The observed delay (and smearing due to the finite size of the reprocessor) of optical bursts from 1636−536 indicate that an approximately Roche-lobe filling accretion disk is the site of the reprocessing [369]. A similar result was obtained from simultaneous observations of X-ray and optical pulsations from 1626−673 [313] (see also Sect. 2.3.5.2).

Since the optical luminosity due to reprocessing generally far exceeds that generated internally in the disk, the structure of the disk is strongly affected by the incident X-rays. In a non-irradiated disk, the internally generated flux (F_i) varies with distance R from the compact star approximately as R^{-3}; thus the (effective) temperature T varies as $R^{-3/4}$ [416]. Since the flux due to reprocessing (F_r) varies as R^{-2}, the relative importance of X-ray heating increases with R. When reprocessing dominates, T varies as $R^{-3/7}$ [488]; and the shape of the disk is also affected – the disk thickness h varies as $R^{9/7}$ (compared with $h(:)R^{9/8}$ for a non-irradiated disk [416]).

2.3.2   *Average optical properties of LMXBs*

We discuss the following optical properties of LMXBs: intrinsic color indices $(B-V)_0$ and $(U-B)_0$, absolute magnitude M_V and the ratio of optical to X-ray flux. The basic data for these quantities come from Ch. 14. For SXTs, we have taken data at maximum, when they are similar to the persistent LMXBs. For eclipsing sources, the data refer to out-of-eclipse observations. In addition, we discuss the temperature of the optical emitter and the kinematic properties of LMXBs.

2.3.2.1   *Color indices*

We obtained the intrinsic colors $(B-V)_0$ and $(U-B)_0$ (see Figure 2.13) using $E(B-V)$ values from Ch. 14 and the standard relation $E(U-B) = 0.72 \times E(B-V)$. In estimating their average values, we have excluded a small number of sources (indicated in Figure 2.13). They are: 0921−631, 2023+338, Her X-1 and Cyg X-2 the orbital periods of which indicate they have an evolved companion which may contribute significantly to the optical emission; 2129+471, the optical emission

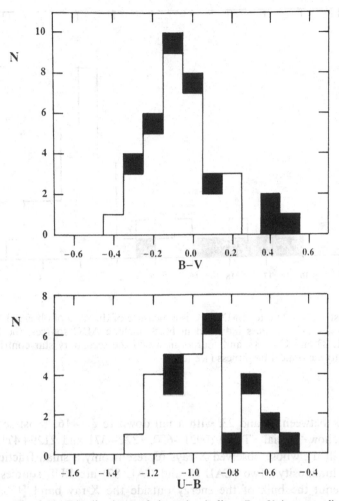

Fig. 2.13. Distributions of the color indices $B - V$ (top panel) and $U - B$ (bottom panel) for optical counterparts of LMXBs. The sources indicated in black include the LMC sources CAL 83 and CAL 87, sources in which the secondary star contributes significantly to the optical brightness, sources whose optical emission in contaminated by a nearby star, and a source for which there is some doubt about the optical identification (see text).

of which is contaminated by a nearby star (see Sect. 2.6); 0042+327 the optical identification of which is in doubt [289]; and CAL 83 and CAL 87, which may not be accreting neutron stars (see [464]). The average values of $(B - V)_0$ and $(U - B)_0$ are -0.09 ± 0.14 (1 s.d.) and -0.97 ± 0.17, respectively, similar to those of a flat energy distribution (i.e. $F_\nu = $ constant), for which (using [211]) $U - B = -1.0$ and $B - V = 0.2$.

2.3.2.2 *Ratio of X-ray to optical luminosity*

As for HMXBs we use as a parameter characterizing the ratio of X-ray to optical flux the quantity $\xi = B_0 + 2.5 \log F_X(\mu\text{Jy})$. The ξ distribution (see Figure 2.14)

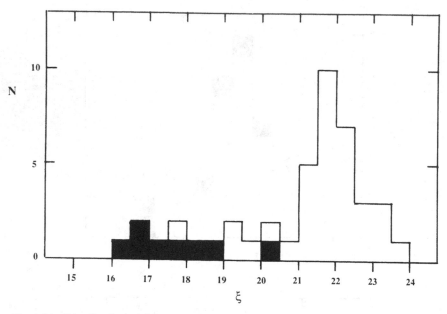

Fig. 2.14. Distribution of ξ for LMXBs; ξ is a measure of the ratio of observed X-ray to optical fluxes. The sources indicated in black include ADC sources, the LMC sources, CAL 83 and CAL 87, and sources in which the secondary star contributes significantly to the optical brightness (see text).

shows a strong peak between 21 and 23, with a tail down to $\xi \sim 16$. Most sources in this tail are somehow 'special'. Three (0921−630, 1822−371 and 2129+471) are ADC sources (see Ch. 1), whose observed X-rays represent only a small fraction of the intrinsic X-ray luminosity. Two (CAL 83 and CAL 87) are EUS sources (see Sect. 2.3.5), which emit the bulk of the energy outside the X-ray band [170,360], and may not be accreting neutron stars (see [464]). The companions of GX 1+4, 2023+338, Cyg X-2 and Her X-1 contribute significantly to the optical brightness; that of Her X-1 also shows a pronounced heating effect due to the tilt of the accretion disk [154]. We do not include these sources in the average value of ξ. This leaves two sources (1354−645 and 2127+119) in the low-ξ tail. It is unlikely that the low ξ value for 2127+119 can be explained by assuming it is an ADC source [126,485]; it may related to a very low metallicity [136].

For the remaining 31 LMXBs, the average value $\bar{\xi} = 21.8 \pm 1.0$ (1 s.d.); if we include the two sources in the low-ξ tail, $\bar{\xi} = 21.4 \pm 1.7$. Some of the scatter is due to errors in the reddening corrections and to non-simultaneity of most X-ray and optical observations. The different projections of the disks will contribute a scatter of ± 0.35 mag [470]. Thus, for a large majority of LMXBs the X-ray to optical flux ratios are distributed within a range of less than ± 1 mag (r.m.s.). Taking into account that the optical energy distribution of LMXBs is approximately flat (Sect. 2.3.2.1.), we find (using [231]) for the average ratio of the X-ray (2–11 keV) to optical (3000−7000 Å) flux a value of ~ 500.

Table 2.3. *Absolute visual magnitudes of LMXBs*

Source	M_V	Remark
CAL 83	−2.5	$d = 50$ kpc [139]
CAL 87	0.0	$d = 50$ kpc [139]
LMC X-2	−0.3	$d = 50$ kpc [139]
2127+119	1.0	$d = 10$ kpc [485]
1850−087	5.6	$d = 6.5$ kpc [106]
0748−676	1.4	radius expansion burst [169]
1636−536	1.3	radius expansion burst [275]
1735−444	2.2	radius expansion burst [107]
1905+000	4.4	radius expansion burst [73]
1916−053	5.3	radius expansion burst [427]
2127+119	1.6	radius expansion burst [485]
0620−000	0.7	secondary spectrum [306]
Cen X-4	0.9	secondary spectrum [74]
Aql X-1	0.9	secondary spectrum [443]
Sco X-1	0.0	Z source [194]
GX 349+2	0.0	Z source [194]
Cyg X-2	−2.0	Z source [194]

2.3.2.3 *Absolute visual magnitude*

For several groups of LMXBs, we can estimate the absolute visual magnitude (see Table 2.3).

(1) *LMXBs in stellar systems at known distances.* These include CAL 83, CAL 87 and LMC X-2 in the Large Magellanic Cloud, and the globular cluster sources 2127+119 (in M15) and 1850−087 (in NGC 6712). Distances to these stellar systems have been determined by optical methods (see Table 2.3 for references).

(2) *Sources with Eddington limited X-ray bursts.* The peak luminosity of an X-ray burst that causes photospheric radius expansion equals the Eddington limit (see Ch. 4); this is a reasonably good standard candle, so the distance to the burst source can be estimated. We have taken for the Eddington limit a value of 3×10^{38} erg s^{-1} (appropriate for helium-rich material, a 1.4 $M_\odot$ neutron star and a moderate (factor 1.2) redshift correction).

(3) *Soft X-ray transients.* Several SXTs in quiescence have shown the late-type spectrum of the secondary. If the orbital period is not known, one has to make an assumption about the structure of the secondary, e.g. that it is a main-sequence star. Then its absolute magnitude $M_{V,2}$ is known directly (see, e.g., [29]), and its M_V during outburst is given by $M_V = M_{V,2} + V_{max} - V_{min}$. The magnitude at minimum may have to be corrected for some remaining contaminating emission related to mass transfer (e.g. [74]). If the orbital period is known, an estimate of the secondary radius can be made (it will be somewhat dependent on the component masses). With the reasonable assumption that spectral type uniquely determines the V band surface brightness of a star, one can then make the required correction to the standard spectral type-$M_{V,2}$ relation.

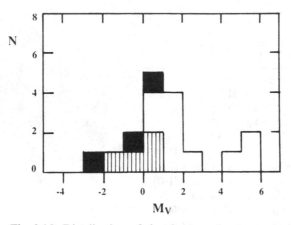

Fig. 2.15. Distribution of the absolute visual magnitudes for LMXBs. Sources indicated in black are located in the Large Magellanic Cloud. Burst sources are indicated in white; the vertical stripes indicate galactic LMXBs that have not shown X-ray bursts.

(4) *Z sources.* Theoretical considerations [195,262] indicate that the X-ray luminosity of Z sources in the 'normal-branch' state [194] is very close to the Eddington limit (see also [371]). We have used a standard-candle value of 2×10^{38} erg s^{-1} (for hydrogen-rich accreting material).

In the following discussion we will not include the EUS sources CAL 83 and CAL 87 (see above), or Cyg X-2, which has an evolved component that contributes significantly to the optical brightness [88]. The optical counterparts of burst sources are systematically fainter than those of the non-bursting sources (see Figure 2.15). This is likely to be due to a systematic difference in their respective X-ray luminosities. The Z sources and LMC X-2 radiate close to the Eddington limit, and the high mass of the X-ray source in 0620−000 [306] allows even sub-Eddington luminosities to reach very high values; the persistent X-ray luminosities of burst sources are generally between one and ten per cent of the Eddington limit [481].

Three burst sources (1850−087, 1905+000 and 1916−053) are much fainter than the others. A possible explanation may be based on the following consideration. For a given (relative) disk shape, the average temperature of the disk will roughly scale with X-ray luminosity L_X and orbital period P according to T (:) $L_X^{1/4} a^{-1/2}$ (:) $L_X^{1/4} P^{-1/3}$ (a is the binary separation). In the temperature range encountered in LMXB disks (see Sect. 2.3.2.4), the visual surface brightness S_V varies approximately as S_V (:) T^α, with $\alpha \sim 2$. For the visual luminosity L_V, one therefore expects L_V (:) $S_V a^2$ (:) $L_X^{1/2} P^{2/3}$, i.e., for a given X-ray luminosity, the optical luminosity of the disk decreases as the disk gets smaller because the disk is hotter, and a relatively larger fraction of the reprocessed emission appears in the UV. This decrease may be amplified if the assumption of a constant relative disk shape is invalid, e.g. if a smaller disk is simply the inner (and thinner) part of a (concave) larger disk. For the burst sources in Table 2.3 with known period, we have calculated the quantity $\Sigma = (P/1 \text{ hr})^{2/3} \gamma^{1/2}$, where $\gamma = L_X/L_{Edd}$ (see [481]). The average of $\log \Sigma$ for the four burst sources with $M_V \sim$

1.5 and known P is -0.2 ± 0.2 (1 s.d.); the value for 1916–05 is -1.1. Based on this simple analysis, one would expect that the optical brightness of 1916–053 is weaker by 2.3 mag; this accounts reasonably well for the results in Table 2.3. On the basis of the foregoing, one would expect that 1850–087 and 1905+000 have very short orbital periods as well.

2.3.2.4   Temperature

An estimate of the temperature T of the optical emitter in LMXBs may be obtained from an energy balance consideration of X-ray reprocessing, put in the form $R^2 T^4 = \epsilon\, R_X^2 T_X^4$. Here, T and T_X are the (effective) temperatures of the optical emitter and the X-ray source, respectively, and R and R_X are their sizes; ϵ is an efficiency factor that includes the X-ray albedo and geometric factors. With $R_X \sim 10$ km, $R \sim 1$ light second, $T_X \sim 10^7$ K and $\epsilon \sim 0.1$, we obtain $T \sim 10^{4.5}$ K. This result is confirmed by observational estimates.

Blackbody fits to the shape of the UV continuum of Sco X-1, 1735–444 and 0620–000 yield temperatures in the range 25 000–30 000 K [187,522,523]; more detailed models for X-ray irradiated accretion disks [28,488] confirm these estimates. Observed X-ray/optical co-variability can be used to measure a temperature if the optical emitter is assumed to be a blackbody [134]; then, $d\log F_{opt}/d\log F_X = (h\nu/4kT)$ $[1 - exp\,(-h\nu/kT)]^{-1}$. From simultaneous X-ray/optical burst observations, quiescent temperatures in the range from 30 000 to 60 000 K have been derived for 1636–536 [303,369]. Similar values have been inferred from the X-ray and optical decay rates of SXT [134,340,470]. From the variation of the *UBV* colors during a burst Lawrence *et al.* [268] obtained a non-burst temperature of 25 000 K.

2.3.2.5   Kinematic properties of LMXBs

Based on their own radial-velocity data collected for many LMXBs, and on published data, Cowley *et al.* [92,94] estimated center-of-mass velocities for 19 LMXBs with known orbital periods. From a solar-motion analysis of their results, they found that the LMXBs as a group have a high velocity dispersion ($\sigma \sim 130$ km s^{-1}) and a small galactic rotation velocity ($V_{rot} \sim 85$ km s^{-1}), intermediate between the corresponding values for the metal-poor globular clusters (galactic halo population) and the metal-rich globular clusters (thick-disk population). Since the age difference between these two groups of globular clusters is relatively small, Cowley *et al.* concluded from this that LMXBs are among the oldest objects in the Galaxy, with ages in excess of 10^{10} years. However, we caution that, in this interpretation of the kinematic data, the possibility that LMXBs may obtain a substantial kick velocity at the formation of the neutron star is not taken into account.

2.3.3   Optical spectra of LMXBs

As viewed from the compact source, a typical accretion disk has a half thickness of $\sim$10–15° [296,470], and it therefore intercepts about one-quarter of the flux from the central source, or about 10^{37} erg s^{-1}. Thus, much of the disk is intensely heated, and it greatly outshines a typical K/M dwarf secondary, which has an intrinsic luminosity of $\lesssim 10^{33}$ erg s^{-1}. Moreover, the secondary is often not an important site for reprocessing X-rays because of its modest solid angle as viewed from the compact source ($\Delta\Omega/4\pi \lesssim 0.02$), and because the secondary is largely shielded by the accretion disk.

Thus, except for a few systems (e.g. those that contain giant companions), the optical/UV spectrum of an LMXB is the spectrum of an X-ray illuminated accretion disk. Typically it consists of a blue continuum ($T_{eff} \sim 25\,000–30\,000\,K$; see Sect. 2.3.2.4) and a few high-excitation emission lines, most notably He II $\lambda4686$ and a blend at $\lambda\lambda4640–50$. Balmer emission lines are also frequently observed. Numerous examples of such spectra can be found in the literature ([42,92,94,301,423,425,470]; for additional references, see Table 2.4 and Ch. 14).

2.3.3.1   Spectrum of Sco X-1

Because few LMXBs are much brighter than $V \sim 18$, most spectra are of moderate or low quality. A spectrum of Sco X-1, which is by far the brightest persistent LMXB ($V \sim 12.5$), is shown in Figure 2.16. This spectrum contains numerous lines of the Balmer and Paschen series of H, several lines of He I (e.g. $\lambda\lambda5876,6678,7065$) and He II (most notably $\lambda4686$), broad components of emission due to multiplets of Fe II, and the O III and N III Bowen lines (see Sect. 2.3.3.4). Interestingly, absorption lines in the range 3580–3770 Å may also be present [405]; absorption lines are only rarely observed in the spectra of active LMXB.

This spectrum of Sco X-1 illustrates two features that are common to almost all spectra of active LMXBs. First, the object is blue ($B - V \sim 0.2$, $U - B \sim -0.8$); the continuum is flat, i.e. $F_\nu \sim$ constant (see Sect. 2.3.2.1). Secondly, He II $\lambda4686$ and N III/C III $\lambda\lambda4640–50$ are the most intense, high-excitation lines present in the commonly observed blue spectral band (~ 4000–5000Å).

2.3.3.2   Equivalent widths of prominent emission features

The profiles of emission lines contain information on the emission region that may be uncovered with, e.g., a Doppler tomographic analysis [293]. In view of the faintness of the optical counterparts of most LMXBs, however, the statistical quality of their spectra is often insufficient for a detailed analysis; in general, the only quantitative information available is the equivalent width.

In Table 2.4 we have collected equivalent widths of the three most prominent emission lines found in blue spectra of LMXBs: Hβ, He II $\lambda4686$ and the C III–N III $\lambda\lambda4640–50$ blend. The table is based on two previous compilations [333,474] and more recent literature. (For near IR spectra, see [97].) The average values of the equivalent widths of Hβ, $\lambda4686$ and $\lambda4640$ are 2.6 ± 3.7Å, 5.5 ± 3.8Å, and 2.8 ± 2.1Å, respectively (the errors are 1 s.d. of the distributions). Their largest observed values are 28 Å, 20 Å and 8 Å, respectively. Very few individual values exceed 6 Å.

2.3.3.3   X-ray heated gas: nebulae, stars and disks

A salient characteristic of the spectrum of Sco X-1 and all LMXB spectra is the absence of forbidden lines. This immediately distinguishes the photoionized nebula near a compact X-ray source from familiar photoionized nebulae such as H II regions, planetary nebulae, nova shells and supernova remnants. The spectra of these latter, low-density ($n_e \sim 10^4$ cm^{-3}) plasmas are dominated by forbidden lines [352].

The spectra most like the spectra of LMXBs are those of SyI galaxies and quasars

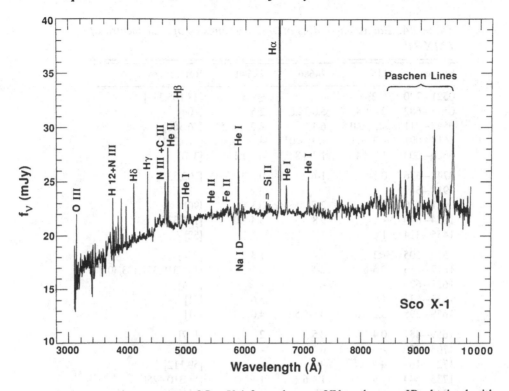

Fig. 2.16. A composite spectrum of Sco X-1 from the near UV to the near IR obtained with the Lick 3 m Shane telescope (adapted from [405]). Three gratings were used: the resolution is 4 Å in the blue, 8 Å in the red and 16 Å in the infrared. All the data are binned at 2 Å. The spectrum has not been corrected for reddening, which is small ($E_{B-V} \approx 0.2$ mag).

(AGNs). Here we ignore the 'narrow-line' component in AGN spectra, which arises in an extended, low-density region and is similar to the spectrum of an H II region or a planetary nebula. We consider the 'broad-line' region (BLR), which emits both very broad, permitted emission lines and a relatively weak, featureless blue continuum. The BLR consists of dense clouds ($n \sim 10^9$ cm^{-3}), which are located relatively near ($R \sim 10^{17}$ cm) to a luminous ionizing source ($L_X \sim 10^{46}$ erg s^{-1}) [352, 527]. At the surface of such a cloud, the large value of the ionization parameter, $\xi \equiv L_X/nR^2 \sim 10^3$, establishes that the gas is highly ionized [240].

Now consider an LMXB containing a uniform disk with a radius of $\sim 10^{11}$ cm, half thickness $\sim 15°$, total mass $\sim 10^{-9} M_\odot$ [377] and therefore a mean gas density $n \sim 10^{15}$ cm^{-3}. Near the surface of the inner disk ($R \sim 10^{10}$ cm) and with $L_X \sim 10^{38}$ erg s^{-1}, one finds $\xi \sim 10^3$. Thus, the value of the scaling parameter, ξ, which uniquely determines the state of the gas (ignoring the moderate spectral differences between AGNs and LMXBs) is comparable for the inner disk region in LMXBs and for the BLR in AGNs. Not surprisingly, therefore, there are a number of similarities between the optical/UV spectra of LMXBs and the BLRs in AGN: for example, the absence of forbidden lines, large line widths due to high bulk velocities, and the presence of Fe II permitted lines [352].

Table 2.4. *Equivalent widths of emission lines in optical spectra of LMXBs[a]*

Source	Hβ	λ4686	λ4640	References
0521−720	(−2)-8	4-6	≲0.5	[38,105,333]
0543−682	2.4-6.4	5.0-12.8	2.5	[104]
0547−711	<3; ≲0.5	6-17	<3; ≲0.5	[96,360]
0614+091	<3; <1	<3; <0.9	4.3-7.5	[110,281]
0620−003	1.9-4.4	1.7-4.8	0.3-1.1	[350]
0748−676	0.5-5	1-6	0.5-2	[103,333]
0918−549			+?	[175]
0921−630	0-4	2-7	0.3-5	[90]
1254−690	<3	9.8	5.4-8.0	[171,335]
1455−314	1	1.5	1.5	[52]
1556−605	<1	4.4	1.3	[336]
1617−155	0.5-6.2	2-5	3-9	[101,310,331,403,405]
1627−673	-	-	-	[374]
1636−536	1.9	4.6	5.6	[51]
1658−298	3.2-3.6	4.5-6.5	4.0-5.9	[51]
1659−487	0.4-1.0	2.5	2	[177]
1705−250	-	6	-	[172]
1728−169	+?	+	+	[59,112]
1735−444	1.4	1.4-6.2	1-9	[51,310,425]
1755−338	-	-	-	[311]
1822−371	(−1.3)-1.0	1.8-2.5	0.4-2.8	[62,89,171]
1822−000			+?	[175]
1837+049	0	+?		[445]
1908+005	1.5	0.9-2.8	1.3-2.8	[61]
1957+113	-	-	-	[292]
2000+251		12:		[66]
2023+338	7-28	10-20	3-6	[55]
2127+119	<0	∼8	<0.6	[64,333,345]
2129+470		2	2.3	[204]
2142+380	2.5-7	1.2-4.3	0.4-2.5	[333,484]

[a] Equivalent widths are given in angstroms; negative values indicate absorption lines. + : present; - : not detected; +? : probably present; a blank indicates lack of information.

A number of photoionization codes have been developed with applications to X-ray binaries. Most codes are built around a simple geometry: a spherically symmetric, constant-density nebula, which surrounds a central, compact source of continuum X-rays (e.g. [240] and references therein). The ionization structure and emissivity of the gas are controlled primarily by photoionization, recombination and fluorescence. Excitation and ionization by electron collisions are less important. Such models have proved useful in understanding gas dynamic effects near a compact object and in modeling certain optical/UV spectral features, especially in the case of HMXBs

[240]. However, in an LMXB, the photoionized plasma, which is the surface of an accretion disk, and/or the inner face of the secondary, is irradiated from the outside by the compact source. Therefore, the nebular model has limited applicability to LMXBs.

Models for the UV/optical radiation emitted by an X-ray heated secondary star are applicable to a few LMXBs, e.g. Her X-1, Cyg X-2 and 2129+470 [277]. On the other hand, models for the UV/optical appearance of X-ray illuminated accretion disks are widely applicable. Among the most successful applications of such disk models to date are IUE UV and Ginga X-ray studies of Cyg X-2 [488] and Sco X-1 [490]. We comment briefly on the latter study.

Sco X-1 has an orbital period of 18.9 hours and an orbital separation of $\sim 3 \times 10^{11}$ cm. Vrtilek *et al.* [490] computed the disk temperature vs. radius for an assumed steady accretion disk with the mass transfer rate ($\dot{M}$) and the outer disk radius as free parameters; the mass and radius of the compact object were held fixed at $1.4 M_\odot$ and 10^6 cm. The thickness of the disk vs. radius was determined self-consistently and iteratively, taking into account both the absorbed X-ray flux and the viscous dissipation in the disk. The UV spectra were obtained by summing stellar spectra at the same effective temperature. These composite model spectra were fitted to several dozen spectra of the UV continuum (1200–3200 Å), thereby determining values of $\dot{M}$ in the range $(0.4 - 1.1) \times 10^{-8} M_\odot \text{ yr}^{-1}$. A single value was found for the best-fit outer disk radius, 6×10^{10} cm.

A total of nine UV lines were definitely detected with IUE, including familiar resonance lines (e.g. C IV $\lambda 1545$) and the strongest Bowen fluorescence line O III $\lambda 3133$ (see Sect. 2.3.3.4). A first attempt at interpreting the emission lines is described (details to be given in [392]). It is plausibly assumed that the emission lines are formed in a hotter layer above the photosphere of the accretion disk. The authors also assume that the X-rays, which strike the disk at grazing angles, are due to thermal bremsstrahlung at 10^8 K. Using the parameters from the continuum fits mentioned above, they calculated the ionization and thermal balances in the irradiated surface layer of the disk and computed the fluxes in the emergent UV emission lines (for $6.3 \times 10^9 \text{ cm} < R < 6.3 \times 10^{10}$ cm). Good agreement was found between the overall line fluxes computed for the model and the IUE spectra, which sampled a broad range of Sco X-1's behavior. The authors stress: 'No strong lines are predicted by the model that do not exist in the observations, and no strong lines are detected in the observations that are not predicted by the model.' Thus, the model of X-rays reprocessed in the outer disk of Sco X-1 can reasonably account for the UV observations.

2.3.3.4 *Bowen fluorescence in Scorpius X-1*

Another quantitative approach to interpreting the spectrum of an LMXB was made by Schachter *et al.* [405]. They obtained high-quality spectra of Sco X-1 (Figure 2.15), and focussed their attention on interpreting the intensities of the He II $\lambda 4686$ line and the O III and N III Bowen fluorescence lines.

In 1934, Bowen [39,40] suggested a resonance fluorescence mechanism for the creation of certain intense permitted emission lines of O III and N III in planetary nebulae. The mechanism works because of coincidences in the wavelengths of EUV

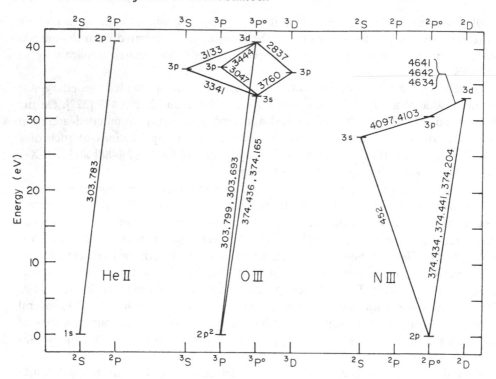

Fig. 2.17. Grotrian diagrams of He II, O III and N III, including only the lines which play an important role in the Bowen fluorescence process.

resonance lines of He II, O III and N III. The relevant levels and most important transitions of the ions are shown in Figure 2.17. The He II $\lambda303.783$ resonance line (hereafter, He II Lyα) pumps the resonance line of O III $\lambda303.799$. Usually, the excited O III decays to the ground state by emitting a $\lambda304$ photon; however, 1.8 per cent of the time the O III emits one or two Bowen lines ($\lambda\lambda3133$ and $\lambda3444$ are the strongest) followed by other resonance lines. One of these, O III $\lambda374.436$, may in turn pump the resonance lines of N III $\lambda\lambda374.434, 374.441$. As before, the excited N III usually decays to the ground state by emitting a $\lambda374$ photon. In 1.3 per cent of the cases, however, the N III will emit one or two Bowen lines from the following multiplets: N III $\lambda\lambda4634$–4642 and N III $\lambda\lambda4097$–4103.

McClintock *et al.* [309] proposed that one of the dominant emission features in the optical spectra of LMXBs – a broad blend near $\lambda4640$ – is due to the Bowen mechanism and is composed of the N III $\lambda\lambda4634$–4642 Bowen lines plus possible contributions from C III and O II. This hypothesis was confirmed a few years later via the detection of the O III $\lambda3444$ Bowen line in HZ Her [290]. (The $\sim \lambda4100$ Bowen blend mentioned above is difficult to observe because it is blended with Hδ.)

In the case of planetary nebulae, He II emission is due solely to recombination, and therefore the intensity of He II Lyα is simply proportional to the intensity of He II $\lambda4686$; the proportionality constant depends weakly on the temperature in the nebula [239]. In order to compare theory and observation, one usually computes the yield (= efficiency) of the Bowen mechanism. The yield y_B (or y'_B) is defined to be the fraction of the He II Lyα photons that eventually produce O III (or N III) Bowen line emission [239]:

$$y_B = \kappa I(\lambda 3444)/I(\lambda 4686)$$

$$y'_B = 2.4 I(\lambda\lambda 4634 - 4642)/I(\lambda 3444),$$

where $\kappa \approx 1$ ([239] and references therein). For a model calculation of a typical planetary, NGC 7027, Kallman & McCray [239] find mean Bowen yields of $\bar{y}_B = 0.42$ and $\bar{y}'_B = 0.39$. This is in reasonable agreement with the observed mean values and dispersions for a sample of 19 nebulae: $\bar{y}_B = 0.37 \pm 0.04$ and $\bar{y}'_B = 0.61 \pm 0.08$ [239]. The success of the theory allows one to place constraints on the ion abundances, emission measures, and optical depths.[*]

In X-ray binaries, the column densities of ionized gas are much greater than in planetary nebulae, and the optical depths in resonance lines are so large that the escape probability for the EUV pump photons is negligible. As a result, a representative model of an accreting X-ray source ($L_X = 10^{38}$ erg s^{-1}; $kT = 10$ keV; $n = 10^{11}$ cm^{-3}) gives large values for the mean Bowen yields: $\bar{y}_B = 0.95$; $\bar{y}'_B = 0.70$ [239].

The only observational, quantitative study of Bowen fluorescence in LMXBs to date is the work of Schachter *et al.* [405] on Sco X-1. Their overall spectrum is shown in Figure 2.16, and detailed spectra and line intensities can be found in their paper. They observed the two O III Bowen lines that are predicted to be the strongest optical lines in a static nebula: $\lambda 3133$ and $\lambda 3444$. Also, they possibly observed the O III $\lambda 3407$ line, which would imply that large velocity gradients may be present in the nebula [117]. They extracted the intensities of the N III Bowen lines ($\lambda 4634$ and $\lambda\lambda 4641,4642$) from the blend near $\lambda 4640$, and measured the intensities of the O III Bowen lines and He II $\lambda 4686$. The following are four conclusions of their study. (1) The intensity ratio, $I(\lambda 3133)/I(\lambda 3444)$, agrees with theory. (2) The observed value of the oxygen Bowen yield, $\bar{y}_B \approx 0.55$, is in reasonable agreement with the model calculations of Kallman & McCray (see above). (3) The anomalously high value of the nitrogen Bowen yield, $\bar{y}'_B \approx 3.5$, may be due to the presence of large velocity gradients; support for this suggestion comes from the possible detection of the O III $\lambda 3407$ line [117]. (4) Charge transfer, a mechanism that may modify the Bowen line intensities [432], is a small effect (< 9 per cent) in Sco X-1.

In this study by Schachter *et al.*, several assumptions are made that require future scrutiny. For example, the assumption of Case B He II recombination may be invalidated by the high He II opacity. Also, both He II $\lambda 4686$ and He II Lyα are assumed to be produced solely by recombination, although electron collisions may be more important than recombination (cf. [239]). In this context, Schachter *et al.*

---

[*] Kastner & Bhatia [244,245] argue that the $\lambda 4640$ and $\lambda 4100$ multiplets of N III emitted by planetary nebulae (and X-ray binaries) are not produced by the Bowen process. (They agree, however, that the O III lines are pumped by the Bowen process.) More recently, Ferland [140] has supplied an alternative model for producing the N III lines in planetary nebulae, which is based on continuum fluorescence rather than the Bowen process. These conclusions are inconsistent with the substantial NIII Bowen yield found by Kallman & McCray [239] for a model of NGC 7078 (see above). This inconsistency is not addressed by Kastner & Bhatia [244,245] or by Ferland [140]. Given these uncertainties, we have chosen to ignore the recent results and to complete our discussion using Bowen's model for the N III line strengths and Kallman and McCray's [239] conclusions.

[405] derive the following constraints on the size and electron density of the Bowen emission-line region (ELR):

$$4 \times 10^{11} < R\,(\text{cm}) < 1 \times 10^{14}$$

$$3 \times 10^6 < n_e\,(\text{cm}^{-3}) < 2 \times 10^{10}$$

(Here we have simplified the expressions given by Schachter *et al.* [405] by inserting the nominal values for several parameters; e.g. $D = 700$ pc.) The minimum radius of the ELR is thus a few times larger than the expected Roche lobe radius of the primary ($\sim 10^{11}$ cm). If this is true, most of the emitting gas in the ELR is freely escaping from the system and has to be continually replenished. Alternatively, Schachter *et al.* suggest that the ELR is a low-density fringe of a denser feature. Finally, it is possible that Schachter *et al.* have made an invalid assumption (see above) that has led to an erroneously large size for the ELR.

2.3.3.5 Metallicity

It is reasonable to suppose that the strength of the $\lambda\lambda4640{-}50$ blend, which is composed of the N III Bowen lines and possibly some lines of C III and O II [309], might be positively correlated with metallicity. Motch and Pakull [333] have made this idea useful by normalizing the equivalent width (EW) of $\lambda\lambda4640{-}50$ by the EW of He II $\lambda4686$, $R \equiv$ EW $(\lambda\lambda4640{-}50)/EW(\lambda4686)$, and by tabulating R for 18 LMXBs. Four systems that are located in low-metallicity environments (LMC X-2, CAL83 and CAL87 in the LMC, and 2127+119 in M15) have low values of R, ranging from $R < 0.12$ to $R < 0.36$. The values of R for 12 galactic systems are much greater: $< R > = 0.99 \pm 0.37$ (r.m.s.). The two remaining LMXBs, Cyg X-2 and 0921-63, which are probable halo-population systems with low metallicity [90], have intermediate values of $R : R = 0.46$ and $R = 0.33$, respectively. Thus, Motch and Pakull make a good observational case that R is a useful indicator of metallicity.

Furthermore, the observational evidence is supported by theoretical considerations. For example, photoionization models of compact X-ray sources (Sect. 2.3.3.4) show that the He II $\lambda4686$ line and the N III–C III–O II $\lambda\lambda4640{-}50$ blend originate approximately in a common region [240]. Moreover, the intensity of He II $\lambda4686$ is intimately tied to the intensity of the N III $\lambda\lambda4634{-}42$ Bowen blend via the intensity of He II Lyα (Sect. 2.3.3.4). It is not too surprising, therefore, that Motch and Pakull's study of individual LMXBs has shown there is a good correlation between the EW variations of He II $\lambda4686$ and the $\lambda\lambda4640{-}50$ blend. This correlation holds both for variations with orbital phase and for random variations.

Motch and Pakull mention two possible applications. First, metallicity can also be inferred for several LMXBs that exhibit periodic dips in their X-ray intensity curves (if the dips are attributed to photoelectric absorption by cold matter.) A comparison of the two methods might be fruitful. Secondly, model calculations indicate that the behavior of an X-ray burster depends upon the CNO abundance (see Ch. 4). It may be possible to test this hypothesis by making refined measurements of the metallicity indicator, R.

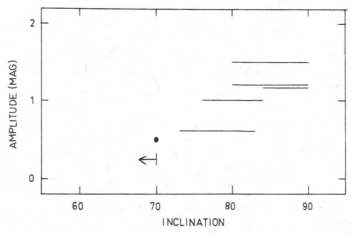

Fig. 2.18. Dependence of the amplitude of optical light curves of LMXBs on the orbital inclination [482]. Eclipsing systems are represented by horizontal lines, which indicate the allowed range of inclination. Three sources which show dips but not eclipses are assumed to have $i = 70°$. Systems which do not show eclipses or dips are likely to have smaller inclinations. Systems in which the companion star contributes significantly to the optical brightness have not been included in this figure.

2.3.4   Optical light curves of low-mass X-ray binaries

2.3.4.1 Introduction

Many LMXBs show an orbital modulation of their optical light (see [471] for an extensive review). This indicates that the distribution of reprocessing material is not axi-symmetric around the X-ray source. The model for the light curves of HMXBs (see Sect. 2.2.2.1) can be applied to LMXBs as well, albeit with different system parameters [335,471]. Since the secondary stars contribute little to the optical brightness, ellipsoidal variations are not observed (only when the X-rays are absent do they appear; see Sect. 2.3.4.4 and 2.3.6).

The amplitude, A, of the light curve depends on the orbital inclination, i (see Figure 2.18). For low inclinations (no X-ray eclipses or 'dips', see Ch. 1) the light curves are approximately sinusoidal, with $A \sim 0.2 - 0.3$ mag (Sect. 2.3.4.3). At somewhat higher i (i.e. X-ray dips but no eclipses), $A \sim 0.5$ mag. For these non-eclipsing systems, minimum light coincides with the superior conjunction of the compact star; this indicates that the optical light curve reflects the X-ray heating of the secondary, inasfar as this is not in the X-ray shadow of the disk (see Sect. 2.3.4.2). The same is likely to be the case for the light curves of the low-inclination systems for which the amplitude of the light curve predominantly reflects the ratio of solid angles (as seen from the X-ray source) of the secondary and the disk.

When X-ray eclipses occur, A reaches ~ 1.5 mag. The light curves of these systems can be decomposed into the sine wave that is also observed for the low-i systems, and a cusp superposed on the minimum of the sine wave. The cusps are due to the eclipse of the accretion disk.

The dependence of A on i reflects the fact that as i decreases the relative brightness

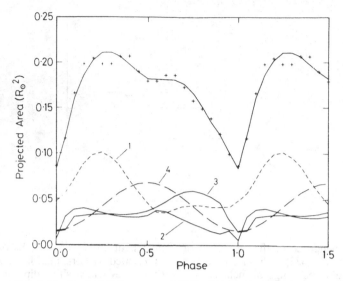

Fig. 2.19. Optical light curve of 1822−371 and best-fit theoretical light curve consisting of four components (see text), which are also shown individually [298].

of the disk increases (larger projection factor, less self shielding), and the importance of the variable component (eclipses of the disk, X-ray heating of the companion star) decreases.

In Sect. 2.3.4.2 and 2.3.4.3, we describe some representative LMXB light curves separately for high- and low-inclination systems. In Sect. 2.3.4.4, we discuss the light curves of LMXBs during X-ray off states.

2.3.4.2   High-inclination systems

1822-371 and CAL 87.   The optical light curve of the ADC source 1822−371 (see Ch. 1) contains a broad minimum, preceded by a slow ingress due to absorption by a non-axisymmetric outer disk rim [300]. Mason & Cordova [298] were able to successfully model the optical, UV, and IR light curves using an azimuthally structured disk, with contributions from (1) the inward facing side of the rim of the disk, (2) the disk, (3) the outer side of the rim, and (4) the X-ray heated side of the secondary (see Figure 2.19).

The light curve of CAL 87 [46, 360] is similar to that of 1822−371 (see Figure 2.19), which suggests that the disk structure inferred for 1822−370 may be representative for LMXBs in general.

1254-690 and 1755-338.   The X-ray intensity curves of 1254−690 and 1755−338 show dips, but not eclipses [86,511]. Their optical light curves [302,335] are sinusoidal, with minima ∼0.2 cycles after the dips (see Figure 2.20). In view of the phase relation between dips and eclipses when both are observed (see Ch. 1), this indicates that minimum light coincides with superior conjunction of the X-ray source. The same is likely to be true for the light curves of low-inclination systems which show neither dips nor eclipses in their X-ray intensity curves.

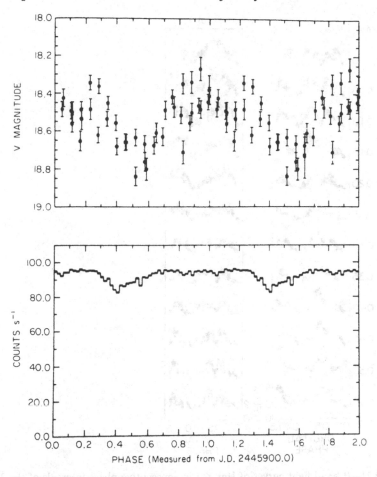

Fig. 2.20. Orbital variations of the *V* band magnitude and the X-ray intensity of 1755−338, showing the phase relation between the approximately sinusoidal optical light curve and the X-ray dips [302].

Her X-1, 2129+470 and 0748−676. The optical light curve of Her X-1 shows a single minimum that coincides with the X-ray eclipses [5,11,15,33,154,279]. The light curve is dominated by X-ray heating of the companion and eclipses of the accretion disk. The average X-ray intensity of Her X-1 shows a long-period (35 day) on−off cycle [155], which also appears in the optical light curve in the form of a feature that continuously drifts toward earlier orbital phase during the 35 day cycle [41,71,154] (see Figure 2.21). This 35 day X-ray and optical variation can be explained by precession of a tilted accretion disk [154,413]. Her X-1 shows extended off states (see Sect. 2.3.4.4).

The light curves of the ADC source 2129+470 [314,444] and the transient 0748−676 [103,337,408,482] are similar to those of Her X-1 (see Figure 2.22), but with reduced amplitude. 2129+470 was not detected in 1983 [378], and has remained in this off state since (see Sect. 2.3.4.4 and 2.6).

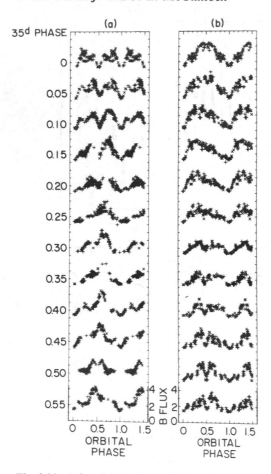

Fig. 2.21. *B* band light curves of Her X-1 in consecutive phase intervals of the 35 day precessional period [154]. (a) The curves in the left column have been folded modulo the 1.7 day orbital period, and data have been combined according to the symmetry rule $B(-\varphi, -\psi) = B(\varphi, \psi)$, where φ and ψ are the orbital and precessional phases, respectively. (b) The curves in the right column have been sampled along a diagonal in the (φ, ψ) plane and represent light curves that would be measured by an observer moving around the binary system in 35 days together with the precession of the disk.

2.3.4.3   Low-inclination systems

Sinusoidal light curves.   LMXBs with low inclinations show approximately sinusoidal optical light curves, which are due to X-ray heating of the secondary star. Their amplitudes are mainly determined by the orbital inclination and by the ratio of solid angles (as seen from the X-ray source) of the accretion disk and the secondary (insofar as it is not shielded by the disk). Such sinusoidal curves have been observed for CAL 83 [428], Sco X-1 [7,168], 1636−536 [368,410,424,483] GX 339−4 in the high state [48], GX 9+9 [406], 1735−444 [85,426,460] and 1957+115 [334,446].

Sco X-1 and Cyg X-2.   From an analysis of archival photographic plates, Gottlieb *et al.* [168] found a 0.2 mag brightness variation of Sco X-1 with a period

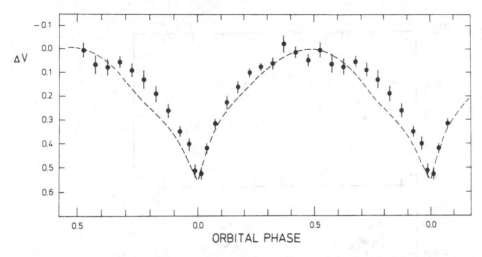

Fig. 2.22. Average *V* band light curve of 0748−676, compared with that of 2129+470, after scaling the amplitude of the latter (dashed curve) [482].

of 18.9 hours (Figure 2.23). Superposed on this modulation are much larger irregular variations which make it difficult to detect the orbital light curve from short stretches of data. The sinusoidal light curve is recognizable in both the active and inactive states [7].

The secondary of Cyg X-2 is a giant star whose optical brightness is comparable to that of the accretion disk [88]. In line with this, Cyg X-2 shows a double-waved light curve, with minima that occur at the conjunctions [88,165]; superposed on this ellipsoidal component are irregular brightness increases, due to variable heating of the disk (see Sect. 2.3.5.2).

2.3.4.4 *Ellipsoidal variability during X-ray off states*

Soft X-ray transients in quiescence. Several quiescent SXT (A0620−00, Cen X-4, GS 2000+251, N Mus 1991, and GS 2023+338) have shown double-waved optical light curves [45,74,306,308,317,409,494]. Some accretion-related optical emission from quiescent SXTs has been inferred from the presence of emission lines in their optical spectra (see Sect. 2.3.6), and from non-orbital optical brightness variations [74]. Such emission is also recognizable in some light curves, e.g. by their unequal maxima, which indicate a contribution from a 'bright spot', analogous to the orbital hump in the light curves of many CVs (see Figure 2.24 and Ch. 8).

Remarkably enough, during low-state observations of 2129+471 and 1543−475, the expected ellipsoidal variations were not detected [70,75,447]; this may indicate that these systems are triples (see Sect. 2.6).

Her X-1 'off' states. During long intervals, Her X-1 is in an optically low state [236], indicating that the X-ray emission responsible for the heating of the secondary is then very weak. In some low-state data, the light curve is double-waved [236,504], reflecting ellipsoidal variability. At other times, the light curve showed a

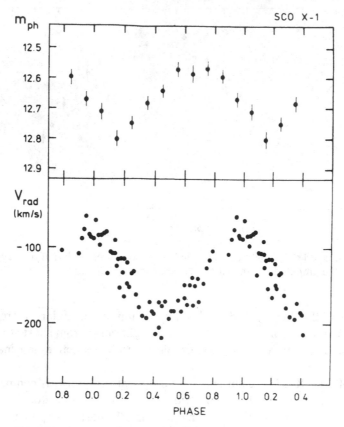

Fig. 2.23. Average photographic light curve and He II λ 4686 radial-velocity curve of Sco X-1 [168].

single deep minimum, which can be ascribed to the occultation of luminous matter near the compact star [211,212] (see Figure 2.25).

2.3.5   Correlated optical/X-ray brightness variations of LMXBs

Many LMXBs show correlated optical and X-ray brightness variations, on time scales between seconds and many days. Most of this co-variability can be readily explained by X-ray reprocessing (see Sect. 2.3.1). Informative examples of optical/X-ray co-variability are optical bursts and optical pulsations, and irregular optical/X-ray variations driven by variations in the accretion rate. We will discuss these in turn.

2.3.5.1   Optical bursts

Soon after the discovery of X-ray bursts (see Ch. 4), attempts were made to detect coincident optical bursts. After initial unsuccessful attempts on 1837+049 [2,22,441], the first simultaneous optical/X-ray burst was detected from 1735−444 [176] (see Figure 2.26). The fluence in the optical burst was $\sim 2 \times 10^{-5}$ that in the X-ray burst; this is too large by $\sim$ six orders of magnitude to explain the optical

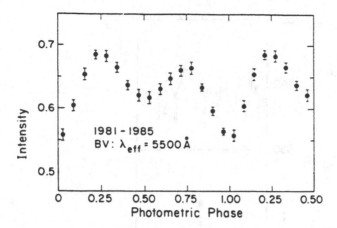

Fig. 2.24. Average quiescent optical light curve of the soft X-ray transient A0620—00 [306].

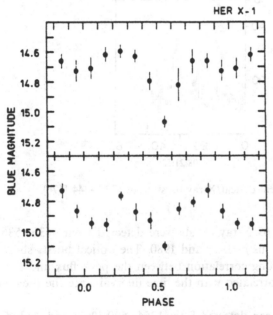

Fig. 2.25. Average blue light curve of Her X-1 during extended low states according to [211] (upper panel) and [236] (lower panel).

emission as the low-energy tail of the blackbody X-ray burst emission. The optical burst was delayed by ~ 3 s [312].

Optical bursts, in several cases correlated with X-ray bursts, have since been detected from 1837+049 [183] and 1636—536 [268,303,369,370,453,457]. All, except one, coincident optical bursts were delayed relative to the X-ray burst. The apparent zero delay of one optical burst from 1636—536 [457] may be caused by a clock problem (H. Pedersen, private communication).

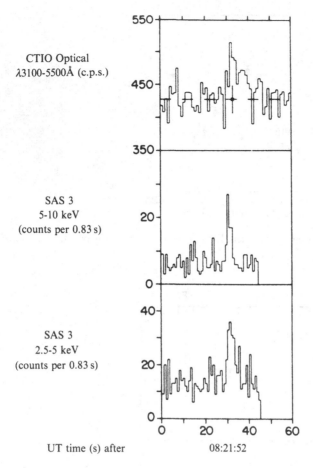

Fig. 2.26. First coincident optical/X-ray burst, from 1735−444 [176].

A total of 41 optical and 69 X-ray bursts were detected from 1636−536 during about ten weeks of observations in 1979 and 1980. The optical bursts show a large variety of profiles, with a strong correlation between the peak flux and the fluence, both of which, in turn, are correlated with the time interval since the previous burst [370].

Optical bursts have also been detected from 1254−690 [299] and Aql X-1 [477] (see also [470]); no simultaneous X-ray observations were made.

The optical bursts are delayed (and also smeared) as a result of a range of travel-time differences between the X-rays that directly go to the observer, and those that first hit the reprocessor, and then continue to the observer as optical photons (see Figure 2.27). Delay and smearing due to radiative processes are not very important [369]. Thus, the varying X-ray burst signal serves as a probe: it illuminates the surroundings of the neutron star, which are heated, and reveals their presence by optical radiation. For example, if the reprocessing is in the accretion disk, a quantitative analysis of the correlated optical and X-ray bursts can yield estimates of the size, thickness and average temperature of the disk.

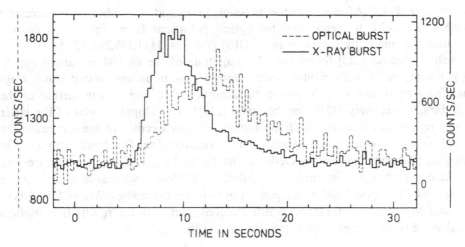

Fig. 2.27. Coincident optical/X-ray burst from 1636−536, showing the delay and smearing of the optical signal [453].

Based on this reprocessing model, Pedersen *et al.* analyzed three coincident optical/X-ray bursts from 1636−536 [369]. Independent of detailed geometric assumptions, they find that the reprocessor has a linear size of ∼1.5 light seconds (lt s). This puts a limit on the size of the disk, which is consistent with a binary separation corresponding to the 3.8 hour orbital period [368,424,483]. One of the events was observed in three optical passbands. Analysis of this time-resolved color information shows that X-ray reprocessing into optical light is reasonably described by a single-temperature blackbody model [268].

Matsuoka *et al.* [303] found that for five (of six) optical/X-ray bursts the delay times are consistent with a single value near 2.5 s; for one the delay was significantly smaller. This suggests that the reprocessing region is either moving or varying in size. A natural explanation would be that a significant fraction of the reprocessing occurs on the secondary star. One would then expect a correlation between the apparent size of the optical emitter and the delay time (both vary in phase with the orbital cycle). However, Matsuoka *et al.* did not find such a correlation (see also [483]). Also, the ratio of optical to X-ray burst fluences (and peak fluxes) does not follow the average optical light curve [483]. These results suggest that in 1636−536 substantial long-term changes occur in the spatial distribution of reprocessing material.

2.3.5.2 *Optical pulsations*

Optical pulsations arise from reprocessing of beamed X-rays in the accretion disk and the secondary; similarly to coincident optical/X-ray bursts, they can be used to probe the reprocessing material and constrain the system parameters. Optical pulsations have been detected from Her X-1 and 1627−673. The detection of IR pulsations reported for 2259+587 [325] has not been confirmed in later observations [114].

Her X-1.   After a possible detection [261] and a number of unsuccessful searches (see [470] for references) for optical pulsations from Her X-1, a definite detection was made by Davidsen *et al.* [109] (see also [111,179,291,322,346]). Middleditch & Nelson [323] found that the amplitude of the optical pulsations (up to 0.3 per cent) varies with orbital phase. Optical pulsations are present which originate from reprocessing of X-rays in the accretion disk and on the surface of the secondary, respectively [323] (see, however, [238]). The Doppler shifts of the latter do not represent the orbital motion of the secondary's center of mass because the optical pulsations come preferentially from the vicinity of the first Lagrange point. By assuming that the companion exactly fills its Roche lobe, and that the reprocessing time is constant across its surface, Middleditch & Nelson were able to correct for this and use the optical and X-ray periods to estimate the masses: $M_X = 1.30 \pm 0.14$ $M_\odot$, and $M_2 = 2.11 \pm 0.11$ $M_\odot$, which are consistent with the results from optical radial-velocity variations [99,100,255] (see also [12]).

1627−673.   Doppler shifts in the (7.68 s) X-ray pulse arrival times of 1627−673 have not been observed ($a_X \sin i < 10.5 \, 10^{-3}$ lt s). Using $i = 18(+18, -7)°$ [324], the corresponding upper limit on the secondary mass ranges between 0.03 and $0.09 M_\odot$ [273]. Optical pulsations [226,374] with ~ 3 per cent amplitude occur at the X-ray pulse period (these are interpreted as reprocessing of X-rays in the accretion disk). In addition there is also a weaker (~ 0.5 per cent) pulsation with a slightly longer period [324]. These are believed to be due to reprocessing at the surface of the companion star, which orbits the neutron star with a period of ~ 2492 s in the same direction as the neutron star rotation.

The two pulsations show a varying phase difference, the average value of which corresponds to the light travel-time from the X-ray source to the companion; the variation of the phase difference is due to the variation of the distance of the companion to the observer. Making a correction for the fact that the site of the reprocessing does not coincide with the center of mass of the companion, Middleditch *et al.* [324] derive $a_c \sin i = 0.36 \pm 0.10$ lt s, $a = 1.14 \pm 0.40$ lt s, $i = 18(+18, -7)°$ and $M_X = 1.8 \, (+2.9, -1.3) \, M_\odot$.

2.3.5.3   *X-ray/optical co-variability driven by variations in the accretion rate*

Among the many examples of X-ray/optical co-variability driven by variations in the accretion rate, we mention here: (i) the soft X-ray transients, which coincident with the X-ray outburst show an optical nova (they are discussed in Sect. 2.3.6), and (ii) the correlated 1000 s X-ray and optical flares observed from 1627−673 [313]. These examples of correlated X-ray and optical brightness variations are well described by the reprocessing of X-rays in an accretion disk.

Until recently, X-ray reprocessing did not appear to explain the irregular X-ray and optical variability of Sco X-1. Extensive X-ray and optical observations of Sco X-1 (some of them simultaneous) were made in the decade after its optical identification [403], which showed that its behavior is complicated [43,50,227,332,510,517]. The *B* magnitude of Sco X-1 varies between 13.6 and 12.2 on a time scale of hours to days, with little change in the optical/X-ray colors [332,517]. The optical brightness histogram shows long-term changes [328].

Sco X-1 shows 'active' and 'inactive' states. Optical flares and flickering are observed only when $B < 12.8$; also, X-ray active states occur only when Sco X-1 is above this optical threshold [50,328]. The radio behavior appears to be correlated with this optical threshold: radio flares appear only when Sco X-1 is in the optically faint state [43,50,332] (see Ch. 7). In the active state, the X-ray and optical brightness (averaged over minutes) often show a correlation, roughly according to $F_{opt} (:) F_X^{0.5}$ (consistent with reprocessing of X-rays in a medium with a temperature of $\sim 30\,000$ K, see Sect. 2.3.2.4). In the inactive state, relatively large optical brightness variations can occur without a corresponding change in X-rays.

Some understanding of this behavior has come with the recognition of two groups of LMXBs with different (but in each group correlated) X-ray variability and spectral properties. These are called the Z and atoll sources, after the shapes of the tracks they follow in X-ray color–color diagrams [194] (see Ch. 6). Sco X-1 is a Z source. Priedhorsky *et al.* [386] proposed that the active and inactive states of Sco X-1 are correlated with the flaring and normal branches of the Z track, respectively.

Coordinated X-ray/UV(IUE) observations of the Z source Cyg X-2 [194] showed that its UV brightness increased as it moved from the horizontal branch via the normal branch to the flaring branch. Using a model for the X-ray heated accretion disk model, Vrtilek *et al.* [488] found that the accretion rate varies by a factor of 2.5 along the Z track, and concluded that the X-ray flux is not a good tracer of the accretion rate. This suggests that the angular distribution of the X-rays is sensitive to geometric changes in the inner disk region, as also suggested by theoretical considerations [20,262]. The UV and optical flux are expected to be better tracers of the accretion rate, since the reprocessing material around the X-ray source subtends a large solid angle, and is therefore less affected by these geometric effects.

Co-ordinated X-ray/UV/optical observations of Sco X-1 [7,202,490] have confirmed the results for Cyg X-2 (see Figure 2.28).

2.3.6 Soft X-ray transients

Transient LMXBs are characterized by episodic periods of outburst at X-ray, optical and radio frequencies, which are separated by long intervals of quiescence. Our working definition of a transient source is a source that has been observed to vary in X-ray intensity by a factor >100 (with known binary eclipses discounted). There are several types of transient sources (see Sect. 2.2.1.1, 2.3.4.4 and 2.6.1); however, in this section we consider only the soft X-ray transients (SXTs), a subclass of LMXBs that are sometimes referred to as X-ray novae. For the X-ray properties of SXTs, see Ch. 3 and [512]. For a comparison of SXTs and dwarf novae, see [382,474].

An SXT outburst is due to a sudden dramatic increase in the rate of mass accretion onto the compact primary. The rise time of the X-ray flux is of the order of days, and its subsequent decline occurs on a time scale of months [512]. As illustrated in Figure 2.29, an optical counterpart brightens by $\Delta V \sim 7$ mag, which greatly facilitates its optical identification. During the decline phase, the optical flux always falls much slower than the X-ray flux; the relative decay rates can be used to estimate the average temperature of the reprocessing region [134]. Temperatures around $30\,000$ K have been reported, in agreement with values found by other techniques for persistent

98   *J. van Paradijs and J. E. McClintock*

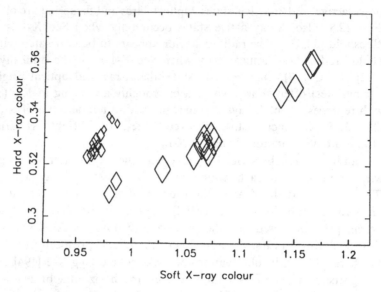

Fig. 2.28. Variation of the *B* band magnitude of Sco X-1 along the normal and flaring branches in an X-ray color diagram of this Z-type LMXB. The size of the symbol is a measure of the *B* magnitude. The magnitude range represented in this diagram is from 13.1 (smallest symbol) to 12.5 (largest symbol) [7].

LMXBs (see Sect. 2.3.2.4; [474]). Several SXTs are recurrent on time scales that range from about 1 to 60 years, and it is likely that all SXTs are recurrent.

During outburst, SXTs are very similar to persistent LMXBs in their X-ray luminosities, X-ray spectra, frequency of type I burst behavior, absence of periodic pulsations and optical spectra. Thus, it is entirely reasonable to assume that SXTs are a subgroup of LMXBs in which accretion occurs spasmodically. It is believed that the cause of the sudden increase in mass accretion onto the compact object is either a mass loss instability that arises in the secondary star [19,184] or an instability in the accretion disk [138,321,456].

Near the peak of the outburst, the optical spectra of SXTs are indistinguishable from the spectra of persistent LMXBs. The spectrum consists of a flat continuum ($F_v \approx$ constant) and weak emission lines (Balmer, He II $\lambda4686$ and N III $\lambda\lambda4634$–42; see Sect. 2.3.3). During decline, the strengths of the high-excitation lines decrease, whereas the Balmer equivalent widths increase substantially. In quiescence, the secondary itself often becomes plainly visible: ellipsoidal variability (see Sect. 2.2.2) and K-star absorption lines are frequently observed. Consequently, it has been possible to determine precise and reliable radial-velocity curves for the secondary stars in four systems: A0620−00, GS2023+338, GRS1124−68, and Cen X-4. Based on these dynamical data, the first three systems contain massive primaries ($M_X > 3M_\odot$), and therefore are black-hole candidates (see Sect. 2.4 and Ch. 3). Cen X-4, on the other hand, which is a type I X-ray burst source, has the mass expected for a neutron star ($M_X \sim 1.4M_\odot$).

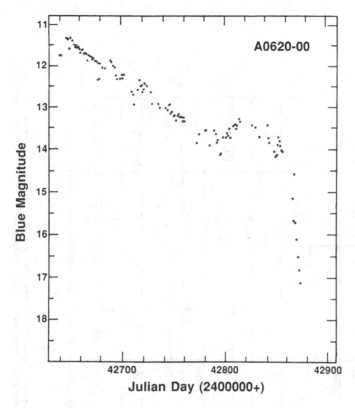

Fig. 2.29. The light curve of the X-ray nova A0620−00 for the 1975 outburst (adapted from [500]). Since outburst, the approximate quiescent level (1978–92) has been $V \sim 18.3, B \sim 20$ [307,349].

In Table 2.5, we summarize some key data for the 12 known optically identified SXTs.

A new SXT, J0422+32, is not included in the table, but is discussed below. Several transient LMXBs have been excluded from the table because their optical identifications are not secure (e.g. 1704+240 and 1742−289) or because their behavior does not conform to the behavior of SXTs, which is described above (e.g. GX339−4, 2129+470, and EXO 0748−676). The nature of the compact object – black hole or neutron star – is given in column 3, and the supporting evidence is indicated in column 4. The surest evidence for a black hole is dynamical evidence that reveals a massive primary ($M_X > 3M_\odot$; Sect. 2.4); such evidence is available for 3 of the 12 systems (entries 1–3). Weaker evidence for a black hole is provided by an $\sim$ 1–10 keV X-ray spectrum that is 'ultrasoft' ($kT \sim 1$ keV) and/or has a power-law tail at energies $\gtrsim 20$ keV (see Ch. 3); five additional compact sources have one or both of these spectral characteristics and therefore are probable black holes (entries 4–8). A 13th SXT, J0422+32, which was recently discovered by GRO/BATSE [355], is a probable black hole because its X-ray spectrum is similar to Cyg X-1 and extends to $\sim$ 600 keV [49,437]. At X-ray maximum ($\sim$3 Crab), it was identified with a $V \sim 12.6$ mag star [58], which has the optical and UV characteristics expected for an

Table 2.5 SOFT X-RAY TRANSIENT SOURCES

#	X-ray/Optical Names	Primary[a]	Evidence[b]	P_{orb}(h)	$F_x^{max}(\mu Jy)$[c]	Sp. Type	E_{B-V}(mag)	Mag (quiesc.) / Mag (outburst)	References[d]
Black Hole Candidates									
1	A0620-00 / N Mon '75 / V616 Mon	BH	Dynamical Xspec	7.8	50,000	K5V	0.35	V = 18.3 / V = 11.2	see Sect. 2.3.6.1
2	1124-684 / N Mus '91	BH	Dynamical Xspec	10.4	3,000	K2V	0.3	V = 20.5 / V = 13.6	see Sect. 2.3.6.1
3	2023+338 / V404 Cyg	BH	Dynamical Xspec	155.4	20,000	K0IV	1.0:	V ~ 19 / V = 12.7	see Sect. 2.3.6.1
4	1354-64 / Cen X-2? / BW Cir	BH?	Xspec		120		1.0:	R ~ 22? / V = 16.9	[252]
5	1524-617 / TrA X-1 / KY TrA	BH?	Xspec		950		0.7	R > 21 / B = 17.5	[340]
6	1543-475	BH?	Xspec		15,000	A2V	0.7	V = 16.7 / V = 14.9	see Sect. 2.6.2
7	1705-250 / Nova Oph '77 / V2107 Oph	BH?	Xspec		3,600		0.5	V~ 21? / V = 15.9	[171]
8	2000+251 / QZ Vul	BH?	Xspec	8.3	11,000		1.5	R = 21.2 / R = 16.2	
Neutron Star Primaries									
9	1455-314 / Cen X-4 / V822 Cen	NS	Burster	15.1	20,000	K7V	0.1	V = 18.4 / V = 12.8	see Sect. 2.3.6.2
10	1608-522	NS	Burster		1,100		1.5:	I > 20 / I = 18.2	[175a,372]
11	1658-298 / V2134 Oph	NS	Burster	7.1	80		0.3	V = 18.3	[51,127]
12	1908+005 / Aql X-1 / V1333 Aql	NS	Burster	19.0	1,300	K0V	0.4	V = 19.2 / V = 14.8	see Sect. 2.3.6.2

Footnotes:

a Primary is either a black hole (=BH) or a neutron star (=NS).

b *Dynamical* proof of a massive primary ($M_x \geq 3\ M_\odot$) is the best evidence for a black hole: certain characteristics of the X-ray spectrum (=Xspec) may also indicate a black hole (see Sect. 2.3.6, and references therein). Type I X-ray bursts (=Burster) provide clear evidence for a neutron star; for Cen X-4 the conclusion is corroborated by dynamical evidence (see Sect. 2.3.6.2).

c Flux density averaged over 2–11 keV; F_x(Crab Nebula) = 1060 μJy.

d For additional references and discussion see Ch. 4 & Ch. 14.

Table 2.6 X-ray/Optical Orbital Parameters of X-ray Binaries

	Source	Type[a]	P_{orb}(d)[c]	$a_x\sin i$(lt-s)	$f_x(M/M_\odot)$	K_c(km s^{-1})	$f_c(M/M_\odot)$[b]	$M_X(M_\odot)$	References
	A. Neutron Star Primaries[c]								
1	LMC X-4	H	1.41	26.31±0.03[d]	9.86±0.04	37.9±2.4	0.008	1.38±0.25	[246,272]
2	Cen X-3	H	2.09	39.664±0.007	15.386±0.001	24±6	0.003	1.06(+0.56, -0.53)	[284,394]
3	4U1538-52	H	3.73	52.8±1.8	11.4±1.2	19.8±1.1	0.003	1.3±0.2	[387,395]
4	SMC X-1	H	3.89	53.46±0.05	10.84±0.03			1.6±0.1	
5	Vela X-1	H	8.96	113.0±0.4	19.29±0.21	21.8±1.2	0.010	1.77±0.21	[116]
6	Her X-1	L	1.70	13.1831±0.0003	0.8513±0.0001	83±3	0.10	0.98±0.12	[84]
7	4U1907+09	H	8.38	83±3	8.8±1.0				
8	4U0115+63	H	24.3	140.13±0.16	5.007±0.019				
9	2S1553-54	H	30.6	164±22	5.0±2.1				
10	V0332+53	H	34.3	48±4	0.101±0.025				[404]
11	GX301-2	H	41.5	371.2±3.3	31.9±0.8				[364]
12	EXO2030+375	H	46[e]	240±15[e]	7.1±1.3[e]				[273]
13	4U1626-67	L	0.029	<0.010	<1.3×10^{-6}				[199]
14	4U1700-37	H	3.41			18±3	0.002	1.8±0.4	[199]
15	Cen X-4	L	0.63			146±12	0.20		[95,308]
	B. Black Hole Candidates								
16	LMC X-3	H	1.70			235±11	2.3±0.3	>7[f]	[91]
17	LMC X-1	H	4.23			68±8	0.14±0.05		[224]
18	Cyg X-1	H	5.60			74.6±0.13	0.241±0.013	>7[f]	[156]
19	A0620-00	L	0.32			442±4	2.90±0.08	>3.4[g]	[306,307]
20	Nova Mus '91	L	0.43			409±18	3.07±0.40	>2.9[g]	[393]
21	GS2023+338	L	6.47			210.6±4	6.26±0.31	>5.6[g]	[54]

Footnotes:

[a] H = HMXB and L = LMXB.

[b] For entries 1-15 the errors are large and asymmetric, and are not given. They can be computed easily using the expression for $f_c(M)$ given in the text and the values tabulated here for P_{orb}, K_c, and ΔK_c.

[c] Data in Part A are adopted from Tables 3 and 4 in [342], and from the supplementary references cited above.

[d] August 1989 Ginga observations; see [272] for a summary of the results of two other recent X-ray timing observations of LMC X-4.

[e] Parameters for Model III [364].

[f] Model dependent (see text).

[g] Firm 2σ limits set by the value of the mass function (see text).

SXT in outburst [418,492]. The last four systems listed in Table 2.5 (entries 9–12) are type I X-ray burst sources, which establishes that they contain neutron stars (see Ch. 4). *It is remarkable that at least 3 out of 13 (23 per cent), and possibly as many as 9 out of 13 (69 per cent), of the optically identified SXTs appear to contain black hole primaries.*

Seven SXTs have known orbital periods, which are listed in column 5; all are ~10 hours, except for the remarkable system 2023+338 (see Sect. 2.3.6.1, and Ch. 3). Spectral types for six SXT secondaries are given in column 7. For each of the four systems that contain K dwarfs, the luminosity class can be deduced either from the spectrum or the short orbital period. It is unlikely that the secondary of 1543−475 is an A2V star (see Sect. 2.6.2). In the following subsection we discuss the three black-hole candidates mentioned above. Two neutron star SXTs, Cen X-4 and Aql X-1, are discussed in Sect. 2.3.6.2.

2.3.6.1   *The dynamical black-hole candidates*

Two SXTs, A0620−00 and GS2023+338, are relatively bright in quiescence (Table 2.5), and therefore it has been possible to measure precisely the radial velocities of their secondaries. A less precise result has been obtained for a fainter SXT, GS1124−684. The velocity semi-amplitude, K_c, the mass function, $f_c(M)$, and a lower bound on the mass of the compact X-ray source, M_X, are given for each SXT in Sect. 2.4.1 (Table 2.6). Their primaries are very strong black-hole candidates because their masses exceed the maximum mass of a neutron star (~3 $M_\odot$; see Sect. 2.4.4).

A0620−00 (Nova Mon 1917/1975; V616 Mon).   The transient A0620−00 was first detected on August 3 1975. By mid-August, the source was about 50 times as bright as the Crab Nebula. The peak 1–10 keV X-ray luminosity was 1.0×10^{38} erg s^{-1} [133] for a distance of 870 pc [349]. During outburst, A0620−00 was identified with a blue star of 12th magnitude. Fifteen months later the optical counterpart had returned to its pre-outburst brightness ($V \approx 18.3$; see Figure 2.29). Its quiescent spectrum was found to consist of two components: a K5V stellar spectrum and an accretion-disk component, which is composed of a blue continuum and double-peaked Balmer emission lines [306,341,349,505].

More recent studies have centered on dynamical measurements. Following the determination of the velocity of the secondary [306], three techniques have been applied to yield constraints on the mass ratio ($M_X/M_c \equiv q$). Johnston *et al.* [232] modeled the optical emission line profiles (principally Hα) arising from the accretion disk. Using their determination of the radial-velocity amplitude of the secondary, $K_c = 468 \pm 44$ km s^{-1}, they concluded that $M_X/M_c \gtrsim 13.3$, which implies $M_X > 6.6M_\odot$ for a plausible, assumed value of $M_c = 0.5M_\odot$ (limits are 3σ). Their result strengthens the case for a black hole; however, the result should be viewed with some reserve for two reasons. First, if one uses a more refined value for K_c (Table 2.6), then one finds an extraordinarily large value of the mass ratio, $M_X/M_c \gtrsim 40$ (3σ), which disagrees with the determination quoted below. Secondly, their model for the profile of disk emission lines is known to have shortcomings [203].

Haswell and Shafter [196] inferred the orbital radial velocity variations of the compact star from observations of the Hα disk emission line. They report K_X

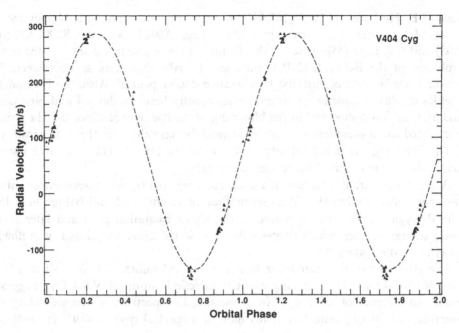

Fig. 2.30. Radial-velocity data for the secondary of V404 Cyg. The smooth curve is a fit to a circular orbit [54].

$= 43 \pm 8$ km s^{-1}, which implies $M_X/M_c = 10.3 \pm 1.9$ (for $K_c = 442$ km s^{-1}). Consequently, the *minimum* masses (for $i = 90°$) are $M_X = 3.50 \pm 0.21$ $M_\odot$ and $M_c = 0.34 \pm 0.08$ $M_\odot$. This result is less model dependent than the result discussed above; nevertheless, the extraction of reliable orbital parameters from observations of the broad, multicomponent Hα line is by no means a straightforward and certain procedure, as Haswell and Shafter acknowledge.

2023+338 (V404 Cyg). The discovery of GS2023+338 was reported on May 22 1989 (see Ch. 3). It was promptly identified with the previously known recurrent nova V404 Cyg [295,493], which had erupted previously in 1938, 1956 and possibly also in 1979 [128,396]. The X-ray light curve is unique: the source was exceptionally variable (it once varied by a factor of 200 in intensity in a few minutes) and its spectrum was generally very hard (Ch. 3). Similarly, the optical flux was highly variable [458,493]. Quasi-periodic oscillations in the mHz range were reported in the optical band [166] and at radio frequencies [190].

During the nova's decline, six groups suggested a total of five different candidate orbital periods in the range 3.05–5.7 hours (see [53,167]). A 10.5 minute period was also reported [491]. Later, in quiescence, the 5.6 day orbital period was discovered by measuring the radial velocities of the G/K secondary (see Figure 2.30; [54]). None of the shorter periods mentioned above appears to be persistent, with the possible exception of the 5.7 hour period, which Casares *et al.* [54] suggest may be the inner period of a triple system (see Sect. 2.6.4).

During decline, the optical spectrum displayed strong, broad and complex emission

lines of H, He I and He II [53,167,493]. The spectrum was unlike the plainer spectra
observed for other X-ray novae in outburst (e.g. A0620−00 and GS2000+25; see
[505] and Fig. 17 in [55]). One notable feature of the outburst spectra was the profile
evolution of the Balmer and Paschen lines. Initially they were single peaked, but
within a few months of outburst they became double peaked. Also, the Hα and Hβ
profiles displayed some peculiarities: an asymmetry between the red and blue peaks
and an emission component in the blue wing of the line. Nonetheless, the Hα profiles
were fitted to an accretion disk model to yield formal values for the disk parameters
[167]. Similarly, the radial velocity variations of the Hα and Hβ lines were used to
derive the 5.7 hour period mentioned above [55].

A second remarkable feature of the outburst was the transient appearance (about
two weeks after outburst) of P Cygni profiles for several He I and Balmer lines [55].
The P Cygni profiles first appeared in the higher excitation lines and later in the
lower excitation lines, which suggests that the ejected mass cooled as it was flowing
away from the system [55].

The strengths of the interstellar lines imply a reddening of $E(B-V) \approx$ 1.0–1.1
mag [65,167,493]. The reddening and the galactic location of V404 Cyg suggest a
minimum distance of 1.5–2 kpc. In quiescence, the spectrum of the secondary star
together with its ellipsoidal variability indicate a spectral type of K0IV [54,494].

GS1124−683 (GRS1121−68; Nova Muscae 1991). GS1124−683 was discovered on January 8 1991 and reached a maximum X-ray intensity of 2.2 Crab on
January 11 (see Ch. 3). The X-ray properties of the source (an intense soft component
plus a high energy tail) suggest strongly that the compact object is a black hole (see
Ch. 3). Especially remarkable was the probable detection on Days 13–14 (following
discovery) of a 511 keV positron–electron annihilation line; comparable line features
have been reported previously only for a few black-hole candidates [163,436].

The optical counterpart was identified on Day 5 as a $V \approx$ 13.5 star [119], which
had brightened by $\Delta V \approx$ 7 mag [393]. The optical behavior of Nova Muscae 1991
in outburst closely resembled that of A0620−00 (and was unlike the behavior of
GS2023+338). The outburst spectrum of Nova Muscae 1991 consisted of a blue
continuum plus weak emission lines of H, He I, He II and N III [119]. Most LMXBs,
including A0620−00 in outburst [505], have similar spectra. The initial light curve of
Nova Muscae 1991 (its decay rate and the amplitude of the outburst) is a close match
to the light curve of A0620−00 [119]. A third similarity between the two sources is
their modest aperiodic variability (Nova Muscae: $\approx$ 0.2 mag on a time scale of ∼1
day [119]).

Far-ultraviolet observations on Day 10 with IUE revealed a hot continuum extending shortward of Lyα and several resonance emission lines [164]. The reddening,
which was inferred from the 2200 Å interstellar feature, is $E(B-V) = 0.20\pm0.05$.
Far-UV observations with the HST Faint Object Spectrograph four months after
outburst showed a power-law spectrum ($f_\nu \propto \nu^{0.3}$) and a reddening of $E(B-V) \sim$
0.29 [67].

Extensive CCD photometry, obtained three to five months after outburst, revealed
a single-humped light curve with a period of 10.5 hours, an amplitude of $\Delta V = 0.2$
mag, and negligible color variations [17]. Photometric observations in the quiescent

state confirm that the orbital period is 10.4 hours [393], and is slightly shorter than the active-state period determined by Bailyn [17]. Spectroscopic observations reveal an absorption-line velocity curve with a semi-amplitude of 409 ± 18 km s^{-1} [393]. The light curve shows ellipsoidal variations with an amplitude of ±0.2 mag in the I band, and ±0.15 mag in a band that approximates $B + V$. The quiescent optical spectrum is very similar to the spectrum of the black-hole candidate A0620−00. It comprises a K0V-K4V stellar component plus broad, double-peaked emission lines and a blue continuum [393].

2.3.6.2 SXTs with neutron star primaries

Type I X-ray bursts have been observed from Cen X-4 and Aql X-1, which establishes that they contain neutron star primaries (Ch. 4). Moreover, the X-ray spectra of these sources (which are harder than most of the SXT black-hole candidates) and the modest value of the mass function for Cen X-4 support a neutron star model. Type I bursts have also been observed from two other SXTs, 1608−522 and 1658−298; however, we choose not to discuss these sources because little is known about their optical counterparts (see Table 2.5).

Cen X-4. The value of the mass function, $f(M) = 0.20\pm0.05\ M_\odot$ [95,308], is consistent with a $\sim1.4\ M_\odot$ neutron star primary and a low-mass secondary viewed at an inclination $i \sim 45°$. Two difficulties with the model arise when one considers the K-dwarf spectral type of the secondary and the rather long orbital period of 15.1 hours [74]. A main sequence K dwarf would fill only about 10 per cent of the volume of its Roche lobe, thereby making it difficult to account for (1) mass transfer and (2) the 0.2 mag (full) amplitude of the light curve [74,95].

Cowley *et al.* [95] propose an alternative model. They invoke an evolved K-type secondary that fills its Roche lobe and has a radius about twice as large as the radius of a main sequence K dwarf. The light variations are attributed to the tidal distortion of the secondary viewed at $i \sim 45°$. This model has been criticized because it implies a distance of at least 2.5 kpc and therefore a peak X-ray burst luminosity that greatly exceeds the Eddington luminosity for a $\sim1.4M_\odot$ neutron star [74, 308].

Chevalier *et al.* [74] reject the model of Cowley *et al.* [95] and present two widely different models. (1) A high-inclination ($i > 80°$) model in which a main sequence secondary fills only a small fraction of its Roche lobe volume. Gas is transferred via a massive wind, and the complex light curve is attributed to the eclipses of three components: the K star, an accretion disk, and an accretion wake, which is located near the neutron star. (2) A model in which the secondary is a peculiar low-density star; the light curve is ellipsoidal and the system is viewed at low inclination ($i \sim 30°$).

The latter model, in which the secondary is a stripped giant with $M_c \sim0.1M_\odot$, is favored by McClintock and Remillard [308]. They show that their light curves can be explained by the tidal distortion of such a secondary plus some 'excess light' from the region where the accretion stream strikes the edge of the disk. On the other hand, the secular variability in the light curve observed by Chevalier *et al.* [74] cannot be readily explained. McClintock and Remillard [308] suggest some differences between Cen X-4 and the black-hole candidate A0620−00 that may be due to the presence of a disk/star boundary layer in Cen X-4 and its absence in A0620−00. These include the detection in quiescence of high-excitation optical lines and X-ray emission from Cen X-4 and their absence in A0620−00.

Aql X-1 (V1333 Aql). The optical counterpart was identified and first studied by Thorstensen *et al.* [443]. This object is known to undergo frequent X-ray and optical outbursts on a time scale of ∼1 year. Recent optical outbursts have been monitored regularly by Ilovaisky and Chevalier [225]. In each of three consecutive years (1989–91), they observed mild but extensive outbursts, which have a very different profile from the major outburst in 1978 when the counterpart reached $V = 14.8$. During mild outbursts, the object is highly variable around a mean magnitude of $V = 17.8$; the quiescent magnitude is $V = 19.2$ [76].

A 19 hour periodic modulation of the V band flux has been reported, and is probably the orbital period of the system [76]. It is comparable to the 15.1 hour orbital period of Cen X-4. The peak-to-peak amplitude of the V band light curve is 0.4 mag. No radial-velocity data have been reported.

2.3.7   *Ultraviolet observations of low-mass X-ray binaries*

Ultraviolet observations have so far been reported for only a handful of LMXBs (see Table 2.2). Their UV/optical spectra show rather flat (F_v constant) continua, whose shape does not change much with source intensity [28,187,488,517]. Superposed on this continuum are emission lines, the strongest of which are due to N V, Si IV, C IV, and He II (see Figure 2.31). Thus, the UV spectra of LMXBs look roughly like their optical spectra (see Sect. 2.3.3). We also note that these UV spectra are similar to those of some cataclysmic variables (see, e.g., [439]).

The UV (and optical) emission of LMXBs are dominated by reprocessing of X-rays in the disk. In view of the large solid angle of the disk as seen from the neutron star, this reprocessed emission probably gives a better estimate of the mass accretion rate than the observed X-ray flux (which may be sensitive to geometric effects). Vrtilek *et al.* [488,490] therefore argued that the increase of the UV flux of the (Z type) LMXBs Sco X-1 and Cyg X-2, which they observed as these sources moved in an X-ray color–color diagram from the horizontal branch, via the normal branch to the flaring branch (see Ch. 6), reflects an increase of the mass accretion rate, even though the X-ray flux does not increase correspondingly (cf. [7]).

The equivalent widths of the UV emission lines do not depend much on the UV flux, i.e. the line fluxes vary together with the continuum flux. This indicates also that the emission lines are powered by reprocessing of X-rays. It is then natural to interpret the lines as emission from an X-ray heated optically thin plasma, as modeled by Hatchett *et al.* [198] and Kallman & McCray [240]. It turns out that these models have difficulty in accounting for the observed ratio of the strengths of the N V λ1240 and C IV λ1550 lines; this has been interpreted as an enhanced nitrogen abundance [28,77]. However, some of the assumptions underlying the models (sphericity, negligible optical depth in the line) may not be justified. A different model, in which the line emission originates in an X-ray illuminated accretion disk (with an inverted temperature profile), has been worked out in detail by Kallman *et al.* [242], who found that the line-forming region is optically thick in the CNO resonance

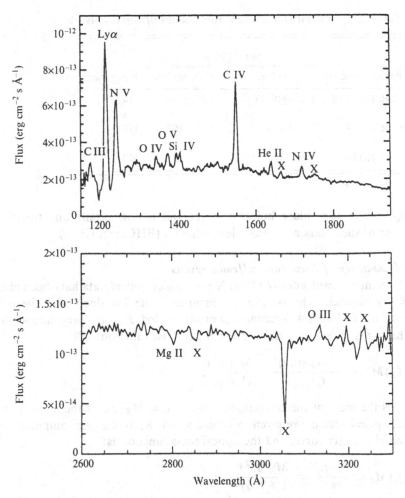

Fig. 2.31. Average ultraviolet spectrum of Sco X-1 [490].

lines; this is supported by the ratio of the two components of the C IV emission line, observed in a high-resolution IUE spectrum of Sco X-1 [242]. The resulting reduced cooling of this line-forming region likely accounts for much of the enhancement of the N V line over the C IV line.

2.4   Neutron star and black-hole masses

Masses have been derived for neutron stars and black holes in about 15 binary systems from radial-velocity and (X-ray or radio) pulse arrival time studies. For thorough and critical reviews of the X-ray pulsar timing data, see [237,342,389]; the binary radio pulsar data have been discussed by Taylor & Weisberg [442]. The emphasis in the discussion that follows is on three topics: (1) our empirical knowledge of neutron star and black hole masses; (2) the importance of systematic errors in optical radial-velocity data; and (3) the maximum mass of a neutron star, which is believed to be about $3M_\odot$ [78]. It is very probable that compact stars more massive

Table 2.7. *Neutron star masses from studies of radio pulsars*

Pulsar name	Mass ($M_\odot$)		Reference
	Pulsar	Companion	
PSR 1913+16	1.442±0.003	1.386±0.003	[442]
PSR 1534+12	1.32±0.03	1.36±0.03	[521]
PSR 1855+09	1.27(+0.23,−0.15)		[401]

than $3M_\odot$, are, in fact, black holes; however, this is not certain, and therefore we usually refer to such stars as black-hole candidates (BHCs; cf. Ch. 3).

2.4.1   A summary of X-ray/optical/radio results

Dynamical results derived from X-ray and/or optical data have been obtained for 21 X-ray binaries. The systems are listed in Table 2.6 along with the relevant dynamical data. Columns 4–6 contain the orbital period, P_{orb}, the projected semimajor axis of the X-ray pulsar, $a_X \sin i$, and the X-ray mass function:

$$f_X(M) \equiv \frac{4\pi^2(a_X \sin i)^3}{GP_{orb}^2} = \frac{M_c \sin^3 i}{(1+q)^2}, \tag{2.1}$$

where M_c is the mass of the companion star and $q = M_X/M_c$ is the mass ratio. The analogous optical results are given in columns 7–8; K_c is the semi-amplitude of the optical radial-velocity curve, and the optical mass function is:

$$f_c(M) \equiv \frac{P_{orb}K_c^3}{2\pi G} = \frac{M_X \sin^3 i}{(1+1/q)^2}. \tag{2.2}$$

In the following, we only consider those optical results that are based on observations of the absorption line spectrum of the companion star. Results based on optical emission lines and on other considerations [93,531] are less secure and are not considered here. A full mass determination requires both mass functions and the orbital inclination, i, to be measured. The best constraint on i is probably obtained from the duration of X-ray eclipses; in principle, optical light curves and polarimetric variations can also provide information on i (see [470]).

For six X-ray pulsar binaries (Table 2.6, entries 1–6), the pulsar is eclipsed by its companion, and the companion's radial velocity curve has been measured. These systems have yielded complete orbital solutions and the masses of six neutron stars (column 9). Seven additional X-ray pulsar binaries, for which useful radial velocity data are unavailable, are listed for completeness (Table 2.6, entries 7–13). These systems are of passing interest here because the value of the X-ray mass function does not provide a useful constraint on M_X. The most precise neutron star masses come from studies of radio pulsars in binary systems. Five masses have been determined; they are listed in Table 2.7.

For the last eight systems listed in Table 2.6 (entries 14–21), only the orbits of the companion stars have been determined. They include six BHCs, the HMXB 4U1700–37 and the neutron-star burst source Cen X-4. The indirect determination of the mass of the neutron star in 4U1700–37 (Table 2.6, entry 14) is based in part on using X-ray light curve data and a model of the stellar wind. We do not consider Cen X-4 (entry 15) further because its orbital inclination is too uncertain to provide a useful constraint on M_X. The six BHCs are discussed in the following subsection.

2.4.2   *Comments on specific models for six black-hole candidates*

A number of possible ways of distinguishing a black hole from a neutron star have been suggested and tried (Ch. 3). The most reliable method, however, remains the determination of the mass of the compact object via a Doppler-velocity study of the absorption-line spectrum of its companion. The maximum mass of a neutron star is believed to be about $3M_\odot$ (see Sect. 2.4.4). Thus the operational definition adopted here for a BHC is a compact object with $M_X \gtrsim 3M_\odot$.

For the discussion at hand, the following fact is crucial: M_X cannot be less than the value of the mass function; i.e., $M_X \geq f_c(M)$. An inspection of Eq. (2.2) reveals that the rock-bottom limit, $M_X = f_c(M)$, corresponds to a system with a zero-mass companion ($M_c = 0$, or $q = \infty$) viewed at the maximum inclination angle ($i = 90°$). Thus, as an example, the optical mass function alone establishes that the compact star in GS2023+338 cannot be less massive than $6.26 \pm 0.31 M_\odot$ (Table 2.6). For Cyg X-1, on the other hand, this rock-bottom limit is an uninteresting $0.24 M_\odot$; consequently, a viable black-hole model for Cyg X-1 requires $M_c \gg 0$ and/or $i \ll 90°$. In the following discussion we comment on published models and probable values of M_X for the six dynamical BHCs (see also Ch. 3).

LMC X-3.   Cowley *et al.* [91] concluded that $14M_\odot > M_X > 7M_\odot$ if the B3V companion is a normal star with $8M_\odot > M_c > 4M_\odot$. However, the brightness of the system varies considerably [98,465,480], and possibly the optical companion is rather faint. This uncertainty, and the possibility that the radial-velocity data may be affected by large systematic errors (see Sect. 2.4.3), led Mazeh *et al.* [304] to conclude that the mass of the compact star could be as small as $M_X = 2.5M_\odot$ (which would imply $M_c = 0.7M_\odot$), and that a 'heavy neutron-star' model cannot be fully excluded. Kuiper *et al.* [258] analyzed the optical light curve, deduced a conservative lower limit of $M_X > 2.8M_\odot$, and concluded that the X-ray star is likely a black hole. Thus, LMC X-3 is a strong BHC; however, the possibility of it being a massive neutron star primary cannot be ruled out.

LMC X-1.   Hutchings *et al.* [224] found that the most probable component masses are $M_X = 6M_\odot$ and $M_c = 20M_\odot$. However, the small and very uncertain mass function, $f_c(M) = 0.14\pm0.05M_\odot$, and the paucity of observations make LMC X-1 the weakest of the BHCs.

Cygnus X-1.   This venerable BHC [36,502] has withstood the challenges posed by several non-black-hole models, including the following: triple-star systems [14,415], undermassive companions (see below), massive accretion disks [259] and

X-rays via magnetic reconnection [13]. If the optical companion is a normal O9.7 supergiant ($M_c \sim 33 M_\odot$), as it appears to be, then $M_X \sim 16 M_\odot$; moreover, M_X is very unlikely to be less than $7 M_\odot$ [8,159]. On the other hand, the mass function of Cyg X-1 is small, and therefore if the companion star is severely undermassive ($M_c \sim 1 M_\odot$) for its spectral type [356,452,461,526], then the compact object might be a neutron star with $M_X \sim 2.5 M_\odot$. Such an extreme model is very unlikely because 20 years of close scrutiny has failed to reveal any evidence that the companion star is abnormal (e.g., [8,156,159,160,347]). We conclude that Cyg X-1 remains a very strong black-hole candidate.

A0620-00.   The value of the mass function gives an absolute lower limit (3σ) on the mass of the compact star: $M_X > 2.7 M_\odot$ ($M_X = 2.7 M_\odot$ corresponds to $M_c = 0$ and $i = 90°$).

A more realistic *minimum* value for the mass of the compact star is $M_X = 4.24 \pm 0.09 M_\odot$, which is based on the following model: a main sequence K5V secondary, $M_c = 0.7 M_\odot$, and the maximum inclination angle that is consistent with the absence of X-ray eclipses, $i = 77°$. Even for an undermassive secondary, $M_c = 0.3 M_\odot$ and $i < 80°$ (no eclipse), one finds $M_X = 3.56 \pm 0.08 M_\odot$. Of course, at a moderate inclination, e.g. $i = 45°$, the mass is quite large, $M_X = 9.1 M_\odot$ ($M_c = 0.5 M_\odot$). There have been attempts to determine the mass ratio, q, by modeling the emission line profiles [232] and by measuring the orbital radial-velocity variations of the Hα emission line [196]. The former approach has serious shortcomings (see Sect. 2.3.6.1). The latter approach appears to be more promising; Haswell & Shafter [196] report $K_X = 43 \pm 8$ km s^{-1}, a result that requires confirmation. Two other studies, which are ongoing, aim to determine q by measuring the rotational broadening of the K dwarf's absorption lines, and i by modeling the light curves [305,307]. Despite the present uncertainties in q and i, the large and secure value of the mass function make A0620−00 a very strong BHC. (Also see Sect. 2.3.6.1 and Ch. 3.)

Nova Muscae 1991.   The value of the mass function is rather uncertain; nevertheless, Remillard *et al.* [393] find that it strongly favors a black-hole model. There were no reported optical or X-ray eclipses during outburst. Therefore, if the secondary is an $0.7 M_\odot$ main sequence star that fills its Roche lobe, then the inclination angle (i) must be less than 77°, which implies that M_X is not less than $M_X = 4.45 \pm 0.46 M_\odot$. Similarly, for an assumed $0.3 M_\odot$ undermassive secondary, one finds $i < 80°$, and the corresponding mass limit is $M_X = 3.75 \pm 0.43 M_\odot$ ($M_X > 2.9 M_\odot$; 2σ). Therefore, neutron star models are excluded at about the 2σ confidence level, even for extreme values of i and M_c. A second reason for favoring black-hole models are the many similarities, both in outburst and quiescence, between Nova Mus 1991 and A0620−00, an established BHC ([393] and references therein). Nova Mus 1991 is a strong BHC. (Also see Sect. 2.3.6.1 and Ch. 3.)

GS 2023+338.   This is the most secure BHC by virtue of its large mass function, $f_c(M) = 6.26 \pm 0.31 M_\odot$ [54]. The recent detection of ellipsoidal variability indicates that the secondary is probably a K0IV star that fills or nearly fills its Roche lobe; the corresponding distance to the system is ~ 3.5 kpc [494]. Wagner *et al.* [494]

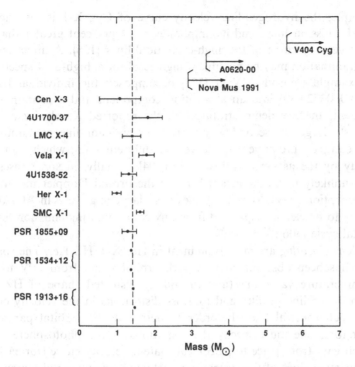

Fig. 2.32. Empirical knowledge of neutron-star and black-hole masses as summarized in Table 2.6 and Table 2.7. The error bars are too small to be shown for four neutron stars located in the binary radio pulsars PSR1913+16 and PSR1534+12 (see Table 2.7). The 2σ lower limits on the masses of three BHCs (plotted at the top of the figure) are based on firm dynamical measurements. The dashed line corresponds to $1.4M_\odot$, which is approximately the expected mass of a neutron star based on current scenarios of neutron star formation.

suggest an orbital inclination of $\sim 60°$ (if the secondary nearly fills its Roche lobe), which corresponds to $M_X \sim 11 M_\odot$; however, this result is rather uncertain. (See also Sect. 2.3.6.1 and Ch. 3.)

The available data on neutron star and BHC masses are summarized in Figure 2.32.

2.4.3 *Sources of systematic errors in radial-velocity data*

There are several systematic effects that may modify the absorption-line velocity amplitude, K, of the companion star and thereby affect the determination of M_X: spectral contamination by emission lines, X-ray heating, tidal distortion, and non-synchronous rotation*. Any of these effects, singly or in combination, may displace the photocenter of the optical companion, as measured in the light of the absorption lines, away from its center of mass. In the following we discuss these effects in turn.

Systematic errors in the absorption-line velocities can be caused by *blending with emission lines* from an accretion disk, accretion stream, or circumstellar material (e.g.

* An additional effect, phase dependent absorption in a stellar wind, has also been discussed [327].

[339]). For example, the hydrogen-line velocity curve of Cyg X-1 is distorted by
the blending of H emission lines, and its amplitude is ~4 per cent greater than the
less-contaminated velocity curve of the 'high-excitation' lines [156]. A direct test for
emission-line contamination may be possible if high-resolution, high-S/N spectra are
available. For example, Penrod & Vogt [373] decomposed the individual Balmer
lines of X Per (4U 0352+30) into an absorption component and a contaminating
emission component, thereby demonstrating that the reported K-velocity of these
lines was spuriously large. A second way to test for emission-line contamination
is to obtain and compare radial-velocity curves for different lines, which should be
affected differently by the assumed emission [156,304]. Finally, in some cases, the
contamination is unlikely to be important because the orbital Doppler modulation
of the narrow absorption (FWHM $\approx$ 2 Å) lines is so large (e.g. 16 Å in A0620−00)
that they are likely to be cleanly separated from any contaminating emission line for
most of the orbital cycle [306].

The effects of *X-ray heating* are most prominent in Her X-1/HZ Her. The spectral
type of HZ Her has been observed to vary with orbital phase from F0V to B0V
[100]. These temperature variations (and the tidally distorted shape of HZ Her)
lead to distortion in the line profiles and to gross distortions in the velocity curves,
which differ for each spectral line. In order to determine the orbital parameters
and masses, therefore, one must specify the conditions in the photosphere of the
companion (which vary from place to place), calculate a velocity curve (for each line
of interest) for various values of the system parameters (K, q, etc.), and match these
curves to the observed velocity curves [223,255]. In the case of Her X-1, Hutchings
et al. [223] conclude that $K = 83\pm3$ km s^{-1}. The estimated uncertainty (which seems
surprisingly small given the severity of the heating effect) is presumably dominated
by uncertainties in the model and systematic effects, and not by measurement errors.
X-ray heating effects may also be important in some HMXBs, despite their low ratio
of X-ray luminosity to optical luminosity [326]; however, see [160,214,258].

Tidal distortion of a companion star is accompanied by variations in effective
temperature and gravity over the star's surface. Consequently, the continuum intensity,
limb darkening coefficient, and line strengths will also vary over the surface. Moreover,
surface elements of a synchronously rotating star may have large velocities relative
to the star's center-of-mass. Thus it is possible that the observed velocity of the star
– a spectrophotometric average over its surface – may deviate significantly from its
center-of-mass velocity. This problem was studied by Sterne [433] and by Wilson &
Sofia [520]. Van Paradijs *et al.* [476] developed models for the distortion and dynamics
of HD77581 (Vela X-1) and concluded that both the K velocity and eccentricity may
be strongly affected by distortion. Hutchings [214] modeled five HMXB primaries,
including HD77581, and found that no observations to date showed evidence for
velocity distortions.

For most of the companion stars in Table 2.6, it is uncertain if they are in (or near)
synchronous rotation, $\Omega = 1$, where Ω is the ratio of the stellar rotational frequency
to the orbital frequency (see discussion and references in [389]). This, however, is an
important consideration. For eclipsing systems, the unknown rotational velocity can
cause an uncertainty in the calculated upper limit for M_X of ~20 to 50 per cent [10].
Also, non-synchronous rotation can magnify the effects of heating and tidal distortion

discussed above. In order to determine a reliable value for Ω, one must determine v sin i and i. One must measure the widths of several lines *versus* orbital phase and employ a model to include the effects of heating and tidal distortion. Such an analysis has been carried out for the non-eclipsing system Cyg X-1; it favors solutions with Ω near unity [159].

2.4.4   Upper bound to the mass of a neutron star

At present, the surest way to establish that a compact object is a black hole is to show that its mass exceeds the maximum allowed mass of a stable neutron star. That maximum mass is a sensitive function of the unknown equation of state (EOS) for nuclear matter. Numerical models covering a very wide range of macroscopic properties have by now been computed using dozens of different EOSs (e.g. [6,108,267,525]). Several analytic models of neutron stars have been explored (e.g. [1,253,363]). Models of neutron stars with quark cores and hypothetical quark stars have been studied [4,267,351,354]. All of the above models are for non-rotating neutron (or quark) stars; the most realistic and sophisticated of these models imply a maximum mass of $\lesssim 2.0 M_{\odot}$, and even the stiffest EOS predicts a maximum mass of only $2.7 M_{\odot}$. Moreover, rotation cannot increase the maximum mass by more than about 20 per cent [142,417,501]. *Therefore, models with the stiffest EOS and maximal rotation have a maximum mass that does not exceed $3.2 M_{\odot}$.*

In attempting to establish the existence of black holes, one might be concerned that an unthought of EOS might yield viable neutron star models with $M >> 3 M_{\odot}$. Fortunately, however, there is a firm bound on the mass of a neutron star that does not depend on the details of the unknown EOS at high densities. The key assumptions are the following: (1) general relativity is the correct theory of gravity; (2) the EOS of nuclear matter is known below $\varrho_o = 5.1 \times 10^{14}$ g cm^{-3} (which is ≈ 1.8 times saturated nuclear density); and (3) the velocity of sound at large wavelengths is less than the velocity of light, $dp/d\varrho < c^2$. These minimal assumptions lead to a strong conclusion: *a firm upper bound to the mass of a non-rotating neutron star is about $3 M_{\odot}$* [78,192,344,397].

2.5   The Magellanic Cloud sources

Nine X-ray binaries are known in the LMC (Table 2.8). Interestingly, this is about the number of sources one might expect to find based on scaling down the total number of galactic sources, ~ 180, by the ratio of the mass of the Galaxy to the mass of the LMC, $M_G/M_{LMC} \sim 23$ [27]. For the SMC, which has one-tenth the mass of the LMC, a similar scaling to the Galaxy predicts only about one source, whereas four are known (Table 2.8).

2.5.1   The high-mass systems

Three of the seven (~ 43 per cent) HMXBs in the Magellanic Clouds (MCs) contain X-ray pulsars, which is the same as the fraction in the Galaxy (45 per cent; see Table 2.8 and Ch. 14). The pulse periods of the three MC pulsars are relatively short; that of the LMC transient, 0535−668, is the second shortest known (and the source is also exceptionally luminous). Two of the seven (~ 29 per cent) HMXBs, LMC X-3 and LMC X-1, contain dynamical black-hole candidates, compared to

Table 2.8. *X-ray binaries in the Magellanic Clouds*

A. Massive systems (HMXBs)

Location	Name(s)	V_{mag} sp. type	L_X^a(erg s^{-1})	P_{orb} P_{pulse}
SMC	0050 − 727 SMC X-3	~14 O9III-Ve	4.7×10^{37}	
SMC	0053 − 739 SMC X-2	16.0 B1.5Ve	6.6×10^{37}	
SMC	0103 − 762	17[b]	2×10^{37}	
SMC	0115 − 737 SMC X-1	13.3 B0Ib	5.3×10^{38}	3.89 d 0.71 s
LMC	0532 − 664 LMC X-4	14.0 O7III-V	3.9×10^{38}	1.40 d 13.5 s
LMC	0535 − 668 'LMC transient'	12.3-14.9 B2III-IVe	1.2×10^{39}	16.7 d 0.069 s
LMC	0538 − 641 LMC X-3	16.4-17.5 B3Ve	2.9×10^{38}	1.70 d
LMC	0540 − 697 LMC X-1	14.5 O7-9III	1.6×10^{38}	4.22 d
LMC	0544 − 665	15.4[b]	1.6×10^{37}	

B. Low-mass systems (LMXBs)

Location	Name(s)	V_{mag}	$L_{\bar{X}}$(erg s^{-1}) kT_{bb}	P_{orb}
LMC	0521 − 720 LMC X-2	18.0–19.0	3×10^{38c} 1.3	
LMC	0527.8 − 6954 [HV2554][e]		$\sim 1 \times 10^{38d}$ ~0.03	
LMC	0543 − 682 CAL 83	16.2-17.3	$\sim 1 \times 10^{38d}$ ~0.03	1.04 d
LMC	0547 − 711 CAL 87	18.8–20	~0.1[f]	0.44 d

Data are from Ch. 14 unless otherwise noted.
[a] The X-ray luminosity derived from the maximum observed value of the flux density (F_X), which is given in Ch. 14. We use $L_X = 2.6 \times 10^{35}$ $F_X(\mu Jy)(D/10\,kpc)^2$ (see [42]) with D(LMC) = 50kpc and D(SMC) = 60kpc [503].
[b] Optical counterpart not completely certain.
[c] Maximum observed luminosity and blackbody temperature ([38] and references therein).
[d] Minimum luminosities and blackbody temperatures [170].
[e] Possible optical counterpart.
[f] Blackbody temperature [360].

only one out of 60 (∼2 per cent) of the galactic HMXBs (Cyg X-1). Even if one would include with Cyg X-1 the less-certain black-hole candidate SS433 [125,532], it would appear that HMXB black-hole candidates are more plentiful in the Clouds [216]. Interestingly, the black-hole candidate LMC X-1 is surrounded by an extended, X-ray photoionized He III nebula [359]. No comparable nebulae have been observed around galactic X-ray binaries. For some additional distinctive properties of the MC X-ray, sources see [216].

The high X-ray luminosities of the MC X-ray sources have been noted by a number of authors. (For references and a comparison of the luminosity distributions of the LMC and galactic X-ray sources, see [233].) Restricting attention to the seven sources in the Clouds with OB-type companions (Table 2.8), one finds a mean (maximum) luminosity of $< \log L > = 38.4 \pm 0.5$ r.m.s. The mean X-ray luminosity of galactic HMXBs is given in Table 2.1: $< \log L > = 36.7 \pm 0.9$ for OB supergiant systems, and the time-averaged luminosity of Be systems is significantly less. Thus, on average, the MC sources are ∼50 times as luminous as their galactic counterparts.

The higher luminosities of the MC sources are believed to be linked to the lower abundances of metals in the Clouds. (In H II regions, e.g., the O/H ratio in the LMC and the SMC is one-third and one-fifth of the solar value, respectively; [400].) One linking mechanism is the effect of X-ray heating of gas as it falls toward a compact object, which depends strongly on the atomic number Z via the photoelectric cross section ($\sigma \propto Z^4$). For spherical accretion, such heating can seriously impede the accretion flow and thereby reduce the limiting luminosity to a value far below the Eddington limit [353]. Clark *et al.* [79] argue that this effect should occur also in more realistic, anisotropic accretion flows through disks and magnetospheres. Thus, compared with a galactic source, a comparable MC source may be more luminous because its low-Z accretion flow is less impeded by heating.

It is probable that a second metallicity-dependent effect also boosts the luminosities of the wind-fed OB-star systems in the MCs. For the case of a compact object accreting the stellar wind of an OB star, the accretion rate ($\dot{M}$) depends sensitively on the stellar-wind velocity (v) at the orbit of the compact object: $\dot{M} \propto v^{-4}$ (e.g., [233]). All available evidence indicates that the terminal wind velocity decreases with Z. This behavior is predicted by the very successful theory of radiation-driven stellar winds [256,257]. Furthermore, UV spectroscopic data show that the SMC has lower terminal wind velocities than the LMC, and the Galaxy has the highest [151,152]. Thus, the lower metallicity in the Clouds implies a lower terminal wind velocity, a higher mass accretion rate, and a more luminous X-ray source, as observed.*

Therefore, there is good circumstantial evidence that links the higher luminosities of the MC sources to the lower metal abundances found there.† Not only are the MC sources exceptionally luminous relative to their galactic counterparts, they are also very luminous in an absolute sense: the maximum luminosities of four sources in Table 2.8 are approximately two to eight times greater than the Eddington limit

---

* The luminosity boost will be reduced somewhat if the stellar mass loss rate is proportional to ∼ $Z^{0.5}$, as predicted by radiation driven wind theory [257]. This prediction, however, is not supported by the existing (marginal) data, which suggests that the mass loss rate is independent of Z [270].

† Note that low metallicity *per se* does not necessarily imply high luminosity. For example, consider galactic LMXBs, which are powered by Roche lobe overflow. The systems in globular clusters are probably low-Z sources and also have rather low luminosities relative to LMXBs in the field (see [487]).

for a $1.4M_\odot$ neutron star (1.6×10^{38} erg s^{-1} for $R = 10$ km and $X = 0.7$). In part, the super-Eddington luminosities may be due to the anisotropic emission expected from strongly magnetized pulsars (although such spectacular luminosities are not seen among the galactic pulsars). Recently, Paczynski [357] suggested that the MC accretors may have very strong magnetic fields that decrease the electron scattering opacity, thereby raising the Eddington limit. Some possible explanations for super-Eddington luminosities of burst sources have been summarized by Van Paradijs & Stollman [473].

2.5.2   Low-mass systems

There are four cataloged LMXBs in the LMC (Table 2.8), and none in the SMC. Greiner *et al.* [170] mention that there are two LMXBs in the SMC, but no further information has been published. One of them, LMC X-2, resembles the LMXBs found in the Galaxy. Both its optical spectrum and its X-ray spectrum are similar to the spectra of galactic LMXBs [38], and it has a characteristically high ratio of X-ray to optical luminosity (L_X/L_{opt} ~600; [47]). The orbital period is highly uncertain: 8.15 hours [47] and 12.5 days [105] have been proposed. The short period would imply kinship with the many LMXBs that have orbital periods of several hours, whereas the long period would suggest that the system is similar to the halo LMXBs, Cyg X-2 and 0921−630, which have giant companions. It seems unlikely that the long period is correct since the expected features of a giant companion are not present in optical or infrared spectra [97,333].

LMC X-2 differs from galactic LMXBs in the following ways, all of which probably can be attributed directly or indirectly to the lower metallicity of the LMC: LMC X-2 is exceptionally luminous in the optical band [38] and at X-ray energies (Table 2.8); the N III–C III $\lambda\lambda 4640$−50 blend is absent in optical spectra [333]; and the ~6.7 keV Fe line is weak (EW < 80 eV, 90 per cent confidence; [38,513]).

The most extraordinary sources in the Clouds are the last three sources listed in Table 2.8. Their spectra (which are best fitted by a blackbody model) are exceedingly soft: $kT_{bb} = 0.03$–0.10 keV (Table 2.8). By comparison, both the common 'soft' ($kT_{bb} \sim 1.5$ keV) LMXB and the so-called 'ultrasoft' ($kT_{bb} \sim 0.3$ keV) LMXBs have hard(!) spectra. We will refer here to these three LMC sources as 'extreme ultrasoft' or 'EUS' sources, in analogy to the well-known nomenclature, extreme ultraviolet (EUV). The relatively short orbital periods, which are known for two of the sources (Table 2.8), are the principal reason for believing that the EUS sources are low-mass systems. A second reason is the absence in their spectra of any stellar absorption features. (Such features are readily observed in good quality spectra of all HMXBs except SS433.)

Consider, for example, the well-studied EUS source CAL 83. Its observed photon spectrum, which is best fitted by a blackbody with $kT_{bb} = 26$ eV, is narrowly sandwiched between ~ 0.2 keV (the IS cutoff) and ~ 0.7 keV (the Wien cutoff) (see Fig. 1 in [170]). The (unattenuated) model spectrum, however, peaks well below this band at ~ 0.1 keV. Therefore, we observe only a small fraction of the source flux, and consequently the model is weakly constrained. EUS sources are severely attenuated by even the modest column depth of the LMC (log $N_H = 20.8$), which may explain their rarity in the Galaxy [170,454]. Nevertheless, it may be possible to detect galactic

EUS sources, if they exist. Scaling by the relative masses of the Galaxy and the LMC, one might possibly expect ∼70 galactic EUS sources (about half the number of presently cataloged LMXBs). Many galactic X-ray binaries have column depths of log N_H ∼ 21.5 (e.g. [489]), which is several times the LMC value. Even with such severe attenuation, however, it may be possible to discover a galactic EUS source in the ROSAT data bank.* Alternatively, it may be possible to discover them optically. Their optical spectra are very similar to the distinctive spectra of ordinary LMXBs (see Sect. 2.3.3, and references therein). Moreover, they can be optically much more luminous than standard LMXBs: M_V(CAL 83) = −2.5, which is ∼2 mag brighter than LMC X-2 and ∼3.5 mag brighter than galactic LMXBs [470] (see Sect. 2.3.2.3).

Another interesting EUS source, CAL 87, is an eclipsing system with an orbital period of 10.6 hours. The spectrum of the secondary, which can be observed near minimum light, appears to be late F [96]. Based on the phase and low amplitude of the velocity curve for the He II λ4686 emission line, Cowley *et al.* [96] conclude that the X-ray source is a probable black hole with $M_X > 6M_\odot$. This result should be viewed with reserve because dynamical results based on emission line data often prove to be unreliable. A very different model for CAL 87 and all EUS sources was recently suggested by Van den Heuvel *et al.* [464]. They argue that EUS sources are not accreting neutron stars or black holes (i.e. not LMXBs), but instead are accreting white dwarfs whose ∼Eddington-limited luminosity is derived from thermonuclear burning of accreted hydrogen.

2.6   Triple-star systems

About 4 per cent of solar-like stars in the solar neighborhood are members of triple systems [131]. Therefore, naively one might expect a significant number of triple systems (∼ ten) among the few hundred X-ray binaries in the Galaxy. In fact, there are no established X-ray triple systems. There are about a dozen candidates, however, and in this section we discuss several of them.

2.6.1   2129+470/V1727 Cyg

The X-ray source 2129+470 was observed to shine persistently throughout most of the 1970s. The 5.2 hour orbital period of the system implies that a main sequence secondary would be a late K dwarf [444]. The X-ray source emits type I X-ray bursts and therefore is a neutron star [147; Ch. 4]. The optical counterpart exhibits a large-amplitude (ΔB = 1.5 mag) light curve reminiscent of HZ Her [314]. The optical light curve and the properties of the X-ray light curve show that the orbital inclination is high ($i \sim 80°$; [315]).

The X-ray source went into a low state in 1983 [378]. It was subsequently detected only recently at a flux level about 100 times lower than observed during the 1970s [148]. This low level of X-ray emission has probably persisted since 1983 [330]. Also in 1983, the optical counterpart faded from a mean V mag of 17.0 to 18.0, about 0.6 mag fainter than minimum light in the earlier high (X-ray 'on') state [447]. Given the 5.2 hour orbital period and the high inclination of the system, observers

---

* Since this chapter was written, seven EUS sources have been detected in the Galaxy (see, e.g., the contribution from P. Kahabka in the proceedings of *IAU Symposium 165, 'Compact Stars in Close Binaries',* 1995).

expected to discover ellipsoidal variability ($\Delta V \sim 0.3$ mag; cf. [32]) and to observe a large velocity amplitude for the secondary ($K \sim 300$ km s^{-1}). Surprisingly, these effects were not observed; no variability at all was detected at the orbital period: $\Delta V < 0.012$ mag [447]; $K \le 5 \pm 9$ km s^{-1} [87,150].

The spectrum of the quiescent counterpart shows no emission lines and is consistent with a late F star [75,87,447]. It is very unlikely that the late FV (or FIV; [87]) star is the secondary because it would greatly overfill its Roche lobe if it were in a 5.2 hour orbit with an $\sim 1.4 M_\odot$ neutron star. Several models have been suggested to account for these unexpected findings. The most plausible, however, invoke another star in the line-of-sight. Because the probability of an unrelated background/foreground star is small ($\sim 10^{-3}$; [447]), the favored model [75,150,447] is a triple system consisting of an LMXB with a 5.2 hour period and an F star in a much larger orbit around the inner binary.

The most direct test of this model, a search for radial velocity variations of the F star, has been initiated by Garcia *et al.* [150]. Their observations at six epochs, which span $\sim$ two years, show variations of ~ 40 km s^{-1} in the systemic velocity [146,150]. Further observations of this kind may be the best way to prove the existence of an LMXB triple system, which could deepen our understanding of LMXBs. For example, the expected slow modulation in the eccentricity of the inner binary of a triple could provide the explanation for the multi-year 'on' and 'off' cycles observed in 2129+470, Her X-1 and some other LMXBs [150]. Also, if 2129+470 were a triple system with an outer period ~ 10–100 days, this would constrain standard evolutionary schemes for LMXBs that invoke a huge common envelope. This envelope, which is believed to engulf the initially wide binary ($P \gtrsim 1000$ days; e.g. [486]) and cause the secondary to spiral in, would also engulf a third star.

2.6.2   1543−475

The soft X-ray transient (SXT) 1543−475 erupted in 1971 [276], 1983 [251] and 1992 [191]. In the 1983 outburst, its spectrum was ultrasoft ($kT = 1.6$ keV), indicating a possible black-hole primary. The optical counterpart was discovered near X-ray maximum at $V = 14.9$ [367]. Based on the optical light curves of other SXTs, one would predict a very high magnitude for the quiescent state, $V \sim 22$ (typically $\Delta V \sim 7$ mag; see Sect. 2.3.6). It was therefore a great surprise to find that the quiescent counterpart had dimmed by only $\Delta V = 1.8$ mag to $V = 16.7$ [70]. A second surprise was the early spectral type of the quiescent counterpart, A2V [70]; the five well-studied secondaries of SXTs are K stars (Table 2.5). Finally, Balmer emission lines, which are present in the spectra of all quiescent SXTs, are absent in the spectra of 1543−475.

Chevalier [70] lists three hypotheses to explain the presence of an A2V star at the position of 1543−475: (1) the A star may be just a very unusual mass donor star; (2) although improbable, the A star may be an interloper in the line-of-sight; or (3), as suggested for 2129+470, the A star may be the outer component of a triple system.

2.6.3   1916−053

Periodic X-ray dips of variable intensity (20 per cent – 100 per cent) are almost always observed for the persistent source 1916−053 [429,430,495,508]. The X-ray dip morphology is highly variable from day to day, and even cycle to cycle; consequently, no one has succeeded in phasing the X-ray observations that are separated by ~ 1 year or more. A recent good determination of the X-ray period, $P_X = 50.00\pm0.08$ min [430], is consistent with earlier determinations, which are summarized by Smale *et al.* [429]. X-ray observers generally believe that the X-ray period is the orbital period of the system. The dips are thought to be the result of obscuration of the central X-ray source by a thickened portion of the disk, which is produced by the impact of the gas stream from the companion [365], or by clouds in an inner 'circularization ring' in the disk [141]. Many such dipping sources are now known [365; Ch. 1]. A $B = 21$ optical counterpart (V1405 Aql) was found by Grindlay *et al.* [178]. It exhibits a sizable optical modulation that is phase coherent over at least an interval of 2.5 years [173]. They suggest that the optical modulation is due to a partial eclipse of the accretion disk by an $\sim 0.1\,M_\odot$ degenerate secondary. An extensive campaign of optical monitoring during 1987–89 yielded an ephemeris for the times of the optical minima with $P_{opt} = 50.459\pm0.003$ min [173]. This period has been confirmed [174]; see also [407]. Remarkably, the optical period is 0.9 per cent longer than the X-ray period.

A triple star model, which is based on the assumption that the optical period is the orbital period, has been proposed to explain the shorter X-ray period [173,178]. The authors invoke a third star of low mass ($M < 1\,M_\odot$) in a 2.5 day retrograde orbit that modulates the eccentricity and therefore the mass transfer rate in the 50 min binary [16]. As before, the X-ray dips are due to absorption by a bulge on the disk (see above). In this picture, however, the eccentricity, mass transfer rate and X-ray dips are modulated at the beat of the orbital period of the inner binary with the orbital period of the third star, which is the 50.00 min X-ray period. An extra feature of the model is its ability to account for the 199 day X-ray modulation observed for 1916–053 (see Sect. 2.6.4; [173]).

White [506] criticizes the triple star model (see also [430]), pointing out that it does not explain how a modulation of the mass transfer rate could cause the X-ray dip period to differ from the orbital period. A modulation of the mass transfer rate by itself might be expected to modulate the depths of the dips, but not their phase. White notes that the disk radius might be affected by modulations in the mass transfer rate. In this case, however, the orbital phase of the stream–disk impact point would remain fixed if averaged over a full orbital cycle of the third star, and the average X-ray period would equal the optical period, which is contrary to observation [506,508]. The triple-star model also predicts a strong 2.5 day modulation of the depth of the dips, which is not apparent in the X-ray data [429,508]

More recent optical observations, conducted on seven nights in September 1990, have revealed complexities in the light curve that were not apparent in the earlier studies [174]. For example, these data contain not only the optical periodicity, but also strong evidence for the X-ray dip period. Moreover, the optical modulation was not always present. The new findings motivate Grindlay [174] to consider two alternative models: disk precession as developed for Her X-1, and tidally-induced precession, as applied to SU UMa CVs [514]. The latter model, in modified form, is discussed and advocated by White [506] and by Smale *et al.* [430].

2.6.4   *Other candidate triples*

A number of other X-ray triple systems have been proposed. For example, triple-star models were considered for Cyg X-1 as an alternative to the standard black-hole model [14]; however, the predicted effects were not observed [3,157,415], and triple models are now only marginally viable. The LMXB GX17+2 is coincident with a 17th magnitude G star to within 0.5 arcsec, and has been suggested as a possible triple system [18]. However, see Van Paradijs & Lewin [533] for a critical evaluation. In Cyg X-3, Molnar [329] found the radio flare period to be about 3 per cent longer than the X-ray binary period, which is 4.79 hours. In one model he suggested a third star as the cause of the beat period between the X-ray and radio periods. The globular cluster source 4U1820–30 has an 11 min orbital period that has steadily *decreased* during the past 15 years. Acceleration by a distant triple companion is one possible explanation given for this unusual behavior [467].

A model for the X-ray nova 2023+338, which invokes a main sequence K0 star orbiting ($P_{outer} = 6.5$ days) a mass exchange binary with $P_{inner} = 5.7$ hours was proposed by Casares *et al.* [54]. The inner period, which is based on Hα emission-line velocities, is not a secure orbital period; moreover, the recent finding that the K0 star is an ellipsoidal variable [494] undercuts the original rationale for suggesting a third star (cf. [54]).

Long-term cycles of ~30–300 days have been confirmed for several massive systems (SS433, Her X-1, LMC X-4, and Cyg X-1), and are suspected in several others (see [383] for a review). Such long cycles, which are presumably due to the precession of some component in the system, have been interpreted in terms of triple star models. For example, a triple model was proposed to account for the 164 day cycle in SS433 [135]. This model invokes an interacting binary with an ~ 1 day period, which orbits the massive companion every 13.1 days and precesses at 164 days due to the tidal influence of the companion. The weakness of the model is the failure to detect any evidence of an ~ 1 day periodicity (see [383] and references therein).

Quasi-periodic cycles, ranging from 0.5 to 2 years, are observed for several LMXBs (e.g. 1916–053, 1820–303, and the Rapid Burster 1730–335 [Sect. 4.6]). They seem to be due to changes in the mass transfer rate rather than aspect variations due to precession [383]. These long cycles may be due to the type of (binary) mass transfer instability that cause the super-outburst cycles of the SU UMa cataclysmic variables [383]. On the other hand, they may be driven by the presence of a third star, as discussed in the model of 1916–053 above.

References

[1]   Abhyankar, K.D. 1991, Bull. Astr. Soc. India 19, 105
[2]   Abramenko, A.N. et al. 1978, MNRAS 184, 27P
[3]   Abt, H.A. et al. 1977, ApJ 213, 815
[4]   Alcock, C. et al. 1986, ApJ 310, 261
[5]   Antokhina, E.A. & Cherepashchuk, A.M. 1990, SvA Lett. 16, 182
[6]   Arnett, W.D. & Bowers, R.L. 1977, ApJS 33, 415
[7]   Augusteijn, T. et al. 1992, A&A 265, 177
[8]   Avni, Y. & Bahcall, J.N. 1975, ApJ 197, 675
[9]   Baade, D. (ed.) 1990, ESO Workshop on Rapid Variability in
       OB-Stars: Nature and Diagnostic Value (ESO, October 1990)
[10]   Bahcall, J.N. 1978, ARA&A 16, 241
[11]   Bahcall, J.N. & Bahcall, N.A. 1973, ApJ 178, L1
[12]   Bahcall, J.N. & Chester, T.J. 1977, ApJ 209, 214
[13]   Bahcall, J.N. et al. 1973, Nature Phys. Sci. 243, 27
[14]   Bahcall, J.N. et al. 1974, ApJ 189, L17
[15]   Bahcall, J.N. et al. 1976, PASP 87, 141
[16]   Bailyn, C.D. 1987, ApJ 317, 737
[17]   Bailyn, C.D. 1992, ApJ 391, 298
[18]   Bailyn, C.D. & Grindlay, J.E. 1987, ApJ 312, 748
[19]   Bath, G.T. 1975, MNRAS 171, 311
[20]   Begelman, M., C. & McKee, C.F. 1983, ApJ, 271, 89
[21]   Bernacca, P.L. & Bianchi, L. 1981, A&A 94, 345
[22]   Bernacca, P.L. et al. 1979, MNRAS 186, 287
[23]   Bessell, M.S. et al. 1975, ApJ 195, L117
[24]   Bianchi, L. & Bernacca, P.L. 1980, A&A 89, 214
[25]   Bianchi, L. & Pakull, M. 1985, A&A 146, 242
[26]   Bieging, J.H. et al. 1989, ApJ 340, 518
[27]   Binney, J. & Tremaine, S. 1987, Galactic Dynamics (Princeton:
       Princeton Univ. Press), 428
[28]   Blair, W.P. et al. 1984, ApJ 278, 270
[29]   Blaauw, A. 1963, Basic Astronomical Data, ed. K. A. Strand, p. 383
[30]   Blondin, J.M. et al. 1990, ApJ 356, 591
[31]   Blondin, J.M. et al. 1991, ApJ 371, 684
[32]   Bochkarev, N.G. et al. 1979, SvA 23, 8
[33]   Bochkarev, N.G. et al. 1988, SvA Lett. 14, 421
[34]   Bochkarev, N.G. et al. 1988, SvA 32, 405
[35]   Bohlin, R.C. 1970, ApJ 162, 571
[36]   Bolton, C.T. 1972, Nat 235, 271
[37]   Bonnet-Bidaud, J.M. et al. 1981, A&A 101, 184
[38]   Bonnet-Bidaud, J.M. et al. 1989, A&A 213, 97
[39]   Bowen, I. 1934, PASP 46, 146
[40]   Bowen, I. 1935, ApJ 81, 1
[41]   Boynton, P.E. et al. 1973, ApJ 186, 617
[42]   Bradt, H.V.D. & McClintock, J.E. 1983, ARA&A 21, 13
[43]   Bradt, H.V. et al. 1975, ApJ 197, 443
[44]   Branduardi, G. et al. 1978, MNRAS 185, 137
[45]   Callanan, P.J. & Charles, P.A. 1991, MNRAS 249, 573
[46]   Callanan, P.J. et al. 1989, MNRAS 241, 37P
[47]   Callanan, P.J. et al. 1990, A&A 240, 346
[48]   Callanan, P.J. et al. 1992, MNRAS 259, 395
[49]   Cameron, R.A. et al. 1992, IAU Circular 5587
[50]   Canizares, C.R. et al. 1975, ApJ 197, 457
[51]   Canizares, C.R. et al. 1979, ApJ 234, 556
[52]   Canizares, C.R. et al. 1980, ApJ 236, L55
[53]   Casares, J. & Charles, P.A. 1992, MNRAS 255, 7
[54]   Casares, J. et al. 1992, Nat 355, 614
[55]   Casares, J. et al. 1991, MNRAS 250, 712
[56]   Cassinelli, J. 1979, ARA&A 17, 725
[57]   Castor, J.I. & Lamers, H.J.G.L.M. 1979, ApJS 39, 481
[58]   Castro-Tirado, A.J. 1992, IAU Circular 5588
[59]   Charles, P.A. et al. 1977, IAU Circular 3096
[60]   Charles, P.A. et al. 1978, MNRAS 183, 813
[61]   Charles, P.A. et al. 1980, ApJ 237, 154
[62]   Charles, P.A. et al. 1980, ApJ 241, 1148
[63]   Charles, P.A. et al. 1983, MNRAS 202, 657
[64]   Charles, P.A. et al. 1986, Nat 323, 417
[65]   Charles, P.A. et al. 1989, in ref. [213], 103
[66]   Charles, P.A. et al. 1991, MNRAS 249, 567
[67]   Cheng, F.H. et al. 1992, ApJ (in press)
[68]   Cherepashchuk, A.M. 1982, SvA Lett. 8, 336
[69]   Cherepashchuk, A.M. & Khruzina, T.S. 1981, SvA 25, 697
[70]   Chevalier, C. 1989, in ref. [213], 341
[71]   Chevalier, C. & Ilovaisky, S.A. 1973, Nature Phys. Sci. 245, 87
[72]   Chevalier, C. & Ilovaisky, S.A. 1977, A&A 59, L9
[73]   Chevalier, C. & Ilovaisky, S.A. 1990, A&A 228, 115
[74]   Chevalier, C. et al. 1989, A&A 210, 114
[75]   Chevalier, C. et al. 1989, A&A 217, 108
[76]   Chevalier, C. & Ilovaisky, S.A. 1991, A&A 251, L11
[77]   Chiapetti, L. et al. 1983, ApJ 265, 354
[78]   Chitre, D.M. & Hartle, J.B. 1976, ApJ 207, 592
[79]   Clark, G.W. et al. 1978, ApJ 221, L37
[80]   Clayton, D.D. 1968, Principles of Stellar Evolution and
       Nucleosynthesis (McGrawHill)
[81]   Coe, M.J. & Payne, B.J. 1985, Ap&SS 109, 175
[82]   Conti, P.S. & Altschuler, W.R. 1971, ApJ 170, 325
[83]   Conti, P.S. & Cowley, A.P. 1975, ApJ 200, 133
[84]   Cook, M.C. & Page, C.G. 1987, MNRAS 225, 381
[85]   Corbet, R.H.D. et al. 1987, MNRAS 222, 15P
[86]   Courvoisier, T.J.-L. et al. 1986, ApJ 309, 265
[87]   Cowley, A.P. & Schmidtke, P. 1990, AJ 99, 678
[88]   Cowley, A.P. et al. 1979, ApJ 231, 539
[89]   Cowley, A.P. et al. 1982, ApJ 255, 596
[90]   Cowley, A.P. et al. 1982, ApJ 256, 605
[91]   Cowley, A.P. et al. 1983, ApJ 272, 118
[92]   Cowley, A.P. et al. 1987, ApJ 320, 296
[93]   Cowley, A.P. et al. 1987, AJ 92, 195
[94]   Cowley, A.P. et al. 1988, ApJ 333, 906
[95]   Cowley, A.P. et al. 1988, AJ 95, 1231
[96]   Cowley, A.P. et al. 1990, ApJ 350, 288
[97]   Cowley, A.P. et al. 1991, ApJ 373, 228
[98]   Cowley, A.P. et al. 1991, ApJ 381, 526
[99]   Crampton, D. 1974, ApJ 187, 345
[100] Crampton, D. & Hutchings, J.B. 1974, ApJ 191, 483
[101] Crampton, D. et al. 1976, ApJ 207, 907
[102] Crampton, D. et al. 1978, ApJ 225, L63
[103] Crampton, D. et al. 1986, ApJ 306, 599
[104] Crampton, D. et al. 1987, ApJ 321, 745
[105] Crampton, D. et al. 1990, ApJ 355, 496
[106] Cudworth, K.M. 1988, AJ 96, 105
[107] Damen, E. et al. 1990, A&A 237, 103
[108] Datta, B. 1988, Fund Cosmic Phys. 12, 151
[109] Davidsen, A. et al. 1972, ApJ 177, L97
[110] Davidsen, A. et al. 1974, ApJ 193, L25
[111] Davidsen, A. et al. 1975, ApJ 198, 653
[112] Davidsen, A. et al. 1976, ApJ 203, 448
[113] Davidson, K. & Ostriker, J.P. 1973, ApJ 179, 585
[114] Davies, S.R. et al. 1989, MNRAS 237, 973
[115] Davis, R. & Hartman, L. 1983, ApJ 270, 671
[116] Deeter, J.E. et al. 1991, ApJ 383, 324
[117] Deguchi, S. 1985, ApJ, 291, 492 (Erratum ApJ 303, 901)
[118] De Jager, C. et al. 1988, A&AS 72, 259
[119] Della Valle, M. et al. 1991, Nat 353, 50
[120] De Loore, C. et al. 1981, A&A 104, 150
[121] De Loore, C. et al. 1984, A&A 141, 279
[122] Densham, R.H. et al. 1983, MNRAS 205, 1117
[123] Dewey, R.J. & Cordes, J.M. 1987, ApJ 321, 780
[124] Doazan, V. et al. 1987, A&A 182, L25
[125] D'Odorico. S. et al. 1991, Nat 353, 329
[126] Dotani, T. et al. 1990, Nat 347, 534
[127] Doxsey, R. et al. 1979, ApJ 228, L67
[128] Duerbeck, H.W. 1987, A Reference Catalogue and Atlas of
       Galactic Novae (Berlin: Springer)

[129] Dupree, A. et al. 1978, Nat 275, 400
[130] Dupree, A. et al. 1980, ApJ 238, 969
[131] Duquennoy, A. & Mayor, M. 1991, A&A 248, 485
[132] Eadie, G. et al. 1975, MNRAS 172, 35P
[133] Elvis, M. et al. 1975, Nat 257, 656
[134] Endal, A.S. et al. 1976, Ap. Lett. 17, 131
[135] Fabian, A.C. et al. 1986, ApJ 305, 333
[136] Fabian, A.C. et al. 1987, MNRAS 225, 29P
[137] Fahlman, G.G. & Walker, G.A.H. 1980, ApJ 240, 169
[138] Faulkner, J. et al. 1983, MNRAS, 205, 359
[139] Feast, M.W. 1991, IAU Symposium 148, 1
[140] Ferland, G.J. 1992, ApJ 389, L63
[141] Frank, J. et al. 1987, A&A 178, 137
[142] Friedman, J.L. et al. 1986, ApJ 304, 115
[143] Friend, D.B. 1990, ApJ 353, 617
[144] Friend, D.B. & Abbott, D.C. 1986, ApJ 311, 370
[145] Friend, D.B. & Castor, J.I. 1982, ApJ 261, 293
[146] Garcia, M.R. 1992, private communication
[147] Garcia, M.R. & Grindlay, J.E. 1987, ApJ 313, L59
[148] Garcia, M.R. & Grindlay, J.E. 1992, IAU Circular 5578
[149] Garcia, M. et al. 1983, ApJ 267, 291.
[150] Garcia, M.R. et al. 1989, ApJ 341, L75
[151] Garmany, C.D. & Conti, P.S. 1985, ApJ 293, 407
[152] Garmany, C.D. & Fitzpatrick, E.L. 1988, ApJ 332, 711
[153] Garmany, C.D. et al. 1981, ApJ 250, 660.
[154] Gerend, D. & Boynton, P. 1976, ApJ 209, 562
[155] Giacconi, R. et al. 1973, ApJ 184, 227
[156] Gies, D.R. & Bolton, C.T. 1982, ApJ 260, 240.
[157] Gies, D.R. & Bolton, C.T. 1984, ApJ 276, L17
[158] deleted
[159] Gies, D.R. & Bolton, C.T. 1986, ApJ 304, 371
[160] Gies, D.R. & Bolton, C.T. 1986, ApJ 304, 389
[161] Gies, D.R. & Bolton, C.T. 1986, ApJS 61, 419
[162] Giovanelli, F. & Graziali, L.S. 1992, Space Sci. Rev. 59, 1
[163] Goldwurm, A. et al. 1992, ApJ 389, L79
[164] Gonzalez-Riestra, R. et al. 1991, IAU Circular 5174
[165] Goranskii, V. & Lyutyi, V.M. 1988, SvA 31, 193
[166] Gotthelf, E. et al. 1991, ApJ 374, 340
[167] Gotthelf, E. et al. 1992, AJ 103, 219
[168] Gottlieb, E.W. et al. 1975, ApJ 195, L33
[169] Gottwald, M. et al. 1986, ApJ 308, 213
[170] Greiner, J. et al. 1991, A&A 246, L17
[171] Griffiths, R.E. et al. 1978, Nat 276, 247
[172] Griffiths, R.E. et al. 1978, ApJ 221, L63
[173] Grindlay, J.E. 1989, in ref. [213], 121
[174] Grindlay, J.E. 1992, in Frontiers of X-ray Astronomy, ed. Y.
      Tanaka & K. Nomoto, (Tokyo Universal Academy Press), 69
[175] Grindlay, J.E. 1981, IAU Circular 3620
[175a] Grindlay, J.E. & Liller, W. 1978, ApJ 220, L127
[176] Grindlay, J.E. et al. 1978, Nat 274, 567
[177] Grindlay, J.E. et al. 1979, ApJ 232, L79
[178] Grindlay, J.E. et al. 1988, ApJ 334, L25
[179] Groth, E.J. 1974, ApJ 192, 517
[180] Gursky, H. et al. 1980, ApJ 237, 163
[181] Haberl, F. & White, N.E. 1990, ApJ 361, 225
[182] Haberl, F. et al. 1989, ApJ 343, 409
[183] Hackwell et al. 1979, ApJ 233, L115
[184] Hameury, J.M. et al. 1986, A&A 162, 71
[185] Hammerschlag-Hensberge, G. & Wu, C.C. 1977, A&A 56, 433
[186] Hammerschlag-Hensberge, G. et al. 1980, A&A 85, 119
[187] Hammerschlag-Hensberge, G. et al. 1982, ApJ 254, L1
[188] Hammerschlag-Hensberge, G. et al. 1984, ApJ 283, 249
[189] Hammerschlag-Hensberge, G. et al. 1990, ApJ 352, 698
[190] Han, X. -H. & Hjellming, R.M. 1990, in: Accretion-Powered
      Compact Binaries, ed. C.W. Mauche (Cambridge: CUP), 25
[191] Harmon, B.A. et al. 1992, IAU Circ. No. 5504
[192] Hartle, J.B. 1978, Phys. Rep. 46, 201

[193] deleted
[194] Hasinger, G. & Van der Klis, M. 1989, A&A 225, 7
[195] Hasinger, G. 1987, A&A 186, 153
[196] Haswell, C.A. & Shafter, A.W. 1990, ApJ 359, L47
[197] Hatchett, S. & McCray, R. 1977, ApJ 211, 552
[198] Hatchett, S. et al. 1976, ApJ 206, 847
[199] Heap, S. & Corcoran, M.F. 1992, ApJ 387, 340
[200] Heemskerk, M. & Van Paradijs, J. 1989, A&A 223, 154
[201] Henrichs, H. et al. 1982, ApJ 268, 807
[202] Hertz, P. et al. 1992, ApJ 396, 201
[203] Horne, K. & Marsh, T.R. 1986, MNRAS 218, 761
[204] Horne, K. et al. 1986, MNRAS 218, 63
[205] Howarth, I.D. 1983, MNRAS 203, 801
[206] Howarth, I.D. & Wilson, R. 1983, MNRAS 202, 347
[207] Howarth, I.D. & Wilson, R. 1983, MNRAS 204, 1091
[208] Howarth, I.D. et al. 1984, MNRAS 207, 287
[209] Howarth, I.D. et al. 1986, in: New Insights in Astrophysics, ESA
      SP-263, p. 475
[210] Huang, S.S. 1972, ApJ 171, 549
[211] Hudec, R. & Wenzel, W. 1976, Bull. Astr. Inst. Czech. 27, 325
[212] Hudec, R. & Wenzel, W. 1986, A&A 158, 396
[213] Hunt, J. & Battrick, B. (eds) 1989, Proc. 23rd ESLAB
      Symposium (Bologna), ESA SP-296
[214] Hutchings, J.B. 1977, ApJ 217, 537
[215] Hutchings, J.B. 1979, PASP 91, 657
[216] Hutchings, J.B. 1984, IAU Symposium 108, 305
[217] Hutchings, J.B. & Crampton, D. 1981, PASP 93, 486
[218] Hutchings, J.B. et al. 1973, ApJ 182, 549
[219] Hutchings, J.B. et al. 1977, ApJ 217, 186
[220] Hutchings, J.B. et al. 1978, ApJ 225, 548
[221] Hutchings, J.B. et al. 1979, ApJ 229, 1079
[222] Hutchings, J.B. et al. 1982, PASP 96, 312.
[223] Hutchings, J.B. et al. 1985, ApJ 292, 670
[224] Hutchings, J.B. et al. 1987, AJ 94, 340
[225] Ilovaisky, S.A. & Chevalier, C. 1992, IAU Circular 5507
[226] Ilovaisky, S.A. et al. 1978, A&A 70, L19
[227] Ilovaisky, S.A. et al. 1980, MNRAS 191, 81
[228] Ilovaisky, S.A. et al. 1984, A&A 210, 251
[229] Iping, R. & Patterson, J.A. 1991, A&A 239, 221
[230] Jackson, J. 1975, MNRAS 172, 483
[231] Johnson, H.L. 1966, ARA&A 4, 193
[232] Johnston, H.M. et al. 1989, ApJ, 345, 492
[233] Johnston, M.D. et al. 1979, ApJ, 233, 514
[234] Johnston, M.D. et al. 1983, MNRAS 183, 11P
[235] Jones, C. & Liller, W. 1973, ApJ 184, L121
[236] Jones, C. et al. 1973, ApJ 182, L109
[237] Joss, P.C. & Rappaport, S.A. 1984, ARA&A 22, 537
[238] Joss, P.C. et al. 1980, ApJ 235, 592
[239] Kallman, T. & McCray, R. 1980, ApJ 242, 615
[240] Kallman, T.R. & McCray, R. 1982, ApJS 50, 263
[241] Kallman, T.R. et al. 1987, ApJ 317, 746
[242] Kallman, T.R. et al. 1991, ApJ 370, 717
[243] Kaper, L. et al. 1990, Nat 347, 652
[244] Kastner, S.O. & Bhatia, A.K. 1990, ApJ 362, 745
[245] Kastner, S.O. & Bhatia, A.K. 1991, ApJ 381, L59
[246] Kelley, R.L. et al. 1983, ApJ 264, 568
[247] Kemp, J.C. et al. 1983, ApJ 271, L65
[248] Kemp, J.C. et al. 1987, SvA 31, 170
[249] Khruzina, T.S. & Cherepashchuk, A.M. 1983, SvA 26, 310.
[250] Khruzina, T.S. & Cherepashchuk, A.M. 1983, SvA 27, 35
[251] Kitamoto, S. et al. 1984, PASJ 36, 799
[252] Kitamoto, S. et al. 1990, ApJ 361, 590
[253] Knutsen, H. 1989, ApSS 162, 315
[254] Kondo, Y. et al. 1983, ApJ 273, 716
[255] Koo, D.C. & Kron, R.G. 1977, PASP 89, 285
[256] Kudritzki, R.P. & Hummer, D.G. 1990, ARA&A 28, 203
[257] Kudritzki, R.P. et al. 1991, IAU Symp. 148, 279

[258] Kuiper, L. et al. 1988, A&A 203, 79
[259] Kundt, W. 1979, A&A 80, L7
[260] LaDous, C. 1990, in Cataclysmic Variables and Related Objects, ed. M. Hack, NASA/CNRS Monograph Series on Non Thermal Phenomena in Stellar Atmosheres, p. 14
[261] Lamb, D.Q. & Sorvari, J.M. 1972, IAU Circular 2422
[262] Lamb, F.K. 1990, in ref. [213], 215
[263] Lamers, H.J.G.L.M. & Groenewegen, M.A.T. 1989, A&AS 79, 359
[264] Lamers, H.J.G.L.M. & Pauldrach, A.W.A. 1991, A&A 244, L5
[265] Lamers, H.J.G.L.M. & Waters, L.B.F.M. 1987, A&A 182, 80
[266] Lang, F. et al. 1981, ApJ 246, L21
[267] Lattimer, J.M. et al. 1990, ApJ 355, 241
[268] Lawrence, A. et al. 1983, ApJ 271, 793
[269] Leibowitz, E.A. 1984, MNRAS 210, 279
[270] Leitherer, C. 1990, in: Intrinsic Properties of Hot Luminous Stars, ed. C.D. Garmany (Astron. Soc. Pacific), p. 315
[271] Lesh, J.R. 1968, ApJS 17, 371
[272] Levine, A. et al. 1991, ApJ 381, 101
[273] Levine, A. et al. 1988, ApJ 327, 732
[274] Lewin, W.H.G. & Van den Heuvel, E.P.J. (eds) 1983, Accretion-Driven Stellar X-ray Sources, (Cambridge Univ. Press)
[275] Lewin, W.H.G. et al. 1987, ApJ 319, 893
[276] Li, F.K. et al. 1976, ApJ 203, 187
[277] London, R. et al. 1981, ApJ 243, 970
[278] Lucy, L.B. 1967, Zeitschr. f. Astroph. 65, 89
[279] Lyutyi, V.M. & Voloshina, I.B. 1989, SvA Lett. 15, 347
[280] MacGregor, K.B. & Vitello, P.A. 1982, ApJ 259, 267
[281] Machin, G. et al. 1990, MNRAS 247, 205
[282] Maeder, A. 1980, A&A 90, 311
[283] Magnier, E. et al. 1993, **MNRAS** (in press)
[284] Makishima, K. et al. 1987, ApJ 314, 619
[285] Maraschi, L. et al. 1976, Nat 259, 292
[286] Maraschi, L. et al. 1980, 241, L23
[287] Maraschi, L. et al. 1981, ApJ 248, 1010
[288] Margon, B. 1984, ARA&A 22, 507
[289] Margon, B. 1992, private communication
[290] Margon, B. & Cohen, J.G. 1978, ApJ 222, L33
[291] Margon, B. et al. 1976, ApJ 208, L35
[292] Margon, B. et al. 1978, ApJ 221, 907
[293] Marsh, T.R., & Horne, K. 1988, MNRAS 235, 269
[294] Marlborough, J.M. 1977, ApJ 216, 446
[295] Marsden, B.G. 1989, IAU Circular 4783
[296] Mason, K. 1989, in ref. [213], 113
[297] Mason, K.O. & Cordova, F.A. 1982, ApJ 255, 603
[298] Mason, K.O. & Cordova, F.A. 1982, ApJ 262, 253
[299] Mason, K.O. et al. 1980, Nat 287, 516
[300] Mason, K.O. et al. 1980, ApJ 242, L109
[301] Mason, K.O. et al. 1982, MNRAS 200 793
[302] Mason, K.O. et al. 1985, MNRAS 216, 1033
[303] Matsuoka, M. et al. 1984, ApJ 283, 774
[304] Mazeh, T. et al. 1986, A&A 157, 113
[305] McClintock, J. 1992, in: Frontiers of X-ray Astronomy, eds. Y. Tanaka & K. Koyama (Tokyo: Universal Academy Pres), p. 333
[306] McClintock, J.E. & Remillard, R.A. 1986, ApJ 308, 110
[307] McClintock, J.E. & Remillard, R.A. 1989, BAAS 21, 1206
[308] McClintock, J.E. & Remillard, R.A. 1990, ApJ 350, 386
[309] McClintock, J.E. et al. 1975, ApJ 198, 641
[310] McClintock, J.E. et al. 1978 ApJ 223, L75
[311] McClintock, J.E. et al. 1978, IAU Circular 3251
[312] McClintock, J.E. et al. 1979, Nat 279, 47
[313] McClintock, J.E. et al. 1980, ApJ 235, L81
[314] McClintock, J.E. et al. 1981, ApJ 243, 900
[315] McClintock, J.E. et al. 1982, ApJ 258, 245
[316] McClintock, J.E. et al. 1984, ApJ 283, 794
[317] McClintock, J.E. et al. 1992, IAU Circular 5499

[318] Mestel, L. 1965, in: Stellar Structure, ed. L.H. Aller & D.B. McLaughlin (Univ. Chicago Press), 465
[319] Meurs, E.J. & Van den Heuvel, E.P.J. 1989, A&A 226, 88
[320] Meurs, E.A. et al. 1992, A&A 265, L41
[321] Meyer, F. & Meyer-Hoffmeister, E. 1981, A&A 104, L10
[322] Middleditch, J. & Nelson, J. 1973, Ap. Lett. 14, 129
[323] Middleditch, J. & Nelson, J. 1976, ApJ 208, 567
[324] Middleditch, J. et al. 1981, ApJ 244, 1001
[325] Middleditch, J. et al. 1983, ApJ 274, 313
[326] Milgrom, M. 1977, A&A 54, 725
[327] Milgrom, M. 1978, A&A 70, 763
[328] Miyamoto, S. & Matsuoka, M. 1977, Space Sci. Rev. 20, 687
[329] Molnar, L.A. 1986, in: Physics of Accretion onto Compact Objects, ed. K. Mason et al. (NY: Springer), 313
[330] Molnar, L. & Neelay, M. 1992, IAU Circular 5595
[331] Mook, D.E. et al. 1972, ApJ 177, L63
[332] Mook, D.E. et al. 1975, ApJ 197, 525
[333] Motch, C. & Pakull, M.W. 1989, A&A 214, L1
[334] Motch, C. et al. 1985, Space Sci. Rev. 40, 239
[335] Motch, C. et al. 1987, ApJ 313, 792
[336] Motch, C. et al. 1989, A&A 219, 158
[337] Motch, C. et al. 1990, in ref. [213], 545
[338] Mouchet, M. et al. 1980, A&A 90, 113
[339] Moulding, M. 1977, A&A 58, 393
[340] Murdin, P. et al. 1977, MNRAS 178, 27P
[341] Murdin, P. et al. 1980, MNRAS 192, 709
[342] Nagase, F. 1989, PASJ 41, 1
[343] Nagase, F. et al. 1992, ApJ 396, 147
[344] Nauenberg, M. & Chapline, G. 1973, ApJ 179, 277
[345] Naylor, T. et al. 1988, MNRAS 233, 285
[345a] Naylor, T. et al. 1992, MNRAS 255, 1
[346] Nelson, J. et al. 1977, ApJ 212, 215
[347] Ninkov, Z. et al. 1987, ApJ 321, 425
[348] Ninkov, Z. et al. 1987, ApJ 321, 438
[349] Oke, J.B. 1977, ApJ 217, 181
[350] Oke, J.B. & Greenstein, J. 1977, ApJ 211, 872
[351] Olive, K.A. 1991, Science 251, 1194
[352] Osterbrock, D.E. 1989, Astrophysics of Gaseous Nebulae and Active Galactic Nuclei (Mill Valley CA: University Science Books)
[353] Ostriker, J.P. et al. 1976, ApJ 208, L61
[354] Overgard, T., & Ostgaard, E. 1991, A&A 243, 412
[355] Paciesas, W.S. et al. 1992, IAU Circular 5580
[356] Paczynski, B. 1973, Annals of the N.Y. Acad. of Sciences 24, 233
[357] Paczynski, B. 1992, Acta Astron. 42, 145
[358] Pakull, M.W. 1982, Proc. Workshop Accreting Neutron Stars, Garching, MPE Report 177, 53
[359] Pakull, M.W. & Angebault, L.P. 1986, Nat 322, 511
[360] Pakull, M.W. et al. 1988, A&A 203, L27
[361] Pakull, M.W. et al. 1993, in preparation
[362] Pallavicini, R. et al. 1981, ApJ 248, 279
[363] Pandey, S.C. et al. 1989, ApSS 159, 203
[364] Parmar, A.N. et al. 1989, ApJ 338, 359
[365] Parmar, A.N. & White, N.E. 1988, Mem. Soc. Astr. Italia 59, 147
[366] Payne, B.J. & Coe, M.J. 1987, MNRAS 225, 987
[367] Pedersen, H. 1983, ESO Messenger 34, 21
[368] Pedersen, H. et al. 1981, Nat 294, 725
[369] Pedersen, H. et al. 1982, ApJ 263, 325
[370] Pedersen, H. et al. 1982, ApJ 263, 340
[371] Penninx, W. 1990, in ref. [213], 185
[372] Penninx, W. et al. 1989, A&A 208, 146
[373] Penrod, G.D. & Vogt, S.S. 1985, ApJ 299, 653
[374] Peterson, B.A. et al. 1980, MNRAS 190, 33P
[375] Petro, L. & Hiltner, W.A. 1974, ApJ 190, 661
[376] Petterson, J.A. 1977, ApJ 218,783
[377] Petterson, J.A. 1983, in ref. [274], 367
[378] Pietsch, W. et al. 1986, A&A 157, 23
[379] Poeckert, R. & Marlborough, J.M. 1978, ApJS 38, 229

[380] Pols, O. et al. 1991, A&A 241, 419
[381] Pounds, K.A. et al. 1975, MNRAS 172, 473
[382] Priedhorsky, W.C. 1986, ApSS 126, 89
[383] Priedhorsky, W.C. & Holt, S.S. 1987, Space Sci. Rev. 45, 291
[384] Priedhorsky, W.C. & Terrell, J. 1983, ApJ 273, 709
[385] Priedhorsky, W.C. et al. 1983, ApJ 270, 233
[386] Priedhorsky, W. et al. 1986, ApJ 306, L91
[387] Primini, F. et al. 1977, ApJ 217, 543
[388] Prinja, R.K. et al 1990, ApJ 379, 734
[389] Rappaport, S.A. & Joss, P.C. 1983, in ref. [274], 1
[390] Rappaport, S.A. & Van den Heuvel, E.P.J. 1982, IAU Symp. 98, 327
[391] Raymond, J.C. 1982, ApJ 258, 240
[392] Raymond, J.C. 1992, ApJ (submitted)
[393] Remillard, R.A. et al. 1992, ApJ 399, L145
[394] Reynolds, A.P. et al. 1992, MNRAS 256, 631
[395] Reynolds, A.P. et al. 1992, MNRAS in press
[396] Richter, G.A. 1989, IAU Inf. Bull. Var. Stars 3362
[397] Rhoades, C.E. & Ruffini, R. 1974, Phys. Rev. Lett. 32, 324
[398] Robinson, E.L. 1976, ARA&A 14, 119
[399] Rose, L.A. et al. 1979, ApJ 231, 919
[400] Russell, S.C. & Dopita, M.A. 1990, ApJS 74, 93
[401] Ryba, M.F. & Taylor, J.H. 1991, ApJ 371, 739
[402] Sadakane, K. et al. 1985, ApJ 288, 284
[403] Sandage, A. et al. 1966, ApJ 146, 316
[404] Sato, N. et al. 1986, ApJ 304, 241
[405] Schachter, J. et al. 1989, ApJ 340, 1049 (Erratum ApJ 362, 379)
[406] Schaefer, B. 1990, ApJ 354, 720
[407] Schmidtke, P. 1988, AJ 95, 1528
[408] Schmidtke, P. & Cowley, A.P. 1987, AJ 92, 374
[409] Schmidtke, P. & Cowley, A.P. 1992, IAU Circular 5451
[410] Schoembs, R. et al. 1987, ESO Messenger 48, 6
[411] Schrader, C.R. et al. 1992, IAU Circular 5591
[412] Schreier, E. et al. 1972, ApJ 172, L79
[413] Schwarzenberg-Czerny, A. 1992, A&A 260, 268
[414] Schwarzschild, M. 1958, Structure and Evolution of the Stars (Dover Publ.)
[415] Shafter, A.W. et al. 1980, ApJ 240, 612
[416] Shakura, N.I. & Sunyaev, R.A. 1973, A&A 24, 337
[417] Shapiro, S.L. & Teukolsky, S.A. 1983, Black Holes, White Dwarfs, & Neutron Stars (New York: Wiley), p. 264
[418] Shrader, C.R. et al. 1992, IAU Circular 5591
[419] Skinner, G.K. 1980, Nat 288, 141
[420] Slettebak, A. 1979, Space Sci. Rev. 23, 541
[421] Slettebak, A. 1982, ApJS 50, 55
[422] Slettebak, A. & Snow, T.P. 1987, Physics of Be Stars, Cambridge University Press
[423] Smale, A.P. & Corbet, R.H.D. 1991, ApJ 383, 853
[424] Smale, A.P. & Mukai, K. 1988, MNRAS 231, 663
[425] Smale, A.P. et al. 1984, MNRAS 207, 29P
[426] Smale, A.P. et al. 1986, MNRAS 223, 207
[427] Smale, A.P. et al. 1988, MNRAS 232, 647
[428] Smale, A.P. et al. 1988, MNRAS 233, 51
[429] Smale, A.P. et al. 1989, PASJ 41, 607
[430] Smale, A.P. et al. 1992, ApJ 400, 330
[431] Stella, L. et al. 1986, ApJ 308, 669
[432] Sternberg, A. et al. 1988, Comm. on Astrophys. 13, 29
[433] Sterne, T.E. 1941, Proc. Nat. Acad. Sci. (U.S.) 27, 168
[434] Stone, R.C. 1979, ApJ 232, 520
[435] Strittmatter, P.A. et al. 1973, A&A 25, 275
[436] Sunyaev, R. et al. 1992, ApJ 389, L75
[437] Sunyaev, R. et al. 1992, IAU Circular 5593
[438] Swank, J.H. et al. 1976, ApJ 209, L57
[439] Szkody, P. & Mateo, M. 1984, ApJ 280, 729
[440] Taam, R.E. et al. 1991, ApJ 371, 696
[441] Takagishi, K. et al. 1978, ISAS Research Note 60
[442] Taylor, J.H. & Weisberg, J.M. 1989, ApJ 345, 434

[443] Thorstensen, J. et al. 1978, ApJ 220, L131
[444] Thorstensen, J. et al. 1979, ApJ 233, L57
[445] Thorstensen, J.R. et al. 1980, ApJ 238, 964
[446] Thorstensen, J.R. et al. 1987, ApJ 312, 739
[447] Thorstensen, J.R. et al. 1988, ApJ 334, 430
[448] Tjemkes, S.A. et al. 1986, A&A 154, 77
[449] Treves, A. et al. 1980, ApJ 242, 1114
[450] Treves, A. et al. 1988, ApJ 335, 142
[451] Treves, A. et al. 1990, ApJ 364, 266
[452] Trimble, V. et al. 1973, MNRAS 162, 1P
[453] Truemper, J. et al. 1985, Sp. Sci. Rev. 40, 255
[454] Truemper, J. et al. 1991, Nat 349, 579
[455] Tuohy, I. & Cruise, A.M. 1975, MNRAS 171, 33P
[456] Tuchman, Y. et al. 1990, ApJ 359, 164
[457] Turner, M.J.L. et al. 1985, Sp. Sci. Rev. 40, 249
[458] Udalski, A., & Kaluzny, J. 1991, PASP 103, 198
[459] Van Amerongen, S. et al. 1986, IAU Inf. Bull. Var. Stars 2901
[460] Van Amerongen, S.F. et al. 1987, A&A 185, 147
[461] Van den Heuvel, E.P.J. 1983, in ref. [274], 303
[462] Van den Heuvel, E.P.J. & Heise, J. 1972, Nat Phys.Sci. 239, 67
[463] Van den Heuvel, E.P.J. & Rappaport, S.A. 1987, in [422], 291
[464] Van den Heuvel, E.P.J. et al. 1992, A&A 262, 97
[465] Van der Klis, M. et al. 1983, A&A 126, 265
[466] Van der Klis, M. et al. 1982, A&A 106, 339
[467] Van der Klis, M. 1992, MNRAS 260, 686
[468] Van Kerkwijk, M.H. et al. 1992, Nat 355, 703
[469] Van Oyen, J. 1989, A&A 217, 115
[470] Van Paradijs, J. 1983, in ref. [274], 189
[471] Van Paradijs, J. 1991, in: Neutron Stars: Theory and Observation, eds J. Ventura & D. Pines (Kluwer), 289
[472] Van Paradijs, J. & Kuiper, L. 1984, A&A 138, 71
[473] Van Paradijs, J. & Stollman, G.M. 1984, A&A 137, L12
[474] Van Paradijs, J. & Verbunt, F. 1984, in: High Energy Transients in Astrophysics, ed. S.E. Woosley (New York: AIP), 49
[475] Van Paradijs, J. & Zuiderwijk, E.J. 1977, A&A 61, L19
[476] Van Paradijs, J. et al. 1977, A&AS 30, 195
[477] Van Paradijs, J. et al. 1981, IAU Circular 3626
[478] Van Paradijs, J. et al. 1983, A&A 124, 294
[479] Van Paradijs. J. et al. 1983, A&AS 55, 7
[480] Van Paradijs, J. et al. 1987, A&A 184, 201
[481] Van Paradijs, J. et al. 1988, MNRAS 231, 379
[482] Van Paradijs, J. et al. 1988, A&AS 76, 185
[483] Van Paradijs, J. et al. 1990, A&A 234, 181
[484] Van Paradijs, J. et al. 1990, A&A 235, 156
[485] Van Paradijs, J. et al. 1990, PASJ 42, 633
[486] Verbunt, F. 1992, in: Frontiers of X-ray Astronomy, ed. Y. Tanaka & K. Koyama (Tokyo: Univeral Academy Press), 57
[487] Verbunt, F. et al. 1984, MNRAS 210, 899
[488] Vrtilek, S.D. et al. 1990, A&A 235, 162
[489] Vrtilek, S.D. et al. 1991, ApJS 76, 1127
[490] Vrtilek, S.D. et al. 1991, ApJ 376, 278
[491] Wagner, R.M. et al. 1990, in: Accretion-Powered Compact Binaries, ed. C.W. Mauche (Cambridge: Camb. U. Press), 30
[492] Wagner, R.M. et al. 1992, IAU Circ. 5589
[493] Wagner, R.M. et al. 1991, ApJ 378, 293
[494] Wagner, R.M. et al. 1992, ApJ 401, L97
[495] Walter, F.M. et al. 1982, ApJ 253, L67
[496] Warwick, et al. MNRAS 190, 243
[497] Waters, L.B.F.M. 1989, in ref. [213], 25
[498] Watson, M.G. & Griffiths, R.E. 1977, MNRAS 178, 513
[499] Watson, M.G. et al. 1982, MNRAS 199, 915
[500] Webbink, R.F. 1978, A provisional Optical Light Curve of the X-ray Recurrent Nova V616 Mon = A0620-00, unpublished report
[501] Weber, F. et al. 1991, ApJ 373, 579
[502] Webster, B.L., & Murdin, P. 1972, Nat 235, 37
[503] Westerlund, B.E. 1991, IAU Symp. 148, 15
[504] Whelan, J. 1973, ApJ 185, L127

[505] Whelan, J.A.J. et al. 1977, MNRAS 180, 657
[506] White, N.E. 1989, A&A Rev. 1, 85
[507] White, N.E. & Carpenter, G.F., 1978, MNRAS 183, 11P
[508] White, N. & Swank, J. 1982, ApJ, 253, L61
[509] White, N.E. & Swank, J.H. 1984, ApJ 287, 856
[510] White, N.E. et al. 1976, MNRAS 176, 91
[511] White, N.E. et al. 1984, ApJ 283, L9
[512] White, N.E. et al. 1984, in: High Energy Transients in
       Astrophysics, ed. S.E. Woosley (New York: AIP), 31
[513] White, N.E. et al. 1986, MNRAS 218, 129
[514] Whitehurst, R. 1988, MNRAS 232, 35
[515] Wickramasinghe, D.T. 1975, MNRAS 173, 21
[516] Wickramasinghe, D.T. & Whelan, J. 1975, MNRAS 172, 175
[517] Willis, A.J. et al. 1980, ApJ 237, 596
[518] Wilson, R.E. 1979, ApJ 234, 1054
[519] Wilson, R.E. & Devinney, E.J. 1971, ApJ 166, 605
[520] Wilson, R.E. & Sofia, S. 1976, ApJ 203, 182
[521] Wolszczan, A. 1991, Nat 350, 688
[522] Wu, C.C. 1979, ApJ 227, 291
[523] Wu, C.C. et al. 1976, A&A 50, 445
[523a] Wu, C.C. et al. 1982, PASP 94, 149
[524] Wu, C.C. et al. 1983, PASP 95, 391
[525] Wu, K., et al. 1991, A&A 246, 411
[526] Young, A. & Wentworth, S.T. 1982, PASP 94, 815
[527] Zombeck, M.V. 1990, Handbook of Space Astronomy &
       Astrophysics (Cambridge: Cambridge U. Press), 195
[528] Zorec, J. & Briot, D. 1991, A&A 245, 150
[529] Zuiderwijk, E.J. 1979, Ph.D. Thesis, University of Amsterdam
[530] Zuiderwijk, E.J. et al. 1974, A&A 35, 353
[531] Zwitter, T. & Calvani, M. 1989, MNRAS, 236, 581
[532] Zwitter, T. et al. 1989, Fund. of Cosmic Phys. 13, 309
[533] **Van Paradijs, J. & Lewin, W.H.G., 1985, MNRAS 142, 361**

3

Black-hole binaries

Yasuo Tanaka
Institute of Space and Astronautical Science, 3-1-1 Yoshinodai, Sagamihara, Kanagawa-ken 229, Japan

Walter H. G. Lewin
Massachusetts Institute of Technology, Center for Space Research, 37-627, Cambridge, MA 02139, USA

3.1   Some historic notes

The story of black-hole X-ray sources began in the early seventies when Webster & Murdin (1972) and Bolton (1972) discovered the 5.6 day orbital period of the supergiant that stood out prominently in the X-ray error box of Cyg X-1. Based on the mass function and on the classification of the bright star, they independently considered the possibility that this binary system contained an accreting black hole (Figure 3.1). Webster & Murdin write, 'it is inevitable that we should also speculate that it might be a black hole'. Bolton writes, 'this raises the distinct possibility that the secondary is a black-hole'. (It is often forgotten that this was prior to the discovery with Uhuru of the binary nature of Cen X-3 (Schreier *et al.* 1972).)

The possibility of a black hole in Cyg X-1 stirred up the emotions in the astrophysical community; numerous 'pro' and 'con' black-hole articles have been written, including some in which equations of state have been suggested which can allow for very massive compact objects which have escaped the black-hole catastrophe (e.g. Bahcall *et al.* 1990). No one, however, has succeeded (or even attempted) to show that such objects can be created. The evidence that accreting stellar black-hole binaries

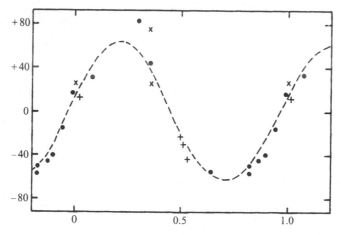

Fig. 3.1. The radial velocity of HD 226868 (Cyg X-1) plotted against phase in the 5.60 day orbital period. From Webster & Murdin (1972).

exist is now very strong. The very high mass of the compact objects, in combination with their X-ray properties, make it likely that they are black holes, though no certainty exists as none of the information that we have obtained to date is a test of general relativity in the strong field limit, and we will refer to them in this article as black-hole candidates (BHCs).

A large fraction of the BHCs are X-ray transients. The story of X-ray transients began on April 4, 1967, when Harries *et al.* (1967), during a rocket observation from Woomera, Australia, detected a very bright X-ray source (they named it CRUX) in a region in the sky where no source had been seen before. CRUX (later named Cen X-2) rose to an unprecedented brightness (outdoing Sco X-1), then diminished. By the end of September 1967, it was again undetectable. Its intensity had increased and subsequently decreased by at least a factor of 50 (Francey *et al.* 1967; Chodil *et al.* 1968; Cooke *et al.* 1968).

In the mid-seventies, it became clear that X-ray binaries could be roughly divided into two groups: the eclipsing pulsars with relatively hard spectra (population I objects) and a group (an old population) which included the galactic bulge sources and the X-ray sources in globular clusters (Tananbaum 1973; Canizares 1975; Jones 1977). Ostriker (1977) added a third group consisting of the two BHCs Cyg X-1 and Cir X-1, which stood out because of their very soft X-ray spectra in their high states. White & Marshall (1984) suggested, on the basis of much more information, that sources with 'ultrasoft' X-ray spectra should all be considered as *potential* BHCs. Their paper was a milestone in the way that the BHC research progressed.

3.2   Introduction

3.2.1   *What is a BHC?*

A neutron star is only roughly three times as large as its Schwarzschild radius, and therefore the accretion flows around a low-magnetic-field neutron star and a stellar-mass black hole have many similarities. A number of X-ray characteristics believed to be typical of black holes at one time, such as very soft spectra, fast variability, high-energy power-law spectrum, are now known to occur in neutron stars as well. Cir X-1 was long believed to be a BHC based on its temporal characteristics and on its soft spectrum (Samimi *et al.* 1979), but we now know that the compact object is a neutron star as it exhibits type I X-ray bursts (Tennant *et al.* 1986). X 0331+53 = V 0332+53 exhibits erratic fast variability with characteristics very similar to that of Cyg X-1 in its low state (Davelaar *et al.* 1983; Tanaka 1983a). However, it displays coherent pulsations with a period of 4.4 s and is therefore almost certainly a neutron star (Stella *et al.* 1985). High-energy (>20 keV) power-law tails were, and still are, considered signatures of a BHC (e.g. Sunyaev *et al.* 1988, 1991a–f, 1992b; Döbereiner *et al.* 1989; Tanaka 1989, 1992a,b; Grebenev *et al.* 1991, 1992a,b, 1993; Barret *et al.* 1992; Harmon *et al.* 1992a–c). However, low-luminosity neutron star LMXBs (most of which are type I bursters), often exhibit a single power-law type spectrum which can extend to 100 keV and beyond (see, e.g., Barret *et al.* 1991). Thus, the hard power-law spectrum alone is not a unique feature of BHCs.

In spite of this, the above characteristics (fast variability, very soft spectra, high-energy tails) are still very useful indicators. A 0620−00, LMC X-1, GS 2023+338, and

GRS/GS 1124—68 were first identified as BHCs on the basis of these characteristics. Subsequent mass determinations lent support (in some cases very strong support) to the earlier suggestions. It is unclear how good a BHC indicator the presence of positron annihilation radiation is; such radiation has been detected from GRS/GS 1124—68 and possibly from 1E 1740.7—2942 (see below).

Models have been proposed to account for the spectral differences between black holes and neutron star binaries. However, the physics of accretion is enormously complex, controversial, and the interpretation is by no means unique. In Sect. 3.4 we discuss the X-ray spectra and their interpretation (see also Ch. 1).

The most reliable black-hole characteristic remains, of course, the mass of the compact object as determined from Doppler shift measurements. As no stable neutron stars are believed to exist with masses in excess of $\sim 3 M_\odot$, any compact object exceeding this mass is usually assumed to be a black hole (Rhoades & Ruffini 1974). This presupposes that general relativity is correct, and that exotic particle physics cannot prevent the black-hole catastrophe (see Sect. 2.4). Kundt & Fischer (1989) have tried to argue that BHCs are accreting neutron stars with more or less massive accretion disks. However, Heemskerk *et al.* (1992) have shown that systems with a disk more massive than the central star are unstable to non-axisymmetric modes with azimuthal wave number $m = 1$.

In an X-ray binary, it is not trivial to distinguish a black hole from a neutron star; none of the information that we have obtained to date from BHCs is a test of general relativity in the strong-field limit. Connors *et al.* (1980) discuss how measurements of X-ray polarization, at least in principle, allow for measurements of the metric near a compact object.

3.2.2   *What is an X-ray transient?*

All X-ray binaries are variable; the degree of variability has historically been used as a dividing line between transients and permanent sources. Different authors have used different criteria; what some call a transient, others do not. There is an emotional bias as well. Bright sources which are always detectable (because they are so bright) are less likely to be called a transient than weak sources which are not always detectable. However, when the variability of a source exceeds many orders of magnitude, the transient behavior is never disputed. Increases in luminosity in excess of a factor of 10^3–10^4 are common. At the peak of the outbursts, luminosities are typically 10^{37} to 10^{38} erg s^{-1}, but they can be as high as $\sim 10^{39}$ erg s^{-1}.

The Tables in Ch. 14 by Van Paradijs contain 124 LMXBs, of which ~ 38 are listed as transients, and 69 HMXBs, of which 30 are listed as transients. About half of these transients were listed in previous reviews on transients (Kaluzienski *et al.* 1977; Cominsky *et al.* 1978; White *et al.* 1984a; Priedhorsky & Holt 1987). Of the 20 BHCs (see Table 3.1), ~ 16 are LMXBs, and three are HMXBs. All 16 LMXB BHCs are listed as transients in the Tables by Van Paradijs; none of the three HMXB BHCs are transients.

3.2.3   *Recurrence*

Many (possibly all) transients are recurrent; they exhibit repeated outbursts. The intervals between outbursts vary from source to source; they can be periodic, quasi-periodic or irregular on time scales from less than one month to tens of years or more (Priedhorsky & Holt 1987). It is known from archival optical data that A 0620−00 (V616 Mon) was in outburst in 1917 and that GS 2023+338 (V404 Cyg) was in outburst in 1938 and 1956 (for references, see Sect. 3.3).

3.2.4 Low-mass X-ray binaries

The LMXB transients have relatively soft spectra; ∼1/3 of them have ultrasoft spectra. The compact objects are neutron stars in most cases (X-ray bursts are often observed; see Ch. 4) and probably a black hole in many cases. It is probably not an accident that almost *all* LMXBs whose compact object is believed to be a black hole are transients, including the strongly variable source GX 339−4 (notice, not everyone would call GX 339−4 a transient). The formation and evolution of these systems are discussed in Ch. 11.

The optical counterparts of LMXB transients are typically ∼10^2 times brighter near transient maximum than in quiescence; this greatly facilitates the search for them (see Ch. 2). The X-ray rise times of these transients are typically a few days (the 1/e decay times are typically ∼ one month), but one or more mini-outbursts have been observed in several cases during decay (see Sect. 3.3). The very bright transients (the best studied and relatively nearby systems) such as, e.g., A 0620−00, GS 1124−68, Cen X-4, Aql X-1, GS 2000+25 have been detectable in X-rays up to ∼ one year after outburst. The 1/e decay times of the optical counterparts are typically 0.5–1 month. Radio emission has been detected from most bright LMXB transients, with decay times ranging from several days to several tens of days. GS 2023+338 (still observable three years after outburst) and GS 1124−68 had a second phase of radio emission with a much longer decay time (see Ch. 7 and references therein).

3.2.5 High-mass X-ray binaries

The HMXB transients have relatively hard spectra, and pulsations are often observed (see Ch. 1); the compact object is probably in most (perhaps all) cases a neutron star, and the companions are almost always ∼10–$20M_\odot$ unevolved Be stars which underfill their Roche lobe (Rappaport & Van den Heuvel 1982). Optical brightening during the outbursts is absent or very modest, though increases up to 1 and 2 mag have been observed (see Ch. 2). For information on radio emission from these systems, see Ch. 7.

3.2.6 Transient mechanisms

Transient outbursts are due to a large surge in the accretion rate onto a compact object. There are many ways that nature can do this, and various mechanisms are at work. We will summarize here briefly the possible outburst mechanisms for LMXBs as so many of the BHCs listed in Table 3.1 are transient LMXBs (see also Priedhorsky & Holt 1987). The outburst mechanism for HMXBs may be qualitatively different; we shall not discuss them here.

There are similarities between the outbursts of LMXBs and the outbursts of dwarf novae (DN) (see, e.g., Lin & Taam 1984; Van Paradijs & Verbunt 1984). The various types of DN show different outburst morphologies. Only the short orbital period

Table 3.1 Summary of Black-Hole Candidates†

Source [1]	Type [2]	Log (Flux) (erg cm⁻² s⁻¹) [3]	Log (Lum) (ergs) [4]	Dist. (kpc) [5]	Mass Funct (sol. mass) [6]	Optical [7]	Orb. Period [8]	Long Period [9]	X-ray Spectrum [10]	Detection >50 keV [11]	QPO >0.04 Hz [12]	Other Wavel. [13]
LMC X-3 0538-641	HMXB	-9.0 (max) (1 - 10 keV)	38.5 (max) (1 - 10 keV)	~50	2.3 ± 0.3 M_x > 7	B3 V V = 17	1.70 d	198 d X & Opt	ultras + high-en pl tail pl ind 2.2 ± 0.1			IR UV
LMC X-1 0540-697	HMXB	-9.2 (max) (1 - 10 keV)	38.3 (max) (1 - 10 keV)	~50	0.14 ± 0.05 M_x > 2.6 M_x = 6	O7-9 III V = 14.5	4.22 d		ultras + high-en pl tail pl ind 2.3 ± 0.1		0.075	UV
0620-003 V616 Mon [1975]	LMXB RT	-5.9 to <-12.2 (1 - 6 keV)	38.1 to <32.0 (1 - 6 keV)	~0.9	3.18 ± 0.16 M_x > 7.3	K5 V V = 12 - 18	7.75 h		ultras; no evidence for high-en tail >30 keV			radio IR UV
1124-684 Nova Muscae [1991]	LMXB T	-6.6 to <-11.0 (1 - 6 keV)	38.4 to <33.0 (1 - 6 keV)	~3	3.1 ± 0.4 M_x > 3.1	K0V - K4V B = 13 - 21	10.4 h		Jan-May '91 ultras + hard pl with ind 2.6-2.2 June-Sept '91 single pl with ind 1.5-1.8 positron annihil. line	up to ~600 keV	3 - 10	radio UV
1354-645 [1987]	LMXB T	-8.5 (peak) (1 - 10 keV)				V = 17			ultrasoft with high-en pl tail; pl ind =2.1			
1524-617 TrA X-1	LMXB RT	-7.9 (peak) (3 - 6 keV)				B=17 - >19.5			ultrasoft high-en tail (1990)	up to ~100 keV?		
1543-475	LMXB RT	-7.0 to <-11.3 (2 - 6 keV)	37.7 to <33.4 (2 - 6 keV)	~4		A1-2 V** V = 15 - 17			ultras, low-en pl ind =4-5 (1983) hi-en pl ind 2-3 (1992)			UV
1630-472	LMXB RT	-7.6 to -9.5 (1 - 50 keV)							soft + high-en pl tail (1984) with index = 2.5 - 1.0			
GX 339-4 1659-487 V821 Ara	LMXB RT *	-7.3 to -8.0 (H) -8.0 to -10.0 (L) <-10.0 (Off) (1 - 20 keV)	37.3 to 38.0 (H) 37.3 to 35.3 (L) <35.3 (Off) (1 - 20 keV)	~4 assumed		V=15.4 - >20	14.8 h		ultras + high-en pl tail (H) pl ind ~2.5 hard pl spectrum (L & Off) pl ind ~1.7	up to ~400 keV	0.05, 0.1 X,L 0.8 X,L 0.05 O,L ~0.13 O,L 6 X,H	IR UV
1705-250 Nova Ophiuchi [1977]	LMXB T	-6.9 (peak) (2 - 18 keV)				B=16.5 - >21			soft, kT =3 keV (2-18 keV) pl with ind ~2.7 (2-120 keV)			

Table 3.1 (cont)

Source [1]	Type [2]	Log (Flux) (erg cm⁻² s⁻¹) [3]	Log (Lum) (ergs) [4]	Dist. (kpc) [5]	Mass Funct (sol. mass) [6]	Optical [7]	Orb. Period [8]	Long Period [9]	X-ray Spectrum [10]	Detection >50 keV [11]	QPO >0.04 Hz [12]	Other Wavel. [13]
1740.7-2942		-8.5 (4 - 300 keV)	37.4 (4 - 300 keV)	~8.5 assumed					hi-en pl with hi-en cutoff pl ind =1.9-2.9 possible positron annihil	up to >500 keV		radio
1741-322 [1977]	LMXB T	-7.6 (peak) (1 - 10 keV)							soft with high-en pl tail with ind = 2.4			UV
1755-338	LMXB	-8.4 (1 - 20 keV)				V = 19	4.4 h?		soft, kT =2 keV			
1758-258	LMXB?	-8.6 (4 - 300 keV)							pl with ind =2 (4 - 300 keV)	up to ~300 keV		
1826-238 [1988]	LMXB? T	-8.9 (1 - 40 keV)							single pl with ind =1.7 (1 - 40 keV)			
1846-031 [1985]	LMXB T		~38 assumed	~10 to match lumin.					ultras + pl tail up to at least 25 keV			
1957+115 V1408 Aql	LMXB	-9.3 (2 - 11 keV)				V = 18.7	9.3 h		<10 keV th br kT=1.7 keV (1978.8) 8-20 keV pl ind 2.1 (1990)			UV
Cyg X-1 1956+350 HD226868	HMXB	see [4] & [5]	36.5 - 37.0 (H) 37.5 (L) (10 - 200 keV)	~2.5	0.24 ± 0.01 Mx > 7 Mx = 16 ± 5	O9.7 Iab V = 9	5.6 d	294 d X & Opt	ultras + high-en pl tail (H) hard pl spectrum (L) pl ind =1.5-2.2 (>20 keV)	up to ~MeV	-0.04	radio IR UV
2000+251 QZ Vul Nova Vulpec. [1988]	LMXB T	-6.5 to <-11.2 (1 - 40 keV)	38.2 to <33.4 (1 - 40 keV)	~2		early K dwarf V = 16 - >21	~8.3 h		ultras + high-en pl with ind 1.9-2.5 up to 300 keV 12/16 appr. single pl with ind 1.7±0.1	up to >100 keV		radio
2023+338 V 404 Cyg [1989]	LMXB RT	-5.9 to -9.4 (1 - 40 keV)	39.3 to 35.8 (1 - 40 keV)	~3.5	6.26 ± 0.31 Mx = 8 - 12	K0 IV V = 12 - 21	6.47 d		pl with ind =1.3-1.6 up to 300 keV	up to >100 keV		radio

systems called SU UMa systems show 'superoutbursts' as well as 'normal outbursts' (see Ch. 8 and Sect. 10.4.3). It has been suggested that the normal outbursts are caused by 'explosive' rapid mass loss of the donor (Bath 1969, 1975) or by a thermal disk instability (Smak 1970, 1971; Osaki 1974; for more references see Mineshige & Wheeler 1989) and that the superoutbursts are due to an irradiation-induced instability in the donor star (Smak 1992 and references therein). Osaki (1989) suggested that the superoutbursts could be due to a combination of the thermal instability and a tidal disk instability (Whitehurst 1988). We refer the reader to Ch. 8 and 10 for a much more detailed discussion of the various instabilities, including the pros and cons (see also Verbunt 1986 and Van Amerongen *et al.* 1990). By analogy, thermal disk instabilities and irradiation-induced instabilities have been proposed for the transient outbursts in LMXBs. (It is perhaps possible that the outbursts in LMXBs are due to the combined mechanisms as proposed by Osaki (1989) for the superoutbursts.)

3.2.7   *Thermal disk instabilities*

Thermal disk instabilities (Smak 1970, 1971; Osaki 1974; Hoshi 1979; Meyer & Meyer-Hoffmeister 1981; Bath & Pringle 1982; Cannizzo *et al.* 1982, 1985; Lin & Taam 1984; Huang & Wheeler 1989; Mineshige & Wheeler 1989) result from the fact that the hydrogen opacity (due to ionization of hydrogen near 6000 K) increases rapidly as a function of disk temperature in a narrow range of temperatures. This leads to a double-valued relation between disk surface density and temperature, such that any perturbation of the disk temperature leads to a thermal runaway. This instability can only occur when the mass transfer rate into the disk is below a critical value. Outbursts due to this instability cannot be very large unless one

---

Comments to Table 3.1: Summary of black-hole candidates
As this article was nearly finished, a very hard (up to ~500 keV) transient J0422+32 was discovered with BATSE (Paciesas *et al.* 1992). This may well be a BHC (see numerous IAU Circ. following No. 5580; Harmon *et al.* 1992b; also references in Ch. 14). In the absence of a spectrum below 10 keV, we decided not to incorporate this source in Table 3.1.
The numbers in square brackets below refer to the columns in the Table.
† For references, see the text and Ch. 14.
[1] The number in brackets indicates the year of the outburst relevant to the numbers in the table.
[2] T and RT stand for transient and recurrent transient, respectively.
* GX 339–4 is listed as a transient in the Tables of Ch. 14. Even though the source is highly variable, not everyone refers to it as a transient.
[3] Approximate values of the logarithm of the flux. (H) & (L) correspond to high & low states. GX 339–4 also has an 'off' state (see text). (Peak) refers to the maximum during the transient outburst; (max) refers to typical high values for LMC X-1 and LMC X-3, which are quite variable; see text.
[4] Approximate values of the logarithm of the luminosity in cases where a distance is known. (H) & (L) correspond to high & low states. (Max) refers to typical high values for LMC X-1 and LMC X-3, which are quite variable; see text.
[6] In addition to the mass function, the lower limits to and the most probable values of the mass of the compact objects (M_x) are listed. See also Table 2.6 and references therein.
[7] Optical counterparts and magnitudes.
** This may not be the donor (see Sect. 3.3.11).
[9] The long-term periods observed in the X-rays (X) and optical (Opt).
[10] All quoted power-law (pl) indices refer to *photon* spectral indices. (H) & (L) correspond to high & low states. Iron lines have been observed in many cases (see Ch. 1) but are not listed here. Years are indicated between brackets in some cases (notice e.g. the recurrent transient 1543–475).
[12] QPO frequencies (in Hz) greater than 0.04 Hz. (H) & (L) correspond to high & low states.
(X) and (O) refer to X-ray and optical QPO. For more information, see the text and Ch. 6.
[13] Other wavelengths in which the sources have been detected; for more information, see Ch. 2 and 7.

artificially and arbitrarily increases the instability (Smak 1984). The recurrence time scale of the outbursts increases with increasing mass of the compact object, and the amplitude of the associated optical outburst is a strong function of the assumed dependence of α (see Ch. 10) on disk radius and disk thickness (Mineshige & Wheeler 1989, 1992). Unlike in the irradiation-induced instability models, in the thermal disk instability models the accretion rate into the disk *between outbursts*, thus the mass transfer rate from the companion, is approximately constant; the outburst is due to sudden accretion onto the compact object triggered by an instability in the disk. It is likely that *during the outburst*, due to X-ray heating of the companion, the mass transfer is higher (irradiation-induced). It is not clear whether the thermal disk instability can play an important role in outbursts of LMXBs as the disk temperature is kept high due to X-ray heating of the disk, and this suppresses the instability.

3.2.8   *Irradiation-induced instabilities*

In these models (see Ch. 10 (especially Sect. 10.4.3) and references therein; Hameury *et al.* 1986, 1988, 1990; Sarna 1990; Frank *et al.* 1992) an outburst is due to a mass loss instability in the donor as a result of X-ray heating. The resulting expansion of the donor's atmosphere increases the mass transfer rate, which leads to a further increase in the X-ray flux. In general, the star can adjust its atmosphere and the mass transfer rate to be in equilibrium under these two effects. But for a certain range of transfer rates the atmosphere is unstable (Hameury *et al.* 1986); the system must jump between low mass transfer states with negligible radiation and high mass transfer states dominated by irradiation. Screening of the donor by the accretion disk forces the system back into the low mass transfer state and ends the instability. The outburst stops when the whole disk has been accreted onto the compact object. A new instability cannot develop immediately as the screening decreases as the mass loss from the donor has changed its envelope structure (size reduction). Once the disk is gone, it takes many years for the X-rays to expand the donor slightly to bring it back again to an unstable situation. These models require a minimum amount of accretion in quiescence, and they require that the illuminating X-ray flux penetrate the photosphere. Thus, a significant fraction of the accretion luminosity must be in the form of high-energy (> 7 keV) X-rays. If the accreting object is a neutron star, a minimum accretion rate of $\sim 10^{12}$ g s^{-1} would be required, and the hard flux is likely to be generated near the neutron star surface (Hameury *et al.* 1986). For black-hole transients, the required accretion rate would be about three orders of magnitude higher, and the high-energy flux would have to come from the disk (Hameury *et al.* 1990). Unlike in the disk instability models, in the irradiation-induced instability models the mass transfer rate into the disk is many orders of magnitude lower during quiescence than during outburst.

It is unclear in the irradiation-induced instability models how the secondary surface can be heated with the very low X-ray luminosity during quiescence. McClintock *et al.* (1983) estimated that in A 0620–00 the temperature increase of the companion's atmosphere due to X-ray heating in quiescence is <10 K, which may be too small to cause the mass transfer instability (Priedhorsky & Holt 1987). However, since the L1 region is cooler than the average temperature of the companion, the very low quiescent X-ray flux irradiating the L1 region (where the donor loses its mass) may be sufficient to get the instability going (Hameury *et al.* 1986). Mineshige *et al.* (1992) also express their concerns whether, in the case of GS 2000+25, the very low accretion rate observed 155 days prior to the outburst was high enough to trigger the outburst if it were irradiation-induced.

Disk instability models and irradiation-induced models cannot be responsible for all outbursts in LMXBs. Aql X-1 (with an orbital period of ∼19 h) and 1630−47 (perhaps a black-hole binary) exhibit outbursts which can recur with an underlying cycle of ∼123 days and ∼600 days, respectively, but most outbursts are 'missed'. There is no easy provision in these models to remember the phase in the cycle even though an outburst is missed. An additional problem is that the intervals between outbursts of 1630−47 can be fairly regular (∼600 days) even after very different size outbursts. This argues in favor of an independent clock, probably intrinsic to the companion star. However, such variations (mass ejection) are not at all understood (for more details, see Priedhorsky & Holt 1987).

3.3   Description of the individual black-hole candidates

In this section we give a description of each of the BHCs listed in Table 3.1 (see also Sect. 2.4). We start with the six cases where a mass function is available. They are discussed in order of increasing right ascension, with the exception of Cyg X-1, which we discuss first for historic reasons. We then discuss GX 339−4 and GS 2000+25, which in our opinion are very strong BHCs as well, though no mass function is known to date. The remaining 12 sources are discussed in order of increasing right ascension.

3.3.1   *Cyg X-1 (1956+350, HD226868, V1357 Cyg)*

Cyg X-1, discovered by Bowyer *et al.* (1965), is the brightest X-ray source in the sky at energies >20 keV (it has been detected up to ∼1 MeV), and it is one of the best studied X-ray sources. Cyg X-1 was the first source to be identified with a binary system in which the X-ray emission results from mass transfer onto a compact object. Both historic discovery papers of the 5.6 day orbital period include the mass function and mention the distinct possibility that the compact object is a black hole (Bolton 1972; Webster & Murdin 1972; see Sect. 3.1). Cyg X-1 made headlines all over the world, and is often referred to as the first observational evidence for the existence of black holes.

The compact object in Cyg X-1 has a mass in excess of ∼$7M_\odot$ and a probable mass of ∼$16M_\odot$ (Gies & Bolton 1982, 1986); the donor is a supergiant (O9.7 Iab) and has a mass in excess of ∼$20M_\odot$ and a probable mass of ∼$33M_\odot$. For references, see Ch. 14 and Oda (1977); for critical discussions on mass estimates, see McClintock (1992), and references therein. The X-ray spectrum in the range ∼0.5 keV to ∼1 MeV is very complex (see, e.g., Barr & Van der Woerd 1990 and the references below), but is often satisfactorily described by two components (plus an iron line): a soft component and a hard component in the range ∼2 keV to a few hundred keV. The source exhibits two distinct spectral states: the common low state, in which the spectrum is relatively hard, and the rare high state, in which the spectrum is ultrasoft accompanied by a high-energy tail.* Approximate luminosities are given in Table 3.1; a division is made into energies < 10 keV and 10–200 keV (Liang & Nolan 1984). Notice that the luminosity in the range 10–200 keV is higher in the low state than it is in the high state.

A transition from the high to the low state occurred in 1975; the low-energy flux (~2–8 keV) decreased by a factor of ~3, and at the same time the high-energy component (the flux near ~100 keV) increased by a factor of ~2 (as one goes up, the other goes down, and vice versa; however, see Ling *et al.* 1983). The X-ray flux varies on all time scales down to a few milliseconds. Shot noise models have been extensively used as a phenomenological model for the aperiodic variability (the power density spectrum >0.1 Hz is approximately of the form $1/f$, see Figure 3.5); the total light curve is the superposition of the individual shots (various shot profiles have been used) which may last from milliseconds to a few seconds (Terrell 1972; Lochner *et al.* 1991 and references therein). X-ray flares (flux increases ~30%) have been observed which lasted from hours to days, and X-ray dips (up to ~10 min duration) have been observed in which the flux near 1 keV decreased by a factor of ~7 and near 10 keV by a factor of ~2 (Kitamoto *et al.* 1984a, 1989b). The spectral changes that occur during these dips cannot be explained by simple absorption by uniform cold matter alone.

It has been suggested that the high-energy X-ray emission is due to Compton scattering of low-energy photons by a hot plasma of ~10^8–10^9 K (Shapiro *et al.* 1976; Sunyaev & Trümper 1979; see also Liang & Nolan 1984; and Sect. 1.5). Miyamoto *et al.* (1989) have shown that the observed time lags between photons of different energies increases from ~2 ms to a few seconds with increasing Fourier period (from 0.1 to 300 s). This indicates that the Comptonization picture is not the complete story (see Miyamoto & Kitamoto 1989).

In the low state, Ling *et al.* (1987) have reported three levels at high energies. Transitions between these levels occur on time scales of days to weeks (Schwartz *et al.* 1991 and references therein). It remains to be seen whether these are distinct levels. Remarkable is a super low state where the 100 keV flux is very low, but the spectrum shows a strong broad bump centered at ~1 MeV. Ubertini *et al.* (1991) list all high-energy observations from 1977 through 1985 (see also Liang & Nolan 1984) and the states in which the source was observed, with the corresponding spectral fits according to several models (for details on the spectral fits and their interpretations, see Sect. 3.4). High-energy spectra have been obtained with GRANAT (Grebenev *et al.* 1993). There is no convincing evidence that Cyg X-1 emits a narrow 511 keV annihilation line (Nolan & Matteson 1983; Ling & Wheaton 1989).

Cyg X-1 exhibits a distinct ~294 day period in X-rays which is uncorrelated with the low and high states (Priedhorsky *et al.* 1983). This period has also been detected in the optical flux (Kemp *et al.* 1987) though not in the radial velocity (Doppler shift) of optical spectral lines (Gies & Bolton 1984). Intermittent quasi-periodic oscillations (QPOs) at 0.04 Hz were observed over a wide range of energies (Angelini & White

---

* More appropriate names would be 'soft' and 'hard' states with respect to the spectrum below 10 keV, instead of 'high' and 'low' states. The luminosity, including the energy up to a few hundred keV, can be higher in the 'low' state than in the 'high' state (see Table 3.1).

1992; Kouveliotou *et al.* 1992a,b; Vikhlinin *et al.* 1992). For details on the power density spectra, see Ch. 6 and Sect. 3.5 below.

In the low state, the source is radio bright, ~15–20 mJy, whereas in the high state the source is radio weak (less than a few millijanskys). For references and many details, see Ch. 7.

3.3.2   *LMC X-3 (0538−641)*

LMC X-3 can be very luminous (3×10^{38} erg s^{-1} in the range <10 keV). It is a non-eclipsing, highly variable (by factors of ~25) massive X-ray binary in the Large Magellanic Clouds. Its orbital period is 1.70 days, its orbital inclination is ~50–60°, the donor is a B3 V star (*V* magnitude ~16.7–17.5). The ellipsoidal optical amplitude (with half the orbital period) can be as high as ~0.16 mag (peak to peak) in *V* but is not always observed. The compact object has a mass probably in excess of ~$7M_\odot$ which makes it a prime candidate as a black hole (Cowley *et al.* 1983; Paczynski 1983; Ebisawa *et al.* 1993). An X-ray and optical period of 198 (possibly 99) days is probably associated with a precessing disk. This period may change slightly from epoch to epoch. The source varies in the UV by a factor ~2 in the range ~120–200 nm; lines were observed in this region in January 1987 when the source was relatively bright. The X-ray spectrum is ultrasoft with a high-energy tail clearly visible at energies above 10 keV. For references, see the Tables in Ch. 14 (see also Treves *et al.* 1988 and references therein).

The modulation of the X-ray and the optical flux in the 198 day period is very strong; at the minimum, the flux is ~25% of the maximum in X-rays and ~40% in the optical (peak to peak ~1 mag in *V*); the X-ray modulation lags that of the optical by ~20 days (Cowley *et al.* 1991b). Below 13 keV there is a strong positive correlation between the X-ray spectral hardness and the intensity; in 1987 through 1990 intensity changes by a factor of ~4 (198 day period) were associated with hardness ratio (4.7–9.3 keV)/(1.2–4.7 keV) changes by a factor of ~1.7. Thus, the X-ray spectrum is softer at low intensity than at high intensity. No correlation between intensity and spectral hardness was observed at energies >13 keV. The high-energy tail varies independently of the low-energy part of the spectrum < 9 keV (Ebisawa *et al.* 1993).

The X-ray spectrum (Treves *et al.* 1988; White *et al.* 1988a; Sunyaev *et al.* 1989; Tanaka 1989, 1992a,b; see also references therein) is similar to that of Cyg X-1 in the high (spectrally soft) state and to the spectra of other BHCs. The ultrasoft component probably comes from the accretion disk. For spectral fits, possible interpretations, and their consequences, see Sect. 3.4.

3.3.3   *LMC X-1 (0540−697)*

LMC X-1 can be very luminous (~2×10^{38} erg s^{-1} in the range 1–10 keV). It is a non-eclipsing variable (by factors of ~8) massive X-ray binary in the Large Magellanic Cloud. If star #32 (Hutchings *et al.* 1987) is the optical counterpart, which seems likely, the donor is an O7-9 III star (V magnitude ~14.5). The orbital period is 4.22 days, and the orbital inclination <60°; this would mean that the compact object has a mass in excess of ~$2.6M_\odot$ (its mass is probably ~$6M_\odot$; Hutchings *et al.* 1987), which makes it a good candidate as a black hole. The weak UV lines and the lack of stellar wind signatures in the visible suggest that the donor may be near the Roche limiting surface. The source, located in the OB association N159, is surrounded by a highly ionized HeIII region (N159F) with a diameter of $\sim$3 pc. The X-ray spectrum is ultrasoft with a high-energy (power-law) tail clearly visible at energies above $\sim$10 keV. For more references see Ch. 14.

The X-ray spectrum of LMC X-1 is similar to that of Cyg X-1 in the high state and to that of other BHCs (see Sect. 3.4). The ultrasoft component probably comes from the accretion disk. The high-energy power-law component can vary substantially, while the soft component remains approximately constant (Ebisawa *et al.* 1989). For more details on spectral fits, possible interpretations, and their consequences, see Sect. 3.4.

QPOs (rms $\sim$3%) at $\sim$0.075 Hz ($\sim$13 s period) were detected in the range 1.2–15.7 keV during one observation, and it has been suggested that the QPOs are associated with the high-energy component (Ebisawa *et al.* 1989). For more details on fast-timing variability and possible interpretations, see Ch. 6.

3.3.4 *0620−003 (V616 Mon, Nova Mon 1975)*

The recurrent transient A 0620−00 was in outburst in 1917 (Eachus *et al.* 1976) and in 1975, when it was discovered on August 3 with Ariel 5 (Elvis *et al.* 1975); after a small precursor, it reached a maximum flux of $\sim$50 Crab ($\sim$1.5–6 keV) by August 12; its luminosity was then $\sim$1.3$\times$10^{38} erg s^{-1} ($\sim$1–6 keV, for a source distance of $\sim$900 pc, for references, see below and Ch. 14). The flux decreased at first with a decay time (1/e) of $\sim$24 days with day to day variability (Matilsky *et al.* 1976). About 40 days after maximum, the decay flattened off, and a distinct second maximum was observed about 55 days after maximum, and a third broad maximum was observed about 200 days after maximum. The X-ray light curve, covering a period of $\sim$160 days, is shown in Figure 3.2; the spectrum became softer throughout this period. The X-ray spectrum is ultrasoft (Rickets *et al.* 1975; White & Marshall 1984); above 10 keV there is a hard tail (White *et al.* 1984a). Coe *et al.* (1976) have reported an X-ray flux above 30 keV; however, the statistical significance of their data is very marginal, and their conclusion that there is an anti-correlation between the low- and high-energy flux seems to be insufficiently supported by their data.

At maximum, the optical counterpart was very bright ($V\sim$12; Boley *et al.* 1976), and the luminosity ratio X-rays/optical was $\sim$2$\times$10^3 (Whelan *et al.* 1976). During the first $\sim$125 days after outburst, the decay in the optical (in both V and B) was $\sim$0.015 mag per day (1/e decay time $\sim$65 days). This decay then flattened off during the next $\sim$70 days followed by a fast decline of $\sim$3 mag in $\sim$3 weeks (Whelan *et al.* 1976 and references therein; Lloyd *et al.* 1979). Since the optical emission is largely due to X-ray heating of the disk, a slower decay in the optical compared to X-rays is expected, as the optical emission is not proportional to the X-ray flux (the optical emission is largely in the Rayleigh Jeans part of the spectrum). Optical spectra obtained when the source had returned to its quiescent brightness ($V \sim$ 18.3, $B \sim$ 19.7; Murdin *et al.* 1980) showed absorption lines characteristic of a K5 V dwarf and emission lines characteristic of an accretion disk (Oke 1977; Whelan *et al.* 1976; Murdin *et al.* 1980) The optical emission is modulated (largely due to the tidal distortion of the K dwarf) with the orbital period of 7.75 h (McClintock *et al.* 1983).

A 0620−00 made headlines when McClintock & Remillard (1986) measured that

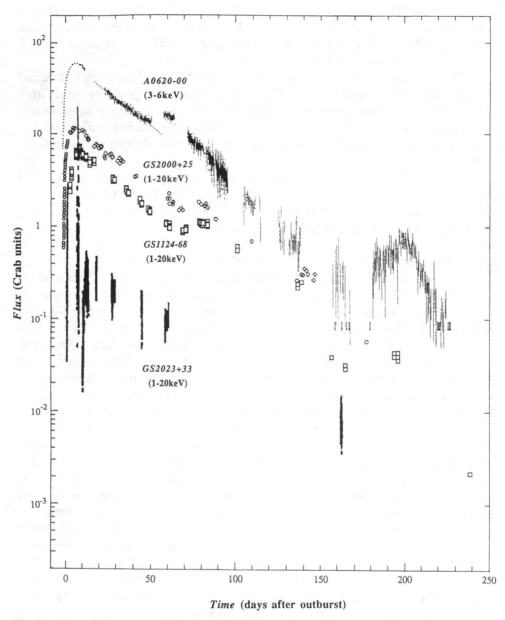

Fig. 3.2. X-ray light curves of four bright BHC transients. The observed fluxes are shown in units of the Crab Nebula in an energy band indicated separately for each source. The dotted curve for A 0620—00 is from Elvis *et al.* (1975); the thin vertical bars indicate statistical errors (Kaluzienski *et al.* 1977). The GS 2000+25 data (open circles) are from Tsunemi *et al.* (1989b) and Takizawa (1991). The GS 2023+33 data (the thick vertical bars indicate actual large excursions in the observed flux) are from (Tanaka 1992a,b). The GS/GRS 1124—68 data (open squares) are from Kitamoto *et al.* (1992) and Ogawa (1992).

the radial-velocity amplitude K of the dwarf companion is 457 ± 8 km s^{-1} with a corresponding mass function of $3.18 \pm 0.16 M_\odot$. They concluded that the compact object is therefore probably a black hole. McClintock & Remillard (1986) tentatively concluded that if the K dwarf fills its Roche lobe, the inclination i of the system is $< 50°$, and the mass of the compact object is $>7.3 M_\odot$ (see also Johnston *et al.* 1989; Haswell & Shafter 1990).

In quiescence (September 1979), the X-ray luminosity was $<10^{32}$ erg s^{-1}, assuming that the X-ray spectrum was similar to earlier measurements (Long *et al.* 1981). De Kool (1988) showed that the absence of a significant X-ray flux in combination with the observed optical emission from the disk during quiescence is in agreement with the standard α disk models (see Ch. 10) if the mass accretion during quiescence is $< 10^{-11} M_\odot$ yr^{-1} (near maximum $\sim 5 \times 10^{-8} M_\odot$ yr^{-1}).

A 0620−00 was bright in the radio ($\sim$100–150 mJy at $\sim$31 cm) on August 16–17, 1975 (Davis *et al.* 1975); the radio flux density decayed rapidly by a factor of $\sim$5 in 2 weeks (1/e time $\sim$5.2 days). The data from August 20 in the range $\sim$6 cm to $\sim$31 cm can be fitted by a spectral index of $\sim$0.7 (Little *et al.* 1976). On August 24, when Davis *et al.* observed a flux of 21 ± 9 mJy ($\sim$31 cm), Bieging & Downes (1975) detected a flux of 28 ± 6 mJy (at $\sim$6.5 cm); on September 7–8, 1975, the radio flux at $\sim$6.5 cm was estimated at < 20 mJy. In June 1977, the source may have been detected at 2 cm (44 ± 14 mJy; Duldig *et al.* 1979). For more information, see Ch. 7.

3.3.5   1124−684 (Nova Muscae, 1991)

An outburst of GS/GRS 1124–68 (Nova Muscae, 1991) was detected on January 8, 1991, with the All Sky Monitor (ASM) (Makino 1991) and independently with the GRANAT/Watch (Lund & Brandt 1991; Sunyaev 1991).

Figure 3.2 shows the X-ray light curve obtained with the Ginga ASM (Kitamoto *et al.* 1992) and the LAC. The flux was less than 50 mCrab (1–6 keV) on January 5, 1991. When the outburst was detected on January 8, the flux was $\sim$0.8 Crab. It increased to a maximum of $\sim$8 Crab on January 15. The peak luminosity is estimated to be $\sim 2.5 \times 10^{38}$ erg s^{-1} for a distance of 3 kpc (see below). The count rate decayed exponentially with a time constant of $\sim$31 days, while the decay time of the bolometric flux was $\sim$40 days. A flux increase by a factor of $\sim$2 occurred between 60 and 75 days after maximum, after which the count rate decayed with a time constant of $\sim$37 days through mid-May, 1991. During the observations in June, the count rate dropped by a factor of $\sim$2 associated with a change in the spectrum (see below). A second count rate increase occurred between June 21 and July 21, 1991 (a similar increase was observed in A 0620−00, see above). During the last observation, on October 16–17, 1991, the source had become undetectable with Ginga ($<10^{-11}$ erg cm^{-2} s^{-1} in the range 1–40 keV, thus <0.2 mCrab).

Figure 3.3 shows sample spectra during the first eight months (Ogawa 1992). Until mid-May, the spectrum was ultrasoft (steadily softening in time) with a hard power-law tail with photon spectral indices between 2.2 and 2.6 (Ogawa 1992). With ROSAT, the spectrum was observed down to 0.4 keV on January 24 and 25 (Greiner *et al.* 1991). The hard power-law component diminished gradually during the first two months, remained very weak in March and April, 1991, but came back in May. The power-law component extended up to $\sim$500 keV with a photon index of $\sim$2.3–2.7

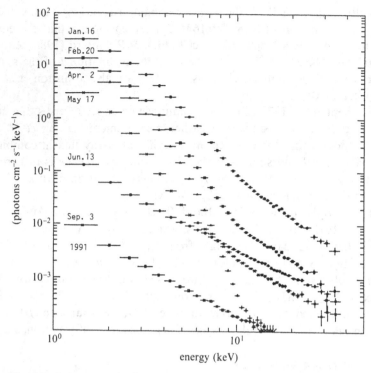

Fig. 3.3. Spectra of GS/GRS 1124—68 corrected for the detector response (Ogawa 1992; Tanaka 1992b). The Ginga observation dates (in 1991) are indicated.

(Goldwurm *et al.* 1992; Sunyaev *et al.* 1992b; Gilfanov *et al.* 1993; see also Grebenev *et al.* 1992b). On January 20–21, the spectrum (Figure 3.4) showed a narrow emission line at 481 ± 22 keV, which is interpreted as a positron annihilation line (Goldwurm *et al.* 1992; Sunyaev *et al.* 1992a; Gilfanov *et al.* 1993). The line appeared only in the last 13 h of a 21 h observation. On a subsequent LAC observation on June 13, 1991, the ultrasoft component was absent, and the spectrum was approximately a single power law. The slope of this power law (photon index 1.5–1.8) was distinctly smaller (thus the spectrum was harder) than when the ultrasoft component was present; this hard power-law spectrum persisted during the remaining observations till September 3, 1991.

Variations on various time scales were observed; Figure 3.5 shows a PDS up to 64 Hz (see also the footnote on p. 165). When the spectrum was ultrasoft with a hard tail, rapid (<1 s) fluctuations were relatively weak with an rms variation averaged over 1–40 keV of a few per cent or less (in the range 8 ms–128 s). The rms increased with energy, which indicates that the fluctuations occurred largely in the power-law component. The power-law component above 40 keV varied up to a factor of ~2 on time scales of several hours (Goldwurm *et al.* 1992; Sunyaev *et al.* 1992a,b). When the spectrum was a single power law, the rms values (30–40%) were independent of photon energy (in the range 8 msec–128 s). Strong QPOs were observed on January 11 and February 14, 1991. The QPO frequency changed between ~3 Hz and 8 Hz

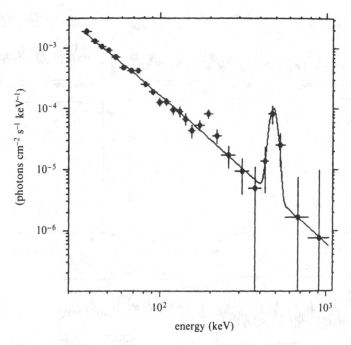

Fig. 3.4. Spectrum of GS/GRS 1124−68 during the last 13.3 h observations with SIGMA on January 20, 1991. The solid line is the best-fit model to a power law plus a Gaussian (Goldwurm *et al.* 1992).

(Takizawa *et al.* 1994). QPO at ∼10 Hz was observed with the ART-P (Grebenev *et al.* 1993).

The optical counterpart was identified with a star of $V \sim 13.5$ (Della Valle *et al.* 1991). The peak brightness probably occurred between January 11 and 12 with $V \sim 13.3$ mag and $B \sim 13.5$ mag (Della Valle *et al.* 1991). Before outburst, $B \sim 21$ and $R \sim 19.4$. The average optical decay rate was ∼0.017 mag per day (about 1.0 mag in 55 days) with variations of ∼0.2 mag on a time scale of ∼1 day (Della Valle *et al.* 1991). This average decay is about twice as slow as that in the X-ray range. A slower decay in the optical compared to that in the X-rays is expected if the optical emission is due to reprocessing of X-rays in the disk (see also Sect. 2.3.5 and 2.3.6.1).

Nova Muscae 1991 was observed with IUE on January 17. The color excess $E(B - V) \sim 0.30$ was obtained in the optical and UV bands (Della Valle *et al.* 1991; Gonzalez-Riestra *et al.* 1991; Shrader & Gonzalez-Riestra 1991). The companion has a K-type spectrum. The distance is between 1 and 5 kpc; ∼3 kpc is a probable value (West 1991).

McClintock *et al.* (1992) determined the mass function of GS/GRS 1124−68 to be $3.1 \pm 0.4 M_{\odot}$; the orbital period is 10.42 ± 0.04 h (Bailyn 1991). They note that the quiescent optical spectrum is strikingly similar to that of V616 Mon (A 0620−00). See also Sect. 2.4.2.

Kesteven & Turtle (1991) detected strong radio emission from the source during outburst. Two major bursts were observed: one started prior to January 17 and the

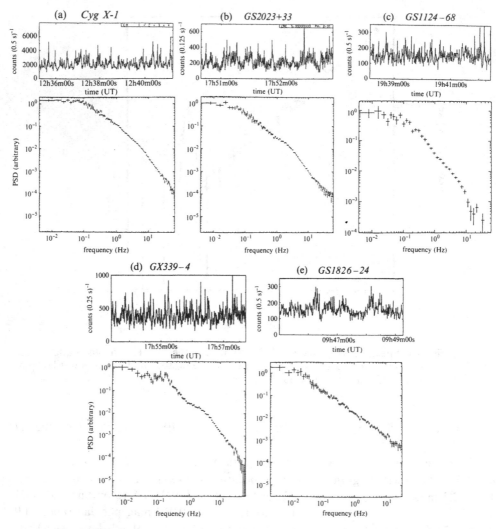

Fig. 3.5. Examples of X-ray light curves (Ginga data) of five sources which show an approximately single power-law spectrum and strong flickering (upper panels) and associated power density spectra (lower panels). From Tanaka (1992b).

other on January 30, 1991. The spectrum is non-thermal with a spectral index of 0.6–0.7. A linear polarization of 5–10% was detected. For detailed information of the radio emission, see Ch. 7.

3.3.6   2023+338 (V404 Cyg)

A very unusual and bright X-ray recurrent transient GS 2023+33 was discovered on May 21, 1989 (Kitamoto *et al.* 1989a; Makino 1989) with the Ginga All Sky Monitor (ASM). Within a few days, the source was optically identified with V404 Cyg (Hurst & Mobberley 1989; Wagner *et al.* 1989).

On May 11, 1989, about ten days before the outburst, the flux (1–6 keV) was <50

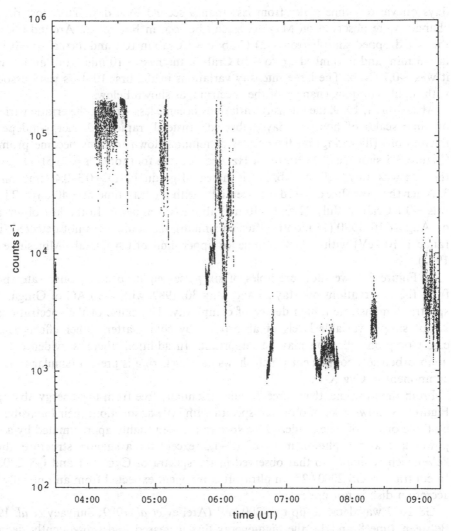

Fig. 3.6. X-ray light curve (Ginga data) of GS 2023+33 (1–30 keV) on May 30, 1989 (Tanaka 1989). The count rates have been corrected for deadtime.

mCrab. On May 21, the source was detected with the ASM at a level of ~0.1 Crab (1–6 keV). Within a day between May 21 and 22, the flux increased from ~0.1 Crab to ~4 Crab (1–6 keV) and from ~1 to ~3.5 Crab (6–20 keV). The observation with the LAC started on May 23. In Figure 3.2 we show the X-ray light curve observed with the LAC and the ASM. The highest flux was ~21 Crab (1–40 keV) observed on May 30. For isotropic emission this corresponds to ~10^{39} erg s^{-1} (1–40 keV) at a distance of ~2.5 kpc (see below). There is an indication that the flux saturated at this level (see Figure 3.6).

In the early phase of the outburst, the source behaved unlike any known transient (its behavior was also very unusual in the optical and the radio, as discussed below). Erratic and large-amplitude intensity variations were observed during the first ~10 days on various time scales from less than a second to a day. The most dramatic changes were observed on May 30, as can be seen in Figure 3.6. Around 06:00 UT the flux dropped sharply from ~21 Crab to ~1 Crab in 10 s and further to ~0.3 Crab in ~3 min, and it climbed up to ~10 Crab in the next ~10 min. Half an hour later it was ~0.1 Crab. The large intensity variations in the first 10 days were associated with highly complex changes of the spectrum, as shown below.

After June 1, 1989, the intensity variations became less strong; the erratic variability on time scales of hours to days subsided. Instead, rapid and energy-independent fluctuations (flickering) on time scales of minutes down to 2 ms became prominent (Figure 3.5 shows a PDS up to 64 Hz; see also the footnote on p. 165). On June 1, the flux was only ~20 mCrab, but it increased gradually to ~0.3–0.4 Crab on June 3. After that, the flux decayed exponentially with a decay time of ~40 days. The flux was ~0.1 Crab on July 22 and ~10 mCrab on November 1. In the last observation, on August 16, 1990 (15 months after maximum), the source was not detected (in the range 1–10 keV) with a 90% confidence upper limit of 0.4 mCrab (Mineshige *et al.* 1992).

In Figure 3.7, we show examples of six pulse-height spectra (count rate spectra) from the observations on May 23 and May 30, 1989, with the LAC of Ginga. These spectra demonstrate a high degree of complexity. The cause of the spectral changes is not simply variable levels of absorption by cold matter; other effects such as, e.g., Compton reflection may be important. In addition, there is evidence that an unabsorbed soft component which shows little flickering is present (similar to the soft component in Cyg X-1).

From time to time, the source became essentially free from low-energy absorption. Figure 3.8 shows a set of observed spectra with little absorption; their intensities vary by three orders of magnitude. These spectra can be roughly approximated by a single power law with a photon index of 1.3–1.5, except for a shallow structure above 7 keV which is similar to that observed in the spectra of Cyg X-1 and GS 2000+25. In contrast to GS 2000+25, an ultrasoft component expected from an optically thick accretion disk is not present.

GS 2023 was detected up to ~300 keV (Aref'ev *et al.* 1989; Sunyaev *et al.* 1991e). Between June 8 and 14, the high-energy flux increased and subsequently decreased by ~50%. In the range 105–200 keV it reached a maximum of ~3 Crab on June 10 (see Figure 5 in Sunyaev *et al.* 1991e), when the luminosity, for a distance of 2.5 kpc, was ~5×10^{37} erg s^{-1} (2–300 keV). At times, the spectra seemed to exhibit complex structures (Figure 3.9). At energies >100 keV, the spectrum became harder as the source became weaker, and the luminosity of the source became dominated by the high-energy X-rays. In June, 1989, the main contribution came from the region 50–100 keV, and in July and August from an even higher energy range (Sunyaev *et al.* 1991e, 1992b). (For possible interpretations, see Sect. 3.4.) The 30–100 keV power density spectrum in the range from 0.01–5 Hz had an approximately $1/f$ shape (Sunyaev *et al.* 1991e).

The optical counterpart of GS 2023+33 is V404 Cyg (Hurst & Mobberley 1989; Marsden 1989; Wagner *et al.* 1989). Previous outbursts of this system were recorded in 1938 (Wachmann 1948) and 1956, and possibly in 1979 (Richter 1987). The behavior in the optical, like that in X-rays, was unlike any known transient. Fluctuations of

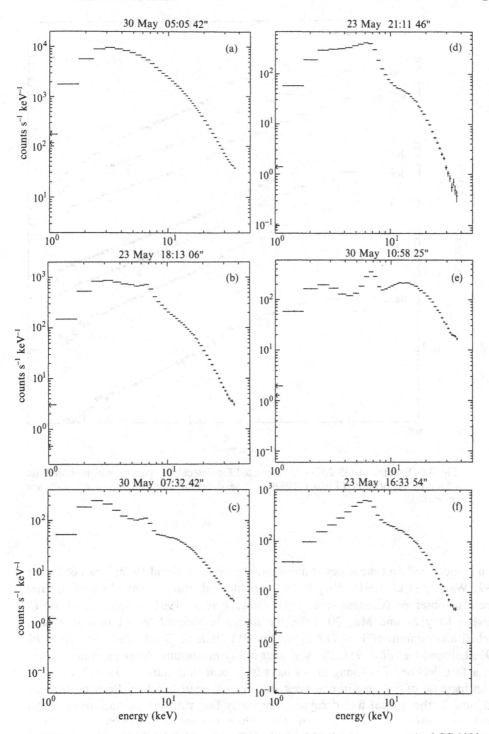

Fig. 3.7. Examples of pulse-height spectra (not corrected for detector response) of GS 2023+33 taken with Ginga on May 23 and 30, 1989. The dates and times are indicated at the top of each panel. The spectrum under (a) corresponds to an approximate single power law (Tanaka 1992b).

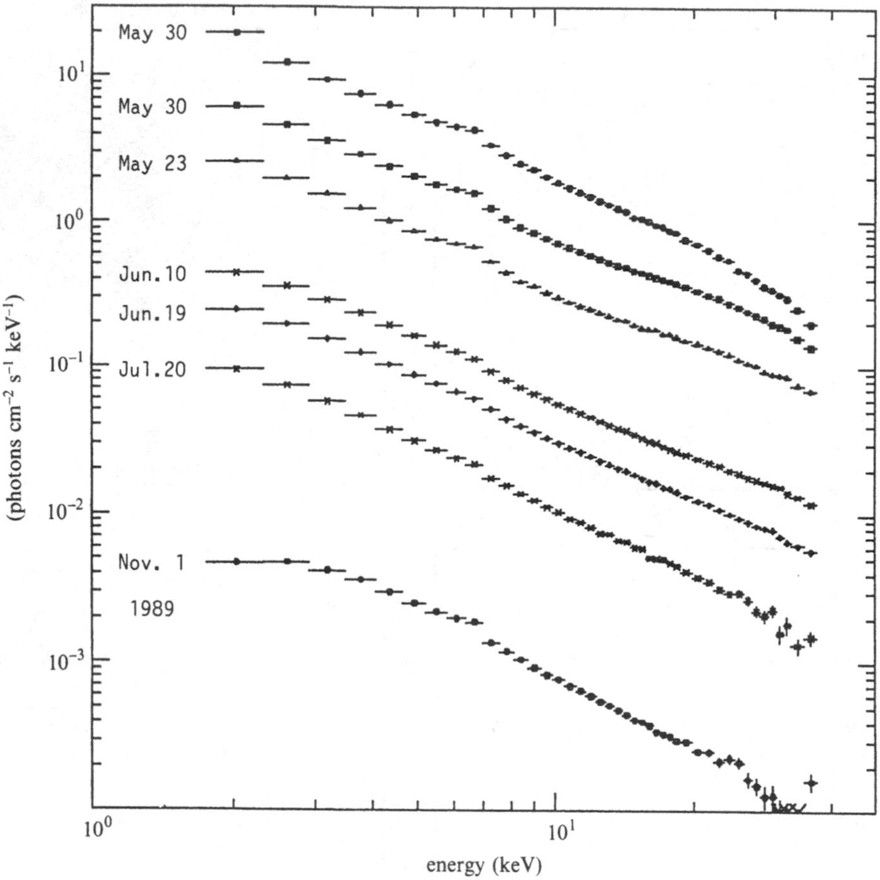

Fig. 3.8. Spectra of GS 2023+33 (corrected for detector response), which show little effect of absorption (Tanaka 1992b). The dates of the Ginga observations are indicated.

~1 mag occurred on time scales of a few minutes (Buie & Bond 1989; Jones & Carter 1989; Wagner *et al.* 1991). Very strong variable and double peaked emission lines were also observed (Casares *et al.* 1991; Gotthelf *et al.* 1991; Wagner *et al.* 1991). Between May 27 and May 30, 1989, the source brightened by ~1 mag in V and reached a maximum of $V \sim 11.6$ near May 30.1 (Buie & Bond 1989; Wagner *et al.* 1991; Leibowitz *et al.* 1991); this was near X-ray maximum. After maximum, there was a fast decline of ~3 mag in ~4 days to a local minimum on June 2, followed by an increase of ~1 mag in one week, and then a steady decline. Between May 30 and June 2, the optical flux decayed surprisingly fast with erratic variations. After mid-June (continuing at least till early November), the optical flux declined according to a power-law function of the form $t^{(-0.89\pm0.08)}$ (Wagner *et al.* 1990). If a correction is made for the presence of a 19 mag star 1.4" from V404 Cyg (Gotthelf *et al.* 1991; Udalski & Kaluzny 1991), the source may have been already brightening on April 13,

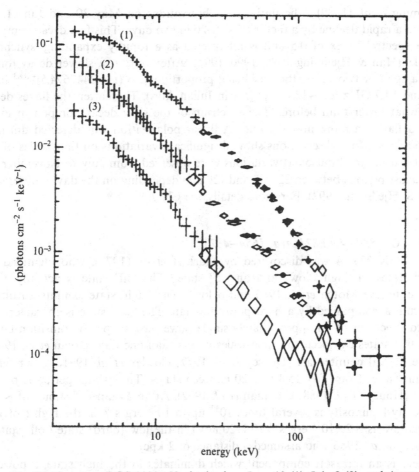

Fig. 3.9. Spectra of GS 2023+33 obtained on June 10 (1), July 8 (2), and August 19, 1989 (3). The crosses indicate data from TTM; the diamonds are data from Hexe; and the circles are from Pulsar X-1 (Sunyaev *et al.* 1991e).

1989 (Szkody & Margon 1989). The brightness increase from quiescence to maximum was ~8.6 magnitudes in *B* and ~7.5 in *V*.

The distance to V404 Cyg is 2–3 kpc (Charles *et al.* 1989; Casares *et al.* 1991; Gotthelf *et al.* 1991). Han & Hjellming (1990) put the source at a distance >3 kpc. There are several reports of periodicities in the optical ranging from ~10 min to ~6 h (Haswell & Shafter 1990; Wagner *et al.* 1990; Leibowitz *et al.* 1991; Udalski & Kaluzny 1991; Casares & Charles 1992). From observations made after the system had returned to quiescence (in July and August, 1991), Casares *et al.* (1992) obtained an orbital period of 6.473 ± 0.001 days and a mass function of 6.26 ± 0.31$M_\odot$. Wagner *et al.* (1992) showed that the donor is a ~1$M_\odot$ K0 IV star at a distance of ~3.5 kpc. The mass of the compact object is probably in the range 8–12$M_\odot$ (Casares *et al.* 1992; Wagner *et al.* 1992).

The radio behavior of GS 2023+33 was very unusual. A strong (~1.1 to ~1.6 Jy, depending on wavelength) variable radio source was observed on May 30, 1989, by

Hjellming *et al.* (1989). The initial two observations on May 30 and June 1, 1989, showed a rapid decline by a factor of ~15–20 in two days. This fast decay component has a spectral index of 0.5 and is interpreted as a rapidly expanding 'synchrotron bubble' (Han & Hjellming 1989, 1990, 1992). After June 1, a slower decay followed over a span of two years, the flux being proportional to $(t - 2447674.5)^{-0.83}$ for 4.9, 8.4, and 14.9 GHz for ~150 days (t is in Julian days). Thereafter, the fluxes decayed somewhat faster than before. The spectrum of the slow decay component evolved from a flat to a more inverted one. A linear polarization was detected during the first 50 days. Most observations showed significant variations on time scales of a few minutes to hours. Sinusoidal variations were observed from July to November 1989, with quasi-periods between 22 min and 120 min depending on the day of observations (Han & Hjellming 1992). For more details, see Ch. 7.

3.3.7   *GX 339–4 (V821 Ara, 1659–487)*

GX 339–4 was discovered by Markert *et al.* (1973), who identified three X-ray states: a 'high', 'low', and an 'off' state. The 'off' state is actually a weak hard state (see Motch *et al.* 1985 and below). The 'high' state exhibits an ultrasoft spectrum accompanied by a hard power-law tail. The 'low' state is characterized by a hard spectrum of an approximately single power-law shape. A transition between these two states can occur on time scales of less than one day (Dolan *et al.* 1987).

The optical counterpart (Doxsey *et al.* 1979; Cowley *et al.* 1991a and references therein) varies from $V \sim 15.4$ to >20 between states. The orbital period is probably 14.8 h (Honey *et al.* 1988; Callanan *et al.* 1992). At an assumed distance of ~4 kpc, the X-ray luminosity is several times 10^{37} up to 10^{38} erg s^{-1} in the high (soft) state and as low as $\sim 2 \times 10^{35}$ erg s^{-1} (0.1–20 keV) in the low (hard) state ('off' state; see Ilovaisky *et al.* 1986 who assumed a distance of 2 kpc).

There is an ultrasoft component which dominates in the high state, a power-law component (observed up to ~430 keV), and an iron absorption edge and emission line (e.g. Nolan *et al.* 1982; Ilovaisky *et al.* 1986; Makishima *et al.* 1986; Dolan *et al.* 1987; Ebisawa 1991; Fishman *et al.* 1991; Miyamoto *et al.* 1991; Tanaka 1991; Grebenev *et al.* 1992a, 1993; Harmon *et al.* 1992c). The photon spectral index changes between 2.4–2.7 in the high (soft) state and is ~1.6 in the low (hard) state (Ebisawa 1991). In the low (hard) state, the ultrasoft component vanishes and rapid fluctuations (flickering) with rms variations as high as 30% have been observed on time scales from tens of milliseconds to tens of seconds (Figure 3.5 shows a PDS up to 64 Hz; see also the footnote on p. 165), similar to Cyg X-1 in the low state (Samimi *et al.* 1979; Maejima *et al.* 1984; Ebisawa 1991; Iga *et al.* 1991; Miyamoto *et al.* 1991; Grebenev *et al.* 1992a, 1993). Such a rapid variability is, in general, not significant during the high (soft) state (Maejima *et al.* 1984; Motch *et al.* 1985; Ilovaisky *et al.* 1986; Ebisawa 1991). However, the power-law component can vary significantly by factors up to ~3 on time scales of a day or less without an appreciable change in the flux of the ultrasoft component (Makishima *et al.* 1986; Miyamoto *et al.* 1991).

QPOs with a mean period of ~1.2 s (Grebenev *et al.* 1991), ~10 s and ~20 s (Motch *et al.* 1983) have been observed in the low (hard) state. Miyamoto *et al.* (1991) observed a strong ~6 Hz QPO during a very bright high (soft) state along

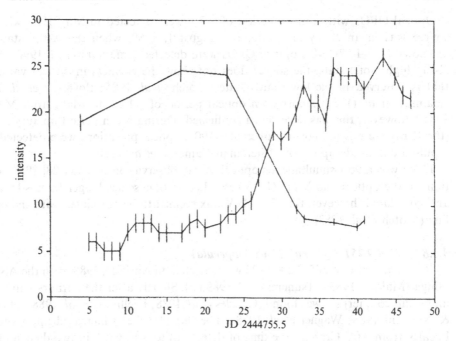

Fig. 3.10. Simultaneous X-ray (Hakucho) and optical (V CCD from ESO) observations of GX 339—4 during the June 1981 X-ray hard (low state) to soft (high state) transition. The V intensity scale is arbitrary. The optical V and the soft X-ray (3–6 keV) fluxes are clearly anti-correlated during the transition. From Motch *et al.* (1985).

with low-frequency and very-low-frequency noise enhanced by 'dips' and 'flip-flops' (for more details, see Ch. 6).

The relationship between the optical and X-ray flux is very poorly understood (Figure 3.10; see also Table 1 in Motch *et al.* 1985). In the high state, V ranges from ~16 to 18; in the low state $V \sim 15.4$–17; and in the off state $V \sim 17.7$–20.2 (Motch *et al.* 1985; Remillard & McClintock 1987). The source underwent an unprecedented sequence of events during 1981. Between February and June, the X-ray state changed from an off (spectrally hard) state (during which the optical counterpart was fainter than $B \sim 21$) to a low (also spectrally hard) state (during which the optical counterpart was extremely bright, $V \sim 15.4$). The optical transition took place between March 8 and May 7, 1981 (Motch *et al.* 1983 and references therein). The transition to the X-ray high (spectrally soft) state occurred near June 23, 1981 (see Figure 3.10; notice the anti-correlation between the optical brightness and the X-ray intensity). On May 28–29, when the optical counterpart was presumably at maximum and the X-ray state low, the optical flux exhibited 20 s QPOs together with fast flaring activity on time scales as short as 10–20 ms, while the X-ray flux showed 20 s and 10 s (second harmonic) QPOs (Motch *et al.* 1983). The rms of the 20 s oscillations was ~40% in both the optical and the X-ray flux. The rms was independent (within ~10%) of X-ray energy (1–17 keV). During the May 1981 hard X-ray state, the spectrum could be represented by a single power law with photon index 1.58 all the way from the IR to X-ray energies, in contrast to the high X-ray state, during which much less visible and IR light was produced (Motch *et al.* 1985).

Optical QPOs with a mean period of $\sim$7 s were detected in May 1982 when the source was in an X-ray low state; on August 1, 1989, when the X-ray state was unknown ($V \sim$17.7), $\sim$8 s optical QPOs were detected (Imamura *et al.* 1990). In the X-ray high (soft) state, the source does not show the richness in optical variability that is observed in the low (hard) state (Motch *et al.* 1985; Ilovaisky *et al.* 1986). Imamura *et al.* (1987) reported an optical period of $\sim$1.13 ms (data from May 29, 1985); however, this has never been confirmed. During a 2 h period on July 5, 1986 (the X-ray state is unknown), coherent $\sim$190 s optical pulsations were detected with a pulsed amplitude up to 50% (Steinman-Cameron *et al.* 1990).

There was a 96 s simultaneous optical/X-ray observation on May 28, 1981. Variations in the optical and X-ray (1–13 keV) flux on time scales longer than $\sim$15 s were anti-correlated; however, the 13–20 keV flux seemed to be correlated with the optical flux (Motch *et al.* 1983).

3.3.8   2000+251 (QZ Vul, Nova Vulpecula)

An outburst of GS 2000+25 was detected on April 23, 1988, with the ASM on Ginga (Makino 1988a; Tsunemi *et al.* 1989a,b). Shortly after the outburst, an optical and radio counterpart was found (Charles *et al.* 1988; Hjellming *et al.* 1988; Okamura & Noguchi 1988; Wagner *et al.* 1988). The flux (1–6 keV) increased approximately linearly from $\sim$0.5 Crab (at the time of detection) to $\sim$10 Crab in two days and then more slowly to a maximum of $\sim$12 Crab near April 28. For a distance of $\sim$2 kpc (see below), the luminosity at the maximum was $\sim$1.5$\times$10^{38} erg s^{-1} (1–40 keV). The X-ray light curve is shown in Figure 3.2. After reaching a maximum, the count rate (1–20 keV) decreased exponentially with a 1/e decay time of $\sim$30 days. In July, $\sim$73 days after maximum, the source brightened by $\sim$50%, and then the count rate continued its exponential decay on the same time scale as before.

The spectrum softened gradually as the count rate decreased (see Figure 3.11). The 1/e decay time of the bolometric flux was $\sim$40 days through September 8 and became gradually longer (see Figure 3.14). On May 15, 1989 (15 months after the outburst started), the source had become undetectable; its luminosity (1–40 keV) was less than $\sim$3$\times$10^{33} erg s^{-1} for a distance of 2 kpc (Tanaka 1989; Mineshige *et al.* 1992).

In Figure 3.11 we show X-ray spectra as obtained with the LAC during the first eight 8 months. The spectrum was ultrasoft with a high-energy (power-law) tail from April 29 (near maximum) through December 7. The photon power-law component varied between 1.9 and 2.5 (Takizawa 1991). Whereas the luminosity of the ultrasoft component decreased smoothly, the flux of the hard power-law component changed enormously and irregularly; the ratio of the luminosity of the hard component to that of the ultrasoft component differed by a factor $\sim$100 between the observations on May 3 and September 9, as can be seen in Figure 3.14. On December 16, the ultrasoft component had disappeared, and only the power-law component was present, and the photon power-law index decreased to 1.7 (Takizawa 1991). Between May 15 and June 8, 1988, the source was observed from the Mir Kvant Module. A high-energy component was detected up to $\sim$300 keV (Sunyaev *et al.* 1988, 1992b; Döbereiner *et al.* 1989; Efremov *et al.* 1989). In Sect. 3.4, we discuss the energy spectra and their interpretations.

Erratic intensity fluctuations were observed on various time scales down to 100

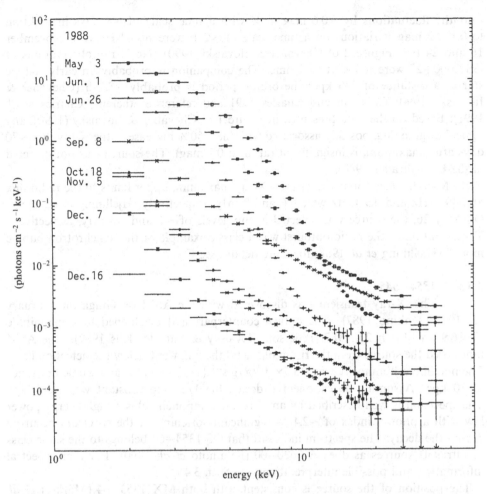

Fig. 3.11. Spectra of GS 2000+25 covering the first eight months after outburst in 1988. The spectra have been corrected for the detector response (Takizawa 1991; Tanaka 1992b). The dates of the Ginga observations are indicated.

ms. The rms variations in the range from 62 ms to 256 s were ~1.8% on May 3–7, when the ultrasoft component was dominant, and ~10% on September 8, when the luminosity of the power-law component was comparable to that of the ultrasoft component (see Figure 3.11). Sinusoidal variations were occasionally observed with quasi-periods ranging from 7 to 10 h (no coherent periodicity) (Takizawa 1991).

The optical counterpart QZ Vul was found shortly after the outburst (Charles *et al.* 1988; Okamura & Noguchi 1988; Wagner *et al.* 1988; Tsunemi *et al.* 1989b). The optical behavior was classic for an LMXB in outburst (see Ch. 2). The source reached a maximum of V ~16.4 mag and B ~17.5 mag; in quiescence before the outburst, the B magnitude was >21 mag. Between the end of July and December 1988, the decay in the optical was approximately exponential with a decay time of ~65 days, which is significantly longer than the X-ray decay time (Pavlenko *et al.* 1989; Charles *et al.* 1991), as is expected (see A 0620−00 above). The decay was not smooth: fluctuations by ∼0.6 mag in V with a time scale of ∼8 days in addition to 0.1–0.2 mag variations on a time scale of ∼2 h were noted between November 16 and 24 (see Figure 2 of Chevalier & Ilovaisky 1990). No X-ray observations of GS 2000+25 were made at that time. The companion is probably an early K-type star at a distance of 2–3 kpc; the orbital period is probably ∼8.3 h (Chevalier & Ilovaisky 1990; Callanan and Charles 1991 and references therein; Charles *et al.* 1991). Based on the data presented in Figure 1 of Chevalier & Ilovaisky (1990), any optical brightening, possibly associated with the ∼50% increase in the X-ray flux ∼70 days after maximum, is insignificant (at most 0.2 mag). The source was not detected in the UV (Shrader 1988).

On May 3, 1988, about 7 days after X-ray maximum, upper limits to the radio flux at 1.49 GHz and 4.9 GHz were 0.6 and 8 mJy, respectively (Hjellming *et al.* 1988). On May 26, the source was detected at flux levels of ∼6 and ∼5 mJy, respectively. The evolution of the radio outburst was a classic example of the 'synchrotron bubble model' (Hjellming *et al.* 1988; for more details, see Ch. 7).

3.3.9   1354–645

This X-ray transient was discovered with the ASM on Ginga on February 13, 1987 (Makino 1987). The optical counterpart had brightened to a magnitude V∼16.9 on March 27, 1987 (Pederson, Ilovaisky & Van der Klis 1987). The ASM monitored the source from the rising phase till the flux went below the detection limit. The maximum luminosity was ∼3.5×10^{37} erg s^{-1} (1–10 keV) for an assumed distance of 10 kpc. Assuming an exponential decay, the 1/e time constant was ∼66 days. The spectrum is well described by an ultrasoft component plus a high-energy power law with a photon index of ∼2.4. A significant softening of the spectrum occurred during the decay. The spectrum indicates that GS 1354−64 belongs to the same class of ultrasoft sources as does A 0620−00 (Kitamoto *et al.* 1990). For more spectral information and possible interpretations, see Sect. 3.4.

The position of the source is consistent with both MX 1353−64 (Markert *et al.* 1977, 1979) and Cen X-2 (e.g. Chodil *et al.* 1968; Francey 1971). However, the characteristics of GS 1354−64 differ from both of these sources. Cen X-2 is one of the strongest soft transient sources (Francey *et al.* 1967; Cooke and Pounds 1971; Cominsky *et al.* 1978; and see Sect. 3.1). The maximum flux of Cen X-2 was about 50 times larger than that of GS 1354−64, while the flux of MX 1353−64 was comparable to GS 1354−64. Cen X-2 appears to belong to the ultrasoft transients but with a spectrum that is variable (power-law indices ranged from 1.15–2.8; see, e.g., Chodil *et al.* 1968). Markert *et al.* (1979) reported that the spectrum of MX 1353−64 was harder than that of the Crab (unlike GS 1354−64). If all of the above detections were made of the same source, then it would have to show at least four states ranging greatly in both intensity and hardness (Kitamoto *et al.* 1990).

3.3.10   1524–617 (TrA X-1, KY TrA)

This soft X-ray transient was discovered with Ariel 5. The X-ray light curve between April 1975 and May 1976 (see Figure 3.12) showed an unusual precursor which lasted ∼1 month and reached an intensity of ∼40% of the maximum flux, which was ∼0.9 Crab (3–6 keV). During the precursor, the spectrum softened during the

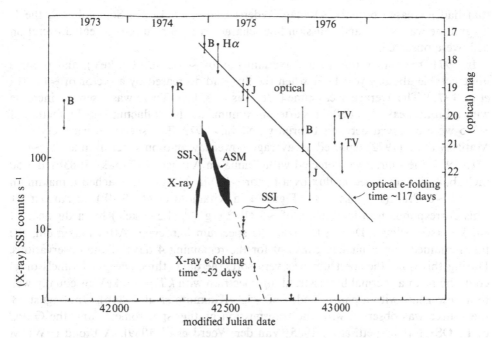

Fig. 3.12. X-ray and optical light curves of A 1524 − 62 from Murdin *et al.* (1977). The X-ray upper limits are at a 3σ level of confidence. ASM and SSI refer to instruments on board Ariel 5. For details on the optical data, see Murdin *et al.* (1977). Notice that the e-folding time of the optical data is ∼117 days, and that of the X-ray data is ∼52 days.

rise but remained approximately constant during its decay (Kaluzienski *et al.* 1975). For the main outburst, the 1/e decay time was ∼2 months (Kaluzienski *et al.* 1975; Pounds 1976). The spectrum softened during the rise, while, during the decay, the spectral hardness remained approximately constant (Kaluzienski *et al.* 1975). There was a distinct correlation between the optical and X-ray flux (see Figure 3.12). The optical counterpart had a *B* magnitude of ∼17.5 near maximum. About two years before outburst, the *B* magnitude was greater than 19.5. The optical (1/e) decay time was ∼117 days (Murdin *et al.* 1977). The decay of the optical light was similar to that of A 0620−00 (see above).

During the August 27–28, 1990 observations with SIGMA, hard X-rays were detected from a source with an error box of 2 arcmin, which included 1524−62 (Sunyaev 1990). For an assumed distance of 2 kpc, the luminosity (30–100 keV) was ∼2.5×10^{35} erg s^{-1} (Barret *et al.* 1992). It is unclear whether this detection was another outburst of A 1524−62; however, the presence of an ultrasoft component at that time is unknown. No outbursts of this source were observed with Ginga.

3.3.11   1543−475

This source is a recurrent transient; outbursts have been recorded in 1971 (Matilsky *et al.* 1972; Belian *et al.* 1973), in 1983 (Kitamoto *et al.* 1984b; Chiapetti *et al.* 1985) and in 1992 (Harmon *et al.* 1992a). On August 31, 1983, 14 days after outburst, a comparison of a red plate with the red ESO Sky Survey showed a star that had increased by only ∼1 mag (Pedersen *et al.* 1983). On September 2, the *V* magnitude was ∼14.9, and emission lines characteristic for an X-ray heated accretion disk were observed.

In 1971, the source rose to a maximum flux of ∼2 Crab (2–6 keV), and became undetectable about a year later when the flux had decreased by a factor of ∼10^3 (Li *et al.* 1976). The average decay time (1/e) was ∼55 days. There was a second increase with a broad peak about 80 days after maximum, and large fluctuations by factors of ∼3–5 were observed between February and July 1972. The spectrum was very soft; Matilsky *et al.* (1972) reported an average power-law photon spectral index of ∼4.

In 1983, the source was detected with Tenma on August 17 (Tanaka 1983b) at ∼50 mCrab, and it increased in flux by a factor ∼100 in 2 days and reached a maximum flux of ∼6 Crab (1.5–7 keV; see Figure 1 in Kitamoto *et al.* 1984b) on August 20. This corresponds to a luminosity of ∼5×10^{38} erg s^{-1} (1.5–7 keV) for a distance of ∼4 kpc (see below). During the rise, the spectrum hardened. After maximum, the flux remained approximately constant for the remaining 4 days of the observations. During this time, the spectrum was very soft; the data (three energy channels only) could be fit to a thermal bremsstrahlung spectrum with $kT = 1.6$ keV or equally well to a power law with a photon index of ∼5.3 (Kitamoto *et al.* 1984b). On August 28, the source was observed with the transmission grating spectrometer and the GSPC of EXOSAT (Chiapetti *et al.* 1985; Van der Woerd *et al.* 1989). A broad (FWHM ∼2.7 keV) line at ∼5.9 keV was observed. The continuum (2–10 keV) could be fit with an unsaturated Comptonization spectrum (Sunyaev & Titarchuk 1980) with kT_e ∼0.84 keV and an optical depth τ ∼27 (with a corresponding y parameter of ∼4.7). The presence of a high-energy tail >10 keV was statistically marginal. The source was undetectable on February 17, 1984, when the flux in the range 0.04–2.0 keV had decreased by a factor of 20 000.

In 1992, an outburst was recorded with GRO (Harmon *et al.* 1992a). The flux above 20 keV increased in 3 days to a maximum on April 4 when the flux was ∼1.1 Crab (20–40 keV), ∼0.54 Crab (50–120 keV) and 0.29 ± 0.04 Crab (120–230 keV). The data could be fit to a power law with photon indices of 2.2 ± 0.1 (on April 19) and 2.9 ± 0.1 (on April 20–21). The combination of an ultrasoft spectrum and a power-law high-energy tail makes this source a BHC. No radio flux has been reported.

The quiescent star is a dwarf of spectral type between A1 V and A2 V; it has a *V* magnitude of ∼16.7, which is only 1.8 mag fainter than at maximum (Chevalier 1989). The visual absorption A_v is ∼2.2, thus the de-reddened *V* magnitude is ∼14.5. The distance is estimated at ∼4 kpc. An A star is an unusual mass donor. Chevalier (1989) considers the possibility that the A star is a chance alignment (thus not the companion of 1543—47), and also that the A star is associated with an LMXB in a triple system (see Sect. 2.6).

3.3.12   1630—472

This LMXB has been several times in outburst. Between December 1970 and March 1976, four transient outbursts were observed with a separation of ∼600 days (Jones *et al.* 1976; see also Priedhorsky & Holt 1987 and references therein). It was the first transient X-ray source which exhibited separate and regularly spaced outbursts.

The source was also found in outburst with Tenma in April 1984 at a flux level of ~0.5 Crab (Tanaka 1984). The spectrum was very soft with a high-energy tail. In subsequent EXOSAT observations (Parmar *et al.* 1986), the flux decayed over a period of ~100 days from ~2.8×10^{38} to 4×10^{36} erg s^{-1} (1–50 keV) for an assumed distance of 10 kpc. The spectrum could be fit well by a soft Wien-like spectrum with a high-energy power-law tail. The relative strength of the soft component (compared with the high-energy tail) decreased by at least a factor of two as the luminosity decreased. During the decay, the power-law component became progressively flatter as the photon index changed from ~2.5 to ~1.0. However, the spectra in the last two observations (July 1 and 29, 1984) could be fit by a single power law with photon indices of ~1.8 to ~1.2.

During the observation on April 11, 1984, flickering was observed with rms variations of ~7% (1–7 keV; mostly the soft component), and ~4% (14–30 keV; mostly the hard component). An autocorrelation analysis of the April 11 data showed variability on time scales down to 50 ms and a characteristic time scale of ~20 s (Parmar *et al.* 1986). No optical counterpart has been found to date.

3.3.13 1705–250 (Nova Ophiuchi 1977)

This X-ray transient was discovered independently by Griffiths *et al.* (1977) and Kaluzienski & Holt (1977a) in August 1977. In September 1977, an optical nova was detected with a magnitude in the blue of $B \sim 16.5$, which was identified with the X-ray source. The source at quiescence was fainter than $B \sim 21$. The optical spectrum shows a continuum with no absorption lines but an emission line identified as $\lambda 4686$ HeII (Griffiths *et al.* 1978). The X-ray light curve of HEAO-1 data of August 1977 showed the fast rise and the slow decay of 'classic' X-ray transients; however, it was double-peaked near the maximum (see Figure 3.13 and Watson *et al.* 1978). The maximum flux (2–18 keV) was ~3.5 times that of the Crab Nebula (Watson *et al.* 1978). The very irregular and variable decline (not exponential in form) lasted at least 30 days with a decay time of roughly 2–3 months. No regular periodicity was observed. Watson *et al.* (1978) found that the post-maximum spectrum could be best fit by a thermal bremsstrahlung model with $kT \sim 3$ keV. In September 1977, a high-energy spectral component up to ~120 keV was found with HEAO-1 (Wilson & Rothschild 1983; Cooke *et al.* 1984), which was best fit with a power law of photon index ~2.4 above ~10 keV. The entire spectrum obtained with HEAO-1 in mid-September is expressed by two components, a soft one ($kT \sim 1.8$ keV, if approximated by a thermal bremsstrahlung spectrum) and a hard power-law component (Wilson & Rothschild 1983; White *et al.* 1984a).

3.3.14 1740.7–2942

This source, located 40 arcmin from the Galactic Center, was first detected with the Einstein Observatory as one of a large group of weak X-ray sources surrounding Sgr A West (the source at the dynamical center of the Galaxy). Subsequent hard X-ray and γ-ray experiments have established that at times this is the dominant source of hard radiation within several degrees of the Galactic Center (see, e.g.,

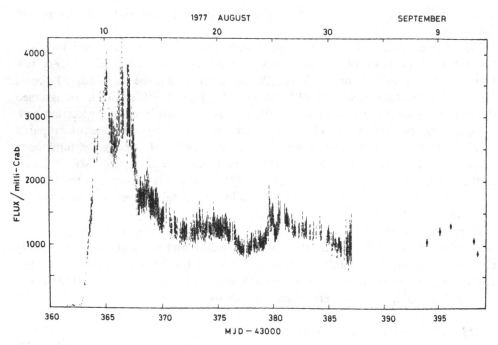

Fig. 3.13. X-ray light curve of H 1705−25 (Nova Ophiuchi 1977). The error bars are ± 1σ. The data are from Ariel V SSI (2–18 keV); the last five data points are from HEAO-1 (0.9–13.3 keV). From Watson *et al.* (1978).

Skinner *et al.* 1987; Kawai *et al.* 1988; Cook *et al.* 1991). Based upon over a decade of observations, the spectrum has been shown to consist normally of a high-energy power-law with a high-energy cutoff and absorption at low energies. The power-law photon index ranged from 1.9 to 2.9 (Bazzano *et al.* 1992 and references therein). The average luminosity (4–300 keV) was ~3×10³⁷ erg s⁻¹ for an assumed distance of 8.5 kpc. Sunyaev *et al.* (1991a,b) reported a variability in the 4–200 keV band of ~50% on a time scale of weeks. Bazzano *et al.* (1992) reported that the flux varied by factors of 3 at 40 keV and 2.5 at 100 keV over the period of ~1 year. Its spectrum was similar to those of GS 2023+338 and Cyg X-1 in its low (hard) state. No strong, soft component (characteristic of A 0620−00, and Cyg X-1 and GX 339−4 in their high states) has been detected (Sunyaev *et al.* 1991a,b).

Observations on October 13–14, 1990, with SIGMA revealed a high-energy spectral feature which appeared as a bump, reaching a maximum intensity at ~500 keV, followed by an ~700 keV cutoff. A day later, the feature was absent (Bouchet *et al.* 1991; Sunyaev *et al.* 1991f). Because of this high-energy feature, it has been suggested that this source may be responsible for the variable 511 keV positron annihilation radiation seen from the Galactic Center region (Lingenfelter & Ramaty 1989; Mandrou *et al.* 1990b; Sunyaev *et al.* 1991c; Bally & Leventhal 1991; Paul *et al.* 1991). However, this is still very uncertain.

Radio observations have revealed filamentary structures up to 50 pc long that are hypothesized to be related to the high-energy radiation from 1740.7−2942 (Gray *et al.* 1991; Reich & Schlickeiser 1992). In addition, a likely radio counterpart has been detected at several frequencies (Leahy 1991; Mirabel *et al.* 1991; Prince *et al.* 1991). From VLA observations at 6 and 20 cm, Mirabel *et al.* (1992) have shown that this radio source is located at the center of an aligned double radio jet whose total extent is ~1 arcmin. The time variations from September 1991 to February 1992 of the compact radio counterpart of 1740.7−2942 appeared to be correlated with variations in high-energy X-rays (Mirabel *et al.* 1992; Schmitz-Frayasse *et al.* 1992).

3.3.15 1741−322

An outburst of this transient was observed with Ariel 5 and HEAO-1 in August 1977 (Doxsey *et al.* 1977a; Kaluzienski & Holt 1977b). The spectrum was very soft (White & Marshall 1984; White *et al.* 1984a). A high-energy tail (power-law photon spectral index ~2.4) was observed up to ~100 keV in September 1977 and March 1978, but not in September 1978 (Cooke *et al.* 1984). We list here the fluxes by day (half day averages in units of the flux from the Crab) in the range 3–6 keV. August 12: 0.1 ± 0.1, August 19: 0.4 ± 0.05, August 31: 0.38 ± 0.09, September 3: 0.60 ± 0.08, September 5: 0.65 ± 0.11 (Kaluzienski & Holt 1977). In September 1977, when the spectrum was somewhat harder than that of the Crab, the flux (1–10 keV) was 0.73 ± 0.1 Crab (Doxsey *et al.* 1977a; for the light curve, see White *et al.* 1984a). The average flux (2–10 keV) in the period July 29–August 6, 1985, was $<4\times10^{-11}$ erg cm^2 s^{-1} (<0.002 Crab; Skinner *et al.* 1990). On March 12–18, 1978, the flux in the range 1–10 keV was $\sim10^{-9}$ erg cm^2 s^{-1} (~0.03 Crab; Wood *et al.* 1978; see also Wood *et al.* 1984).

3.3.16 1755−338

This source is a very bright LMXB and has a soft spectrum whose characteristic temperature is ~2 keV (Jones 1977; White & Marshall 1984). No high-energy tail has ever been observed. White *et al.* (1984b) discovered three equally spaced dips in the X-ray flux with a period of 4.4 ± 0.2 h. Each dip lasted ~30 min, with considerable variability within each dip on time scales between 30 and 600 s. No coherent periodicities between 15 ms and 1250 s were observed. The variations are spectrally independent with a maximum reduction in flux of 40%. The non-dip luminosity (1–20 keV) was $\sim4\times10^{36}$ erg s^{-1} for an assumed distance of 3 kpc (White *et al.* 1984b). The optical counterpart is a blue, 19th magnitude star (McClintock *et al.* 1978). The spectrum showed very weak HeII lines (Cowley *et al.* 1988). Simultaneous optical and X-ray observations during August 1984 showed an optical sinusoidal modulation with the same period as the X-ray dips. The amplitude was ~0.4 mag in *V*. The optical minimum occurred ~0.15 cycles after the center of the X-ray dip (see Figure 2.20). Four X-ray dips were recorded during these observations. The dips, which lasted ~1 h, reached a maximum depth of 30% of the steady intensity and showed an irregular morphology. They were not associated with any detectable change in spectral hardness (Mason *et al.* 1985).

3.3.17 1758−258

This source, located ~40 arcmin from GX 5-1, was discovered with GRANAT in March–April 1990 (Mandrou *et al.* 1990a; Paul *et al.* 1991; Sunyaev *et al.* 1991a).

It may be the origin of the hard X-ray flux previously thought to have originated from GX 5-1 (Skinner 1991). Levine *et al.* (1984) point out in the HEAO-1 A-4 Catalogue that the high-energy flux (>80 keV) from the near vicinity of GX 5-1 may be due to confusion with a nearby hard source.

The spectrum of 1758−258 extended from 4 to 300 keV and could be well fitted by a power law with a photon index of ∼2. Its luminosity was ∼2.2×10^{37} erg s^{-1} (4–300 keV) for an assumed distance of 8.5 kpc. Variations by a factor of ∼2 over several days were observed in the 4–30 keV and 50–170 keV energy bands (Sunyaev *et al.* 1991a; see also Grebenev *et al.* 1993). At energies above 50 keV, this is one of the two brightest sources in the Galactic Center region (Sunyaev *et al.* 1991d). The source does not appear to be a transient (Skinner 1991). No optical counterpart has been found (Mirabel *et al.* 1992).

3.3.18   1826−238

GS 1826−24 was first detected on September 8, 1988, with Ginga (Makino 1988b). The average X-ray flux was ∼26 mCrab in the range 1–40 keV. During September 9–16, 1988, rapid fluctuations (flickering) were observed on time scales down to 2 ms with an rms variation of ∼30% (Figure 3.5 shows a PDS up to 32 Hz; see also the footnote on p. 165). The power density spectrum exhibited a ∼1/f form between 0.02 and 500 Hz. The X-ray spectrum is expressed approximately by a single power law with a photon index of ∼1.7 (Tanaka 1989). The source flux was below 50 mCrab (1–6 keV, Ginga ASM) during the month before the discovery, and in October 1988.

This source is previously uncatalogued (hence possibly a 'classic' transient), and the light curve and the spectral evolution are not known. We list this source here as a BHC because of its similarities with Cyg X-1 and GX 339−4 in the low (hard) state. No optical counterpart has been reported.

3.3.19   1846−031

This transient has an ultrasoft spectrum and a high-energy tail extending up to at least 25 keV (Parmar & White 1985; Parmar *et al.* 1993). Between April and September 1985, the flux decreased by a factor of ∼5. The 1/e decay time was ∼60 days. The spectrum softened as the flux decayed. If the peak luminosity is assumed to be 10^{38} erg s^{-1}, the distance is ∼10 kpc. No optical counterpart was found (Parmar *et al.* 1993).

3.3.20   1957+115 (V1408 Aql)

This is not a typical transient; it changes its intensity by a factor as high as ∼5. White & Marshall (1984) put this source in the ultrasoft group. No single model could describe the spectrum from the 1978 HEAO-1 observations; a thermal bremsstrahlung spectrum of kT ∼1.7 keV can fit the data in the range from 0.5 to 10 keV (White & Marshall 1984). The luminosity was ∼5×10^{36} erg s^{-1} (>0.5 keV) at an assumed distance of 7 kpc. No X-ray periodicity has been observed (Priedhorsky & Terrell 1984).

More recently, at a luminosity of ∼(6–8)×10^{36} erg s^{-1} the spectrum as observed with Ginga in 1990 consisted of a soft and a hard component; the latter could be expressed by a power law with a photon index of ~2.1 (8–20 keV; Yaqoob *et al.* 1993). The spectral hardness increased with overall source intensity, which was mainly due to an increase of the high-energy component while the soft component remained relatively stable. Based on its spectral properties, Yaqoob *et al.* (1993) argue that the compact object is likely to be a *neutron star*.

The optical counterpart is a faint blue star with a *V* mag of ~18.7 (Doxsey *et al.* 1977b; Margon *et al.* 1978). The optical spectrum exhibited narrow emission lines of HeII (Cowley *et al.* 1988). Thorstensen (1987) observed an optical sinusoidal variation with a period of 9.32 ± 0.01 h. The *V* band light curve had an amplitude of ~0.23 mag peak-to-peak but with considerable random variation.

3.4   X-ray spectra – interpretations

The ultrasoft spectral component often observed in BHCs is typically of a thermal nature, but its break is sharper and falls off faster at high energies than an optically thin thermal bremsstrahlung spectrum. The observed spectra are Wien-like, suggesting an optical thickness ≫1 for the emission region.

3.4.1   *Optically thick accretion disk*

Several models for the thermal emission from an optically thick accretion disk have been proposed (e.g. Shakura & Sunyaev 1973; Mitsuda *et al.* 1984; Stella & Rosner 1984; see Sect. 1.5). The disk temperature varies from a low value near the outer edge to a maximum T_{in} at the inner edge of the disk with radius R_{in}. R_{in} may be approximately three times the Schwartzschild radius $3r_g$ ($r_g = 2GM/c^2$), where M is the mass of the compact object. The observed spectrum is therefore the composite of a range of temperatures, hence the name multicolor disk. The temperature versus radius relation is a function of M, the mass accretion rate, and, in some cases, the disk viscosity parameter α (Shakura & Sunyaev 1973); obviously, it also depends on specific assumptions made. Various models for the emission from an optically thick disk were discussed and compared with the observations by White *et al.* (1988b). Some models assume blackbody spectra while others assume modified blackbody spectra. Regardless of the model, an observed spectrum with a high-energy break allows one to measure quite sensitively the maximum *color* temperature T_{in} and a quantity proportional to the associated surface area.

It is important to note that the temperature measured is the color temperature which can be significantly higher than the effective temperature for blackbody temperatures (kT) in excess of ~1 keV (Ebisuzaki *et al.* 1984; London *et al.* 1986). A correct interpretation should be based on a self-consistent solution of the vertical disk structure and radiative transfer to the disk surface, as done by Ebisuzaki *et al.* (1984) and London *et al.* (1986) for a neutron star atmosphere.

3.4.2   *Unsaturated Comptonization*

If low-energy photons are injected into a high-temperature plasma with a Comptonization optical depth ≫ 1, the spectrum of the emergent photons approaches a Wien spectrum (e.g. Sunyaev & Titarchuk 1980). With an appropriate choice of the electron temperature T_e and the Comptonization optical depth τ, the observed ultrasoft spectrum can also be produced (White *et al.* 1988b). Unlike in the case of a multicolor blackbody disk, the X-ray flux is then determined by the rate of injection of low-energy photons from an unspecified source.

Both types of models can explain equally well the shape of the observed ultrasoft spectra (White *et al.* 1988b; Ebisawa *et al.* 1991, 1992; Mitsuda 1992). Therefore, model fitting alone cannot tell which of the two represents the physics. As we will show below, there are persuasive arguments why the multicolor disk approximation is a satisfying representation.

The simplest of the multicolor disk models is given by Mitsuda *et al.* (1984). It assumes blackbody emission from the surface of the disk with an innermost radius R_{in}. If all gravitational potential energy, released as the matter moves through the optically thick disk, is emitted locally as blackbody radiation, the effective temperature $T(r)$ will depend on the radius, r, as $r^{-0.75}$, and the spectrum is given by the surface integral of the blackbody spectra. The two parameters in the model are $R_{in}^2\cos(\theta)/d^2$ (where θ is the inclination angle of the disk and d is the distance to the source) and T_{in} (the blackbody *effective* temperature at R_{in}). Note that the *observed* value T_{in} is the *color* temperature. Since the effective temperature is lower, one should be careful when interpreting the measured values for R_{in}.

The multicolor blackbody disk model (representing the ultrasoft component), including an iron line and disk reflection (Tanaka 1991) plus a power law (to represent the high-energy tail), can satisfactorily fit the observed spectra of the BHCs LMC X-1, LMC X-3, GX 339−4 (Ebisawa 1991), GS 2000+25 (Takizawa 1991), GS 1354−64 (Kitamoto *et al.* 1990) and GS/GRS 1124−68 (Ogawa 1992).

The time histories of the two parameters are shown in Figure 3.14 for the three BHCs GS 2000+25 (Takizawa 1991), GS/GRS 1124−68 (Ogawa 1992) and LMC X-3 (Ebisawa 1991), for which large luminosity changes were observed with Ginga. The observed color temperature T_{in} decreases gradually as the luminosity of the ultrasoft component decreases. The values of $(R_{in}/d)^2(\cos\theta)^{0.5}$ remain remarkably constant, while the bolometric ultrasoft flux (and thus presumably the accretion rate) decrease by more than an order of magnitude. The constancy of $(R_{in}/d)(\cos\theta)^{0.5}$ over a wide range of accretion rates is considered to be characteristic of an optically thick accretion disk. It is plausible that the parameter R_{in} is related to the radius of the innermost stable orbit, $3r_g$ which seems to be the only quantity that is independent of the accretion rate.

High-luminosity LMXBs whose compact objects are probably neutron stars, such as Sco X-1, Cyg X-2 and LMC X-2, exhibit a soft component and a blackbody component with a color temperature (kT) of ~2 keV (Mitsuda *et al.* 1984). This blackbody component is interpreted as emission from the neutron star surface, and the soft component is interpreted as emission from the accretion disk. Since accreting matter does not know whether the central object is a low-magnetic-field neutron star or a black hole, a similar accretion disk structure is expected in these two classes of objects. The soft component of these high-luminosity LMXBs is also fitted well with the multicolor blackbody disk model (e.g. Mitsuda *et al.* 1984; Mitsuda 1992; Tanaka 1992a,b).

Table 3.2 lists the values of $R_{in}(\cos\theta)^{0.5}$ determined for several BHCs and neutron star LMXBs for which distance estimates are available (see Table 3.1). There is a systematic difference by a factor of 3–4 in the values of $R_{in}(\cos\theta)^{0.5}$ between the BHCs

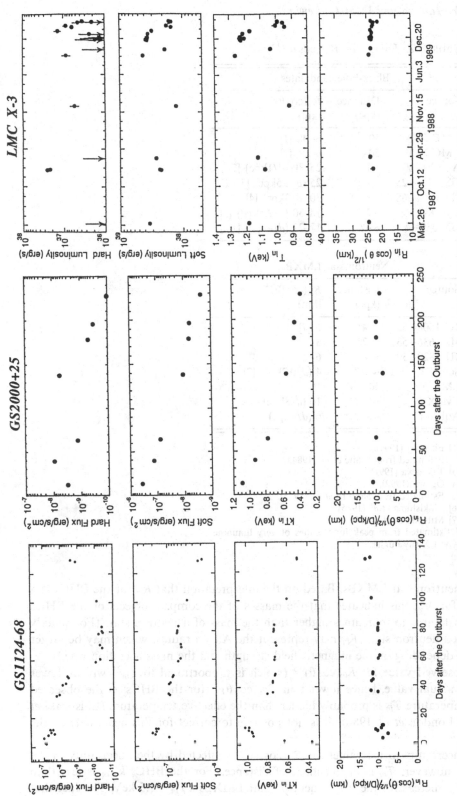

Fig. 3.14. Time histories of the spectral parameters for GS/GRS 1124–68 (Ogawa 1992), GS 2000+25 (Takizawa 1991), and LMC X-3 (Ebisawa 1991) based on a two-component model. The model consists of a power-law and an ultrasoft component described by the multicolor blackbody disk model (see text). From top to bottom, we present the power-law flux (2–30 keV), the bolometric flux of the ultrasoft component, the color temperature kT_{in} at the innermost disk radius R_{in}, and $R_{in}(\cos \theta)^{0.5}(D/\text{kpc})^{-1}$ (see text). For LMC X-3 the distance D is taken to be 50 kpc, and luminosities are shown instead of fluxes. From Tanaka (1992b).

Table 3.2. *Values for* $R_{in}(\cos\theta)^{0.5}$

Black-hole candidates		
Source	Distance (kpc)	$R_{in}(\cos\theta)^{0.5}$ (km)
LMC X-1	50	~ 40 [1]
LMC X-3	50	~ 24 [1]
A 0620−00	~1	25–30 (d/1kpc) [2]
GS 2000+25	~2–3	25 (d/2.5kpc) [3]
GS 1124−68	~3	30 (d/3kpc) [4]
Cyg X-1	~2.5	18–30 (d/2.5kpc) [5]
GX 339−4	~4?	17–22 (d/4kpc) [1,6]

Neutron star LMXBs		
Source	Distance (kpc)	$R_{in}(\cos\theta)^{0.5}$ (km)
4U 1608−52	~ 4[a]	6 [7]
4U 1636−53	~ 7[a]	8
4U 1820−30	~ 8.5	6
Sco X-1	~ 0.7	4 (d/0.7kpc) [7]
LMC X-2	50	10
Cyg X-2	~8	10 (d/8kpc)
Aql X-1	<4[a]	6 (d/4kpc)

[1] Ebisawa (1991).
[2] Estimated from White *et al.* (1984a).
[3] Takizawa (1991).
[4] Ogawa (1992).
[5] Estimated from Sanford *et al.* (1975).
[6] Makishima *et al.* (1986).
[7] Mitsuda *et al.* (1984).
[a] Estimated from peak luminosities of very luminous type I X-ray bursts.

and the neutron star LMXBs. Based on the interpretation that R_{in} for the BHCs is a measure for $3r_g$, this indicates that the masses of the compact objects of the BHCs are at least three to four times higher than the mass of a neutron star. (For weakly magnetized neutron stars, R_{in} may represent the Alfvén radius, which may be larger than $3r_g$, depending on the magnetic field strength and the mass accretion rate.)

The observed value for $R_{in}(\cos\theta)^{0.5}$ (which is proportional to T_{in}^2) will be lower than the actual value, hence lower than $3r_g(\cos\theta)^{0.5}$ for the BHCs as the observed color temperature T_{in} is probably higher than the effective temperature (Ebisuzaki *et al.* 1984; London *et al.* 1986). It is not possible to correct for this accurately as the relation between color and effective temperature is as yet poorly known.

The uncertainty in the values for $R_{in}(\cos\theta)^{0.5}$ is affected by the uncertainty in the distance; however, T_{in} is independent of distance. For the BHCs kT_{in} is less than ~1.3 keV, whereas for the bright neutron star LMXBs kT_{in} ~1.5 keV. The fact that

T_{in} for a given luminosity is lower for the BHCs than for the neutron star LMXBs is consistent with their larger masses. For a similar disk structure, the value for T_{in} would scale as $M^{-0.25}$.

3.4.3 *Is there a surface?*

An obvious difference between a neutron star and a black hole is the presence or absence of a solid surface. One-half of the total gravitational potential energy released will be carried in the form of the kinetic energy of accreting matter. For the case of a neutron star, this energy will be dissipated eventually as thermal emission from an optically thick envelope. The single-color blackbody component (not to be confused with the multicolor blackbody component as mentioned above) observed in many high-luminosity neutron star LMXBs can be interpreted as coming from the neutron star surface (Mitsuda *et al.* 1984; White *et al.* 1988b). In contrast, the BHCs never show such a single-color blackbody component at any luminosity level, which is in support of the idea that the compact objects are black holes (without a solid surface).

The power-law slopes in the range 2–40 keV are $\sim$1.4–1.7 (photon index) for those spectra which can be fit with a single power law, and they are $\sim$2.0–2.5 (determined in the range $\sim$5–40 keV) for those cases where two spectral components (an ultrasoft component and a high-energy power-law component dominating >10 keV) are needed (Tanaka 1992b). Note that the power-law slopes for the latter cases are steeper. (However, we caution the reader that a fit to a two-component spectrum, in which one component is ultrasoft, may artificially increase the power-law slope of the underlying power-law component.) The intensity of the power-law component can vary without a change in its slope, and its relative strength (relative to that of the ultrasoft component) can change by more than an order of magnitude. The high-energy power-law tail extends in some cases up to 100 keV and beyond (see Sect. 3.3).

The origin of the power-law component remains unclear. A possible mechanism is Comptonization in a plasma with temperature $kT > 100$ eV. However, it is difficult to see why the Comptonization y parameter would remain constant (the spectral slope does not change) over a wide range of observed accretion rates.

On the other hand, there is evidence for Comptonization in a few low-luminosity neutron star LMXBs (predominantly bursters). During observations on May 13–18 and June 18–24, 1984, as the luminosity decreased, the soft component of 4U 1608−52 gradually softened, while the blackbody component gradually hardened, showing a power-law like high-energy tail (Mitsuda *et al.* 1989). Simultaneously, the blackbody spectrum of the type I X-ray bursts (see Ch. 4) developed high-energy tails which could be the result of Comptonization of the blackbody radiation in a hot plasma (Nakamura *et al.* 1989). The spectral shape of relatively low-luminosity LMXBs may resemble the two-component (ultrasoft + power law) spectra of the BHCs, except that they exhibit much smaller values for R_{in} (see above). 4U 1957+11 (Yaqoob *et al.* 1993) and 1755 338 could be such cases.

As the luminosity of the neutron star LMXBs decreases further (below $\sim$1/10 of the Eddington limit), the spectra become a single power-law type. Among the sources with a single power-law spectrum, the distinction between a neutron star and a BHC is difficult, except for a possible (but not established) difference in the amplitudes of flickering. (The low-luminosity neutron star LMXBs usually show less flickering.)

The spectrum of GS 2023+33 exhibited a single power law over a wide range of luminosities in the energy range 2–40 keV. It would indicate that an optically thick accretion disk was absent near the compact object even when the luminosity was very high ($\sim 10^{39}$ erg s^{-1}, see Table 3.1). This is difficult to explain. Perhaps it is due to a disk instability in cases of supercritical accretion.

There are some similarities between the X-ray spectra of BHCs and AGNs despite the many orders of magnitude differences in mass and scale.

- The slopes of the power law components cover the same range.
- The power law extends to $\sim$100 keV and beyond.
- The intensity of the power law components show large and irregular changes of several tens to a few hundred per cent in minutes to hours.
- The slope of the power law for a given source remains approximately constant regardless of the source intensity.

These similarities perhaps suggest that the power-law components are produced by the same mechanism for both classes of sources, the BHCs and the AGNs. One should note that a possible ultrasoft component for AGNs, if at all present, would appear at much longer wavelengths than the X-ray band. Hence only the power-law portion would be observed in X rays.

3.5   BHC diagnostics

None of the information that we have obtained to date from BHCs is a test of general relativity in the strong field limit. It is therefore premature to state that any of the BHCs in X-ray binaries are black holes. However, it is generally agreed that a necessary condition for there to be a black hole in an X-ray binary is that the mass of the compact object (i.e. the black hole) is in excess of $\sim 3 M_\odot$ (see Sect. 2.4.4). We will therefore concentrate on the properties of systems which satisfy this criterion.

The first compact object with such a high mass was Cyg X-1; the second was LMC X-1. Subsequently, many sources which share certain characteristics with Cyg X-1 have been called BHCs long before the mass of their compact objects were measured. Apart from Cyg X-1 and LMC X-1, the verdict is now in for four more sources: LMC X-3, A 0620−00, GS 2023+338 and GS 1124−68; the mass of their compact objects are all $> 3 M_\odot$ (see Table 3.1). However, Cir X-1 was also considered to be a BHC, but it turned out later that its compact object is a neutron star as it emits type I bursts.

The characteristics most often used (in the absence of a mass measurement of the compact object) in evaluating the possibility that the binary contains a black hole are:

- ultrasoft spectra;
- high-energy power-law tail above 20 keV;
- two spectral states: a soft spectrum when the source is bright at energies <10 keV (sometimes called the high state) and a hard spectrum when the source is relatively weak at energies below 10 keV (sometimes called the low state);
- millisecond variability and 'flickering' in the hard state.

Below we will examine whether the six BHCs whose compact objects have a mass $>3M_\odot$ (see Table 3.1) exhibit these characteristics and whether these characteristics are present in sources whose compact objects are neutron stars.

3.5.1 *Ultrasoft spectra*

Five of the six sources (except GS 2023+33) have a very soft spectrum when they are luminous ($\geq 10^{38}$ erg s^{-1}). Such soft spectra, called ultrasoft by White & Marshall (1984), are a good indicator for a source to be a BHC; however, this is not conclusive. Neutron star binaries can have very soft spectra as well (e.g. Cir X-1) though such soft spectra in neutron star systems are much less common.

3.5.2 *High-energy power-law tail >20 keV*

Five of the six sources (except GS 2023+33) have a high-energy power-law tail which dominates above 20 keV (note such a hard component at these energies is not established for A 0620–00; see Sect. 3.3). These tails are good BHC indicators (e.g. Cir X-1 does not have a hard tail), but they alone are not conclusive. Many neutron star LMXBs exhibit a high-energy tail as the luminosity (persistent emission) decreases (e.g. the burst sources 1608–52, 1636–53, 1722–40 in the globular cluster Ter 2 and 1728–34; see Sect. 3.2 and 3.4).

GS 2023+33 showed a single power-law spectrum independent of luminosity, from very high luminosities ($\gg 10^{38}$ erg s^{-1}) to many orders of magnitude lower luminosities. However, a single power-law spectrum alone, if the luminosity is not taken into consideration, is not a useful indicator since many low-luminosity neutron star LMXBs (e.g. bursters) exhibit single power-law spectra.

3.5.3 *High-soft and low-hard states*

Of the six sources, Cyg X-1 and GS 1124–68 exhibit both high–soft and low–hard states (see Sect. 3.3). There is an indication that the transition between the two states occurs around a certain threshold luminosity ($\sim 10^{37}$ erg s^{-1}). LMC X-1 and LMC X-3 appear to be almost always above this threshold (they are almost always in the high-soft state). However, the two-state behavior is also seen in the neutron star binary Cir X-1; hence it is not a conclusive indicator.

3.5.4 *Variability, flickering, power-density spectra, time lags*

Cyg X-1, GS 2023+33 and GS 1124–68 exhibit large-amplitude (rms variation $\sim$30%) flickering[*] (over the entire range from $\sim$20 ms to minutes for the Ginga observations) when they are in the single power-law state. While the neutron star LMXBs generally appear to show flickering (see Figure 6.18 for a remarkable similarity in the PDS between the burst source 1608–52 in the island state and Cyg X-1 in

---

[*] The word 'flickering' is very ill-defined; various authors have used it rather loosely though no formal definition has ever been given in terms of frequency ranges and amplitudes. As pointed out in Ch. 6, the mere presence or absence of variability (flickering) is not very meaningful unless the degree of variability is specified (e.g. in terms of rms). As an example, all sources that vary slowly (e.g. aperiodically with a period of an hour, a day or a decade) vary on any time scale, no matter how short, and we will be able to observe variability in these sources on a time scale of, e.g., milliseconds provided that we have sufficient statistics. Consequently, all sources will show excess power in their power density spectra at 1000 Hz (assuming that the time resolution is better than 0.5 ms) and that our statistics is good enough.

the low state) with smaller amplitudes, this distinction is certainly not conclusive as there are exceptions, such as, e.g., Cir X-1. Binary X-ray pulsars also show significant flickering (e.g. 0331+53). We conclude that it is not possible to make an assessment of flickering as a BHC indicator (see also Ch. 6).

The study of BHCs has recently shifted in the direction of temporal studies which include power-density spectra at various spectral states (e.g., Kitamoto 1989; Miyamoto *et al.* 1989; see also references below). QPOs, which have already been extensively studied in neutron stars (for reviews, see Lewin *et al.* 1988; Fortner *et al.* 1989; Hasinger & Van der Klis 1989; Van der Klis 1989a,b, 1991), are now being observed more and more in BHCs. If some of these phenomena are due to the same physical processes, this will make it possible to make some detailed comparisons between neutron stars and black-hole binaries. The observed centroid frequencies in the BHCs are 0.075 Hz for LMC X-1 (Ebisawa *et al.* 1989); 3–10 Hz for 1124–68 (Grebenev *et al.* 1993; Takizawa *et al.* 1994); 0.026, 0.05, 0.1, 0.8, and 6 Hz for GX 339–4 (Motch *et al.* 1985; Ilovaisky *et al.* 1986; Imamura *et al.* 1990; Grebenev *et al.* 1991; Harmon *et al.* 1992c); and 0.04 Hz for Cyg X-1 (Angelini & White 1992; Kouveliotou 1992a,b; Vikhlinin *et al.* 1992).

Of particular interest are the results obtained by Miyamoto (1992) and Miyamoto *et al.* (1992a,b). They found that the characteristics of the three BHCs Cyg X-1, GX 339–4 and GS 2023+338 showed remarkable similarities when these sources were in their spectrally hard states (accompanied by distinct flickering). Their 'normalized' power-density spectra (PDS) above 0.2 Hz, as well as the time lags between high-energy photons (4.5–10 keV) and low-energy photons (1–4.5 keV) were very similar. The time lags are roughly similar over a frequency range from about 0.02 to 5 Hz. Similarities (in the PDS and the time lags) also exist between the BHCs GX 339–4 and GS 1124–68 in their 'high' (spectrally soft) states (Miyamoto *et al.* 1993).

3.5.5   *The best diagnostics*

None of the above characteristics alone are successful indicators to distinguish BHCs from neutron-star binaries. Perhaps it is because most of these spectral as well as temporal characteristics originate from the accretion phenomenon itself, which may not be strongly related to the nature of the compact object. As discussed in Sect. 3.4, the accretion disk structures at a given accretion rate will be similar for both a weakly magnetized neutron star and a stellar-mass black hole. Not a great deal of qualitative differences are expected in the properties of radiation from the accretion disks. Therefore, with respect to X-ray properties, the best diagnostics for BHCs to examine are those which will reflect the difference in the compact objects.

3.5.6   *Presence or absence of a solid surface*

The fundamental difference that distinguishes a black hole from a weakly magnetized neutron star is the absence of a solid surface. A black hole should never produce type I X-ray bursts, and none of the six sources have exhibited such X-ray bursts. However, 'the absence of evidence is not the evidence of absence'. There are many LMXB neutron stars which do not produce X-ray bursts when they are very luminous (see Ch. 4).

On the other hand, as discussed in Sect. 3.4, almost all these luminous LMXBs have a blackbody component which is interpreted to be the emission from the neutron star surface. None of the six BHCs has such a blackbody spectral component. Instead, they have a hard power-law tail, though its origin is still unknown. This spectral difference is fairly distinct, particularly for luminous sources (however, to obtain a luminosity, a distance estimate must be available). Note, however, that this distinction becomes ambiguous when the luminosity is relatively low ($\sim 10^{37}$ erg s^{-1}; see Sect. 3.4).

3.5.7 Detailed examination of spectra

As discussed in Sect. 3.4, it is likely that both the ultrasoft components of the BHCs and the soft component of the bright neutron star LMXBs represent the emission from an optically thick accretion disk. The multicolor disk interpretation of observed spectra allows us to determine the inner radius of the disk, R_{in} (if the source distance is approximately known) and the blackbody temperature, T_{in}, for which no source distance is required. The constancy of R_{in} observed for two of the six BHCs, LMC X-3 and GS 1124–68 (see Figure 3.14), independent of luminosity, suggests that R_{in} is related to $3r_g$ (though the exact value for R_{in} depends on the poorly known relation between T_{color} and T_{eff}; for details see Ch. 4). Of the six sources, five are ultrasoft and their values for R_{in} are all at least three to four times larger than those for the neutron star LMXBs. Although the actual values are model dependent and the distance estimates include uncertainties, the differences in R_{in} between the BHCs and the neutron star LMXBs are quite significant (Table 3.2). This is interpreted to indicate that the masses of the BHCs are at least a factor of three to four larger than the neutron star masses. If the disk of the weakly magnetized neutron star terminates at a larger distance than the neutron star radius, as seems likely, the mass differences become even larger.

For a given luminosity, the more massive a compact object is, the lower is the value for T_{in}. Hence, T_{in}, which is independent of the distance to the source, is also a good parameter to distinguish a BHC from a neutron star LMXB. This may be the reason why the ultrasoft sources are such good BHCs.

Very different boundary conditions at the inner edge of an accretion disk are expected in neutron star and black-hole binaries. With neutron stars, the disk must terminate at the magnetosphere or at the surface of the neutron star (where X-rays are produced). In the case of a black hole, the disk must terminate at a distance of $3r_g$ (see Sect. 3.4). This could explain some important differences (in X-ray luminosity and in the presence of high excitation optical lines) between Cen X-4 (a neutron star system) in quiescence and A 0620–00 in quiescence (McClintock & Remillard 1990).

3.6   How many black-hole binaries are there?

3.6.1 Formation and evolution

If it is assumed that compact objects with a mass in excess of $\sim 3 M_\odot$ are black holes, the total number of stellar mass black holes in our Galaxy could be $\sim 10^8$ (Van den Heuvel 1992). The total number of black-hole X-ray binaries (most become X-ray active at least once in a period of 10 to 100 years) is estimated (based on the statistics of the known BHC transients) to be between a few hundred and a few thousand. At any moment in time, only a small fraction (a few per cent) are X-ray active (Tanaka 1992b; Van den Heuvel 1992).

The formation of an HMXB with an $\sim 10 M_\odot$ black hole or with a neutron star is rather straightforward (e.g. Van den Heuvel 1983). However, the formation of an LMXB with $\sim 10 M_\odot$ black hole requires exceptional conditions (Van den Heuvel 1992 and references therein). It is conceivable that a triple system is required for their formation (Eggleton & Verbunt 1986) or a finely tuned binary spiral-in scenario (Van den Heuvel & Habets 1984; De Kool *et al.* 1987; Romani 1992). The formation of an LMXB containing a neutron star will be a similarly rare event, and it is quite possible that the majority of them are X-ray dormant at any given time (Frank *et al.* 1992 and references therein). In Ch. 11 Verbunt & Van den Heuvel discuss in detail the formation and evolution of LMXBs and HMXBs containing neutron stars as well as black holes.

There are ~ 125 known LMXBs (see Table 14.1), of which $\sim 31\%$ are transients and $\sim 13\%$ are BHCs; 80% of these BHCs are transients (Table 3.1). Of the ~ 69 known HMXBs (Table 14.2) $\sim 43\%$ are transients and only $\sim 4\%$ are BHCs. There may be evolutionary reasons why more LMXBs contain black holes than HMXBs. However, one has to be very careful in arriving at this conclusion; there may be a selection effect in our list of BHCs that favors the LMXBs. We only know of six of the 20 BHCs (Table 3.1) where the mass of the compact object is in excess of $\sim 3 M_\odot$; three of them are LMXBs and three are HMXBs. Some of the sources listed in Table 3.1 will almost certainly turn out to be neutron star LMXBs.

Acknowledgements

We are very grateful to France Córdova, Ken Ebisawa, Andrew King, Shunji Kitamoto, Lori M. Lubin, Jeff McClintock, Fumiyoshi Makino, Christian Motch, Mina Ogawa, Arvind Parmar, Rashid Sunyaev, Ron Taam, Mamoru Takizawa, Jan van Paradijs, Michiel van der Klis, Brian Vaughan, Frank Verbunt, and Nick White who have provided us with invaluable information, help and advice; some of them kindly proofread (and substantially improved) parts of this manuscript. Without the invaluable Tables and references provided by Jan van Paradijs (Ch. 14 of this book), this article may never have been finished. WHGL thanks his friends and colleagues at ISAS for their kind hospitality and NASA for support under grants NAG8-700 and NAG8-216.

References

Angelini, L. & White, N.E. 1992, *IAU Circ. 5580*
Aref'ev, V. *et al.* 1989, in: *Two Topics in X-ray Astronomy*, 23rd ESLAB Symp., Bologna, Italy (ESA, Paris), p. 255
Bahcall, S., Lynn, B.W. & Selipsky, S.B. 1990, *ApJ*, 362, 251
Bailyn, C. 1991, *IAU Circ. 5259*
Bally, J. & Leventhal, M. 1991, *Nature*, 353, 234
Barr, P. & Van der Woerd, H. 1990, *ApJ*, 352, L41
Barret, D. *et al.* 1991, *ApJ*, 379, L21
Barret, D. *et al.* 1992, *ApJ*, 392, L19
Bath, G.T. 1969, *ApJ*, 158, 571
Bath, G.T. 1975, *MNRAS*, 171, 311
Bath, G.T. & Pringle, J. 1982, *MNRAS*, 199, 267

Bazzano, A. *et al.* 1992, *ApJ*, 385, L17

Belian, R.D., Conners, J.P. & Evans, W.D. 1973, Proc. Conf. on Transient Cosmic Gamma- and X-ray Sources, Los Alamos, NM

Bieging, J. & Downes, D. 1975, *Nature*, 258, 307

Boley, F. *et al.* 1976, *ApJ*, 203, L13

Bolton, C.T. 1972, *Nature*, 235, 271

Bouchet, L. *et al.* 1991, *ApJ Lett.*, 383, L45

Bowyer, S., Byram, E.T., Chubb, T.A. & Friedman, H. 1965, *Science*, 147, 394

Buie, M.W. & Bond, H.E. 1989, *IAU Circ. 4786*

Callanan, P.J. & Charles, P.A. 1991, *MNRAS*, 249, 573

Callanan, P.J. *et al.* 1992, *MNRAS*, 259, 395

Canizares, C. 1975, *ApJ*, 201, 589

Cannizzo, J.K., Wheeler, J.C. & Ghosh, P. 1982, in: *Proc. Boulder Conference on Pulsations in Classical and Cataclysmic Variable Stars*, ed. J.P. Cox, (University of Colorado and National Bureau of Standards, Boulder), p. 13

Cannizzo, J.K., Wheeler, J.C. & Ghosh, P. 1985, in: *Cataclysmic Variables and Low-Mass X-ray Binaries*, eds. D.Q. Lamb & J. Patterson (Reidel, Dordrecht), p. 307

Casares, J. & Charles, P.A. 1992, *MNRAS*, 255, 7

Casares, J. *et al.* 1991, *MNRAS*, 250, 712

Casares, J., Charles, P.A. & Naylor, T. 1992, *Nature*, 355, 614

Charles, P.A. *et al.* 1988, *IAU Circ. 4609*

Charles, P.A. *et al.* 1989, in: Two Topics in X-ray Astronomy, 23rd ESLAB Symp., Bologna (ESA, Paris), p. 103

Charles, P.A., Kidger, M.R., Pavlenko, E.P., Prokofieva, V.V. & Callanan, P.J. 1991, *MNRAS*, 249, 567

Chevalier, C. 1989, in: *Two Topics in X-ray Astronomy*, 23rd ESLAB Symp., Bologna, Italy (ESA, Paris), p. 341

Chevalier, C. & Ilovaisky, S.A. 1990, *A&A*, 238, 163

Chiappetti, L. *et al.* 1985, *Space Sci. Rev.*, 40, 207

Chodil, G., Mark. H., Rodrigues, R., Seward, F.D. & Swift, C.D. 1968, *ApJ*, 152, L45

Coe, M.J., Engel, A.R. & Quenby, J.J. 1976, *Nature*, 259, 544

Cominsky, L., Jones, C., Forman, W. & Tananbaum 1978, *ApJ*, 224, 46

Connors, P.A., Piran, T. & Stark, R.F. 1980, *ApJ*, 235, 224

Cook, M.C. *et al.* 1991, *ApJ*, 372, L75

Cooke, B.A. & Pounds, K.A. 1971, *Nat. Phys. Sci.*, 259, 544

Cooke, B.A., Pounds, K.A., Stewardson, E.A. & Adams, D.J. 1968, *ApJ*, 150, L189

Cooke, B.A. *et al.* 1984, *ApJ*, 285, 258

Cowley, A.P. *et al.* 1983, *ApJ*, 272, 118

Cowley, A.P. *et al.* 1988, *ApJ*, 333, 906

Cowley, A.P. *et al.* 1991a, *ApJ*, 373, 228

Cowley, A.P. *et al.* 1991b, *ApJ*, 381, 526

Davelaar, J. *et al.* 1983, *IAU Circ. 3893*

Davis, R.J. *et al.* 1975, *Nature*, 257, 659

De Kool, M. 1988, *ApJ*, 334, 336

De Kool, M. *et al.* 1987, *A&A*, 183, 47

Della Valle, M. *et al.* 1991, *IAU Circ. 5165*, and Proc. Workshop on Nova Muscae 1991, Lyngby, ed. S. Brandt, p. 107

Döbereiner, S. *et al.* 1989, in: *Two Topics in X-ray Astronomy*, 23rd ESLAB Symp., Bologna (ESA, Paris), p. 387

Dolan, J.F. *et al.* 1987, *ApJ*, 322, 324

Doxsey, R. *et al.* 1977a, *IAU Circ. 3113*

Doxsey, R. *et al.* 1977b, *Nature*, 269, 112

Doxsey, R. *et al.* 1979, *ApJ*, 228, 167

Duldig, M.L. *et al.* 1979, *MNRAS*, 187, 567

Eachus, L.J. *et al.* 1976, *ApJ*, 203, L17

Ebisawa, K. 1991, Ph.D. Thesis, ISAS, University of Tokyo

Ebisawa, K., Mitsuda, K. & Inoue, H. 1989, *PASJ*, 41, 519

Ebisawa, K. *et al.* 1991, *ApJ*, 367, 213

Ebisawa, K. *et al.* 1992, in: *Frontiers of X-ray Astronomy*, eds. Y. Tanaka and K. Koyama, Frontiers Science Series-2 (Universal Academy Press, Inc., Tokyo), p. 351

Ebisawa, K. *et al.* 1993, *ApJ*, 403, 684

Ebisuzaki, T., Hanawa, T. & Sugimoto, D. 1984, *PASJ*, 36, 551

Efremov, V.V. *et al.* 1989, in: *Two Topics in X-ray Astronomy*, 23rd ESLAB Symp., Bologna (ESA, Paris), p. 15

Eggleton, P.P. & Verbunt, F. 1986, *MNRAS*, 220, 13

Elvis, M. *et al.* 1975, *Nature*, 257, 656

Fishman, G.J. *et al.* 1991, *IAU Circ. 5395*

Fortner, B., Lamb, F.K. & Miller, G.S. 1989, *Nature*, 342, 775

Francey, R.J. 1971, *Nature Phys. Sci.*, 229, 229

Francey, R.J., Fenton, A.G., Harries, J.R. & McCracken, K.G. 1967, *Nature*, 216, 773

Frank, J., King, A.R. & Lasota, J.P. 1992, *ApJ*, 385, L45

Gies, D.R. & Bolton, C.T. 1982, *ApJ*, 260, 240

Gies, D.R. & Bolton, C.T. 1984, *ApJ*, 276, L17

Gies, D.R. & Bolton, C.T. 1986, *ApJ*, 304, 371

Gilfanov, M. *et al.* 1993, *A&A Suppl.*, 97, 303

Goldwurm, A. *et al.* 1992, *ApJ*, 389, L79

Gonzalez-Riestra, R. *et al.* 1991, *IAU Circ. 5174*

Gotthelf, E. *et al.* 1991, *AJ*, 103, 219

Gray, A.D. *et al.* 1991, *Nature* 353, 237

Grebenev, S.A. *et al.* 1991, *Sov. Astr. Lett.* 17(6), 413

Grebenev, S.A. *et al.* 1992a, in: *Frontiers of X-ray Astronomy*, eds. Y. Tanaka and K. Koyama, Frontiers Science Series-2 (Universal Academy Press, Inc., Tokyo), p. 319

Grebenev, S.A. *et al.* 1992b, *Sov. Astr. Lett.* 18, 5

Grebenev, S.A. *et al.* 1993, *A&A Suppl.*, 97, 281

Greiner, J. *et al.* 1991, in: Proc. Workshop on Nova Muscae 1991, Lyngby, ed. S. Brandt, p. 79

Griffiths, R.E. 1977, *IAU Circ. 3110*

Griffiths, R.E. *et al.* 1978, *ApJ*, 221, L63

Hameury, J-M., King, A.R. & Lasota, J-P. 1986, *A&A*, 162, 71

Hameury, J-M., King, A.R. & Lasota, J-P. 1988, *A&A*, 192, 187

Hameury, J-M., King, A.R. & Lasota, J-P. 1990, *ApJ*, 353, 585

Han, X.-H. & Hjellming, R.M. 1989, *IAU Circ. 4879*

Han, X.-H. & Hjellming, R.M. 1990, in: Proc. 11th North American Workshop on CVs and LMXBs, ed. C.W. Mauche (Cambridge University Press), p. 25

Han, X. & Hjellming, R.M. 1992, *ApJ*, 400, 304

Harmon, B.A. *et al.* 1992a, *IAU Circ. 5504*

Harmon, B.A. *et al.* 1992b, in *Compton Gamma-ray Observatory*, AIP Conf. Proc. 280, eds. M. Friedlander, N. Gehrels and D. Macomb. (AIP, New York), p. 314

Harmon, B.A. *et al.* 1992c, in *Compton Gamma-ray Observatory*, AIP Conf. Proc. 280, eds. M. Friedlander, N. Gehrels and D. Macomb. (AIP, New York), p. 350

Harries, J.R., McCracken, K.G., Francey, R.J. & Fenton, A.G. 1967, *Nature*, 215, 38

Hasinger G. & Van der Klis, M. 1989, *A&A*, 225, 79

Haswell, C.A. & Shafter, A.W. 1990, *ApJ*, 359, L47

Heemskerk, M.H.M., Papaloizou, J.C. & Savronije, G.J. 1992, *A&A*, 260, 161

Hjellming, R.M., Calovini, T.A., Xiao Hong Han & Córdova, F.A. 1988, *ApJ*, 335, L75

Hjellming, R.M., Han, X.-H. & Cordova, F.A. 1989, *IAU Circ. 4790*

Honey, W.B. *et al.* 1988, *IAU Circ. 4532*

Hoshi, R. 1979, *Progr. Theor. Phys.*, 61, 1307

Huang, M. & Wheeler, J.C. 1989, *ApJ*, 343, 229

Hurst, G.M. & Mobberley, M. 1989, *IAU Circ. 4783*

Hutchings, J.B. *et al.* 1987, *AJ*, 94, 340

Iga, S. *et al.* 1991, in: *Frontiers of X-Ray Astronomy*, eds. Y. Tanaka & K. Koyama, Frontiers Science Series-2 (Universal Academy Press, Inc., Tokyo), p. 309

Ilovaisky, S.A. *et al.* 1986, *A&A*, 164, 67

Imamura, J.N. *et al.* 1987, *ApJ*, 314, L11

Imamura, J.N. *et al.* 1990, *ApJ*, 365, 312
Johnston H.M. *et al.* 1989, *ApJ*, 345, 492
Johnston, M., Griffith, R. & Ward, M.J. 1980, *Nature*, 254, 578
Jones, C. 1977, *ApJ*, 214, 856
Jones, C. *et al.* 1976, *ApJ*, 364, 664
Jones, D. & Carter, D. 1989, *IAU Circ. 4794*
Kaluzienski, L.J. & Holt, S.S. 1977a, *IAU Circ. 3104*
Kaluzienski, L.J. & Holt, S.S. 1977b, *IAU Circ. 3099 & 3106*
Kaluzienski, L.J. *et al.* 1975, *ApJ*, 201, L121
Kaluzienski, L.J., Holt, S.S., Boldt, E.A. & Serlemitsos, P.J. 1977, *ApJ*, 212, 203
Kawai, N. *et al.* 1988, *ApJ*, 330, 130
Kemp, J.C. *et al.* 1987, *SvA*, 31, 170
Kesteven, M.J. & Turtle, A.J. 1991, *IAU Circ. 5181*
Kitamoto, S. 1989, in: *Two Topics in X-ray Astronomy*, 23rd ESLAB Symp., Bologna (ESA, Paris),
    p. 231
Kitamoto, S. *et al.* 1984a, *PASJ*, 36, 731
Kitamoto, S. *et al.* 1984b, *PASJ*, 36, 799
Kitamoto, S. *et al.* 1989a, *Nature*, 342, 518
Kitamoto, S. *et al.* 1989b, *PASJ*, 41, 81
Kitamoto, S. *et al.* 1990, *ApJ*, 361, 590
Kitamoto, S. *et al.* 1992, *ApJ*, 394, 609
Kouveliotou, C. *et al.* 1992a, *IAU Circ. 5576*
Kouveliotou, C. *et al.* 1992b, Compton Symposium, Washington Univ., St Louis
Kundt, W. & Fischer, D. 1989, *J. Astrophys. Astr.*, 10, 19
Leahy, D.A. 1991, *MNRAS*, 251, 22P
Leibowitz, E.M., Ney, A., Drissen, L., Grandchamps, A. & Moffat, A.F.J. 1991, *MNRAS*, 250, 385
Levine, A.M. *et al.* 1984, *ApJS*, 54, 581
Lewin, W.H.G., Van Paradijs, J. & Van der Klis, M. 1988, *Space Sci. Rev.* 46, 273
Li, F.K. *et al.* 1976, *ApJ*, 203, 187
Liang, E.P. & Nolan, P.L., 1984, *Space Sci. Rev.*, 38, 353
Lin, D.N.C. & Taam, R.E. 1984, in: AIP Conference Proc. 115 (AIP, New York), p. 83
Ling, J.C. & Wheaton, Wm. A. 1989, *ApJ*, 343, L57
Ling, J.C. *et al.* 1983, *ApJ*, 275, 307
Ling, J.C. *et al.* 1987, *ApJ*, 321, L117
Lingenfelter, R.E. & Ramaty, R. 1989, *ApJ*, 343, 686
Little, A.G. *et al.* 1976, *Nature*, 261, 113
Lloyd, C. *et al.* 1979, *MNRAS*, 179, 675
Lochner, J.C. *et al.* 1991, *ApJ*, 376, 295
London, R.A., Taam, R.E. & Howard, W.M. 1986, *ApJ*, 306, 170
Long, K.S. *et al.* 1981, *ApJ*, 248, 925
Lund, N. & Brandt, S. 1991, *IAU Circ. 5161*
McClintock, J.E. 1992, in: *X-ray Binaries and Recycled Pulsars*, eds. E. van den Heuvel & S.
    Rappaport, NATO ASI Series C: Math. & Phys. Sci. Vol. 377, (Kluwer, Dordrecht), p. 27
McClintock, J.E. & Remillard, R.A. 1986, *ApJ*, 308, 110
McClintock, J.E. & Remillard, R.A. 1990, *ApJ*, 350, 386
McClintock, J.E. *et al.* 1978, *ApJ* 223, L75
McClintock, J.E. *et al.* 1983, *ApJ*, 266, L27
McClintock, J.E., Bailyn, C. & Remillard, R. 1992, *IAU Circ. 5499; ApJ*, 399, L145
Maejima, Y. *et al.* 1984, *ApJ*, 285, 712
Makino, F. 1987, *IAU Circ. 4342*
Makino, F. 1988a, *IAU Circ. 4587 & 4600*
Makino, F. 1988b, *IAU Circ. 4653*
Makino, F. 1989, *IAU Circ. 4782 & 4786*
Makino, F. 1991, *IAU Circ. 5161*
Makishima, K. *et al.* 1986, *ApJ*, 308, 635
Mandrou, P. *et al.* 1990a, *IAU Circ. 5032*
Mandrou, P. *et al.* 1990b, *IAU Circ. 5140*

Margon, B. *et al.* 1978, *ApJ*, 221, 907
Markert, T.H. *et al.* 1973, *ApJ*, 184, L67
Markert, T.H. *et al.* 1977, *ApJ*, 218, 801
Markert, T.H. *et al.* 1979, *ApJS*, 39, 573
Marsden, B.G. 1989, *IAU Circ. 4783*
Mason, K. *et al.* 1985, *MNRAS*, 216, 1033
Matilsky, T. *et al.* 1972, *ApJ*, 174, L53
Matilsky, T. *et al.* 1976, *ApJ*, 210, L127
Meyer, F. & Meyer-Hofmeister, E. 1981, *A&A* 104, L10
Mineshige, S. & Wheeler, J.C. 1989, *ApJ*, 343, 241
Mineshige, S. & Wheeler, J.C. 1992, in: *Frontiers of X-ray Astronomy*, eds. Y. Tanaka & K.
    Koyama, Frontiers Science Series-2 (Universal Academy Press, Inc., Tokyo), p. 329
Mineshige, S. *et al.* 1992, *PASJ*, 44, 117
Mirabel, I.F. *et al.* 1991, *A&A*, 251, L43
Mirabel, I.F. *et al.* 1992, *IAU Circ. 5477; Nature*, 358, 215
Mitsuda, K. 1992, in: Ginga Memorial Symp., eds. F. Makino & F. Nagase, p. 11
Mitsuda, K. *et al.* 1984, *PASJ*, 36, 741
Mitsuda, K. *et al.* 1989, *PASJ*, 41, 97
Mitsuda, K. *et al.* 1992, in: *Frontiers of X-Ray Astronomy*, eds. Y. Tanaka, & K. Koyama, Frontiers
    Science Series-2 (Universal Academy Press, Inc., Tokyo), p. 115
Miyamoto, S. 1992, in: *Frontiers of X-Ray Astronomy*, eds. Y. Tanaka & K. Koyama, Frontiers
    Science Series-2 (Universal Academy Press, Inc., Tokyo), p. 303
Miyamoto, S. & Kitamoto, S. 1989, *Nature*, 342, 773
Miyamoto, S. *et al.* 1989, *Nature*, 336, 450
Miyamoto, S. *et al.* 1991, *ApJ*, 383, 784
Miyamoto, S. *et al.* 1992a, *ApJ*, 391, L21
Miyamoto, S. *et al.* 1992b, in: Ginga Memorial Symp., eds. F. Makino & F. Nagase, p. 37
Miyamoto, S. *et al.* 1993, *ApJ*, 403, L39
Motch, C. *et al.* 1983, *A&A*, 119, 171
Motch, C. *et al.* 1985, *Space Sci. Rev.*, 40, 219
Murdin, P. *et al.* 1977, *MNRAS*, 178, 27P
Murdin, P. *et al.* 1980, *MNRAS*, 192, 709
Nakamura, N. *et al.* 1989, *PASJ*, 41, 617
Nolan, P.L. & Matteson, J.L. 1983, *ApJ*, 265, 389
Nolan, P.L. *et al.* 1982, *ApJ*, 262, 727
Oda, M. 1977, *Space Sci. Rev.*, 20, 757
Ogawa, M. 1992, M.Sc. Thesis, ISAS, Rikkyo Univ.
Okamura, S. & Noguchi, T. 1988, *IAU Circ. 4589*
Oke, J.B. 1977, *ApJ*, 217, 181
Osaki, Y. 1974, *PASJ*, 26, 429
Osaki, Y. 1989, *PASJ*, 41, 1005
Ostriker, J.P. 1977, *Ann. New York Acad. Sci.*, 302, 229
Paciesas, W.S. *et al.* 1992, *IAU Circ. 5580* (see also many following issues of *IAU Circ.*)
Paczynski, B. 1983, *ApJ*, 273, L81
Paul, J. *et al.* 1991, in: *Gamma Ray Line Astrophysics* (AIP Proc. 232), P. Duchouroux, & N.
    Prantzos, (AIP, New York), p.17
Parmar, A.N. & White, N.E. 1985, *IAU Circ. 4051*
Parmar, A.N., Stella, L. & White, N.E. 1986, *ApJ*, 304, 664
Parmar, A.N. *et al.* 1993, *A&A*, 279, 179
Pavlenko, E.P. *et al.* 1989, *Sov. Astron. Lett.*, 15, 262
Pedersen, H. *et al.* 1983, *IAU Circ. 3858*
Pedersen, H., Ilovaisky, S. & van der Klis, M. 1987, *IAU Circ. 4357*
Pounds, K.A. 1976, *Comm. Astr.*, 6, 145
Priedhorsky, W. & Holt, S.S. 1987, *Space Sci. Rev.*, 45, 291
Priedhorsky, W. & Terrell, J. 1984, *ApJ*, 280, 661
Priedhorsky, W.C., Terrell, J. & Holt, S.S. 1983, *ApJ*, 270, 233
Prince, T. *et al.* 1991, *IAU Circ. 5252*

Rappaport, S.A. & Van den Heuvel, E.P.J. 1982, in: Be Stars, IAU Symp. No. 98, eds. M. Jaschek & H.-G. Groth (Reidel, Dordrecht), p. 327

Reich, W. & Schlickeiser, R. 1992, *A&A*, 256, 408

Remillard, R. & McClintock, J.E. 1987, *IAU Circ. 4383*

Rhoades, C.E. & Ruffini, R. 1974, *Phys. Rev. Lett.*, 32, 324

Richter, G.A. 1987, *IBVS*, 3362

Rickets, M.J., Pounds, K.A. & Turner, M.J.L. 1975, *Nature*, 257, 657

Romani, R.W. 1992, *ApJ*, 399, 621

Samimi, J. *et al.* 1979, *Nature*, 278, 434

Sanford, P.W. *et al.* 1975, *Nature*, 256, 109

Sarna, M. 1990, *A&A*, 239, 163

Schmitz-Frayasse, M.C. *et al.* 1992, *IAU Circ. 5472*

Schreier, E., Levinson, R., Gursky, H., Kellogg, E., Tananbaum, H. & Giacconi, R. 1972, *ApJ*, 172, L79

Schwartz, R.A. *et al.* 1991, *ApJ*, 376, 312

Shakura, N.I. & Sunyaev, R. 1973, *A&A*, 24, 337

Shapiro, S.L., Lightman, A.P. & Eardley, D. 1976, *ApJ*, 204, 187

Shrader, C.R. 1988, *IAU Circ. 4605*

Shrader, C.R. & Gonzalez-Riestra, R. 1991, in: Proc. Workshop on Nova Muscae 1991, Lyngby, ed. S. Brandt, p. 85

Skinner, G.K. 1991, in: *Gamma Ray Line Astrophysics* (AIP Proc. 232), eds. P. Duchouroux & N. Prantzos (AIP), p. 358

Skinner, G.K. *et al.* 1987, *Nature*, 330, 544

Skinner, G.K. *et al.* 1990, *MNRAS*, 243, 72

Smak, J.I. 1970, *IAU Coll.* 15, 248

Smak, J.I. 1971, *Acta Astron.*, 21, 15

Smak. J.I. 1984, *Acta Astron.*, 34, 61

Smak, J.I. 1992, *Acta Astron.*, 41, 269

Steinman-Cameron, T. *et al.* 1990, *ApJ*, 359, 197

Stella, L. & Rosner, R. 1984, *ApJ*, 277, 312

Stella, L., White, N.E., Davelaar, J., Parmar, A.N., Blissett, R.J. & Van der Klis, M. 1985, *ApJ*, 288, L45

Sunyaev, R.A. 1990, *IAU Circ. 5104*

Sunyaev, R.A. 1991, *IAU Circ. 5176*

Sunyaev, R.A. & Titarchuk, L.G. 1980, *A&A*, 201, 379

Sunyaev, R.A. & Trümper, J. 1979, *Nature*, 279, 506

Sunyaev, R.A. *et al.* 1988, *Sov. Astron. Lett.*, 14, 327

Sunyaev, R.A. *et al.* 1989, *Sov. Astron. Lett.*, 16, 55

Sunyaev, R.A. *et al.* 1991a, *Sov. Astron. Lett.*, 17, 50

Sunyaev, R.A. *et al.* 1991b, *Sov. Astron. Lett.*, 17, 54

Sunyaev, R.A. *et al.* 1991c, in: *Gamma Ray Line Astrophysics* (AIP Proc. 232), eds. P. Duchouroux, & N. Prantzos (AIP, New York)

Sunyaev, R.A. *et al.* 1991d, in: *Frontiers in X-ray Astronomy*, eds. Y. Tanaka & K. Koyama, Frontiers Science Series-2 (Universal Academy Press, Inc., Tokyo), p. 241

Sunyaev, R. A. *et al.* 1991e, *Sov. Astron. Lett.*, 17, 123

Sunyaev, R. A. *et al.* 1991f, *ApJ Lett.*, 383, L49

Sunyaev, R. *et al.* 1992a, *ApJ*, 389, L75

Sunyaev, R. *et al.* 1992b, in: *Frontiers of X-ray Astronomy*, eds. Y. Tanaka & K. Koyama, Frontiers Science Series-2 (Universal Academy Press, Inc., Tokyo), p. 697

Szkody, F., & Margon, B. 1989, *IAU Circ. 4794*

Takizawa, M. 1991, M.Sc. Thesis, ISAS, Univ. Tokyo

Takizawa, M. *et al.* 1994, *PASJ*, to be submitted

Tanaka, Y. 1983a, *IAU Circ. 3801*

Tanaka, Y. 1983b, *IAU Circ. 3854*

Tanaka, Y. 1984, *IAU Circ. 3936*

Tanaka, Y. 1989, in: *Two Topics in X-ray Astronomy*, 23rd ESLAB Symp., Bologna, Italy (ESA, Paris), p. 3

Tanaka, Y. 1991, in: *Iron Line diagnostics in X-Ray Sources*, Lecture Notes in Physics 385, eds. A. Treves, G.C. Perrola & L. Stella (Springer, Berlin), p. 98

Tanaka, Y. 1992a, in: *X-ray Binaries and Recycled Pulsars*, eds. E.P.J. Van den Heuvel & S.A. Rappaport (Kluwer Academic Publishers Dordrecht), p. 37

Tanaka, Y. 1992b, in: Ginga Memorial Symp., eds. F. Makino, & F. Nagase, p. 19

Tananbaum, H.D., 1973, IAU Symp. No. 55, eds. H. Bradt & R. Giacconi, p. 9

Tennant, A.F., Fabian, A.C. & Shafer, R.A. 1986, *MNRAS*, 219, 871

Terrel, N.J. 1972, *ApJ*, 174, L35

Thorstensen, J.R. 1987, *ApJ*, 312, 739

Treves, A. *et al.* 1988, *ApJ*, 325, 119

Tsunemi, H., Kitamoto, S., Manabe, M., Miyamoto, S., Yamashita, K. & Nakagawa, M. 1989a, *PASJ*, 41, 391

Tsunemi, H., Kitamoto, S., Okamura, S. & Roussel-Dupré, D. 1989b, *ApJ*, 337, L81

Ubertini, P. *et al.* 1991, *ApJ*, 366, 544

Udalski, A. & Kaluzny, J. 1991, *PASP*, 103, 198

Van Amerongen, S. *et al.* 1990, *MNRAS*, 242, 522

Van den Heuvel, E.P.J. 1983, in: *Accretion Driven Stellar X-ray Sources*, eds. W.H.G. Lewin & E.P.J. van den Heuvel (Cambridge University Press), p. 303

Van den Heuvel, E.P.J. 1992, Proc. European ISY Conf. '92, Symp. No.3, ESA Publ., ESTEC, Noordwijk.

Van den Heuvel, E.P.J. & Habets, G.M.H.J. 1984, *Nature*, 309, 698

Van der Klis, M. 1989a, *ARA&A*, 27, 517

Van der Klis, M. 1989b, in: *Timing Neutron Stars*, eds. Oegelman & Van den Heuvel (Kluwer Academic Publishers, Dordrecht), p. 27

Van der Klis, M. 1991, in: Frontiers of X-ray Astronomy, eds. Y. Tanaka & K. Koyama, Frontiers Science Series-2 (Universal Academy Press, Inc., Tokyo), p. 139

Van der Woerd, H. *et al.* 1989, *ApJ*, 344, 320

Van Paradijs, J. & Verbunt, F. 1984, in: *High-Energy Transients in Astrophysics*, ed. S. Woosley (AIP, New York), p. 49.

Verbunt, F. 1986, in: *Physics of Accretion onto Compact Objects*, eds. K. Mason, M.G. Watson & N.E. White (Springer Verlag, Berlin), p. 59

Vikhlinin, A. *et al.* 1992, *IAU Circ. 5576*

Wachmann, A.A. 1948, *Erg. Astron. Nachr.*, 11, No.5, E42

Wagner, R.M., Henden, A.A., Bertram, R. Starrfield, S.G. 1988, *IAU Circ. 4600*

Wagner, R.M., Starrfield, S. & Cassatella, A. 1989, *IAU Circ. 4783*

Wagner, R.M. *et al.* 1990, in: *Proc. 11th North American Workshop on CVs and LMXBs*, ed. C.W. Mauche (Cambridge University Press), p. 29

Wagner, R.M. *et al.* 1991, *ApJ*, 378, 293

Wagner, R.M. *et al.* 1992, *ApJ*, 401, L97

Watson, M.G. *et al.* 1978, *ApJ*, 221, L69

Webster, B.L. & Murdin, P. 1972, *Nature*, 235, 37

West, R.M. 1991, in: Proc. Workshop on Nova Muscae 1991, Lyngby, ed. S. Brandt, p. 143

Whelan, J.A.J. *et al.* 1976, *MNRAS*, 180, 657

White, N.E. & Marshall, F.E. 1984, *ApJ*, 281, 354

White, N.E., Kaluzienski, J.L. & Swank, J.L. 1984a, in: *High Energy Transients in Astrophysics*, ed. S.E. Woosley (AIP, New York), p. 31

White, N.E. *et al.* 1984b, *ApJ*, 283, L9

White, N.E. *et al.* 1988a, *ApJ*, 324, 363

White, N.E., Stella, L. & Parmar, A. 1988b, *ApJ*, 324, 363

Whitehurst, R. 1988, *MNRAS*, 232, 35

Wilson, C.K. & Rothschild, R.E. 1983, *ApJ*, 274, 717

Wood, K.S. *et al.* 1978, *IAU Circ. 3203*

Wood, K.S. *et al.* 1984, *ApJS*, 56, 507

Yaqoob, T., Ebisawa, K. & Mitsuda, K. 1993, *MNRAS*, 264, 411

4

X-ray bursts

Walter H. G. Lewin
Massachusetts Institute of Technology, Physics Department,
Center For Space Research, MIT 37-627, Cambridge, MA 02139, USA
Astronomical Institute 'Anton Pannekoek', University of Amsterdam,
and Center for High Energy Astrophysics, Kruislaan 403,
1098 SJ Amsterdam, The Netherlands
Institute of Space and Astronautical Sciences, 3-1-1 Yoshinodai,
Sagamihara-shi, 229 Japan

Jan Van Paradijs
Astronomical Institute 'Anton Pannekoek', University of Amsterdam,
and Center for High Energy Astrophysics, Kruislaan 403,
1098 SJ Amsterdam, The Netherlands
Institute of Space and Astronautical Sciences, 3-1-1 Yoshinodai,
Sagamihara-shi, 229 Japan

Ronald E. Taam
Department of Physics and Astronomy, Northwestern University,
2145 Sheridan Road, Evanston, IL 60208, USA

4.1   Introduction

4.1.1   *Progress during the past decade*

Since the last comprehensive reviews on X-ray bursts were written about a decade ago by Lewin and Joss (1981, 1983), the number of type I burst sources, all of which are low-mass X-ray binaries (LMXBs), has only increased from ∼32 to ∼40 (see Ch. 14). However, lots more data have become available on the type I bursts (thermonuclear flashes on the surface of a neutron star) and on the type II bursts from the Rapid Burster (spasmodic accretion), and much theoretical work has been done. (i) Photospheric radius expansion of the neutron star atmosphere during very luminous type I X-ray bursts has been well established and has led to the possibility, at least in principle, of measuring both the mass and radius of neutron stars, thereby gaining information about the equation of state of neutron star matter. (ii) A 4.1 keV absorption line was detected in several type I burst spectra from different sources, the origin of which is unclear. (iii) It appears that burst sources are largely (but not exclusively) 'atoll' sources (see Ch. 6), and that several type I burst properties depend on the spectral state of the source. (iv) The evolutionary connection between LMXBs and millisecond radio pulsars has come into focus during the past decade (Ch. 5), and (v) new ideas have evolved about the magnetic-field decay of neutron stars in LMXBs (Ch. 12). (vi) Quasi-periodic oscillations (QPOs) were detected in type II bursts and in the persistent emission of the Rapid Burster (Sect. 4.6.10). (vii) Several new peculiarities have been detected in the type II bursts and in the persistent emission from the Rapid Burster. (viii) Much theoretical work was done during the past decade on neutron star model atmospheres, on the interpretation of the burst spectra, and on the theory of the thermonuclear flashes themselves. (ix) Recent theoretical work suggests the possibility for the significant depletion of CNO elements by nuclear spallation reactions, which may have an important impact on models for type I bursts. In spite of much new information on the Rapid Burster, little progress was made during the past decade in understanding its bizarre behavior.

This review is a shortened version of an article we wrote recently for *Space Science Reviews* (Lewin, Van Paradijs and Taam 1993); we will refer to it here as LVT93.

4.1.2   Brief history – highlights

X-ray bursts were discovered in 1975 independently by Grindlay *et al.* (1976) and Belian, Conner and Evans (1976). In February 1976, two additional burst sources were found by Lewin *et al.* (1976a) within a few degrees of the galactic center; the presence of a third burst source was suspected and searched for in early March 1976. It was found (Lewin 1976b), and another extraordinary source, only ~0.5° away, was discovered which produced about 1000 bursts per day (Lewin *et al.* 1976b); it was later called the Rapid Burster. The Rapid Burster was found to be located in a previously unknown, highly reddened globular cluster (Liller 1977). The activity of the Rapid Burster stopped by mid-April 1976. About a year later, White *et al.* (1978) observed the source to be active again. Many active periods of the Rapid Burster (each lasting several weeks) have been observed since then (see Figure 4.18). Within a year, over 20 burst sources were found, largely due to observations with SAS-3 and OSO-8.

Maraschi and Cavaliere (1977), and independently Woosley and Taam (1976), were the first to discuss the possibility that the X-ray bursts were due to thermonuclear flashes on the surface of accreting neutron stars. Dr L. Maraschi suggested this in early February, 1976, when she was visiting MIT at the time that the galactic center region was observed with SAS-3; many X-ray bursts were then observed (Lewin 1976a; Lewin *et al.* 1976a). (Several years earlier, before the discovery of X-ray bursts, Hansen and Van Horn (1975) had made pioneering investigations into thermonuclear instabilities in the surface layers of accreting neutron stars.) However, after the discovery of the Rapid Burster in early March, 1976, it became clear that the rapidly repetitive bursts could not be due to thermonuclear flashes; if they were, a very high flux of persistent X-ray emission due to the release of gravitational potential energy should be present, and this was not observed. Thus the thermonuclear-flash model could not explain all burst phenomena, and that was not in its favor (see, e.g., Lewin and Joss 1977).

A break came in the fall of 1977 when Hoffman, Marshall and Lewin (1978a) discovered that the Rapid Burster emits two very different kinds of bursts. They introduced the classification of type I and type II bursts and suggested that the rapidly repetitive type II bursts are due to accretion instabilities and that the type I bursts are due to thermonuclear flashes. Detailed calculations by Joss (1978) and Taam and Picklum (1979) strengthened the idea that thermonuclear flashes on the surface of neutron stars produce type I bursts. Effects of nonthermal equilibrium and thermal inertia effects were stressed by Taam (1980), and compositional inertia effects were stressed by Ayasli and Joss (1982) and Woosley and Weaver (1985). More recent work by Bildsten *et al.* (1992) points out the possibility for a reduction of the CNO nuclei due to the interaction of the accretion flow with the neutron star atmosphere, and calls into question the pure helium flash model for the accretion of hydrogen rich matter.

In the summer of 1977, 1735−444 was the first burst source to be optically identified by McClintock, Canizares and Backman (1978). This was also the first burst source from which, in the summer of 1978, a simultaneous optical and X-ray burst was observed (Grindlay *et al.* 1978; McClintock *et al.* 1979).

It was long suspected that burst sources are low-mass X-ray binaries. Direct evidence for this was obtained in 1980 and 1981: (i) the quiescent optical counterparts of the transient (burst) sources Cen X-4 and Aql X-1 were found to be low-luminosity late-type stars (Thorstensen, Charles and Bowyer 1978; Van Paradijs *et al.* 1980; Koyama *et al.* 1981), (ii) an orbital period of ∼3.8 h was found for the burst source 1636−536 (Pedersen, Van Paradijs and Lewin 1981).

Photospheric radius expansion of the neutron star atmosphere, due to radiation pressure during a strong X-ray burst, became well established in 1984 (see Sect. 4.3.4). It led to the possibility, at least in principle, of obtaining both the mass and the radius of the neutron star and thereby gaining information about the equation of state of neutron star matter (Sect. 4.4).

In that same year, a 4.1 keV absorption line was found in a type I burst from 1636−536 (Waki *et al.* 1984; Sect. 4.3.3). Since that time, absorption lines at 4.1 keV have been observed in at least seven bursts from three different sources. Recent work by Pinto, Taam and Laming (1994) suggests that the line shift is due to a combination of Doppler and gravitational redshifts.

It was found that for the three burst sources that were extensively observed with EXOSAT, the energy in a type I burst is largely determined by the interval since the previous burst (Lewin *et al.* 1987), thus by the amount of time that accretion occurred, *independent* of the accretion rate. The accretion rates differed by a factor of ∼30 (Van Paradijs, Penninx and Lewin 1988a). This means effectively that the time-averaged luminosity in bursts is independent of the time-averaged luminosity in the persistent emission between the bursts (thus of the accretion rate). This is a puzzling result which perhaps suggests that thermal and compositional inertia effects are operating to determine the ignition conditions of the nuclear fuel (Sect. 4.3.8).

Some burst characteristics (among them the burst durations) are strongly correlated with the spectral state (Van der Klis *et al.* 1990); this may indicate that the nuclear fuel composition in the bursts changes as the accretion rate changes (Sect. 4.3.8).

4.1.3 *Burst nomenclature*

Hereafter, type I bursts may be simply called *bursts* and type-I burst sources may be called *burst sources*. However, whenever we discuss type II bursts the 'II' will always be mentioned.

4.2 Characteristics of burst sources

The ∼40 known burst sources form a subset of the low-mass X-ray binaries (LMXBs); the number of known LMXBs is ∼124 (see Ch. 14).

4.2.1   Characteristics of LMXBs

LMXBs are bright X-ray sources ($>10^{34}$ erg s^{-1}); they accrete through Roche lobe overflow. Observationally, they distinguish themselves from the $\sim$70 known high-mass X-ray binaries (HMXBs, with donor stars $>10M_\odot$) in our galaxy by the following characteristics (see also Ch. 1, 2 and 14):

- Their star-like optical counterparts are faint (typically $M_v >0$), in contrast to the luminous HMXBs which are more than two orders of magnitude brighter. Many LMXBs have no known optical counterpart due to severe interstellar absorption.
- Their optical spectra (except for transients in quiescence and systems in which the companion is a late-type giant star) are devoid of normal stellar absorption features.
- The ratio of their X-ray to optical luminosities ranges from $\sim$$10^2$ to $\sim$$10^4$ (except for transients in quiescence). For HMXBs this ratio ranges from $\sim$$10^{-3}$ to $\sim$$10^1$.
- Their X-ray spectra are substantially softer than the spectra of the HMXBs.
- They rarely show periodic X-ray pulsations such as are often observed from highly magnetized, rotating neutron stars in HMXBs.
- Many of them produce X-ray bursts. In contrast, no X-ray bursts have been observed from any of the HMXBs or any LMXB that shows pulsations.

All LMXBs (except two or three[*]) are believed to have the above characteristics.

4.2.2   Old population

The LMXBs are concentrated towards the galactic center (Figure 4.1). About 17 are found in globular clusters (two of which in ω Cen (F. Verbunt, private communication), and four in 47 Tuc (Verbunt *et al.* 1993)); nine of these are known burst sources (Ch. 14). Their locations outside regions of active star formation identify the LMXBs and burst sources as members of an old population (Figure 4.1). The low mass of the donor stars supports the idea that the LMXBs, of which the burst sources are a subset, are an old population.

Neutron stars in LMXBs almost certainly have a weak magnetic field compared to the neutron stars in the massive (Population I) binary systems (Ch. 12). The absence of X-ray pulsations from LMXBs and the fact that type I X-ray bursts never occur in systems which show X-ray pulsations, are generally used as arguments for the weak magnetic fields (Joss and Li 1980).

4.3   Type I X-ray bursts

4.3.1   Introduction

Type I X-ray bursts (hereafter called X-ray bursts or simply bursts) have been observed from $\sim$40 sources so far (Ch. 14). In this section, we discuss the

---

[*] The following two LMXBs have hard spectra and they show pulsations: Her X-1 (donor mass $\sim$$2.0M_\odot$; Nagase 1989) and 1626−673 (donor mass $<0.06M_\odot$; Levine *et al.* 1988). GX 1+4 has a hard spectrum and it shows pulsations (Lewin, Ricker and McClintock 1971; White *et al.* 1976); if the optical identification suggested by Doxsey *et al.* (1977) is correct, GX 1+4 is an LMXB.

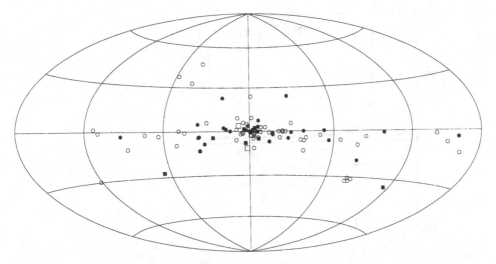

Fig. 4.1. Sky distribution in galactic coordinates of LMXBs. The galactic longitude is zero at the center and increases toward the left. Filled symbols represent sources from which type I bursts have been observed; open symbols are the nonbursting sources. Squares and circles indicate sources in globular clusters and outside globular clusters, respectively.

observed properties of X-ray bursts and their relation to properties of the persistent emission.

4.3.2 Burst profiles

X-ray bursts show a large variety in profiles. Profiles of bursts from a given source may look quite similar and have characteristic shapes (see, e.g., Hoffman, Lewin and Doty (1977b) for 1728−337, and Haberl *et al.* (1987) for 1820−303), but this is by no means a general phenomenon (see Figure 4.2).

Burst rise times vary from less than a second to ∼10 s, and decay times are in the range of ∼10 s to minutes. In general, burst profiles depend strongly on photon energy; decays are much shorter at high photon energies than at low energies (see Figure 4.3). This energy dependence of the burst profile corresponds to a softening of the burst spectrum during the decay, which is the result of the cooling of the neutron star photosphere (see Sect. 4.3.3 and 4.4).

For the transient source 1608−522, Murakami *et al.* (1980a) found that the burst rise times were anti-correlated with the persistent X-ray flux. In general, the burst rise times do not depend in an obvious way on other burst properties, except that most bursts which show evidence for photospheric radius expansion (Sect. 4.3.4) have rise times less than ∼1 s (see, e.g., Lewin *et al.* 1987).

Approximately linear relations between the (bolometric) burst fluence, E_b, and the (bolometric) peak burst flux, F_{max}, have been found in several sources (1735−444: Lewin *et al.* 1980; 1636−536: Ohashi *et al.* 1982; Lewin *et al.* 1987; Ser X-1: Sztajno *et al.* 1983; 1728−337: Basinska *et al.* 1984; see Figure 4.4). For very strong bursts, the (E_b, F_{max}) relation saturates (e.g. Figure 4.4b): the peak luminosity cannot exceed the Eddington limit, and an increase in E_b corresponds to a longer burst. For the

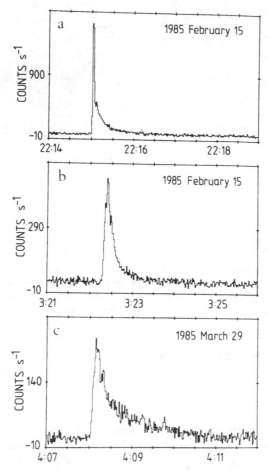

Fig. 4.2. Three bursts (1.5–15 keV) from 0748–676, with very different durations. The horizontal axes are UT (in hours:minutes). Burst (a) showed radius expansion during its initial phase. The count rates have not been dead-time corrected. (From Gottwald *et al.* 1986.)

transient source 1608−522 a linear (E_b, F_{max}) relation was only found when the source was bright (Figure 4.4a). When it was faint, F_{max} was approximately constant (at a value below the highest peak fluxes reached when the source had a high persistent flux); E_b then varied by a substantial factor (Figure 4.4a).

In about a dozen sources, bursts have been observed with double-peaked profiles which are particularly distinct at high energies (examples are shown in Figure 4.5). These burst profiles are caused by photospheric radius expansion due to radiation pressure (Sect. 4.3.4). These bursts, which all have very high peak fluxes, should not be confused with the relatively weak bursts with multi-peaked (bolometric) profiles observed from 1636−536 (Sztajno *et al.* 1985; Van Paradijs *et al.* 1986; Figure 4.6); the latter profiles likely reflect a genuine variation in the rate at which thermonuclear energy is generated or released.

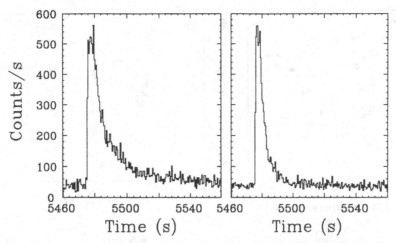

Fig. 4.3. One X-ray burst from 1702—429 observed with EXOSAT in the 1.2–5.3 keV band (left), and the 5.3–19.0 keV band (right). Time is in seconds since April 8, 1986, UT 03:31:31. The softening of the X-ray burst spectrum during decay is apparent as a relatively long tail in the low-energy burst profile. (Courtesy T. Oosterbroek.)

In some cases, a relation has been observed between burst profiles and the persistent emission. Clark *et al.* (1977) observed a gradual decrease of the decay time of bursts from 1820—303, as the persistent X-ray flux increased by a factor of ~5. Murakami *et al.* (1980a) observed 1608—522 during an active phase in April, 1979, when the persistent flux was relatively high; the majority of the bursts then had a fast rise (< 2 s) and a fast decay (< 15 s). Two months later, when the persistent flux had decreased by a factor of ~5, the bursts had a slower rise and decay. Also, the maximum observed burst flux was lower at that time (by a factor of ~3) than it was during the early part of the active phase (Figure 4.4a). Their findings are supported by later observations of 1608—522 by Nakamura, Inoue and Tanaka (1988). A striking example of a relation between burst profiles and the level of the persistent emission is the case of 0748—673, for which Gottwald *et al.* (1986) found that both the fluence (E_b) and the peak flux (F_{max}) of the bursts increased as the persistent flux increased by about an order of magnitude; the ratio E_b/F_{max}, which is a measure of the effective duration of the burst, decreased by a factor of ~3. Van Paradijs, Penninx and Lewin (1988a) found from a statistical study that an anti-correlation between burst duration and the persistent X-ray luminosity is a global property of X-ray burst sources (Sect. 4.3.8).

4.3.3 *Burst spectral analysis*

It has become standard procedure to study the spectra of X-ray sources by assuming a particular shape of the incident photon spectrum and folding this spectrum through the detector response matrix to obtain a predicted count-rate spectrum. The effect of the energy dependent interstellar absorption is taken into account (for details see LVT93). Whether the assumed incident spectrum describes the observations satisfactorily is then assessed by some quality estimator, generally a

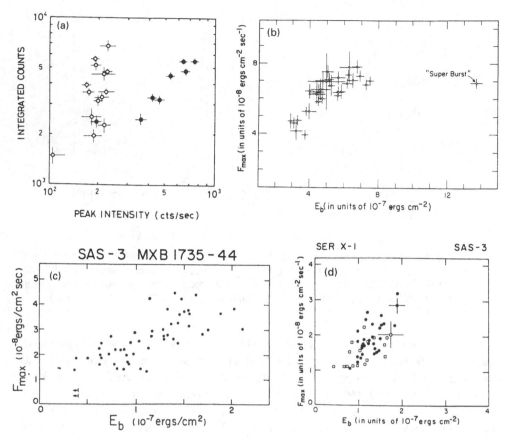

Fig. 4.4. Relationship between peak burst flux and burst fluence. (a) 1608−522 (Murakami *et al.* 1980a). Filled circles indicate bursts observed during a period when the persistent X-ray flux was high; the burst peak fluxes were then generally high, and variable, and strongly correlated with the burst fluence. Open circles indicate bursts observed when the persistent flux was low, the bursts then had low peak fluxes which did not vary much from burst to burst. (b) 1728−337 (Basinska *et al.* 1984). For weak bursts there is a linear relationship between burst peak flux and fluence, which saturates for stronger bursts. Panels (c) and (d) show that for 1735−444 and 1837+049 there are linear relationships between burst peak flux and burst fluence. (From Lewin *et al.* 1980; Sztajno *et al.* 1983.)

least-squares fitting of the predicted count-rate spectrum to that observed. This fitting leads to a determination of parameters of the spectral model, and an estimate of the quality of the fit in terms of a minimum χ^2 value.

Swank *et al.* (1977) and Hoffman, Lewin and Doty (1977a,b) found that the time-dependent spectra of X-ray bursts are well described by a blackbody spectrum. In making spectral fits, it is customary to subtract the pre-burst persistent emission from the total signal. Since the burst emission may affect the accretion flow, it is not obvious that this is correct. However, it is unclear what the effect of the burst on the persistent flux is. For instance, for bursts which at their peaks reach the Eddington limit (Sect. 4.3.4), one might expect a temporary suppression of the accretion; however, it has been argued (Walker and Meszaros 1989) that, instead of a

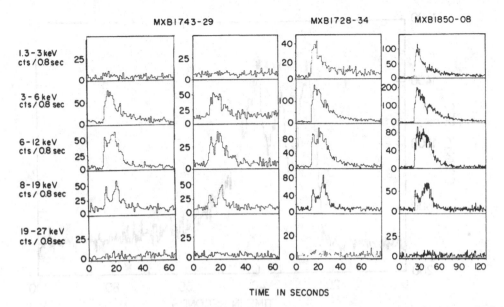

Fig. 4.5. X-ray bursts from three different sources showing energy-dependent double-peaked profiles, indicative of photospheric radius expansion near burst maximum. The X-ray spectrum of MXB 1743−29 (located near the galactic center) is highly cut off below 3 keV due to interstellar absorption. From Hoffman, Cominsky and Lewin (1980).

suppression, the accretion rate is then enhanced due to increased angular momentum loss of the inflowing matter by radiation drag.

The best-fit blackbody temperature can be used to determine the bolometric flux, F_{bol}, in the burst by multiplying the flux by a bolometric correction factor, which is a function of the blackbody temperature alone. From the bolometric flux and the blackbody temperature, one can determine the apparent blackbody radius, R_{bb}, of the burst emitting region (identified with the surface of a neutron star) through:

$$R_{bb} = d(F_{bol}/\sigma T_{bb}^4)^{1/2}. \tag{4.1}$$

Here, T_{bb} is the blackbody temperature, and d is the source distance (see also Eq. (4.2)).

Thus, the result of a typical spectral analysis of an X-ray burst is the variation with time of blackbody temperature, bolometric flux and blackbody radius (Figure 4.7). A useful way to display these results is on a diagram of log F_{bol} versus log kT_{bb}; to adhere to astronomical convention (Hertzsprung–Russell diagram), temperature is plotted horizontally, and increases to the left (Figure 4.8). During burst decay, one finds often that R_{bb} is approximately constant, i.e. in the HR diagram the burst is then represented by points located along a straight line ('cooling track') with slope 4 (notice, this is not the case in Figure 4.8). As we will discuss in more detail in Sect. 4.3.4, in some bursts photospheric radius expansion occurs due to strong radiation pressure. During the expansion phase, the luminosity remains approximately constant. In the HR diagram, the burst is then represented by a nearly horizontal track that runs to the right from the upper part of the cooling track (Figure 4.8).

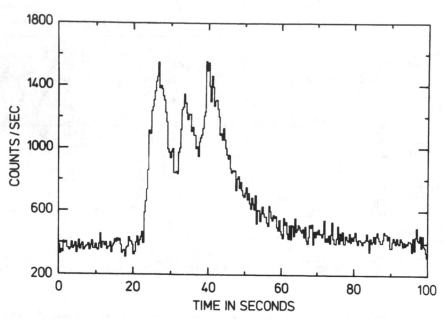

Fig. 4.6. Triple-peaked burst observed with EXOSAT from 1636−536 (from Van Paradijs *et al.* 1986).

It was pointed out by Van Paradijs and Lewin (1985) that if the neutron star in an X-ray burst source is sufficiently hot to give a significant blackbody (or blackbody-like) contribution to the persistent X-ray flux, blackbody radii obtained from this standard burst analysis contain systematic errors which can be very large near the end of the burst when the burst flux reaches very low values. In this case, the burst emission is given by the difference between the emission of two blackbodies; this difference does not have a blackbody energy distribution. Van Paradijs and Lewin showed that when the burst flux is small the blackbody temperature assigned to its spectrum is directly related to the quiescent (prior to the burst) blackbody temperature T_0 (it looks somewhat like a blackbody with $T_{bb} \sim 1.3 T_0$) independent of the burst flux. As a consequence, during the final decay part of the burst, the apparent blackbody radius (Eq. (4.1)) becomes artificially small, and goes to zero as the burst flux goes to zero. This effect was found to be important in a burst observed from the luminous source GX 17+2 (Sztajno *et al.* 1986).

Although blackbody spectra in general provide a good fit to observed burst spectra, evidence for significant deviations from a blackbody has been obtained in some cases (see, e.g., Figure 4.8). Blackbody temperatures kT_{bb} in excess of 3 keV have been observed for several sources; this is substantially higher than the Eddington temperature, in spite of the fact that burst luminosities are not expected to exceed the Eddington limit (Sect. 4.3.4); this indicates that the spectra emitted during bursts mimic those of blackbodies at temperatures higher than the effective temperature. According to Nakamura *et al.* (1989), burst spectra from 1608−522 show high-energy tails when the persistent flux is relatively low. They interpret these tails as a result of Comptonization of the burst emission in a hot plasma surrounding

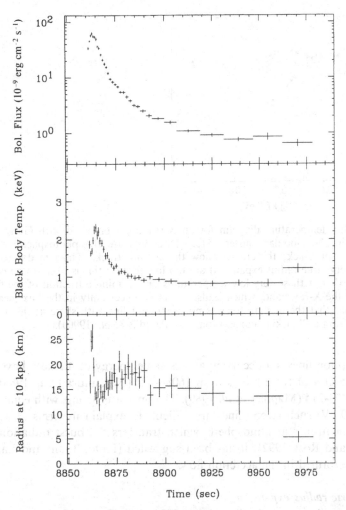

Fig. 4.7. Variation of the bolometric flux, blackbody temperature and blackbody radius during a burst observed with EXOSAT from 1636−536, as obtained from time-resolved blackbody fittings of burst spectra (see text). (Courtesy T. Oosterbroek.)

the neutron star. From a detailed analysis of possible high-energy tails in burst spectra observed from 1636−536, Damen *et al.* (1990) found no evidence for such Comptonization. Systematic deviations from a blackbody spectrum, in the form of temperature dependent 'bumps' have been found in burst spectra observed during one very long X-ray burst from 2127+119 in M15 (Van Paradijs *et al.* 1990a; Figure 4.8); their interpretation is presently unclear.

A particularly interesting non-Planckian feature in burst spectra was found by Waki *et al* (1984), who detected absorption lines in X-ray spectra of four (out of 13) X-ray bursts from 1636−536. These lines were observed only during part of these four bursts. Three of the lines occurred at an energy of 4.1 keV, the fourth at 5.7 keV. Waki *et al.* interpreted these lines as redshifted Lyman α lines of helium-like

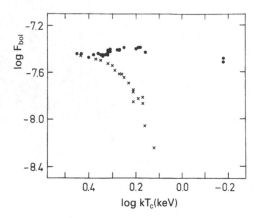

Fig. 4.8. Flux–temperature diagram for an X-ray burst observed with Ginga from 2127+119 in the globular cluster M15. Dots indicate the photospheric expansion/contraction track; the crosses show the cooling track. The two dots on the far right (very large radius expansion) are too low in flux; this is likely to be caused by the fact that at these very low temperatures only a minute fraction of the flux is sampled in the X-ray band, which leads to a large uncertainty in the flux measurement. Notice that the cooling track is not a straight line with slope 4; this reflects deviations from a Planckian curve. From Van Paradijs *et al.* (1990a).

iron atoms. Absorption lines, all occurring at this same energy of 4.1 keV, have also been detected in spectra of three bursts from 1608−522 (Nakamura *et al.* 1988) and one burst from 1747−214 (Magnier *et al.* 1989). The lines are strong, with equivalent widths up to ~500 eV; such strong lines are difficult to explain in terms of atomic transitions in the neutron star atmosphere, which transfers the burst radiation (see also Day, Fabian and Ross 1992); it has been suggested (Pinto, Taam and Laming 1992) that they are formed in the accretion flow.

4.3.4   *Photospheric radius expansion*

Strong burst-like events, lasting up to 1500 s, have been observed (Hoffman *et al.* 1978b; Lewin, Vacca and Basinska 1984; Tawara *et al.* 1984a,b; Van Paradijs *et al.* 1990a) which, because of their peculiar profiles and very long duration, were not initially recognized as type I X-ray bursts.

The events (see Figure 4.9) start with a brief increase of the X-ray intensity (the precursor), which rises rapidly (within a second) and lasts for a few seconds; it is followed by a return to the persistent flux level for a time interval lasting between ~5 and ~10 s. After this, the main part of the event starts: a relatively slow increase of the X-ray intensity first appears in the low-energy band and progressively later X-rays become visible at higher energies. Thus, during the first part of this main event, the X-ray spectrum gradually becomes harder. After the blackbody temperature has reached a maximum value (in excess of kT ~2.5 keV), the main event starts to decay with a gradual decrease of the X-ray flux accompanied by a softening of the spectrum, corresponding to the cooling of a blackbody of approximately constant size, similar to what is observed in a 'normal' type I X-ray burst.

Tawara *et al.* (1984a) and Lewin, Vacca and Basinska (1984) independently pro-

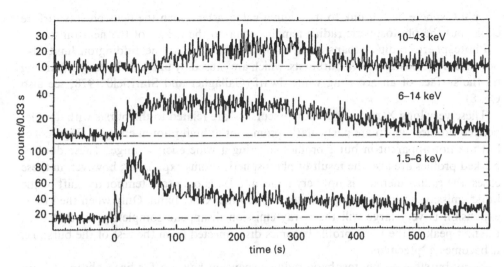

Fig. 4.9. Profile of a very long X-ray burst observed on February 7, 1977, with SAS-3. Before the main event, a precursor occurred which was separated from the main event by ~7 s in the (1.5–6 keV) band and ~60 s in the (10–43) keV band. This profile is the result of very strong radius expansion during the initial phase of the burst (see text). Adapted from Lewin, Vacca and Basinska (1984).

posed that the precursor and the main event are both part of one very energetic type I X-ray burst during which the luminosity becomes so high (i.e. reaches the Eddington limit) that temporarily the atmosphere of the neutron star, through which the burst radiation is transported, expands due to radiation pressure, possibly through the formation of a stellar-wind outflow of material from the neutron star. After the luminosity decreases below the Eddington limit, the photosphere contracts. The observed variation of the blackbody radius during the main parts of these events supports this picture.

Detailed calculations of the effect of a (super-) Eddington luminosity on the structure of a neutron star atmosphere have shown that, during the expansion and contraction of the atmosphere, the luminosity always remains very close (to within 1% or less) to the Eddington limit; the excess luminosity is transformed very effectively into kinetic and potential energy of the extended atmosphere (Hanawa and Sugimoto 1982; Taam 1982; Wallace *et al.* 1982; Ebisuzaki *et al.* 1983; Kato 1983; Paczynski 1983a,b; Paczynski and Anderson 1986; Paczynski and Proszynski 1986). This theoretical result can be used to constrain the mass–radius relation of neutron stars (LVT93).

As the photospheric radius, R_{ph}, increases at a near constant luminosity L, the effective temperature, T_e, will decrease according to $T_e = (L/4\pi\sigma)^{1/4} \times (R_{ph})^{-1/2}$. The end of the observed precursor signals that the radius has become so large and, correspondingly, the temperature so low, that X-rays are no longer emitted. As the photospheric radius slowly starts decreasing (and the temperature starts increasing), X-rays will again be emitted, first at low energies, and later on, as the temperature continues to increase, at higher energies. The radius decrease stops when the photosphere has shrunk to its original value, after which the neutron star surface cools, but the photospheric radius remains equal to the radius of the neutron star.

Photospheric radius expansion at a luminosity close to the Eddington limit also occurs during classical nova outbursts, which are caused by unstable hydrogen burning on the surface of an accreting white dwarf (Gallagher and Starrfield 1978; see also Ch. 8).

These very long X-ray bursts with 'precursors' are related to the bursts with double-peaked profiles, which are particularly prominent at high photon energies (Figure 4.5) but are not apparent in burst profiles covering a wide energy range. These double-peaked profiles are also the result of photospheric radius expansion; however, in these cases, the radius increase is not very large, and leads only to a temporary shift of the burst emission to lower-energy photons within the X-ray band. Only when the radius expansion is very large will the X-ray emission disappear completely; in that case, the first peak of the burst profile becomes disconnected from the rest of the burst, i.e. it becomes a 'precursor'.

X-ray bursts with photospheric radius expansion have so far been observed from 13 sources. Five of these (1724−307, 1746−370, 1820−303, 1850−087 and 2127+119) are located in globular clusters, and for them distances can be estimated using optical observations of the cluster. This allows us to evaluate whether the peak luminosities of bursts with radius expansion can be considered standard candles (cf. Van Paradijs 1978; Lewin 1984). We find that, except for the bursts from 1746−370, these peak luminosities are 3.0 ± 0.6 (1 standard deviation)$\times 10^{38}$ erg s^{-1} (for assumed isotropic burst emission). The low value for 1746−370 (1.0×10^{38} erg s^{-1}) has been discussed in detail by Sztajno *et al.* (1987).

Sugimoto, Ebisuzaki and Hanawa (1984) found that the distribution of burst peak fluxes for 1636−536 is bi-modal, with a gap between $F_{max} \sim 4.0 \times 10^{-8}$ and $\sim 6.5 \times 10^{-8}$ erg cm^{-2} s^{-1}. Below the gap, the F_{max} values range over a factor ~ 4; the peak fluxes above the gap are consistent with a single value. All the bursts above the gap (but none below the gap) show evidence for photospheric radius expansion. These results have been confirmed by subsequent observations (see Fujimoto *et al.* 1987b, 1988; Damen 1990). Sugimoto *et al.* proposed that the peak flux above the gap and the upper bound to the distribution below the gap correspond to the Eddington limit for hydrogen-poor matter, and for matter with cosmic composition, respectively; these Eddington limits are expected to be in the ratio ~ 1.7, consistent with the observed ratio of the corresponding peak fluxes. They suggested that during the radius expansion the upper hydrogen-rich layers are ejected from the neutron star surface, thereby exposing the hydrogen-poor layers which were involved in the nuclear processing. There is a finite chance that a radius expansion burst has a peak flux that lies in the gap (Fujimoto *et al.* 1987b), but this requires some fine tuning of the conditions in the outer layers. (Damen (1990) found that one of 61 bursts observed from 1636−536 with EXOSAT has a peak flux inside the gap.) The distribution of peak fluxes for 1636−536, in particular the existence and location of the gap, has not been observed to change over the years; this indicates that any anisotropy of the burst emission remains constant.

The double-peaked radius expansion bursts should not be confused with the small number of bursts observed from 1636−536 which showed two (in one case even three) peaks in their bolometric profiles (see Sect. 4.3.2, especially Figure 4.6). These bursts had peak luminosities substantially less than the Eddington limit, and they showed no evidence for radius expansion. Melia (1987) suggested that these multi-peaked profiles are caused by diffuse scattering of the burst radiation in a burst-induced accretion disk corona. However, Penninx, Van Paradijs and Lewin (1987) showed this is not the case; it is likely that these profiles reflect variations in the rate of generation or release of nuclear energy.

4.3.5 *Burst intervals*

Burst intervals can be regular or irregular on time scales of hours to days; they range from ~5 min to days; burst activity can stop altogether for periods from days to months. Very regular burst behavior has been observed from, e.g., 1658−298 (Lewin 1977), 1820−303 (Clark *et al.* 1977; Haberl *et al.* 1987) and 1323−619 (Parmar *et al.* 1989).

The burst occurrence rate is sometimes (but not always) related to the level of persistent X-ray emission. Clark *et al.* (1977) found that the persistent flux of 1820−303 increased by a factor of ~5 while the burst intervals gradually decreased by ~50%. The persistent flux continued to increase, and the bursts stopped completely. According to Priedhorsky and Terrell (1984), the X-ray flux of 1820−303 shows a long-term (176 days) periodic variation; X-ray bursts have only been observed when the source was in a low state (Vacca, Lewin and Van Paradijs 1986; Haberl *et al.* 1987; Stella, Priedhorsky and White 1987). The transient source 1658−298 produced bursts at very regular intervals of ~2.5 hours when the persistent flux was less than $\sim 5 \times 10^{-11}$ erg cm^{-2} s^{-1} (Lewin, Hoffman and Doty 1976). When the persistent flux was $\sim 2 \times 10^{-9}$ erg cm^{-2} s^{-1}, no bursts were observed (Lewin *et al.* 1978; Share *et al.* 1978). Similar behavior has been observed for the bright source GX 3+1, which was found in a burst active state, when its persistent X-ray flux was relatively low (Makishima *et al.* 1983).

Bursts from the transient sources Cen X-4, 1608−522, Aql X-1 and 0748−673 have only been observed when these sources were in the bright state (Matsuoka *et al.* 1980; Murakami *et al.* 1980a; Koyama *et al.* 1981; Gottwald *et al.* 1986; Czerny, Czerny and Grindlay 1987; Nakamura *et al.* 1989). However, this does not contradict the above-mentioned anti-correlation between burst occurrence and persistent luminosity. In the case of Cen X-4 and Aql X-1, X-ray bursts occurred when the outburst had already decayed by a substantial factor. In the case of 0748−673 and 1608−522 the persistent flux remained high for a long time after the transient outburst started; when the type I X-ray bursts were observed, the persistent X-ray luminosity was at least a factor of ten below the Eddington limit.

In several well studied sources, large variations in the observed burst intervals occurred that were apparently unrelated to variations in the rate of the accretion that feeds the bursts (e.g., Ser X-1: Li *et al.* 1977, Sztajno *et al.* 1983; 1735−444: Lewin *et al.* 1980; Van Paradijs *et al.* 1988b; 1728−337: Basinska *et al.* 1984; 1636−536: Lewin *et al.* 1987). Very irregular burst behavior tends to occur in sources with relatively high persistent X-ray luminosities (Sect. 4.3.7); this may be related to the fact that these persistent luminosities are close to a critical value (near half the Eddington luminosity), above which X-ray bursts do not occur (as expected from some thermonuclear-flash models – Sect. 4.5). In 1735–444, there is a tendency for the bursts to come in clusters (Lewin *et al.* 1980; Van Paradijs *et al.* 1988b), and it is perhaps possible that in this relatively luminous source the bursts themselves create conditions in the accreted material that quench further burst activity for some time.

Burst intervals as short as $\sim$10 min have been observed from 0748–673 (Gottwald *et al.* 1986), 0836–429 (Aoki *et al.* 1992), 1608–522 (Murakami *et al.* 1980b), 1636–536 (Ohashi 1981; Pedersen *et al.* 1982b), 1705–440 (Langmeier *et al.* 1987), 1743–28 (Lewin *et al.* 1976c), and 1745–248 (Inoue *et al.* 1984). According to current models of thermonuclear flashes (Sect. 4.5), these intervals are much too short to replenish, through accretion, a sufficient amount of nuclear fuel to account for the second burst. Thus, one requires the presence of a reservoir of nuclear fuel which, having survived the previous thermonuclear flash, can be prematurely rekindled (Sect. 4.5; Hanawa and Fujimoto 1984; Fujimoto *et al.* 1987b, 1988).

4.3.6   *Burst energy and burst intervals*

For the simplest type of thermonuclear-flash models, one would expect that there is a correlation between the time interval since the previous burst and the integrated burst energy: the longer this time interval is, the larger is the amount of nuclear fuel available for the burst.

Hoffman, Lewin and Doty (1977b) were the first to find a possible correlation (for 1728–337) between the burst fluence, E_b, and the time interval, $\Delta\tau$, since the previous burst: the lower the observed burst frequency was, the larger was the fluence. This was confirmed by later, more extensive, observations of this source (Basinska *et al.* 1984). Lewin *et al.* (1980) found evidence for a correlation between the average burst interval and the burst fluence for 1735–444. Hoffman, Cominsky and Lewin (1980) suggested that the very strong bursts (with double-peaked profiles; see Figure 4.5) came after long waiting times. For optical bursts from 1636–536, Pedersen *et al.* (1982b) found a clear correlation between the fluence and the interval since the previous optical burst.

Extensive studies of the $(E_b, \Delta\tau)$ relation have been made with EXOSAT for 1636–536, 1735–444 and 0748–673 (for references, see above). For all three sources, one finds globally that the longer the burst interval, the higher is the burst fluence; because of the correlation between burst fluence and peak flux, the same is also true for the latter quantity (Figure 4.10). The scatter around these relations is, however, substantial. For time intervals below a value that ranges between $\sim$1 h for 1735–444 and $\sim$6 h for 1636–536, the $(E_b, \Delta\tau)$ relation is approximately linear; for longer intervals, the burst fluence does not increase much. Gottwald *et al.* (1987) pointed out that for 0748–673 the linear relation between E_b and $\Delta\tau$ leads to a positive value for E_b ($2 \pm 0.5 \times 10^{-8}$ erg cm^{-2}) at $\Delta\tau = 0$; this reflects the occurrence of X-ray bursts after very short waiting times (Sect. 4.3.5).

Bursts that come after very long waiting times tend to show photospheric radius expansion, and some of the flattening of the $(E_b, \Delta\tau)$ relation may therefore be explained in terms of the energy consumed in mass ejection. The flattening may also be explained by a growing leak of nuclear fuel, e.g. due to steady nuclear burning during the intervals between bursts (Fujimoto *et al.* 1987b; Lewin *et al.* 1987). For 1636–536 it may be possible to explain the energy leak simply in terms of continuous

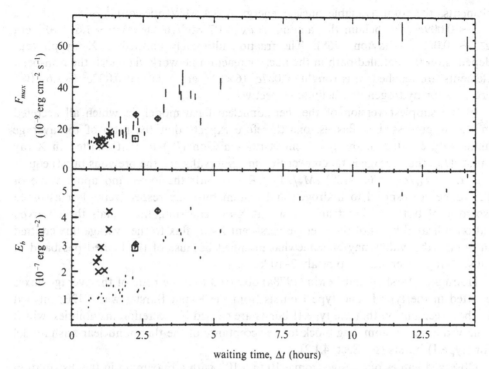

Fig. 4.10. Dependence of burst peak fluxes (top panel) and burst fluences (bottom panel) on the waiting time since the previous burst, for 1636−536. Crosses and diamonds indicate bursts that occurred during 'island' states in 1983 and 1985, respectively; the square indicates a burst that had a triple-peaked profile (see Figure 4.6). Adapted from Damen (1990).

stable hydrogen burning through the CNO cycle (Fujimoto *et al.* 1987b). However, this explanation is unlikely to work for 1735−444 (Van Paradijs *et al.* 1988b). For this source, the α-value (Sect. 4.3.7) for bursts which come after an interval of less than an hour (α ∼ 250) are consistent with the values expected for thermonuclear flashes in which only helium participates. The very high α-values for the bursts that occurred after very long time intervals indicate that prior to these bursts both hydrogen and helium are burning stably (cf. Van Paradijs *et al.* 1979a). The rapid transition to very long burst intervals (in which helium is stably burnt) is not accounted for by present models of thermonuclear flashes on the surface of a neutron star (cf. Fujimoto *et al.* 1987a,b).

4.3.7   *Burst energetics and the persistent flux*

The fuel for the thermonuclear flashes that give rise to X-ray bursts was accreted on the surface of the neutron star prior to the burst. The accretion of one gram of material leads to the emission of an amount of energy equal to $\epsilon_G = GM_*/R_*$, where M_* and R_* are the mass and radius of the neutron star. For 'canonical' values $M_* = 1.4 M_\odot$ and $R_* = 10$ km, one finds $\epsilon_G = 1.8 \times 10^{20}$ erg g^{-1} = 180 MeV(nucleon)$^{-1}$ = $0.2c^2$. The amount of energy liberated in thermonuclear burning depends on the composition of the fuel. In the transformation of pure hydrogen into iron-peak elements, the total available nuclear energy is 8.4 MeV(nucleon)$^{-1}$ $\sim 8 \times 10^{18}$ erg g^{-1} $\sim 0.009c^2$; for helium this amount is $\epsilon_N = 1.7$ MeV(nucleon)$^{-1}$ $\sim 1.6 \times 10^{18}$ erg g^{-1} $\sim 0.002c^2$ (Clayton 1968). The fraction ultimately emitted in X-ray photons depends on the detailed path in the nuclear-reaction network via which the iron-peak elements are reached; it is roughly $0.007c^2$ (6×10^{18} erg g^{-1}) and $0.002c^2$ ($\sim 1.6 \times 10^{18}$ erg g^{-1}) for hydrogen and helium, respectively.

In the simplest version of the thermonuclear-flash model, in which all accreted matter is processed in flashes, one therefore expects that the ratio of the average luminosity emitted in the persistent X-ray emission (L_p) to that emitted in X-ray bursts (L_b; the average is taken over the time interval since the previous burst) equals $L_p/L_b = \epsilon_G/\epsilon_N \sim (25\text{--}100)(M_*/M_\odot)/(R_*/10 \text{ km})$; the lower and upper value of the range correspond to hydrogen and helium burning, respectively. For assumed isotropy of both the burst and persistent X-ray emission, this equals the observed ratio (indicated by α) of the average persistent X-ray flux to the average flux emitted in bursts (this value may be somewhat modified because of the limited passband of most X-ray observations, typicaly 2–10 keV).

Hoffman, Marshall and Lewin (1978a) observed that the ratio of the average fluxes emitted in the type II and type I bursts from the Rapid Burster was ~ 120; this led to their recognition that the type II bursts are caused by accretion instabilities, which removed a major stumbling block to the acceptance of the thermonuclear-flash model for (type I) bursts (see Sect. 4.1.2).

Observed values of α range from ~ 10 to $\sim 10^3$, with a maximum in the distribution of α-values near 10^2, roughly as expected for the thermonuclear-flash model. The α-ratio varies with the persistent luminosity of the source (Sect. 4.3.8).

The observed values of α can differ substantially from the above estimates for several reasons. Prior to the occurrence of an X-ray burst, steady burning of the accreted fuel may occur, in which case the observed value for α may reach arbitrarily high values (see, e.g., Van Paradijs et al. 1988b). In case part of the nuclear fuel survives an X-ray burst, fuel is available for a subsequent burst, and low values for α are observed (Sect. 4.3.5).

4.3.8   Dependence of burst properties on accretion rate

As mentioned in Sect. 4.3.2 and 4.3.5, in a number of sources a correlation has been observed between burst profiles (and intervals) and the persistent X-ray flux F_p. A striking example is the transient source 0748−673, for which Gottwald et al. (1986, 1987) obtained the following results:

(1) The burst frequency was anti-correlated with F_p; it varied from 0.75 h^{-1} to 0.12 h^{-1} as F_p varied from $\sim 3 \times 10^{-10}$ to $\sim 10^{-9}$ erg cm^{-2} s^{-1}. A similar anti-correlation has been observed in some other transients (e.g., 1658−298, see Lewin et al. 1978), but not in others (e.g., 1608−522, see Murakami et al. 1980a). This anti-correlation may be similar to that observed in some luminous persistent sources, such as 1820−303 (Clark et al. 1977) and GX 3+1 (Makishima et al. 1983); in these sources burst activity ceased altogether when the persistent flux was above a certain 'critical' level.

(2) The average ratio, α, of the total energy emitted in the persistent flux to that emitted in bursts (Sect. 4.3.7) increased strongly with F_p, from $\sim$12 to $\sim$75 over the range given above.

(3) Over the above range in F_p, the average burst fluence E_b increased by a factor of two, the maximum peak flux, F_{max}, increased by a factor of $\sim$10, and the average burst duration $\tau = E_b/F_{max}$ decreased from $\sim$25 s to $\sim$6 s.

(4) The blackbody radius of the burst emitting region depends strongly on F_p.

An anti-correlation between burst duration τ and the persistent flux was also found for 1608−522 by Murakami *et al.* (1980a). Van der Klis *et al.* (1990) found for bursts from 1636−536 and 1705−440 that τ is strongly correlated with the spectral state of the persistent X-ray emission (as indicated by the position in the X-ray color–color diagram; see Figure 4.11). They found a similar result for the blackbody radii obtained for these bursts, and they interpreted this as evidence that it is the mass accretion rate that controls the burst duration and the blackbody radius (e.g. through its influence on the structure of the outer envelope of the neutron star, or the angular dependence of the X-ray emission); if their interpretation is correct, the observed persistent X-ray flux is not always a good measure of the accretion rate.

Van Paradijs, Penninx and Lewin (1988a) investigated the relation between important burst parameters and the persistent X-ray luminosity, which they took to be a measure of the accretion rate. Guided by the notion that the peak luminosity of X-ray bursts that show photospheric radius expansion equals the Eddington luminosity (Sect. 4.3.4), they assumed this peak luminosity is a standard candle (see also Van Paradijs 1978; Van Paradijs *et al.* 1979a); this assumption is probably correct to within a factor of two for most (if not all) sources. This makes it possible to define a relative distance scale and leads to the following expression for the persistent X-ray luminosity L_p:

$$L_p = L_0(\zeta_p/\zeta_b)\gamma. \tag{4.2}$$

Here, ζ_p and ζ_b are anisotropy factors for the persistent and burst emission, respectively, and γ is the ratio of the observed persistent flux, F_p, to the net peak flux (not including the persistent emission), F_{re}, of bursts with radius expansion. Their results have been summarized in Figures 4.12 and 4.13, which show the γ-dependence of the burst duration $\tau (= E_b/F_{max})$ and the α-value, respectively.

From Figure 4.12, it appears that τ is anti-correlated with γ over the entire observed range of γ. It varies between 30 s and a few minutes at $\log \gamma \sim -2$, and is $\sim$5 s at $\log \gamma \sim -0.5$. The average (γ, τ) relation is rather tight and encompasses both the bursts with radius expansion (Sect. 4.3.4) and the more common X-ray bursts quite well. There is no evidence for a clear separation of bursts into two groups, one with long and another with short durations. In the framework of thermonuclear-flash models, this decrease of τ with γ indicates that hydrogen becomes less important in the energetics of the bursts as the mass accretion rate increases (see Sect. 4.5).

The average values of α are strongly correlated with γ, with $\log \alpha$ varying from $\sim$1 to >3 for $\log \gamma$ varying from ~ -2 to -0.5 (Figure 4.13); this confirms the

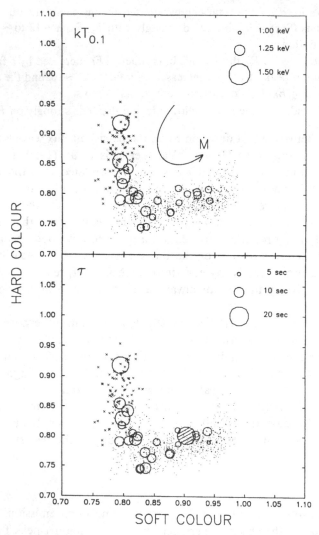

Fig. 4.11. X-ray color–color diagrams (small symbols) for 1636−536, with X-ray bursts superposed (large symbols). Each color measurement corresponds to 200 s of data. The hard color is the ratio of the count rates in the 6.1–20.5 keV and 4.5–6.1 keV bands; the soft color is that of the 2.9–4.5 and 0.9–2.9 keV bands. Crosses and dots indicate 'island' and 'banana' states, respectively, as determined from the X-ray variability characteristics. The position of a large circle indicates the average colors of the persistent emission just before and after the burst; the size of the circle indicates the temperature $kT_{0.1}$ when the burst flux is 10% of the Eddington flux (top panel) or the burst duration (bottom panel). The hatched circle in the lower panel indicates the burst with a triple-peaked profile; apart from this burst there is a strong correlation between the source state and both the temperature $kT_{0.1}$ and burst duration. The direction in which $\dot{M}$ is believed to increase is indicated. From Van der Klis *et al.* (1990).

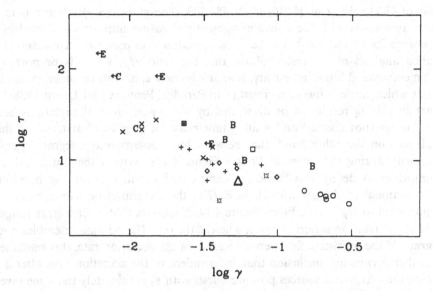

Fig. 4.12. Relationship between burst duration τ and the persistent X-ray luminosity, represented by the ratio, γ, of the persistent X-ray flux to the Eddington flux for burst sources that produced bursts with photospheric radius expansion. The symbols represent different sources as follows: 0748–676 ($\times$); 1516–569 ($\square$); 1608–522 (B); 1636–536 ($+$); 1715–321 (E); 1724–307 (C); 1728–337 ($\diamond$); 1735–444 ($\bigcirc$); 1746–370 ($\triangle$); 1820–303 ($\bowtie$). From Van Paradijs, Penninx and Lewin (1988a).

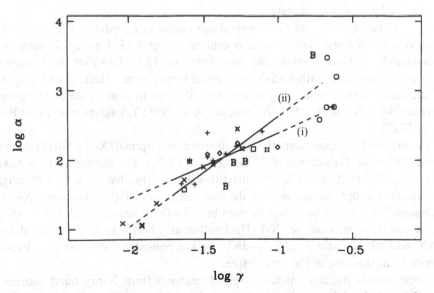

Fig. 4.13. Relation between log γ and log α for burst sources that have produced X-ray bursts with radius expansion. The symbols are explained in the caption of Figure 4.12. The figure contains two straight lines. (i) represents a least-squares fit to the data in the interval $-1.7 < \log \gamma < -1.0$. (ii) is the long axis of the dispersion ellipse of a bivariate (log α, log γ) distribution of the same data points. From Van Paradijs, Penninx and Lewin (1988a).

result of Gottwald *et al.* (1986) for 0748−673. The global (α, γ) relation is in qualitative agreement with the above-mentioned decreasing importance of hydrogen in the energetics of the bursts as the mass accretion rate increases. However, a more detailed analysis of the results shows that the ratio α/γ, which is proportional to the time-averaged burst luminosity, is approximately constant, i.e. independent of the rate at which nuclear fuel is accreted (Van Paradijs, Penninx and Lewin 1988a). This rather surprising result can be illustrated by a comparison of the relations between burst energy (not fluence) and waiting time since the previous burst, $\Delta\tau$, on the one hand and, on the other hand, that between burst energy and integrated persistent luminosity during that interval $\Delta\tau$ (see Figure 4.14). Within the framework of the assumptions made by Van Paradijs, Penninx and Lewin (1988a), the burst energy is proportional to the quantity $U_b = E_b/F_{re}$; the integrated persistent luminosity is proportional to $U_p = \gamma\Delta\tau$. From Figure 4.14, it appears that, over a large range in γ, the $(U_b, \Delta\tau)$ relation is fairly unique, whereas the (U_b, U_p) relation resembles a scatter diagram. If the persistent flux were a measure of the accretion rate, this would lead to the rather surprising conclusion that, independent of the accretion rate, after a given waiting time $\Delta\tau$, burst sources produce bursts with approximately the same (average) energy. This would suggest that continuous stable burning of a sizeable fraction of the accreted nuclear fuel occurs not only in luminous X-ray burst sources (see Sect. 4.3.6), but in less luminous sources as well.

4.3.9   *Observations at other wavelengths*

4.3.9.1   *Optical bursts*

Between 1977 and 1980, several campaigns of coordinated optical and X-ray observations of burst sources were organized by the SAS-3 group. During the 1977 'burstwatch', optical observations were made of 1837+049 (Ser X-1) when X-ray bursts were detected with SAS-3. No optical bursts were detected, with upper limits to the ratio of the fluence E_{opt} in any optical burst to that in the X-ray burst (E_X) between $\sim 5\times 10^{-5}$ and $\sim 10^{-4}$ (Abramenko *et al.* 1978; Takagishi *et al.* 1978; Bernacca *et al.* 1979).

During the 1978 campaign, the first simultaneous optical/X-ray burst was detected, from 1735−444 (Grindlay *et al.* 1978; see Figure 2.26). The fluence in the optical burst was $\sim 2\times 10^{-5}$ that in the X-ray burst; this is too large by ~ 6 orders of magnitude to explain the optical emission as the low-energy tail of the blackbody X-ray burst emission. The optical burst was delayed by ~ 3 s (McClintock *et al.* 1979). An optical burst was detected from Ser X-1 (Hackwell *et al.* 1979), which was delayed by 1.4 ± 0.5 s, and whose ratio E_{opt}/E_X $(\sim 3\times 10^{-6})$ was consistent with the non-detection of bursts from this source the year before.

These results indicated that all optical emission from X-ray burst sources is the result of reprocessing of X-rays in material, within a few light seconds from the X-ray source, which is enhanced during an X-ray burst (for reviews, see also Ch. 2). Plausible sites for this reprocessing are an accretion disk around the neutron star and the hemisphere of the secondary star facing the X-ray source (insofar as it is not shielded by the accretion disk). The optical signal will be delayed (and also smeared) as a result of a range of travel-time differences between the X-rays that go directly

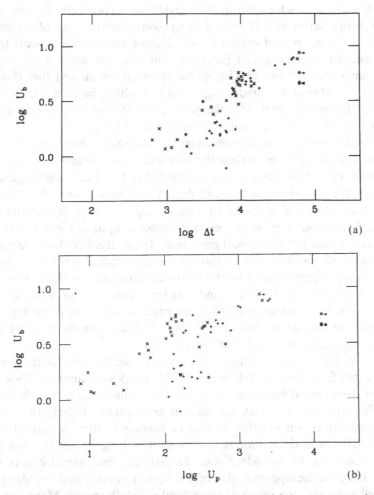

Fig. 4.14. (a) Dependence of the parameter U_b, which is a measure of the integrated burst energy (not fluence) on the waiting time Δt since the previous burst, for 0748–676 (×), 1636–536 (+), and 1735–444 (◊), which have log γ values in the range between −2.0 and −0.6. (b) Dependence of the parameter U_b on $U_p = \gamma \Delta t$ for the same sources as in (a). U_p is a measure of the integrated persistent X-ray luminosity (not flux) in the interval since the previous burst. From Van Paradijs, Penninx and Lewin (1988a).

to the observer and those that first hit the disk, become transformed into optical photons, and then go to the observer (see Figure 2.27). Delay and smearing due to radiative processes are not very important (Pedersen *et al.* 1982a). According to this picture, the X-ray bursts serve as a probe; they illuminate the surroundings of the neutron star, which reveal their presence by optical radiation. By comparing the changes in the X-ray and optical signals, one can deduce information about these surroundings.

Very extensive simultaneous optical/X-ray burst observations have been made of 1636–536, which was regularly monitored with Hakucho and ESO telescopes during periods of ~6 weeks and ~1 month in 1979 and 1980, respectively. In this period, a total of 69 X-ray bursts and 41 optical bursts were detected, ten of which were detected both in the X-ray and optical bands. During none of the optical (X-ray) bursts for which an X-ray (optical) burst was not observed were X-ray (optical) observations simultaneously made. The optical observations showed that there is a strong correlation between E_{opt} and $F_{opt,max}$, both of which, in turn, are positively correlated with the time interval since the previous burst; two of the optical bursts occurred at an interval of only 5.5 min (Pedersen *et al.* 1982b).

During the 1979 observations, five coincident optical/X-ray bursts were detected (Pedersen *et al.* 1982a); for three of them the data were of sufficient quality to make a detailed analysis. These observations were interpreted in terms of a simple geometric model in which the optical emission (both during and outside bursts) is that of a (single-plate) blackbody, and is caused by reprocessing of X-rays in material in the vicinity of the X-ray source. (One of these simultaneous optical/X-ray bursts, from 1636−536, was observed in three optical passbands. The analysis of this time-resolved color information shows that X-ray reprocessing into optical light is reasonably described by a single-temperature blackbody model (Lawrence *et al.* 1983b).) From their analysis, Pedersen *et al.* (1982a) found that the reprocessor has a linear size of ~1.5 light seconds, independent of detailed assumptions about its geometry. These results put limits on the size of the accretion disk which are consistent with a binary separation corresponding to the 3.8 h orbital period.

Matsuoka *et al.* (1984) analyzed the six best cases of the ten coincident optical/X-ray bursts detected from 1636−536 during the 1979 and 1980 campaigns. They found that for five of these optical bursts the delay times were consistent with a single value near 2.5 s. However, for one burst, the delay is significantly smaller; this suggests that the reprocessing region is either moving or varying in size. A natural way to account for variations in delay time is to assume that a significant fraction of the reprocressing occurs in the secondary star. In that case, one would expect to see a correlation between the apparent size of the optical emitter and the delay time (both of which vary with the phase of the orbital cycle); however, Matsuoka *et al.* (1984) did not find such a correlation. Using an orbital ephemeris for 1636−536, based on optical brightness variations, covering the period between 1980 and 1988, Van Paradijs *et al.* (1990b) find no evidence for a relation between delay time and orbital phase. Also, the variation with orbital phase of the ratio of optical to X-ray burst fluences (and peak fluxes) does not follow the average optical light curve. These results suggest that substantial long-term changes occur in the spatial distribution of reprocessing material in the binary system.

Mason *et al.* (1980) observed an optical burst from 1254−690; X-ray bursts from this source were observed later with EXOSAT (Courvoisier *et al.* 1986). An optical burst was detected from the recurrent transient burst source Aql X-1 when it was in outburst (Van Paradijs, Pedersen and Lewin 1981).

4.3.9.2 *Radio and infrared observations*

Radio observations were made of five X-ray burst sources during the 1977 coordinated burst campaign; no radio bursts were detected (Johnson *et al.* 1978; Ulmer *et al.* 1978; Thomas *et al.* 1979).

Kulkarni *et al.* (1979) and Jones *et al.* (1980) reported the detection of infrared bursts from 1730–335 (Rapid Burster). It is unclear whether these events are real (cf. Apparao and Chitre 1980). Sato *et al.* (1980) did not detect infrared bursts from this object. Radio bursts from the Rapid Burster were reported; no X-ray bursts were seen during simultaneous Hakucho observations (Hayakawa 1981). Based on extensive X-ray, radio and infrared observations of the Rapid Burster made during 1979 and 1980, Lawrence *et al.* (1983a) concluded that (i) it is unlikely that the previously reported radio bursts from this source are real; (ii) the reported infrared bursts are harder to dismiss, but difficult to reconcile with the many null observations. For more information on radio emission from burst sources, see Ch. 7.

4.3.10 Rapid variability during X-ray bursts

In a number of cases, rapid variability (time scales less than a second) has been observed during X-ray bursts; in some cases, the variations were coherent (i.e. coherence limited by the length of the burst). It is unlikely that these variations can be explained by a single mechanism (see Livio and Bath (1982) for a discussion of models).

Mason *et al.* (1980) found a coherent brightness modulation during an optical burst they observed from 1254–690, at a frequency of 36.40 Hz. We note that the high frequency of this modulation makes it unlikely that the optical burst was associated with a type I X-ray burst; if the optical burst originates from reprocessing of X-rays in a region a few light seconds across the high-frequency oscillation is expected to be washed out by light travel time effects.

Sadeh *et al.* (1982) reported the detection of 12 ms oscillations just before and during one of four X-ray bursts from 1728–337, which they observed with HEAO-1. The period of the oscillations showed a drift from 12.254 ms before the burst to 12.244 ms during the burst. We note that the significance level of this (drifting) periodic signal given by Sadeh *et al.* (99.7%) is likely to be overestimated since they did not take into account the presence of irregular variability in the signal (as found during this same burst by Hoffman *et al.* 1979).

Murakami *et al.* (1987) detected oscillations with a period of 0.65 s during the phase of photospheric radius expansion of a strong X-ray burst they observed from 1608–522. The oscillation of the low-energy X-ray count rate is anti-correlated with that at high-energy X-rays. Murakami *et al.* conclude that these oscillations are caused by oscillations in the photospheric radius which occur at an approximately constant bolometric luminosity. Possibly, similar oscillations at a much longer period ($\sim$10 s), which lasted for only a few cycles, were found during the phase of photospheric radius expansion of a burst from 2127+119 (Van Paradijs *et al.* 1990a).

Schoelkopf and Kelley (1991) detected a 7.6 Hz oscillation at a 4.1σ confidence level during an X-ray burst from Aql X-1 observed with Einstein. They proposed that the oscillations reveal the spin period of the neutron star through a relatively small non-uniformity of the surface brightness across the neutron star.

4.4 Mass–radius relation of neutron stars

Information on masses and radii of neutron stars can, in principle, be obtained from observations of X-ray bursts. If identifiable emission or absorption lines were detected from the surface of a neutron star, we would have a direct measurement of the gravitational redshift from the surface, and thereby a measurement of the ratio of the mass M and the radius R_* of the neutron star (see Eq. (4.4)).

Information on mass and radius can also be obtained from the burst continuum X-ray spectra (Goldman 1979; Van Paradijs 1979; Hoshi 1981; Marshall 1982). We mention here only the basic idea; all details and limitations can be found in LVT93.

If the burst spectra are approximately Planckian, the observed flux, F_∞, will vary approximately as the fourth power of the *observed* blackbody temperature $T_{c\infty}$ (the subscript c indicates 'color'; the subscript ∞ indicates that the observation is made at a very large distance from the neutron star). For an assumed uniform spherical blackbody emitter of radius R_∞, at a distance d, we have:

$$L_{b\infty} = 4\pi (R_\infty)^2 \ \sigma (T_\infty)^4 = 4\pi d^2 F_{b\infty}. \tag{4.3}$$

Here, L stands for luminosity, and the subscripts b and e stand for 'burst' and for 'effective', respectively. The quantity $F_{b\infty}$ is the bolometric flux. If the blackbody temperature is measured (from the shape of the X-ray spectrum) at a given time in the burst, and also the associated bolometric flux, one can find R_∞ if the source distance is known. Since the energy of each photon, and also the rate at which photons arrive, undergo gravitational redshift, the observed value $T_{c\infty}$ and $L_{b\infty}$ differ from the corresponding values at the neutron star surface. As a result,

$$R_\infty = R_*(1 + z_*). \tag{4.4}$$

Here, $1 + z_*$ is the gravitational redshift factor given by

$$1 + z_* = [1 - 2GM/(R_* c^2)]^{-1/2}. \tag{4.5}$$

Here, M is the gravitational mass of the neutron star, and R_* is its radius as measured by a local observer on the neutron star surface. Thus, a measurement of R_∞ leads to information about the radius R_* and the mass of the neutron star, and thereby of the equation of state of neutron star matter.

Various methods have been used. When the bursts are so energetic that there is substantial photospheric radius expansion (and thus the gravitational redshift varies with time), knowledge of the source distance is not even required (LVT93).

In practice, there are several major stumbling blocks; they are all described in LVT93. The main factor which presently prevents the derivation of useful constraints on masses and radii of neutron stars is our limited understanding of the spectra of X-ray bursts. The observed blackbody temperature T_c is larger than T_e by a factor of ~ 1.5, but present models of neutron star atmospheres are inadequate to allow for a sufficiently accurate correction. Progress in this respect may depend on the availability of improved spectral resolution of X-ray burst observations (such as expected from the Japanese satellite ASCA), and from detailed theoretical studies of neutron star atmospheres, including the possible effects of magnetic fields.

4.5 Theory of type I X-ray bursts

4.5.1 Introduction

Theoretical models for X-ray bursts involve the accretion of matter onto a compact object. The mechanisms deemed responsible for the outburst are classified into two types. One mechanism derives the energy for the outburst from the nuclear energy stored in the accreted matter, and the other mechanism relies on the release of gravitational potential energy associated with the accretion process. There is now compelling evidence that neutron stars are involved in the X-ray burst phenomenon and that the type I X-ray bursts are produced by a thermonuclear shell flash instability in the star's surface layers. The mechanism involving an instability in the accretion flow, although not responsible for type I X-ray bursts, is very likely responsible for the type II bursts seen in the Rapid Burster.

The basic physical picture of the thermonuclear-flash model involves the accretion of nuclear fuel, in the form of hydrogen and helium, from the neutron star's stellar companion. As the accreted layer is built up on the neutron star surface, the high pressures exerted by the weight of this matter lead to its fusion to iron nuclei and eventually to neutrons at densities $\gtrsim 10^{11}$ g cm^{-3}. This transformation to iron is likely to be explosive in a wide range of circumstances because the degree of electron degeneracy in these layers is high. For temperature sensitive nuclear reaction rates, these layers are susceptible to a thermal instability driven by the exothermic reactions. The nuclear energy released is eventually transported to the surface, which can give rise to the X-ray burst.

The thermonuclear model has been remarkably successful in reproducing the basic features of the X-ray burst phenomenon. Of note are the short rise time scales ($\sim$ 1 s), the recurrence time scales ($\sim$ hours), the energetics ($\sim 10^{39}$–10^{40} erg), the ratio of burst to accretion luminosities, and the spectral softening during burst decay. It is this success that provided the strongest theoretical evidence that neutron stars are involved in the phenomenon. Furthermore, the theory suggested that the magnetic fields of these neutron stars are weak ($\lesssim 10^{10}$–10^{11} G) for, otherwise, the thermonuclear instabilities would be suppressed, for the range of mass accretion rates inferred for X-ray burst sources, when matter is funnelled to the magnetic poles (Joss 1978; Taam and Picklum 1978; Joss and Li 1980).

During the past decade, new theoretical insights have been gleaned from studies of the neutron star's accreted envelope in response to repeated thermonuclear shell flashes (Fujimoto *et al.* 1985; Woosley and Weaver 1985), of nuclear burning when nonthermal equilibrium effects are included (Fushiki and Lamb 1987), of the possibility that heavy elements can be destroyed in the accreted matter via nuclear spallation reactions (Bildsten, Salpeter and Wasserman 1992), and of X-ray bursts as probes of the thermal state of the neutron star interior (Fushiki *et al.* 1992). We review these more recent investigations and discuss their impact on the interpretation of the observational data.

4.5.2 Nuclear processes on accreting neutron stars

The nuclear reactions that take place in the accreted layer of a neutron star have been discussed in detail by Taam (1985). Hydrogen burning by itself cannot produce an X-ray burst since the rate at which protons are burned is limited by the β decays associated with the weak interaction process of transforming a proton into a neutron (Joss 1977; Lamb and Lamb 1978). Although the fusion of protons to helium by the pp chain and the CNO cycle cannot provide a sufficiently rapid energy release, it can profoundly affect the nuclear burning development in the accreted layer, since the energy liberated from hydrogen burning can heat the neutron star envelope to such temperatures that helium and other nuclear fuels (not inhibited by the weak interaction process) can ignite and burn rapidly as first pointed out by Taam and Picklum (1978, 1979). In this regime, the nuclear burning development is described by nonstandard nuclear flows (see Wallace and Woosley 1981, 1985; Hanawa, Sugimoto and Hashimoto 1983) and involves a series of (p,γ) reactions followed by positron decays, which skews the nuclear flow toward the proton drip line (called the rapid proton or rp process). Modifications to this flow occur at the high densities and temperatures characteristic of a neutron star envelope, where (α,p) reactions can compete with the positron decays. In the latter case, the evolution to iron nuclei is accomplished via a series of (α,p) and (p,γ) reactions.

In contrast to hydrogen, the ignition of helium can give rise to a thermonuclear instability leading to an X-ray burst since helium burns via the highly temperature sensitive triple-alpha reaction. This burning occurs under conditions of high electron degeneracy, and the reactions are enhanced by strong electron screening. In contrast to hydrogen burning, helium burning involves charged-particle reactions and does not require the operation of weak interaction processes.

4.5.3   *Envelope structure*

For a given mass and radius of the neutron star, the envelope structure is determined by the composition of the accreted matter and the main energy generating processes.

For most binary X-ray sources, the companion is a main sequence-like star, and the matter which is transferred will be rich in both hydrogen and helium. In others, such as 1820−303, no hydrogen is expected to be present in the accreted matter. This latter circumstance can arise if the neutron star is a member of an ultra-compact binary in which mass is transferred from either a degenerate dwarf or from a helium rich star. In all cases, a trace of heavy elements will be transferred. For hydrogen burning, the CNO element abundance is particularly important. The mass abundance of these nuclei in the transferred matter can be expected to lie in the range from ~ 0.0001 to 0.01 since X-ray burst sources are located in globular clusters as well as in the field of our galaxy. Recently, it has been suggested by Bildsten, Salpeter and Wasserman (1992) that the CNO abundances in the neutron star envelope can be reduced by a factor $\sim 10^3$ as a result of their spallation in the accretion flow. Much less depletion would be expected, however, if the accretion disk extends to the neutron star surface in which case the flow would not be radial or if radiative deceleration of the accreting plasma is sufficient to reduce the infall velocities below free fall.

The complete description of the envelope structure is given once the thermal structure is known. In the absence of accretion, the thermal structure is determined by the photon and neutrino luminosities emitted from the core region and, hence, is a direct function of the neutron star age. On the other hand, in the presence of accretion, the thermal structure is influenced by compressional heating. Here, gravitational energy is released as a given mass element is compressed to higher densities as a result of the increased weight of the overlying matter. This heating is more efficient at higher accretion rates since the energy generation would be localized. A more important factor in determining the thermal structure is the energy released by the nuclear burning reactions. This is most effective if the burning is in steady state since then the surface luminosity, L, will be equal to the nuclear luminosity, L_{nuc}. If the burning is not steady and the energy is released rapidly, as in the X-ray burst event, then only a fraction of the energy ($< 10\%$) will be conducted into the core region. In this case, the thermal structure of the envelope will continually evolve as a consequence of successive thermonuclear flashes (Taam 1980; see also Sect. 4.5.4.1).

4.5.4 *Steady versus nonsteady-state models*

The structure and stability of nuclear burning shells in accreting neutron stars can be most simply studied within the steady-state approximation where matter is assumed to burn at the same rate at which it is accreted (see Fujimoto, Hanawa and Miyaji 1981; Taam 1981b; Hanawa and Fujimoto 1982). Stable burning of both hydrogen and helium occurs at accretion rates $\gtrsim \dot{M}_{Edd}$ (where $\dot{M}_{Edd}$ is the Eddington mass accretion rate) for neutron stars at high temperatures ($T \gtrsim 5 \times 10^8$ K), where the degree of electron degeneracy is low and for $\dot{M} \lesssim 10^{-6}\dot{M}_{Edd}$ at low temperatures ($T \lesssim 5 \times 10^6$ K) when the nuclear burning occurs in the pycnonuclear regime. Unstable nuclear burning is classified into three categories, each delineated by a range of mass accretion rates. At high mass accretion rates ($\dot{M}_{Edd} \gtrsim \dot{M} \gtrsim \dot{M}_{c1}$), a helium flash develops in the presence of a hydrogen rich environment; at intermediate accretion rates, hydrogen burning leads to the formation of an underlying unstable helium burning layer; at low accretion rates ($10^{-6}\dot{M}_{Edd} \lesssim \dot{M} \lesssim \dot{M}_{c2}$), the instability in the hydrogen rich layer triggers a helium flash. Here, $\dot{M}_{c1}$ and $\dot{M}_{c2}$ depend on the metal abundance of the accreted matter and range from 5×10^{-11} to $5 \times 10^{-10}M_{\odot}yr^{-1}$ and from 10^{-12} to $5 \times 10^{-11}M_{\odot}yr^{-1}$, respectively, for the metal abundance, $Z \sim 0.0004$–0.02.

For the intermediate mass accretion rate regime, where hydrogen burning leads to the formation of a helium layer devoid of hydrogen, all the helium at the base of the accreted layer is burned to heavy elements via a series of helium capture reactions. The nuclear energy released in these reactions results in the ratio of persistent emission to time-averaged burst emission, α, $\gtrsim 100$, with recurrent time scales no less than 10 h. Shorter recurrence time scales must either involve a combined hydrogen–helium burning phase or involve the accretion of pure helium at a high rate. In the latter case, the helium burning shell is unstable for accretion rates in the range (10^{-14}–$10^{-8})M_{\odot}yr^{-1}$. Because of the rapid release of nuclear energy, the rise time scales associated with these flashes can be short ($\lesssim 1$ s). We note that if the CNO elements are significantly depleted by nuclear spallation reactions, as suggested by Bildsten, Salpeter and Wasserman (1992), this pure helium flash regime is not expected, thus restricting the diversity of nuclear shell burning behavior that the observations suggest is required (e.g., Tillett and MacDonald 1992). Specifically, the explanation of the gap in the peak fluxes in MXB 1636–536 (Ohashi 1981) of a factor of ~ 1.7 would be difficult to quantify in such a model along the ideas of Sugimoto, Ebisuzaki and

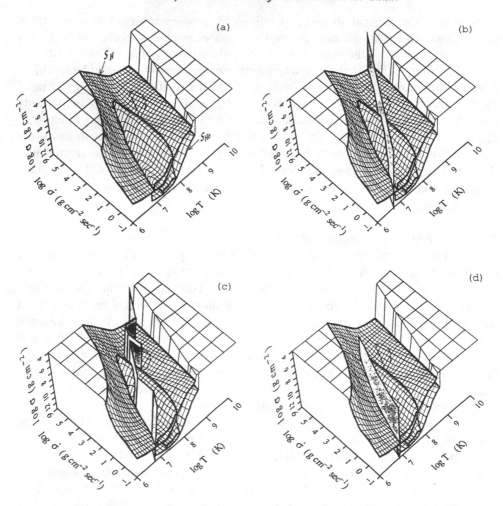

Fig. 4.15. (a) The hydrogen and helium fuel surfaces are denoted by S_H and S_{He} respectively. The ignition surfaces lie just above the fuel surfaces. The plane cutting (a) corresponds to the conditions for a steady state in (b), to the thermal equilibrium case in (c), and the compressional heating case in (d). From Fushiki and Lamb (1987).

Hanawa (1984) since the surface composition would not change significantly after mass loss for significant CNO depletion.

In the high and low mass accretion rate regimes, the nuclear burning involves a series of complex nuclear reactions leading up to iron nuclei. Because of the weak interactions, the nuclear burning development is slower, and the rise time of the burst is correspondingly longer than for a pure helium flash. This aspect of the nuclear burning development may be responsible for the differences in bursts characterized by a fast rise time (< 2 s) and slow rise time (see Murakami *et al.* 1980a; Sect. 4.3.2). In addition, it is possible that not all of the accreted fuel will be consumed (see Sect. 4.5.5). As a result, the α-values are smaller than for the pure helium flash case, but can be no smaller than ~ 20.

By relaxing the steady-state assumption, Fushiki and Lamb (1987) pointed out that the temperature of the envelope can also be an important independent parameter and, therefore, that the behavior of the nuclear burning shells is a function of three parameters. To elucidate the behavior, the concepts of a nuclear ignition surface and a fuel surface were introduced. The former corresponds to a surface at which the nuclear burning is unstable, and the latter to where the nuclear fuel is exhausted or where the luminosity due to compressional heating and nuclear burning equals that radiated from the neutron star surface. These surfaces are illustrated in Figure 4.15(a), and various cuts are shown in Figures 4.15(b)–(d), corresponding to the steady state, thermal equilibrium, and the compressional heating case (see Fushiki and Lamb 1987). It is immediately seen that the steady-state approximation is overly restrictive (since the temperature at the bottom of the accreted layer is determined by the mass accretion rate) and severely limits the possible range of available parameter space and nuclear burning shell behaviors that are accessible. The steady-state description, for those situations in which bursts occur, is found to be adequate only in the case in which stable hydrogen burning leads to helium shell flashes. This implies that the critical mass accretion rates ($\dot{M}_{c1}$, $\dot{M}_{c2}$) delineating the different unstable nuclear burning shell behaviors are similar in the steady-state and nonsteady-state models. However, the steady-state prescription is inadequate in the phase space corresponding to the occurrence of a combined hydrogen–helium flash. The relaxation of the steady-state assumption allows the possibility for additional behavior at high temperatures (corresponding to young, hot neutron stars) and low mass accretion rates where a series of hydrogen flashes could occur which eventually leads to a helium flash. This regime is not accessible to steady-state models since the temperatures required are greater than that obtainable in the steady-state approximation.

The results obtained from these analyses, for a hydrogen rich composition, are also known to depend upon the abundance of the CNO nuclei, Z_{CNO}, in the accreted matter, as these nuclei are catalysts for hydrogen burning via the CNO cycle. The possibility that the metal abundance in the accreted envelope of the neutron star is different than that transferred from the stellar companion (see Bildsten, Salpeter and Wasserman 1992) makes the structure and stability of the nuclear burning shells an intrinsically four-dimensional problem depending on the column accretion rate, $\dot{\sigma}$, the ignition density, σ, T, and Z_{CNO}. The difficulty in determining the properties of these shells is further compounded by the fact that the ignition conditions in the burning shell depend not only on the composition of the accreted matter, but also on the composition profile (of hydrogen, helium and the heavy elements) resulting from the previous outburst.

4.5.4.1 *Thermal state and relation to burst types*

From the analyses in both the steady- and nonsteady-state approximations the behavior of the nuclear burning shells depends critically on the thermal state of the neutron star. This state depends on the previous history of the star, which may reflect that it is relatively young (for a hot neutron star), the effects of compresssional heating associated with the mass accretion process, and the history of the nuclear burning that has previously taken place (on an older neutron star).

Of particular importance is the relationship between the thermal state of the envelope and the interior. This subject has been addressed in the study by Fujimoto *et al.* (1984, 1987a), in which it is found that the thermal structure of the envelope is effectively decoupled from the interior provided that the mass accretion rate and abundance of the CNO nuclei are high ($\dot{M} \gtrsim 0.01 \dot{M}_{Edd}$, $Z_{CNO} \sim 0.01$). If these conditions are not satisfied, the nuclear burning takes place deep in the conductive region of the neutron star and the envelope thermal structure is significantly influenced by the interior.

The steady-state models can approximately describe the thermal state of accreting neutron stars in which the nuclear burning in the envelope is stable. This description is justifiable at very high mass accretion rates, where both hydrogen and helium are stable, at intermediate accretion rates, where the hydrogen burning shell is stable, and at very low accretion rates, where both burning shells are stable in the pycnonuclear regime. Such models may apply to those neutron stars in nonbursting binary X-ray sources. On the other hand, the mass accretion rate and thermal regimes characterizing the coexistence of hydrogen and helium burning in the neutron star envelope are not well described by the steady-state assumption. In this case, a small amount of energy is conducted into the core region amounting to $L_{core} \lesssim 0.1 L_{nuc}$ (Ayasli and Joss 1982; Fujimoto *et al.* 1985; Woosley and Weaver 1985), indicating that the temperatures characterizing the accreted envelopes are significantly lower (by about a factor of two) in comparison with the temperatures obtained from the steady-state assumption ($L = L_{nuc}$). Although the nuclear burning behaviors are the same as for the steady-state case in this regime, the ignition conditions, and therefore the resulting burst properties, differ. The soft X-ray transients which also exhibit X-ray bursts, such as Aql X-1 and 1608−522, are in such a category where it is likely that the neutron stars in these sources are cool (Fushiki *et al.* 1992).

The calculation of the specific properties of X-ray bursts requires the determination of the relationship between the mass accretion rate and thermal state of the neutron star. This aspect of the theory cannot be addressed within the steady-state or nonsteady-state approximations. To provide understanding of this important relationship, we now turn to the full-scale time dependent numerical calculations.

4.5.5   *Numerical calculations*

The comparison of X-ray burst observations with the early numerical calculations of an individual burst provided persuasive evidence for the viability of the thermonuclear-flash model. Although the model was successful in reproducing the gross properties of the X-ray bursts (see Sect. 4.5.1), detailed comparisons between observations and theory were lacking. Progress was made in this direction by Ayasli and Joss (1982), who incorporated the fully general relativistic equations of stellar structure in their calculations. They demonstrated that these effects could significantly modify the X-ray burst properties, especially for neutron star models based on a soft nuclear-matter equation of state. Notwithstanding these effects, theoretical models were still incomplete since the thermal state of the neutron star was assumed *a priori* and not calculated as part of the global analysis. As emphasized in Sect. 4.5.4.1, the burst properties are sensitive to thermal-inertia effects (Taam 1980) reflecting the heating associated with the previous outbursts. Thus, long-term studies of the neutron star envelope subject to a number of repeated thermonuclear flashes were required for a complete description of the model. The investigations of Woosley and Weaver (1985) are especially noteworthy in this respect, illustrating that the burst properties are not only influenced by thermal-inertia effects, but also by the compositional profile left from the previous outburst. That is, not all of the accreted fuel is burned in the X-ray burst event (see Ayasli and Joss 1982; Hanawa and Fujimoto 1984; Woosley and Weaver 1985). This result has important implications for the interpretation of the observed burst characteristics. The presence of this residual fuel significantly modifies the burst properties expected from the steady-state and nonsteady-state analyses described in Sect. 4.5.4, and may have important implications for the regularity, or lack of regularity, of the burst behavior (e.g., 1735−444; see Sect. 4.3.5). In addition, the possibility of inverted molecular weight gradients and temperature profiles in the immediate post burst state have consequences for bursts which are produced with short recurrent time scales ($\sim$ 5–10 min; see Sect. 4.3.5 and 4.5.5.2).

4.5.5.1 Limit cycle behavior

The first demonstration of a limit cycle behavior was presented in the paper by Woosley and Weaver (1985) in which 12 successive combined hydrogen–helium flashes were followed over a time interval of $\sim$ 30 h. The bursting behavior was regular, and the bursts were characterized by similar light curves and recurrence time scales ($\sim$ 2.4 h). From their results of other calculations, there is some indication of erratic behavior for $Z_{CNO} \lesssim 0.01$ and for $\dot{M} \lesssim 10^{-9} M_\odot yr^{-1}$. A complicating issue is related to the fact that a significant fraction of hydrogen remains after a shell flash. In these cases, hydrogen burning contributes to the quiescent luminosity level during the burst inactive state (Van Paradijs, Penninx and Lewin 1988a; see also Sect. 4.3.8), which in turn influences the evolution toward either a limiting cyclic or erratic state.

Regular bursting behavior is also expected in situations where helium ignites in a region where hydrogen is absent. Such behavior is expected for $\dot{M}_{c1} > \dot{M} > \dot{M}_{c2}$ for the accretion of hydrogen rich matter and for $\dot{M}_{Edd} > \dot{M} > 10^{-6} \dot{M}_{Edd}$ for the accretion of pure helium matter. Note, however, that for helium flashes the recurrence time scales are long ($\sim$ 1 day) unless the accretion rates are high ($\gtrsim 10^{-9} M_\odot yr^{-1}$).

For those circumstances in which helium is ignited at high densities ($> 10^6$ g cm^{-3}), the nuclear luminosity generated is sufficient to cause photospheric expansion of the neutron star. The surface luminosity reaches the Eddington limit and a wind driven from the neutron star results (Hanawa and Sugimoto 1982; Taam 1982; Wallace, Woosley and Weaver 1982; Paczynski and Anderson 1986; Paczynski and Proszynski 1986; Joss and Melia 1987; see also Sect. 4.3.4). A long lasting Eddington limited phase ($\sim$ minutes) associated with a pure helium flash underlying a hydrogen rich region is likely to be responsible for the rapid transients which are accompanied by a presursor. Although the bolometric luminosity is nearly constant, the X-ray precursor can be understood to reflect variations of the neutron star photosphere (see Sect. 4.3.4). The identification of these transients with a deep seated helium flash is also consistent with the short rise time (< 1 s) seen for bursts which show evidence for radius expansion (Lewin *et al.* 1987). Further support for this interpretation is provided by the X-ray bursts emitted by the ultra-compact binary 1820−303 (see

Sect. 4.3.4). In this source, all bursts show evidence for radius expansion, and it is likely that the companion to the neutron star is transferring nearly pure helium rich material (see Sect. 4.5.3).

4.5.5.2   Bursts with short time intervals

X-ray bursts with short time intervals have been reported from a number of sources (e.g. MXB 1743–28, Lewin *et al.* 1976c; 0748–676, Gottwald *et al.* 1986; 1745–248, Oda 1982). These bursts have similar burst properties to other type I X-ray bursts, but are of an enigmatic variety since $\alpha \lesssim 10$ (and in several cases $\alpha < 2$). This small value of α cannot be explained within the framework of the simple picture of the thermonuclear model outlined in the steady- or nonsteady-state analyses since there is insufficient time to accumulate mass to initiate another outburst if all matter accreted in a given burst is burned. As a consequence, a number of mechanisms have been suggested for these bursts which rely on the storage of nuclear fuel. Among them are ideas related to the pooling of fuel over the different locations of the neutron star surface (Fujimoto, Hanawa and Miyaji 1981; Nozakura, Ikeuchi and Fujimoto 1984) and others wherein the second burst is attributed to the occurrence of mixing during the post burst phase (Woosley and Weaver 1985; Fujimoto *et al.* 1987b).

Fujimoto *et al.* (1987b) have suggested a mechanism involving elemental mixing and dissipative heating associated with hydrodynamical instabilities caused by the redistribution of angular momentum in the accreted envelope. However, no detailed calculation exists, at present, to test the viability of this hypothesis.

The most detailed study of the mixing hypothesis was carried out by Woosley and Weaver (1985), who make use of the fact that an inverted molecular-weight gradient forms as a natural consequence of the nuclear-burning development during the main outburst. This situation is Rayleigh–Taylor unstable; however, during the post burst phase, an inverted temperature gradient is present which stabilizes this layer. As the envelope cools during the post burst phase, the inverted temperature gradient is relieved and the mixing of matter results. The mixing of the residual carbon from the burnt out helium burning region with the residual hydrogen from the overlying layer leads to a rekindling of the fuels, with the consequence that a second burst can be produced. The time delay from the main burst to the onset of mixing is determined by the cooling time scale of the envelope, and is typically a few minutes. A light curve of the main burst and of the recurrent burst is shown in Figure 4.16. A characteristic of these bursts is that they are much weaker (by a factor of five) than the main bursts. The energetics and the recurrence time of the weaker X-ray burst are reminiscent of the bursts seen, for example, from MXB 1743–28 (Lewin *et al.* 1976c) and 0748–676 (Gottwald *et al.* 1986; see also Sect. 4.3.5), but inconsistent with the repetitive bursts exhibiting similar properties seen in 1745–248 (Oda 1982). Since the amplitude of the second burst is sensitive to the temperatures achieved in the nuclear-burning regions and the extent over which an inverted molecular weight gradient exists, and since the delay time of the second burst depends upon the depth of the accreted layer, a larger burst might be produced to yield burst properties more in common with those seen in the main burst if a larger fraction of the burning region or a deeper layer can be involved.

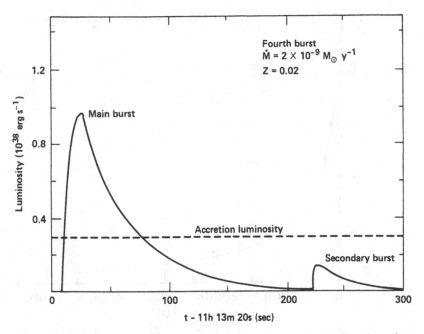

Fig. 4.16. The light curve of the main burst and a recurrent burst at short time intervals caused by convective mixing during the post burst phase. From Woosley and Weaver (1985).

4.5.5.3 *Probe of interior structure*

The properties of the neutron star can, in principle, be constrained by using results obtained from the thermonuclear-flash model. Provided that the first X-ray burst emitted by an object from a transient source is observed, then complications associated with thermal and compositional inertia effects are not essential for determining the properties of the nuclear burning shells. The observations of the unusually long X-ray burst from the recurrent soft X-ray transient Aql X-1 may provide such a testing ground.

Fushiki *et al.* (1992) interpret the long X-ray burst tail (lasting for more than 2500 s) of the first burst observed by Czerny, Czerny and Grindlay (1987) as an extended phase of hydrogen burning accelerated by electron capture processes in regions of high densities ($\sim 10^7$ g cm^{-3}). The light curve obtained from the numerical calculation is shown in Figure 4.17. In this interpretation, the neutron star envelope is out of thermal equilibrium. Consequently, only the first X-ray burst emitted by Aql X-1 exhibits such a prolonged X-ray phase. The subsequent bursts, on the other hand, do not exhibit long X-ray tails, in agreement with observations, since the ignition densities are reduced as a result of the thermal-inertia effects in the neutron star envelope. The properties of these subsequent bursts vary from burst to burst, and are more similar to those of typical type I X-ray bursters.

During the X-ray inactive phase, the envelope of the neutron star relaxes to its nonbursting thermal state. It is likely that for all recurrent soft X-ray transients which exhibit X-ray bursts (see Sect. 4.3.5) the core temperature is low and that

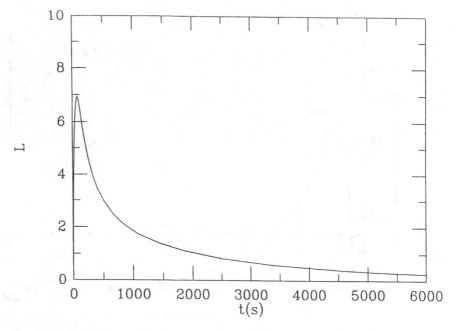

Fig. 4.17. The light curve for the first X-ray burst produced after the onset of accretion. Note the long, relatively flat, X-ray tail extending out to 6000 s. The luminosity, L, is in units of 10^{37} erg s^{-1}. From Fushiki *et al.* (1992).

the description of the bursts must involve models out of thermal equilibrium. As a consequence, the burst patterns of these sources are expected to exhibit very irregular behavior not correlated with the properties of the persistent emission.

4.5.6   Outlook

A foundation for the interpretation of X-ray bursts has been provided by the thermonuclear-flash model, which has been remarkably successful in reproducing the gross properties of the observed bursts. Yet a number of outstanding discrepancies remain between observation and theory. Among them are the erratic correlations between persistent emission and burst behavior exhibited by such sources as 1735—444 (Lewin *et al.* 1980; Van Paradijs *et al.* 1988b), Ser X-1 (Li *et al.* 1977; Sztajno *et al.* 1983), 1608—522 (Murakami *et al.* 1980a) and 0748—676 (Gottwald *et al.* 1986), and the nearly identical bursts separated by 10 min seen in 1745—248 (Oda 1982).

The comparison of observations with theory has now progressed beyond the general considerations of burst properties. The next step must involve the confrontation of data with detailed numerical models in which successive bursts are followed. In addition, theoretical efforts should also be directed toward the construction of more realistic multi-dimensional models in which the ignition, propagation and evolution of nuclear burning fronts are investigated.

4.6   The Rapid Burster (1730−335)

4.6.1   Introduction

The Rapid Burster (RB) was discovered by Lewin *et al.* (1976b). It is located in the highly reddened globular cluster (Liller 1977), at a distance of ∼10 kpc (Kleinmann, Kleinmann and Wright 1976). Its behavior is unlike any other source. When the RB is active (see Figure 4.18), it can produce X-ray bursts in quick succession with intervals as short as ∼7 s (see Figure 4.19). Hoffman, Marshall and Lewin (1978a) discovered that the RB produces two very different types of bursts, which led to the classification of type I (thermonuclear flashes on the surface of a neutron star) and type II bursts (spasmodic accretion).

We would like to remind the reader who feels encouraged to unravel the mysteries of the RB that, in August 1983, this binary system behaved like other 'normal' LMXBs; instead of type II bursts, a persistent flux of X-rays was observed (Kunieda *et al.* 1984b; Barr *et al.* 1987). Thus, the RB with all its idiosyncrasies is almost a 'normal' LMXB.

Hereafter, whenever we quote luminosities or burst energies, we have always assumed a source distance to the RB of 10 kpc (Kleinmann, Kleinmann and Wright 1976) and isotropic emission.

4.6.2   Type I and type II bursts

The distinct spectral softening during burst decay is very characteristic for a type I burst (Sect. 4.3.3, 4.3.4). The type II bursts lack this signature; they are due to a yet unknown accretion instability (see Sect. 4.7).

Type I bursts from the RB have been detected at intervals of ∼1.5 to 4 h (Hoffman, Marshall and Lewin 1978a; Marshall *et al.* 1979; Kunieda *et al.* 1984b; Kawai 1985; Barr *et al.* 1987; Kaminker *et al.* 1990); type I burst intervals from other sources are typically in the same range. Type II bursts from the RB can have intervals as short as ∼7 s and as long as ∼1 h.

The type II burst mechanism behaves like a relaxation oscillator (Lewin *et al.* 1976b); the fluence in a type II burst is approximately linearly proportional to the interval to the *next* type II burst. This is very characteristic for the type II bursts from the RB; there is no other source like it!

4.6.3   Normal and abnormal accretion

The most common mode of activity of the RB is type II bursts in combination with type I bursts. The only type II burst active period during which no type I bursts were observed was August–September 1985 (Stella *et al.* 1988a,b; Lubin *et al.* 1991a,b; Tan *et al.* 1991).

On one occasion, near August 5, 1983, when the source became active, no type II bursts were detected at first. Instead, a strong persistent X-ray flux and type I bursts (with intervals of ∼1.5 h) were observed for at least several days but probably for ∼10 days (Kunieda *et al.* 1984b; Kawai 1985; Barr *et al.* 1987; Kaminker *et al.* 1990).

The average luminosity of the persistent emission was ∼ 1.3×10^{37} erg s^{-1} (∼1−9

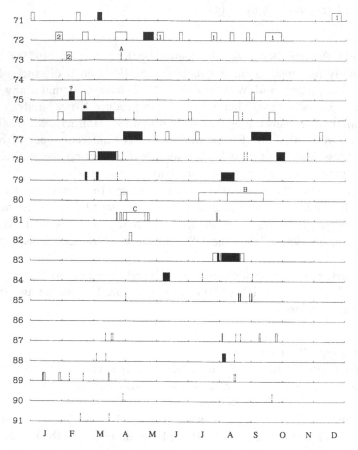

Fig. 4.18. Record of observations of the Rapid Burster. The horizontal scale is one year; the months are indicated. Burst active periods are indicated by black squares; no activity was observed during the periods indicated by open squares. A '1' or '2' in an open square indicate that one or two type I bursts were observed which were probably from 1728–337. The vertical dashed lines indicate observing periods of less than one day during which no bursts were detected. The periods marked 'A' and '?' represent times during which the Rapid Burster may have been active. During the time marked 'B', intermittent observations were made with ~3-day intervals. During the time marked 'C', the sensitivity was low, and small bursts from the Rapid Burster could have been missed. The * indicates the time of discovery of the Rapid Burster. For details covering the period prior to 1981, see Figure 2.18 in Lewin and Joss (1983); for details of the observations in 1981 and thereafter, see LVT93. This figure was prepared by Eugene Magnier. We are thankful for the large amount of unpublished information provided by our Japanese colleagues: T. Dotani, N. Kawai, K. Koyama, H. Tsunemi, and S. Yamauichi.

keV), which is comparable to the time-averaged type II burst luminosities in the early phase of type II burst activity. Near August 19, 1983, type II bursts were observed (possibly already on August 16; see Kaminker *et al.* 1990), and they continued till the burst activity ceased on August 31, 1983.

24-minute snapshots from 8 orbits on March 2/3,1976

|100 s|

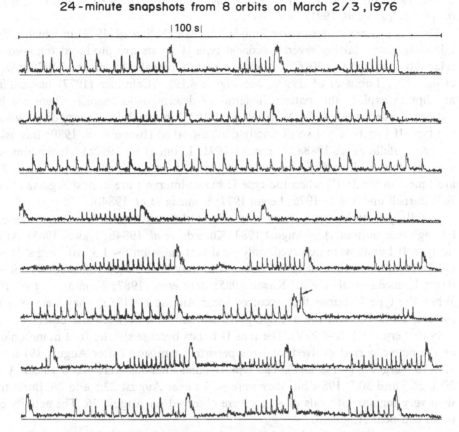

Fig. 4.19. Discovery with SAS-3 of type II bursts from the Rapid Burster in mode I (third curve from top is in mode II). Notice the flat tops in the large bursts; the 'ringing' during burst decay of the large bursts is clearly visible. The burst indicated with an arrow is a type I burst from 1728–337. 'Anomalous' bursts are visible in the top, the fifth and the bottom curves. It is possible that these are type I bursts from the Rapid Burster. From Lewin (1977).

4.6.4 *Type II burst patterns, evolution, persistent emission, average luminosities*

Type II burst intervals vary from ~7 s (Lewin 1977; Marshall *et al.* 1979) to ~1 h (Inoue *et al.* 1980; Kunieda *et al.* 1984a; Stella *et al.* 1988a; Tan *et al.* 1991). The type II burst patterns vary a great deal.

The *time-averaged* type II burst luminosity (~2–15 keV) decreases (though not monotonically; Marshall *et al.* 1979) during the active periods, which last typically several weeks. The time-averaged type II burst luminosity was ~ 1.5×10^{37} erg s^{-1} near March 2, 1976, and it had decreased to ~5×10^{36} erg s^{-1} by April 3–4. In August, 1979, Kunieda *et al.* (1984a) observed a turn on in activity; the time-averaged type II burst luminosity increased during the first ~3 days (from ~7×10^{36} erg s^{-1} to ~4×10^{37} erg s^{-1}) and then decreased by about a factor of two during the following week. The time-averaged type II burst luminosity was ~3×10^{37} erg s^{-1} on August

28–31, 1985; on September 13, 1985, it had decreased to $\sim 4 \times 10^{36}$ erg s^{-1} (Stella *et al.* 1988a; Tan *et al.* 1991).

There is a burst pattern (called mode I by Marshall *et al.* 1979) in which a dozen relatively short (lasting several seconds) type II bursts are produced followed by a relatively long burst (~ 30–60 s long); this pattern repeats (Lewin *et al.* 1976b; Dotani *et al.* 1990; Lubin *et al.* 1992a; see Figure 4.19). (Celnikier (1977) has made an attempt to explain this pattern in terms of deterministic chaos.) There are burst patterns (called mode II by Marshall *et al.* 1979) in which the energy distribution of the type II bursts is not so distinctly double-peaked (Inoue *et al.* 1980; Basinska *et al.* 1980; Stella *et al.* 1988a; Tan *et al.* 1991; Lubin *et al.* 1992a). (It remains to be seen whether a clear distinction between mode I and II is always possible.) There are times (in mode II) when the type II bursts intervals are almost regular (Mason, Bell-Burnell and White 1976; Lewin 1977; Kunieda *et al.* 1984b).

A full cycle (from turn-on to turn-off) of the active period was only observed (though intermittently) in August 1983 (Kunieda *et al.* 1984b; Kawai 1985). At first, no type II bursts were observed, only persistent emission ($\sim 1.3 \times 10^{37}$ erg s^{-1}; ~ 1–9 keV) in combination with type I bursts (this may have lasted continuously for ~ 10 days; Kunieda *et al.* 1988b; Kawai 1985; Barr *et al.* 1987; Kaminker *et al.* 1990). When the type II bursts first occurred (near August 18, 1983), they were very long (several minutes), and the average persistent emission between the bursts was near $\sim 8 \times 10^{36}$ erg s^{-1} (~ 1–9 keV). The type II bursts became shorter (not monotonically) as the active period evolved, and the persistent emission (after August 19) became $< 5 \times 10^{36}$ erg s^{-1} (~ 1–9 keV). The type II burst intervals were ~ 100 s near August 20.5, 24.5 and 30.5, 1983, but they were ~ 25 s near August 23.5 and 26. Burst trains with very regular intervals of ~ 16 s were observed on August 26. The activity came to an end near August 31.

A turn-on was observed in the August 1979 active period; at first short (< 30 s) type II bursts were observed, the bursts then became longer (~ 40 s to ~ 700 s) for about one week, after which they became shorter again; the turn-off was not observed (Kunieda *et al.* 1984a).

4.6.5   *Type II burst profiles – energies and luminosities*

Type II bursts vary in duration from ~ 2 s to ~ 11 min. Their peak luminosities range from $\sim 5 \times 10^{37}$ to $\sim 4 \times 10^{38}$ erg s^{-1}, and the integrated burst energies range from $\sim 1 \times 10^{38}$ erg to $\sim 7 \times 10^{40}$ erg. When the type II burst duration exceeds ~ 15 s, the type II bursts in general have approximately flat tops, whose flux levels, however, can vary greatly from burst to burst. Thus, these flat tops are not the result of an Eddington limit.

Type II bursts which are shorter than ~ 50 s almost always show successive peaks during their decay which we call 'ringing'. Kawai (1985), Tawara *et al.* (1985) and Kawai *et al.* (1990) discussed data in which the type II bursts (all with durations in the range ~ 2 s to ~ 50 s) have nearly identical profiles, though the characteristic time scale of the profiles and peak fluxes differ from burst to burst. When one explores the entire range of type II burst durations, the profiles are not time-scale-invariant (Tan *et al.* 1991). There is a gradual change in the type II burst profiles as the bursts

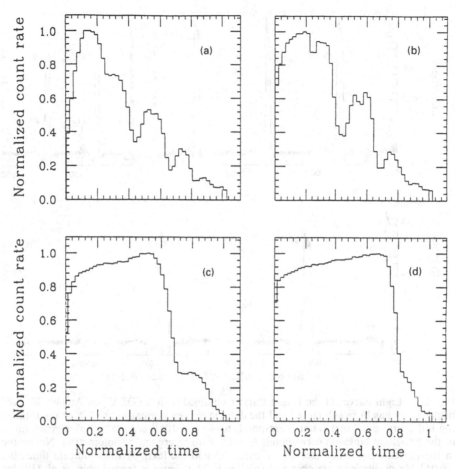

Fig. 4.20. Composite type II burst profiles (1–22 keV). The lengths and heights of the profiles are normalized to unity. (a) 342 bursts with durations from ~2 to 17 s; (b) 169 bursts with durations from ~4 to 27 s; (c) 49 bursts with durations from ~40 to 200 s; (d) 39 bursts with durations from ~90 to 680 s. From Tan *et al.* (1991).

get longer (Figure 4.20). As the bursts become long and approximately flat-topped, the peak fluxes become lower (Kunieda *et al.* 1984a; Stella *et al.* 1988a).

Type II bursts of very low fluence ($<1.1 \times 10^{-8}$ erg cm^{-2}; this corrresponds to a total burst energy of less than ~1.3×10^{38} erg) can have erratic profiles, showing a series of flares. On September 13, 1985, the very low-fluence bursts occur 'too early'; the burst intervals to the burst *prior to* the low-fluence bursts are significantly shorter, while at the same time the burst intervals *after* the low-fluence bursts appear to be longer than one would expect on the basis of the observed relaxation oscillator relation between burst fluence and the interval to the *following* burst. It appears in these cases as if the accretion instability responsible for the type II bursts occurs prematurely, with only a relatively small amount of matter being accreted on the neutron star (Lubin *et al.* 1991b).

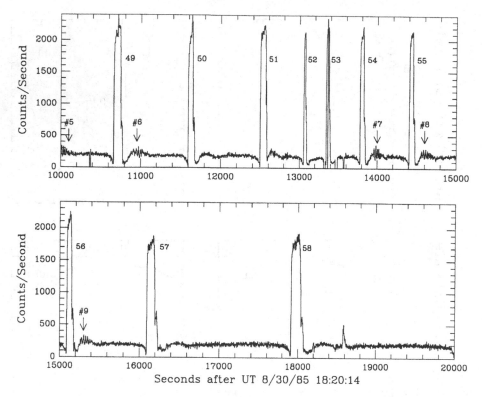

Fig. 4.21. Light curves of the Rapid Burster obtained with EXOSAT on August 30, 1985; the background has been subtracted and the data have been normalized to 7/8 of the total area of the full array; one burst count corresponds to $\sim 1.6 \times 10^{-11}$ erg cm^{-2} (bolometric), one count in the persistent emission corresponds to $\sim 7.8 \times 10^{-12}$ erg cm^{-2} (bolometric). Notice the dips in the persistent emission (see also Figure 4.25 and the text). Arrows indicate times that the ~ 0.042 Hz oscillations are observed (see text). This figure is from Lubin *et al.* (1992b). An expanded light curve of the interval between bursts #54 and #56 is shown in Figure 4.22.

4.6.6   *Radio and infrared bursts?*

Radio and infrared bursts have been reported from the Rapid Burster. However, their believability is in question (see Sect. 4.3.9.2).

4.6.7   *Persistent emission after energetic bursts*

Persistent X-ray emission is often observed after type II bursts with a relatively high burst fluence (this is the case for most bursts which last longer than ~ 30 s); persistent emission is not observed after low-fluence type II bursts. This persistent X-ray flux emerges gradually after the type II bursts, and it decreases gradually before the occurrence of the next type II burst; this is often referred to as the 'dips' in the persistent emission (Figures 4.21, 4.22, 4.25; Marshall *et al.* 1979; Van Paradijs *et al.* 1979b; Kunieda *et al.* 1984a; Stella *et al.* 1988a; Lubin *et al.* 1992b). The disappearance of this persistent emission is not the result of photo-electric absorption (Stella *et al.* 1988a). No satisfying explanation exists for this behavior.

In March 1976, the mean flux in this persistent emission after the long type II

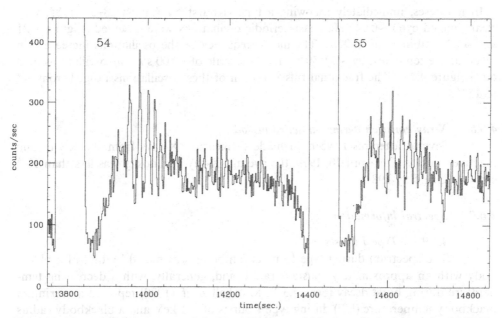

Fig. 4.22. Expanded light curves of the Rapid Burster which show the 'naked-eye' ~0.042 Hz quasi-periodic oscillations (see text and Figure 4.21); the background has been subtracted and the data have been normalized to 7/8 of the total area of the full array (for conversion from counts to flux, see Figure 4.21). Notice the increasing frequency as the oscillations die out. Adapted from Lubin *et al.* (1992b).

bursts was ~5% of the time-averaged type II burst flux (Van Paradijs *et al.* 1979b). Persistent emission after high-fluence type II bursts was observed with HEAO-1 in March 1978 (Hoffman *et al.* 1978c) and with Hakucho in August 1979. During the latter observations, between August 8 and 15, 1979, when long (~40 to ~700 s) type II bursts were observed, the average persistent luminosity was ~ 1.8×10^{37} erg s^{-1}; this was ~50% of the time-averaged type II burst luminosity (Kunieda *et al.* 1984a). On August 28–31, 1985, all type II bursts (observed with EXOSAT) were in excess of ~1 min, and persistent emission was observed after each burst (Figures 4.21, 4.22 and 4.25). The average persistent luminosity was ~2.2×10^{37} erg s^{-1}, which is comparable to the average type II burst luminosity of ~2.8×10^{37} erg s^{-1} during the same period.

No persistent emission has been observed to date after type II bursts with durations less than ~20 s. On July 17, 1984, and September 10, 1985, when the type II bursts durations varied between ~2 s and ~17 s, the average persistent luminosity between the bursts was <2.5×10^{36} erg s^{-1}(2σ upper limit); at those times the time-averaged type II burst luminosity was ~8×10^{36} erg s^{-1} (Tan *et al.* 1991). On September 13, 1985, when the type II bursts were all shorter than 30 s, the time-averaged type II burst luminosity had decreased to ~4×10^{36} erg s^{-1}, and the 2σ upper limit to the persistent emission between the bursts was ~2×10^{36} erg s^{-1}.

The persistent flux between the type II bursts, when observed, varies a great deal. The light curves can, at times, be carbon copies of each other, including such fine details as 'humps', sharp 'glitches' and 'bumps' (Lubin *et al.* 1993; LVT93).

In ten cases, immediately following a type II burst on August 28–31, 1985, very clear ('naked eye') $\sim$0.042 Hz quasi-periodic oscillations were observed (Figures 4.21 and 4.22; Lubin *et al.* 1992b). The mean frequency of the oscillations increased (in seven of the ten cases) by $\sim$30–50% on a time scale of $\sim$100 s as the oscillations died out (Figure 4.22). The fractional rms variation of these oscillations ranged from $\sim$5 to 15%.

4.6.8   *X-ray emission during an active period*

Several attempts have been made to measure X-rays from the RB during inactive periods. On April 10, 1979, the average X-ray luminosity was less than 10^{34} erg s^{-1} (Grindlay 1981).

4.6.9   *Spectral information*

4.6.9.1   *Type I bursts*

The spectrum during type I bursts can be approximated by that of a blackbody with an approximately constant radius and, generally, with a decreasing temperature during burst decay (Sect. 4.3.3). Marshall *et al.* (1979) reported a maximum blackbody temperature (kT) during type I bursts of $\sim$2 keV and a blackbody radius of 9 ± 2 km; Barr *et al.* (1987) reported a blackbody radius of $\sim$10 km. (We caution that the interpretation of blackbody temperatures and radii is not at all straightforward; see LVT93.) Marshall *et al.* (1979) found an average type I burst peak luminosity of $\sim$2$\times$10^{38} erg s^{-1} (active period of September–October 1977); Barr *et al.* (1987) reported peak type I burst luminosities (2–15 keV) of 9.6$\times$10^{37} erg s^{-1} (August 7, 1983) and 7.2$\times$10^{37} erg s^{-1} (August 14/15, 1983). If the anomalous bursts as reported by Ulmer *et al.* (1977) are type I bursts, as argued by Lubin *et al.* (1991b), their peak luminosities are $\sim$5$\times$10^{36} erg s^{-1}.

4.6.9.2   *Type II bursts*

The burst peak luminosities in type II bursts range from $\sim$5$\times$10^{37} to $\sim$4$\times$10^{38} erg s^{-1}, and the integrated burst energies range from $\sim$1$\times$10^{38} to $\sim$ 7$\times$10^{40} erg. The spectra during type II bursts have been approximated by that of a blackbody by various authors. Very roughly, the blackbody temperature remains constant throughout the type II bursts; the blackbody temperatures (kT) are $\sim$1.8 keV. Studied in more detail, one finds that there is some temperature evolution during the type II bursts correlated with the ringing during burst decay; blackbody temperatures (kT) vary from $\sim$1.5 to $\sim$2 keV (Kawai *et al.* 1990; Tan *et al.* 1991). Tan *et al.* (1991) have shown that the temperature evolution in type II bursts is different for bursts with different profiles (and thus different burst durations). The roughly constant blackbody temperature throughout the type II bursts implies that the size of the emitting region decreases as the bursts decay. The largest blackbody radii are observed early on in the type II bursts. Marshall *et al.* (1979) report an average blackbody radius of $\sim$16 $\pm$ 2 km during the first 15 s of the type II bursts. Kawai *et al.* (1990) and Tan *et al.* (1991) report maximum radii from $\sim$10 to $\sim$15 km. Marshall *et al.* (1979) and Kawai *et al.* (1990) concluded that the maximum blackbody radii measured during type II bursts are larger than the blackbody radii measured during type I bursts.

Kunieda *et al.* (1984a), using Hakucho data from August 1979, and Tan *et al.* (1991), using EXOSAT data from August 1985, have shown that there is a correlation between the color temperature, T, and the luminosity, L, as observed during the flat tops of relatively long type II bursts; they find that $L \propto T^6$. This led to the suggestion by Tan *et al.* (1991) that the radiation during a type II burst comes from a photosphere whose radius is larger for higher type II burst accretion rates. This could explain in a natural way why the blackbody radii during the early part of type II bursts are larger than those observed during type I bursts. Using the Ginga data of August 1988, Lubin *et al.* (1992a) found that the above relation between L and T holds approximately for type II bursts but not for type I bursts. (We again caution that the interpretation of blackbody temperatures and radii is not at all straightforward; see LVT93.)

Stella *et al.* (1988a) concluded that blackbody fits to the type II bursts are unacceptably poor; they showed that an unsaturated Comptonization model with an iron emission line at ∼6.7 keV gives acceptable fits. The iron line flux during the type II bursts was ∼7.3×10^{-3} photons cm^{-2} s^{-1}.

Dotani (1990) has fitted the type II burst spectra to a Comptonized blackbody (Kawai 1985) with a fixed electron temperature. He reports blackbody temperatures of ∼ 1.5 and ∼1.7 keV and optical depths of 0.22 ± 0.03 and 0.27 ± 0.06 for the initial peak and the following peak in the type II bursts, respectively.

4.6.9.3 Persistent emission and type I bursts

In early August 1983, the RB was active, but it emitted no type II bursts; however, persistent emission and type I bursts were observed (Kunieda *et al.* 1984b; Kawai 1985; Barr *et al.* 1987). The average luminosity of the persistent emission was ∼1.3×10^{37} erg s^{-1} (∼1–9 keV). Barr *et al.* (1987) showed that blackbody fits to the data were unacceptably poor. Both power law spectra (photon index $\Gamma = 2.1 \pm 0.2$) and thermal-bremsstrahlung spectra ($kT = 9 +3/-2$ keV) gave acceptable fits.

From August 8–15, 1979, when long (∼40 to ∼700 s) type II bursts were observed, the time-averaged persistent emission between the bursts was $(1.8 \pm 0.3) \times 10^{37}$ erg s^{-1} (∼50% of the time-averaged luminosity in type II bursts observed during the same time).

4.6.10 Quasi-periodic oscillations

4.6.10.1 Introduction

Quasi-periodic oscillations (QPOs) of ∼2 Hz were discovered (with Hakucho) in two out of 63 long type II bursts (Tawara *et al.* 1982). Several years later, QPOs were observed during many (not all) type II bursts and often (not always) during the persistent emission between the type II bursts (Figures 4.23, 4.24, 4.25; Stella *et al.* 1988a,b; Lubin *et al.* 1991a; Dotani 1990; Dotani *et al.* 1990). No QPOs have been observed to date in any type I burst from the Rapid Burster. It is presently unclear whether the QPOs observed in the RB have any relation to other forms of QPOs observed in many bright LMXBs (see Ch. 6).

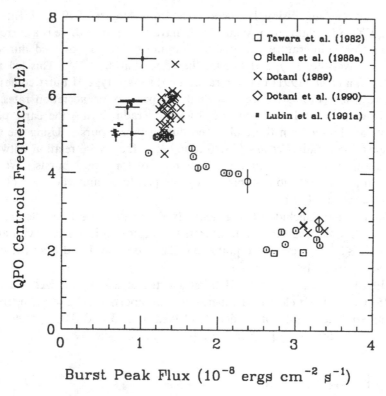

Fig. 4.23. Type II peak burst flux versus the QPO centroid frequency. Adapted from Lubin *et al.* (1991a).

4.6.10.2 Type II bursts

Tawara *et al.* (1982) discovered ~2 Hz QPOs with a strength of 10% (fractional rms variation) in two of 63 long (~40–600 s) type II bursts. The upper limits to the strength in the other 61 were ~4%.

During August 28–29, 1985, 41 type II bursts were detected which lasted between ~1.5 min and ~11 min; the shortest bursts had higher peak fluxes than the longer bursts (Stella *et al.* 1988a). QPOs with frequencies between 2 and 5 Hz were observed in 23 (of the 41) type II bursts. The fractional rms variation ranged from <2% (90% confidence level) to ~20% for the longest and most energetic type II bursts (these energetic bursts are the least luminous bursts and have QPO frequencies near 5 Hz). Lubin *et al.* (1991a) made a study of the QPOs in 784 short (<30 s) and low-luminosity type II bursts observed with EXOSAT, and found that QPOs were only observed in bursts within a certain range of durations. Combining all results, the QPO centroid frequencies in the type II bursts range from ~2 to ~7 Hz, and they are strongly anti-correlated with the average burst peak flux (Figure 4.23).

During the August 1988 active period, relatively short type II bursts (<35 s) were observed (Dotani 1990; Dotani *et al.* 1990). Dotani *et al.* (1990) found evidence that the presence of QPOs is correlated with the ringing in the burst profiles. The QPO is very strong during the initial peak in the burst profile, absent in the second peak,

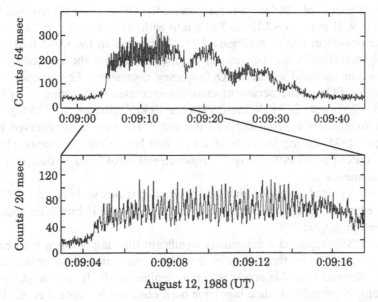

Fig. 4.24. Light curve of a type II burst from the Rapid Burster observed with Ginga on August 12, 1988. Notice the distinct 'ringing' during burst decay and the 'naked-eye' ~5 Hz QPO (bottom panel). Adapted from Dotani *et al.* (1990).

and strong again at the onset of the third peak; average fractional rms variations are ~18%, < 2% and ~13%, respectively. The strength of the 5 Hz QPO was very large, and the individual oscillations could be seen in the burst profiles (Figure 4.24). Dotani *et al.* (1990) showed that, during the large majority of the type II bursts, the oscillations can be described as a periodic modulation without phase jumps whose frequency changes gradually by up to ~25%. The frequency drifts can explain the width of the QPO peaks in the power density spectra.

The oscillations can be described by changes of the temperature of a blackbody emitter with constant apparent area, but they can equally well be described by changes in the photospheric radius and associated temperature changes (Dotani *et al.* 1990). Lewin *et al.* (1991) have given some arguments why the latter may be more likely (see also Lubin *et al.* 1992a).

The fractional rms variation of the QPOs in the type II bursts increases with increasing photon energies. It increases from ~8% near 3 keV to ~13% near 10 keV (Stella *et al.* 1988b; Dotani *et al.* 1990). No significant time lags (between high-energy and low-energy photons) were observed in the QPOs in the type II bursts (Stella *et al.* 1988b; Dotani *et al.* 1990).

4.6.10.3 *Persistent emission*

In early August 1983, when the RB showed only persistent emission and type I bursts, no QPOs were observed in the persistent emission with an upper limit of ~10% in the fractional rms variation (Barr *et al.* 1987). During the active period in August 1985, QPOs were occasionally observed in the persistent emission after relatively short (1–2 min) type II bursts. The fractional rms variation could be as high as ∼35% (Stella *et al.* 1988a). In several cases, the QPO frequency evolved from ∼4 Hz after a type II burst to ∼2 Hz in 3 to 6 min without a general correlation with the flux of the persistent emission (Figure 4.25). Variations in the QPO frequencies were positively correlated with the spectral hardness such that the spectrum of the persistent emission softened as the QPO frequency decreased. The fractional rms variation of the QPOs in the persistent emission increases with increasing photon energies, from ∼20% near 2–3 keV to ∼35% near 10 keV (Stella *et al.* 1988a).

QPOs with frequencies ranging between 0.4 and 1 Hz were also observed occasionally (August 1985) during intervals of 1 to 3 min before type II bursts. On one occasion, two QPO peaks (with centroid frequencies of ∼0.44 and ∼ 0.88 Hz) were observed simultaneously.

Dotani (1990) reported an upper limit of 9% fractional rms variation for QPOs in the persistent emission observed after some of the longest type II bursts (in excess of ∼25 s) observed in August 1988.

Stella *et al.* (1988b) reported a marginally significant time lag between high-energy and low-energy photons in the QPOs in the persistent emission; a reanalysis of the same data showed that this result was not significant (R. E. Rutledge, private communication). No significant time lags have been observed by Dotani *et al.* (1990).

'Naked eye' QPOs with centroid frequencies of ∼0.042 Hz have been observed in the persistent emission immediately following ten type II bursts (Lubin *et al.* 1992b); they are described in Sect. 4.6.7 (Figures 4.21 and 4.22). In eight out of ten cases ∼4 Hz QPOs were observed; therefore, there is a high probability that the occurrence of the ∼0.042 Hz and of the ∼4 Hz QPOs are related (Lubin *et al.* 1992b).

4.7   Models for the Rapid Burster

4.7.1   Introduction

Numerous theoretical models have been advanced for the explanation of the Rapid Burster since its discovery in 1976. Since the time-averaged type II burst emission can be comparable to or greater than the persistent emission, a common hypothesis in all models is that the bursts are caused by the sudden release of energy associated with the accretion of matter onto the neutron star surface. Despite the continued monitoring of the Rapid Burster over the last decade, which has revealed a plethora of new observational results, and despite a number of theoretical attempts to model this enigmatic source, the nature of the instability responsible for the type II bursts has yet to be established.

4.7.2   Instability picture

The fundamental difficulty in making significant progress in understanding the behavior of the Rapid Burster has been the lack of an identification for the proper gating mechanism. A simple mathematical representation of the behavior has been proposed by Celnikier (1977) in a form similar to that encountered in the population dynamics field. Physical descriptions have invoked viscous or thermal instabilities in an accretion disk (Taam and Lin 1984; Hayakawa 1985; Meyer 1986), magnetospheric substorms resulting from magnetic reconnection (Davidson 1982), interactions between an accretion disk and a magnetosphere of a rotating neutron

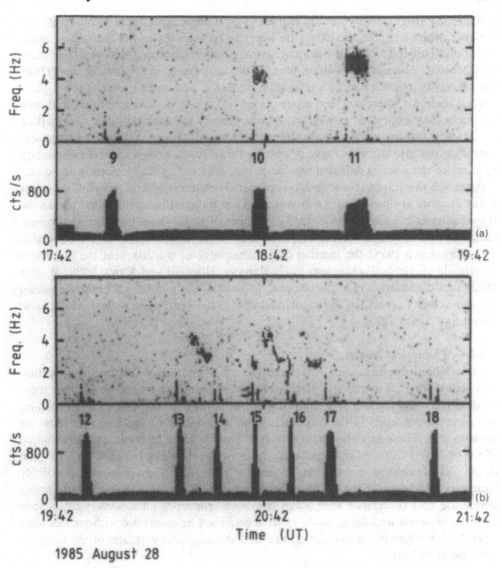

Fig. 4.25. (a) Power density spectra and (b) light curves (1–15 keV count rate) of Rapid Burster EXOSAT data from August 28, 1985. The shade of grey in the power density spectra indicates the power; the darker the shade, the higher is the power. Notice the evolution in the QPO frequency in the persistent emission. Relatively high-frequency QPOs (~4 Hz) occur when the spectral hardness in the persistent emission is relatively high; the QPO frequency is lowest (~2.5 Hz) just before a type II burst. From Stella *et al.* (1988a).

star (Baan 1977, 1979; Lamb *et al.* 1977; Michel 1977; Horiuchi, Kodonaga and Tominatsu 1981; Singh and Duorah 1983; Hanami 1988; Hanawa, Hirotani and Kawai 1989; Spruit and Taam 1993), and instabilities in the mass flow in the disk associated with radiation drag effects (Milgrom 1987; Walker 1992). In all models, mass is accumulated up to a critical level above which instabilities develop to produce the intermittent accretion.

Accretion disks are a common ingredient in models since there is observational evidence which suggests that they are likely to be relevant in the theoretical interpretation. This follows from the approximate linear relationship between the energy of the burst, E_b, and the waiting time to the next burst, Δt (Lewin *et al.* 1976b). This behavior mimics that of a relaxation oscillator, and can be understood if there exists a reservoir where matter is accumulated and stored, suddenly released, and then replenished at a constant rate. Most workers identify the accretion disk surrounding the neutron star as this temporary storage medium. Accretion disks are an attractive candidate because the wide range of observed time scales finds a natural explanation in terms of the viscous diffusion time scale over different spatial regions in the disk.

Although the Rapid Burster exhibits properties characteristic of 'normal' low-mass X-ray binaries at times (Sect. 4.6.3), it is unique in its burst behavior. What makes the Rapid Burster unique in this regard? A number of possibilities have been proposed, including the alignment of the magnetic axis of the neutron star with its rotation axis (Hayakawa 1985), the location of the inner edge of the disk near the innermost marginally stable orbit (Milgrom 1987; Hanawa, Hirotani and Kawai 1989; Walker 1992), and the location of a magnetospheric boundary at a point where the Keplerian angular velocity is close to the angular velocity of the neutron star (Baan 1977, 1979; Spruit and Taam 1993).

4.7.3   *Theoretical models*

Nonmagnetic models of the type II burst phenomenon involve the operation of viscous, thermal or radiation feedback instabilities in an accretion disk. In these models, the mass transfer rate from the companion star to the disk can be constant, but physical processes in the disk lead to the modulation of the flow rate onto the neutron star. These models can, in principle, provide for a theoretical understanding of the observed E_b *vs.* Δt relation, since, for the more energetic bursts, a larger region of the disk is involved, and the time required to replenish the mass is correspondingly increased. Although many of the gross properties of the type II bursts (duration, amplitude and recurrence time scale) can be reproduced, all models yield strictly periodic behavior and the E_b *vs.* Δt relation could not be quantified without requiring a secular variation in the mass input rate into the disk or a variation of the viscosity from burst to burst.

A failing of such models is their generality, since they would also be applicable to other low-mass X-ray binary systems as well as to the Rapid Burster. Thus, it is unclear what essential ingredient distinguishes the Rapid Burster from all other X-ray sources in these models. One possible difference may be related to the strength of the neutron star's magnetic field. To prevent the hot inner disk (which is susceptible to the instabilities described by Lightman and Eardley (1974), Taam and Lin (1984), Meyer (1986), Milgrom (1987) and Walker (1992)) from extending down to the neutron star surface in other sources, a magnetosphere may be sufficiently large that the inner region is disrupted. Although the existence of such a field alleviates the problem, it does not remove it since, in these models, instability of a different type may be expected (see below).

Theoretical models incorporating the effects of magnetic fields are more attractive for explaining the type II burst phenomenon. In most models, the magnetosphere acts as the gate with the matter entering the magnetosphere as a consequence of large scale Rayleigh–Taylor instabilities (Baan 1977, 1979; Lamb *et al.* 1977; Singh and Duorah 1983). The accretion geometry can be either spherical (Baan 1977, 1979; Lamb *et al.* 1977; Singh and Duorah 1983) or disk-like (Baan 1977, 1979; Horiuchi, Kodonaga and Tomimatsu 1982; Hanami 1988; Hanawa, Hirotani and Kawai 1989; Spruit and Taam 1993). Of these models, those involving the angular momentum of the accreted matter are the most promising. Here, the accretion disk is disrupted by the magnetosphere surrounding a rotating neutron star, and the intermittent phases of accretion are presumed to result from the action of a plasma Rayleigh–Taylor instability (Kruskal and Schwarzschild 1954), perhaps accompanied by the occurrence of tearing-mode instability (Furth, Killeen and Rosenbluth 1963). However, the model as applied to the Rapid Burster by Hanami (1988) requires magnetic field strengths $\sim 10^{15}$ G.

Models involving a weak magnetic field ($\sim 10^8$ G) have been proposed by Hanawa, Hirotani, and Kawai (1989) in order to explain the lack of type II burst activity at luminosities $\sim 10^{37}$ erg s^{-1} (Kunieda *et al.* 1984b; Barr *et al.* 1987; see also Sect. 4.6.1). During this phase, they suggest that the magnetosphere shrinks to become smaller than the neutron star or the radius of the innermost stable orbit of the disk. At lower luminosities, the mass flow is assumed to be regulated in a nonsteady manner by the transfer of angular momentum from the accreting plasma to the neutron star via unspecified magnetic torques in the burst phase. The model, in its present version, is difficult to quantify due to the lack of physical understanding of the precise form for the magnetic torque.

A different viewpoint is adopted in the investigations of Baan (1977, 1979) and Spruit and Taam (1993). Specifically, they exploit the variations in the effective gravity at the inner edge of the disk as a means for modulating the mass accretion rate. In this scenario, the mass flow rate through the magnetosphere of a rotating neutron star is inhibited or reduced when the magnetopause is located outside the corotation point as a result of the centrifugal barrier which forms when the magnetosphere rotates more rapidly than the accreting matter (Illianarov and Sunyaev 1975). In this phase, matter accumulates outside the magnetosphere in the inner region of the disk. Eventually the surface density at the boundary rises to the point where it is sufficient to drive the magnetosphere inside the corotation radius, and matter accretes onto the neutron star as a result of the Rayleigh–Taylor instabilities. To provide for an explanation of the most energetic type II bursts ($\sim 10^{40}$ erg) the model by Baan (1977, 1979) requires magnetic moments comparable to those found for X-ray pulsars, which, based on the existence of type I bursts and the lack of coherent pulsations in the Rapid Burster, argues against such an interpretation.

In a recent study, Spruit and Taam (1993) have identified an instability associated with the magnetosphere–accretion disk interaction which operates in concert with the centrifugal-barrier gating mechanism. The burst profiles resulting from the nonlinear evolutionary calculations of the instability (Spruit and Taam 1993) are illustrated in Figures 4.26 and 4.27 for variations in the magnetospheric boundary surface density function and for variations in the average mass accretion rate, respectively. A variety of burst profiles and time scales can be produced. As can be seen, the length of the cycle can be expected to vary and can, in principle, be very long. Since the variation

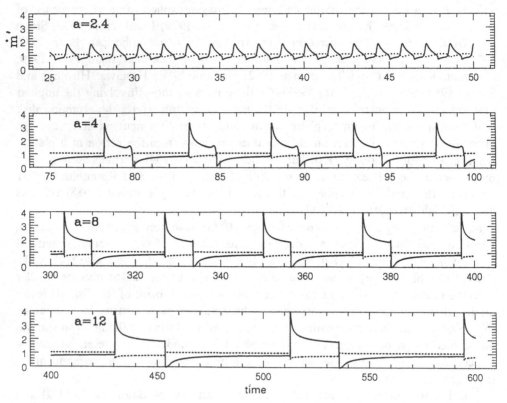

Fig. 4.26. The variation of the mass accretion rate and fastness parameter as a function of time for a steepness parameter of the boundary function ranging from 2.4 to 12. The solid curve corresponds to the accretion rate and the dashed curve to the fastness parameter. The accretion rate is normalized to the rate which places the magnetosphere at the corotation point in the steady-state case, and the unit of time is normalized to the viscous diffusion time from the inner edge of the disk.

of the magnetopause about the corotation point is essential for the instability, Spruit and Taam (1993) suggest that the Rapid Burster is near equilibrium spin for some phase in its evolution. It is argued that this parameter distinguishes the Rapid Burster from other low-mass X-ray binary systems. Although several properties of the Rapid Burster can be simulated, the model is unable to reproduce the correlation between the burst energy with the time interval to the next burst and the remarkable self similarity observed in the profiles of medium sized bursts (Kawai *et al.* 1990; Tan *et al.* 1991; see also Sect. 4.6.5). The failure to explain the E_b *vs.* Δt relation stems from the fact that, in its present form, the model yields strictly periodic bursts at a given mass accretion rate.

4.7.4   *Areas for future work*

The theoretical models developed to date are based upon simple considerations of the accretion physics. Since all models are based upon parameterizations of rather complex physical processes, detailed numerical work is clearly needed to

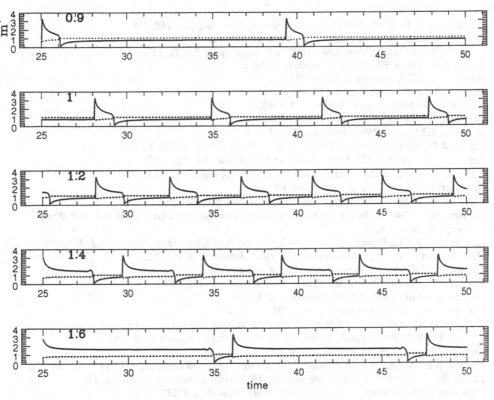

Fig. 4.27. Same as for Figure 4.26 except that the dependence of the burst properties on the average mass accretion rate ranging from 0.9 to 1.6 is illustrated. The accretion flow in the disk is steady for average accretion rates less than 0.82 and greater than 1.7 in this simulation.

place our understanding of the interactions between the many physical processes on a firm foundation. Specific areas for future work should involve multi-dimensional simulations of accretion disks, with the goal of identifying the mechanism responsible for the effective viscosity and elucidating the relevance of viscous and thermal instabilities. Equally important for further investigation are large scale studies of the Rayleigh–Taylor and Kelvin–Helmholtz instabilities in the accretion disk context.

Acknowledgements

This work was supported in part by the National Aeronautics and Space Administration under Grants NAGW-2526, NAGW-2935 and NAG8-700, and by the National Science Foundation under Grant PHY89-04035. This manuscript was written in part at the Institute for Theoretical Physics in Santa Barbara and at the Aspen Center for Physics. WHGL thanks L. M. Lubin for her valuable contributions. We thank T. Oosterbroek, L. M. Lubin, R. Rutledge and E. I. Faverey for their assistance in preparing many figures.

References

Abramenko, A.N. *et al.*: 1978, *Monthly Notices Roy. Astron. Soc.* **184**, 27P.

Aoki, T. *et al.*: 1992, *Publ. Astron. Soc. Japan* **44**, 641.

Apparao, K.V.M. and Chitre, S.M.: 1980, *Astrophys. Space Sci.* **72**, 127.

Ayasli, S. and Joss, P.C.: 1982, *Astrophys. J.* **256**, 637.

Baan, W.A.: 1977, *Astrophys. J.* **214**, 245.

Baan, W.A.: 1979, *Astrophys. J.* **227**, 987.

Barr, P. *et al.*: 1987, *Astron. Astrophys.* **176**, 69.

Basinska, E.M. *et al.*: 1980, *Astrophys. J.* **241**, 787.

Basinska, E.M. *et al.*: 1984, *Astrophys. J.* **281**, 337.

Belian, R.D., Conner, J.P., and Evans, W.D.: 1976, *Astrophys. J.* **206**, L135.

Bernacca, P.L., *et al.*: 1979, *Monthly Notices Roy. Astron. Soc.* **186**, 287.

Bildsten, L., Salpeter, E.E., and Wasserman, I.: 1992, *Astrophys. J.* **384**, 143.

Celnikier, L.M.: 1977, *Astron. Astrophys.* **60**, 421.

Clark, G.W., *et al.*: 1977, *Monthly Notices Roy. Astron. Soc.* **179**, 651.

Clayton, D.D.: 1968, *Principles of Stellar Evolution and Nucleosynthesis* (McGraw-Hill Book Company, New York).

Courvoisier, T.J.-L., Parmar, A.N., Peacock, A., and Pakull, M.: 1986, *Astrophys. J.* **309**, 265.

Czerny, M., Czerny, B., and Grindlay, J.E.: 1987, *Astrophys. J.* **312**, 122.

Damen, E.: 1990, Ph.D. Thesis, University of Amsterdam.

Damen, E., *et al.*: 1989, *Monthly Notices Roy. Astron. Soc.* **237**, 523.

Damen, E., *et al.*: 1990, *Astron. Astrophys.* **233**, 121.

Davidson, G.T.: 1982, *Astrophys. J.* **255**, 705.

Day, C.S.R., Fabian, A.C., and Ross, R.R.: 1992, *Monthly Notices Roy. Astron. Soc.* **257**, 471.

Dotani, T.: 1990: Ph.D. Thesis, University of Tokyo (*ISAS Research Note 418*).

Dotani, T., *et al.*: 1990, *Astrophys. J.* **350**, 395.

Doxsey, R.E., Apparao, K.M.V., Bradt, H.V., Dower, R.G., and Jernigan, J.G.: 1977, *Nature* **270**, 586.

Ebisuzaki, T., Hanawa, T., and Sugimoto, D.: 1983, *Publ. Astron. Soc. Japan* **35**, 17.

Fujimoto, M.Y., Hanawa, T., and Miyaji, S.: 1981, *Astrophys. J.* **247**, 267.

Fujimoto, M.Y., Hanawa, T., Iben, I., Jr., and Richardson, M.B.: 1984, *Astrophys. J.* **278**, 813.

Fujimoto, M.Y., *et al.*: 1985, in S. E. Woosley (ed.), *High Energy Transients in Astrophysics*, AIP Conf. Proc. 115, (AIP Press, New York), p. 302.

Fujimoto, M.Y., Hanawa, T., Iben, I., Jr., and Richardson, M.B.: 1987a, *Astrophys. J.* **315**, 198.

Fujimoto, M.Y., Sztjano, M., Van Paradijs, J., and Lewin, W.H.G.: 1987b, *Astrophys. J.* **319**, 902.

Fujimoto, M.Y., Sztajno, M., Lewin, W.H.G., and Van Paradijs, J.: 1988, *Astron. Astrophys.* **199**, L9.

Furth, H.P., Killeen, J., and Rosenbluth, M.N.: 1963, *Phys. Fluids* **6**, 459.

Fushiki, I. and Lamb, D.Q.: 1987, *Astrophys. J.* **323**, L55.

Fushiki, I., Taam, R.E., Woosley, S.E., and Lamb, D.Q.: 1992, *Astrophys. J.* **390**, 634.

Gallagher, J.S. and Starrfield, S.: 1978, *Ann. Rev. Astron. Astrophys.* **16**, 171.

Goldman, I.: 1979, *Astron. Astrophys.* **78**, L15.

Gottwald, M., Haberl, F., Parmar, A.N., and White, N.E.: 1986, *Astrophys. J.* **308**, 213.

Gottwald, M., Haberl, F., Parmar, A.N., and White, N.E.: 1987, *Astrophys. J.* **323**, 575.

Grindlay, J.E.: 1981, in R. Giacconi (ed.), *X-ray Astronomy with the Einstein Satellite* (Reidel, Dordrecht), p. 79.

Grindlay, J.E., *et al.*: 1976, *Astrophys. J.* **205**, L127.

Grindlay, J.E., *et al.*: 1978, *Nature* **274**, 567.

Haberl, F., Stella, L., White, N.E., Priedhorsky, W.C., and Gottwald, M.: 1987, *Astrophys. J.* **314**, 266.

Hackwell, J.A., *et al.*: 1979, *Astrophys. J.* **233**, L115.

Hanami, H.: 1988, *Monthly Notices Roy. Astron. Soc.* **233**, 423.

Hanawa, T. and Fujimoto, M.Y.: 1982, *Publ. Astron. Soc. Japan* **34**, 495.

Hanawa, T. and Fujimoto, M.Y.: 1984, *Publ. Astron. Soc. Japan* **36**, 119.

Hanawa, T. and Sugimoto, D.: 1982, *Publ. Astron. Soc. Japan* **34**, 1.

Hanawa, T., Hirotani, K., and Kawai, N.: 1989, *Astrophys. J.* **336**, 920.

Hanawa, T., Sugimoto, D., and Hashimoto, M.: 1983, *Publ. Astron. Soc. Japan* **35**, 491.

Hansen, C.J. and Van Horn, H.M.: 1975, *Astrophys. J.* **195**, 735.

Hayakawa, S.: 1981, *Space Sci. Rev.* **29**, 221.
Hayakawa, S.: 1985, *Physics Reports* **121**, 317.
Hoffman, J.A., Lewin, W.H.G., and Doty J.: 1977a, *Astrophys. J.* **217**, L23.
Hoffman, J.A., Lewin, W.H.G., and Doty J.: 1977b, *Monthly Notices Roy. Astron. Soc.* **179**, 57P.
Hoffman, J.A., Marshall, H.L., and Lewin, W.H.G.: 1978a, *Nature* **271**, 630.
Hoffman, J.A., Cominsky, L., and Lewin, W.H.G.: 1980, *Astrophys. J.* **240**, L27.
Hoffman, J.A., *et al.*: 1978b, *Astrophys. J.* **221**, L57.
Hoffman, J.A., *et al.*: 1978c, *Nature* **276**, 587.
Hoffman, J.A., *et al.*: 1979, *Astrophys. J.* **233**, L51.
Horiuchi, R., Kodonaga, T., and Tominatsu. A.: 1981, *Prog. Theor. Phys.* **66**, 172.
Hoshi, R.: 1981, *Astrophys. J.* **247**, 628.
Illianarov, A.F. and Sunyaev, R.A.: 1975, *Astron. Astrophys.* **39**, 185.
Inoue, H., *et al.*: 1980, *Nature*, **283**, 358.
Inoue, H., *et al.*: 1984, *Publ. Astron. Soc. Japan* **36**, 855.
Johnson, A.M., *et al.*: 1978, *Astrophys. J.* **222**, 664.
Jones, A.W., Selby, M.J., Mountain, C.M., Wade, R., Magro, C.S., and Munoz, M.P.: 1980, *Nature* **283**, 550.
Joss, P.C.: 1977, *Nature* **270**, 310.
Joss, P.C.: 1978, *Astrophys. J.* **225**, L123.
Joss, P.C. and Li, F.K.: 1980, *Astrophys. J.* **238**, 287
Joss, P.C. and Melia, F.: 1987, *Astrophys. J.* **312**, 700
Kaminker, A.D., *et al.*: 1990, *Astrophys. Space Sci.* **173**, 189.
Kato, M.: 1983, *Publ. Astron. Soc. Japan* **35**, 33.
Kawai, N.: 1985, Ph.D. Thesis, University of Tokyo (*ISAS Research Note 302*).
Kawai, N., *et al.*: 1990, *Publ. Astron. Soc. Japan*, **42**, 115.
Kleinmann, D.E., Kleinmann, S.G., and Wright, E.L.: 1976, *Astrophys. J.* **210**, L83.
Koyama, K., *et al.*: 1981, *Astrophys. J.* **247**, L27.
Kruskal, M. and Schwarzschild, M.: 1954, *Proc. Roy. Soc.* **A223**, 348.
Kulkarni, P.V., Ashok, N.M., Apparao, K.V.M., and Chitre, S.M.: 1979, *Nature* **280**, 819.
Kunieda, H., *et al.*: 1984a, *Publ. Astron. Soc. Japan* **36**, 215.
Kunieda, H., *et al.*: 1984b, *Publ. Astron. Soc. Japan* **36**, 807.
Lamb, D.Q. and Lamb, F.K.: 1978, *Astrophys. J.* **220**, 291.
Lamb, F.K., Fabian, A.C., Pringle, J.E., and Lamb, D.Q.: 1977, *Astrophys. J.* **217**, 197.
Langmeier, A., Sztajno, M., Hasinger, G., Trümper, J., and Gottwald, M.: 1987, *Astrophys. J.* **323**, 288.
Lawrence, A., *et al.*: 1983a, *Astrophys. J.* **267**, 301.
Lawrence, A., *et al.*: 1983b, *Astrophys. J.* **271**, 793.
Levine, A., *et al.*: 1988, *Astrophys. J.*, **327**, 732.
Lewin, W.H.G.: 1976a, *IAU Circular* No. 2918.
Lewin, W.H.G.: 1976b, *IAU Circular* No. 2922.
Lewin, W.H.G.: 1977, *Annals New York Acad. Sci.* **302**, 210.
Lewin, W.H.G.: 1984, in S.E. Woosley (ed.), *High Energy Transients in Astrophysics*, AIP Conf. Proc. 115 (AIP, New York), p. 249.
Lewin, W.H.G. and Joss, P.C.: 1977, *Nature* **270**, 211.
Lewin, W.H.G. and Joss, P.C.: 1981, *Space Sci. Rev.* **28**, 3.
Lewin, W.H.G. and Joss, P.C.: 1983, in W.H.G. Lewin and E.P.J. van den Heuvel (eds.), *Accretion Driven Stellar X-ray Sources*. (Cambridge University Press) p. 41.
Lewin, W.H.G., Hoffman, J.A., and Doty, J.: 1976, *IAU Circular* 2994.
Lewin, W.H.G., Ricker, G.R., and McClintock, J.E.: 1971, *Astrophys. J.* **169**, L17.
Lewin, W.H.G., Vacca, W.D., and Basinska, E.M.: 1984, *Astrophys. J.* **277**, L57.
Lewin, W.H.G., Van Paradijs, J., and Taam, R.E.: 1993, *Space Sci. Rev.* **62**, 223 (LVT93).
Lewin, W.H.G., *et al.*: 1976a, *Monthly Notices Roy. Astron. Soc.* **177**, 93P.
Lewin, W.H.G., *et al.*: 1976b, *Astrophys. J.* **207**, L95.
Lewin, W.H.G., *et al.*: 1976c, *Monthly Notices Roy. Astron.Soc.* **177**, 83P.
Lewin, W.H.G., *et al.*: 1978, *IAU Circular* 3190.
Lewin, W.H.G., *et al.*: 1980, *Monthly Notices Roy. Astron. Soc.* **193**, 15.
Lewin, W.H.G., *et al.*: 1987, *Astrophys. J.* **319**, 892.

Lewin, W.H.G., Lubin, L.M., Van Paradijs, J., and Van der Klis, M.: 1991, *Astron. Astrophys.* **248**, 538.

Li, F.K., *et al.*: 1977, *Monthly Notices Roy. Astron. Soc.* **179**, 21P.

Lightman, A.P. and Eardley, D.M.: 1974, *Astrophys. J.* **187**, L1.

Liller, W.: 1977, *Astrophys. J.* **213**, L21.

Livio, M. and Bath, G.T.: 1982, *Astron. Astrophys.* **116**, 286.

Lubin, L.M., *et al.*: 1991a, *Monthly Notices Roy. Astron. Soc.* **249**, 300.

Lubin, L.M., *et al.*: 1991b, *Monthly Notices Roy. Astron. Soc.* **252**, 190.

Lubin, L.M., *et al.*: 1992a, *Monthly Notices Roy. Astron. Soc.* **256**, 624.

Lubin, L.M., *et al.*: 1992b, *Monthly Notices Roy. Astron. Soc.* **258**, 759.

Lubin, L.M., *et al.*: 1993, *Monthly Notices Roy. Astron. Soc.* **261**, 149.

McClintock, J.E., Canizares, C., and Backman, D.E.: 1978, *Astrophys. J.* **223**, L75.

McClintock, J.E., *et al.*: 1979, *Nature* **279**, 47.

Magnier, E., *et al.*: 1989, *Monthly Notices Roy. Astron. Soc.* **237**, 729.

Makishima, K., *et al.*: 1983, *Astrophys. J.* **267**, 310.

Maraschi, L. and Cavaliere, A.: 1977, in: E.A. Müller (ed.), *Highlights in Astronomy*, (Reidel, Dordrecht), Vol. 4, Part I, p. 127.

Marshall, H.L.: 1982, *Astrophys. J.* **260**, 815.

Marshall, H.L., Ulmer, M.P., Hoffman, J.A., Doty, J., and Lewin, W.H.G.: 1979, *Astrophys. J.* **227**, 555.

Mason, K., Bell-Burnell, J., and White, N.E.: 1976, *Nature* **262**, 474.

Mason, K.O., Middleditch, J., Nelson, J.E., and White, N.E.: 1980, *Nature* **287**, 516.

Matsuoka, M., *et al.*: 1980, *Astrophys. J.* **240**, L137.

Matsuoka, M., *et al.*: 1984, *Astrophys. J.* **283**, 774.

Melia, F.: 1987, *Astrophys. J.* **315**, L43.

Meyer, F.: 1986, in D. Mihalas and K.H.A. Winkler (eds.), *Radiation Hydrodynamics in Stars and Compact Objects* (Springer Verlag, Berlin), p. 249.

Michel, F.C.: 1977, *Astrophys. J.* **216**, 838.

Milgrom, M.: 1987, *Astron. Astrophys.* **172**, L1.

Murakami, T., *et al.*: 1980a, *Astrophys. J.* **240**, L143.

Murakami, T., *et al.*: 1980b, *Publ. Astron. Soc. Japan* **32**, 543.

Murakami, T., Inoue, H., Makishima, K., and Hoshi, R.: 1987, *Publ. Astron. Soc. Japan* **39**, 879.

Nagase, F.: 1989, *Publ. Astron. Soc. Japan*, **41**, 1.

Nakamura, N., Inoue, H., and Tanaka, Y.: 1988, *Publ. Astron. Soc. Japan* **40**, 209.

Nakamura, N., *et al.*: 1989, *Publ. Astron. Soc. Japan* **41**, 617.

Nozakura, T., Ikeuchi, S., and Fujimoto, M.Y.: 1984, *Astrophys. J.* **286**, 221.

Oda, M.: 1982, in R.E. Lingenfelter, H.S. Hudson, and D.M. Worrall (eds.), *Gamma Ray Transients and Related Astrophysical Phenomena*, AIP Conf. Proc. 77, (AIP Press, New York), p. 319.

Ohashi, T.: 1981, Ph. D. Thesis, University of Tokyo (*ISAS Research Note* 141).

Ohashi, T., *et al.*: 1982, *Astrophys. J.* **258**, 254.

Paczynski, B. 1983a, *Astron. J.* **264**, 282.

Paczynski, B. 1983b, *Astron. J.* **267**, 315.

Paczynski, B. and Anderson, N.: 1986, *Astrophys. J.* **302**, 1.

Paczynski, B. and Proszynski, M.: 1986, *Astrophys. J.* **302**, 519.

Parmar, A.N., Gottwald, M., Van der Klis, M., and Van Paradijs, J.: 1989, *Astrophys. J.* **338**, 1024.

Pedersen, H., Van Paradijs, J., and Lewin, W.H.G.: 1981, *Nature* **294**, 725.

Pedersen, H., *et al.*: 1982a, *Astrophys. J.* **263**, 325.

Pedersen, H., *et al.*: 1982b, *Astrophys. J.* **263**, 340.

Penninx, W., Van Paradijs, J., and Lewin, W.H.G.: 1987, *Astrophys. J.* **321**, L67.

Pinto, P.A., Taam, R.E., and Laming, J.M.: 1994, *Astrophys. J.* in press.

Priedhorsky, W. and Terrell, J.: 1984, *Astrophys. J.* **284**, L17.

Sadeh, D., *et al.*: 1982, *Astrophys. J.* **257**, 214.

Sato, S., Kawara, K., Kobayashi, Y., Maihara, T., Okuda, H., and Jugaku, J.: 1980, *Nature* **286**, 668.

Schoelkopf, R.J. and Kelley, R.L.: 1991, *Astrophys. J.* **375**, 696.

Share, G., *et al.*: 1978, *IAU Circular* 3190.

Singh, L.M. and Duorah, H.L.: 1983, *Astrophys. Space Sci.* **92**, 143.

Spruit, H.C. and Taam, R.E.: 1993, *Astrophys. J.* **402**, 593

Stella, L., Kahn, S.M., and Grindlay, J.E.: 1984, *Astrophys. J.* **282**, 713.
Stella, L., Priedhorsky, W., and White, N.E.: 1987, *Astrophys. J.* **312**, L17.
Stella, L., *et al.*: 1988a, *Astrophys. J.* **324**, 379.
Stella, L., *et al.*: 1988b, *Astrophys. J.* **327**, L13.
Sugimoto, D., Ebisuzaki, T., and Hanawa, T.: 1984, *Publ. Astron. Soc. Japan* **36**, 839.
Swank, J.H., *et al.*: 1977, *Astrophys. J.* **212**, L73.
Sztajno, M., Basinska, E.M., Cominsky, L., Marshall, F.J. and Lewin, W.H.G.: 1983, *Astrophys. J.* **267**, 713.
Sztajno, M., *et al.*: 1985, *Astrophys. J.* **299**, 487.
Sztajno, M., *et al.*: 1986, *Monthly Notices Roy. Astron. Soc.* **222**, 499.
Sztajno, M., *et al.*: 1987, *Monthly Notices Roy. Astron. Soc.* **226**, 39.
Taam, R.E.: 1980, *Astrophys. J.* **241**, 351.
Taam, R.E.: 1981b, *Ap. Space Sci.* **77**, 257.
Taam, R.E.: 1982, *Astrophys. J.* **258**, 761.
Taam, R.E.: 1985, *Ann. Rev. Nuc. Part. Sci.* **35**, 1.
Taam, R.E. and Lin, D.N.C.: 1984, *Astrophys. J.* **287**, 761.
Taam, R.E. and Picklum, R.E.: 1978, *Astrophys. J.* **224**, 210.
Taam, R.E. and Picklum, R.E.: 1979, *Astrophys. J.* **233**, 327.
Takagishi, K., *et al.*: 1978, *ISAS Research Note* 60.
Tan, J., *et al.*: 1991, *Monthly Notices Roy. Astron. Soc.* **251**, 1.
Tawara, Y., Hayakawa, S., Kunieda, H., Makino, F., and Nagase, F.: 1982, *Nature* **299**, 38.
Tawara, Y., *et al.*: 1984a, *Astrophys. J.* **276**, L41.
Tawara, Y., Hirano, T., Kii, T., Matsuoka, M., and Murakami, T.: 1984b, *Publ. Astron. Soc. Japan* **36**, 861.
Tawara, Y., Kawai, N., Tanaka, Y., Inoue, H., Kunieda, H., and Ogawara, Y.: 1985, *Nature* **318**, 545.
Thomas, R.M., *et al.*: 1979, *Monthly Notices Roy. Astron. Soc.* **187**, 299.
Thorstensen, J., Charles, P., and Bowyer, S.: 1978, *Astrophys. J.* **220**, L131.
Tillett, J.C. and MacDonald, J.: 1992, *Astrophys. J.* **388**, 555.
Ulmer, M.P., Lewin, W.H.G., Hoffman, J.A., Doty, J., and Marshall, H.: 1977, *Astrophys. J.* **214**, L11.
Ulmer, M.P., *et al.*: 1978, *Nature* **276**, 799.
Vacca, W.D., Lewin, W.H.G., and Van Paradijs, J.: 1986, *Monthly Notices Roy. Astron. Soc.* **220**, 339.
Van der Klis, M., *et al.*: 1990, *Astrophys. J.* **360**, L19.
Van Paradijs, J.: 1978, *Nature* **274**, 650.
Van Paradijs, J.: 1979, *Astrophys.J.* **234**, 609.
Van Paradijs, J. and Lewin, W.H.G.: 1985, *Astron. Astrophys.* **157**, L10.
Van Paradijs, J., Pedersen, H., and Lewin, W.H.G.: 1981, *IAU Circular* 3626.
Van Paradijs, J., Joss, P.C., Cominsky, L., and Lewin, W.H.G.: 1979a, *Nature* **280**, 375.
Van Paradijs, J.A., Cominsky, L., and Lewin, W.H.G.: 1979b, *Monthly Notices Roy. Astron. Soc.* **189**, 387.
Van Paradijs, J., Verbunt, F., Van der Linden, T., Pedersen, H., and Wamsteker, W.: 1980, *Astrophys. J.* **241**, L161
Van Paradijs, J., *et al.*: 1986, *Monthly Notices Roy. Astron. Soc.* **221**, 617.
Van Paradijs, J., Penninx, W., and Lewin, W.H.G.: 1988a, *Monthly Notices Roy. Astron. Soc.* **233**, 437.
Van Paradijs, J., Penninx, W., Lewin, W.H.G., Sztajno, M., and Trümper, J.: 1988b, *Astron. Astrophys.* **192**, 147.
Van Paradijs, J., Dotani, T., Tanaka, Y., and Tsuru, T.: 1990a, *Publ. Astron. Soc. Japan* **42**, 633.
Van Paradijs, J., *et al.*: 1990b, *Astron. Astrophys.* **234**, 181.
Verbunt, F., *et al.*: 1993, *Proceedings of Cospar Meeting*, Washington, D.C., September 1992.
Waki, I., *et al.*: 1984, *Publ. Astron. Soc. Japan*, **36**, 819.
Walker, M.A.: 1992, *Astrophys. J.* **385**, 651.
Walker, M.A. and Meszaros, P.: 1989, *Astrophys. J.* **346**, 844.
Wallace, R.K. and Woosley, S.E.: 1981, *Astrophys. J. Suppl.* **45**, 389.
Wallace, R.K., Woosley, S.E.: 1985, in S.E. Woosley (ed.) High Energy Transients in Astrophysics, AIP Conf. Proc. 115 (AIP Press, New York), p. 319.
Wallace, R.K., Woosley, S.E. and Weaver, T.A.: 1982, *Astrophys. J.* **258**, 696.

232   *Walter H. G. Lewin, Jan Van Paradijs and Ronald E. Taam*

White, N.E., Mason, K.O., Huckle, H.E., Charles, P.A., and Sanford, P.W.: 1976, *Astrophys. J.* **209**, L119.
White, N.E., *et al.*: 1978, *Monthly Notices Roy. Astron. Soc.* **184**, 1P.
Woosley, S.E. and Taam, R.E.: 1976, *Nature* **263**, 101.
Woosley, S.E. and Weaver, T.A.: 1985, in S. E. Woosley (ed.) *High Energy Transients in Astrophysics*, AIP Conf. Proc. 115 (AIP Press, New York), p. 273.

5

Millisecond pulsars

D. Bhattacharya

Raman Research Institute, Bangalore 560080, India

5.1   Overview

The discovery of the 1.6-millisecond pulsar PSR 1937+21 in 1982 [3] marked the beginning of a new era in pulsar astrophysics. This pulsar was unique in having a spin period very close to the minimum attainable by a neutron star and a magnetic field nearly three orders of magnitude lower than most known pulsars. In addition to generating a substantial amount of discussion regarding the structure of rapidly rotating neutron stars, the magnetospheric properties and radiation mechanisms, the most important thing this discovery did was to highlight the presence of a class of pulsars whose origin is closely tied to the evolution of interacting binary systems. In a decade since then, more millisecond pulsars have been discovered, and their connection with binary systems has been further strengthened.

The fact that a short-period pulsar could be produced due to the spin-up of a neutron star in a binary system had already been recognized with the discovery of the first binary pulsar PSR 1913+16 [23]. Along with this also came the realization that the magnetic field of the neutron star would dictate to what extent it could be spun up – the lower the field strength, the faster the possible final spin [66, 69]. The possible existence of an entire population of spun-up pulsars (also called 'recycled' pulsars) had also been envisaged before the discovery of millisecond pulsars [56], but this idea gained a widespread support and attention only after PSR 1937+21 was discovered.

The short period of PSR 1937+21 and the absence of any sign of its being a young object (such as there being an associated supernova remnant) immediately led to the conjecture that it is a recycled pulsar [55]. The magnetic field strength predicted on the basis of this hypothesis was confirmed with the measurement of its spin-down rate, lending much credence to the proposed origin of this pulsar [1]. A difficulty that remained with this explanation was that, unlike, PSR 1913+16, this pulsar had no companion! However, the discovery of a 6-ms binary millisecond pulsar (PSR 1953+29) soon afterwards [11] supported the idea of the origin of millisecond pulsars from binary systems [44, 64], and suggestions were made that PSR 1937+21 must have lost its companion in some way.

Among the many suggestions of how the companion of the 1.6-ms pulsar could have been lost, the most important one was made by Ruderman *et al.* [60] – that the companion could have been vaporized by the strong radiation emerging from the millisecond pulsar. This idea received a dramatic support with the discovery of PSR 1957+20, a 1.6-ms pulsar whose companion is in a state of being vaporized [20]! The discovery of this pulsar demonstrated that the presence of a millisecond pulsar as a companion may very significantly influence the evolution of a normal star, as well as that of the binary system as a whole. A great deal of interest was thus generated in the evolution of irradiated stars and of the binaries containing them. This is one of the active areas of research at present.

To spin-up a neutron star to periods as short as a few milliseconds requires a rather large quantity of mass ($\sim 0.1 M_\odot$) to be transferred on to it. At an Eddington accretion rate ($\dot{M}_{\text{Edd}} \simeq 10^{-8} \ M_\odot \ \text{yr}^{-1}$), this would take $\sim 10^7$ yr to achieve. The duration of the mass transfer phase in most of the *massive* binaries would be much less than this. Therefore, the view commonly held is that millisecond pulsars are descendants of *low-mass* X-ray binaries, in which the donor stars have mass $\lesssim 1 M_\odot$. The nature of the companions of the binary millisecond pulsars also seems to support this conclusion. This has led several groups to look for millisecond pulsars in globular clusters, which are much richer in LMXBs than the galactic disk. These surveys have already discovered more than 30 pulsars in globular clusters, of which the majority have periods shorter than 10 ms.

While the connection with LMXBs motivated the search for millisecond pulsars in globular clusters, the unexpectedly large number of discoveries has posed a new problem. Statistical studies suggest that the birth rate of millisecond pulsars is much larger than the birth rate of detectable low-mass X-ray binaries in the globular clusters, and this is perhaps true also in the galactic disk [31, 32] (Sect. 11.5.2). The solution of this problem is not yet entirely clear – perhaps the active lifetimes of LMXBs are not as long as is usually assumed, or a fraction of millisecond pulsars originate from binary systems which do not manifest themselves as strong X-ray sources, or both. Several suggestions have been made in this regard, and more are forthcoming.

In the sections to follow we shall discuss all of the above points in greater detail, as well as several related issues.

5.2   The general characteristics of millisecond pulsars

The term *millisecond pulsars* usually refers to pulsars with spin periods $\lesssim 10$ milliseconds. The limit at ~ 10 ms is, however, rather arbitrary, and need not include all pulsars with similar characteristics and evolutionary history. We shall, therefore, regard this as a loose definition and include pulsars with somewhat longer spin periods too in the discussion whenever necessary.

Apart from their short spin periods, the other distinguishing character of the millisecond pulsars is their weak magnetic fields. In all cases where the spin-down rate ($\dot{P}$) of millisecond pulsars has been measured, the derived dipole field strength ($B \propto \sqrt{P\dot{P}}$) turns out to be $\lesssim 10^9$ G, about three orders of magnitude lower than the derived magnetic fields of most known pulsars. The magnetic field–period diagram of Figure 5.1 shows the uniqueness of this population: millisecond pulsars occupy the lower left corner of the diagram, whereas most of the normal pulsars are located at the upper right.

An overwhelmingly large fraction of the millisecond pulsars are members of binary systems. Among the millisecond pulsars known in the galactic disk, $\sim 90\%$ are in binaries, whereas among all known radio pulsars the binary fraction is as small as $\sim 5\%$. The fraction of binaries among the millisecond pulsars in globular clusters is

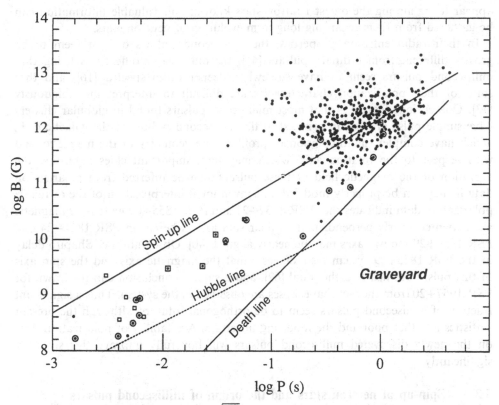

Fig. 5.1. The derived magnetic fields ($\propto \sqrt{P\dot{P}}$) of known radio pulsars plotted against their spin periods P. Pulsars in binaries are shown encircled. Squares denote pulsars in globular clusters; binaries among them are shown with dots inside the squares. The 'death line' corresponds to a polar cap voltage below which the pulsar activity is likely to switch off. The 'Hubble line' represents a spin-down age $\tau \equiv P/2\dot{P}$ of 10^{10} yr for a pulsar. The 'spin-up line' represents the minimum period to which a neutron star can be spun up in an Eddington-limited accretion. Pulsars in the lower left corner are the millisecond pulsars. This figure was updated on July 11, 1993. Notice the difference between this figure and Figures 12.1 and 12.4, which were not updated; this field is moving very rapidly!

somewhat smaller, $\sim$ 50%, than that in the disk, but is still much larger than that among the ordinary disk pulsars. This clearly suggests an intimate link between the evolution of binary systems and the origin of millisecond pulsars. Except in a few cases in globular clusters, the orbits of binaries containing millisecond pulsars are very nearly circular. In practically all known cases, the companion stars of millisecond pulsars are of low mass, with the most probable mass being $\lesssim 0.3 M_\odot$.

By all appearance, the neutron stars that function as millisecond pulsars are very old. The occurrence of millisecond pulsars in globular clusters is in itself one such indication, since the massive stars capable of producing neutron star remnants existed only in the early phase of evolution of globular clusters. The surface temperature of the white dwarf companion of PSR 1855+09, a 5.4-ms pulsar, suggests that it is at least 10^9 yr old [12, 30]. This is a lower limit on the age of the neutron star, since it must have been formed before the white dwarf. Thus the millisecond pulsars appear to be among the oldest neutron stars known, and valuable information can be gathered from them regarding long-term evolution of neutron stars.

In their radio emission properties, the millisecond pulsars do not seem to be grossly different from ordinary pulsars [41], the only major differences being that millisecond pulsars seem to have somewhat steeper radio spectra [16], and that some of their polarization characteristics are difficult to interpret unambiguously [73]. One curious feature is that most millisecond pulsars found in globular clusters have simple pulse profiles, in contrast to the millisecond pulsars in the galactic disk, which have complex, multi-component profiles. The geometry of the magnetic field with respect to the rotation axis, which may have important clues regarding the evolution of the magnetic field of these pulsars, can be inferred from polarization data if they can be properly modeled. A conventional interpretation of the available polarization data indicates that PSR 1937+21 and PSR 1855+09 have their magnetic axes oriented nearly perpendicular to their spin axes, whereas in PSR 1821−24 and PSR 1953+29 the two axes may be nearly aligned [40]. Observation of Shapiro delay in the PSR 1855+09 system also suggests that the magnetic axis and the spin axis of this pulsar are nearly orthogonal [62], and a similar conclusion can be drawn for PSR 1957+20 from the fact that eclipses are observed in the system. Thus a significant fraction of millisecond pulsars seem to be orthogonal rotators, although the present statistics is rather poor and the modeling uncertain. Availability of polarization data on the newly discovered millisecond pulsars will hopefully improve this situation significantly.

5.3   Spin-up of neutron stars and the origin of millisecond pulsars

Millisecond pulsars are believed to acquire their rapid spin via spin-up during the phase of accretion from their companions. In most of the pulsating X-ray sources in which the neutron star accretes through an accretion disk, the spin period is seen to decrease continuously; the shortest period observed in such a system is 69 ms, which is still decreasing (cf. the X-ray pulsar A0538−66) [42], (see Ch. 1). It seems reasonable to suppose that if such a spin-up is allowed to proceed far enough, a millisecond-period neutron star may be the result.

The extent to which spin-up can proceed is determined by several factors: the magnetic field of the neutron star; the accretion rate; and the duration of the accretion phase. The matter being accreted, having left the donor star (secondary), settles into an accretion disk and proceeds towards the accretor (the neutron star) by gradually losing angular momentum through viscous exchange due to differential rotation in the disk. At any radius, the matter in the accretion disk rotates at the local Keplerian velocity. Significant interaction of this matter with the neutron star occurs when the matter arrives at the magnetospheric boundary of the neutron star. At this point, the neutron star's magnetic field is strong enough to dominate the dynamics and causes the disk matter to corotate with the neutron star. If, in doing so, it needs to accelerate the disk matter to a higher angular velocity, the neutron star would lose spin angular momentum, whereas if corotation would demand that the disk matter must be slowed to a sub-Keplerian speed, then a part of the excess angular momentum carried by the disk matter would be transferred to the neutron star via magnetic torques. Given enough time, therefore, an equilibrium state will be achieved with the corotation speed at the magnetospheric boundary equalling the Keplerian speed at the same point. The spin period of the neutron star in this state would equal the period of Keplerian motion at the magnetospheric boundary. This is the so-called 'equilibrium period':

$$P_{eq} = \frac{2\pi}{\Omega_K(r = R_{mag})} = 2\pi \left(\frac{R_{mag}^3}{GM}\right)^{1/2}, \tag{5.1}$$

where Ω_K denotes the Keplerian velocity and R_{mag} is the radius of the magnetosphere. M is the mass of the neutron star and G is the gravitational constant.

While the correct effective radius R_{mag} of the magnetosphere is difficult to estimate and is model-dependent, a fairly good estimate of P_{eq} is obtained by setting $R_{mag} = R_A$, the Alfvén radius, defined as the distance from the neutron star at which the pressure due to the neutron star's magnetic field equals the ram pressure of the infalling matter, assuming both to be spherically symmetric:

$$R_A = \left(\frac{B_s^2 R_s^6}{\dot{M}\sqrt{2GM}}\right)^{2/7}, \tag{5.2}$$

where B_s is the dipole field strength at the surface of the neutron star of radius R_s and $\dot{M}$ is the mass accretion rate. Combining Eqs. (5.1) and (5.2), one finds

$$P_{eq} = 1.9 \text{ ms } B_9^{6/7} \left(\frac{M}{1.4M_\odot}\right)^{-5/7} \left(\frac{\dot{M}}{\dot{M}_{Edd}}\right)^{-3/7} R_6^{16/7}, \tag{5.3}$$

where B_9 is the surface dipole field of the neutron star in units of 10^9 G, R_6 is the neutron star radius in units of 10^6 cm and $\dot{M}_{Edd}$ is the maximum possible rate of spherical accretion (called the Eddington rate) on the neutron star, beyond which the pressure of the accretion-generated radiation would stop the accretion flow. The value of $\dot{M}_{Edd}$ for a neutron star is $\sim 10^{-8}$ $M_\odot$ yr^{-1}.

According to this picture, therefore, for a neutron star to be spun-up to periods as short as a few milliseconds, its surface dipole field strength must necessarily be low: $\lesssim 10^9$ G. In all observed millisecond pulsars, this does happen to be the case.

Even if the condition of low magnetic field and a near-Eddington accretion rate is fulfilled, the neutron star may still not achieve the ultra-rapid spin unless the accretion phase lasts long enough. The matter being accreted brings with it angular momentum, a substantial amount of which must be deposited on the neutron star before it can spin at a millisecond period. The maximum rate of angular momentum accretion is, roughly speaking, the product of the accretion rate and the specific Keplerian angular momentum at the magnetospheric boundary. At this rate of angular momentum enhancement, the neutron star could reach the equilibrium spin period after a total mass of $\Delta M \sim 0.1 M_\odot (P_{eq}/1.5 \text{ ms})^{-4/3}$ has been accreted. At the Eddington rate, the spin-up to a millisecond period would thus require a sustained accretion for at least $\sim 10^7$ yr. If the donor star has a mass well above the mass of the neutron star, such a long duration of heavy mass transfer is very unlikely to occur: soon after the beginning of the Roche lobe overflow, mass transfer would become unstable, a common envelope will form, and evolution will last for $\lesssim 10^4$ yr. Because of this reason, the progenitors of millisecond pulsars are thought to be the low-mass X-ray binaries, in which the donor is a star less massive than the neutron star. In these binaries, as mass transfer proceeds, the orbit tends to expand, keeping the mass transfer stable. The slow evolution of the donor star on account of its low mass ensures a prolonged mass transfer phase, enough to spin the neutron star up to millisecond periods. In the following section the evolution of low-mass X-ray binaries is briefly sketched.

A related question of considerable importance is what happens to the spin of the neutron star if the accretion continues much longer than that necessary to achieve millisecond rotation? Magnetic field strength permitting, would the neutron star continue to spin-up to sub-millisecond periods? The answer to this depends on the equation of state of dense matter which composes the neutron star. The stability of the star against disruption by centrifugal forces requires that the rotation period be larger than a certain P_{min}, which is a function of the equation of state. A variety of such equations of state have been constructed using different ways of extrapolation from known laboratory physics. No observational handle is, however, available at present to clearly distinguish between them. For the equations of state so far proposed, P_{min} lies in the range 0.5–1.5 ms, the 'harder' the equation of state, the larger being the value of P_{min} [18, 65]. An additional effect that may limit the spin rate of the neutron star is the following. As the star is being spun-up, it may become unstable against physical deformations and the consequent emission of gravitational radiation as P_{min} is approached. An equilibrium may set in between the angular momentum coming in through the accretion flow and that going out in gravitational radiation [45, 85]. This, in effect, may raise the value of P_{min} by $\sim$ 10–30% [17, 35], unless viscosity is able to suppress this instability [36].

The fastest pulsar known at present has a spin period of 1.56 ms. Observations, however, do not preclude existence of even faster pulsars, since none of the pulsar surveys have been conducted with sufficient sensitivity for sub-millisecond periods. Search for sub-millisecond pulsars is a computationally demanding task, but advances in computing technology may soon make it possible to conduct such searches routinely. Discovery of a sub-millisecond pulsar will, for the first time, provide meaningful constraints on the neutron star equation of state, and hence considerable attention is likely to be devoted to this area in the near future.

5.4   Evolution of LMXBs: the standard model

One of the important issues surrounding millisecond pulsars is their evolutionary connection to low-mass X-ray binaries (LMXBs). As mentioned above, LMXBs seem to be the most likely candidates to provide the right conditions for the spin-up of neutron stars to millisecond periods. In this section we briefly describe the evolution of LMXBs, referring the reader to other, more comprehensive, reviews [9, 74, 79] and Ch. 11 for details. In the standard model of evolution, LMXBs can be divided into three broad classes according to their initial orbital period P_{b0}.

Wide systems: $P_{b0} \gtrsim 1$–2 d. In these systems, the secondary star fills its Roche lobe after evolving away from the main sequence, i.e. in the sub-giant phase or the giant phase of evolution. Since the donor star is less massive than the accretor, the orbit widens as mass is transferred. However, due to nuclear evolution, the secondary expands too, and sustains the mass transfer as a result. The mass transfer and consequent expansion of the orbit continues till only the degenerate core of the donor is left. At this point, the binary detaches and mass transfer stops. The expected end-product of this evolution is a 0.2–$0.4 M_\odot$ white dwarf in a wide circular orbit, with orbital period P_b in the range tens of days to years. The increase in the orbital period from the initial state (beginning of mass transfer) to the final configuration is typically by an order of magnitude, with some variation over the range of systems. The wider the original binary, the later in the evolutionary phase of the secondary does the system come into contact. Since the rate of expansion of a star due to nuclear evolution progressively increases on the giant branch, in an initially wider orbit the mass transfer occurs at a higher rate in comparison to the system where the orbit was initially smaller. The duration of the mass transfer phase is thus correspondingly smaller in wider orbits. For a $1 M_\odot$ donor of solar composition, the average rate of mass transfer in wide-orbit LMXBs is $\langle \dot{M} \rangle \approx 8 \times 10^{-10} (P_{b0}/1 \text{ d}) \ M_\odot \text{ yr}^{-1}$, and the corresponding X-ray lifetime is $\tau_x \approx 10^9 (1 \text{ d}/P_{b0})$ yr [14, 79, 86].

Close systems: $P_{b0} \lesssim 12$ h. These systems are brought into contact by loss of orbital angular momentum due to gravitational radiation and magnetic braking while the secondary is still on the main sequence. Magnetic braking causes loss of orbital angular momentum in the following way. Due to tidal interaction in the binary system, the secondary is brought into synchronous rotation with the orbit (spin period = orbital period). Stellar wind from the secondary escapes from the system and exerts a spin-down torque on the secondary via the secondary's magnetic field. As the secondary spins down, tidal effects attempt to spin it up again at the cost of the orbital angular momentum. Thus the magnetic torque exerted on the secondary by its wind results in the loss of angular momentum from the orbit [84]. After the system is brought into contact, mass transfer ensues. Since the secondary is on the main sequence, on loss of mass its radius decreases. In absence of angular momentum loss, this would terminate mass transfer as the secondary would sink into the Roche lobe – since, due to mass transfer, its radius would decrease and at the same time the orbit would expand as a result of the donor being less massive than the accretor. However, angular momentum loss drives the components closer together and the binary stays in contact. In this process, the orbit continues to shrink as mass transfer proceeds. If mass transfer continues uninterrupted, these systems would pass through a minimum orbital period of ~ 80 min, following which the secondary becomes degenerate, its mass–radius relation reverses (i.e. it expands on loss of mass) and the orbit begins to expand again. If this course of evolution is strictly followed, mass transfer is unlikely to finish in a Hubble time. However, several effects may intervene and turn off the mass transfer, as we shall discuss in the following section.

Intermediate systems: 12 h $\lesssim P_{b0} \lesssim 1$–$2$ d. In these systems, both processes, namely the nuclear evolution of the secondary and angular momentum loss from the system, are important. The outcome depends rather strongly on the assumed rate of magnetic braking. The expected end product in this case would be a low-mass degenerate dwarf (core of a sub-giant) in circular orbit around the neutron star, with orbital period between several hours and a few days [52, 53].

It is evident from the above that the wide LMXBs provide the ideal conditions for the spin-up of a neutron star to millisecond periods, and the first binary millisecond pulsar to be discovered, namely PSR 1953+29, fits this description perfectly. Its 117-d orbital period, near-perfect circular orbit (eccentricity $= 3.3 \times 10^{-4}$), and low mass of the companion (0.2–$0.4 M_\odot$) are all exactly as expected of a wide LMXB product. The initial orbital period of this system would have been ~ 12 d, and the average rate of mass transfer $\sim 6 \times 10^{-9} M_\odot$ yr^{-1}, a little below the Eddington rate, consistent with the spin period of the neutron star ($P = 6.13$ ms at present, but would have been shorter just after the end of the mass transfer) [11, 44, 64]. The orbital characteristics of the millisecond pulsar discovered next, PSR 1855+09, also were fully consistent with the wide-LMXB evolution scenario (see [53] and Ch. 11 for detailed evolutionary models).

The discovery of these millisecond pulsars in wide binaries represented major successes of the wide-LMXB evolution theory [86] and the hypothesis that millisecond pulsars originate from LMXBs [1]. But, as discussed above, in addition to the right mass-transfer conditions, spin-up to millisecond periods also requires the magnetic field of the neutron star to be low. One would obviously ask if there is evidence for this in the low-mass X-ray binary systems. It turns out that the absence of detectable X-ray pulsations in LMXBs has for a long time been attributed to the weak field strength of the accreting star. If the magnetic field strength of the neutron star is low, the incoming matter is not well collimated and accretes onto a large fraction of the stellar surface, heating it uniformly. The absence of pronounced polar hot spots would explain the absence of detectable pulsations. Further, many LMXBs are seen to undergo X-ray bursts, which are thought not to occur if the magnetic field of the neutron star is strong (see Ch. 4). Thus there seems to be reasonable evidence that most LMXBs harbor low-field neutron stars.

Given the low field strength and the continuing accretion, it is likely that in many LMXBs the neutron star is already spinning at millisecond periods. No direct evidence of this can be obtained unless pulsations in LMXBs are detected. Some indirect evidence of rapid spin of the accreting stars in LMXBs has, however, emerged in recent years through the quasi-periodic oscillation (QPO) phenomenon [77] (see also Ch. 6). These oscillations (presently called HBOs), with frequencies in the range 15–55 Hz, are best explained as the 'beat frequency' of the Keplerian motion at the magnetospheric boundary and the spin rate of the neutron star [2]. At this frequency, blobs of matter moving in Keplerian orbits at the edge of the magnetosphere encounter the open field line cusps above the magnetic poles of the rotating neutron star, where they can enter the magnetosphere, fall to the neutron star surface, and produce X-ray emission. If this picture of the origin of HBOs is correct, then the observed variation of the HBO frequency with the X-ray luminosity (and hence accretion rate) yields a spin period of ~ 3–20 ms for the neutron star in the LMXB GX 5–1 [34], thus indicating that spin-up to millisecond periods is indeed possible in LMXBs.

5.5   Statistics

Despite the above arguments in favor of LMXBs being the progenitors of millisecond pulsars, this hypothesis has been constantly challenged on the basis of statistical estimates. If the LMXBs are indeed the progenitors of millisecond pulsars, in a steady state the birth rate of these two classes of objects must match, and here lies the controversy.

The first attempts to compute the birth rate of millisecond pulsars were made after the discovery of PSR 1855+09 [8, 76]. This immediately led to interesting results. The prevalent opinion at that time regarding the evolution of neutron star magnetic fields was that, starting from a high value $\sim 10^{12}$ G, the field strength decreases continuously due to ohmic dissipation, with a time constant of $\lesssim 10$ Myr. The crude statistics based on three millisecond pulsars, however, already showed that if LMXB birth rates and millisecond pulsar birth rates are to match, such a field evolution could not be correct, not at least for the millisecond pulsars. The equality of the two birth rates demanded active lifetimes in excess of 10^9 yr for millisecond pulsars, whereas, according to the above model for field decay, their active lifetimes could not be longer than a few times 10^7 yr. Thus, millisecond pulsar fields had to be long-lived, a conclusion which in the following years has led to a series of studies causing fundamental changes in our ideas of neutron star magnetic field evolution (see Ch. 12 for a review).

A much more detailed analysis of the statistics of millisecond pulsars was undertaken following the above preliminary results, carefully taking into account the selection effects which limit our ability to detect millisecond pulsars [31]. In this method, the estimate of the total number of millisecond pulsars in the galaxy is obtained by assuming that, for every observed pulsar i, the galaxy contains S_i pulsars with similar characteristics, where $1/S_i$ is the fraction of the galactic volume within which pulsars with characteristics similar to those of pulsar i could have been detected in the present surveys.

The result of this exercise was striking: the estimated total number of active millisecond pulsars in the galaxy turned out to be $\sim 10^5$, a number too large to be accommodated as the descendants of LMXBs, even allowing an active lifetime equal to a Hubble time for a millisecond pulsar. The total number of LMXBs known in the galaxy is ~ 125 (Ch. 14). A ratio of $\sim 10^3$ between the number of millisecond pulsars and that of LMXBs meant that if the LMXB–millisecond pulsar link has to survive, the lifetimes of LMXBs must be $\lesssim 10^7$ yr, a factor of 10–100 smaller than the predictions from the standard models. Thus originated the suggestion that LMXBs may not be the real progenitors of millisecond pulsars. While they may contribute to the millisecond pulsar population, most of the millisecond pulsar population originates in a different way [31].

Since then, several more millisecond pulsars have been discovered, but the estimated size of the millisecond pulsar population has remained more or less the same. After a revision of the distance to the very near millisecond pulsars, it was argued that the total number of millisecond pulsars may be smaller than the above estimate by about an order of magnitude [26], but the discovery since then of at least half a dozen close-by low-luminosity millisecond pulsars has brought the estimated number back to $\sim 10^5$ [4]. Thus, it is becoming increasingly clear that low-mass X-ray binaries, at least according to the standard evolutionary model, would be hopelessly inadequate to generate the galactic population of millisecond pulsars.

This statistical problem is, however, not the only one facing the standard model of LMXB evolution, particularly in the context of generating millisecond pulsars. Another obvious problem, recognized quite early, was the absence of a companion to PSR 1937+21. It seemed extremely unlikely that after settling into an orbital configuration similar to that of PSR 1953+29 or PSR 1855+09 the companion of the pulsar could have been lost in some way, unless an encounter with a third star 'ionized' the binary – an event of near-vanishing probability in the galactic disk. Suggestions were made that PSR 1937+21 could be a descendant of a *close* LMXB, in which the secondary has been tidally disrupted after going through the period minimum [54], but there were many difficulties with this. First of all, whether such a tidal disruption is a physical possibility was itself doubtful [50, 81]. In addition, it was pointed out that during this course of evolution the mass transfer rates would be very low, $\lesssim 10^{-3} \dot{M}_{Edd}$, and the equilibrium period of the neutron star would be much longer than that required for it to eventually function as a radio pulsar [24].

A related problem has been the absence of LMXBs with X-ray luminosities below $\sim 10^{35}$ erg s^{-1}. This problem is due to the fact that the standard model makes no definite prediction about how the mass transfer in the close systems ($P_{b0} \lesssim 12$ h) should end. If the full course of standard evolution is followed, the mass-transfer rate in such a system should continue to drop and the system should become progressively fainter in X-rays as it crosses the period minimum. Further, the evolution would also become slower and slower, and as a result the sources would 'pile up' at low X-ray luminosities, $\lesssim 10^{35}$ erg s^{-1}. This, however, is not observed [37]. If this is not due to observational selection, it indicates that either most LMXBs are 'born wide', or that mass transfer is somehow prevented from entering the low-$\dot{M}$ phase. The former seems unlikely, since many LMXBs are *known* to have $P_b < 10$ h.

Prevention of the low-$\dot{M}$ phase also seems to be indicated by the paucity of LMXBs with orbital periods less than ~ 2 h. To illustrate this point, it is instructive to compare the distribution of orbital periods (see Figure 1.1) of low-mass X-ray binaries and cataclysmic variables (CVs). The latter are systems similar to LMXBs, but the accreting compact objects in them are white dwarfs rather than neutron stars (for a review, see Ch. 8). The main difference in the period distributions of these two species is, however, that a large number of CVs have orbital periods between 1.3 h and 2 h, where LMXBs are virtually absent. In fact, the distribution of the orbital periods of CVs shows a 'gap' between 2 h and 3 h. This is thought to occur due to a sudden loss of magnetic braking as the donor becomes completely convective [57, 67]. Above the period gap, magnetic braking drives the mass transfer in a time scale shorter than the thermal time scale of the secondary, causing it to depart from thermal equilibrium and inflate. When magnetic braking disappears, the time scale for orbital decay (which is now driven by gravitational radiation alone) lengthens by an order of magnitude, during which time the secondary can cool and shrink within the Roche lobe. The system comes into contact again when the orbital period has been reduced to ~ 2 h.

LMXBs, on the other hand, become rare at periods below the gap. This indicates that most LMXBs are unable to emerge from the period gap and resume mass transfer. The low-$\dot{M}$ phase of standard evolution, which lies below the period gap, apparently does not occur in most LMXBs.

This has led to a number of suggestions for modifying the standard model of LMXB evolution; these are briefly discussed in the next section.

5.6 Evolution of LMXBs: beyond the standard model

The inadequacy of the standard model of LMXB evolution was demonstrated clearly with the discovery of the 1.6-ms pulsar PSR 1957+20, the companion of which is being vaporized by the wind of the pulsar [20]! Indeed, the existence of such systems was anticipated in a remarkable paper by Ruderman *et al.* [60]. This turns out to be a missing link in the evolutionary path from binary millisecond pulsars to singles. Timing data show that the orbital period of this binary is decreasing on a very short time scale ($\sim$ 30 Myr) due to heavy loss of mass and associated angular momentum from the system [63]. This suggests that the $\sim 0.01 M_\odot$ secondary will evaporate entirely, leaving a solitary millisecond pulsar. The effect of the pulsar radiation, and of the high-energy radiation in the mass transfer phase on the secondary star, adds a new dimension to the evolution of LMXBs, and calls for a major modification of the standard evolutionary scenario.

An interesting suggestion to provide a common answer to the problems mentioned in the previous section, as well as to integrate the effect of pulsar radiation into the scheme of evolution, was made by Van den Heuvel and Van Paradijs [75]. According to this hypothesis, close LMXBs continue their magnetic braking-driven large $\dot{M}$ till they enter the period gap. At this point, the binary detaches, and the neutron star, which has by this time been spun-up to period of a few milliseconds, switches on as a pulsar. The radiation and e^+e^- wind from the pulsar impinges on the companion and erodes it away, leaving a solitary millisecond pulsar at the end. In this way, the absence of low-luminosity systems and LMXBs below the period gap, as well as the origin of solitary millisecond pulsars, can be explained. It would also reduce the active X-ray lifetime of these LMXBs, though it is not clear that this reduction would be large enough to solve the millisecond pulsar birth rate problem.

A different school of thought attributes the short X-ray lifetimes of LMXBs to an accelerated evolution caused by the irradiation of the secondary by high-energy radiation *during* the X-ray phase. According to this model, irradiation of the secondary drives a strong wind, which results in a near-Eddington mass transfer onto the neutron star, and probably a large quantity of mass loss from the system. This would explain the absence of low-luminosity LMXBs, and may also shorten LMXB lifetimes enough to bring the derived birth rates of LMXBs and millisecond pulsars into agreement [70]. However, no observational evidence for strong winds from LMXBs exists except in one case (AC 211), and the X-ray luminosity of many LMXBs is well below the Eddington limit, contrary to the prediction of this model. In this model, the orbital evolution would be strongly affected by mass loss, and it has been suggested that systems will shrink at first, but will begin to widen again at periods $\gtrsim$ 3 h, explaining the paucity of systems below the so-called period gap. This explanation is somewhat doubtful, however, considering the fact that even in cataclysmic variables, for which the evolution is not expected to be significantly affected by the irradiation of the secondary, the 'gap' begins at almost exactly the same period. Further, this model by itself does not provide a way to form single millisecond pulsars. The evolution ends with a $\sim 0.4 M_\odot$ main-sequence star underfilling its Roche lobe, and one has to invoke the radiation from the pulsar to eventually vaporize it away [29, 71].

An important effect of the illumination of the secondary has recently been noted by Podsiadlowski [49] and Harpaz and Rappaport [22]. If the illuminating flux is above a certain critical value, the radius of a low-mass ($\lesssim 2 M_\odot$) secondary would increase considerably due to a change in the ionization structure of its envelope. This would cause enhanced mass transfer, expansion of the orbit, consequent reduction in $\dot{M}$ and illuminating flux, and finally a detachment of the orbit when the illuminating flux falls below the critical value causing the radius of the secondary to shrink again. If the neutron star has been spun up to a short enough spin period in the meantime, it may be in a position to ablate the companion away without further recurrence of the LMXB phase. During the contact phase of the binary, the mass transfer rate sustained by this mechanism is close to the Eddington limit, or even larger. Due to rapid orbital expansion, the duration of such a contact phase is only $\sim 10^7$ yr, thus shortening the active lifetime of the LMXB.

A similar effect can also occur if there are large variations in $\dot{M}$ of LMXBs due to other reasons (see, e.g., Hamuery, King and Lasota [21]). It is possible that at low $\dot{M}$, LMXBs become transient X-ray sources [87]. These sources have bright (10^{37-38} erg s^{-1}) outburst phases separated by long, quiescent phases in which the mass accretion rate on the neutron star is very low ($\lesssim 10^{-12} M_\odot$ yr^{-1}). If the neutron star has achieved a short enough spin period, such a drop in $\dot{M}$ may move the magnetospheric boundary outside the speed-of-light cylinder. This would allow the neutron star to turn on as a pulsar, following which its wind and radiation pressure may halt any further matter inflow and also evaporate the companion away [61, 68]. However, recent attempts to detect radio pulsations from the quiescent LMXBs Cen X-4 and 4U2129+47 have proved unsuccessful [33]; and the detection of X-ray emission from Cen X-4 in the quiescent state [78] suggests that the neutron star in this binary does not have sufficient spin-down power to arrest the accretion flow even in the low-$\dot{M}$ phase. These observations cast doubt on the ability of soft X-ray transients to produce millisecond pulsars.

A more radical suggestion has been that millisecond pulsars form due to accretion-induced collapse, and then become single by evaporating their companions, without going through an intermediate LMXB phase [5]. This would sever the evolutionary connection between LMXBs and millisecond pulsars (though, in some cases, the pulsar may turn into an LMXB if not enough spin-down power is available to completely evaporate the companion away [10, 28]), and would render much of the above discussion meaningless. However, it is not clear why a neutron star born in accretion-induced collapse should necessarily have a low magnetic field as in a millisecond pulsar. Further, it appears that for an accretion-induced collapse to occur rather special conditions are necessary [43]; so the importance of this route for pulsar formation is very uncertain (see [7, 9] for related discussions).

A more likely way to hide some progenitors of millisecond pulsars is by having a phase of super-Eddington mass transfer, during which the emerging X-rays may be screened. $\dot{M} > \dot{M}_{Edd}$ would result if the initial orbital period is very large, and also if the secondary mass is slightly above the stable regime. As pointed out by Coté and

Pylyser [14], this might make an important contribution towards reconciling the birth rates of LMXBs and millisecond pulsars.

Discussion of the models for the origin of millisecond pulsars could not be complete without mentioning the enigmatic PSR 1257+12, which has two (or more) planet-sized companions in circular orbits [90]. There is no clear understanding as yet of how such a system could originate. An initial suggestion involved planets residing around a massive binary: one of the stars evolved, underwent supernova explosion producing the neutron star, which later spiralled into the other star along with the planets, giving rise to the presently observed system [88]. If this view is correct, then it would argue for some millisecond pulsars originating from massive, rather than low-mass, binaries. It has, however, recently been argued that a more plausible way to form the system is to have the planets condense from an excretion disk around a binary containing a neutron star and a very low-mass secondary, created by pulsar-wind-induced mass loss from the secondary which was destroyed in the process [6] (see also Sect. 11.3.4).

5.7   Pulsars in globular clusters

Following the suggestion that millisecond pulsars descended from LMXBs, the attention of pulsar astronomers was turned to the globular clusters because of the well-known overabundance of LMXBs in them [27]. This has met with an extraordinary success. Not only have a large number (> 30) of pulsars been discovered, but also the variety of their properties and their distribution has produced a wealth of information about the globular clusters themselves.

Not all the pulsars discovered in globular clusters are millisecond pulsars; in fact, the distribution of periods extends all the way from 1.7 ms to 1 s. But a majority of them, about 70%, have periods in the millisecond range [38]. About half of these pulsars are in binary systems, while the rest are single. Several pulsars among these have been of particular interest. PSR 2127+11A, a single 110-ms pulsar in the cluster M15, was the first one to show a *negative* period derivative, i.e. the pulsar is apparently spinning up. This indicates that the pulsar is undergoing an accelerated motion towards the Earth. Clearly this is due to the action of the strong gravitational potential of the cluster as a whole. In fact, several more cluster pulsars have now been shown to have negative period derivatives, and these can yield useful information about the mass distribution inside the clusters [46].

Another pulsar in the same cluster, PSR 2127+11C, has a neutron star companion, and is located many core radii away from the center of the cluster. It is very unusual to find a neutron star so far away from the cluster core, especially a binary one, if the standard distribution of stellar mass in the cluster potential is followed. Obviously an encounter, or some violent event connected with the formation of this system, moved it out of the cluster core. It is likely that after this 30-ms pulsar was spun-up to its present period in a binary system, it underwent a close encounter with another neutron-star binary and the present system formed as a result [48]. This system is remarkably similar in characteristics to the first binary pulsar PSR 1913+16, and its discovery raises the possibility that similar systems might be released from the globular clusters into the galactic halo.

Systems similar to the eclipsing millisecond pulsar PSR 1957+20 have also been found in the globular clusters. The first one, PSR 1744—24A in the cluster Ter 5, has a spin period of 11 ms and is in a 1.8-h binary orbit with a $\sim 0.09 M_\odot$ evaporating secondary. Estimates of the mass loss rate suggest, however, that this secondary will not entirely evaporate due to ablation [72]. If so, it is quite plausible that this system will turn into an LMXB some time in the future [28].

The second eclipsing pulsar found in globular clusters, namely PSR 1718−19 in NGC 6342, is quite a remarkable one [39]. Unlike the other eclipsing pulsars, this has a long spin period (1 s) and a high magnetic field ($\sim 10^{12}$ G). The $\sim 0.1 M_\odot$ companion in a 6-h binary orbit receives a rather modest amount of flux from the pulsar, but this is apparently enough to cause the secondary to overflow its Roche lobe and blow so much ionized wind as to attenuate the pulsar's observed intensity when our line of sight passes through it. It is possible that this will finally become a solitary long-period pulsar. The existence of such a high-field pulsar in a globular cluster also argues strongly against the spontaneous decay of the magnetic fields of isolated pulsars.

Apart from the individual pulsars, the pulsar population as a whole in the globular clusters has also taught us several new things and has posed a number of new questions. Before the globular cluster pulsars were discovered, it had been shown that close tidal encounters of neutron stars with other stars in globular clusters could comfortably explain the formation of low-mass X-ray binaries in the cluster system [80]. Binary formation due to tidal capture proceeds as follows: as a solitary neutron star passes close by a solitary normal star, the tidal forces in the gravitational encounter deform the normal star at the cost of the relative kinetic energy of the stars. If the deformed star is able to dissipate a part of the potential energy of the tidal bulge, and if the amount of energy lost in this way exceeds the initial total (positive) energy of the two-star system, then a bound binary system forms. It turns out that, for a capture to occur, the distance of closest approach between the two stars must be less than or of the order of three times the radius of the extended star [15].

It is clear that the rate of neutron star binary production via tidal capture must be proportional to the rate of stellar collisions in a globular cluster. This rate is, in turn, proportional to $n_n n r_c^3 / \sigma_c$, where n and n_n are the densities of normal stars and neutron stars in the cluster core, respectively, r_c is the core radius and σ_c is the velocity dispersion in the cluster core. Assuming that both n_n and n are proportional to the core mass density ρ_c, one finds that the collision rate is proportional to $\rho_c M_c / \sigma_c$, where $M_c \propto \rho_c r_c^3$ is the core mass.

If tidal capture is the way to form X-ray binaries and thence the recycled pulsars in globular clusters, then the X-ray binaries and the recycled pulsars should be distributed among the clusters according to their collision rates. However, of the observed recycled pulsars in globular clusters, too many lie in low-density clusters where the probability of tidal capture is very small. The observed distribution is, in fact, better fit by a cluster weighting proportional to $\rho_c^{0.5} M_c$ rather than $\rho_c M_c$ of the collision–number model [25]. An estimate of the relative number of pulsars between clusters based on their continuum radio flux also leads one to the same conclusion [19].

It is, therefore, apparent that processes other than tidal capture must contribute to the production of recycled pulsars in clusters. One of the most promising routes appears to be collisions of neutron stars with primordial binaries in clusters. The evidence for existence of primordial binaries in clusters is rather recent [51], and it is now thought that clusters may contain $\sim 10\%$ of stars in primordial binaries. The importance of these binaries lies in the fact that they present much larger collision cross sections than the single stars. Thus, collision with primordial binaries may dominate neutron-star binary production except in the very densest of clusters, where wide binaries are destroyed by stellar encounters [47]. The encounter between a neutron star and a primordial binary could result in a neutron star binary if the neutron star replaces one of the original components of the binary system, or if, during the course of the encounter, the neutron star physically collides with one of the stars, destroying it in the process.

One important indicator of the formation route of the recycled pulsars in globular clusters is, of course, their birth rate, and this has received a considerable amount of attention since the discovery of cluster pulsars. The most detailed attempt [25, 47] estimates the total number of pulsars in the cluster system by first estimating what fraction of the 'production rate volume' has actually been covered in the present surveys, taking into account the known selection effects, and dividing the observed number by this fraction. This leads to an estimate of ~ 2500 pulsars in the whole globular cluster system. The birth rate this yields, $\sim 2 \times 10^{-7}$ yr^{-1}, is consistent with that obtained on the basis of the timing ages ($\equiv P/2\dot{P}$) of observed pulsars [47]. However, there is still room for disagreement. For example, one group [89] finds the total number of cluster pulsars to be only ~ 200 based on the radio continuum luminosity of the clusters, a number that has not been satisfactorily reconciled with the above estimate.

The large number of millisecond pulsars inferred in the clusters has serious implications. If cluster LMXBs have to supply the high pulsar birth rate of $\sim 2 \times 10^{-7}$ yr^{-1}, then the active lifetimes of these LMXBs must not exceed $\sim 5 \times 10^7$ yr. This is sufficient to spin a neutron star up to millisecond periods, but only if the accretion rate is near the Eddington rate. However, none of the 12 observed LMXBs in globular clusters are undergoing near-Eddington accretion, as indicated by the fact that their luminosities are 10 to 100 times smaller than the Eddington value [83]. Thus, one may be forced to look for millisecond pulsar progenitors in systems unlike the observed LMXBs in globular clusters. One of the suggestions has been that there may be two kinds of LMXBs in globular clusters: one kind, with high birth rates of $\gtrsim 10^{-7}$ yr^{-1} and short lifetimes of $\lesssim 10^6$ yr, undergoes heavy accretion and produces the majority of the spun-up pulsars with $P \gtrsim 5$ ms, whereas another kind, with birth rates of $\lesssim 10^{-8}$ yr^{-1} and lifetimes of $\gtrsim 10^9$ yr, result in the fastest pulsars and the observed LMXBs [25, 47]. The former channel may result from physical collisions with normal stars, and the latter from tidal capture.

A model for the origin of globular cluster pulsars must also explain the high abundance of singles among them. If these pulsars are indeed spun-up, then their companions must have in some way disappeared. The mechanism originally proposed for this was encounters with a third star, which may either 'ionize' the binary or replace the neutron star as a binary member [59, 82]. Detailed computation of probabilities, however, show that these effects are important for only the long orbital periods ($\gtrsim 100$ d) and in the densest of clusters [58]. This would not be sufficient to account for the birth rate of the single pulsars. One must therefore invoke additional mechanisms, such as vaporization of the companion (cf. PSR 1744—24A, PSR 1718—19) and spin-up due to accretion of the debris of a normal star after a direct physical collision [9, 47].

One more question that presents itself is whether the globular clusters have enough primordial neutron stars produced in the early history of the cluster to explain the abundance of recycled pulsars. Assuming that stars more massive than $8M_\odot$ left neutron stars, and a Salpeter mass function ($dN/dM \propto M^{-2.35}$) for the upper main-sequence, one would arrive at a total number of neutron star progenitors $\sim 3 \times 10^5$ in the whole globular cluster system (a total mass of $5 \times 10^7 M_\odot$ for the cluster system has been assumed). Measurement of velocities of pulsars in the disk shows that they are high-velocity objects, with a mean speed ~ 200 km s^{-1}, much higher than the escape velocity from a typical globular cluster (~ 25 km s^{-1}). Only the slowest $\lesssim 15\%$ of the original neutron stars are thus likely to be retained in the globular clusters, bringing the number of neutron stars available for recycling to $\lesssim 4.5 \times 10^4$. A fraction of these will eventually undergo close gravitational interactions with other stars and binaries and produce recycled pulsars.

Although the above number seems adequate to explain the total inferred number of pulsars in the cluster system, it must be remembered that the Salpeter mass function has been used to arrive at this result. If, on the other hand, mass functions steeper than M^{-3} are used, the total number of retained neutron stars falls below ~ 6000, making it difficult to produce > 2500 active recycled pulsars out of them. There has been considerable argument in the literature regarding the possible upper main-sequence mass function in globular clusters. It has been noted that mass loss in supernovae and stellar winds could unbind a cluster if the lost mass exceeds one-half the total original mass, and a mass function as flat as Salpeter could contain enough high-mass stars to lead to cluster disruption [13]. This has led to the speculation that a large fraction of neutron stars in globular clusters originates in accretion-induced collapse of heavy white dwarfs [5]. However, this conclusion is very sensitive to the assumed lower limit to the mass of stars in clusters. While a value of $0.4M_\odot$ for the lightest stars only allows mass functions steeper than $M^{-3.5}$ if the cluster has to survive, a lower cutoff mass of $0.1M_\odot$ can accommodate the Salpeter mass function [25]. This is still a rather open question, and will need careful determination of the lower cutoff mass of globular cluster stars to be settled beyond doubt.

5.8   Future prospects

It is clear from the above discussion that while millisecond pulsars, both in the galactic disk and in globular clusters, have taught us much since their discovery, several questions still remain unresolved. This will continue to drive millisecond pulsar astronomy in the coming years. First of all, to understand the population well enough, the sample of millisecond pulsars has to be enlarged, especially in the galactic disk. A step towards this has already been taken with the starting of a number of new surveys (see [4] for a review). Some major new telescopes, such as the 110-m telescope in the USA and the Giant Metrewave Radio Telescope in India, are nearing completion, and, along with the resurfaced Arecibo telescope, these will be the instruments of the coming decade for millisecond pulsar astronomy. Not only will they be able to significantly improve on the sensitivities of the present surveys, they will also enable one to do more precision timing and polarization studies. Advances in computing technology will make possible faster data recording and analysis, and pulsar surveys with acceleration search (for detecting short-period binaries) as well as searches for sub-millisecond pulsars will come within the realm of routine activities. One thus expects to see a lot of exciting pulsar discoveries in the next ten years.

Detailed, high-resolution optical studies of globular clusters, probing their mass distribution and binary content, would also be immensely helpful in understanding the origin of pulsars in globular clusters.

At the theoretical front, as we continue to grapple with pulsar electrodynamics, better modeling of the emission geometry and polarization characteristics of millisecond pulsars will become possible. One of the most important ingredients of the evolutionary schemes leading up to millisecond pulsars, namely the evolution of neutron star magnetic fields, will also have to be better understood. Evolutionary models of LMXBs must be refined, taking into account the effect of radiation and wind from the neutron star on the secondary. Finally, it will also have to be properly investigated whether low-field, rapidly spinning pulsars can originate via routes not involving LMXBs. Could a newly born neutron star behave as a low-field millisecond pulsar? One would eagerly await the neutron star in supernova 1987A to make its appearance. The low energy input from it into the supernova ejecta leaves only two alternatives – either it has a very long spin period or a very low magnetic field. Should it prove to be a low-field millisecond pulsar, pulsar astronomy will see a new revolution.

References

[1] Alpar, M. A., Cheng, A. F., Ruderman, M. A., and Shaham, J., 1982, *Nature*, **300**, 728.
[2] Alpar, M. A. and Shaham, J., 1985, *Nature*, **316**, 239.
[3] Backer, D. C., Kulkarni, S. R., Heiles, C. E., Davis, M. M., and Goss, W. M., 1982, *Nature*, **300**, 615.
[4] Bailes, M. and Johnston, S., 1993, *Review of Radio Science 1990–1992* (URSI/Oxford University Press), p. 677.
[5] Bailyn, C. D. and Grindlay, J. E., 1990, *Astrophys. J.*, **353**, 159.
[6] Banit, M., Ruderman, M. A., Shaham, J., and Applegate, J. H., 1993, *Astrophys. J.*, **415**, 779.
[7] Bhattacharya, D., 1991, in Ventura, J. and Pines, D., eds. *Neutron Stars: Theory and Observation* (Kluwer Academic Publishers, Dordrecht), page 103.
[8] Bhattacharya, D. and Srinivasan, G., 1986, *Curr. Sci.*, **55**, 327.
[9] Bhattacharya, D. and Van den Heuvel, E. P. J., 1991, *Physics Reports*, **203**, 1.
[10] Bisnovatyi-Kogan, G. S., 1989, *Astrofizika*, **31**, 567.
[11] Boriakoff, V., Buccheri, R., and Fauci, F., 1983, *Nature*, **304**, 417.
[12] Callanan, P. J., Charles, P. A., Hassal, B. M. J., Machin, G., Mason, K. O., Naylor, T., Smale, A. P., and Van Paradijs, J., 1989, *Mon. Not. R. Astr. Soc.*, **238**, 25P.
[13] Chernoff, D. and Weinberg, M., 1990, *Astrophys. J.*, **351**, 121.
[14] Coté, J. and Pylyser, E. H. P., 1989, *Astron. Astrophys.*, **218**, 131.
[15] Fabian, A. C., Pringle, J. E., and Rees, M. J., 1975, *Mon. Not. R. Astr. Soc.*, **172**, 15P.
[16] Foster, R. S., Backer, D. C., and Fairhead, L., 1992, in Van den Heuvel, E. P. J. and Rappaport, S. A., eds. *X-ray Binaries and Recycled Pulsars* (Kluwer Academic Publishers, Dordrecht), page 115.
[17] Friedman, J. L., Imamura, J. N., Durisen, R. H., and Parker, L., 1988, *Nature*, **336**, 560.
[18] Friedman, J. L., Ipser, J. R., and Parker, L., 1984, *Nature*, **312**, 255.
[19] Fruchter, A. S. and Goss, W. M., 1990, *Astrophys. J.*, **365**, L63.
[20] Fruchter, A. S., Stinebring, D. R., and Taylor, J. H., 1988, *Nature*, **333**, 237.
[21] Hameury, J. M., King, A. R., and Lasota, J. P., 1986, *Astron. Astrophys.*, **162**, 71.

[22] Harpaz, A. and Rappaport, S. A., 1991, *Astrophys. J.*, **383**, 739.

[23] Hulse, R. A. and Taylor, J. H., 1975, *Astrophys. J.*, **191**, L51.

[24] Jeffrey, L. C., 1986, *Nature*, **319**, 384.

[25] Johnston, H. M., Kulkarni, S. R., and Phinney, E. S., 1992, in Van den Heuvel, E. P. J. and Rappaport, S. A., eds. *X-ray Binaries and Recycled Pulsars* (Kluwer Academic Publishers, Dordrecht), page 349.

[26] Johnston, S. and Bailes, M., 1991, *Mon. Not. R. Astr. Soc.*, **252**, 277.

[27] Katz, J., 1975, *Nature*, **253**, 698.

[28] Kluźniak, W., Czerney, M., and Ray, A., 1992, in Van den Heuvel, E. P. J. and Rappaport, S. A., eds. *X-ray Binaries and Recycled Pulsars* (Kluwer Academic Publishers, Dordrecht), page 425.

[29] Kluźniak, W., Ruderman, M., Shaham, J., and Tavani, M., 1988, *Nature*, **334**, 225.

[30] Kulkarni, S. R., Djorgovski, S., and Klemola, A. R., 1990, *Astrophys. J.*, **367**, 221.

[31] Kulkarni, S. R. and Narayan, R., 1988, *Astrophys. J.*, **335**, 755.

[32] Kulkarni, S. R., Narayan, R., and Romani, R. W., 1990, *Astrophys. J.*, **356**, 174.

[33] Kulkarni, S. R., Navarro, J., Vasisht, G., Tanaka, Y., and Nagase, F., 1992, in Van den Heuvel, E. P. J. and Rappaport, S. A., eds. *X-ray Binaries and Recycled Pulsars* (Kluwer Academic Publishers, Dordrecht), page 99.

[34] Lewin, W. H. G., Lubin, L. M., Tan, J., Van der Klis, M., Van Paradijs, J., Penninx, W., Dotani, T., and Mitsuda, K., 1992, *Mon. Not. R. Astr. Soc.*, **256**, 545.

[35] Lindblom, L., 1992, in Pines, D., Tamagaki, R., and Tsuruta, S., eds. *The Structure and Evolution of Neutron Stars* (Addison-Wesley, New York), page 122.

[36] Lindblom, L. and Mendel, G., 1992, in Pines, D., Tamagaki, R., and Tsuruta, S., eds. *The Structure and Evolution of Neutron Stars* (Addison-Wesley, New York), page 227.

[37] Long, K. S. and van Speybroeck, L. P., 1983, in Lewin, W. H. G. and Van den Heuvel, E. P. J., eds. *Accretion Driven Stellar X-ray Sources* (Cambridge University Press), page 117.

[38] Lyne, A. G., 1992, in Van den Heuvel, E. P. J. and Rappaport, S. A., eds. *X-ray Binaries and Recycled Pulsars* (Kluwer Academic Publishers, Dordrecht), page 79.

[39] Lyne, A. G., Biggs, J. D., Harrison, P. A., and Bailes, M., 1993, *Nature*, **361**, 47.

[40] Lyne, A. G. and Manchester, R. N., 1988, *Mon. Not. R. Astr. Soc.*, **234**, 477.

[41] Manchester, R. N., 1992, in T. H. Hankins, J. M. Rankin and Gil, J., eds. *Magnetospheric Structure and Emission Mechanism of Radio Pulsars* (Pedagogical University of Zielona Gora Press, Zielona Gora, Poland), 204.

[42] Nagase, F., 1989, *Publ. Astr. Soc. Japan*, **41**, 1.

[43] Nomoto, K. and Yamaoka, H., 1992, in Van den Heuvel, E. P. J. and Rappaport, S. A., eds. *X-Ray Binaries and Recycled Pulsars* (Kluwer Academic Publishers, Dordrecht) page 189.

[44] Paczynski, B., 1983, *Nature*, **304**, 421.

[45] Papaloizou, J. and Pringle, J. E., 1978, *Mon. Not. R. Astr. Soc.*, **184**, 501.

[46] Phinney, E. S., 1992, *Philos. Trans. R. Soc. London Ser. A.*, **341**, 39.

[47] Phinney, E. S. and Kulkarni, S. R., 1994, *Ann. Rev. Astron. Astrophys.*, in press.

[48] Phinney, E. S. and Sigurdsson, S., 1991, *Nature*, **349**, 220.

[49] Podsiadlowski, P., 1991, *Nature*, **350**, 136.

[50] Priedhorsky, W. C. and Verbunt, F., 1988, *Astrophys. J.*, **333**, 895.

[51] Pryor, C., McClure, R. D., Fletcher, J. M., and Hesser, J. E., 1989, *Astron. J.*, **98**, 596.

[52] Pylyser, E. H. P. and Savonije, G. J., 1989, *Astron. Astrophys.*, **208**, 52.

[53] Pylyser, E. H. P. and Savonije, G. J., 1988, *Astron. Astrophys.*, **191**, 57.

[54] Radhakrishnan, V. and Shukre, C. S., 1985, in Srinivasan, G. and Radhakrishnan, V., eds. *Supernovae, Their Progenitors and Remnants* (Indian Academy of Sciences, Bangalore), page 155.

[55] Radhakrishnan, V. and Srinivasan, G., 1982, *Curr. Sci.*, **51**, 1096.

[56] Radhakrishnan, V. and Srinivasan, G., 1984, in Hidayat, B. and Feast, M. W., eds. *Proc. 2nd Asia-Pacific Regional Meeting of the IAU (1981)* (Tira Pustaka, Jakarta), page 423.

[57] Rappaport, S., Verbunt, F., and Joss, P. C., 1983, *Astrophys. J.*, **275**, 713.

[58] Rappaport, S. A., Putney, A., and Verbunt, F., 1989, *Astrophys. J.*, **345**, 210.

[59] Romani, R., Kulkarni, S., and Blandford, R. D., 1987, *Nature*, **329**, 309.

[60] Ruderman, M., Shaham, J., and Tavani, M., 1989, *Astrophys. J.*, **336**, 507.

[61] Ruderman, M., Shaham, J., Tavani, M., and Eichler, D., 1989, *Astrophys. J.*, **343**, 292.

[62] Ryba, M. F. and Taylor, J. H., 1991, *Astrophys. J.*, **371**, 739.
[63] Ryba, M. F. and Taylor, J. H., 1991, *Astrophys. J.*, **380**, 557.
[64] Savonije, G. J., 1983, *Nature*, **304**, 422.
[65] Shapiro, S. L., Teukolsky, S. A., and Wassermann, I., 1984, *Astrophys. J.*, **272**, 702.
[66] Smarr, L. L. and Blandford, R. D., 1976, *Astrophys. J.*, **207**, 574.
[67] Spruit, H. C. and Ritter, H., 1983, *Astron. Astrophys.*, **124**, 267.
[68] Srinivasan, G. and Bhattacharya, D., 1989, *Curr. Sci.*, **58**, 953.
[69] Srinivasan, G. and Van den Heuvel, E. P. J., 1980, *Astron. Astrophys.*, **108**, 143.
[70] Tavani, M., 1991, *Astrophys. J.*, **366**, L27.
[71] Tavani, M., Kluźniak, W., Ruderman, M., and Shaham, J., 1989, *Ann. N. Y. Acad. Sci.*, **571**, 427.
[72] Thorsett, S. E. and Nice, D. J., 1991, *Nature*, **353**, 731.
[73] Thorsett, S. E. and Stinebring, D. R., 1990, *Astrophys. J.*, **361**, 644.
[74] Van den Heuvel, E. P. J., 1992, in Van den Heuvel, E. P. J. and Rappaport, S. A., eds. *X-ray Binaries and Recycled Pulsars* (Kluwer Academic Publishers, Dordrecht), page 233.
[75] Van den Heuvel, E. P. J. and Van Paradijs, J. A., 1988, *Nature*, **334**, 227.
[76] Van den Heuvel, E. P. J., Van Paradijs, J. A., and Taam, R. E., 1986, *Nature*, **322**, 153.
[77] Van der Klis, M., 1989, *Ann. Rev. Astr. Astrophys.*, **27**, 517.
[78] Van Paradijs, J., Verbunt, F., Shafer, R. A., and Arnaud, K. A., 1987, *Astron. Astrophys.*, **182**, 47.
[79] Verbunt, F., 1990, in Kundt, W., ed. *Neutron Stars and Their Birth Events* (Kluwer Academic Publishers, Dordrecht), page 179.
[80] Verbunt, F. and Hut, P., 1987, in Helfand, D. J. and Huang, J. H., eds. *The Origin and Evolution of Neutron Stars* (D. Reidel, Dordrecht), page 187.
[81] Verbunt, F. and Rappaport, S. A., 1988, *Astrophys. J.*, **332**, 193.
[82] Verbunt, F., Van den Heuvel, E. P. J., Van Paradijs, J., and Rappaport, S. A., 1987, *Nature*, **329**, 312.
[83] Verbunt, F., Van Paradijs, J., and Elson, R., 1984, *Mon. Not. R. Astr. Soc.*, **210**, 899.
[84] Verbunt, F. and Zwaan, C., 1981, *Astron. Astrophys.*, **100**, L7.
[85] Wagoner, R. V., 1984, *Astrophys. J.*, **278**, 345.
[86] Webbink, R. F., Rappaport, S. A., and Savonije, G. J., 1983, *Astrophys. J.*, **270**, 678.
[87] White, N. E., Kaluzienski, J. L., and Swank, J. H., 1984, in Woosley, S., ed. *High Energy Transients in Astrophysics* (American Institute of Physics), page 31.
[88] Wijers, R. A. M. J., Van den Heuvel, E. P. J., Van Kerkwijk, M. H., and Bhattacharya, D., 1992, *Nature*, **355**, 593.
[89] Wijers, R. A. M. J. and Van Paradijs, J. A., 1991, *Astron. Astrophys.*, **241**, L37.
[90] Wolszczan, A. and Frail, D. A., 1992, *Nature*, **355**, 145.

6

Rapid aperiodic variability in X-ray binaries

M. van der Klis

Astronomical Institute Anton Pannekoek, University of Amsterdam, and
Center for High-Energy Astrophysics,
Kruislaan 403, 1098 SJ Amsterdam, The Netherlands

6.1   Introduction

The subject of this chapter is rapid aperiodic variability in the X-ray emission of X-ray binaries. It will be discussed below what the word 'rapid' means in this context. The variations we are talking about are neither periodic, such as pulsations, dips or eclipses (Ch. 1), nor do they consist of easily recognized isolated events, such as X-ray bursts (Ch. 4). They include the types of variability that are usually called quasi-periodic oscillations (QPOs) and noise, and also sometimes flickering, irregular flaring, fluctuations, etc. Such variability occurs in all types of X-ray binaries: black-hole candidates (Sect. 6.2), low-magnetic-field neutron star systems (Sect. 6.3) and accreting pulsars (Sect. 6.4.4).

Together with X-ray spectroscopy, the study of rapid (periodic and aperiodic) X-ray variability is one of the very few ways to obtain direct information about the physical circumstances in the vicinity of accreting compact objects. The rapid aperiodic variability is thought to originate in the irregular nature of the inner accretion flow. As the flow is in close physical interaction with the compact object, some of the properties of the variability can be expected to reflect properties of the compact object. This is the main reason to study rapid aperiodic variability. Some of the information on the compact object can be extracted without a full understanding of the causes of the variability, such as, for example, turbulent and magnetic phenomena in the accretion disk.

There are a number of parameters of neutron stars and black holes that can *a priori* be expected to influence the rapid X-ray variability. The *mass M* of the compact object is one obvious parameter; gravity not only determines the accretion flow but also affects the paths of the emitted photons. Neutron stars are probably confined to a relatively narrow mass range, but black holes could occur over a wide range of masses. The *magnetic moment μ* of the compact object is important for the accretion flow when electromagnetic forces on the plasma are of the same order as the gravitational forces (or larger). Neutron stars have a wide range of magnetic moments from maybe close to zero to more than 10^{30} G cm^3 (corresponding to a surface field of $B \gtrsim 10^{12}$ G); for fields $\gtrsim 10^9$ G, strong effects on the accretion flow are expected, and it is probably only for fields $<<10^6$ G that magnetic forces can be neglected (F.K. Lamb, 1993, private communications). Black holes in X-ray binaries are unlikely to carry a large amount of electrical charge, so that electromagnetic forces there are probably small. The compact-object *spin rate ω* can be expected to affect the accretion flow in two distinct ways, through gravity and through electromagnetic forces. In a rapidly spinning black hole, relativistic frame dragging will be important for the innermost flow and for the emerging radiation. A neutron star produces little frame dragging (as long as $\omega \ll c/R$). A spinning neutron star with a non-aligned magnetic field produces periodic effects in the accreting plasma that lead to observable phenomena (pulsations, QPOs). Non-aligned magnetic fields cannot occur in black holes. The *radius R* of the compact object is not strictly an independent parameter, as it is directly coupled either (in neutron stars) through the equation of state (EOS) or (in black holes) through gravitation theory to the mass M. However, as the EOS of bulk nuclear density matter is not known, the independent determination of R neutron stars can help to constrain the EOS.

Apart from these basic properties of the compact object itself, the *accretion rate* $\dot{M}$ at which matter arrives in the vicinity of the compact object will be important for the flow dynamics as well as for the emerging radiation. The same may apply to the magnitude and the orientation of the *specific angular momentum j* of the accreting matter. Observationally, $\dot{M}$ and j can be distinguished from other parameters because they can change rapidly with time.

A first aim of the study of the phenomenology of X-ray binaries is to distinguish between systems containing black holes and those with neutron stars. In spite of early hopes, it is not easy to distinguish between neutron stars with a low magnetic field strength and black holes on the basis of their rapid X-ray variability (Sect. 6.5). This is expected from first principles. A neutron star is only roughly three times as large as its Schwarzschild radius (at $2R_g$, where the gravitational radius $R_g \equiv GM/c^2$). Therefore the accretion flows around a low-magnetic-field neutron star and a stellar-mass black hole in an X-ray binary system will have many similarities, including similar characteristic time scales. Rapid X-ray variability properties once believed to be typical of black holes are now known to occur in neutron stars too. The relatively subtle observational distinctions between low-magnetic-field neutron stars and black holes are presently the subject of much research. They will be discussed throughout this chapter.

As a second goal, one would like to map the observed phenomena for each compact object type into the parameter space defined by M, μ, ω, $\dot{M}$, and j. For accreting neutron stars one might expect $\dot{M}$, μ and ω to underly most of the observational differences, and for black holes $\dot{M}$, M and ω may be most crucial. For all objects, we should remember that accidental observational circumstances such as, for example, the binary inclination i, and relative details, such as the chemical composition of the accreting matter and the type of the companion star, may further complicate the picture.

For neutron stars, the distinction between pulsars and bursters provides some of the mapping along the μ axis of this parameter space (Ch. 1), and the difference between Z and atoll sources may reflect a more subtle difference in magnetic field strength (Sect. 6.3). $\dot{M}$ mapping has recently become clear (at least in rank) in the bright low-mass X-ray binaries (LMXBs), and turned out quite different than expected (Sect. 6.3). For pulsars, ω is known, and variations in ω provide indications for j (Ch. 1). So-called horizontal-branch QPOs in principle allow estimates of ω in some bright LMXBs (Sect. 6.3) and perhaps of μ in pulsars (Sect. 6.4). Strong indications now exist about $\dot{M}$ mapping in black-hole candidates (Sect. 6.2), but very little can be said as yet about the effects of M and ω of black-hole candidates on the rapid variability.

The faster the aperiodic variability is, and the larger its amplitude, the more likely it is that it carries information about the compact object, where the energy is produced and the time scales are short. In the immediate vicinity of the compact object, the strong gravity will cause rapid bulk motion of the accreting matter. The dynamical time scale $\tau_d \equiv (r^3/GM)^{0.5}$ at the surface of a $1.4M_\odot$ neutron star is ~ 0.1 ms, and at the innermost stable orbit (at $6R_g$) of a $3M_\odot$ black hole ~ 0.2 ms. If the effective gravity is reduced by radiation pressure, the time scales are longer.

The fastest detected variations last a few milliseconds, but, as discussed in Sect. 6.2.4, the associated time scales are longer than that. In any case, a range of fluctuation phenomena is seen in neutron stars as well as in black-hole candidates on time scales of several tens of milliseconds. In this chapter I will concentrate on aperiodic variability faster than 1 s. This corresponds to the time scale for matter moving solely under the influence of gravity within a distance of several thousand kilometers from a stellar-mass compact object. Variability observed on these time scales might, for example, also reflect the flow time through a ~ 100 km radius radiation-pressure-limited spherical inflow for luminosities within 1% of the Eddington limit (Lamb 1991) or the flow time through the inner few 10 km of an $\alpha \sim 1$ Shakura–Sunyaev (1973) accretion disk. Of course, variability whose amplitude is only a small fraction of the total emission is not guaranteed to have causes as global as this. It could also be related to smaller-scale phenomena further away from the compact object, such as turbulent cells or magnetic loops.

In practice, it turns out that most types of rapid aperiodic variability cover a (sometimes wide) range of time scales. To properly study such variations down to <1 s, it is sometimes necessary to also include slower variations in the analysis. The meaning of the word 'rapid' in the title of this chapter is therefore somewhat relative: emphasis is on variations faster than 1 s, but variability up to $\sim 10^3$ s will be discussed when necessary. Mostly, I shall deal with variations in X-ray emission, as the bulk of the X-rays is certain to originate close to the compact object. However, I shall also mention rapid aperiodic optical variability that is clearly related to the X-ray variations.

Aperiodic X-ray intensity variations bear a close relation both in general aspect and in probable underlying physics to rapid rotation-frequency fluctuations in accreting pulsars (Ch. 1). These are caused by fluctuations in the torque exerted by the accreting matter upon the neutron star, and perhaps originate in similar irregularities in the accretion flow that cause the X-ray intensity variations. These spin fluctuations will not be further discussed in this chapter.

6.1.1   *History*

The discovery by Oda *et al.* in 1971 of rapid aperiodic X-ray intensity variations in the black-hole candidate Cyg X-1 (Figure 6.1) initiated the first wave of interest in aperiodic variability in X-ray binaries. Soon after Oda's discovery, the accretion-powered pulsars were discovered (Giacconi *et al.* 1971), and in the subsequent 15 years the number of papers about aperiodic intensity variations constituted only a trickle compared with the stream of papers on X-ray pulsations and, later,

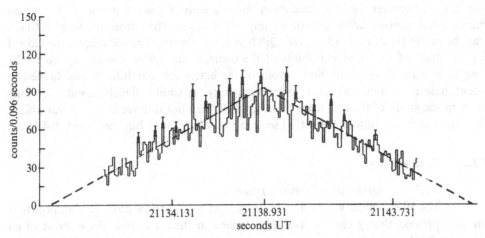

Fig. 6.1. Discovery of the LS noise in Cyg X-1 (Oda *et al.* 1971). Shown is one scan of *UHURU* over Cyg X-1; the dashed line indicates the expected response to a constant intensity source.

bursts. Most studies of aperiodic variations were done on Cyg X-1 (Sect. 6.2); later, some attention was also paid to Cir X-1 (Sect. 6.4.1) and GX 339–4 (Sect. 6.2). Strong emphasis was put on applying shot noise models to the data. As noted by Bradt *et al.* in their 1982 review of rapid X-ray variability, the pulsations and the bursts seemed much more amenable to interpretation than did the aperiodic variability. However, obviously bursts and pulsations do not provide information on black holes, and no pulsations have so far been detected from accreting low-magnetic-field neutron stars. Rapid quasi-periodic oscillations (QPOs) now provide a diagnostic that is intermediate in interpretability between noise and pulsations.

A revival of the study of aperiodic variability in X-ray binaries came with the *EXOSAT* (1983–86) and *Ginga* (1987–91) X-ray satellites (Turner *et al.* 1981, 1989; Makino *et al.* 1987) whose large collecting area, long observations of bright sources, and capability to follow up on the discovery of X-ray transients provided a flood of new information on rapid X-ray variability.

The work with *EXOSAT* on bright low-mass X-ray binaries (LMXBs) containing low-magnetic-field neutron stars (Sect. 6.3) led to the insight that the basic phenomenology of these sources consists of correlated changes in the properties of the rapid X-ray variability and the X-ray spectrum. This made it possible to map the source behaviour as a function of $\dot{M}$. The discovery of two different types of rapid QPOs played an important part in this, and the properties of the QPOs suggested specific models for the interaction of the accretion flow with the magnetic field and with the emerging radiation. Two subclasses of bright LMXBs (Z and atoll sources) were identified, each showing several different modes of correlated X-ray spectral and rapid X-ray variability behaviour. Possibly underlying these subclasses is a difference in magnetic field strength. One of the surprises was that *no* simple relation holds in these sources between X-ray intensity and accretion rate.

The *Ginga* observations of persistent as well as newly discovered transient blackhole candidates (Sect. 6.2) suggested that similar correlations between spectral and variability characteristics and accretion rate as seen in low-magnetic-field neutron stars existed in black holes. One of the important new contributions made with *Ginga* was the discovery of rapid (3–10 Hz) QPOs in these objects. The techniques developed for the study of the rapid variability of the neutron star systems were applied to the long-known rapid aperiodic fluctuations of the black-hole candidates, and helped to disentangle black-hole candidate phenomenology. A similar development may take place in the study of the accreting pulsars, which turn out to have irregular variability that has much in common with that seen in the other X-ray binaries (Sect. 6.4.4).

6.1.2   Data analysis

6.1.2.1   Description of random processes

The phenomena which are the subject of this chapter have in common that random-process theory can be fruitfully applied in their analysis. As is most often the case in the study of random processes, what is of greatest interest is not the details of any individual fluctuation, but the average properties of the variability, which are most likely to contain information about the physical processes causing them. Statistical methods are therefore extensively applied in the analysis of the rapid aperiodic variability of X-ray binaries. In many cases, especially those of the most rapid variations, this approach is a requirement, as the individual fluctuations cannot be identified in the light curves.

After initial attempts in 1971–2 at interpreting the then newly discovered rapid variability in Cyg X-1 in terms of transient fast periodicities (Sect. 6.2.5), Terrell (1972) pointed out that the observed variations were consistent with a random shot noise model, where the light curve is made up of randomly occurring discrete and identical events, the 'shots'. In the subsequent years, the average properties of the rapid intensity variations were nearly always reported in terms of shot noise model parameters determined from the autocorrelation function (ACF) of the time series. As any random process (up to its second-order statistics, of which the ACF and the power spectrum are examples), can be modelled with pure random shot noise (Doi 1978), this is a valid way to report the average properties of the variations. However, it is important to note that a succesful fit of a shot noise model to the data does not imply that the underlying physical process must actually consist of shots.

A shot noise model more complicated than 'pure' shot noise was introduced by Sutherland *et al.* (1978) and was used in most later work. It involves shot noise plus a constant intensity level. This model requires a higher-order statistic in addition to the ACF (namely, the third moment of the time series) to estimate its parameters. The third moment is difficult to measure, which in practice often causes the parameters of this model to be ill-constrained. If only the results of fits to such a shot noise model are presented, then the well-defined second-order statistics tend to get washed out.

In more recent years, observers have preferred to report the average parameters of the variations in the form of Fourier spectra (power spectra and cross-spectra), which provide an estimate of the amplitude and the phase of the fluctuations at each Fourier frequency. One of the advantages of this approach is that it does not bias thinking towards any particular random process. Shot noise can be treated as one of the possible time-series models that can generate the observed Fourier spectra, and, when they are measurable, higher-order statistics such as the third moment can be used as additional constraints. Other second-order statistics, such as the autocorrelation function, the cross-correlation function, and the variation function, have simple mathematical relations to the Fourier spectra (e.g. Maejima *et al.* 1984).

In some cases, attempts have been made to apply chaos theory to the description of the aperiodic variability (Miyamoto *et al.* 1988b; Atmanspacher *et al.* 1989; Morfill *et al.* 1989; Unno *et al.* 1990). So far, these attempts have not led to much physical insight beyond what was obtained with conventional methods.

6.1.2.2 *Fourier spectra*

The ways in which average properties of random fluctuations can be measured using Fourier techniques are discussed in detail in Van der Klis (1989b). The basic technique is to divide the data into time segments, to calculate power spectra of these segments and to average the power spectra. The data segments should be sufficiently short to allow the detection of changes in the variability as the source changes state and sufficiently long to measure relevant low-frequency variability components. The average power spectra (e.g. Figures 6.3, 6.9, 6.10 and 6.15) can show various different features that are indicative of the presence of aperiodic variability in the X-ray intensity.

The most common of these is broad-band noise (noise for short), with power spread over a large range, usually several decades in frequency. 'Red noise' is a term used for noise whose power decreases monotonically as a function of Fourier frequency. Broad-band noise indicates the presence in the data of random fluctuations covering a range of time scales; in red noise the slower fluctuations have the larger amplitudes. In a limited frequency range, red noise power spectra can often be satisfactorily described with a power law $P \propto v^{-\gamma}$ (e.g. Figure 6.9, 'UB'). However, towards the lowest frequencies such power laws must turn over, as no infinite amplitudes occur in the time series. Towards higher frequencies, the power law must be steeper than $\gamma = 1$ to avoid infinite integrated power. Another commonly occurring type of broad-band noise has a power spectrum that becomes progressively steeper towards higher frequencies (e.g. Figures 6.3 and 6.17). The cut-off frequency of such 'band-limited noise' can be characterized in various different ways, for example by fitting broken or exponentially cut-off power laws to it and reporting, respectively, v_{break} or v_{cut}. Sometimes, the power spectrum of band-limited noise has a clear flat top below a certain frequency (see, e.g., Figure 6.4), and this frequency, v_{flat}, is used to characterize the cut-off frequency of the noise. These different methods can lead to appreciable differences in the reported cut-off frequency for the same power spectral shape.

A well-defined resolved peak in the power spectrum (e.g. Figure 6.10) is evidence for the presence of quasi-periodic oscillations (QPOs), fluctuations with a preferred frequency (the centroid frequency of the peak). Usually, the term QPO is reserved for approximately symmetric peaks whose relative width (FWHM/centroid frequency)[*] does not exceed 0.5, and broader features are termed noise, even if the power spectrum has a local maximum (this is sometimes called 'peaked' noise); however, there are no hard and fast rules for this. The width of a QPO peak is a measure for the coherence

---

[*] FWHM: full width at half maximum.

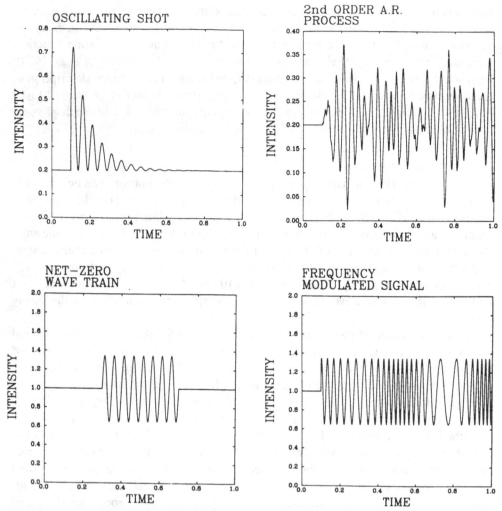

Fig. 6.2. Examples of time series that cause a QPO peak in the power spectrum (Van der Klis 1989b).

time, τ_{coh}, of the quasi-periodic signal (for a Lorentzian QPO peak, and defining τ_{coh} as the e-folding time of the ACF, FWHM $= (\pi\tau_{coh})^{-1}$). The reason that coherence is lost after this time may be, for example, frequency shifts or phase jumps in a persistent signal, or a finite lifetime of an otherwise strictly periodic signal.

A time series is not completely specified by its power spectrum (the power spectrum gives only the amplitudes of the Fourier components of the series, not the phases). Therefore, various different types of random processes, which are very different in the time domain, can all produce a similar power spectrum (for example, a QPO peak; see Figure 6.2.).

The Fourier power is a measure of the variance of the time series around its mean. The strength of the QPO and noise components as measured from power spectra is usually expressed in terms of fractional root-mean-square (rms) amplitude (sometimes called fractional rms variation). This dimensionless quantity is defined as $\sigma_{I_x}/\overline{I_x}$, where σ_{I_x} is the square root of the variance in the X-ray intensity caused by the variability in the frequency range of interest, and $\overline{I_x}$ is the mean X-ray intensity. It can be thought of as some kind of mean amplitude of the intensity variations, expressed as a fraction of the average intensity. Fractional rms amplitudes of less than 1% are detectable in bright sources (with total count rates of several thousand counts per second).

In the graphical presentation of power spectra, various approaches have been used. For the study of QPO and noise components in X-ray binaries, the present standard is to use log–log plots of power density versus Fourier frequency, rebinned (where possible) into equal logarithmic frequency intervals, with the flat power spectral component due to the counting statistics (Poisson) noise subtracted and with the power normalized to fractional rms amplitude squared per hertz (units Hz^{-1}). This normalization, used by Belloni and Hasinger (1990a) and Miyamoto *et al.* (1991) differs from the previously popular Leahy *et al.* (1983) normalization by a factor I_x (in counts per second): divide Leahy normalized powers by I_x to obtain fractional rms normalized powers. The advantage of this normalization is that fractional rms amplitudes can be directly estimated from the level of the power spectrum; its disadvantage that the significance of an excess power cannot.

An important diagnostic is the photon energy dependence of the rapid variability. Two quantities are necessary to characterize this dependence: the amplitude of the variability as a function of photon energy, and its phase as a function of photon energy. These quantities usually depend on Fourier frequency. The first quantity, variously called the 'rms spectrum', the 'X-ray spectrum' or the 'energy spectrum' of the variability is measured from power spectra in different photon energy bands. The second quantity is variously discussed in terms of 'time lags' or 'phase lags' of the variability between photon energy bands. It is measured by constructing cross-correlation functions or cross-spectra between the data in the two bands (see Van der Klis *et al.* 1987b; Ch. 2 in Lewin *et al.* 1988; and Vaughan *et al.* 1994 for descriptions of these techniques).

6.2 Black-hole candidates

The aim of this section is to describe the properties of the rapid X-ray variability in black-hole candidates. I refer to Ch. 3 for a discussion of the body of information that suggests that the objects discussed here may be black holes, and for a detailed description of their X-ray spectral and optical properties.

The existence of X-ray spectral states which correlate with the properties of the rapid X-ray variability ('high' and 'low' states) has long been recognized in the black-hole candidate Cyg X-1. On the basis of observations of other black-hole candidates, an 'off' state and a 'very high' state were added. The transient black-hole candidate GS 1124–68 (Nova Muscae 1991) in its decay subsequently went through the very high, high and low states (Ch. 3; Miyamoto *et al.* 1992a; Kitamoto *et al.* 1994), strongly suggesting that the states relate directly to $\dot{M}$, and that the accretion rate decreases in this order. On the basis of this and similar evidence, in the following I shall discuss the properties of the rapid X-ray variability of the black-hole candidates within a general framework of $\dot{M}$-driven states. This approach is motivated by the fact that in the bright low-magnetic-field neutron star systems (Sect. 6.3) the same approach has been very succesful. It should be noted, however, that this classification of black-hole candidate phenomenology is preliminary. Cases of observations that may not fit the picture will be noted below.

6.2.1   *The spectral states*

The concept of an *off state* was introduced by Markert *et al.* (1973) in the black-hole candidate GX 339—4 for cases when the source dropped below the ~5 mCrab (1–6 keV) sensitivity of the *OSO*-7 detector. In later observations (Ilovaisky *et al.* 1986), the source has been detected at levels down to 1.5 mCrab (2–10 keV), with no evidence that the X-ray spectrum at these levels is different from that in the low state (below). For a level of 4 mCrab (1–37 keV), Iga *et al.* (1992) report that up to 1 Hz the power spectrum of GX 339—4 is very similar to that at a level of 40 mCrab. There is therefore no need to distinguish a separate off state; the data are adequately described by a low state which encompasses a large (factor of ten or more) range in X-ray intensity.

The difference between the high and the low states was first noticed in the X-ray spectra of Cyg X-1 (Tananbaum *et al.* 1972). In the *low state* (LS) the X-ray spectrum is a flat power law, $CE^{-\alpha}$ photons cm^{-2} s^{-1} keV^{-1} with photon spectral index[*] $\alpha \sim 1.5$–2. Strong rapid aperiodic variability with fractional rms amplitudes of several tens of per cent that does not strongly depend on photon energy (Oda *et al.* 1971) is characteristic of this state. In the *high state* (HS) the 1–10 keV flux is an order of magnitude higher than in the low state due to the presence of a soft (sometimes called 'ultrasoft'; $kT \sim 1$ keV) component. The low-state power-law component is, in the HS, sometimes still visible at higher energies ('sticking out' from under the soft component; see, e.g., Figure 3.11), with usually (but not always) a somewhat steeper slope ($\alpha \sim 2$–3) than in the LS; sometimes the power law does not 'stick out' and could be part of the soft component itself, whose shape is not certain. This seems to be normally the case in the *very high state* (VHS), which has similar X-ray spectral properties as the high state (at a two to eight times higher 1–10 keV luminosity) and is distinguished from the HS by its broad-band noise properties and 3–10 Hz QPOs. In the following, I shall refer to the power-law component as the *hard component* and to the (ultra)soft component as the *soft component*. See Ch. 3 for more information about black-hole candidate spectral decomposition and spectral models.

Rapid variability similar to that seen in the LS can sometimes also be detected in the HS and the VHS and is often assumed to occur only in the hard component (see Sect. 6.2.3.1). As will become apparent below, there is no observational evidence that contradicts the idea that there is a continuous range of behaviour from the extreme low state to the high state, the only parameters changing along the way being the strength (and maybe spectral index) of two X-ray spectral components: a rapidly variable hard component and a much less rapidly variable soft component.

In the literature, one usually refers to sources as being in the 'low state' when X-ray spectroscopy or the 1–10 keV intensity indicate that the soft spectral component is negligible compared to the 1–10 keV luminosity of the hard component. The 'high

---

[*] *Photon* spectral indices (symbol α) are used throughout this chapter. Subtract one to obtain energy index.

state' conversely encompasses all observations where this is not the case and therefore ranges from cases where the soft component is relatively weak to cases where it is the only spectral component seen. It seems clear that a more quantitative approach to specifying source state, for example by quoting the luminosity ratio of the two components, is preferable. However, when the hard component is steep, the spectral decomposition into the soft and the hard components is model-dependent. The soft component is virtually constrained to the <10 keV band. At higher photon energies, the X-ray flux is often higher in the 'low' state than in the 'high' state (see also Ch. 3).

Cyg X-1 is usually in the low state and has only occasionally been observed in the high state (see Oda 1977; Liang and Nolan 1984). In GX 339–4, high and low states both occur, with no evidence that one state is more frequent than the other (Ilovaisky *et al.* 1986 and references therein); on one occasion, the source was observed in the very high state (Miyamoto *et al.* 1991). LMC X-1 and LMC X-3 appear to be always in the high state (White and Marshall 1984; Ch. 3). Of the transient black-hole candidates for which information about the rapid X-ray variability is available, 4U 1630–47 (Parmar *et al.* 1986) and EXO 1846–031 (Parmar *et al.* 1994) showed evidence for high-state properties, and GS 1826–24 (Tanaka 1989) for low-state properties similar to those in Cyg X-1, and GS 2000+25 (Tanaka 1989; Inoue 1991; Ch. 3) for both. In GS 1124–68 (Nova Muscae 1991; see Miyamoto *et al.* 1992a; Takizawa *et al.* 1994a,b) VHS behaviour was seen during the first 70 days of its outburst; later, HS and LS behaviour occurred. The exceptional transient GS 2023+33 (Inoue 1989, 1991; Kitamoto *et al.* 1989; Terada *et al.* 1992) exhibited a power-law X-ray spectrum at epochs throughout its entire decay, perhaps indicating that it was always in a 'low' state, even at its peak intensity of 17 Crab (see Ch. 3 for a full account of the remarkable X-ray spectral behaviour of this exceptional source). The reports about the 1–20 keV rapid X-ray variability of GS 2023+33 (see below) have so far only covered the latter part of its decay, when it was similar to the other sources in the low state. GRO J0422+32 may be another example of a black-hole candidate with low-state properties when it is very bright (Denis 1994).

6.2.2 *Power spectrum*

Broad-band power spectra have been reported from black-hole candidates in all three states. Table 6.1 lists the main references and indicates for each source in which state(s) the power spectra were obtained.

6.2.2.1 *Noise in the low state (LS)*

The power spectra of black-hole candidates in the LS are quite similar to each other. Their shape can be described as follows (Nolan *et al.* 1981; Makishima 1988; Miyamoto and Kitamoto 1989; Belloni and Hasinger 1990b; see Figure 6.3): below a frequency ν_{flat} at roughly 0.1 Hz (usually between 0.04 and 0.4 Hz), the power spectrum is approximately flat; it can be described by a power law $P \propto \nu^{-\gamma}$ with an index γ of 0–0.3. Above ν_{flat}, the spectrum steepens to $\gamma=1$–1.7. In the range from 0.1 to 10 Hz, the power spectral slope sometimes varies, creating one or more 'shoulders' or 'wiggles', but on the whole the spectrum becomes gradually steeper. Eventually, at frequencies $\gtrsim$10 Hz, power falls off steeply following a power law of index 1.5–2. Although there are differences in details (sometimes three distinct

Table 6.1. *Published power spectra of black-hole candidates*

Source	LS	HS	VHS	References
Cyg X-1	•			Terrell (1972), Brinkman *et al.* (1974), Weisskopf *et al.* (1975), Oda *et al.* (1976), Nolan *et al.* (1981), Makishima (1988), Miyamoto and Kitamoto (1989), Belloni and Hasinger (1990a), Lochner *et al.* (1991), Sunyaev *et al.* (1991), Negoro *et al.* (1992)
GX 339−4	•		•	Motch *et al.* (1983, 1985), Maejima *et al.* (1984), Belloni and Hasinger (1990b), Grebenev *et al.* (1991a, 1993), Miyamoto *et al.* (1991, 1992b), Dotani (1992), Iga *et al.* (1992)
LMC X-1		•		Ebisawa *et al.* (1989)
LMC X-3		•		Treves *et al.* (1990)
GS 1826−24	•			Tanaka (1989)
GS 2023+33	•[a]			Sunyaev *et al.* (1991, 1992), Terada *et al.* (1992), Miyamoto *et al.* (1992b)
GS 2000+25	•	•[b]		Kitamoto (1989), Sunyaev *et al.* (1991)
GS 1124−68	•	•	•	Grebenev *et al.* (1991b, 1993), Dotani (1992), Miyamoto *et al.* (1992a, 1993), Takizawa *et al.* (1993a,b)
GRO J0422+32	•[a]			Kouveliotou *et al.* (1992), Denis (1994)

[a] Source was very bright; see text. [b] Uncertain.

slopes can be distinguished in the power spectrum, the strength of the shoulder(s) can differ, and also v_{flat} is variable, see below) in all cases an obvious roll-over is detected from rather flat at frequencies $\lesssim 0.1$ Hz to quite steep $\gtrsim 10$ Hz. Roughly, the overall shape resembles a Lorentzian centred on zero frequency with a HWHM[*] of ∼0.5 Hz. The amplitude of the fluctuations varies between 30 and 50% (rms) (e.g. Belloni and Hasinger 1990a). In the following, I shall refer to this band-limited power spectral component as the 'LS noise'. The power spectrum of the optical flux of GX 339−4 in the low state shows broad-band noise similar to that in the X-ray power spectrum (Motch *et al.* 1983).

The low-state power spectra are similar not only in shape, but also in strength: the fractional rms amplitude of the LS noise as a function of Fourier frequency appears approximately the same for frequencies above v_{flat}. This was reported to hold not only between various low-state epochs in Cyg X-1 (Belloni and Hasinger 1990a), but even between the sources Cyg X-1, GX 339−4, GS 2023+33 (Miyamoto *et al.* 1992b) and GS 1124−68 (Miyamoto *et al.* 1992a, Terada *et al.* 1992). This similarity is well illustrated (Figure 6.4) by superimposing power spectra normalized to fractional rms amplitude squared (Sect. 6.1.2.2). In Cyg X-1 in the low state, v_{flat} varies between 0.04 and 0.4 Hz, and the associated 'saturation level' below

[*] HWHM: half width at half maximum; I shall consistently use this quantity to describe the width of zero-centred Lorentzians, and FWHM for other Lorentzians. The HWHM is larger than v_{flat}: for the LS power spectral shapes described here, usually by a factor between 1.5 and 10.

BLACK−HOLE−CANDIDATE POWER SPECTRA

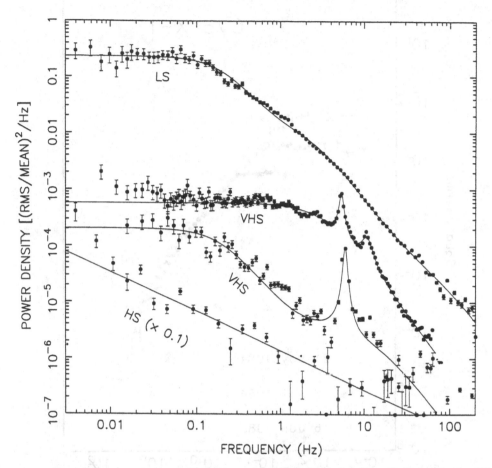

Fig. 6.3. Typical power spectra of black-hole candidates in the LS (Cyg X-1) and in the HS and VHS (GS 1124−68). The HS power spectrum was shifted down by one decade for clarity. (*Ginga* data, courtesy B. Vaughan.)

ν_{flat} goes up and down in anticorrelation to it (Belloni and Hasinger 1990a). A tendency towards a positive correlation between ν_{flat} and I_x was reported in Cyg X-1, GX 339−4 and GS 2023+33 (Nolan *et al.* 1981; Miyamoto *et al.* 1992b). However, in other observations (Cyg X-1, Belloni and Hasinger 1990a; GS 1124−68, Miyamoto *et al.* 1992a), no simple correlation with X-ray intensity was seen.

6.2.2.2 *Noise in the high state (HS)*

The power spectral shape typical of the HS seems to be a power law $P \propto \nu^{-\gamma}$ with an index $\gamma \sim 1$ and a fractional amplitude of a few per cent (Figure 6.3). Such a shape was observed in LMC X-3, LMC X-1, GS 2000+25 and GS 1124−68 in the HS (Ebisawa *et al.* 1989; Treves *et al.* 1990; Miyamoto *et al.* 1992a; Kitamoto *et al.* 1994). I shall refer to this power spectral component as 'HS noise'.

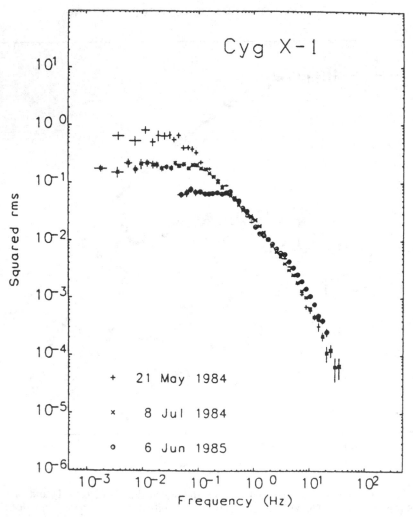

Fig. 6.4. The variable flat-top effect in Cyg X-1 in the LS (Belloni and Hasinger 1990a).

In various cases, LS noise was seen in the high state (Terrell 1972; Sutherland *et al.* 1978; Page *et al.* 1981; Treves *et al.* 1988; Miyamoto *et al.* 1992a). It seems likely that this happens when, in the high state, the hard component with its associated LS noise is strong enough to dominate the power spectra. Because the amplitude of the LS noise is large compared to that of the HS noise, a small admixture of the hard component is sufficient to make this happen. Only when the hard component is practically absent is the weak HS noise in the soft component revealed. The energy dependence of the LS noise in the high state is a strong argument that this description is correct (Sect. 6.2.3.1). In GX 339—4 in the HS, Belloni and Hasinger (1990b) observed a broad-band noise component that looks like LS noise, but with ν_{flat} ~10 Hz and a fractional amplitude of only ~3.6% (rms). This would fit in with a positive correlation between I_x and ν_{flat} (see Sect. 6.2.2.1). Perhaps GX 339—4 was in the VHS (below) during this observation.

6.2.2.3 Noise in the very high state (VHS)

The very high state (VHS) is characterized by the presence of 3–10 Hz QPOs (Sect. 6.2.2.4). Only two occurrences of this state have been reported, one in the early stages of the outburst of the transient GS 1124—68, at 1–10 keV flux levels of up to about eight times those in the high state (with *Ginga*: Dotani 1992; Miyamoto *et al.* 1992a, 1993; Takizawa *et al.* 1994a,b; and *GRANAT*: Grebenev *et al.* 1993), and once in GX 339—4 at levels two to three times higher than in the high state (with *Ginga*: Miyamoto *et al.* 1991, 1993; Dotani 1992). The broad-band noise in the VHS is intermediate in fractional amplitude between those of the high and low states (it can be as weak as the HS noise); amplitudes vary between 1 and 15% (rms). The power spectrum (Figure 6.3) sometimes resembles a three to eight times weaker version of the LS noise, but with ν_{flat} between 1 and 10 Hz (in the low state this is ~0.1 Hz). At other times it resembles HS noise, stronger by a factor of one to two. Transitions between the two noise types have been observed to occur within less than a second (Takizawa *et al.* 1994b). Sharp-edged 10–20% intensity dips with lengths of 10 s to several minutes sometimes contribute to the low-frequency part (<0.3 Hz) of the VHS power spectra. When dips are numerous, dip transitions are sometimes called 'flip-flops'. The strong LS-like noise switches on and off, sometimes within a second, on entry and exit, respectively, of a dip (Miyamoto *et al.* 1991; Takizawa *et al.* 1994b).

Relatively subtle X-ray spectral variations take place in the VHS on time scales of minutes to days. They can be usefully parametrized using X-ray hardness–intensity and colour–colour diagrams (see also Sect. 6.3). The power spectra correlate with these X-ray spectral variations. In GX 339—4, the strength of the LS-like noise (>0.3 Hz, so excluding the effect of the dips) increased from 2 to 10% (rms) when the X-ray colours, following a clear branch in an X-ray colour-colour diagram (Figure 6.5), increased by a factor of two to three (Miyamoto *et al.* 1991). In GS 1124—68, a more complex correlation between X-ray colours and noise parameters involving two intersecting branches was reported (Takizawa *et al.* 1994b).

It has been suggested that the LS-like noise in the VHS appears when the hard component is stronger (Miyamoto *et al.* 1993) and is therefore associated with this component. Note, however, that the X-ray spectral decomposition in the VHS is rather model-dependent.

6.2.2.4 QPOs

QPOs with frequencies varying between 3 and 10 Hz were seen during the two VHS episodes described in the previous section (Figure 6.3). They occurred together with both types of VHS noise, but were less evident in dips (when the noise is LS-like) (Takizawa *et al.* 1994b). In GX 339—4 (Miyamoto *et al.* 1991), QPO frequencies were usually near 6 Hz. Second harmonics to these peaks were seen in most cases, and a possible subharmonic near 3 Hz in one case. Sometimes the QPOs were directly visible in the data (Figure 6.6). In GS 1124—68 (Dotani 1992; Grebenev *et al.* 1993; Miyamoto *et al.* 1993; Takizawa *et al.* 1994a,b) the QPOs had frequencies between 3

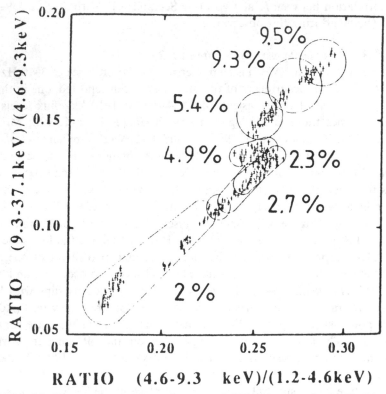

Fig. 6.5. X-ray colour–colour diagram of GX 339–4 in the VHS. The fractional amplitude of the LS-like noise in the various sections of the diagram is indicated. (After Miyamoto *et al.* 1991.)

and 10 Hz. The frequency increased with X-ray intensity at different rates depending on the presence or absence of LS-like noise. Strong second harmonics and possible subharmonics were detected.

Considerably slower QPOs have been observed from black-hole candidates in the high and low states. The first glimpse of this may have been the observation by Frontera and Fuligni (1975) of sporadic 0.06 Hz peaks in the 30–200 keV flux of Cyg X-1. More recently, 0.04 Hz QPOs were reported from this source in various bands between 1 and 300 keV (Angelini *et al.* 1992; Kouveliotou *et al.* 1992; Vikhlinin *et al.* 1992a). None of these reports included an estimate of the significance of the QPO peak. The peak reported by Angelini *et al.* (1992) is very broad (0.07 Hz), but that reported by Kouveliotou *et al.* (1992) may be more narrow.

In GX 339–4 in the LS, QPOs in the X-ray flux near 0.05 and near 0.1 Hz were reported by Motch *et al.* (1983). Only the 0.1 Hz QPOs were significant (at the 3σ level), although confidence in the existence of the 0.05 Hz peak was boosted by the simultaneous detection of 0.05 Hz QPOs in the optical band. Further optical QPOs in GX 339–4 at 0.12 and 0.14 Hz were also reported (Motch *et al.* 1985; Imamura *et al.* 1990). Highly significant 0.8 Hz QPOs were detected in GX 339–4 in the 2–60 keV band (Grebenev *et al.* 1991a). In LMC X-1 (Ebisawa *et al.* 1989), a narrow 0.08 Hz

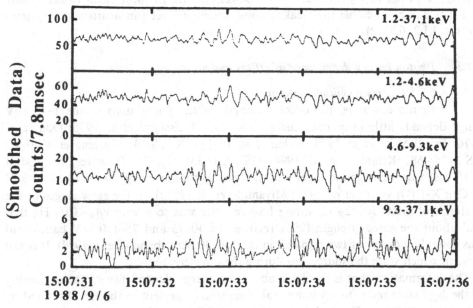

Fig. 6.6. The 6 Hz QPOs in GX 339—4 in the VHS show up directly in these (slightly smoothed) light curves (Miyamoto *et al.* 1991).

Table 6.2. *QPOs in black-hole candidates[a]*

Source	State	Frequency (Hz)	FWHM (Hz)	Fractional amplitude (% rms)	References (remarks)
GX 339—4	VHS	5.9–6.6	0.6–2.5	3.5–4.7	Miyamoto *et al.* (1991)
		3.1	0.5	1.6	(possible subharmonic)
		11.4–13.8	1.5–4.3	1.4–3.0	(second harmonics)
GS 1124—68	VHS	3.0–7.6	0.4–3.3	0.8–3.6	Takizawa *et al.* (1994a)
		2.7–4.3	1.6	0.9–2.1	(subharmonics)
		6.2–15.6	1.7–5.1	0.5–2.7	(second harmonics)
GX 339—4	LS	0.8	0.15	7.0	Grebenev *et al.* (1991a)
GX 339—4	LS	0.11	—	~1–15	Motch *et al.* (1983)
LMC X-1	HS	0.08	0.009	2.9	Ebisawa *et al.* (1989)
		0.14	<0.07	1.8	(second harmonic)

[a] Verifiably significant and narrow X-ray QPO peaks only; see text for more reports.

QPO peak with a harmonic were seen in the 1–16 keV range. Broad QPO peaks near 0.25 and 0.04 Hz were reported from GRO J0422+32 (Kouveliotou *et al.* 1992; Vikhlinin *et al.* 1992b).

All in all, there are so far three cases of verifiably narrow, significant X-ray QPOs in black-hole candidates in the LS and HS, and two in the VHS (see Table 6.2).

The QPO peaks in Cyg X-1 and GRO J0422+32 are in most cases rather broad, and they rather resemble the 'peaked noise' sometimes seen in neutron star systems (Sect. 6.3.1.2, 6.3.2.2).

6.2.3   *Photon energy dependence of QPOs and noise*

6.2.3.1   *rms amplitude spectra*

In the *low state*, the overall strength of the fluctuations in the 1–15 keV range depends little on photon energy (Cyg X-1: Weisskopf *et al.* 1975; Oda *et al.* 1976; Priedhorsky *et al.* 1979; Nolan *et al.* 1981; GX 339−4: Maejima *et al.* 1984; GS 1124−68: Kitamoto *et al.* 1994; GS 2023+33: Ch. 3). However, the fastest ($\lesssim 10$ ms) variations get stronger with increasing photon energy from 10 to 60 keV in Cyg X-1 (Ogawara *et al.* 1977; Miyamoto *et al.* 1988a). In the early stages of the outburst of GRO J0422+32, strong LS-like noise was seen with $v_{flat} \sim 0.03$ Hz that had about the same strength (20% rms) in the 40–75 and 75–150 keV bands, but was roughly a factor of two stronger in the 150–300 keV band (Denis 1994). It is not clear in which state the source was during these observations.

The information that is available about the energy dependence of the variability in the *high state* refers to LS noise that is sometimes present in the HS. As noted by Oda *et al.* (1976), in Cyg X-1 in the high state LS-like noise is observed that seems energy-dependent; its fractional amplitude is larger at higher energies, where the hard component dominates the flux, than at lower energies, where the soft component dominates. This has been the main argument for the idea that the LS noise is a property of the hard component, and that this variability when seen in the HS is the same as that in the LS, but suppressed in the 1–10 keV range by the presence of the much less variable soft component. However, Oda *et al.* (1976) did not quantify the effect. GS 1124−68 shows a similar effect (large amplitude increase above 10 keV; Miyamoto *et al.* 1992a). No information exists about the energy dependence of the HS noise (Sect. 6.2.2.2). In an observation of 4U 1630−47 (Parmar *et al.* 1986) in the high state, the amplitude was higher in the 1–7 keV band (7% rms) than in the 14–30 keV band (4%), indicating that a mixture of the HS and LS noise may have contributed to the variations in the 1–7 keV band.

The broad-band noise, as well as the QPOs seen in the *very high state* of GX 339−4, had energy spectra similar to the X-ray spectrum of the hard component (Miyamoto *et al.* 1991), so that the amplitude of the fluctuations as a fraction of the total flux increased towards higher photon energy. Such an increase was also seen in the QPOs of GS 1124−68 in the VHS (Figure 6.7; Miyamoto *et al.* 1992a; Takizawa *et al.* 1994a). The LS-like noise in the VHS is strongest near 4 keV and decreases toward both higher and lower photon energies (Takizawa *et al.* 1994a).

6.2.3.2   *Time lags*

The possibility and relevance of the detection of time lags between the rapid fluctuations in different photon energy bands were recognized early on (Brinkman *et al.* 1974; Weisskopf *et al.* 1975). Asymmetries in the cross-correlation function that indicated that the hard photons lag the soft ones and interpreted as due to Compton scattering delays were reported by several authors (Priedhorsky *et al.* 1979; Nolan *et*

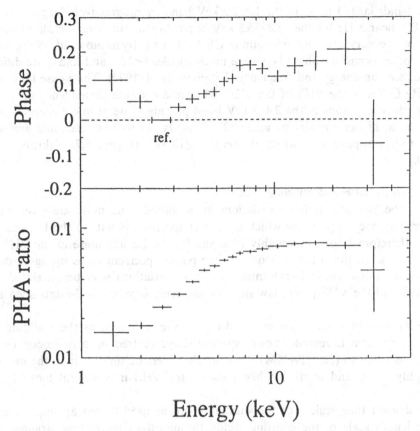

Fig. 6.7. Phase lag (top) and fractional rms amplitude (bottom) spectra of the VHS QPOs in GS 1124−68 (Takizawa *et al.* 1993a).

al. 1981; Page *et al.* 1981; Page 1985; see also Maejima *et al.* 1984). However, it was not until Fourier cross-spectral techniques were applied to the data that the basic phenomenology became clear. In Cyg X-1 in the *low state*, the variations in harder bands lag those in softer bands by an amount depending on both Fourier frequency and photon energy (Miyamoto *et al.* 1988a; Kitamoto *et al.* 1989). The phase lag between the 1.2–4.7 and 4.7–9.3 keV bands increases from 0.3° at 0.01 Hz to 6° at 30 Hz (the implied time lag decreases with frequency). In higher photon energy bands, the lags with respect to the 1.2–4.7 keV band gradually increase, by up to a factor of three in the 14–35 keV band. These results showed that the time delays cannot be due to Compton scattering, which causes similar time lags at all but the lowest Fourier frequencies (Miyamoto *et al.* 1988a). In GX 339−4 and GS 2023+33 in the LS, hard phase lags in the 0.6–6° range are also seen between the 1.2–4.6 and 4.6–9.2 keV bands, but no clear dependence on Fourier frequency is detected (Miyamoto *et al.* 1992b).

In GX 339−4 in the *very high state*, the LS-like noise in the 0.1–10 Hz range showed phase lags which depended on Fourier frequency and had a maximum near ∼3 Hz (near ν_{flat} and clearly below the QPO frequency). The fluctuations in the harder bands lagged those in the 1.2–2.3 keV band by progressively larger amounts, up to 80° near 3 Hz for the 13.8–36.8 keV band (Miyamoto *et al.* 1993). Something similar was observed on one occasion in GS 1124−68 (Miyamoto *et al.* 1993), but on several other occasions phase lags were much smaller (<15°) and had quite different dependences on energy and frequency (Kitamoto *et al.* 1994). The phase lags of the 3–10 Hz QPOs in the VHS of GS 1124−68 have a complex dependence on photon energy: the oscillations in the 2.4–3 keV band precede those at lower energies by up to 6° and at higher energies by up to 12°. Phase lag as well as fractional amplitude of the QPOs appear to saturate at energies $\gtrsim$10 keV (Figure 6.7; Takizawa *et al.* 1994a).

6.2.4 *Observed variability time scales*

The *longest* correlated variations in the broad-band noise are given by the Fourier frequency v_{flat}, below which the power spectrum is flat. In the LS, this time scale is therefore $1/v_{flat}$, or roughly 10 s, and for the LS-like noise in the VHS it is roughly 0.1 s. At much lower frequency, the power spectrum turns up again due to the long-term changes in X-ray intensity. These variations may be unrelated. The weak HS and the VHS power-law noise components have no well-established turn-over.

The *characteristic* time scale of broad-band noise is defined as the e-folding time of its ACF. For a Lorentzian power spectral shape centred on zero frequency, this quantity is equal to $(2\pi \cdot \text{HWHM})^{-1}$, so the characteristic time scale in the low state is roughly 0.3 s, and in the LS-like noise of the VHS it is several tens of milli-seconds.

The *shortest* time scale τ of the variability can be used to put an upper limit τc on the length scale of the emitting region through light travel time arguments. τ is defined as $I_x/\dot{I}_x$; it is a measure of how steeply I_x can vary with time. Note that τ is not identical to the shortest time interval δt in which variability can be *detected*, which of course depends on the statistical quality of the data. Measurements of the highest Fourier frequency where significant power in excess of the counting statistics noise exists, the smallest lag-time interval including zero lag within which the ACF is inconsistent with being flat, or the shortest time interval in which significant deviations from Poisson statistics are found using a chi-squared or similar test (all of which occur in the literature), do not in themselves provide an estimate of τ but only of δt. In a smoothly varying aperiodic source, progressively smaller values for δt are found as the statistics improve (for infinite statistics δt=0), but this does not mean that τ becomes shorter. To estimate τ, the amplitude of the variability must be specified in addition to δt. One can, for example, determine the fractional rms amplitude r of the variability as a function of δt. This can be converted to a time scale τ as defined above using $\tau = \xi\delta t/r$, where ξ is a factor that depends on assumptions about how I_x could have varied to produce the observed variance. For a sinusoidal variation with period δt, $\xi = \frac{1}{2}\sqrt{2}/\pi \sim 0.2$; for a linear variation during the interval δt, $\xi = \frac{1}{3}\sqrt{3} \sim 0.6$.

The observation giving the fastest statistically convincing time scale in a black-hole candidate was reported in Cyg X-1 in the low state (Meekins *et al.* 1984). Using a chi-squared method, a fractional amplitude of 12% (rms) was detected for time scales in the 0.3–3 ms range, implying that τ was $\lesssim 25\xi$ ms. However, this result was only significant at the 2.8σ level of confidence. With high confidence, Meekins *et al.* detected 28% (rms) over 1–10 ms ($\tau \lesssim 35\xi$ ms). The fractional amplitudes derived by Meekins *et al.* (1984) are considerably larger (by a factor of four) than what seems to be usual judging from published power spectra of Cyg X-1 (Sect. 6.2.2.1). The reason for this is not known.

Occasional 'millisecond bursts' lasting $\sim$1 ms and with submillisecond rise and fall times have been reported in Cyg X-1 (Oda *et al.* 1974; Rothschild *et al.* 1974, 1977; Giles 1981). The significance of these events was evaluated under the assumption that the count rates in the short (0.2–0.4 s) data segments that were searched for excesses are distributed according to the Poisson distribution appropriate to the mean count rate in the segment. This assumption is incorrect, as strong fluctuations on time scales of 0.1 s are known to occur in the LS noise, and this strongly affects the significance of the reported events (Press and Schechter 1974). Attempts to evaluate the significance of the claimed bursts using a model for the LS noise (Weisskopf and Sutherland 1978; Giles 1981) were inconclusive: there is a wide range of possibilities for such models, and each has different effects in the shortest time-scale regimes. Two different questions play through the discussion about this subject in the literature, namely: (i) do the millisecond bursts exist as a phenomenon that is independent from the LS noise, and (ii) do changes in the X-ray flux as fast as suggested by the claimed rise and fall times of the millisecond bursts really occur in the data. Question (i) is unanswerable without further specifying what is meant in a statistical sense by 'independent'. Two time-scale shot models (below) are an example of such further specification, but this solution is by no means unique. The answer to question (ii) might be affirmative, even if the millisecond bursts have no independent reality. The 'bursts' could, for example, be due to the occasional accidental pile-up of sharp-edged random shots. However, in my opinion, no convincing quantitative evidence has up to now been presented for submillisecond variations in black-hole candidates.

6.2.5   Shot noise model

The fact that the longest correlation time scale in the low state data is $\sim$10 s is consistent with the idea that the LS noise is made up of a superposition of individual finite events ('shots'). Visual inspection of high time resolution X-ray light curves of black-hole candidates in the low state (Figure 6.8) and some theoretical ideas (Sect. 6.2.6.2) seem to support this. This has motivated the continued application of shot noise models to the data. (The shot noise model was originally proposed (Terrell 1972) in response to the idea (Oda *et al.* 1971) that transient periodicities were causing the observed fluctuations in Cyg X-1.) Shot parameters were often derived from fits to the ACF of the data (but the power spectrum or the variation function can be used as well; e.g. Maejima *et al.* 1984). As its power spectrum is roughly a zero-centred Lorentzian (Sect. 6.2.2.1), the ACF of the LS noise is approximately exponential, and such a shape, interpreted as the average shot shape, was indeed reported in much of the early work on Cyg X-1 (Terrell 1972; Weisskopf *et al.* 1975, 1978; Rothschild *et al.* 1977; Sutherland *et al.* 1978; Priedhorsky *et al.* 1979). Reported decay times for Cyg X-1 in the low state range from 0.24 to 1.4 s (power spectral Lorentzian HWHM 0.1–0.6 Hz).

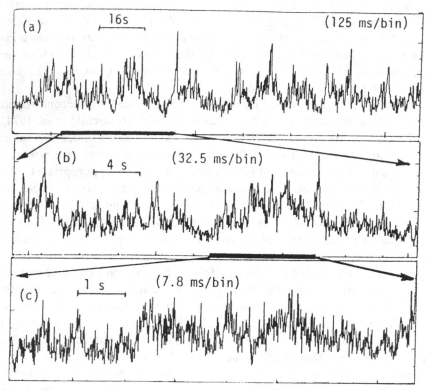

Fig. 6.8. Light curves of Cyg X-1 in the LS in various time resolutions (Makishima 1988).

Evidence for more complicated ACFs corresponding to the more complicated power spectral shape described in Sect. 6.2.2.1 was, however, reported as early as 1977 (Canizares and Oda 1977; Ogawara *et al.* 1977; Nolan *et al.* 1981; Meekins *et al.* 1984; Belloni and Hasinger 1990a; Lochner *et al.* 1991). In a shot noise model, such ACF shapes can be explained by assuming that all shots individually have this more complicated shape (but are still identical), or by allowing several different shot shapes or a distribution of shot shapes, and these solutions have all been considered by authors cited above. Shot noise models are very good at reproducing the properties of any random process (Doi 1978). Therefore, the case for shot noise in black-hole candidates in the low state is undecided.

The fractional rms amplitude of the noise is, in a shot noise model, proportional to $f/\sqrt{\lambda}$, where λ is the shot rate and f is the fraction of the flux that is in the shots (Sutherland *et al.* 1978). The earliest work assumed $f \equiv 1$, and this usually led to LS shot rates of $\lambda = 10$–30 s^{-1}. Later analyses also included determination of f using the third moment of the data, and found $f \sim 25\%$ and $\lambda \sim 1$ s^{-1}, consistent with a similar fractional amplitude. Derived HS shot rates are much higher (Terrell 1972; Sutherland *et al.* 1978), which reflects the much lower fractional amplitude in the high state.

6.2.6 Discussion of the black-hole candidates

6.2.6.1 Phenomenology

The data on black-hole candidates are consistent with the idea that these sources form a phenomenologically more or less homogeneous group, and that the instantaneous properties of a given source are strongly correlated to its instantaneous accretion rate.

The long known high and low states appear to be the consequence of the interplay of two emission components: a soft component that cuts off strongly above 10 keV and a hard component that (in the 1–30 keV band) approximately follows a power law with photon index $\alpha=1.5$–3. The soft component shows little variability; its power spectrum is a power law with an index of $\gamma \sim 1$ and a fractional amplitude of a few per cent (rms) (the HS noise). The hard component exhibits the strong rapid variability for which the black-hole candidates are famous; its power spectrum is approximately flat below a frequency, ν_{flat}, of roughly 0.1 Hz (the LS noise). The hard component varies in approximate but *not* strict anticorrelation to the soft component. Many of the properties of the rapid variability in the HS and the LS can be understood as the consequence of a superposition of these two spectral components and their associated variability.

At the highest 1–10 keV luminosities, in the very high state, the properties of the broad-band noise alternate on a time scale of sometimes less than a second between being similar to HS noise and being similar to LS noise (the LS-like noise has a much higher ν_{flat} (1–10 Hz) than in the LS). It is not yet clear what relation these components have to those seen in the LS and the HS, but it is possible that the LS-like noise in the VHS is the same phenomenon as that in the LS. The VHS in addition shows 3–10 Hz QPOs with a strong harmonic content. The outburst properties of the transient black-hole candidates (Ch. 3) suggest that the luminosity of the soft component correlates with the accretion rate, which is thought to be lowest in the LS, higher in the HS, and highest in the VHS. Perhaps the properties of the VHS have to do with near-Eddington accretion rates, as they are in some respects similar to those of neutron stars near L_{Edd} (Van der Klis 1994a). In Sect. 6.5 we further compare the phenomenology of the black-hole candidates with that of the low-magnetic-field neutron stars.

6.2.6.2 Models

A large variety of physical mechanisms has been considered for the LS noise, including turbulence and other hydrodynamic effects, disk instabilities, magnetic phenomena, plasma effects and radiation pressure feedback (see Fabian 1986, McClintock 1988 and references therein). The models have been mostly aimed at explaining the observed time scales (which are long compared to dynamical and radiative time scales in the inner disk), and, in some cases, energy dependencies. They have not yet been confronted quantitatively with the new information from the *Ginga* satellite. In particular, the power spectral shape variations as a function of $\dot{M}$ and the time lag spectra described in Sect. 6.2.2.1 and Sect. 6.2.3.2 would seem to provide strong model constraints. On an empirical level, Miyamoto and Kitamoto (1989) have proposed that the shots in X-ray intensity that produce the LS noise arise through the passage of inhomogeneities (of unspecified origin) through the innermost part of the accretion disk, where the radiation is produced; this scenario would lead to a population of shots whose X-ray spectrum hardens as the shot progresses. Eyeball shot superposition techniques (Negoro *et al.* 1992) illustrate this 'hardening shot' model but provide no additional evidence for it.

A phenomenon that seems particularly suggestive of interpretation is the variable flat-top effect described in Sect. 6.2.2.1 (Figure 6.4). It seems as if in terms of fractional amplitudes, variations on long time scales (below ν_{flat}) are sometimes removed from the LS noise without any change in the faster variations. One would expect such a phenomenon to correlate with $\dot{M}$, but the observations are inconclusive about this point. It seems possible that a correlation of ν_{flat} with $\dot{M}$ exists, but that because the correlation between I_x and $\dot{M}$ is not strict, a correlation of ν_{flat} with I_x is only apparent when I_x varies over a large range. If the VHS version of the LS noise is really the same phenomenon as in the LS, then there is no doubt that ν_{flat} increases when $\dot{M}$ does.

6.3   Z and atoll sources

Z and atoll sources (Hasinger and Van der Klis 1989, hereafter HK89; see Table 6.3) are thought to be X-ray binaries containing low-magnetic-field neutron stars, as many of the atoll sources and at least one Z source show X-ray bursts (Ch. 2). Their X-ray spectra are soft, and no pulsations have been detected from them (Ch. 1, 4). Together, the two groups encompass all LMXBs that are persistently brighter than 100 μJy (see Ch. 14). Exceptional sources that are not listed in Table 6.3, but that are perhaps related to Z and atoll sources, are Cir X-1 (considered transient by some) and Cyg X-3 (probably an HMXB); these sources are discussed in Sect. 6.4. None of the neutron star soft X-ray transients have been classified as either Z or atoll as none of them have been studied with the required sensitivity and/or time resolution, but some faint or highly variable LMXBs have been classified as 'probably atoll' (see Table 6.3).

Early studies of the rapid time variability of bright LMXBs (Friedman *et al.* 1969; Lampton *et al.* 1970; Boldt *et al.* 1971) indicated a distinct lack of variations on time scales $\lesssim 1$ s, quite contrary to what was seen in Cyg X-1 (Sect. 6.2) and Cir X-1 (Sect. 6.4.1). There was some evidence in Sco X-1 for transient 1–10 Hz power with amplitudes of 1–2% (Angel *et al.* 1971) and later variability at levels of 5–10% was detected in this range in several other bright LMXBs (Parsignault and Grindlay 1978; Petro *et al.* 1981; Stella *et al.* 1984). On longer time scales, the X-ray spectral hardness H was often observed to vary in correlation with X-ray intensity I_x, and X-ray hardness versus intensity diagrams (HIDs) were routinely used to describe this correlation (see, e.g., Mason *et al.* 1976; White *et al.* 1976, 1978, 1980; Parsignault and Grindlay 1978; Charles *et al.* 1980; Holt 1980; Ponman 1982; Sztajno *et al.* 1983; Basinska *et al.* 1984; Shibazaki and Mitsuda 1984; see Sect. 1.5.2). Sometimes, complex HID patterns were seen (e.g. White *et al.* 1976; Branduardi *et al.* 1980; Shibazaki and Mitsuda 1984)

With *EXOSAT*, it was found that some of the bright LMXBs show QPOs in the 6–60 Hz range with amplitudes of a few per cent (Van der Klis *et al.* 1985a; Hasinger *et al.* 1986; Middleditch and Priedhorsky 1986) as well as noise components

Table 6.3. *Z and atoll sources*

Source	$I_{2-10keV}$ (μJy)	P_{orb} (h)	Type	Phenomena[a]
Sco X-1	14000	18.9	Z	QPOs
GX 5−1	1250	—	Z	QPOs
GX 349+2	825	—	Z	QPOs
GX 17+2	700	—	Z	QPOs,(bu)
GX 340+0	500	—	Z	QPOs
Cyg X-2	450	236	Z	QPOs,(bu)
GX 9+1	700	—	atoll	—
GX 3+1	400	—	atoll	(Bu)
GX 13+1	350	—	atoll	(bu)
GX 9+9	300	4.2	atoll	—
4U 1820−30	250	0.2	atoll	(Bu)
4U 1705−44	280[b]	—	atoll	Bu
4U 1636−53	220	3.8	atoll	Bu
4U 1735−44	160	4.6	atoll	Bu
4U 1728−33	150	—	atoll	Bu
4U 1608−52	< 1–110	—	atoll	Bu
Ser X-1	225	—	probably atoll	Bu
4U 1702−42	45	—	probably atoll	Bu
4U 2129+12	6	17.1	probably atoll	Bu
4U 1850−08	7	—	probably atoll	Bu

[a] Bu: regular X-ray bursts; (Bu): occasional episode(s) of regular X-ray bursts; (bu): occasional X-ray bursts.
[b] Ignoring long-term off (Ch. 14).

with frequencies up to ∼100 Hz and amplitudes up to 20% (Van der Klis *et al.* 1985a, 1987a; Hasinger *et al.* 1989; HK89). The relatively low amplitudes of these fluctuations (the strongest occur only in the relatively fainter sources, see below) had, up to then, prevented their detection. The extensive studies that were undertaken after this revealed strong correlations between X-ray spectral shape and the rapid X-ray variability.

The introduction of X-ray 'colour–colour diagrams' (CDs) to track the X-ray spectral shape variations (Hasinger 1988a; HK89; Hasinger *et al.* 1989; Schulz *et al.* 1989) helped to uncover these correlations. It was found that in both HIDs and CDs the sources trace out characteristically shaped patterns, and that the properties of the rapid X-ray variability are strongly correlated to the instantaneous position of a source in its pattern. This almost certainly arises because changes in the accretion rate $\dot{M}$ cause changes in the accretion flow, which simultaneously affect the X-ray spectrum and the rapid X-ray variability. Consistent with this, motion through CD and HID patterns is one-dimensional: a source does not jump through the diagram but always moves smoothly, following the pattern. The HID and CD patterns of the same source are usually rather similar to one another (see Schulz *et al.* 1989; Penninx *et al.* 1990, 1991; Lewin *et al.* 1992; Kuulkers *et al.* 1994), mainly because the 'soft

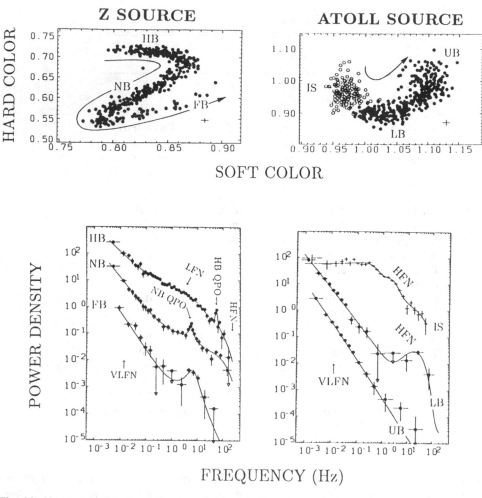

Fig. 6.9. X-ray colour–colour diagrams (top) and power spectra (bottom) typical of Z sources (left) and atoll sources (right). The direction in which $\dot{M}$ is inferred to increase is indicated by arrows in the colour diagrams. Power spectral normalization is arbitrary. Soft colour is approximately (3–5) keV/(1–3) keV, hard colour (6.5–18) keV/(5–6.5) keV. (After HK89.)

colour' (Figure 6.9) is well correlated to the X-ray intensity. Observable parameters, such as the optical and UV continuum and line flux and the X-ray burst properties, correlate with inferred $\dot{M}$. The X-ray intensity does *not* correlate to $\dot{M}$: positive as well as negative correlations are seen. This result was unexpected, as accretion theory says that the energy release rate is proportional to $\dot{M}$; redirection and reprocessing of the produced radiation by the accreting matter must be the cause of this lack of correlation.

The Z and atoll sources are distinguished from each other on the basis of their correlated spectral/timing behaviour. Their names derive from the shapes of the patterns they display in HIDs and CDs (Figure 6.9). *Z sources* display a three-branched rough Z shape; *atoll sources* follow a curved and (due to observational effects) often fragmented branch. The most characteristic aspect of the rapid X-ray variability in Z sources is the occurrence of two types of QPO (Sect. 6.3.2.2) and in atoll sources the occurrence of strong (up to 22%) so-called high-frequency noise (HFN, see Sect. 6.3.1.2). Contrary to the case of the black-hole candidates (Sect. 6.2.1), the X-ray spectral variations of the bright LMXBs are usually rather subtle. This is why HIDs and CDs, which are sensitive to small changes in spectral hardness in broad photon energy bands, work well in parametrizing the X-ray spectral variations (cf. the VHS in black-hole candidates, Sect. 6.2.2.3). Because of this subtleness of the X-ray spectral variations, it is necessary to consider the spectral and timing properties *in correlation* in order to determine the state of a Z or atoll source in a given observation, or the type of an unclassified source; attempts to perform these tasks on the basis of only one of these two kinds of information often fail.

The correlated spectral and timing properties of LMXBs were previously reviewed by Van der Klis (1986, 1987, 1989a,c, 1991, 1992), Hasinger (1987b, 1988a,b, 1991, 1992), Lewin *et al.* (1988), and Stella (1988a,b). For reviews of LMXB QPO models, see Lewin (1986) and Lamb (1988). An extensive overview of observations and theory of Z sources is given by Van der Klis and Lamb (1994). The survey with *EXOSAT* of bright LMXBs that led to the Z/atoll classification is presented in HK89.

6.3.1   Atoll sources

6.3.1.1   X-ray spectral states

The patterns that atoll sources trace out in CDs (HK89; Figure 6.9) are often dominated by observational effects. Ignoring these, atoll sources are characterized by an upwardly curved branch in the CD (with conventional choices for the energy bands, see Figure 6.9) that, at its right hand end, sometimes curves back up and to the left. Motion along this branch is usually much faster at the right hand end of the branch than at the left end, which causes the fragmented nature of many of the observed atoll patterns, with a 'banana' at the right and one or more 'islands' at the left: observations do not usually last long enough to see an atoll source move much when it is in an island state (motion there can take weeks or months; in the banana the time scale is hours to days). For this reason, it is unclear what pattern the sources trace out in this part of the diagram in the long run (for hints, see Mitsuda *et al.* 1989; Inoue 1994). The 'banana' part of the pattern is sometimes further subdivided into a 'lower' and an 'upper' banana (LB and UB, see Figure 6.9).

Invariably the X-ray intensity is lowest in the 'island state' and increases when the source moves to the right along the 'banana'. Combined with the X-ray burst properties (Sect. 6.3.1.3 and Ch. 4) and optical evidence (below), this leaves little doubt that $\dot{M}$ increases from the island to the left of the banana branch and then from left to right along the banana. As mentioned above, usually the islands are found to the left in the CDs, implying that their $\lesssim 5$ keV spectrum is soft. However, there are at least two examples where the overall spectrum became hard in the island state (4U 1705 44: Langmeier *et al.* 1987, 1989; and 1U 1608 52: Mitsuda *et al.* 1989; Inoue 1992). In these cases, the X-ray spectrum had a power-law shape with $\alpha \sim 1.5$–2. In 4U 1705–44, this spectrum produced an island in the right hand part of the CD (HK89). As noted above, information about the rapid X-ray variability is needed in addition to the CD to determine source state; in these two cases, strong high-frequency noise was seen, confirming that these were island states (Sect. 6.3.1.2).

HK89 identified ten atoll sources. Four of these (GX 3+1, GX 9+1, GX 9+9, GX 13+1) displayed only a banana branch (reports about a low state of GX 3+1 suggest that this source is sometimes in the island state: Inoue *et al.* 1981; Makishima *et al.* 1983). Four other sources exhibited both island and banana states (4U 1636−53, 4U 1705−44, 4U 1820−30 and 4U 1735−44), and two were only seen in the island state (4U 1608−52 and 4U 1728−33). 4U 1608−52 was observed in the banana state with *Tenma* (Mitsuda *et al.* 1989; Inoue 1992). A number of probable atoll sources have since been added (4U 2129+12 [AC211, M15]: Van Paradijs *et al.* 1990a; 4U 1702−42: Oosterbroek *et al.* 1991; 4U 1850−08 [NGC 6712]: Kitamoto *et al.* 1992b; Ser X-1: Jongert *et al.* 1994; see also Sect. 6.4.1 on Cir X-1). Note that, for faint sources, counting statistics noise can make a banana branch, with typically 10–30% change in the X-ray colours overall, hard to identify. Without a quantitative evaluation of the smallest colour changes to which the analysis is sensitive, the property of 'no motion' in an X-ray CD is meaningless.

In 4U 1735−44, the optical and X-ray flux are correlated along the banana branch (Corbet *et al.* 1989). This is in accordance with a positive correlation of accretion rate with X-ray intensity on this branch, as the optical flux is probably X-ray radiation that was reprocessed in the accretion disk.

6.3.1.2   *Rapid X-ray variability*

The power spectra of the atoll sources (Figure 6.9) can be described in terms of two rapid variability components (HK89), for historical reasons called the very-low-frequency noise (VLFN) and the high-frequency noise (HFN). VLFN has a power law shape $P \propto v^{-\gamma}$, with $\gamma = 1$–1.5; HFN can be described by a functional shape $P \propto v^{-\gamma} e^{-v/v_{cut}}$, with $\gamma = 0$–0.8 and $v_{cut} = 0.3$–25 Hz. Note that in atoll sources, despite its name, HFN does not usually extend to particularly high frequencies.

The properties of HFN and VLFN correlate strongly with the position of the source in the X-ray colour–colour diagram. At the lowest inferred $\dot{M}$, in the island state, the HFN can be very strong (up to 22%, but usually 10% rms); when $\dot{M}$ increases, the HFN fractional amplitude decreases via 6–<3% (rms) in the LB to usually <2% (rms) in the UB. The VLFN has its lowest fractional amplitudes ($\lesssim 1\%$ rms) in the island state and gradually increases in strength to usually ∼2.5% at the left end of the LB and ∼4% further up the banana. The two cases where the island state X-ray spectrum was very hard (Sect. 6.3.1.1) were also the cases where the HFN became strongest. These cases may be the more extreme island state episodes, perhaps those with the lowest $\dot{M}$. Evidence was recently reported for changes in the shape of the HFN in the atoll source (HK89) 4U 1608−52 reminiscent of those seen in black-hole candidate LS noise (Yoshida *et al.* 1993), providing a further indication for a strong similarity between neutron star and black hole accretion (see Sect. 6.5). The data of Yoshida *et al.* (1993) on 4U 1608−52 also show QPO peaks near 0.4 Hz and 2 Hz.

The interplay between HFN and VLFN leads to a variety of power spectral shapes (Figure 6.9). The HFN is sometimes 'peaked' showing a broad and usually asymmetric local maximum between ∼10 and ∼60 Hz (occasionally described as

QPOs; Lewin *et al.* 1987; Stella *et al.* 1987a; Dotani *et al.* 1989; Makishima *et al.* 1989). Sometimes additional 'wiggles' or 'shoulders' occur, reminiscent of those described in the black-hole candidates (Sect. 6.2.2.1).

The energy dependence of the VLFN was studied in GX 3+1 by Lewin *et al.* (1987) and Makishima *et al.* (1989); the amplitude of the noise increased by a factor of three to ten between 2 and 16 keV. The energy dependence of the peaked HFN in 4U 1820−30 and GX 3+1 has been studied by Mitsuda (1989) and Dotani *et al.* (1989). No time lags were detected; the upper limit was 5 ms (for an HFN peak near 10 Hz, so the upper limit to the phase lag was ~20°). The spectrum of the HFN was harder than the average spectrum. Its *absolute* rms amplitude as a function of photon energy roughly fitted a blackbody spectrum with a temperature of 2.2 keV.

6.3.1.3 X-ray bursts

The duration τ_{burst} of an X-ray burst correlates with its characteristic temperature T_{burst} (Damen *et al.* 1989). In a given source, neither T_{burst} nor τ_{burst} correlate well with the persistent X-ray intensity I_x (although taking together many burst sources that cover a wide range in luminosity, a rough correlation with L_x does exist: Van Paradijs *et al.* 1988b). In 4U 1636−53, τ_{burst} and T_{burst} correlate much better with accretion rate $\dot{M}$ as parametrized by position in the CD pattern than with I_x (Van der Klis *et al.* 1990). Irrespective of the details of the mechanism that causes τ_{burst} and T_{burst} to correlate (see Ch. 4), this suggests that $\dot{M}$ determines CD position as well as τ_{burst} and T_{burst} (see Van der Klis *et al.* 1990 for possible ways in which this could arise), and supports the interpretation of atoll source state in terms of $\dot{M}$. Similar results have been obtained for the atoll source 4U 1705−44 (Van der Klis 1989c). The properties of 4U 1735−44 and 4U 1820−30 are also in accordance with this correlation (Van der Klis *et al.* 1994).

6.3.2 Z sources

6.3.2.1 X-ray spectral states

Z sources produce Z-shaped patterns in X-ray CDs and HIDs. The three branches of the Z (see Figure 6.9) are called the horizontal branch (HB), normal branch (NB), and flaring branch (FB), for various historical reasons (see Van der Klis 1989a). The transition region between the HB and the NB is sometimes called the 'apex' (or 'hard apex') and that between the NB and the FB, the 'antapex' (or 'soft apex'). The branches are little broader than expected from the measurement errors (HK89). Between sources, there are intrinsic differences in the shape of the Z. In particular, the slope of the HB is quite variable from source to source, and also the slope and extent of the FB differs considerably between sources. The HB of GX 5−1 bends upward at its low-intensity end in both HIDs (Shibazaki and Mitsuda 1984) and CDs (Lewin *et al.* 1992; Kuulkers *et al.* 1994). There is some evidence for a change in the rapid X-ray variability of GX 5−1 there (Kuulkers *et al.* 1993). Most Z sources have by now been observed to exhibit all three states (HK89; Penninx *et al.* 1991; Kuulkers *et al.* 1993). GX 349+2 has only been observed on the FB and in the adjacent NB–FB transition region (HK89), and Sco X-1 has not displayed full HB properties.

Irregular motion of a source along the Z is detectable on time scales of minutes (White *et al.* 1976; Makishima and Mitsuda 1985; Hertz *et al.* 1992) and considerable fractions of the Z are covered on time scales of hours (see Lewin *et al.* 1992). The transition between the HB and the NB, as well as that between the NB and the FB, can take place in less than 1000 s (Priedhorsky *et al.* 1986; Hasinger 1988a). As a Z source moves along the Z, it never jumps between branches, but always follows the Z pattern (Branduardi *et al.* 1980; Shibazaki and Mitsuda 1984; Priedhorsky *et al.* 1986; Hertz *et al.* 1992), in accordance with the idea that the pattern is produced by variations in $\dot{M}$. The properties of the optical and UV emission (Sect. 6.3.2.3; see also Sect. 2.3.5.3) and the QPO models (Sect. 6.3.2.4) suggest that $\dot{M}$ increases in the sense HB→NB→FB. In the HB there is usually a positive correlation between $\dot{M}$ and I_x, in the NB a negative one, and in the FB both types of correlation occur, depending on source.

A *rank number* is sometimes defined to describe source position in the Z (Hasinger *et al.* 1990; Hertz *et al.* 1992; Dieters *et al.* 1993). The HB–NB transition point is usually assigned rank number 1.0, and the NB–FB transition point 2.0. Other loci in the Z are interpolated or extrapolated according to their coordinates in the CD or HID; sometimes the observed 'end points' of the Z are assigned rank numbers 0 and 3. Rank number, thus defined, is not necessarily proportional to any physical quantity, but it is expected to be a monotonically increasing function of the mass transfer rate to the neutron star.

In some cases, the Z pattern itself moves in the CD. In Cyg X-2, the Z can shift by factors up to two in I_x and 20% in the colours on time scales of weeks or less (the exact time scale is unknown; Branduardi *et al.* 1980; Vrtilek *et al.* 1986; Hasinger 1987a; Hasinger *et al.* 1990). In the other sources, the changes are smaller ($\lesssim 10\%$; Penninx *et al.* 1990; Kuulkers *et al.* 1993) or none were detected (Van der Klis *et al.* 1987a, 1991; Van Paradijs *et al.* 1988a; Langmeier *et al.* 1990; Lewin *et al.* 1992).

6.3.2.2   Rapid X-ray variability

Power spectra of Z-source X-ray intensity variations show several distinct components (Figure 6.9). Three of these components (VLFN, LFN and HFN) are broad noise components. Z source VLFN is similar in shape to VLFN in atoll sources, and has a similar dependence on accretion rate (it increases with $\dot{M}$). Both LFN and Z source HFN have a similar shape to that of atoll source HFN. Despite the use of the same names for some of these components in Z and atoll sources, there is no formal evidence that the same physical processes underly any of them. The other variability components of the Z sources are QPO components. For these components, extensive modelling has been attempted (Sect. 6.3.2.4).

Very-low-frequency noise (VLFN) and high-frequency noise (HFN).   VLFN and HFN are seen in all Z sources, in all states. They are the 'background continuum' above which other components sometimes appear (Figure 6.9). VLFN (Van der Klis *et al.* 1987a) dominates the power spectra at low frequencies. It extends down to frequencies below 1 mHz. It can be fit by a power law with an index γ of 1.5 to 2.0, and tends to steepen when $\dot{M}$ increases (HK89). The amplitude of the VLFN component increases from < 0.1–3.8% (rms) on the HB (depending on the source) via 0.6–4.0% on the NB to 1.2–8% on the FB (Van Paradijs *et al.* 1988a; HK89; Hasinger *et al.* 1989, 1990; Langmeier *et al.* 1990; Lewin *et al.* 1992; Kuulkers *et al.* 1994). HFN (HK89; Hasinger *et al.* 1989) is unambiguously detected only at frequencies above all other components and in Sco X-1 can be detected up to at least 200 Hz (Dieters *et al.* 1994). It fits the functional shape already discussed in Sect. 6.3.1.2 ($\nu^{-\gamma}e^{-\nu/\nu_{cut}}$), but with $30 < \nu_{cut} < 100$ Hz and $\gamma\sim0$. HFN fractional amplitude depends rather weakly on $\dot{M}$, usually (but not always) showing a weak anticorrelation with $\dot{M}$, from 3–10% (rms) on the HB to <2–5% on the NB and 1.5–4% on the FB (HK89; Hasinger *et al.* 1990; Kuulkers *et al.* 1994).

No analyses of the photon-energy dependence of the HFN or of time lags in the VLFN have been reported. Measurements of the energy spectrum of the VLFN suggest that VLFN and source motion along the Z are the same thing; the energy dependence is that expected if the VLFN intensity variations take place as motion along the spectral branch (Van der Klis 1986, 1991; Lewin *et al.* 1992; see also Van der Klis *et al.* 1987c; Van Paradijs *et al.* 1988a).

Horizontal-branch QPOs (HBOs) and low-frequency noise (LFN). Horizontal branch QPOs (Van der Klis *et al.* 1985a), or HBOs, produce a Lorentzian peak in the power spectrum (see Figure 6.10). LFN (low-frequency noise; Van der Klis *et al.* 1985a) is a broad-band noise component that is strongest at frequencies below the HBO peak but extends above it (see Figure 6.10). As HBOs and LFN appear and disappear together (the ratio of the amplitudes of the LFN and the HBOs is typically in the range 1–2.5), they are thought to be physically related. HBOs and LFN are strongest on the HB but are also seen on the NB. HBOs have been seen on the NB *simultaneously* with normal-branch oscillations (NBOs), proving that the two oscillations are different phenomena. LFN is seen to appear as a bulge near 2 Hz on top of the much broader HFN in Sco X-1 when it moves from the NB into the HB–NB transition region, showing that HFN and LFN are different phenomena (HK89; Hasinger *et al.* 1989; Dieters *et al.* 1994). HBOs and LFN have been detected in all Z sources that have a full HB. They have been studied by Van der Klis *et al.* (1985a, 1987a,b), Elsner *et al.* (1986), Hasinger *et al.* (1986, 1989, 1990), Hasinger (1987a,b, 1991), Stella *et al.* (1987b), Collmar *et al.* (1988), Dotani (1988), HK89, Mitsuda *et al.* (1988, 1991), Langmeier *et al.* (1990), Penninx *et al.* (1990, 1991), Lewin *et al.* (1992), Kuulkers *et al.* (1994), and Vaughan *et al.* (1994).

The HBO frequency ν_H has been observed to vary between 13 and 55 Hz. ν_H and I_x are strictly correlated with position on the HB; both increase towards the NB. The resulting relation between ν_H and I_x (Van der Klis *et al.* 1985a) is usually consistent with a linear one. The ratio of the relative changes in ν_H and I_x on the HB ($\frac{\Delta\nu}{\nu}/\frac{\Delta I}{I}$) is larger in GX 17+2 (4–5) than in other Z sources (1.5–2), probably because in GX 17+2 the HB has a relatively steep slope in the HID. Measured values of ν_H on the NB are all consistent with the maximum value reached at the right (high intensity) end of the HB, i.e. there is no evidence for further change in ν_H as a source moves from the HB–NB junction down the NB (see Figure 6.11). The relative width of the HBO peak (its FWHM divided by ν_H) can have values between 0.1 and 0.4; in GX 5−1 it remains approximately constant (near 0.2) over a range in HBO

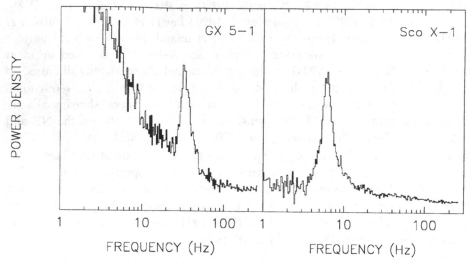

Fig. 6.10. Power spectra showing HBOs and LFN (left) and NBOs (right). Power spectral normalization is arbitrary and power scale is linear (Van der Klis 1989a).

frequency of more than a factor of three. Second harmonics to the HBO peak have been observed on a few occasions; they can be as strong as the first harmonic.

The LFN can, in combination with the VLFN, produce a variety of shapes in the power spectrum, reminiscent of those seen from atoll sources and including both monotonic and 'peaked' broad-band noise shapes. Peaked LFN may be associated with a steep HB and weak HBOs (HK89). Various functional shapes have been used to describe the LFN; a cut-off power law of the form $v^{-\gamma}e^{-v/v_{cut}}$ (the same functional dependence as the HFN) with $\gamma \sim 0$ and $v_{cut}=40$–80 Hz is often acceptable. Z source LFN and HFN are different power spectral components with different dependences on $\dot{M}$ (Van der Klis 1994b).

HBOs and LFN have hard X-ray spectra relative to the time-averaged flux; their rms amplitudes increase by factors of two to six from ~ 2 to ~ 15 keV. HBOs at higher photon energies slightly lag those at lower energies ('hard lags'; Hasinger 1987a; Van der Klis *et al.* 1987b; Mitsuda *et al.* 1988, 1991) and the LFN shows soft lags. The most sensitive time lag measurements to date were done in GX 5−1 (Vaughan *et al.* 1994). In this source, the hard lag gradually increases from ~ 2 to ~ 10 keV and then remains approximately constant up to ~ 18 keV (Figure 6.12). The time lag between the 2 keV and the >10 keV oscillations decreases from 3.8 to 1.0 ms as the HBO frequency increases from 13 to 25 Hz, so that the *phase* lag decreases by a factor of two. Time lags in the second harmonic are significantly different from those seen at the first harmonic. The variation with energy of the lag in the first harmonic agrees with the variation predicted by a simple model that attributes the lag to Comptonization, but, as Vaughan *et al.* (1994) note, neither the fact that the time lags are different between harmonics nor the energy spectrum of the HBOs are consistent with this model. The LFN shows soft lags by up to several tens of

GX 5−1   Cyg X−2

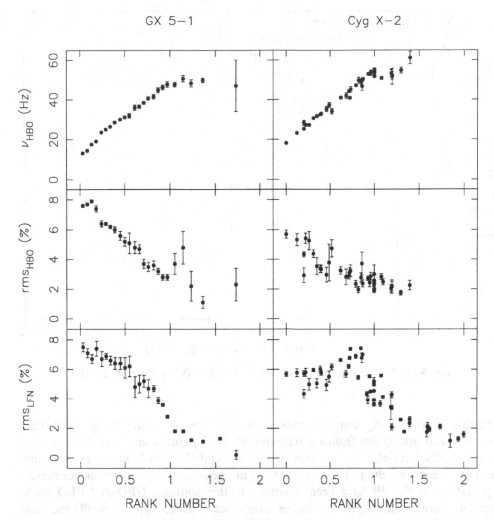

Fig. 6.11. HBO frequency ν_{HBO}, and HBO and LFN fractional rms amplitude as a function of position (see Sect. 6.3.2.1) along the Z in two Z sources. Rank number 1 corresponds to the HB–NB transition. (After Lewin *et al.* 1992 and G. Hasinger 1992, private communication.)

milliseconds (depending on the Fourier frequency range), which also indicates that a simple propagation delay is not sufficient to explain everything seen.

Normal- and flaring-branch QPOs (NBOs and FBOs). Normal-branch QPOs (Middleditch and Priedhorsky 1986), or NBOs, have frequencies between 4.5 and 7 Hz and amplitudes typically between 1 and 3% (rms) in the 1–10 keV band. They are strongest near the middle of the NB. NBOs have been seen in all Z sources which show a full NB. Flaring-branch QPOs (Van der Klis *et al.* 1985b, 1987c; Priedhorsky *et al.* 1986), or FBOs, occur on a small part (∼10% of the total extent) of the FB nearest the NB. Their frequencies increase from ∼6 Hz near the NB–FB junction to ∼20 Hz up the FB. With increasing frequency the fractional amplitude of the FBOs

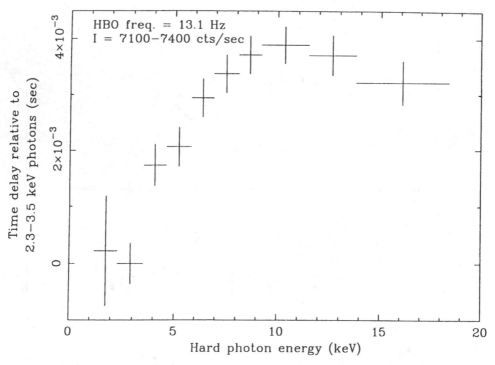

Fig. 6.12. Time lag spectrum of HBOs in GX 5−1 (Vaughan *et al.* 1994).

remains approximately constant while its width increases, until the peak becomes too broad to distinguish from the background (HFN) continuum (Priedhorsky *et al.* 1986). FBOs have clearly been seen in Sco X-1 and GX 17+2; broad excesses that may be due to an FBO peak moving rapidly in frequency as the source moves rapidly up and down the FB have been reported in other sources. NBO and FBO peaks both fit Lorentzian shapes. No higher harmonics of either type of oscillation have been detected. NBOs and FBOs have been studied by Middleditch and Priedhorsky (1986), Priedhorsky *et al.* (1986), Hasinger (1987a, 1991), Norris and Wood (1987), Stella *et al.* (1987b), Van der Klis *et al.* (1987a,c), Dotani (1988), Mitsuda *et al.* (1988), Van Paradijs *et al.* (1988a), Hasinger *et al.* (1989, 1990), Mitsuda (1989), Mitsuda and Dotani (1989), Langmeier *et al.* (1990), Penninx *et al.* (1990, 1991), Hertz *et al.* (1992), Lewin *et al.* (1992), Dieters *et al.* (1994) and Kuulkers *et al.* (1994).

In Sco X-1, the NBO frequency joins smoothly to the FBO frequency as the source moves from the NB to the FB (Priedhorsky *et al.* 1986; Dieters *et al.* 1994; Figure 6.13), suggesting that the two types of oscillations are physically related. The increase from 6 Hz to ∼10 Hz occurs in a very short segment of the Z (unresolved by present observations, Dieters *et al.* 1994; see Figure 6.13) located just at the vertex of the NB–FB junction.

The properties of the NBO vary with photon energy (Dotani 1988; Mitsuda 1989; Mitsuda and Dotani 1989; Lewin *et al.* 1992). In Cyg X-2, the amplitude has a minimum and the phase changes rapidly by ∼150° near 5 keV (Figure 6.14). In

Sco X−1

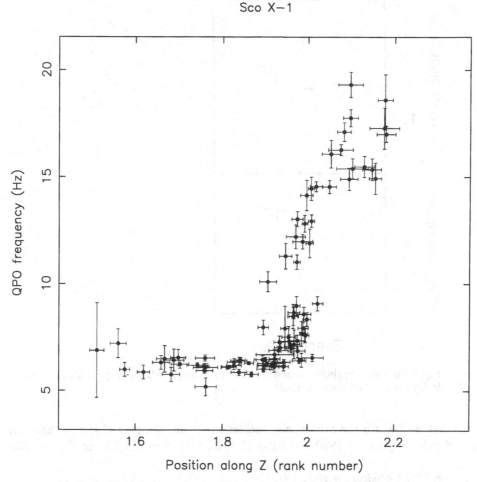

Fig. 6.13. N/FBO frequency as a function of position (see Sect. 6.3.2.1) along the Z in Sco X-1. Rank number 2 corresponds to the NB–FB transition. (Dieters *et al.* 1994.)

GX 5−1, the amplitude increases rapidly above 2.5 keV, accompanied by a substantial phase change. These variations can be interpreted as 'pivoting' of the X-ray spectrum around points near to these energies (Lamb 1989).

6.3.2.3 Optical and UV emission

Three Z sources have known optical counterparts: Sco X-1 (Sandage *et al.* 1966; Gottlieb *et al.* 1975), Cyg X-2 (Giacconi *et al.* 1967; Cowley *et al.* 1979), and GX 349+2 (Cooke and Ponman 1991; Penninx and Augusteijn 1991). The optical and UV properties of Cyg X-2 (Hasinger *et al.* 1990; Van Paradijs *et al.* 1990b; Vrtilek *et al.* 1990) and Sco X-1 (Wood *et al.* 1989; Vrtilek *et al.* 1991; Augusteijn *et al.* 1992) are clearly correlated to source state. Both the continuum and emission-line fluxes increase monotonically in the sense HB→NB→FB, indicating that the X-ray heating

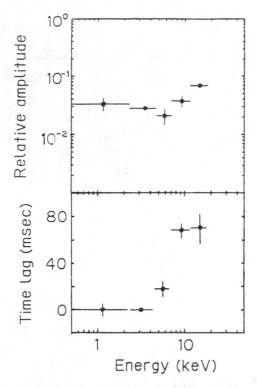

Fig. 6.14. Time lag (bottom) and fractional rms amplitude (top) spectra of the NBOs in Cyg X-2 (after Mitsuda 1988).

of the disk and hence the rate of mass accretion by the compact object increases in this sense (Hasinger *et al.* 1990; Vrtilek *et al.* 1990, 1991; Sect. 2.3.5.3, see Figure 2.28).

6.3.2.4　*Models for the QPOs*

QPOs are more amenable to interpretation than broad-band noise, and a profusion of models was proposed for the QPOs seen in Z sources (see Lewin 1986, Lamb 1988, and references therein). As the observations leave no doubt that at least two different types of QPOs occur in the Z sources, we need at least two models.

For the explanation of the HBOs, the *beat frequency model* (Alpar and Shaham 1985; Lamb *et al.* 1985; Alpar 1986; Lamb 1986; Shaham 1986, 1987; Elsner *et al.* 1987, 1988; Shibazaki and Lamb 1987; Shibazaki *et al.* 1987, 1988; White and Stella 1988; Yongheng and Keliang 1988; Norris *et al.* 1989, 1990; Mitsuda *et al.* 1991; Ghosh and Lamb 1992; these references include those dealing with the 'oscillating shot' time series description inspired by the beat frequency model) stands strong. In this model, plasma in near-Keplerian orbits at the inner edge of an accretion disk interacts with a neutron star magnetosphere. Inhomogeneities ('clumps') enter the magnetosphere more readily at some point(s) (e.g. near to the magnetic poles) than at others. As the magnetosphere spins (with the star) at a rate v_S, a clump of plasma circulating with a frequency v_K will periodically pass by a point of easier entry into the magnetosphere with a frequency $v_B = v_K - v_S$, the beat frequency between

Keplerian disk and neutron star spin frequencies. The clump's accretion, modulated at v_B by its interaction with the field, contributes a wave train or 'oscillating shot' to the X-ray intensity I_x; the time it takes for the clump to accrete corresponds to the duration of the shot, and the beat frequency v_B is the HBO frequency. As the shots are positive-definite, HBO and LFN strength are predicted to be similar, as observed. In principle, the beat frequency model can be used to constrain the magnetic field and spin rate of neutron stars and the relation of magnetospheric radius to accretion rate in Z sources. However, as we cannot be sure about the relation between position in the HB and $\dot{M}$ no definite conclusions have been reached. Ghosh and Lamb (1992), under the assumption that $\dot{M}$ is proportional to I_x on the HB, conclude that, by considering all Z source results together, it is possible to constrain inner disk models. They find spin periods for the neutron stars in Z sources between 3 and 20 ms, depending on source and on inner disk model.

Strong points of the beat frequency model are that it can relatively easily explain the observed HBO frequencies and amplitudes and the fact that HBO frequency is so strongly dependent on accretion rate. When the model was proposed, it was not then clear that I_x is not a good measure for $\dot{M}$ in Z sources, and nothing much was thought of the fact that the model requires $\dot{M}$ to be positively correlated with I_x in the HB; in retrospect, this is a prediction of the model that was verified by the simultaneous X-ray–UV observations of Z sources (Sect. 6.3.2.3). Weak points are that the model predicts the presence of a neutron star with a magnetosphere that is not symmetric around the rotation axis and therefore with millisecond pulsations (which have not been observed), and that no predictions can be made for some of the statistical aspects of the time series, because the properties of the accretion flow inhomogeneities are not completely specified by the model. This latter point is the reason that detailed statistical investigations into the time series properties of the HBOs, and LFN have not led to conclusive results (Norris *et al.* 1990; Lamb 1991; Mitsuda *et al.* 1991; Van der Klis and Lamb 1994). There are convincing arguments that the observational upper limits on pulsations that have been obtained are not in contradiction with the model, as, in particular, scattering can strongly suppress the pulse amplitudes (Lamb *et al.* 1985; Lamb 1991 and references therein).

In most models for the N/FBOs, *radiation pressure* plays the key role. As a group, Z sources are near the Eddington luminosity L_{Edd}, so, when $\dot{M}$ increases, at some point one expects to see the effects of radiation pressure. The NBO frequency is similar in each Z source, suggesting that it is determined by some parameter not varying much between sources, e.g. L_{Edd} (Hasinger 1987a). Two possible consequences of approaching L_{Edd}: (i) a transition to a thick accretion disk and a spherical accretion regime (Hasinger 1987a, 1988a,b; Van der Klis *et al.* 1987c; Lamb 1988, 1989; Hasinger *et al.* 1989), and (ii) a decrease in inflow velocity (Hasinger 1987a; Lamb 1989), have been exploited in explanations for spectral and temporal behavior in NBs and FBs. An important clue is that in Cyg X-2 the NBOs apparently occur by quasi-periodic pivoting of the X-ray spectrum around a point near 5 keV (Sect. 6.3.2.2), suggesting oscillations in the Compton scattering optical depth. It has been proposed that, near to the Eddington luminosity, part of the accretion will occur in an approximately spherically symmetric 'radial flow' with a radius of typically a few 100 km, within which matter falls in at relatively low velocities compared to free fall due to the reduction of the effective gravity by radiation pressure, and that in such a radial flow a radiation pressure feed-back loop can be set up that causes ~6 Hz oscillations in the optical depth of the flow (Fortner *et al.* 1989; Lamb 1989; see also Miller and Lamb 1992). Alpar *et al.* (1992) recently proposed a model in which sound waves in a thick disk, where the effective gravity is reduced by radiation pressure, cause the NBOs (cf. Hasinger 1987a; Van der Klis *et al.* 1987c; see also Van der Klis and Lamb 1994).

A comprehensive model explaining the various types of QPO and the X-ray spectral properties of the Z sources has been proposed by Lamb (1991); see also Van der Klis and Lamb (1994). This model uses the beat frequency model to explain the HBOs and the spherical-flow model for the N/FBOs.

6.3.3   Discussion of Z and atoll sources

In this section, we consider the differences and similarities between Z and atoll sources. For a comparison between the Z and atoll sources and the black-hole candidates, see Sect. 6.5.

Prompted by QPO models along the lines of those discussed in the previous section, HK89 proposed as a working hypothesis for explaining the Z/atoll phenomenology that the neutron stars in Z sources have *both* higher magnetic field strengths B than atoll sources *and* reach higher accretion rates $\dot{M}$. Their higher field ($>10^{9-10}$ G) explains why the (magnetospheric) HBO/LFN phenomenon is displayed by Z sources only, and their higher maximum accretion rate ($\sim \dot{M}_{Edd}$) explains why the same is true for the (near-Eddington) N/FBOs. Of the seven brightest persistent X-ray binaries, six are Z sources (Table 6.3), supporting a difference in average accretion rate between the two groups. However, the idea that Z and atoll sources differ *only* in $\dot{M}$ is unattractive as there is no evidence that Z sources at low $\dot{M}$ in any way tend to become similar to atoll sources at high $\dot{M}$ or *vice versa*. So, the conclusion that another parameter in addition to $\dot{M}$ causes the differences between Z and atoll sources is more or less directly suggested by the phenomenology. However, the conclusion that this other parameter is B is model dependent and subject to our understanding of HBOs and the suppression of millisecond pulsations in LMXBs (Sect. 6.3.2.4). The correlation between $\dot{M}$ and B that is implied by the HK89 working hypothesis could originate in a different evolutionary history of Z and atoll systems (below).

Predictions from the hypothesis are that an atoll source that becomes bright will show Z source NB/FB properties (N/FBOs and spectral branches), but never HBOs, and that a Z source that becomes faint will show millisecond pulsations. The properties of Cir X-1 (Sect. 6.4.1) fit the first prediction. The second prediction has not yet been tested as no Z source has ever been seen to go dim.

The Z/atoll classification groups four of the 'bright bulge' sources (or 'GX' sources) together with Sco X-1 and Cyg X-2 into a 'high luminosity' group, and puts four other GX sources together with classical X-ray burst sources such as 4U 1636−53 into a 'low luminosity' group (atoll sources). This is different from what was usual in earlier classifications of bright LMXBs that did not consider rapid aperiodic variability (Parsignault and Grindlay 1978; Ponman 1982; White and Mason 1985; White *et al.* 1988; Schulz *et al.* 1989), where in nearly all cases the sources GX 9+1, GX 9+9, GX 3+1 and GX 13+1 were classified among the 'high luminosity' sources.

Possibly, Z sources have longer orbital periods (>10 h) than atoll sources (<10 h), but the number of sources for which the orbital period has been measured is too small to reach a firm conclusion about this. If true, this would indicate that there is an evolutionary difference between Z sources (which would then all have evolved companions) and atoll sources (dwarf companions). Proposals have been made that long orbital periods (i.e. the presence of an evolved companion star) should correlate with high luminosities (Webbink *et al.* 1983), and with the presence of a neutron star magnetic field strong enough to form a magnetosphere (10^{9-10} G) and consequently produce HBOs (Van der Klis *et al.* 1985a; see also Lewin and Van Paradijs 1985). The source classifications that include GX 9+9, GX 9+1, GX 3+1 and GX 13+1 among the 'high luminosity' sources are inconsistent with these ideas as none of these sources show HBOs, and GX 9+9 has an orbital period of 4.2 h. The Z/atoll classification, on the other hand, is consistent with these proposed correlations (HK89). Evolutionary calculations (Pylyser and Savonije 1988a,b) suggest that high (Z source) accretion rates require evolved companions, but that *some* systems with an evolved companion could have a low accretion rate, so that atoll sources with long periods are not excluded (Van der Klis 1992). Indeed, the probable atoll source in M15, AC 211, has a 17.1 h period (Table 6.3).

6.4   Other sources

In this section, the rapid X-ray variability of some peculiar X-ray binaries (Cir X-1, the Rapid Burster, Cyg X-3) and of accretion powered pulsars is discussed. The main aim of the descriptions below is to see to what extent these sources fit within the patterns of phenomenology of black-hole candidates and/or Z/atoll sources. Many X-ray binaries are faint, and so their rapid X-ray variability cannot be measured at the level of sensitivity required to compare them with the bright sources. These faint X-ray binaries will not be further discussed here. Many of them are X-ray burst sources and therefore most resemble atoll sources (many of which are also burst sources).

6.4.1   Cir X-1

Cir X-1 is the classical example of a source that was considered a black-hole candidate on the basis of its strong rapid variability (Toor 1977; see also, e.g., Samimi *et al.* 1979), but subsequently found to be a neutron star, as it showed at least three (maybe more) type I X-ray bursts (Tennant *et al.* 1986b; see Ch. 4). The source has a ~17 d period, at phase 0.0 (see Stewart *et al.* 1991 for an ephemeris) of which 'something' usually happens: a steep drop, a steep rise, and/or one or more flares in X-ray intensity have all been observed at these times (Tennant 1988a), and often radio flares occur (Ch. 7). The X-ray spectral properties (Inoue 1994) are similarly varied during these events. It seems likely (see Murdin *et al.* 1980 and references therein; Inoue 1989) that the source is in a highly eccentric 17 d orbit and that complex processes near periastron during near- and super-Eddington mass transfer to the compact object underly this behaviour. Absorption and scattering by matter in the line of sight can be expected to play an important role. The rapid X-ray variability of Cir X-1 can help in understanding this source, as it provides an additional diagnostic of what is going on near the compact object. Note, however, that scattering by matter in the line of sight could suppress the amplitude of the rapid X-ray variability by temporal smearing.

The type I X-ray bursts were seen near phase 0.0, when the luminosity and probably $\dot{M}$ were relatively low (Tennant 1988a). The rapid variability and motion in the CD were at that time similar to those of an atoll source in the banana branch (Oosterbroek et al. 1994). However, the source did not exhibit evidence for strong HFN similar to that of an atoll source in the island state when the X-ray intensity decreased further. On another occasion, near phase 0.5, bursts were seen that could not with certainty be classified as type I (Tennant et al. 1986a). The X-ray intensity was roughly similar to that in the other observation where bursts were seen. In this observation, the power spectrum shows a strong power law (Oosterbroek et al. 1994).

QPOs with frequencies in the 5–21 Hz range were observed on two occasions, in both cases near phase 0.0 (Tennant 1987; Makino et al. 1992). On both occasions, the X-ray intensity showed a large step upward close to phase 0.0 and decreased after this. The QPOs occurred 7 h and 2.5 d after the step in the first and second case, respectively. The X-ray intensity had by that time dropped by a factor of two to five from the level just after the step. In the CD, two connected spectral branches were seen during these episodes, and the QPO frequency depended on source position in these branches. Variable strong noise reminiscent of that seen in black-hole candidates in the very high state was also observed, although the noise shape varied within a few 10^3 s rather than within a second as sometimes occurs in black-hole candidates (Figure 6.15; Tennant 1988a; Oosterbroek et al. 1994). On another occasion the X-ray intensity near phase 0.0 showed a strong flare lasting several hours (Tennant 1988b). During this flare, the power spectrum exhibited a strong noise component that approximately fitted a zero-centred Lorentzian with a HWHM of 0.015 Hz and a fractional amplitude of 21% (rms; Tennant et al. 1988b; Oosterbroek et al. 1994), similar to (but somewhat narrower than) black-hole candidate LS-noise (Sect. 6.2.2.1). During part of this time, a narrow 1.4 Hz 2–4% (rms) QPO peak was seen as well.

What makes the rapid X-ray variability of Cir X-1 so reminiscent of that seen in black-hole candidates is that large-amplitude HFN (or LS-noise) occurs when the source is very bright. Atoll sources also show strong HFN, but only when they get into the (faint) island state. They are never observed to reach luminosity levels as high as Cir X-1. Z sources probably reach similar luminosities to that of Cir X-1, and then also, like Cir X-1, show 6–20 Hz QPOs but they do not show such strong noise. Possibly, Cir X-1 is an atoll source (neutron star with a magnetic field too weak to form a magnetosphere) that differs from the other atoll sources only in that on occasion it reaches very high accretion rates (Van der Klis 1991). When it gets bright, it displays Z source NB/FB phenomena because its accretion approaches the Eddington limit; when it becomes faint it displays atoll source banana properties. See Sect. 6.5 for a discussion of how its behaviour may also fit in with that of the black-hole candidates.

Millisecond X-ray bursts (cf. Sect. 6.2.4) were reported in Cir X-1 by Toor (1977). One burst was very significant. It produced 44 counts in 20 ms over an average rate of 10 counts per 20 ms. Even in the presence of intrinsic source fluctuations with an amplitude of several 10%, as are known to occur sometimes in Cir X-1, it seems

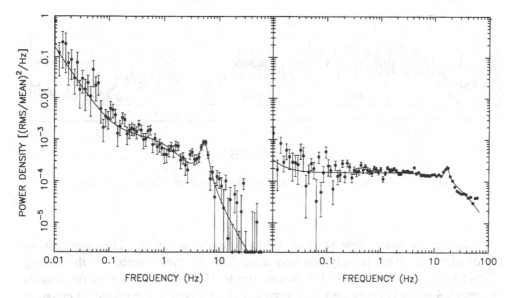

Fig. 6.15. Power spectra of Cir X-1 in a very high state (*EXOSAT* data, courtesy T. Oosterbroek).

hard to attribute this burst to a statistical fluctuation. The associated time scale is $\tau \lesssim 12$ ms.

6.4.2 The Rapid Burster

The intricate phenomenology of the Rapid Burster is described in Sect. 4.6.10; in this section I concentrate on QPO and noise properties that might relate to those seen in other X-ray binaries.

Not much is known as yet about the properties of the broad-band noise of the Rapid Burster. For a comparison to other burst sources, it would obviously be of great interest to know the properties of the broad-band noise (if any) in the persistent emission at a time when the source behaves like a normal type I burst source (as it sometimes does, see Sect. 4.6.10), but low count rates make this difficult (Rutledge *et al.* 1994). During type II bursts, broad-band noise is relatively weak (1.3–6%; Stella *et al.* 1988a) compared to the QPOs (below). Variations are seen on time scales somewhat shorter than the burst duration, but these seem to be deterministic ('ringing' and the time-scale-invariant burst profile, see Sect. 4.6.10). In the persistent emission between type II bursts, weak power law noise with index $\gamma \sim 0.7$ was detected by Dotani *et al.* (1990); <6 to 26% (rms) noise of unspecified shape below 1 Hz was reported by Stella *et al.* (1988a).

QPOs with frequencies between 2 and 7 Hz and fractional amplitudes between 2 and 20% (rms) are sometimes observed in type II bursts (Tawara *et al.* 1982; Stella *et al.* 1988a; Dotani *et al.* 1990; Lubin *et al.* 1991). There is a strong anticorrelation between burst frequency and burst peak flux (Lubin *et al.* 1991). Sometimes the QPOs are directly visible in the light curves (Figure 6.16). They can be described as a periodic phenomenon with no phase jumps, whose frequency drifts by up to

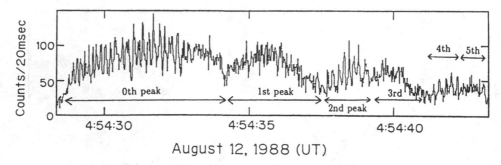

Fig. 6.16. QPOs in a type II burst of the Rapid Burster (Dotani *et al.* 1990).

25% during the burst (Tawara *et al.* 1982; Dotani *et al.* 1990). The amplitude of the oscillations varies strongly and may depend on the cycle number of the ringing (Figure 6.16; see also Sect. 4.6.10). Strong (up to 35% rms) QPOs with frequencies between 0.4 and 4 Hz have also been occasionally observed in the persistent emission *between* type II bursts; in some cases, the QPO frequency evolved in correlation with the average X-ray spectral hardness (Stella *et al.* 1988a). Additionally, oscillations with frequencies of 0.039–0.056 Hz are seen in the persistent emission between type II bursts; they may preferentially occur together with 4 Hz QPOs (Lubin *et al.* 1992).

The energy spectra of the QPOs in these various cases were found to be hard: the fractional amplitude increased by a factor of 1.5–2 between 3 and 10 keV (Stella *et al.* 1988b; Dotani *et al.* 1990). No significant time lags have been seen in the QPOs (Stella *et al.* 1988b; Dotani *et al.* 1990; Sect. 4.6.10).

The QPO and noise properties of the Rapid Burster are as complex and confusing as any of its other properties. It may be significant that, like in the Z sources, strong HFN, such as that seen in the atoll sources in the island state, or strong LS-like noise, such as that seen in the black-hole candidates in the low and very high states, have not been reported from the Rapid Burster (although the nature of the 26% rms noise detected by Stella *et al.* (1988b) is not clear). This could indicate that, like Z sources, the Rapid Burster has a magnetic field strong enough to form a small magnetosphere. The QPO property that seems most suggestive of interpretation is the anticorrelation between QPO frequency and type II burst peak flux. This strongly suggests that, as in all other cases discussed so far, there is a strong dependence of QPO properties on $\dot{M}$. If some kind of magnetic channeling explains the type II bursts, and the width of the channel determines the burst peak flux, then the properties of an instability in the flow through this channel that cause the QPOs might very well also depend on channel width (Lewin *et al.* 1991). For example, the flow through the channel might locally reach the Eddington limit, limiting the accretion rate and causing a flat topped burst, and a model similar to the Fortner *et al.* (1989) NBO model (Sect. 6.3.2.4) might explain the QPOs. No convincing example of such a model has so far been proposed. The 0.039–0.056 Hz QPOs in the persistent emission between type II bursts are in the same frequency range as those seen in some black-hole candidates in the HS and LS (Sect. 6.2.2.4).

6.4.3 Cygnus X-3

After a 20-year career as 'enigmatic' X-ray source (see Bonnet-Bidaud and Chardin 1988 for a review), the nature of the compact object (neutron star or black hole) and the energy source (rotation or accretion) of Cyg X-3 are still unknown (see also Sect. 1.6.3). It does seem clear now, however, that the compact object is embedded in a dense H-poor wind, likely emanating from a Wolf–Rayet companion (Van Kerkwijk *et al.* 1992; see also Van den Heuvel and de Loore 1973). Scattering in this wind can be expected to strongly suppress rapid X-ray variability amplitudes by smearing, and it therefore explains the reported lack of short time scale variability (Kitamoto *et al.* 1992a). Accurate measurements of (or upper limits to) the rapid X-ray variability that *is* transmitted by the wind can provide an independent constraint on the wind properties (Berger and Van der Klis 1994). Slow (0.02–0.001 Hz) QPO trains are occasionally seen in Cyg X-3 (Van der Klis and Jansen 1985).

6.4.4 Accretion powered pulsars

Many (perhaps all) accretion powered pulsars (Ch. 1) show, in addition to their periodic modulation, *aperiodic* rapid X-ray variability (Figure 6.17) that is reminiscent of atoll source HFN or black-hole candidate LS noise (Frontera *et al.* 1985, 1987; Stella *et al.* 1985; Makishima 1988; Angelini 1989; Angelini *et al.* 1989, 1991; Belloni and Hasinger 1989, 1990b; Koyama *et al.* 1989; Nagase 1989; Soong and Swank 1989; Makishima *et al.* 1990; Shinoda *et al.* 1990; Takeshima *et al.* 1991; Robba *et al.* 1992; Takeshima 1992; see also Orlandini and Morfill 1992). The fractional amplitude of the noise varies between several and 30% (rms), and, when fitted to a broken power law, ν_{break} is between 0.001 and a few hertz. Sometimes an additional power-law component is seen in the power spectrum at lower frequencies.

Note that V0332+53, whose aperiodic variability was correctly noted to resemble that of Cyg X-1 (Tanaka *et al.* 1983) and was subsequently found to be a pulsar (Stella *et al.* 1985), is not exceptional in either the strength or the time scale of its aperiodic variability; the only thing that is maybe exceptional about this pulsar is its low pulse amplitude, which made the aperiodic variability stand out more clearly. The history of the rapid aperiodic variability in pulsars is truly remarkable. For 15 years, pulsations were studied in pulsars and strong rapid aperiodic variability in black-hole candidates. During all that time, rapid aperiodic variability very similar to that in black-hole candidates was actually present in the pulsars too, but was practically ignored. When noticed at all, the aperiodic variations in pulsars were usually described in terms of 'pulse shape variations'. The application of broad-band power spectral techniques (Sect. 6.1.2.2) finally brought out the similarity between pulsar aperiodic variability and that seen in other accreting compact objects.

Takeshima (1992) reports a strong correlation between ν_{break} and pulse frequency among a sample of ten pulsars. This would seem to suggest that there is a relation between the Kepler frequency ν_K at the magnetospheric radius and the time scale of the noise, as, under the assumption that the pulsars on average spin at roughly their equilibrium periods (see Ch. 1), ν_K is predicted to be, on average, roughly equal to the pulse frequency. However, the relative constancy of ν_{break} in EXO 2030+375 over a large range in X-ray intensity (below) is in contradiction with this. A difference in the power spectra between disk-fed and wind-fed pulsars (Nagase 1989; Soong and

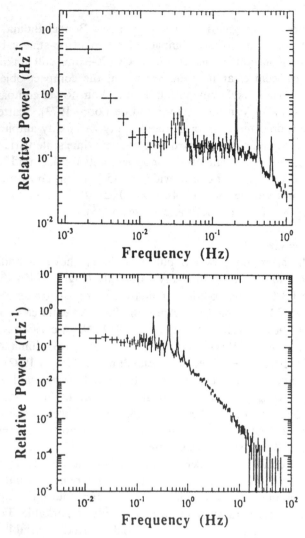

Fig. 6.17. Power spectra showing the noise, QPOs and periodic modulation in the X-ray pulsar Cen X-3 (Takeshima 1992).

Swank 1989), with the wind-fed sources showing power-law and the disk-fed sources flat-topped power spectra may be nothing else than a consequence of the lower pulse frequencies of the wind-fed sources, leading to a break below the frequencies usually covered by the power spectra. The fact that the power spectrum of 4U 1700−37 is apparently a steep power law down to a frequency of 0.001 Hz (Doll and Brinkmann 1987) would then suggest a long (>1000 s) pulse period for this object (which has all the characteristics of an accreting pulsar, except for pulsations).

In Her X-1, the strength of the noise depends on the phase of the 35 d cycle. This can be understood as an effect of scattering in the disk that suppresses the noise by smearing (Belloni and Hasinger 1989, 1990b). In EXO 2030+375, the noise strength

Table 6.4. *QPOs in pulsars*

Source	ν_{pulse} (Hz)	ν_{QPO} (Hz)	FWHM (Hz)	$(\nu_K)^a$ (Hz)	References
Cen X-3	0.21	0.035	0.01	0.24	Nagase (1989), Soong and Swank (1989), Takeshima *et al.* (1991), Takeshima (1992)
EXO 2030+375	0.024	0.19–0.21	~0.045	0.21–0.23	Angelini *et al.* (1989)
4U 1626−67	0.13	0.041	~0.02	0.17	Shinoda *et al.* (1990)

^{a} According to beat frequency model (see text).

may depend on whether the pulsar is spinning up or spinning down: the >0.2 Hz noise amplitude dropped from 20% to 7% when the source went from spin-up to probable spin-down (Parmar *et al.* 1989; Belloni and Hasinger 1990b); there was little change in the cut-off frequency of the noise.

The power spectra of nearly all pulsars show various bumps and wiggles, and the distinction between 'QPO' peaks (in the sense that the relative peak width (Sect. 6.1.2.2) is less than 0.5) and other power spectral features seems somewhat artificial in these sources. Clear QPOs on time scales within an order of magnitude of the pulse frequency were reported from the pulsars Cen X-3 (Figure 6.17), EXO 2030+375 and 4U 1626−67 (see Table 6.4). They could be due to the beat frequency mechanism proposed for Z source HBOs (Sect. 6.3.2.4; the possibility that such QPOs could turn up in pulsars was mentioned before they were discovered; Alpar and Shaham 1985). Using the beat-frequency-model expression $\nu_B = \nu_K - \nu_S$, and identifying ν_{QPO} with the beat frequency ν_B and ν_{pulse} with the spin frequency ν_S, this leads to values for ν_K of ~0.2 Hz in each case (even though the pulse frequencies differ by an order of magnitude between EXO 2030+375 and the other two sources). With a neutron star mass of $1.4 M_\odot$, this means that the magnetospheric radius $r_M \sim 5 \times 10^8$ cm, which is not unreasonable (using the expression of Ghosh and Lamb (1979) for the relation between r_M and the surface magnetic field strength, and assuming $\dot{M} \sim 10^{17}$ g s^{-1}, we have $B \sim 7 \times 10^{12}$ G). In EXO 2030+375, a marginally significant (98% confidence) positive correlation was found between I_x and ν_{QPO} that was consistent with that predicted by the beat frequency model; as in this pulsar ν_S is low, the difference between ν_B and ν_K is small, and a model with $\nu_{QPO} = \nu_K$ is also consistent with the data.

6.5 Overview and outlook

Rapid X-ray variability is a property of all types of accreting stellar-mass compact objects. Strong (several 10%) and rather similar band-limited noise is seen in black-hole candidates in the low state, some low-magnetic-field neutron stars (atoll sources) in the low state, and accretion powered pulsars. Weaker noise (a few per cent) occurs in black-hole candidates in the high state and in all bright low-magnetic-field neutron stars (Z sources, and atoll sources in the high state). QPOs with frequencies

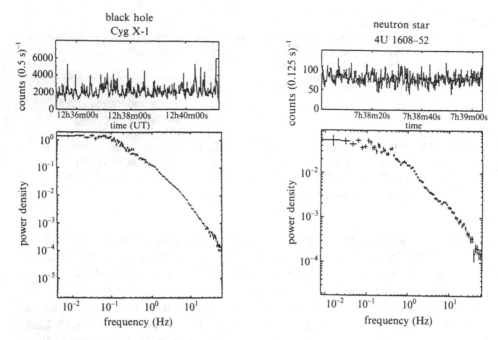

Fig. 6.18. Comparison of the power spectra of a black hole in the LS and an atoll source in the island state. Power spectral normalization is arbitrary. (After Inoue 1992.)

between 0.01 and 50 Hz occur in all these types of sources. There are at least two types of QPOs. One type may be magnetospheric (beat frequency model) and occurs in Z sources and possibly also in accretion powered pulsars. The other type is observed at high $\dot{M}$ (it may be related to near-Eddington accretion rates) and occurs in Z sources and possibly also in black-hole candidates and in the low-magnetic-field neutron star Cir X-1. The $\lesssim 1$ Hz QPOs in the black-hole candidates, and in the Rapid Burster, Cir X-1 and 4U 1608−52, display a large variety of properties. They may well comprise more than one physical phenomenon.

Detailed comparative studies of the relation of these various rapid X-ray variability components with each other and with other source properties are needed to find out to what extent the same physical processes underly the similar phenomena observed in magnetic and non-magnetic neutron stars and black-hole candidates. Several such studies are still underway, but some general results have already emerged.

In the low-magnetic-field neutron stars (Z and atoll sources and the Rapid Burster and Cir X-1) and also in the black-hole candidates, the properties of the rapid X-ray variability are strongly dependent on $\dot{M}$. Comparative studies suggest that differences in $\dot{M}$ as well as in B underly the observational differences between Z and atoll sources. A peculiar source such as Cir X-1 seems to find a niche in the Z/atoll scheme as an exceptional case of a neutron star that has a low B but can reach $\dot{M}_{Edd}$. As we shall see shortly, this may also provide a link to the black-hole candidates, which have $B=0$ and can probably also reach $\dot{M}_{Edd}$.

There are striking similarities between the rapid X-ray variability properties of the black-hole candidates and those of the low-magnetic-field neutron stars. At low accretion rates, the power spectra of black-hole candidate LS noise and atoll source HFN are nearly identical in shape (Inoue 1992; see Figure 6.18 – the shape is roughly a zero-centred Lorentzian) and have similarly high amplitudes (several 10% rms). At higher accretion rates, black-hole candidates (in the HS) and atoll sources (in the banana branch) both have a power-law ($\gamma \sim 1$–1.5) shaped power spectrum with fractional rms amplitudes of only a few per cent. Z sources show noise properties similar to this, but never seem to reach the low accretion rates where noise stronger than $\sim 10\%$ would be expected. Neutron star 6–20 Hz N/FBOs (in Z sources and Cir X-1) and black-hole candidate 3–10 Hz QPOs (in GX 339–4 and GS 1124–68) are also similar. Both appear at the very highest accretion rates, and the (similar) QPO frequencies depend on source position in branch structures which are seen in the colour diagram simultaneously with the QPOs. The broad flat tops ($\nu_{flat} \sim 1$–10 Hz) and rapid noise-shape changes of black-hole candidates in the VHS (Sect. 6.2.2.3) are similar to those seen in Cir X-1 when it becomes very bright (Sect. 6.4.1). (Some *differences* between source types are mentioned below.)

On the basis of these clues, it has been suggested (Van der Klis 1994a) that perhaps there exist similar $\dot{M}$ driven states in Z and atoll sources and black-hole candidates, and that three distinct rapid-variability regimes are common to neutron stars and black holes (Figure 6.19). In a given source, $\dot{M}$ determines the state, but the critical values of $\dot{M}$ at which a source changes state would depend on other source parameters as well (e.g. on compact object mass or inclination).

In such a scheme, the *very high state* of the black-hole candidates (3–10 Hz QPOs; branch structure in the CD) would correspond to the NB/FB of Z sources (6–20 Hz QPOs; two branches). The analogy with the Z sources suggests that the black-hole candidates are also accreting near to the Eddington limit in this state. Atoll sources do not usually attain such high accretion rates, but Cir X-1 in its high state fits in (6–20 Hz QPOs; two branches). Note, however, that Z sources in the NB/FB do not show the strong flat-topped noise with rapid shape changes seen in black-hole candidates and Cir X-1 in the VHS. The *high state* of the black-hole candidates would correspond to the HB of the Z sources and the banana branch of the atoll sources. Only Z sources show the 13–55 Hz HBOs here, as they have a magnetosphere, which is absent in the other source types. Otherwise this state is characterized by little variability ($\gamma \sim 1$–1.5 power law of a few per cent rms). The *low state* of the black-hole candidates (LS noise) would correspond to the island state of atoll sources (strong HFN). The similarity of the atoll sources with the black-hole candidates in the low state is largest in the faintest observed atoll island states. Z sources have never been seen to attain accretion rates that are this low; if they did then one would expect to see millisecond pulsations in them. The reason why Cir X-1 exhibits both black-hole candidate VHS properties and low-magnetic-field neutron star (atoll source) banana properties is, in this interpretation, that the source is the only case of a neutron star, which, like a black hole, has no magnetosphere, that sometimes reaches $\dot{M}_{Edd}$.

The X-ray spectral properties of neutron stars and black-hole candidates may also have similarities across compact object type that conform to this scheme. In particular, flat power-law 1–20 keV spectra similar to those of black-hole candidates

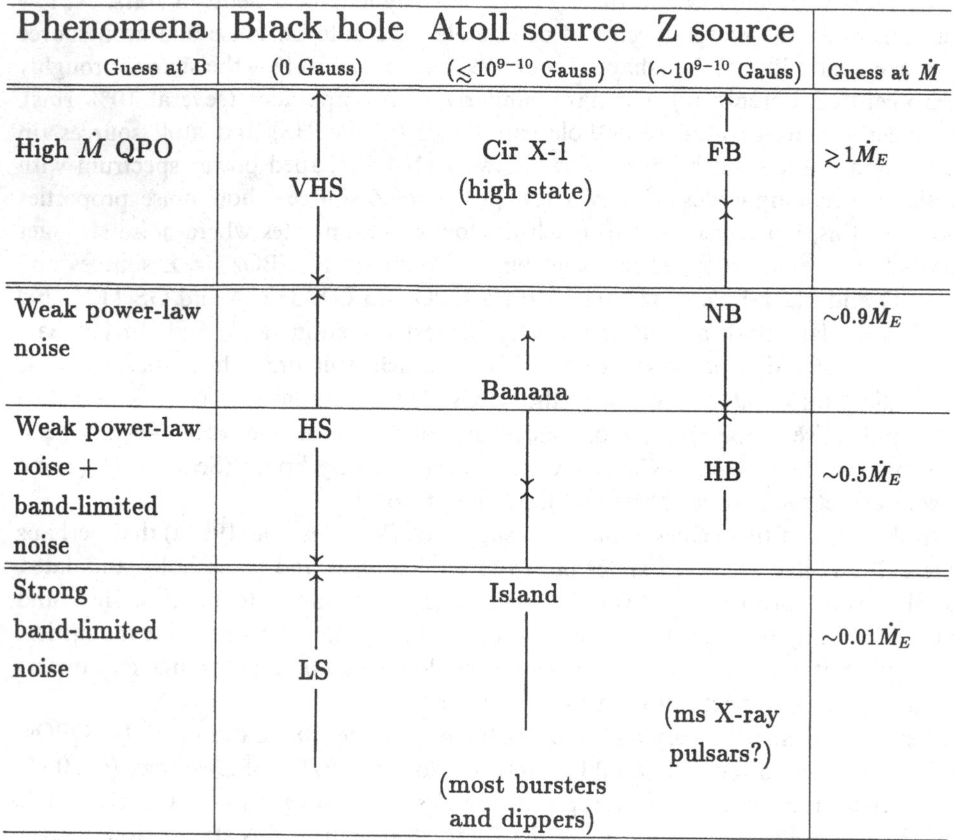

Phenomena Guess at B	Black hole (0 Gauss)	Atoll source ($\lesssim 10^{9-10}$ Gauss)	Z source ($\sim 10^{9-10}$ Gauss)	Guess at $\dot{M}$
High $\dot{M}$ QPO	VHS	Cir X-1 (high state)	FB	$\gtrsim 1\dot{M}_E$
Weak power-law noise		Banana	NB	$\sim 0.9\dot{M}_E$
Weak power-law noise + band-limited noise	HS		HB	$\sim 0.5\dot{M}_E$
Strong band-limited noise	LS	Island (most bursters and dippers)	(ms X-ray pulsars?)	$\sim 0.01\dot{M}_E$

VHS: very high state; HS: high state; LS: low state; FB: flaring branch; NB: normal branch; HB: horizontal branch.

Fig. 6.19. Possible classification scheme for neutron star and black hole accretion phenomenology. There are three states that are common to black holes and neutron stars; $\dot{M}$ determines which state a given source is in. The values of the accretion rate (in terms of the Eddington rate $\dot{M}_E$) at which state transitions occur are indicated. These values are intended as rough indications only, as other source parameters can affect the values of $\dot{M}$ at which state transitions occur and so the transitions may occur at different values of $\dot{M}/\dot{M}_E$ for different source types. (Van der Klis 1994a.)

in their low state occur in atoll sources when they become faint (Sect. 6.3.1.1; see also Barret and Vedrenne 1994).

In a further extension of this scheme, it has been proposed (Van der Klis 1994b) that two classes can be distinguished among the six band-limited noise components identified in accreting compact objects. Black-hole candidate LS and VHS noise, atoll source HFN and Z source LFN comprise one class, characterized by a decrease in fractional rms amplitude and an increase in cut-off frequency when $\dot{M}$ increases, and pulsar noise and Z source HFN comprise the other class, characterized by a weak dependence of noise properties on $\dot{M}$ but a strong dependence on neutron star spin rate. The correlation between pulse frequency and noise break frequency observed in pulsars (Sect. 6.4.4) combined with the 50–100 Hz cut-off frequencies observed in

Z source HFN (Sect. 6.3.2.2) then implies neutron star spin rates of 50–100 Hz in Z sources, consistent with theoretical ideas (Sect. 6.3.2.4). Perhaps the first class of noise components only occurs in accreting compact objects which are surrounded by an accretion disk that has an 'inner' (radiation pressure dominated) part (and therefore does not occur in pulsars where a large magnetosphere disrupts the inner disk), and perhaps the second class only occurs in accreting compact objects which have a magnetosphere (and is therefore absent in black-hole candidates).

It might seem disappointing that the famous black-hole candidate 'shot noise' turns out to be so common, and might in fact be a general inner disk phenomenon. However, similarities between neutron star and black hole accretion are expected from first principles (Sect. 6.1), and the occurrence of similar phenomena in the accretion flows around neutron stars and black holes should therefore not be surprising. Such similarities can in fact be very useful. Compact object mass can reasonably be expected to be a parameter governing the disk phenomena that likely dominate the observed characteristics, and eventually, therefore, order-unity differences should emerge from quantitative comparisons between neutron star and black-hole candidate phenomenology.

What, if any, might be the unique properties of black hole rapid X-ray variability? As we have seen, strength and power spectral shape of LS, HS and VHS broad-band noise are not exceptional. Perhaps black-hole candidate LS noise can be stronger (50% rms) than atoll source HFN ever gets (maximum reported is 22%). The ranges in cut-off frequency of the noise components seen in black-hole candidates and the various types of accreting neutron stars fully overlap.

The black-hole candidate slow LS/HS QPOs are not unique either (1.4 Hz QPOs in Cir X-1, 0.04–0.06 Hz in the Rapid Burster, 0.4 and 2 Hz in 4U 1608–52), nor are the VHS 3–10 Hz QPOs (6–20 Hz N/FBOs in Z sources and Cir X-1; 2–7 Hz in the Rapid Burster). 'Candidate' unique black-hole characteristics of the rapid X-ray variability are the frequency dependent time lags in the LS noise (Sect. 6.2.3.2), the large harmonic content of the VHS 3–10 Hz QPOs (Sect. 6.2.2.4), and the subsecond shape changes of strong broad-band noise in the VHS (Sect. 6.2.2.3), none of which have yet been reported from neutron stars. The variable flat-top effect in the LS noise of black-hole candidates (Sect. 6.2.2.1) has been removed from this list by the recent detection of this effect in 4U 1608–52 (Yoshida *et al.* 1993). The ~180° phase lags seen in the NB QPOs of some Z sources (Sect. 6.3.2.2) might be unique to neutron stars as they have not, thus far, been observed in black-hole candidates. Black-hole candidates may be on average brighter than neutron stars when they exhibit similar phenomena (LS noise versus strong HFN; VHS QPOs versus N/FBOs; flat power-law X-ray spectra), but to what extent this translates into higher luminosities, and this into higher accretion rates, needs careful evaluation. A difference in L_x may be one of the expected order-unity differences mentioned above, as energy release per gram accreted matter is expected to be higher from black holes than from neutron stars.

What the *EXOSAT* and *Ginga* observations of the rapid X-ray variability in X-ray binaries have shown is that when observed with sufficient sensitivity (<1% rms), and for sufficiently long intervals of time to cover all $\dot{M}$ levels (observations of transients are very important for this), correlations emerge between rapid X-ray variability and X-ray spectral properties that guide our interpretative efforts. An example is the realization of the lack of correlation between $\dot{M}$ and I_x in Z sources, which, without correlative studies, would not have been noticed. Especially when the X-ray spectral variations are subtle (such as in Z and atoll sources, and in black-hole candidates in the VHS), HIDs and CDs can help to find out what is going on. The high throughput (attainable count rates) of the *EXOSAT* and *Ginga* large-area proportional counters, resulting in sensitivities to broad-band spectral changes as well as rapid X-ray variability at the $\lesssim 1\%$ level, was crucial for the progress that was made. New instruments with improved spectral resolution will help to sort out what exactly is going on when the sources are tracing out their patterns in CDs and HIDs and also to estimate $\dot{M}$ as a function of position in the pattern. In combination with good time resolution, such instruments will help to identify the relation between the various X-ray spectral (see Ch. 1) and power spectral components by constructing energy spectra of the variability and comparing these with the time-averaged X-ray spectral components. However, the greatest progress is still to be expected from increased throughput. We have not yet reached the time scales relevant to the bulk flow of matter in the innermost regions near accreting stellar mass compact objects ($\lesssim 0.1$ ms). As it looks now, count rates $\sim 10^5$ counts s^{-1} range are necessary to begin to probe this regime. *XTE* will probably be the first mission that will provide such count rates (on Sco X-1); larger detector areas (in the 10 m^2 class) will be required to attain such rates for most of the sources discussed in this chapter.

Acknowledgements

I gratefully acknowledge the help of those who kindly shared with me results of their research in advance of publication: D. Barret, M. Denis, T. Dotani, G. Hasinger, C. Kouveliotou, S. Kitamoto, F. K. Lamb, K. Mitsuda, F. Nagase, A. N. Parmar and T. Takeshima. I am grateful to S. Dieters, E. Kuulkers, T. Oosterbroek, and B. Vaughan for generating the data for some of the figures that were produced for this chapter, and I acknowledge many stimulating discussions with them, and with F. K. Lamb, W. H. G. Lewin and J. Van Paradijs. This work was supported in part by the Netherlands Organization for Scientific Research (NWO) under grant PGS 78-277.

References

Alpar, M.A., 1986, *MNRAS* 223, 469.
Alpar, M.A. and Shaham, J., 1985, *Nature* 316, 239.
Alpar, M.A., Hasinger, G., Shaham, J. and Yancopoulos, S., 1992, *A&A* 257, 627.
Angel, J.R.P., Kestenbaum, H. and Novick, R., 1971, *ApJ* 169, L57.
Angelini, L., 1989, in Proc. 23rd ESLAB Symp. on *Two Topics in X-ray Astronomy*, Bologna, Italy, 13–20 September, 1989 (ESA SP-296), p. 81.
Angelini, L., Stella, L. and Parmar, A.N., 1989, *ApJ* 346, 906.
Angelini, L., Stella, L. and White, N.E., 1991, *ApJ* 371, 332.
Angelini, L., White, N.E. and Stella, L., 1992, *IAU Circ. 5580.*
Atmanspacher, H., Scheingraber, H. and Voges, W., 1989, in *Timing Neutron Stars*, H. Ögelman and E.P.J. Van den Heuvel (eds.), Kluwer, Dordrecht (NATO ASI Series C 262), p. 219.
Augusteijn, T., Karatasos, K., Papadakis, M., Paterakis, G., Kikuchi, S., Brosch, N., Leibowitz, E., Hertz, P., Mitsuda, K., Dotani, T., Lewin, W.H.G., Van der Klis, M. and Van Paradijs, J., 1992, *A&A* 265, 177.

Barret, D. and Vedrenne, G., 1994, in Proc. INTEGRAL Meeting, Les Diablerets, Switzerland, 1993, *ApJS*, in press..

Basinska, E.M., Lewin, W.H.G., Sztajno, M., Cominsky, L.R. and Marshall, F.J., 1984, *ApJ* 281, 337.

Belloni, T. and Hasinger, G., 1989, in Proc. 23rd ESLAB Symp. on *Two Topics in X-ray Astronomy*, Bologna, Italy, 13–20 September, 1989 (ESA SP-296), p. 283.

Belloni, T. and Hasinger, G., 1990a, *A& A* 227, L33.

Belloni, T. and Hasinger, G., 1990b, *A& A* 230, 103.

Berger, M. and Van der Klis, M. (1994), submitted to *A& A*.

Boldt, E.A., Holt, S.S. and Serlemitsos, P.J., 1971, *ApJ* 164, L9.

Bonnet-Bidaud, J.-M. and Chardin, G., 1988, *Phys. Rep.* 170, 325.

Bradt, H.V., Kelley, R.L. and Petro, L.D., 1982, in Proc. NATO ASI Galactic X-ray Sources, Sounion, Greece, P.W. Sanford *et al.* (eds.), Wiley, Chichester, p. 89.

Branduardi, G., Kylafis, N.D., Lamb, D.Q. and Mason, K.O., 1980, *ApJ* 235, L153.

Brinkman, A.C., Parsignault, D.R., Schreier, E., Gursky, H., Kellogg, E.M., Tananbaum, H. and Giacconi, R., 1974, *ApJ* 188, 603.

Canizares, C.R. and Oda, M., 1977, *ApJ* 214, L119.

Charles, P.A. *et al.*, 1980, *ApJ* 237, 154.

Collmar, W., Kendziorra, E. and Staubert, R., 1988, *Adv. Space Res.* 8, (2) 414.

Cooke, B.A. and Ponman, T.J., 1991, *A& A* 244, 358.

Corbet, R.H.D., Smale, A.P., Charles, P.A., Lewin, W.H.G., Menzies, J.W., Naylor, T., Penninx, W., Sztajno, M., Thorstensen, J.R., Trümper, J. and Van Paradijs, J., 1989, *MNRAS* 239, 533.

Cowley, A.P., Crampton, D. and Hutchings, J.B., 1979, *ApJ* 231, 539.

Damen, E., Jansen, F., Penninx, W., Oosterbroek, T., Van Paradijs, J. and Lewin, W.H.G., 1989, *MNRAS* 237, 523.

Denis, M., 1994, in Proc. INTEGRAL Meeting, Les Diablerets, Switzerland, 1993, *ApJS*, in press..

Dieters, S., *et al.*, 1994, in prep.

Doi, K., 1978, *Nature* 275, 197.

Doll, H. and Brinkmann, W., 1987, *A& A* 173, 86.

Dotani, T., 1988, Ph.D. Thesis, University of Tokyo.

Dotani, T., 1992, in *Frontiers of X-ray Astronomy*, Y. Tanaka (ed.), Universal Academy Press, Tokyo, p. 151.

Dotani, T., Mitsuda, K., Makishima, K. and Jones, M.H., 1989, *PASJ* 41, 577.

Dotani, T., Mitsuda, K., Inoue, H., Tanaka, Y., Kawai, N., Tawara, Y., Makishima, K., Van Paradijs, J., Penninx, W., Van der Klis, M., Tan, J. and Lewin, W.H.G., 1990, *ApJ* 350, 395.

Ebisawa, K., Mitsuda, K. and Inoue, H., 1989, *PASJ* 41, 519.

Elsner, R.F., Weisskopf, M.C., Darbro, W., Ramsey, B.D., Williams, A.C., Sutherland, P.G. and Grindlay, J.E., 1986, *ApJ* 308, 655.

Elsner, R.F., Shibazaki, N. and Weisskopf, M.C., 1987, *ApJ* 320, 527.

Elsner, R.F., Shibazaki, N. and Weisskopf, M.C., 1988, *ApJ* 327, 742.

Fabian, A.C., 1986, in *Proc. The Physics of Accretion onto Compact Objects*, Lecture Notes in Physics 266, Springer, p. 229.

Fortner, B., Lamb, F.K. and Miller, G.S., 1989, *Nature* 342, 775.

Friedman, H., Fritz, G., Henry, R.C., Hollinger, J.P., Meekins, J.F. and Sadeh, D., 1969, *Nature* 221, 345.

Frontera, F. and Fuligni, F., 1975, *ApJ* 198, L105.

Frontera, F., Dal Fiume, D., Morelli, E. and Spada, G., 1985, *ApJ* 298, 585.

Frontera, F., Dal Fiume, D., Robba, N.R., Manzo, G., Re, S. and Costa, E., 1987, *ApJ* 320, L127.

Ghosh, P. and Lamb, F.K., 1979, *ApJ* 234, 296.

Ghosh, P. and Lamb, F.K., 1992, in *X-Ray Binaries and Recycled Pulsars*, E.P.J. Van den Heuvel and S. Rappaport (eds.), Kluwer, Dordrecht (NATO ASI Series C 377), p. 487.

Giacconi, R., Gorenstein, P., Gursky, H., Usher, P.D., Waters, J.R., Sandage, A., Osmer, P. and Peach, J.V., 1967, *ApJ* 148, L129.

Giacconi, R., Gursky, H., Kellogg, E., Schreier, E. and Tananbaum, H., 1971, *ApJ* 167, L67.

Giles, A.B., 1981, *MNRAS* 195, 721.

Gottlieb, E.W., Wright, E.L. and Liller, W., 1975, *ApJ* 195, L33.

Grebenev, S.A., Sunyaev, R.A., Pavlinskii, M.N. and Dekhanov, A., 1991a, *Sov. Astron. Lett.* 17(6), 413.

Grebenev, S.A., Sunyaev, R.A. and Pavlinsky, M.N., 1991b, in Proc. Workshop on Nova Muscae, Lyngby, Denmark, May 14–16, 1991.

Grebenev, S., et al., 1993, *A& AS* 97, 281.

Hasinger, G., 1987a, *A& A* 186, 153.

Hasinger, G., 1987b, IAU Symp. 125, p. 333.

Hasinger, G., 1988a, in *Physics of Neutron Stars and Black Holes*, Y. Tanaka (ed.), Universal Academy Press, Tokyo, p. 97.

Hasinger, G., 1988b, *Adv. Space Res.* 8, (2) 377.

Hasinger, G., 1991, in *Particle Acceleration near Accreting Compact Objects*, J. Van Paradijs, M. Van der Klis and A. Achterberg (eds.), North-Holland, Amsterdam, p. 23.

Hasinger, G., 1992, in *X-Ray Binaries and Recycled Pulsars*, E.P.J. Van den Heuvel and S. Rappaport (eds.), Kluwer, Dordrecht (NATO ASI Series C 377), p. 61.

Hasinger, G. and Van der Klis, M., 1989, *A& A* 225, 79. [HK89]

Hasinger, G., Langmeier, A., Sztajno, M., Trümper, J., Lewin, W.H.G. and White, N.E., 1986, *Nature* 319, 469.

Hasinger, G., Priedhorsky, W.C. and Middleditch, J., 1989, *ApJ* 337, 843.

Hasinger, G., Van der Klis, M., Ebisawa, K., Dotani, T. and Mitsuda, K., 1990, *A& A* 235, 131.

Hertz, P., Vaughan, B., Wood, K.S., Norris, J.P., Mitsuda, K., Michelson, P.F. and Dotani, T., 1992, *ApJ* 396, 201.

Holt, S.S., 1980, in *X-ray Astronomy*, R. Giacconi and G. Setti (eds.), Reidel, Dordrecht, p. 237.

Iga, S., Miyamoto, S. and Kitamoto, S., 1992, in *Frontiers of X-ray Astronomy*, Y. Tanaka (ed.), Universal Academy Press, Tokyo, p. 309.

Ilovaisky, S.A., Chevalier, C., Motch, C. and Chiapetti, L., 1986, *A& A* 164, 67.

Imamura, J.N., Kristian, J., Middleditch, J. and Steiman-Cameron, T.Y., 1990, *ApJ* 365, 312.

Inoue, H., 1989, in Proc. 23rd ESLAB Symp. on *Two Topics in X-ray Astronomy*, Bologna, Italy, 13–20 September, 1989 (ESA SP-296), p. 783.

Inoue, H., 1991, ISAS RN 469, paper presented at the Texas/ESO-CERN Symp., Brighton, UK, December 1990.

Inoue, H., 1994, ISAS RN 518, to appear in *Accretion Disks in Compact Stellar Systems*, J.C. Wheeler (ed.).

Inoue, H., et al., 1981, *ApJ* 250, L71.

Jongert, H., et al., 1994, *A& A* , in prep.

Kitamoto, S., 1989, in Proc. 23rd ESLAB Symp. on *Two Topics in X-ray Astronomy*, Bologna, Italy, 13–20 September, 1989 (ESA SP-296), p. 231.

Kitamoto S. and Miyamoto, S., 1989, in *Timing Neutron Stars*, H. Ögelman and E.P.J. Van den Heuvel (eds.), Kluwer, Dordrecht (NATO ASI Series C 262), p. 267.

Kitamoto, S., Tsunemi, H., Miyamoto, S., Yamashita, K., Mizobuchi, S., Nakagawa, M., Dotani, T. and Makino, F., 1989, *Nature* 342, 518.

Kitamoto, S., Mizobuchi, S., Yamashita, K. and Nakamura, H., 1992a, *ApJ* 384, 263.

Kitamoto, S., Tsunemi, H. and Roussel-Dupre, D., 1992b, *ApJ* 391, 220.

Kitamoto, S., Hayashida, K., Ogawa, M. and Ebisawa, K., 1994, in prep.

Kouveliotou, C., Finger, M.H., Fishman, G.J., Meegan, C.A., Wilson, R.B., Paciesas, W.S., Minamitani, T. and Van Paradijs, J., 1992, in Proc. St. Louis Meeting on Gamma Ray Astronomy, St. Louis, USA, preprint.

Koyama, K., Kawada, M., Tawara, Y., Ushimaru, N., Dotani, T. and Takizawa, M., 1989, *ISAS RN* 429.

Kuulkers, E., Van der Klis, M., Oosterbroek, T., Asai, K., Dotani, T., Lewin, W.H.G. and Van Paradijs, J., 1994, *A& A*, in press.

Lamb, F.K., 1986, in *The Evolution of Galactic X-ray Binaries*, J. Truemper, W.H.G. Lewin and W. Brinkmann (eds.), Kluwer, Dordrecht (NATO ASI Series C 167), p. 151.

Lamb, F.K., 1988, *Adv. Space Res.* 8, (2) 421.

Lamb, F.K., 1989, in Proc. 23rd ESLAB Symp. on *Two Topics in X-ray Astronomy*, Bologna, Italy, 13–20 September, 1989 (ESA SP-296), p. 215.

Lamb, F.K., 1991, in *Neutron Stars: Theory and Observation*, J. Ventura and D. Pines (eds.), Kluwer, Dordrecht (NATO ASI Series C 344) p. 445.

Lamb, F.K., Shibazaki, N., Alpar, M.A. and Shaham, J., 1985, *Nature* 317, 681.

Lampton, M., Bowyer, C.S. and Harrington, S., 1970, *ApJ* 162, 181.

Langmeier, A., Sztajno, M., Hasinger, G. and Trümper, J., 1987, *ApJ* 323, 288.
Langmeier, A., Hasinger, G. and Trümper, J., 1989, *ApJ* 340, L21.
Langmeier, A., Hasinger, G. and Trümper, J., 1990, *A& A* 228, 89.
Leahy, D.A., Darbro, W., Elsner, R.F., Weisskopf, M.C., Sutherland, P.G., Kahn, S. and Grindlay, J.E., 1983, *ApJ* 266, 160.
Lewin, W.H.G., 1986, in *Proc. The Physics of Accretion onto Compact Objects*, Lecture Notes in Physics 266, Springer, p. 177.
Lewin, W.H.G. and Van Paradijs, J., 1985, *A& A* 149, L27.
Lewin, W.H.G., *et al.*, 1987, *MNRAS* 226, 383.
Lewin, W.H.G., Van Paradijs, J. and Van der Klis, M., 1988, *Space Sci. Rev.* 46, 273.
Lewin, W.H.G., Lubin, L.M., Van Paradijs, J. and Van der Klis, M., 1991, *A& A* 248, 538.
Lewin, W.H.G., Lubin, L.M., Tan, J., Van der Klis, M., Van Paradijs, J., Penninx, W., Dotani, T. and Mitsuda, K., 1992, *MNRAS* 256, 545.
Liang, E.P. and Nolan, P.L., 1984, *Space Sci. Rev.* 38, 353.
Lochner, J.C., Swank, J.H. and Szymkowiak, A.E., 1991, *ApJ* 376, 295.
Lubin, L.M., Stella, L., Lewin, W.H.G., Tan, J., Van Paradijs, J., Van der Klis, M. and Penninx, W., 1991, *MNRAS* 249, 300.
Lubin, L.M., Lewin, W.H.G., Rutledge, R.E., Van Paradijs, J., Van der Klis, M. and Stella, L., 1992, *MNRAS* 258, 759.
McClintock, J., 1988, *Adv. Space Res.* 8, (2) 191.
Maejima, Y., Makishima, K., Matsuoka, M., Ogawara, Y., Oda, M., Tawara, Y. and Doi, K., 1984, *ApJ* 285, 712.
Makino, F. and the ASTRO-C Team, 1987, *Astrophys. Lett. Comm.* 25, 233.
Makino, Y., Kitamoto, S. and Miyamoto, S., 1992, in *Frontiers of X-ray Astronomy*, Y. Tanaka (ed.), Universal Academy Press, Tokyo, p. 167.
Makishima, K., 1988, in *Physics of Neutron Stars and Black Holes*, Y. Tanaka (ed.), Universal Academy Press, Tokyo, p. 175.
Makishima, K. and Mitsuda, K., 1985, in Proc. of the Japan–U.S. Seminar on Galactic and Extragalactic Compact X-Ray Sources, Tokyo, Y. Tanaka and W.H.G. Lewin (eds.), (ISAS, Tokyo), p. 127.
Makishima, K., *et al.*, 1983, *ApJ* 267, 310.
Makishima, K., Ishida, M., Ohashi, T., Dotani, T., Inoue, H., Mitsuda, K., Tanaka, Y., Turner, M.J.L. and Hoshi, R., 1989, *PASJ* 41, 531.
Makishima, K., *et al.*, 1990, *PASJ* 42, 295.
Markert, T.H., Canizares, C.R., Clark, G.W., Lewin, W.H.G., Schnopper, H.W. and Sprott, G.F., 1973, *ApJ* 184, L67.
Mason, K.O., Charles, P.A., White, N.E., Culhane, J.L., Sanford, P.W. and Strong, K.T., 1976, *MNRAS* 177, 513.
Meekins, J.F., Wood, K.S., Hedler, R.L., Byram, E.T., Yentis, D.J., Chubb, T.A. and Friedman, H., 1984, *ApJ* 278, 288.
Middleditch, J. and Priedhorsky, W.C., 1986, *ApJ* 306, 230.
Miller, G.S. and Lamb, F.K., 1992, *ApJ* 388, 541.
Mitsuda, K., 1989, in Proc. 23rd ESLAB Symp. on *Two Topics in X-ray Astronomy*, Bologna, Italy, 13–20 September, 1989 (ESA SP-296), p. 197.
Mitsuda, K. and Dotani, T, 1989, *PASJ* 41, 557.
Mitsuda, K., Dotani, T. and Yoshida, A., 1988, in *Physics of Neutron Stars and Black Holes*, Y. Tanaka (ed.), Universal Academy Press, Tokyo, p. 133.
Mitsuda, K., Inoue, H., Nakamura, N. and Tanaka, Y., 1989, *PASJ* 41, 97.
Mitsuda, K., Dotani, T., Yoshida, A., Vaughan, B. and Norris, J.P., 1991, *PASJ* 43, 113.
Miyamoto, S. and Kitamoto, S., 1989, *Nature* 342, 773.
Miyamoto, S., Kitamoto, S., Mitsuda, K. and Dotani, T., 1988a, *Nature* 336, 450.
Miyamoto, S., Kitamoto, S., Mitsuda, K. and Dotani, T., 1988b, in *Physics of Neutron Stars and Black Holes*, Y. Tanaka (ed.), Universal Academy Press, Tokyo, p. 227.
Miyamoto, S., Kimura, K., Kitamoto, S., Dotani, T. and Ebisawa, K., 1991, *ApJ* 383, 784.
Miyamoto, S., Iga, S., Terada, K., Kitamoto, S., Hayashida, K. and Negoro, H., 1992a, in Ginga Memorial Symp., F. Makino and F. Nagase (eds.), ISAS, Tokyo, p. 37.
Miyamoto, S., Kitamoto, S., Iga, S., Negoro, H. and Terada, K., 1992b, *ApJ* 391, L21.

Miyamoto, S., Iga, S., Kitamoto, S. and Kamado, Y., 1993, *ApJ* 403, L39.

Morfill, G.E., Atmanspacher, H., Demmel, V., Scheingraber, H. and Voges, W., 1989, in *Timing Neutron Stars*, H. Ögelman and E.P.J. Van den Heuvel (eds.), Kluwer, Dordrecht (NATO ASI Series C 262), p. 71.

Motch, C., Ricketts, M.J., Page, C.G., Ilovaisky, S.A. and Chevalier, C., 1983, *A& A* 119, 171.

Motch, C., Ilovaisky, S.A., Chevalier, C. and Angebault, P., 1985, *Space Sci. Rev.* 40, 219.

Murdin, P., Jauncey, D.L., Haynes, R.F., Lerche, I., Nicolson, G.D., Holt, S.S. and Kaluzienski, L.J., 1980, *A& A* 87, 292.

Nagase, F., 1989, in Proc. 23rd ESLAB Symp. on *Two Topics in X-ray Astronomy*, Bologna, Italy, 13–20 September, 1989 (ESA SP-296), p. 45.

Negoro, H., Miyamoto, S. and Kitamoto, S., 1992, in *Frontiers of X-ray Astronomy*, Y. Tanaka (ed.), Universal Academy Press, Tokyo, p. 313.

Nolan, P.L, Gruber, D.E., Matteson, J.L., Peterson, L.E., Rothschild, R.E., Doty, J.P., Levine, A.M., Lewin, W.H.G. and Primini, F.A., 1981, *ApJ* 246, 494.

Norris, J.P. and Wood, K.S., 1987, *ApJ* 312, 732.

Norris, J.P., Hertz, P., Wood, K.S., Vaughan, B.A., Michelson, P.F., Mitsuda, K. and Dotani, T., 1989, in Proc. 23rd ESLAB Symp. on *Two Topics in X-ray Astronomy*, Bologna, Italy, 13–20 September, 1989 (ESA SP-296), p. 557.

Norris, J.P., Hertz, P., Wood, K.S., Vaughan, B.A., Michelson, P.F., Mitsuda, K. and Dotani, T., 1990, *ApJ* 361, 514.

Oda, M., 1977, *Space Sci. Rev.* 20, 757.

Oda, M., Gorenstein, P., Gursky, H., Kellogg, E., Schreier, E., Tananbaum, H. and Giacconi, R., 1971, *ApJ* 166, L1.

Oda, M., Takagishi, K., Matsuoka, M., Miyamoto, S. and Ogawara, Y., 1974, *PASJ* 26, 303.

Oda, M., Doi, K., Ogawara, Y., Takagishi, K. and Wada, M., 1976, *Ap. Space Sci.* 42, 223.

Ogawara, Y., Doi, K., Matsuoka, M., Miyamoto, S. and Oda, M., 1977, *Nature* 270, 154.

Oosterbroek, T., Penninx, W., Van der Klis, M., Van Paradijs, J. and Lewin, W.H.G., 1991, *A& A* 250, 389.

Oosterbroek, T., Van der Klis, M., Kuulkers, E., Van Paradijs, J. and Lewin, W.H.G., 1994, *A&A*, in prep.

Orlandini, M. and Morfill, G.E., 1992, *ApJ* 386, 703.

Page, C.G., 1985, *Space Sci. Rev.* 40, 387.

Page, C.G, Bennetts, A.J. and Ricketts, M.J., 1981, *Space Sci. Rev.* 30, 369.

Parmar, A.N., Stella, L. and White, N.E., 1986, *ApJ* 304, 664.

Parmar, A.N., White, N.E., Stella, L., Izzo, C. and Ferri, P., 1989, *ApJ* 338, 359.

Parmar, A.N., Angelini, L., Roche, P. and White, N.E., 1994, in prep.

Parsignault, D.R. and Grindlay, J.E., 1978, *ApJ* 225, 970.

Penninx, W. and Augusteijn, T., 1991, *A& A* 246, L81.

Penninx, W., Lewin, W.H.G., Mitsuda, K., Van der Klis, M., Van Paradijs, J. and Zijlstra, A.A., 1990, *MNRAS* 243, 114.

Penninx, W., Lewin, W.H.G., Tan, J., Mitsuda, K., Van der Klis, M. and Van Paradijs, J., 1991, *MNRAS* 249, 113.

Petro, L.D., Bradt, H.V., Kelley, R.L., Horne, K. and Gomer, R., 1981, *ApJ* 251, L7.

Ponman, T., 1982, *MNRAS* 201, 769.

Press, W.H. and Schechter, P., 1974, *ApJ* 193, 437.

Priedhorsky, W., Garmire, G.P., Rothschild, R., Boldt, E., Serlemitsos, P. and Holt, S., 1979, *ApJ* 233, 350.

Priedhorsky, W., Hasinger, G., Lewin, W.H.G., Middleditch, J., Parmar, A., Stella, L. and White, N., 1986, *ApJ* 306, L91.

Pylyser, E.H.P. and Savonije, G.J., 1988a, *A& A* 191, 57.

Pylyser, E.H.P. and Savonije, G.J., 1988b, *A& A* 208, 52.

Robba, N.R., Cusumano, G., Orlandini, M., Dal Fiume, D. and Frontera, F., 1992, *ApJ* 401, 685.

Rothschild, R.E., Boldt, E.A., Holt, S.S. and Serlemitsos, P.J., 1974, *ApJ* 189, L13.

Rothschild, R.E., Boldt, E.A., Holt, S.S. and Serlemitsos, P.J., 1977, *ApJ* 213, 818.

Rutledge, R.E., *et al.*, 1994, in prep.

Samimi, J., *et al.*, 1979, *Nature* 278, 434.

Sandage, A.R., Osmer, P., Giacconi, R., Gorenstein, P., Gursky, H., Waters, J., Bradt, H., Garmire, G., Sreekantan, B.V., Oda, M., Osawa, K. and Jugaku, J., 1966, *ApJ* 146, 316.

Schulz, N.S., Hasinger, G. and Trümper, J., 1989, *A& A* 225, 48.

Shaham, J., 1986, *ApJ* 310, 780.

Shaham, J., 1987, IAU Symp. 125, p. 347.

Shakura, N.I. and Sunyaev, R.A., 1973, *A& A* 24, 337.

Shibazaki, N. and Lamb, F.K., 1987, *ApJ* 318, 767.

Shibazaki, N. and Mitsuda, K., 1984, in *High Energy Transients in Astrophysics*, S.E. Woosley (ed.), AIP Conf. Proc. 115, p. 63.

Shibazaki, N., Elsner, R.F. and Weisskopf, M.C., 1987, *ApJ* 322, 831.

Shibazaki, N., Elsner, R.F., Bussard, R.W., Ebisuzaki, T. and Weisskopf, M.C., 1988, *ApJ* 331, 247.

Shinoda, K., Kii, T., Mitsuda, K., Nagase, F., Tanaka, Y., Makishima, K. and Shibazaki, N., 1990, *PASJ* 42, L27.

Soong, Y. and Swank, J.H., 1989, in Proc. 23rd ESLAB Symp. on *Two Topics in X-ray Astronomy*, Bologna, Italy, 13–20 September, 1989 (ESA SP-296), p. 617.

Stella, L., 1988a, in *X-ray Astronomy with EXOSAT*, R. Pallavicini and N.E. White (eds.), Memorie della Società Astr. Italiana 59, p. 185.

Stella, L., 1988b, *Adv. Space Res.* 8, (2) 367.

Stella, L., Kahn, S.M. and Grindlay, J.E., 1984, *ApJ* 282, 713.

Stella, L., White, N.E., Davelaar, J., Parmar, A.N., Blissett, R.J. and Van der Klis, M., 1985, *ApJ* 288, L45.

Stella, L., White, N.E. and Priedhorsky, W., 1987a, *ApJ* 315, L49.

Stella, L., Parmar, A.N. and White, N.E., 1987b, *ApJ* 321, 418.

Stella, L., Haberl, F., Lewin, W.H.G., Parmar, A.N., Van Paradijs, J. and White, N.E., 1988a, *ApJ* 324, 379.

Stella, L., Haberl, F., Lewin, W.H.G., Parmar, A.N., Van der Klis, M. and Van Paradijs, J., 1988b, *ApJ* 327, L13.

Stewart, R.T., Nelson, G.J., Penninx, W., Kitamoto, S., Miyamoto, S. and Nicolson, G.D., 1991, *MNRAS* 253, 212.

Sunyaev, R.A., *et al.*, 1991, *Sov. Astron. Lett.* 17 (2), 123.

Sunyaev, R.A., *et al.*, 1992, in *Frontiers of X-ray Astronomy*, Y. Tanaka (ed.), Universal Academy Press, Tokyo, p. 325.

Sutherland, P.G., Weisskopf, M.C. and Kahn, S.M., 1978, *ApJ* 219, 1029.

Sztajno, M., Basinska, E.M., Cominsky, L.R., Marshall, F.J. and Lewin, W.H.G., 1983, *ApJ* 267, 713.

Takeshima, T., 1992, Ph.D. Thesis, University of Tokyo.

Takeshima, T., Dotani, T., Mitsuda, K. and Nagase, F., 1991, *PASJ* 43, L43.

Takizawa, M., Dotani, T., Ebisawa, K. and Mitsuda, K., 1994a, in prep.

Takizawa, M., Dotani, T., Ebisawa, K., Ogawa, M. and Aoki, T., 1994b, in prep.

Tanaka, Y., 1989, in Proc. 23rd ESLAB Symp. on *Two Topics in X-ray Astronomy*, Bologna, Italy, 13–20 September, 1989 (ESA SP-296), p. 3.

Tanaka, Y., *et al.*, 1983, *IAU Circ. 3891*.

Tananbaum, H., Gursky, H., Kellogg, E., Giacconi, R. and Jones, C., 1972, *ApJ* 177, L5.

Tawara, Y., Hayakawa, S., Kunieda, H., Makino, F. and Nagase, F., 1982, *Nature* 299, 38.

Tennant, A.F., 1987, *MNRAS* 226, 971.

Tennant, A.F., 1988a, *Adv. Space Res.* 8, (2) 397.

Tennant, A.F., 1988b, *MNRAS* 230, 403.

Tennant, A.F., Fabian, A.C. and Shafer, R.A., 1986a, *MNRAS* 219, 871.

Tennant, A.F., Fabian, A.C. and Shafer, R.A., 1986b, *MNRAS* 221, 27P.

Terada, K., Miyamoto, S., Kitamoto, S., Tsunemi, H. and Hayashida, K., 1992, in *Frontiers of X-ray Astronomy*, Y. Tanaka (ed.), Universal Academy Press, Tokyo, p. 323.

Terrell, N.J., 1972, *ApJ* 174, L35.

Toor, A., 1977, *ApJ* 215, L57.

Treves, A., Belloni, T., Chiapetti, L., Maraschi, L., Stella, L., Tanzi, E.G. and Van der Klis, M., 1988, *ApJ* 325, 119.

Treves, A., Belloni, T., Corbet, R.H.D., Ebisawa, K., Falomo, R., Makino, F., Makishima, K., Maraschi, L., Miyamoto, S. and Tanzi, E.G., 1990, *ApJ* 364, 266.

Turner, M.J.L., Smith, A. and Zimmermann, H.U., 1981, *Space Sci. Rev.* 30, 513.

Turner, M.J.L., Thomas, H.D., Patchett, B.E., Reading, D.H., Makishima, K., Ohashi, T., Dotani, T., Hayashida, K., Inoue, H., Kondo, H., Koyama, K., Mitsuda, K., Ogawara, Y., Takano, S., Awaki, H., Tawara, Y. and Nakamura, N., 1989, *PASJ* 41, 345.

Unno, W., Yoneyama, T., Urata, K., Masaki, I., Kondo, M. and Inoue, H., 1990, *PASJ* 42, 269.

Van den Heuvel, E.P.J. and de Loore, C., 1973, *A& A* 25, 387.

Van der Klis, M., 1986, in *Proc. The Physics of Accretion onto Compact Objects*, Lecture Notes in Physics 266, Springer, p. 157.

Van der Klis, M., 1987, IAU Symp. 125, p. 321.

Van der Klis, M., 1989a, *ARA& A* 27, 517.

Van der Klis, M., 1989b, in *Timing Neutron Stars*, H. Ögelman and E.P.J. Van den Heuvel (eds.), Kluwer, Dordrecht (NATO ASI Series C 262), p. 27.

Van der Klis, M., 1989c, in Proc. 23rd ESLAB Symp. on *Two Topics in X-ray Astronomy*, Bologna, Italy, 13–20 September, 1989 (ESA SP-296), p. 203.

Van der Klis, M., 1991, in *Neutron Stars: Theory and Observation*, J. Ventura and D. Pines (eds.), Kluwer, Dordrecht (NATO ASI Series C 344) p. 319.

Van der Klis, M., 1992, in *X-Ray Binaries and Recycled Pulsars*, E.P.J. Van den Heuvel and S. Rappaport (eds.), Kluwer, Dordrecht (NATO ASI Series C 377), p. 49.

Van der Klis, M., 1994a, in Proc. INTEGRAL Meeting, Les Diablerets, Switzerland, 1993, *ApJS*, in press..

Van der Klis, M., 1993b, *A& A* 283, 469.

Van der Klis, M. and Jansen, F.A., 1985, *Nature* 313, 768.

Van der Klis, M. and Lamb, F.K., 1994, in prep.

Van der Klis, M., Jansen, F., Van Paradijs, J., Lewin, W.H.G., Van den Heuvel, E.P.J., Trümper, J.E. and Sztajno, M., 1985a, *Nature* 316, 225.

Van der Klis, M., Jansen, F., White, N., Stella, L. and Peacock, A., 1985b, *IAU Circ. 4068.*

Van der Klis, M., Jansen, F., Van Paradijs, J., Lewin, W.H.G., Sztajno, M. and Trümper, J., 1987a, *ApJ* 313, L19.

Van der Klis, M., Hasinger, G., Stella, L., Langmeier, A., Van Paradijs, J. and Lewin, W.H.G., 1987b, *ApJ* 319, L13.

Van der Klis, M., Stella, L., White, N., Jansen, F. and Parmar, A.N., 1987c, *ApJ* 316, 411.

Van der Klis, M., Hasinger, G., Damen, E., Penninx, W., Van Paradijs, J. and Lewin, W.H.G., 1990, *ApJ* 360, L19.

Van der Klis, M., Kitamoto, S., Tsunemi, H. and Miyamoto, S., 1991, *MNRAS* 248, 751.

Van der Klis, M., *et al.*, 1994, *A& A* , in prep.

Van Kerkwijk, M.H., Charles, P.A., Geballe, T.R., King, D.L., Miley, G.K., Molnar, L.A., Van den Heuvel, E.P.J., Van der Klis, M. and Van Paradijs, J., 1992, *Nature* 355, 703.

Van Paradijs, J., Hasinger, G., Lewin, W.H.G., Van der Klis, M., Sztajno, M., Schulz, N. and Jansen, F., 1988a, *MNRAS* 231, 379.

Van Paradijs, J., Penninx, W. and Lewin, W.H.G., 1988b, *MNRAS* 233, 437.

Van Paradijs, J., Dotani, T., Tanaka, Y. and Tsuru, T., 1990a, *PASJ* 42, 633.

Van Paradijs, J., Allington-Smith, J., Callanan, P., Charles, P.A., Hassall, B.J.M., Machin, G., Mason, K.O., Naylor, T. and Smale, A.P., 1990b, *A& A* 235, 156.

Vaughan, B., Van der Klis, M., Lewin, W.H.G., Wijers, R.A.M.J., Van Paradijs, J., Dotani, T. and Mitsuda, K., 1994, *ApJ* 421, 738.

Vikhlinin, A., *et al.*, 1992a, *IAU Circ. 5576.*

Vikhlinin, A., *et al.*, 1992b, *IAU Circ. 5608.*

Vrtilek, S.D., Kahn, S.M., Grindlay, J.E., Helfand, D.J. and Seward, F.D., 1986, *ApJ* 307, 698.

Vrtilek, S.D., Raymond, J.C., Garcia, M.R., Verbunt, F., Hasinger, G. and Kürster, M., 1990, *A& A* 235, 162.

Vrtilek, S.D., Penninx, W., Raymond, J.C., Verbunt, F., Hertz, P., Wood, K., Lewin, W.H.G. and Mitsuda, K., 1991, *ApJ* 376, 278.

Webbink, R.F., Rappaport, S. and Savonije, G.J., 1983, *ApJ* 270, 678.

Weisskopf, M.C. and Sutherland, P.G., 1978, *ApJ* 221, 228.

Weisskopf, M.C., Kahn, S.M. and Sutherland, P.G., 1975, *ApJ* 199, L147.

Weisskopf, M.C., Sutherland, P.G., Katz, J.I. and Canizares, C.R., 1978, *ApJ* 223, L17.

White, N.E. and Marshall, F.E., 1984, *ApJ* 281, 354.

White, N.E. and Mason, K.O., 1985, *Space Sci. Rev.* 40, 167.

White, N.E. and Stella, L., 1988, *MNRAS* 231, 325.
White, N.E., Mason, K.O., Sanford, P.W., Ilovaisky, S.A. and Chevalier, C., 1976, *MNRAS* 176, 91.
White, N.E., Mason, K.O., Sanford, P.W., Johnson, H.M. and Catura, R.C., 1978, *ApJ* 220, 600.
White, N.E., Charles, P.A. and Thorstensen, J.R., 1980, *MNRAS* 193, 731.
White, N.E., Stella, L. and Parmar, A.N., 1988, *ApJ* 324, 363.
Wood, K.S., Hertz, P., Norris, J.P., Vaughan, B.A., Michelson, P.F., Mitsuda, K. and Dotani, T., 1989, in Proc. 23rd ESLAB Symp. on *Two Topics in X-ray Astronomy*, Bologna, Italy, 13–20 September, 1989 (ESA SP-296), p. 689.
Yongheng, Z. and Keliang, H., 1988, *Vistas in Astronomy* 31, 411.
Yoshida, K., Mitsuda, K., Ebisawa, K., Ueda, Y., Fujimoto, R. and Yaqoob, T., 1993, *PASJ* 45, 605.

7

Radio properties of X-ray binaries

R. M. Hjellming and X. Han
National Radio Astronomy Observatory*, Socorro, NM 87801-0389, USA

7.1   Introduction

X-ray binaries are the most efficient of the stellar systems in our galaxy at producing strong, highly variable radio emission. The X-ray binary Sco X-1 was the first to be shown to have such radio behavior, and it is now commonly found in both normal X-ray binaries and X-ray transients. All X-ray binary radio emission is apparently due to the capability of X-ray binaries to accelerate relativistic electrons which then radiate by the synchrotron radio emission mechanism when these relativistic electrons interact with magnetic fields. In some cases one sees the radiating relativistic plasma in the form of jets or extended nebulosities; in other cases one can infer the extended or jet properties of the radiating regions.

In radio astronomy, quasars and galactic nuclei with jets and extended lobes are largely believed to result from processes rooted in the accretion environments of these highly energetic objects. In this context, radio-emitting X-ray binaries are local paradigms or 'Rosetta stones' for the behavior of accretion disk environments. Because X-ray binaries are nearby and 'bright' at many wavelengths, multi-wavelength studies are possible, allowing one to obtain more complete information, and hopefully more complete understanding, of the special phenomena associated with binary star accretion disk environments.

7.1.1   *Overview of the radio-emitting X-ray binaries*

The first X-ray binary shown to be a radio source was Sco X-1 (Ables 1969; Hjellming and Wade 1971a; Hjellming 1988). Initially the two radio lobes about 1.2′ NE and SW of Sco X-1 were believed to be related to the X-ray source, but recent work by Fomalont and Geldzahler (1991) has shown that these lobes are probably background extra-galactic sources because they do not share the observed optical and radio proper motion, $(0.0148'' \pm 0.0011'')$ yr^{-1}, of Sco X-1 itself. Now many X-ray binaries and transient X-ray sources are known to have radio counterparts, and some are definitely associated with extended radio emission, generally in the form of asymmetric or jet-like emission on size scales ranging from milliarcseconds to several arcseconds. In Table 7.1 we list the currently known X-ray binaries that have been found to be radio sources, together with brief summaries of their properties and references to recent or original work on the objects.

* The National Radio Astronomy Observatory is operated by Associated Universities, Inc., under a cooperative agreement with the National Science Foundation.

7.1.2   Relativistic electron acceleration in X-ray binaries

Synchrotron emission requires relativistic electrons interacting with magnetic fields. Essentially all radio emission from X-ray binaries has short time scales, ranging from minutes to days. Therefore relativistic electrons must be produced on short time scales in physical environs closely related to X-ray emitting accretion disk environments.

For nearby radio sources, the emission mechanism for the radio photons may be thermal or non-thermal. An emission process is thermal if the radiating particles are in thermal equilibrium, that is if their velocity distribution obeys a Maxwellian distribution for a temperature T_e. In radio astronomy, the most common thermal emission mechanism is bremsstrahlung where the radiating electrons are interacting with ions, so the electron temperature and electron density completely determine the appropriate emission and absorption processes. A thermal distribution of electrons interacting with magnetic fields at sub-relativistic speeds will produce thermal emission described as gyro-resonance radiation. The brightness of these two thermal emission processes are limited by the maximum temperature of the plasma. Neither can produce the extremely bright emission seen in most cosmic radio sources, including X-ray binaries.

The two best understood mechanisms for non-thermal radio emission involve either accelerating the electrons to very high energies with non-Maxwellian velocity distributions, or allowing plasma emission processes to provide emission feeding on waves in the plasma (Zhelenyakov 1970; Hjellming 1988). The latter are difficult to relate to observable emission, because the theory is complex, so it is the practice in radio astronomy to invoke plasma radiation processes only when all other alternatives fail. In practice, plasma radio emission, which is best known from the solar case, tends to be related to very high time variability, hence there is an obvious observable symptom: very short time scale variations. Some of the radio-emitting X-ray binaries are being found with variations on time scales as short as tens of minutes; however, as we will discuss, these observations do not yet force one to abandon incoherent synchrotron emission as the radio emission mechanism.

The most efficient known mechanism for production of intense radio emission from astronomical sources is the so-called synchrotron emission mechanism, in which highly relativistic electrons interacting with magnetic fields produce intense radio emission which tends to be linearly polarized. The observed radio emission can be reasonably explained by assuming there are simple (or complex) spatial distributions of relativistic electrons, usually with a power-law energy distribution, interacting with ordered, or turbulent, distributions of magnetic fields. Because synchrotron emission explains most, if not all, aspects of radio emission from X-ray binaries, the presence of intense non-thermal radio emission is taken as *de facto* evidence for the presence of highly relativistic electrons mixed with magnetic fields. In some cases, linear polarization is observed, further supporting this conclusion. Particularly when one sees flaring, or highly time-variable, radio emission, this implies the need for acceleration of electrons in the environment of the object. Thus, radio emission is usually taken as evidence for particle acceleration. However, before addressing the question of production of relativistic electron plasma, let us summarize some of the characteristics of synchrotron emission.

Table 7.1. *Summary of radio-emitting X-ray binaries*

Name	Main properties	References
SS433	Radio flaring; radio, optical, X-ray jets with $v = 0.26c$	Hjellming and Johnston 1981; Margon 1984; Watson *et al.*1986
Cyg X-1	Periodic variable, $P = 5.6$ days, rare flaring, radio changes as X-ray state changes	Braes and Miley 1971; Hjellming and Wade 1971b; Tananbaum *et al.* 1972; Han and Hjellming 1993
LSI+61°303	Periodic outbursts, $P = 26.5$ days	Gregory and Taylor 1978; Taylor and Gregory 1982
Cir X-1	Periodic outbursts, $P = 16.6$ days	Whelan *et al.* 1977; Preston *et al.* 1983
Cyg X-3	Giant flare events, periodic variable on low level, $P = 4.8$ hours, jet with $v \sim 0.2 - 0.4c$	Braes and Miley 1972; Gregory *et al.* 1972; Geldzahler *et al.* 1983; Spencer *et al.* 1986
Sco X-1	Variable, radio changes with the X-ray spectral state changes; 'Z-source'	Ables 1969; Hjellming and Wade 1971a; Priedhorsky *et al.* 1986; Hjellming *et al.* 1990c
Cyg X-2	Weak, variable, radio changes with X-ray spectral state changes; 'Z-source'	Hjellming and Blankenship 1973; Hjellming *et al.* 1990a
GX 17+2	Weak, variable, radio changes with X-ray spectral state changes; 'Z-source'	Hjellming and Wade 1971b; Hjellming 1978; Geldzahler 1983; Grindlay and Seaquist 1986; Penninx *et al.* 1988
GX 5-1	Weak, variable, radio changes with X-ray spectral state changes	Geldzahler 1983; Tan *et al.* 1991; Grindlay and Seaquist 1986
GX340+0	Weak, variable, radio changes with X-ray spectral state changes	Penninx *et al.* 1993
GX349+2	Weak, variable, radio	Cooke and Ponman 1991
Cen X-4	Transient radio and X-ray source	Hjellming *et al.* 1988
A0620−00	Transient radio and X-ray source	Owen *et al.* 1976; Hjellming *et al.* 1988
GS2000+35	Transient radio and X-ray source	Hjellming *et al.* 1988
Aql X-1	Transient radio and X-ray source	Hjellming *et al.* 1990c
V404 Cyg	Transient radio, X-ray and optical source; second stage radio source	Kitamoto 1990; Wagner *et al.* 1990; Han and Hjellming 1994
GRS1124−683 (Nova Muscae 1991)	Transient radio, optical, X-ray; second stage radio source	Kesteven and Turtle 1991
GRO J0422+32	Transient radio, optical, X-ray; second stage radio source	Han and Hjellming 1992b
GX 13+1	Weak, variable radio; 'atoll' source	Grindlay and Seaquist 1986; Garcia *et al.* 1988
NGC 6624	weak, in globular cluster	Grindlay and Seaquist 1986
NGC 7078	weak, in globular cluster	Machin *et al.* 1990
NGC 6712	weak, in globular cluster	Machin *et al.* 1990

7.1.3   Radio emission and absorption processes

The details of synchrotron radio emission theory can be found in Ginzburg and Syrovatskii (1965); however, let us summarize some of the important points. The radio emissivity of an electron or ion interacting with magnetic fields is inversely proportional to the fourth power of its mass; therefore, relativistic electrons completely dominate the radio emission from any relativistic plasma. Synchrotron emission over the observed frequency ranges of a few hundred megahertz to a few tens of gigahertz require particle energies in the range of mega-electron-volts to several tens of mega-electron-volts, and it is common practice, consistent with the constraints of the observations, to assume a power law for the electron energy distribution. Thus, if there are $N(E)dE$ electrons with energies between E and $E+dE$, then $N(E) = K E^{-\gamma}$, where γ is the energy spectral index, generally between 2 and 3, and K is a constant reflecting the density of relativistic electrons. Most applications of synchrotron emission theory assume an average over finite volumes which have random directions for particle motion and orientation of magnetic field. Under these circumstances, all the emission and absorption properties can be described by simple emission and absorption coefficients, whereby the emission coefficient is given by

$$\epsilon_v = \epsilon_0 (K/K_0)(H/H_0)^{(\gamma+1)/2}(v/v_0)^{-(\gamma-1)/2} \tag{7.1}$$

and the absorption coefficient k_v is related to ϵ_v by a so-called source function, which in this case is

$$\frac{\epsilon_v}{k_v} = \frac{\epsilon_0}{\kappa_0}(H/H_0)^{-1/2}(v/v_0)^{5/2}, \tag{7.2}$$

where H is the strength of the magnetic field. Because the Rayleigh–Jeans limit is valid for all but the shortest radio wavelengths, the source function can be expressed as a source temperature, $T_s = (\lambda^2 \kappa_v)/(\epsilon_v k_B)$, where k_B is the Boltzmann constant. Then, for the cases where the energy distribution is a power law, and Eqs. (7.1) and (7.2) are valid, $T_s = T_\gamma H_{mG}^{-1/2} v_{GHz}^{1/2}$ (H in milligauss, v in gigahertz), where $T_\gamma \approx 2.5 \times 10^{12} e^{-0.65\gamma}$ for $1.5 < \gamma < 3.5$.

Assuming that this simple model applies, observations can be used to determine one or more (or a combination) of the parameters (H, K, γ, D and d). The apparent solid angle subtended by such an emitting region is $\Omega_s = \pi\Theta_s^2/d^2$, where $\Theta_s = D/d$. Defining the optical depth of the source by $\tau_v = k_v L$, whereby L is the size of the emission region, the observed radio flux density for such a source is

$$\begin{aligned} S_v &= \frac{2 k v^2 j_v}{c^2 k_v} (1 - e^{-\tau_v})\Omega_s \\ &= 2.5 \, 10^{12} e^{-0.65\gamma} H_{mG}^{-1/2} v_{GHz}^{1/2} (1 - e^{-\tau_v})(\pi\Theta_s^2/d^2), \end{aligned} \tag{7.3}$$

indicating that the maximum brightness temperatures ($2.5 \times 10^{12} e^{-0.65\gamma} H_{mG}^{-1/2} v_{GHz}^{1/2}$) will be of the order of 10^{12} or less for frequencies of 1 GHz or more and fields of the order of milligauss. It also indicates the brightness temperatures one will tend to have for optically thick emission regions.

If the angular size of the source, Θ_s, is known (e.g. if the source is resolved), then, if one also knows d or D, one can determine some of the interesting parameters of the source. One of the combinations of parameters determined for some X-ray binaries is

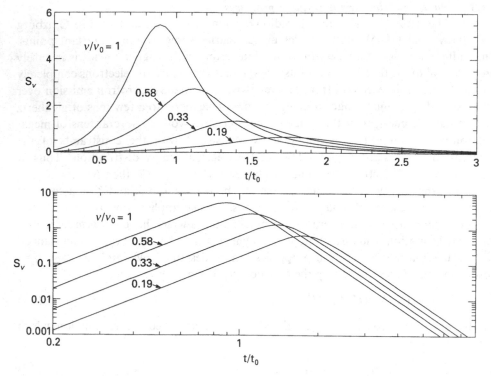

Fig. 7.1. Plots of flux density as a function of time in linear–linear and log–log forms for synchrotron 'bubble' models with frequencies differing by the indicated fractions.

$\Theta_s H^{-1/4}$, which is usually found to be of the order of a few milliarcseconds gauss$^{-1/4}$. This indicates size scales of the order of milliarcseconds since other considerations estimate the fields to be of the order of milligauss.

7.1.4   *Evolving synchrotron bubble events*

The discussion in the previous section applies to radio sources at a particular instant in time. Transient events involve other considerations because of the scaling of the variables with time.

The simplest type of radio event involving relativistic particles, which occurs frequently in Cyg X-3 and transient X-ray sources, is the synchrotron 'bubble event' (Figures 7.1, 7.2, and 7.6). The theory for these events, as examples of expanding, spherical bubbles of relativistic plasma, treated as uniform layers with respect to the line of sight to an observer, was first discussed by Kellermann (1966) and Van der Laan (1966). A recent discussion by Hjellming and Johnston (1988) adds a correction factor for spherical geometry, and provides a formulation that can be directly compared with the two-dimensional twin jet version, with the same level of modeling of expanding relativistic plasmas that we will summarize in a coming section. In this formulation we assume that a spherical 'bubble' of relativistic plasma at time t has a radius r, an angular radius $\theta = r/d$ (where d is the distance), a uniform magnetic field (with a random distribution of orientations with respect to the observer) of strength H, and a power-law spectrum of $N(E) = KE^{-\gamma}$. All variables scale with time, so one assumes that at a time t_0 the size is r_0 (or θ_0), the magnetic field is H_0, each particle has an energy E_0, and the fixed number of relativistic electrons is set by K_0; then each particle loses energy according to $E = E_0(r/r_0)^{-1}$, an effect which has its greatest impact on the upper and lower energy cut-offs (E_1 and E_2) assumed for the power-law spectrum. Conservation of magnetic flux implies that $H = H_0(r/r_0)^{-2}$, and conservation of particles between the evolving cut-off energies gives $K = K_0(r/r_0)^{-(\gamma+2)}$, a result which is independent of the values of cut-off energies. The optical depth for a radius $a \leq r$ is then (cf. Eqs. 7.1 and 7.2)

$$\tau_\nu(a) = \tau_\nu' \left[1 - \left(\frac{a}{r} \right)^2 \right]^{1/2}, \tag{7.4}$$

where

$$\tau_\nu' = \tau_0 \left(\frac{r}{r_0} \right)^{-(2\gamma+3)} \left(\frac{\nu}{\nu_0} \right)^{-(\gamma+4)/2} \tag{7.5}$$

and $\tau_0 = 0.019 g(\gamma)(3.5 \times 10^9)^\gamma K_0 H_0^{(\gamma+2)/2} 2 r_0$, with ν_0 being a fixed reference frequency. Since the source is homogeneous at any point in time, the radio flux density equation can be integrated over the spherically symmetric volume of the source to give

$$S_\nu = 1.42 \, e^{-0.65\gamma} \, H_{0,mG}^{-1/2} \, \nu_{GHz}^{5/2}$$
$$\Theta_{mas}^3 \left\{ 1 - \exp \left[-\tau_0 \left(\frac{r}{r_0} \right)^{-(2\gamma+3)} \left(\frac{\nu}{\nu_0} \right)^{-(\gamma+4)/2} \right] \right\} \xi_{sph}(\tau'). \tag{7.6}$$

The function $\xi_{sph}(\tau)$ varies between 0.67 and 1, and can be described by the power-law expansion $\xi_{sph}(\tau) \simeq 0.66584 + 0.09089\tau - 0.009989\tau^2 + 0.0005208\tau^3 - 0.00001268\tau^4 + 0.000000115\tau^5$ for $\tau < 20$ and $\xi_{sph}(\tau) = 1$ for $\tau > 20$.

Figure 7.1 shows plots of flux density as a function of time, in linear–linear and log–log form, showing the optically thick and thin portions of the light curve, and the different maxima and time delays for a range of frequencies.

If the spherical bubbles of relativistic plasma expand freely with constant velocity, one can assume the radius scales with time according to $t/t_0 = r/r_0$. However, if there is significant deceleration due to interaction with surrounding material, the mapping of spatial into time variations is different: $t/t_0 = (r/r_0)^{2.5}$ or $t/t_0 = (r/r_0)^4$ for the phases where energy or momentum, respectively, are conserved.

7.2 Transient radio emission

7.2.1 *The single radio event transients*

The transient X-ray binaries A0620−00, Cen X-4, GS2000+35, Aql X-1, GS2023+338 (V404 Cyg), and GRS1124−683 (Nova Muscae 1991) have each exhibited transient synchrotron bubble events at radio wavelengths following a change in each system that produced a transient X-ray source. Figure 7.2 shows the data and synchrotron bubble model fits for these six objects (Han 1993; Hjellming *et al.* 1994). The basis for the models that fit the data is discussed in the previous section. These transient radio events indicate that one of the side effects of the sudden appearance

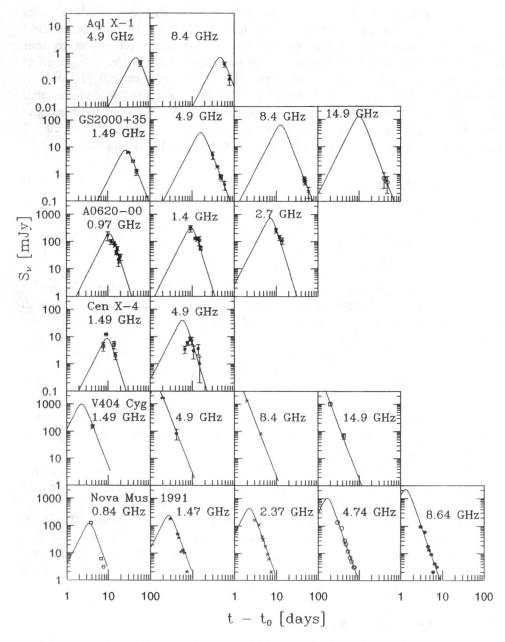

Fig. 7.2. Data and models for synchrotron bubble events in Aql X-1, GS2000+35, A0620−00, Cen X-4, GS2023+338 (V404 Cyg), and GRS1124−683 (Nova Muscae 1991).

of a transient accretion disk environment in these systems is a sudden, effectively single event, production of a compact volume of mixed relativistic electrons and magnetic fields. The bubble subsequently expands adiabatically with all the observational signatures of synchrotron bubble events.

The standard model for a single synchrotron bubble event is the occurrence of a dynamical event that produces an outward-moving shock, and relativistic electrons are accelerated in the shock (Dickel *et al.* 1989). In the context of X-ray transients this is probably the result of the gas dynamical adjustments that occur during the formation of the hot, X-ray-emitting accretion disk.

7.2.2 *Secondary radio emission from transients*

Two of the transient events discussed in the previous section, GS2023+338 (V404 Cyg) and GRS1124−683 (Nova Muscae 1991), were followed by the appearance of a 'second stage' radio source which varied on time scales of hours to days. Figure 7.3 shows the 4.9 GHz radio light curve of GS2023+338 for two years after the outburst, together with the optical and X-ray light curves of this object (Han and Hjellming 1992a). During its slow decay phase, GS2023+338 fluctuated between flat and inverted radio spectra. Figure 7.3 indicates that the same slow decay occurred in all three wavelength regions, indicating that their emitting regions must exist in the same evolving structure around the stellar system. Han (1993) has suggested a layered or 'onion' model that explains the coupling of the apparent 'photospheres' based upon the assumption that the radio emission is generated in the outer portions of a wind from the accretion disk environment, and that the wind is contracting in such a way that the apparent 'photospheres' observed at radio, optical and X-ray wavelengths are decreasing in radius according to $(t - t_0)^{-0.3}$.

GRS1124−683 had extensive ranges of optically thick and thin spectral variation a few days after its radio transient event, but was not observed after that time, so it is unknown whether it had a similar, fluctuating slow decay. In 1992, a third transient, GRO J0422+32, was observed too late to detect a possible fast transient, but it did show a GS2023+338-like, late rise in radio flux followed by a fluctuating decay proportional to t^{-1} – making this object even more similar to V404 Cyg.

7.3   Coupled radio–optical–UV–X-ray state changes

7.3.1 *The 'Z'-source X-ray binaries*

Recent observations have shown that some low-mass X-ray binaries have state changes that are coupled in the sense that changes in behavior in one wavelength region are correlated with state changes in other wavelength regions. Based upon the results of extensive multi-wavelength observing campaigns on Sco X-1, Priedhorsky *et al.* (1986) suggested that the X-ray, optical and radio states of Sco X-1 are related in a predictable way. The nature of this relationship has been more definitively established in recent years through multi-wavelength campaigns involving ground-based observations at radio and optical wavelengths simultaneously with satellite observations in the ultraviolet (IUE) and the X-ray (Ginga). X-ray observations with sensitive measurements in four independent energy ranges and high timing resolution have provided the key classification system for X-ray states; the intensities in the four energy ranges allow one to define 'hard' and 'soft' colors (e.g. $H_1 = I(9 - 18\text{keV})/I(6 - 9\text{keV})$ and $H_2 = I(4 - 6\text{keV})/I(1 - 4\text{keV})$). For the X-ray binaries Sco X-1, Cyg X-2, GX17+2, GX5−1, GX349+2, and GX340+0 plots of H_1 vs. H_2 (called color–color diagrams) indicate branches (Z-shaped; see Figure 6.9), which are

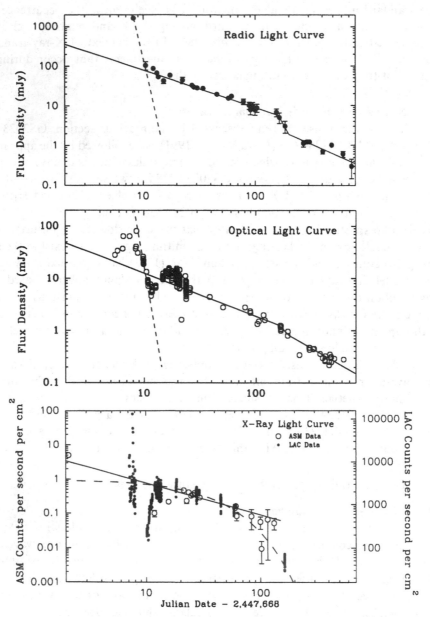

Fig. 7.3. The radio, optical and X-ray light curves of GS2023+338 (V404 Cyg) are plotted on a log–log scale as a function of time since outburst, indicating similar slow decay characteristics at all three wavelengths.

correlated with the X-ray binary's time fluctuation properties (QPOs and noise), and are called 'horizontal', 'normal' and 'flaring' branches (Hasinger and Van der Klis 1989). Not all six Z sources exhibit all three branches (see also Ch. 6). Figure 7.4 shows the so-called 'Z-diagram' of GX17+2 obtained during a multi-wavelength campaign by Penninx *et al.* (1988), with the three branches indicated in the diagram on the left,

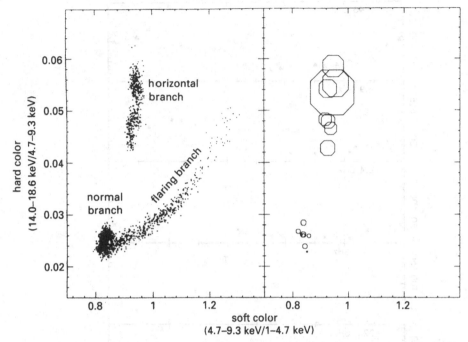

Fig. 7.4. X-ray color–color diagram for GX17+2 from Penninx *et al.* (1988), showing the X-ray data points on the left and the radio fluxes on the right with a symbol size proportional to the flux density.

and a diagram on the right showing the radio fluxes with symbols proportional to the flux densities. The observing campaigns for GX17+2 (Penninx *et al.* 1988), Cyg X-2 (Hjellming *et al.* 1990a), and Sco X-1 (Hjellming *et al.* 1990c) show that for these Z sources the radio emission is strongest on the horizontal branch, decreases as the source proceeds down the normal branch, and is weakest when the source is on the flaring branch. Simultaneous radio–X-ray observations of the Z source GX 5−1 (Tan *et al.* 1991) have not shown a similar correlation. During most of the time, GX 5−1 was on the horizontal branch and the radio emission was quiet (~ 0.6 mJy), and when (once) a radio flare occurred (~ 3 mJy), the X-ray source was on the normal branch.

Penninx (1989) has suggested that the average radio luminosity of the Z sources is approximately the same on the normal branch. The two Z sources that had not yet been detected at radio wavelengths at the time of that prediction, GX 349+2 and GX 340+0, were later detected with approximately the expected radio brightnesses (Cooke and Ponman 1991; Penninx *et al.* 1993).

7.3.2 *Periodic radio variations in X-ray binaries*

7.3.2.1 *Cir X-1*

The radio outbursts of the low-mass X-ray binary Cir X-1 occur with a period of 16.6 days, probably the binary orbit of the system. The radio flares occur near a transition in the X-ray brightness (either a high–low transition as often observed

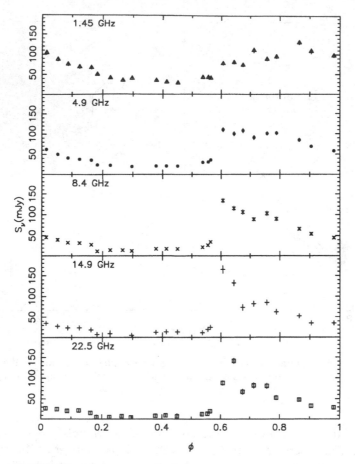

Fig. 7.5. Four frequency radio light curves for LSI+61°303 showing radio flux variations during the last half of one periodic event and the first half of another event in August–September 1991 (Molnar *et al.* 1994).

in the 1970s, or a low–high transition at the end of the 1980s), which has the same periodicity of 16.6 days (Stewart *et al.* 1991). The brightness, the occurrence of the radio flares, and the IR brightness increase appear to be related with the type of X-ray transition (high–low versus low–high). It has been suggested that the orbit of the neutron star around the mass-donor star is eccentric, in which case the transition might occur near periastron; still, how and what causes the transition is unclear (cf. Stewart *et al.* 1991). Haynes *et al.* 1986 have imaged the region around Cir X-1 at 843 MHz, showing that there is a 3′ by 5′ synchrotron radiation nebulosity surrounding the X-ray binary with more compact structures near the center.

7.3.2.2   *LSI+61°303*

Another system with periodic radio behavior is the radio-, optical-, X-ray-, and (probably) γ-ray-emitting Be binary LSI+61°303. Gregory and Taylor (1978) first determined the period of this system to be 26.5 days from the radio flaring data,

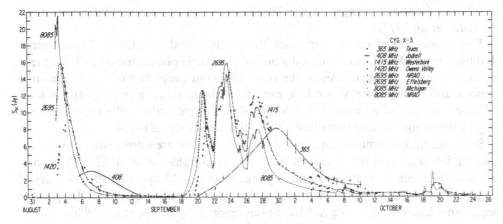

Fig. 7.6. Large radio flares of Cyg X-3 in 1972.

and showed that sudden flaring occurred at a binary phase between 0.5 and 0.6. Observations such as those obtained by Molnar *et al.* (1994), as shown in Figure 7.5, indicate that between phase 0.2 and 0.5 the radio source is decaying with an optically thin synchrotron spectrum, develops an optically thick spectrum during the flaring between phases of 0.5 and 0.65, then evolves to a flat spectrum before going into optically thin decay. The time profile of the true 'repeated event' is probably obtained by plotting phases 0.5 to 1.0, then phases 0 to 0.5.

7.3.3 *Cyg X-3 flares*

Figure 7.6 shows the sequence of multi-frequency radio observations in 1972 which showed that Cyg X-3 exhibits extremely strong radio flares (see Gregory *et al.* 1972, and 20 papers following that article). These and subsequent observations have established that the major flares in Cyg X-3 behave like the synchrotron 'bubble' events discussed earlier in this chapter. Irrespective of whether the relativistic plasma is suddenly created with a mixture of magnetic fields in shell, sphere or conical sheath environments, the subsequent behavior has time variation characteristics of one (or multiple simultaneous) regions of relativistic plasma, expanding at high speeds, with adiabatic expansion dominating the energy losses of the relativistic electrons; however, unlike the X-ray transient events, it is common for other energy losses to affect the decay events in Cyg X-3 (Marscher and Brown 1975; Marti *et al.* 1992).

The 'typical' synchrotron 'bubble' event involves an optically thick rise of the radio flux, with weaker and later peaks as radio frequency decreases, followed by an optically thin power-law decay proportional to $v^{-(\gamma-1)/2} t^{-2\gamma}$, where γ is the energy spectral index of a power-law spectrum of relativistic electrons. If energy losses are dominated by adiabatic expansion, all events will look like Figure 7.1. However, if other energy loss mechanisms become important, the optically thick rise, the peaks, and initial pre-power-law decays can be more complicated (Marscher and Brown 1975). The linear polarization characteristic of synchrotron radio emission is often seen, and sometimes the presence of large quantities of co-existing thermal plasma is inferred from the changes in the Faraday rotation of linear polarization angle (Seaquist *et al.* 1974).

The appearance of radio events with the observational signatures of synchrotron 'bubbles' establishes the sudden production of relativistic plasma mixed with magnetic fields. However, the signatures of the size scales, and properties of the acceleration regions and their geometry, are lost, except for the qualitative result that the scale of time variation of an event is proportional to the size scale of the region when it begins behaving like an adiabatically expanding relativistic plasmoid.

Some strong synchrotron flare events in Cyg X-3 have been precursors of outward moving jets with proper motions of $0.17''$ yr^{-1} (Geldzahler *et al.* 1983; Johnston *et al.* 1986; Spencer *et al.* 1986). Assuming the distance is 12 kpc, the 'jet' velocities of Cyg X-3 are consistent with the highly stable $0.26c$ velocities of SS433. Larger size scale radio lobes around Cyg X-3 have been reported by Strom *et al.* (1989).

7.4   Radio jets and extended radio emission

7.4.1   *SS433 – the paradigm for relativistic particle jets*

SS433 is a weak X-ray source at the center of a very large complex of radio (W50) emission (Königl 1983; Baum and Elston 1985), extended X-ray emission (Watson *et al.* 1986), and optical filaments (Margon 1984). The most important characteristic of SS433 is the essentially continuous ejection of matter at a velocity of $0.26c$, which is seen in the X-ray regime through Doppler shifts of iron lines, in the optical regime through Doppler-shifted emission lines of hydrogen and helium (Margon 1984), and in the radio regime through radio jets that move outward from the stellar system with a proper motion of $3.0''$ yr^{-1} (Hjellming and Johnston 1981, 1985, 1988; Vermeulen 1989).

Figure 7.7 shows a sequence of VLA radio images of SS433 taken at roughly monthly intervals in 1982 using 4.9 GHz for the images on the left and 14.9 GHz for the images on the right. The resolution at 4.9 GHz is a factor of 3.3 greater than that for the 14.9 GHz images, and one sees radio emission developing as a result of continuous ejection into the projected 'cork-screw' pattern of precessing twin jets. The optical and X-ray Doppler shifts, and radio jet proper motions, have a period of 162.5 days, which is believed to be the precession period of a thick accretion disk surrounding the compact object in the SS433 binary system.

The work of Vermeulen *et al.* (1987) and Vermeulen (1989) has shown that high resolution VLB observations of knots in the SS433 jet not only exhibit the same jet kinematics near the central object, but provide the first direct evidence of a phenomenon theoretically inferred for many jet sources: *in situ* particle acceleration in the jets. Sequences of images made with the European VLBI Network by Vermeulen (Figure 7.8) show that the radio emission does not originate *ab initio* from a central region, but rather exhibits a brightening during motion very close to the center, followed by a peak preceding the steady decay that is the dominant factor in the evolution of the ejecta in the radio images on VLA resolution scales of $0.1''$ or more (Figure 7.7).

Figure 7.9 is a schematic of the known geometry of the SS433 binary system with mass transfer from one star to an accretion disk surrounding the compact neutron

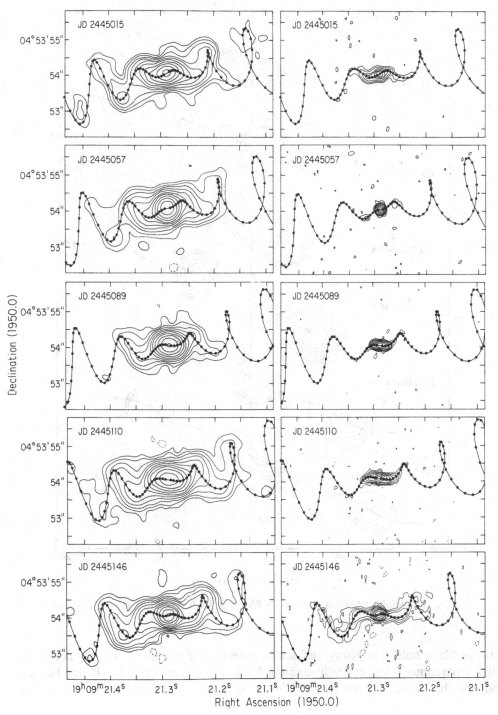

Fig. 7.7. Sequence of VLA images of SS433 made over roughly monthly intervals in 1982, with 4.9 GHz images on the left and 14.9 GHz images on the right, showing mainly the effects of the factor of three difference in resolution (Hjellming and Johnston 1985).

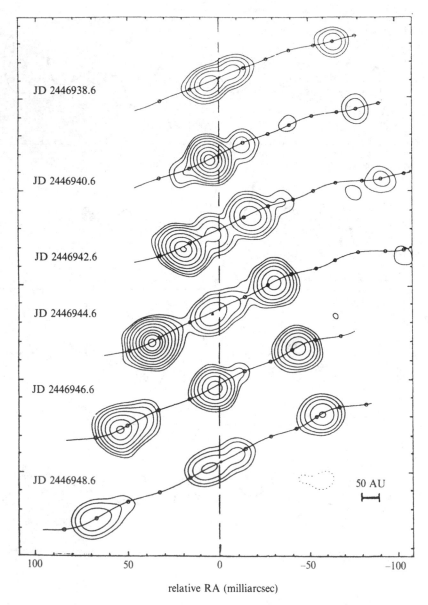

Fig. 7.8. High resolution sequence of SS433 at VLB resolution (Vermeulen 1989).

star or black hole. The X-ray 'jets' are an extension of the X-ray emitting accretion disk environment, the optically emitting knots are recombining 10^4 K gas further outside the jet moving outward at $0.26c$, and the radio emission begins to appear a few light days further out, at angular scales of a few milliarcseconds, as seen in Figure 7.8, and continues out to size scales of several arcseconds, as seen in Figure 7.7.

Hjellming and Johnston (1988) have shown that a conical jet model of laterally expanding sheaths of relativistic plasma, that is initially decelerated by interaction with

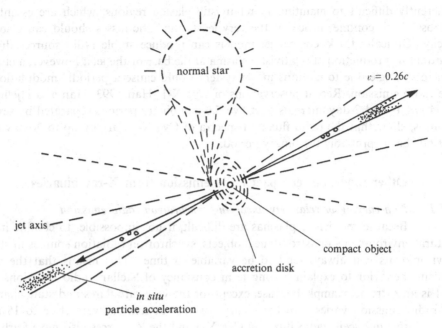

normal star

$v = 0.26c$

jet axis

compact object

accretion disk

in situ particle acceleration

Fig. 7.9. Schematic picture of SS433 (Hjellming and Johnston 1988).

surrounding gas, has optically thin decay characteristics which match the principal decay behavior of the SS433 radio jets. This model is based upon the idea that the driving force behind the production of relativistic particles is the lateral expansion of a hot, X-ray-emitting gas jet with an initial internal sound speed of the order of 2000 km s^{-1}, with a geometry of a conical shock sheath in which the particle acceleration occurs. This hydrodynamical acceleration mechanism may be one of the principal scenarios for the production of relativistic electrons in this and other X-ray binaries. However, even though models of this type have the correct behavior for SS433, it is possible that, even with this object, other scenarios, involving non-emitting jets of relativistic particles being ejected on the axis of the accretion disk environment, will produce the same result. Particle acceleration in a sheath surrounding laterally expanding shocks is consistent with the most commonly discussed mechanisms for particle acceleration of cosmic rays and strong radio sources, but the observations of SS433, where more detailed data are available than for any other radio-emitting X-ray binary, do not yet rule out other models.

7.4.2   Cyg X-1

The radio counterpart of the high-mass X-ray binary Cyg X-1 usually is a relatively stable radio source of $\sim$ 15 mJy with a very flat radio spectrum. The radio component was first identified with the optical and X-ray object because of an X-ray state change that resulted in an increase from an unobservable to stable radio source between March 22 and March 31, 1972 (Tanabaum *et al.* 1972; Hjellming 1973). Except for a radio flaring event at the time of an X-ray flare in 1975, Cyg X-1 has since maintained a radio flux level near 15 mJy. This type of stability is inherently difficult to maintain in relativistic plasma regions, which are essentially impossible to confine; hence, at the very least, adiabatic losses should cause source decay. Optically thick, conical jet models can produce stable radio sources due to continuous production of relativistic plasma at the base of the jets; however, a change in aspect angle due to motions in the system would cause a periodic modulation of the radio emission. Recent observations of Cyg X-1 (Han 1993; Han and Hjellming 1994,) at 1- and 2-day intervals over two 5.6-day binary periods separated by several months, show that the radio flux and spectra of Cyg X-1 vary by up to 20% with a period that is probably the binary period.

7.5   Other models of compact radio emission from X-ray binaries

7.5.1   *Non-varying or relatively unvarying synchrotron radio emission*

Because relativistic plasmas are difficult, if not impossible, to confine in the natural environments of astronomical objects, synchrotron radiation sources in stellar environments will always tend to be variable in time. This means that the most difficult behavior to explain is long term constancy of 'stellar' radio emission. Cyg X-1 is an extreme example because, except for the very rare flares and state changes, its radio emission varies with the binary period, but remains very close to 15 mJy. The relative quiescent radio fluxes in Cyg X-3 and the Z sources also have fairly flat spectra. Flat radio spectra can be produced by a source which is inhomogeneous, that is has a range of optical depths and apparent surface brightness. These facts, and the fact that the radio emission has angular scales less than 0.1″, allow one to eliminate two of the three obvious possibilities and to conclude that it is likely that inhomogeneous, probably jet-like, structures are the sources of relatively steady radio emission.

The larger the size scale of the synchrotron radio source, the longer the time scale of variation due to expansion effects. The extended SS433 jets have time scales for variations of several tens of days, and this is related to these jets being resolved and imageable. The next most obvious reason for source stability is indicated by the relatively flat radio spectra: continuous ejection of relativistic electron plasma, at the same rate with the same properties, will produce a radio-emitting structure that can be constant in time. There are two obvious types of ejecta that will have these properties: a spherical 'wind' of relativistic plasma, and conical jets resulting from continuous ejection along opposite axes, as is seen directly in the images of SS433. The decay of the ejecta limits the size scales over which the jets can be observed. There are three obvious scenarios that can achieve the desired effects: (a) sequential ejection of spherical bubbles of relativistic plasma along ejection axes, where each bubble segment behaves as discussed in the previous section, but the sum over a continuous history of ejected bubbles is a segmented structure with constant emission properties as a whole; (b) continuous ejection of plasma along twin axes with constant axial velocity and only lateral expansion perpendicular to these axes; and (c) a spherically symmetric, continuous 'wind' of relativistic plasma.

It is important to note that all scenarios involving non-spherically symmetric ejection, where the emitting regions range from optically thick to thin extremes from inner to outer regions, have the potential of showing periodic modulation of the strength of the observed radio emission. This is because changes in angle with respect to the observer can change the apparent flux of a structure that would have no variation if there were no changes in this angle. SS433 is an example of a twin jet precessing with a 162.5-day period; however, the SS433 radio emission is always from optically thin regions, so this modulation effect has not been seen. The observable characteristics of optically thick, asymmetric structures changing with a periodicity based upon precession and/or orbital motion would be: broad/flat radio spectra; spectral variations as a function of the precessional and/or orbital period.

The periodic emission of Cyg X-1 and LSI +61° 303, as discussed earlier, may be due to periodic aspect changes in jets. However, periodic initiation of flare events when the compact object is closest to the hot, stellar wind of its companion is also possible (Gregory and Taylor 1978; Stewart *et al.* 1991).

7.5.2 *Multiple synchrotron bubble ejection (episodic/periodic)*

If we assume that, at every Δt interval in time, two spherically symmetric plasmoids are ejected from an object, each plasmoid will expand with the optical depth evolution described by the equations for expanding spherical bubbles. However, the continual replacement of newer, younger bubbles results in a continuous structure that, in outline, would look like twin cones. In the extreme, where successive bubbles do not interact, one can sum over the radio properties of each bubble (allowing for multiple bubbles along individual lines of sight), and predict the total radio emission as a function of Δt and the angle with respect to the line of sight. Rough calculations based upon this model have been made by Penninx (1990), who summed the time variations of multiple bubble events to determine the time variations of a sequence of bubbles. Assuming free expansion of each bubble, the resulting radio spectral index varies between 0.79 and 0.95 for $\gamma = 2$ and 3, respectively. If there is significant deceleration satisfying $t/t_0 = (r/r_0)^2$ or $t/t_0 = (r/r_0)^3$, then the radio spectral indices can range between 0.36–0.56 ($\gamma = 2$–3) and −0.07–0.16 ($\gamma = 2$–3), respectively. This indicates that a constant and fairly flat spectrum can be obtained from a continuous sequence of ejected synchrotron bubbles, but requires that each separate bubble be significantly decelerated.

Several obvious effects can make this more complicated. If, after a finite time, successive bubbles 'collide' because of internal expansion, shock-like structures will occur, requiring a much more complicated model. In addition, if the ejection paths precess, as seen for SS433, the angle with respect to the observer will vary over each segment ejected during a precession period. Finally, modulation of the relativistic electron content and magnetic field at fixed distances along the jet add a dimension of time variability that is very difficult to model.

However, there is one simple case of continuous ejection where the theory of the behavior of the radio emission is almost as simple as the theory of isolated synchrotron bubbles: continuously ejected conical jets (Hjellming and Johnston 1988; Marti *et al.* 1992).

7.5.3 *Continuous ejection with conical jet geometry*

If we assume that there is no acceleration or deceleration of relativistic plasma ejected along twin-jet axes, the expansion effects that determine the simple properties of synchrotron bubbles are changed from three-dimensional to two-dimensional in character. Let z be the coordinate of the jet axis and r be a radius perpendicular to this axis. At any point z_2, the radius of the cross-section of the jet will be r_2. Then the energy of each particle scales as $E = E_0(r_2/r_0)^{-2/3}$, the fields scale as $H = H_0(r_2/r_0)^{-1}$, and the relativistic electron concentration scales as $K = K_0(r_2/r_0)^{-2(\gamma+2)/3}$. As shown by Hjellming and Johnston (1988), the differential contribution to the observed flux density as a function of z along the twin jets is

$$\frac{dS_v}{dz} = \left(\frac{S_0}{r_0}\right)\left(\frac{v}{v_0}\right)^{5/2}\sin(\Theta)\left(\frac{r_2}{r_0}\right)^{3/2}\left[1 - \exp(-\tau_v')\right]\xi_{con}(\tau_v'), \qquad (7.7)$$

where

$$\tau_v' = \left[\frac{\tau_0}{\sin(\Theta)}\right]\left(\frac{v}{v_0}\right)^{-(\gamma+4)/2}\left(\frac{r_2}{r_0}\right)^{-(7\gamma+8)/6} \qquad (7.8)$$

and, analogous to the geometry correction for the spherical case, one has a geometry correction factor which is unity for $\tau > 20$, but can be approximated, for smaller optical depths, by $\xi_{con}(\tau) = 0.78517 + 0.06273\tau - 0.007242\tau^2 + 0.0003905\tau^3 - 0.00000973\tau^4 + 0.00000009\tau^5$.

The total flux density for twin-conical jets is obtained by numerically integrating Eqn. (7.7) along the jet. If Θ is constant, this is a simple calculation; however, if Θ varies with z because of precessional and/or orbital motion, the details of the variations of Θ with z must be included. If the lateral expansion occurs with constant velocity, one can assume r_2 scales with z_2 according to $r_2/r_0 = z_2/z_0$. However, if there is significant deceleration due to interaction with surrounding material, the mapping of z into r_2 is different: $r_2/r_0 = (z/z_0)^{1/2}$ or $r_2/r_0 = (z/z_0)^{1/3}$ for the phases where energy and momentum, respectively, are conserved.

7.5.4   *Continuous ejection with spherical geometry*

The 'wind-like' ejection of relativistic plasma in a spherically symmetric manner is another way to obtain a broad spectrum, inhomogeneous radio source. However, this model has not yet been developed in the context of X-ray binaries.

7.6   Acceleration of relativistic electrons

The previous discussion of spherically symmetric and twin-jet models for synchrotron radio emission addresses the question of what determines the observed radio flux for these geometries – if a specific geometric and energy distribution of relativistic electrons and fields is assumed. It did not, however, address the question of how the relativistic electrons are accelerated. There are two extremes that can be discussed for the environments we are considering. At one extreme, gas dynamical expansions accelerate particles in shocks or trapping regions, and in this case the problem is a two-stage sequence of acceleration of gaseous flows from the binary systems followed by particle acceleration. At the other extreme, the particle acceleration could occur in the magnetospheres of either the accretion disk or the compact object itself.

7.6.1 Gas dynamical acceleration

The acceleration of relativistic electrons, with concomitant generation of magnetic fields, has long been considered to be a natural by-product of the interaction environment of moving shocks, particularly for supernova explosions which expand into the interstellar medium. This acceleration is a result of turbulent acceleration processes behind shocks (Dickel *et al.* 1989). Single synchrotron bubble events in X-ray binaries and transients could be explained by completely analogous processes resulting from sudden changes in the state of the X-ray binary involved. In this scenario, a spherical shock expands from the central region, and there is a sudden phase of relativistic electron acceleration behind/in the spherical shock, which then can either move with the shock, if the sound speed of the relativistic electrons is set by the thermal sound speed, or they can rapidly expand in front of the shock with sound speeds up to the $c/(3^{1/2})$ characteristic of a pure relativistic plasma. The resulting radio emission behaves according to the synchrotron bubble models already discussed.

The analogous situation in a twin-jet geometry, as suggested by SS433 (Hjellming and Johnston 1988), is based upon the assumption that there is a hot, X-ray twin jet of 10^7 K gas emitted perpendicular to an accretion disk, and the ~ 2000 km s^{-1} sound speed in these jets cause lateral expansion that produces a shock geometry in a conical sheath around the expanding, moving gas jet. The same turbulent acceleration, or trapped particle acceleration, mechanisms invoked for spherical shocks in the supernova remnant environment produce the relativistic electrons. Once the relativistic plasma has been produced, the properties of the radio-emitting regions can be described with variations on the previously discussed conical jet model.

While the radio outbursts of X-ray transients (Hjellming *et al.* 1988) have been attributed to the super-Eddington accretion rate, as indicated by the maximum X-ray fluxes, a super-Eddington accretion rate has been observed in Cir X-1 (assuming spherical accretion and isotropic X-ray emission) without a large radio outburst (Stewart *et al.* 1991). A small radio outburst could still have taken place.

7.6.2 Magnetospheric acceleration

Pulsars, which are rapidly rotating neutron stars with very strong magnetic fields, are known to accelerate relativistic electrons in regions near the polar caps of the neutron star magnetosphere (Lyne and Graham-Smith 1990). The acceleration process is not well understood, but can be considered to be empirically well established. Less spectacular, but possibly also analogous, are the particle acceleration processes that occur in the Earth's magnetosphere and the interaction regions between the Earth's magnetic field and the solar wind. These phenomena, plus the fact that the neutron star known in many X-ray binaries could be magnetized, makes it reasonable to consider magnetospheric acceleration for the relativistic electrons in radio-emitting X-ray binaries. In addition, the accretion disk itself is likely to have a magnetic field that threads the accretion disk, leading to the possibility of buoyant magnetic flux tubes. In addition, the Poynting flux along the axis of each accretion disk is a potential means of axial particle acceleration. Thus we are discussing objects where magnetospheres associated with either, or both, compact object and accretion disk may accelerate relativistic particles.

The main difference between pulsars and X-ray binaries which accelerate relativistic electrons is that the latter are 'cocooned' by denser gas environments because of either the wind from a high-mass companion or the mass transfer from a low-mass companion that has filled its Roche lobe. Thus, initially coherent beams of particles may interact with the surrounding medium, either to absorb the energy from the beamed particles, or to randomize the motions in mixtures of relativistic plasma, thermal gas and magnetic fields.

We do not have any developed models for the transmutation of coherent beams of particles into moving relativistic plasmoids, nor do we know whether this is possible. However, there is one observational signature that one might expect: erratic time variations. It is therefore very interesting that the strongest and most erratic time variability of most Z sources occurs when the X-ray spectrum is 'hardest' (horizontal branch) and the strongest radio flares occur. On this branch, the accretion rate is lower than in the other branches, resulting in a somewhat smaller gas pressure, which subsequently results in a somewhat larger neutron-star magnetosphere. It is unclear whether the hard X-ray emission in this branch is coming from the surface of the neutron star, or from Comptonization by a hot electron cloud (see, e.g., Schulz and Wijers 1989). The increase of time variability, hardening of the X-ray spectrum and, on average, increase of radio emission may be a signature of a situation asymptotically approaching that of the coherent beamed emission of a magnetized neutron star, but where the radiating regions are relativistic plasmoids activated by 'pulsar-like' coherent beams.

So far, no 'radio pulsar' type of emission (steep spectra; $\alpha \sim -4$) has been found from the X-ray binaries. One might have expected a radio pulsar when a low-mass X-ray binary shows no indication of mass transfer, and when the characteristics of the neutron star and its environment are similar to those of the millisecond pulsars (see Ch. 5).

7.7   Conclusions

Radio emission from various high- and low-mass X-ray binaries, which can be shown to be due to synchrotron emission processes because of time variation and/or linear polarization, is the most direct evidence that relativistic electrons are accelerated in the environment of these systems. In addition, bulk motion of relativistic plasma, with velocities of the order of $0.26c$ is seen or inferred in some X-ray binaries. Whether gas dynamical effects are primary, with acceleration of electrons to relativistic energies as a by-product of shock and trapping environments, or whether magnetospheres associated with either (or both) accretion disk and compact object are the primary sources of particle acceleration, is still uncertain. The evidence for SS433, a high-mass X-ray binary, argues for gas dynamical acceleration of particles because the radio emission does not appear until a finite distance has occurred in the jet flows. However, the Z-source low-mass binaries have a correlation of strongest, most variable radio emission when the accretion environment is at its lowest levels, indicating that magnetospheric processes may be primary, and that increases in the accretion environment obscure the magnetospheric processes.

The coupling of radio and X-ray behavior in the Z sources, during major state changes in other X-ray binaries, and in X-ray transients like V404 Cyg, may indicate more coupling than expected between these very different emitting environments with very different emission mechanisms. Campaigns to study multi-wavelength relationships in time, and very high resolution imaging (with trans- and inter-continental radio telescope arrays), are the most likely future sources of important new information about the radio emission from these systems.

References

Ables, J. G.: 1969, *Proc. Astron. Soc. Australia*, **1**, 237.

Baum, S.A. and Elston, R.: 1985, in *The Crab Nebula and Related Supernova Remnants*, eds. M.C. Kafatos and R.B.C. Henry (Cambridge University Press), p. 251.

Braes, L. L. E. and Miley, G. K.: 1971, *Nature*, **232**, 246.

Braes, L. L. E. and Miley, G. K.: 1972, *Nature*, **237**, 506.

Cooke, B. A. and Ponman, T. J.: 1991, *Astron. Astrophys.*, **244**, 358.

Dickel, J. R., Eilek, J. E., Jones, E. M. and Reynolds, S. P.: 1989, *Astrophys. J. Suppl.*, **70**, 497.

Fomalont, E.B. and Geldzahler, B.J.: 1991, *Astrophys. J.*, **383**, 289.

Garcia, M. R., Grindlay, J. E., Molnar, L. A., Stella, L., White, N. E. and Seaquist, E. R.: 1988, *Astrophys. J.*, **328**, 552.

Geldzahler, B. J.: 1983, *Astrophys. J.*, **264**, L49.

Geldzahler, B. J. *et al.*: 1983, *Astrophys. J.*, **273**, L65.

Ginzburg, V.L. and Syrovatskii, S.L. 1965, *Ann. Rev. Astron. Astrophys.*, **3**, 297.

Gregory, P. C. and Taylor, A. R. 1978, *Nature*, **272**, 704.

Gregory, P. C. *et al.* 1972, *Nature Phys. Sci.*, **239**, 114.

Grindlay, J.E. and Seaquist, E.R.: 1986, *Astrophys. J.*, **310**, 172.

Han, X.: 1993, *A Study of the Time Variable Radio Emission of X-ray Binaries and Transient X-ray Sources*, Ph. D. Thesis, New Mexico Institute of Mining and Technology, Socorro, New Mexico.

Han, X. and Hjellming, R. M.: 1992a, *Astrophys. J.*, **400**, 304.

Han, X. and Hjellming, R. M.: 1992b, *IAU Circ.* 5593.

Han, X. and Hjellming, R. M.: 1994, *Astrophys. J.*, in press.

Hasinger, G. and Van der Klis, M.: 1989, *Astron. Astrophys.*, **225**, 79.

Haynes, R. F., Komesaroff, M. M., Little, A. G., Jauncey, D. L., Caswell, J. L., Milne, D. K., Kesteven, M. J., Wellington, K. J. and Preston, R. A.: 1986, *Nature*, **324**, 233.

Hjellming, R. M.: 1973, *Astrophys. J.*, **182**, L29.

Hjellming, R. M.: 1978, *Astrophys. J.*, **221**, 225.

Hjellming, R. M.: 1988, in *Galactic and Extra-Galactic Radio Astronomy*, eds. G. L. Verschuur and K. I. Kellermann, (Springer-Verlag, New York) p. 381.

Hjellming, R. M. and Blankenship, L.C.: 1973, *Nature Phys. Sci.*, **243**, 81.

Hjellming, R. M. and Johnston, K. J.: 1981, *Astrophys. J.*, **246**, L141.

Hjellming, R. M. and Johnston, K. J.: 1985, in *Radio Stars*, eds. R. M. Hjellming and D. M. Gibson, (Reidel, Dordrecht), pp. 309–23.

Hjellming, R. M. and Johnston, K. J.: 1988, *Astrophys. J.*, **328**, 600.

Hjellming, R. M. and Wade, C. M.: 1971a, *Astrophys. J.*, **164**, L1

Hjellming, R. M. and Wade, C. M.: 1971b, *Astrophys. J.*, **168**, L21.

Hjellming, R. M., Calovini, T., Han, X.-H. and Córdova, F. A.: 1988, *Astrophys. J.*, **335**, L75

Hjellming, R. M., Han, X., Córdova, F. A. and Hasinger, G.: 1990a, *Astron. Astrophys.*, **235**, 147.

Hjellming, R. M., Han, X. and Roussel-Dupre, D.: 1990b, *IAU Circ.* 5112.

Hjellming, R. M. *et al.*: 1990c, *Astrophys. J.*, **365**, 681.

Hjellming, R. M., Han, X. and Roussel-Dupre, D.: 1994, *Astrophys. J.*, in preparation.

Johnston, K. J. *et al.*: 1986, *Astrophys. J.*, **309**, 707

Kellermann, K. I.: 1966, *Astrophys. J.*, **146**, 621.

Kesteven, M. J. and Turtle, A. J.: 1991, *IAU Circ.* 5181.

Kitamoto, S.: 1990, in *Accretion-Powered Compact Binaries* ed. C.W. Mauche, (Cambridge University Press), p. 21.

Königl, A.: 1983, *Monthly Notices Roy. Astron. Soc.*, **205**, 471.

Lyne, A. G. and Graham-Smith, F.: 1990, *Pulsar Astronomy*, (Cambridge University Press), pp. 184–210.

Machin, G., Lehto, H. J., McHardy, I. M. and Callanan, P. J.: 1990, *Monthly Notices Roy. Astron. Soc.*, **246**, 237.

Margon, B.: 1984, *Ann. Rev. Astron. Astrophys.*, **22**, 507.

Marscher, A. P. and Brown, R. L.: 1975, *Astrophys. J.*, **200**, 719.

Marti, J., Paredes, J. M. and Estalella, R.: 1992, *Astron. Astrophys.*, **258**, 309.

Molnar, L., Hjellming, R.M. and Taylor, A.R.: 1994, in preparation.

Owen, F. N., Balonek, T. J., Dickey, J., Terzian, Y. and Gottesman, S.: 1976, *Astrophys. J.*, **203**, L15.

Penninx, W.: 1989, *23rd ESLAB Symp. on Two Topics in X-ray Astronomy*, eds. J. Hunt and B. Battrick, Bologna, Italy 13–20 September, 1989, (ESA SP-296), pp. 185–96.

Penninx, W.: 1990, *X-ray and Radio Studies of Low-mass X-ray Binaries*, Ph.D. Thesis, University of Amsterdam, Amsterdam, Netherlands.

Penninx, W., Lewin, W. H. G., Zijlstra, A. A., Mitsuda, K., Van Paradijs, J. and Van der Klis, M.: 1988, *Nature*, **336**, 146.

Penninx, W., G. A. A., Van Paradijs, J., Van der Klis, M., Lewin, W. H. G. and Dotani, T.: 1993, *Astron. Astrophys.*, **267**, 92.

Preston, R. A., Morabito, D. D., Wehrle, A. E., Jauncey, D. L., Batty, M. J., Haynes, R. F. and Wright, A. E.: 1983, *Astrophys. J.*, **268**, L23.

Priedhorsky, W., Hasinger, G., Lewin, W. H. G., Middleditch, J., Parmar, A., Stella, L. and White, N. E.: 1986, *Astrophys. J.*, **306**, L91.

Schulz, N. S., and Wijers, R. A. M. J.: 1989, *23rd ESLAB Symp. on Two Topics in X-ray Astronomy*, eds. J. Hunt and B. Battrick, Bologna, Italy 13–20 September, 1989, ESA SP-296, pp. 601–06.

Seaquist, E. R. *et al.*: 1974, *Nature*, **251**, 394.

Spencer, R. E., Swinney, R. W., Johnston, K. J. and Hjellming, R. M.: 1986, *Astrophys. J.*, **309**, 694.

Stewart, R. T., Nelson, G. J., Penninx, W., Kitamoto, S., Miyamoto, S. and Nicolson, G. D.: 1991, *Monthly Notices Roy. Astron. Soc.*, **253**, 212.

Strom, R., Van Paradijs, J. and Van der Klis, M.: 1989, *Nature*, **337**, 234.

Tan, J., Lewin, W. H. G., Hjellming, R. M., Penninx, W., Van Paradijs, J. and Van der Klis, M.: 1991, *Astrophys. J.*, **385**, 314.

Tananbaum, H. *et al.*: 1972, *Astrophys. J. (Lett.)*, **177**, L5.

Taylor, A. R. and Gregory, P. C.: 1982, *Astrophys. J.*, **255**, 210.

Van der Laan, H.: 1966, *Nature*, **211**, 1131.

Vermeulen, R.: 1989, *Multi-wavelength Studies of SS433*, Ph.D. Thesis, Rijksuniversiteit Leiden, Netherlands.

Vermeulen, R. C., Schilizzi, R. T., Icke, V., Fejes, I. and Spencer, R. E.: 1987, *Nature*, **328**, 309.

Wagner, R.M. *et al.*: 1990, in *Accretion-Powered Compact Binaries*, ed. C.W. Mauche (Cambridge University Press), p. 29.

Watson, M. G., Stewart, G. C., Brinkmann, W. and King, A. R.: 1986, *Monthly Notices Roy. Astron. Soc.*, **222**, 261.

Whelan, J. A. J., Mayo, S. K., Wickramasinghe, D. T., Murdin, P. G., Peterson, B. A., Hawarden, T. G., Longmore, A. J., Haynes, R. F., Goss, W. M., Simons, L. W., Caswell, J. J., Little, A. G. and McAdam, W. B.: 1977, *Monthly Notices Roy. Astron. Soc.*, **181**, 259.

Zheleznyakov, V. V.: 1970, *Radio Emission of the Sun and Planets* (Pergamon Press, Oxford).

8

Cataclysmic variable stars

France Anne-Dominic Córdova
The Pennsylvania State University, Department of Astronomy & Astrophysics,
525 Davey Laboratory, University Park, PA 16802, USA

8.1   Introduction

8.1.1   *What is a cataclysmic variable?*

It was on the basis of the continual eruptive behavior of dwarf novae and classical novae that the Gaposchkins invented *c.* 1960 the term by which all accreting white-dwarf stars in close binary systems are known, the 'cataclysmic variables'. Although the first dwarf nova, U Geminorum, was discovered in 1855 (Hind 1856), the binarity of the cataclysmic variables and the picture of transfer of material via an accretion disk onto a degenerate star was not worked out until a century later. Kraft (1990) gives a delightful reminiscence of the period from 1950 to 1965, when observational astronomers and theoretical physicists formed our 'modern' picture of these stars.

A cataclysmic variable star (CV) is a semi-detached binary system comprising an accreting degenerate star and its mass-donating companion star. The latter is usually, but not always, a late-type star near or on the main sequence. The binary orbital periods for such a system are most commonly between $\sim$ 80 minutes and several hours. The entire binary could fit easily inside a star as large as our Sun. A few cases are known where the companion is a giant star or another degenerate dwarf. A compendium of binary orbital parameters for $\sim$ 180 CVs and many other related binaries can be found in Ritter (1990) and in Ch. 15. The space density of each of the principal subclasses of CV (i.e., dwarf novae, novalike objects, and classical novae) is estimated to be a few $\times$ 10^{-7} pc^{-3} by Patterson (1984) and by Downes (1986). This estimate is 100 times lower than that derived by Hertz *et al.* (1990) from the detection of CVs in an X-ray survey of the Galactic plane.

One of the main distinctive features between the subclasses is the geometry of the accretion flow. In some CVs (called 'polars' or AM Herculis stars), the white dwarf has such a strong magnetic field ($B \sim$ few $\times$ 10 MG) that the rotation of this star is phase-locked with the orbit of the binary; material flows along (presumably) dipolar field lines from companion to degenerate star. In other magnetic CVs, the white dwarf is spinning more rapidly than the binary orbital period. These are usually called the 'intermediate polars' or DQ Herculis stars, although the latter designation is reserved by some researchers only for the CVs with the faster coherent periods, namely DQ Herculis itself and AE Aquarii. Astronomers argue over the strength of the magnetic field in intermediate polars (i.e., whether it is of the same strength as the fields in the AM Her stars or an order of magnitude less), and there is, consequently, a debate

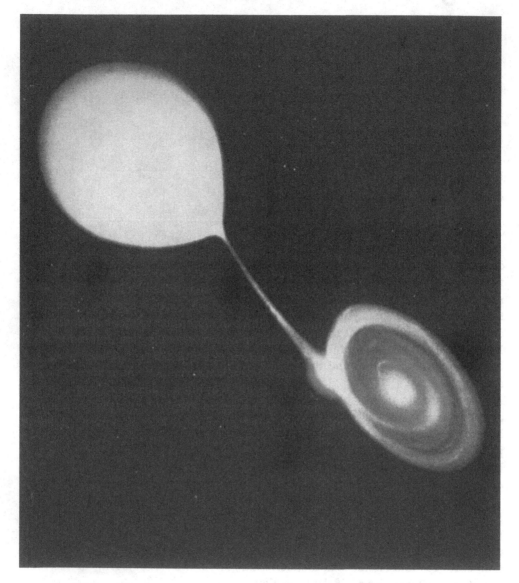

Fig. 8.1. Depiction of a disk-accreting dwarf nova, by Dana Berry (STScI).

over whether or not partial disks can form in these binaries (see Ch. 10). Both
subclasses of magnetic CVs exhibit long-term high- and low-mass-accretion states,
and some intermediate polars (e.g. GK Persei, EX Hydrae, TV Columbae) show
irregular, small-amplitude outbursts in luminosity on a more frequent time scale.

By far the longest-known, and therefore best-studied, CVs are the dwarf novae
and the classical novae. In the former subclass at least, the companion star overfills
its Roche lobe and the high angular momentum material forms a disk around the
degenerate star. Figure 8.1 is a depiction of such a disk-accreting CV. In these systems,

the white dwarf's magnetic field is not strong enough to keep at least an outer disk from forming. The transferred material will impact the disk at its outer edge, causing a luminosity enhancement, or 'bright spot', in a specific range of orbital phases in the photometric light curve; sometimes the bright spot manifests itself as an '*S* curve' in the emission line radial velocity curve, due to its Doppler variations with orbital phase. An example of a light curve with a bright spot hump is shown in Sect. 8.3.4, and an *S* curve is illustrated in Sect. 8.2.1.2. While sometimes prominent at optical wavelengths, the bright spot is 'hot' only relative to the quiescent outer disk: Wood *et al.* (1989) measure the bright-spot temperature in OY Carinae to be 15 000 K, in contrast to the quiescent outer disk temperature of 4000 K. The orbital bright spot is not discernable during states of high mass accretion. A typical dwarf nova brightens by 2–5 magnitudes on a fairly regular, though not periodic, time scale of weeks to months. There are a number of subtypes of dwarf novae, each of which has a different characteristic outburst morphology; representative outburst light curves for each subtype are illustrated in Sect. 8.3.1. Among the dwarf novae no AM Her star has yet been discovered, yet at least one dwarf nova (SW Ursa Majoris) shows in quiescence alone the pulsed signature that is characteristic of the intermediate polars (Shafter, Szkody, and Thorstensen 1986).

A classical nova outburst represents a much different phenomenon. The optical brightness increases by up to 20 magnitudes; by definition, only one such outburst is witnessed. Estimates of the recurrence time of a classical nova range between 3300 and 10^4 years (Patterson 1984; Downes 1986). While at one time classical novae were thought to all be disk-accreting, more recent observations (of, e.g., GK Persei and V1500 Cygni) reveal that strong magnetic fields are present in some systems, preventing the formation of, at least, an inner disk. There is ongoing research on the effect of the type of accretion flow (magnetic or disk) on the classical nova outburst. It is possible that most CVs, magnetic and disk-accreting, produce a classical nova explosion when enough accreting material has accumulated to produce a thermonuclear runaway in the envelope of the degenerate star, and at least the details (e.g. rise and decay times) may depend on the magnetic field strength.

Another type of CV is the recurrent nova, whose outbursts on the time scale of decades make it appear intermediate between the dwarf novae and classical novae. This subclass comprises a mixture of systems; some have giant companions, rather than main-sequence companions, and are long-period binaries. Closely associated with the latter are the symbiotic stars, in which the accreting object is usually a main-sequence star rather than a white dwarf. The outbursts of the various kinds of recurrent novae are thought to be powered by different mechanisms depending on the nature of the accreting object, that is by accretion events onto main-sequence stars or thermonuclear outbursts on white-dwarf stars.

The novalike objects, so named for their spectral and photometric similarities to the classical novae and dwarf novae, but having no observed outbursts, are a mixed bag of subtypes. This nomenclature developed before magnetism was discovered in CVs. Many novalike objects look like dwarf novae in permanent outburst, ie luminous disk-accreting systems like UX Ursae Majoris, but others, such as AM Herculis and TT Arietis, are magnetic systems that have long-duration high and low states. Another subtype, the VY Sculptoris stars, have high and low states, but do not have the classical signatures of a magnetic field (polarization or coherent pulsations). As our knowledge of individual systems grows, more subtypes are proposed, each representing a new behavior that we do not understand. The most recent example is the SW Sextantis stars (Sect. 8.2.1.4). These exhibit significant displacements of the phase of the emission lines with respect to the eclipse light curve and have emission lines without the classical rotational feature (double-line profile) associated with a disk. In addition, their line profiles suggest substantial flux (radial inflow or outflow) perpendicular to the disk. It is postulated that magnetic fields may influence this behavior. More than one-third of all novalikes show coherent optical or X-ray pulsations, which may signify that magnetic accretion is important in these systems.

In reviews of the 1970s and early 1980s, the mass-accretion rate was held to be the single most important variable determining the properties of CVs; today we find that magnetic fields, as well as the mass ratio of the component stars, may be at least as important in defining the behavior of these complex systems.

8.1.2   CVs and their applications

Studies of cataclysmic variables are integral to diverse areas of research concerned with the behavior, structure, and evolution of a broad range of accreting objects. For example:

- Disks are ubiquitous among systems small and large, new and old: they were important in the formation of the early solar nebula out of which planets were formed, and disks or rings presently accompany the larger outer planets. Disks have been observed in newly forming stars and the relatively young Be–neutron star binaries. Disks may provide much of the light of active galactic nuclei (AGN). There is probably no handier place to study the physics of accretion disks than in CVs. This is because of their relatively high space density (allowing a broad sample to be investigated), their proximity to Earth (allowing an unimpeded view of the high-energy emission from the disk's center), the fact that most of the disk luminosity emerges in the relatively accessible optical and ultraviolet wavelengths (in contrast to the neutron star binaries whose disks must be examined at much higher, less accessible, energies from Earth), and their binarity (allowing a determination of most of the salient physical parameters of the system, including the masses and sizes of the individual stars and the orientation of the disk to the line of sight). The study of the spectrum of the dwarf nova's disk as a function of accretion rate has led to models for the evolution of disks and various ideas about the source of the disk's viscosity. The effect of tides on a disk can be examined in at least one subclass of CVs, the 'SU UMa stars' (see Sect. 8.3.1.5), in which the presence of the companion star is thought to cause the disk to precess during high-mass-accretion states, resulting in regular variations in the disk's luminosity.
- The various kinds of luminosity outbursts of CVs have given us whole new fields of inquiry concerning instabilities in disks and mass transfer instabilities from red stars. These may be applicable to soft X-ray transients associated with neutron star or black-hole binaries and may also be relevant to transient luminosity enhancements in AGN.

- The study of CVs yields many clues to the nature of features in the light curves of accreting, low-mass neutron star X-ray binaries (LMXBs): the broad, phase-related dips in the light curve; the rapid, transient, quasi-coherent oscillations; the short bursts of radiation; and the unpredictable transient outbursts of long duration.

- Radial outflow in the form of high-velocity winds make CVs a laboratory for examining theories of the radiation-driven, line-accelerated winds in OB stars, as well as production of X-rays in these systems. Studies of CV winds may also be applicable to the winds or jets of young stars, symbiotic stars, and possibly the He-peculiar main-sequence stars (Sect. 8.2.2.7).

- CVs give us a different view on white-dwarf structure from that provided by the study of isolated white dwarfs. CVs may provide a way for the white dwarf to grow to the Chandrasekhar limit under certain conditions of white-dwarf mass and accretion rate. They then may suffer an accretion-induced collapse and become low-mass X-ray binaries or, in a competing view, supernovae (cf., Wheeler 1990; Nomoto and Kondo 1991).

- CVs can place vital constraints on evolution in globular clusters; they have been implicated, by means of accretion-induced collapse, in the formation of millisecond pulsars in globular clusters (Bailyn and Grindlay 1990). This work, as well as detections of low-luminosity X-ray sources in globulars (Hertz and Grindlay 1983) and scenarios for capture of stars by white dwarfs in the cores of globulars, have given renewed vitality to searches for CVs in these clusters, with some recent success (Cool *et al.* 1993).

- CVs may contribute substantially to the hard X-ray 'Galactic Ridge' (Hertz and Grindlay 1984), whose unresolved emission can be characterized as optically thin thermal bremsstrahlung with 6.7 keV iron line emission (Koyama *et al.* 1986, using Japan's *Tenma* satellite).

- By virtue of the discovery of subclasses of strongly magnetic CVs, these objects now provide a critical tool for examining the effect on the accretion flow of the presence of a strong magnetic field. Even CVs with weaker magnetic-fields provide clues to magnetically related phenomena like outbursts and winds in other magnetic systems (e.g. planets with variable radio outbursts; the young T Tauri stars, which manage bipolar outflows without the benefit of a hot, radiative source; and the magnetic He-peculiar stars).

- Novae ejecta provide clues about the evolutionary history of white dwarfs prior to their eruptions. Novae explosions may be the origin of odd isotopes of abundant elements in the ISM. Novae provide a place to study gas dynamics and the formation and growth of grains in the novae ejecta. They provide standard candles for the extragalactic distance scale because they are luminous at maximum and their rates of decline are correlated well with their absolute magnitudes at maximum (Livio 1992b).

8.1.3 *The scope of this review*

It would take a (large!) volume, rather than one chapter in a book, to satisfactorily review the wealth of information that has been discovered both observationally and theoretically about CVs. The early excellent reviews of Robinson (1976) and Warner (1976) pre-dated the high-energy observations that have yielded so much information about the inner disk and accreting star, the discovery of magnetic CVs, and the variety of disk instability models that were developed in the 1980s. Because of the rapid expansion of the field, it was natural that subsequent reviews concentrated on particular aspects of CVs. Córdova and Mason (1983) emphasized the high-energy (X-ray and ultraviolet) behavior of CVs, and Córdova and Howarth (1987) provided a comparative overview of the ultraviolet properties of CVs and neutron-star binaries, as viewed with the *International Ultraviolet Explorer* satellite. Patterson (1984) formed a compilation of binary data on CVs, and his interpretation of this material greatly influenced subsequent work in the field. The more fundamental and nonvariable aspects of CVs are summarized by Wade (1985), while Wade and Ward (1985) provide a wide-ranging overview of observed behavior of all types of CVs. Much of the optical work summarized by Horne (1991) is based on his seminal attempts to image the disks in these binaries, using eclipses to map the brightness temperature of the disk and Doppler tomography to attempt to map the locations of the emission lines.

An invaluable catalog of the binary parameters of CVs, LMXBs and related systems has been compiled by Ritter (1990), and Duerbeck (1987) has assembled a comprehensive atlas of $\sim$ 200 classical novae. A published conference proceedings, *Accretion-Powered Compact Binaries*, provides a multi-author collection of recent work in the field (Mauche 1990). Accretion disk theory and structure is explained to some extent in most of the above-mentioned reviews, but nowhere more elegantly than by Pringle (1981), who starts with early solar nebular theory. More recent theoretical developments concerning accretion disk physics are detailed in an excellent tutorial by Livio (1993b) and by Andrew King in this volume (Ch. 10).

My intention here is to give an overview of many of the most important aspects of CVs that have been discovered in the past 15 years and to point the interested reader to the considerable literature for detailed information. My approach in Sect. 8.2 is a multispectral one: I show how observations and modeling of a variety of variable spectral and photometric behavior at all wavelengths contribute to our understanding of the emissions of all the components of the system. In Sect. 8.3, I review many of the variable phenomena observed in CVs along with the models developed to explain these. These phenomena include the dwarf nova outburst, the flickering, the rapid oscillations and high-velocity winds observed in luminous CVs, and the humps, dips, and bursts observed in all manner of CVs. I briefly review in Sect. 8.4 the change in the field wrought by the discovery of large magnetic fields in CVs. I discuss angular momentum losses and CV evolution and lifetimes in Sect. 8.5. Finally, in Sect. 8.6, I indicate many directions for future work.

Although I make a number of mentions of the magnetic CVs, including devoting one section to their behavior, I do not discuss them in great detail because the polar subclass, in particular, has been reviewed comprehensively by Cropper (1990). An early review of intermediate polars is given by Warner (1983); the X-ray properties of these objects are reviewed by Osborne (1988) and modeled by Mason, Rosen, and Hellier (1988), and Norton and Watson (1989). I make only a few comparative references to classical novae. There exists an extensive literature on novae that is reviewed by Shara (1989); a book of reviews on novae edited by Bode and Evans (1989); and a recent conference proceedings, *Physics of Classical Novae*, edited by

Casatella and Viotti (1990). Starrfield (1990) provides a multiwavelength overview of classical novae in outburst.

8.2 What multiwavelength observations reveal

The era of the 1950s was 'a time before those terrible nanometers, when all that was right and holy in the world lay between 3500 and 7000 Å'. Robert Kraft, 1990.

Cecelia Payne-Gaposchkin titled her Russell Prize Lecture, published in 1977, 'Fifty years of novae', By novae she meant the four then-known species of novae: dwarf novae, recurrent novae, classical novae and novalike objects. The discoveries of the last 15 years have significantly altered our world view of the behavior and emissions of cataclysmic variables. Many of these findings owe their origin to investigations in regions of the electromagnetic spectrum that were previously inaccessible. Other revelations have come about because of the introduction of new observing capability and analysis methods in the 'right and holy' optical waveband.

8.2.1 *Optical and UV studies of the accretion disk and white dwarf*

8.2.1.1 *The steady-state disk*

There is ample evidence (from, for example, the total luminosity of CVs and the high-energy radiation that is produced) that accretion is taking place. There is also evidence, especially from some eclipsing systems, that there exists an extended disk around the accreting star. Steady-state disk theory assumes that viscosity (of an unspecified form and magnitude) drives the mass transport from a Roche lobe filling star onto the white dwarf by means of a disk (see detailed discussions in Bath and Pringle 1985 and Ch. 10). The viscosity adjusts itself to provide a steady mass flux, hence the disks are called 'steady'. They are called 'α'-disks because the efficiency of the momentum transport is characterized by the parameter α. For turbulent viscosity, $\alpha = \nu/c_s H$, where ν is the effective viscosity, c_s is the sound speed in the disk, and H is the disk's half-thickness.

In the simplest version of the steady-state model, the spectrum of the radiation and the disk temperature are not dependent on α, but the time scale of a mass transfer event, in which the transport changes significantly, is. Thus, variable events such as the outbursts of dwarf novae offer an opportunity to model the magnitude of α.

The original steady-disk theory of Shakura and Sunyaev (1973) assumes radial, thin disk elements that radiate as blackbodies of varying annular radius. The effective disk temperature can be shown to be

$$T(R) = T_*(R/R_{WD})^{-3/4}[1 - (R_{WD}/R)^{1/2}]^{1/4},$$

where T_*, the maximum temperature in the disk, is given by

$$T_* = 4.1 \times 10^4 (\dot{m}/10^{16}\text{g s}^{-1})^{1/4}(M_{WD}/M_\odot)^{1/4}(R_{WD}/10^9\,\text{cm})^{-3/4}\ \text{K}.$$

The total accretion luminosity generated by the flow of material from the companion star to the white dwarf is

$$L_{accr} = \frac{GM_{WD}\dot{m}}{R_{WD}}.$$

It can be shown (e.g. Bath and Pringle 1985) that the luminosity of the steady-state disk is one-half of the total luminosity if $\Omega_* \ll \Omega_{R(WD)}$. The balance of the accretion energy is deposited in the boundary layer (BL), i.e.

$$L_{BL} = L_{disk} \left\{ 1 - \left[\frac{\Omega_*}{\Omega_{R(WD)}} \right] \right\}^2 ,$$

where Ω_* is the white dwarf's angular velocity, and $\Omega_{R(WD)}$ is the Keplerian angular velocity at the white dwarf's surface.

Thus, for a slowly rotating white dwarf, the BL luminosity is approximately equal to the disk luminosity. The temperature of an optically thick BL is of order $10^5 - 10^6$ K (Pringle 1977).

Thus, in steady-disk theory, the disk makes a contribution to all parts of the spectrum between the near infrared and the far ultraviolet, while the BL can contribute approximately an equal amount of radiation in the extreme ultraviolet (EUV)/soft X-ray spectral region. Other sources of luminosity in the system are the companion star, which contributes chiefly to the infrared portion of the spectrum, the optical bright spot (a few times hotter than the disk's outer edge), and various nonsteady-state sources like winds and optically thin regions above the disk and BLs.

8.2.1.2   *Imaging the accretion disk: eclipse mapping and Doppler tomography*

Eclipses of the disk and its central regions by the companion star have proved extremely useful for measuring basic parameters of the system, i.e. the mass ratio, inclination, and size of the disk and bright spot. From the duration of the ingress and egress of the white-dwarf eclipse, the masses of both stars can be measured, assuming a mass–radius relation for the white dwarf, and the radius of the companion star can then be derived (see, for example, Horne, Wood, and Stiening 1991). Yet the eclipses of CVs can be used also in an extremely powerful way to yield images of the 'two-dimensional' disk.

The techniques of *eclipse mapping* and *Doppler tomography* have been used in the last few years to attempt to construct two-dimensional images of the surface brightness and emission line regions, respectively, of accretion disks. Eclipse mapping, which makes use of high-speed (~ 1 s time resolution), multicolor photometry of eclipsing systems, consists of computing an eclipse light curve for any given distribution of brightness on the face of the disk and then selecting, with maximum entropy techniques, the most nearly axisymmetric map that fits the eclipse data. This technique was developed by Horne (1985) and has been applied to several CVs in high- and low-mass-accretion states. Its application to systems accreting at high rates has confirmed the steady-state optically thick disk model in these cases, e.g. for the dwarf nova Z Cha during a decline from optical outburst (Horne and Cook 1985), and the novalike star RW Trianguli (Horne and Stiening 1985). In particular, the surface temperature satisfies the relationship that it is proportional to the radius to the $-3/4$ power (see Ch. 10).

Eclipse mapping of quiescent dwarf novae, however, yields brightness temperature profiles that cannot be fit by the steady-state, optically thick disk model. The profiles of these disks look flat, revealing a disk that is optically thin at all radii. This has been demonstrated for Z Cha by Wood *et al.* (1986); for OY Car by Wood *et al.* (1989); and for HT Cas by Wood, Horne, and Vennes (1992). The latter authors showed that HT Cas's disk emission is characteristic of a partially ionized, optically thin gas with a flat brightness temperature profile of 5000–7000 K. It is unlikely that viscous dissipation is heating the gas because the viscosity parameter would be too high ($\alpha = 10$–200).

While it is possible to measure a mass accretion rate for the high-state CVs (and this number is of order $5 \times 10^{-8} M_\odot$ yr^{-1}), it is not possible to estimate the mass accretion rate for the low-state CVs because the steady-disk model does not apply. The mass injection rate into the bright spot can be estimated, however, and this is $\sim 5 \times 10^{-11} M_\odot$ yr^{-1}.

Figure 8.2 is a quartet of figures illustrating the steps involved in using eclipses to map the surface temperature distribution of OY Carinae's accretion disk during quiescence. The final figure shows that the computed temperature distribution of OY Car's quiescent disk does not fit a steady-disk model; a comparison of the data with the illustrated models for different mass accretion rates shows well the difficulty in estimating the mass accretion rate for quiescent CV disks.

A technique borrowed from medical imaging, Doppler tomography, allows imaging of the accretion flow in two dimensions. The velocity profile of a line observed at a given binary phase measures a projection of the accretion disk's velocity field onto the line of sight. The rotating binary therefore supplies projections along different lines of sight. For an essentially two-dimensional, disk-like flow, the velocity field can be recovered fully from data consisting of the velocity profiles as a function of binary phase. A nice example is the work of Marsh *et al.* (1990) for U Gem. Figure 8.3 shows their result: (a) the observed intensity of the H β and He II λ4686 lines as a function of orbital phase and velocity relative to line center; (b) the Doppler images fitted to these data; and (c) the fits attempting to duplicate (a), as computed from the images. One can easily see the prominent S wave in the narrow emission line components. The bright spot is the dominant contributor to the line flux, although there is an additional Hβ component from a ring, which is the disk emission, and possibly also from the (irradiated?) face of the companion star. The disk is not detected in He II λ4686. It should be noted that, although a number of CVs are thought to have prominent bright spots because of an orbital hump at the apparently correct phase, few actually show, as U Gem does, S wave emission from their bright spots (cf., Livio 1993a).

Doppler maps of the quiescent accretion disk yield a Balmer line emissivity profile that follows an $R^{-3/2}$ power law, i.e. proportional to the local Keplerian rotational frequency. Horne and Saar (1991) note that the correlation between emission line surface brightness and rotation also holds for the Hα and Ca II H and K emission from the active chromospheres of rotating stars. They therefore suggest that a similar dynamo mechanism may power the lines in both classes of object. This fits into other work which favors the idea that emission lines arise in a disk chromosphere, one idea that gets out of the long-standing problem of how to generate emission lines when the underlying disk is optically thick.

8.2.1.3 Modeling the disk spectrum

The mapping of the disk's temperature structure that is described above was done with optical data alone. Data have been combined over a wider energy interval to measure the disk's overall spectrum, to test models for the multifrequency

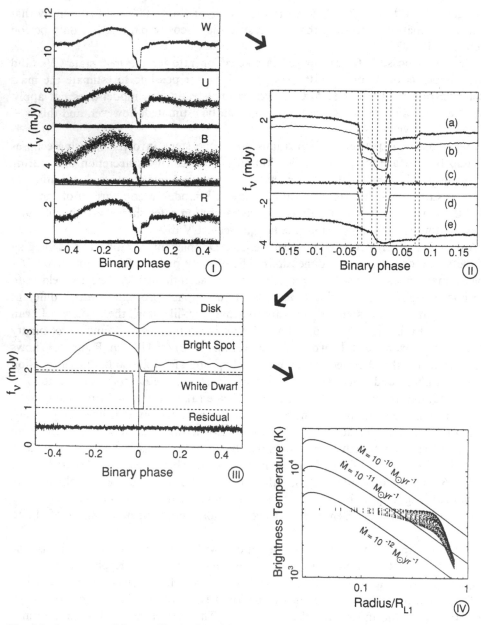

Fig. 8.2. A quartet of figures illustrating the steps involved in using eclipses to map the surface temperature distribution of OY Carinae's accretion disk during quiescence. (I) Averaged white-light light curves of one orbit of this eclipsing dwarf nova. Each light curve is offset by 3 mJy. (II) The extraction of the white-dwarf eclipse light curve from the mean light curve. (a) Original curve; (b) smoothed light curve, offset by −0.5 mJy; (c) derivative of smoothed light curve with spline fit, offset by 1 mJy; (d) reconstructed white-dwarf eclipse, offset by −2.5 mJy; (e) original light curve minus white-dwarf curve, offset by −4 mJy. The vertical dotted lines show the contact phases of the white dwarf and bright spot. (III) The decomposition of the mean white-light light curve. The separated components of the disk, bright spot, and white dwarf are depicted; the dotted lines show the zero levels for each. (IV) The brightness temperature of the disk, shown with models of steady-state optically thick disks for three different mass transfer rates. All figures adapted from Wood *et al.* (1989).

Hβ   HeII 4686

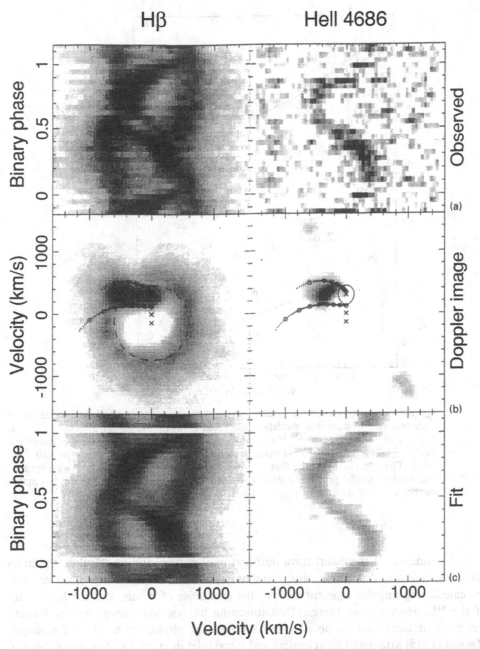

Fig. 8.3. Doppler tomography of two emission lines in U Geminorum during quiescence. (a) Observed spectra, plotted as a function of binary orbital phase and velocity relative to line center. (b) Doppler images fitted to these data. (c) Fits computed from the images. The reference circles have a radius of 600 km s^{-1}. From Marsh *et al.* (1990).

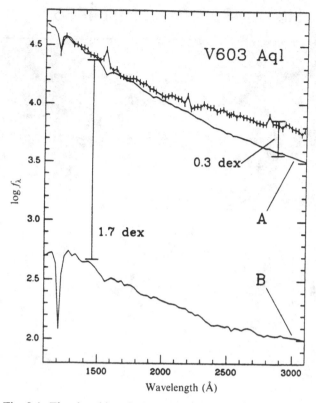

Fig. 8.4. The dereddened ultraviolet spectrum of the ex-nova V603 Aquilae together with two steady-state disk models constructed from model stellar atmospheres. The flux density is in units of 10^{-13} erg cm^{-2} s^{-1} $Å^{-1}$. The main difference between the models is a factor of ~ 100 in mass accretion rate (the top curve is for $\dot{m} \sim 10^{-8} M_\odot$ yr^{-1}). The top curve fits the flux at 1460 Å, but is too 'blue' at longer wavelengths. The bottom model provides a good match to the colors of the CV, but fails to match the object's flux by a large amount. From Wade (1988).

time dependence of the dwarf nova outburst light curves, and to calculate the mass accretion rate and disk luminosity. The mass accretion rate is an important variable for calculating angular momentum loss, the energetics of winds, and the luminosity of the BL, among other things. Disk modeling has yielded ambiguous results and has made it clear that we do not yet have the disk physics in hand. For example, Hassall (1985) attempted to fit optical and ultraviolet data on EK TrA during one of its dwarf nova outbursts. A good fit is obtained with stellar atmosphere models for either the optical or the ultraviolet, but not both simultaneously. Hassall found that the addition of some recombination radiation from an optically thin gas at 10^4 K made the fit a reasonable one at the Balmer jump. This illustrates not that the selected model was the right one, but that no simple blackbody or stellar atmosphere model will fit the entire disk continuum. Woods, Drew, and Verbunt (1990) find that the UV continuum flux can vary much more markedly than the optical flux, suggesting that they come from different regions in the binary, and this is not represented in any of the usual spectral models.

Wade (1988) has compared the UV fluxes and UBV colors of nine novalike stars (thought to be steady-state disk-accreting systems) with disk models comprising a sum of Planck functions or stellar atmospheres. On a color–color diagram, most of the observations lie closer to Kurucz stellar atmosphere models than to the Planck disk models. Accretion rates near 10^{-10} or $10^{-9} M_\odot$ yr^{-1} are suggested on the basis of color by the Kurucz models, yet rates near $10^{-8} M_\odot$ yr^{-1} are suggested for the same systems by the blackbody disk models, illustrating the importance of the model in determining the parameters of the system. Wade finds, however, that the stellar atmosphere model cannot explain both the observed fluxes and colors, and concludes that this model does not encompass the physics of CV disks. The problem is illustrated in Figure 8.4, which shows two model stellar atmosphere fits to the classical nova V603 Aql: Model A, which fits the flux at 1460 Å, gives an accretion rate of $10^{-8} M_\odot$ yr^{-1}, while Model B, which fits the colors of the source, gives an accretion rate of $10^{-10} M_\odot$ yr^{-1}. While the Planck models do, in a statistical sense, pass the criteria established by Wade for fitting the colors and fluxes, they cannot account for the absorption Balmer jump observed in one system. In addition, blackbody models, Wade points out, are unphysical in a disk where there must exist a temperature gradient in the vertical direction to drive radiative losses. This gradient must be shallower than a stellar disk atmosphere since the Planck disk is a better representation of the data than the stellar atmosphere disk. Wade notes that, given the lack of knowledge of the true disk spectrum, one cannot rely on model fitting to derive accurately the mass accretion rate and luminosity of any system, even when the distance to the system is known well. Presently there is a concerted effort to make new, self-consistent disk models, taking into account the effects of a possibly variable gravity, irradiation of the disk, energy deposition into the disk's upper layers, and the contribution of the BL radiation. Shaviv and Wehrse (1991) have published a model to calculate the continuum radiation from disks using a method that solves the disk equations in a self-consistent way. Their model successfully fits the continua of two novalike objects, IX Vel and RW Sex, over a broad wavelength range (see also Hubeny 1989).

8.2.1.4 *When a disk is not present: the link between novae, magnetic CVs, and SW Sex stars*

Astronomers performing Doppler tomography on CVs assume that a disk exists and that it is axisymmetric. Yet Williams (1989) gives a number of examples of emission line CVs that, when observed through eclipse, fail to show evidence for the rotational signature characteristic of a disk. The lines appear to have a velocity component radial to the disk, in addition to some rotational component. All of Williams's examples are classical novae or novalike objects; among the latter are some objects in the SW Sex subclass. Williams explains their emission lines as originating in polar accretion columns (radial infall) or winds. Williams reviews the recent evidence that indicates that strong magnetic fields may be of importance to the nova phenomenon. For example, DQ Her (Nova Her 1934) is probably an intermediate polar, and V1500 Cyg (Nova Cygni 1975) is an asynchronous magnetic system. The novae, magnetic CVs, and SW Sex stars show strong He II emission

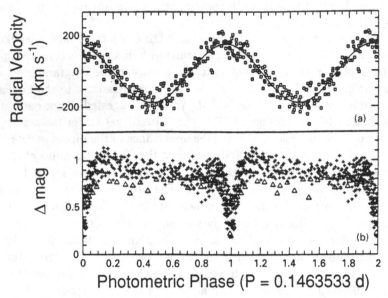

Fig. 8.5. (a) Radial velocity for Hα line versus orbital phase (repeated once for clarity) for the SW Sex star, PG20027+260. (b) V-band light curve, differenced with a comparison star. If the line velocities represented the motion of the white dwarf, the zero crossing toward negative velocity would occur at phase 0 and its integer increments. From Thorstensen *et al.* (1991).

compared to the average nonmagnetic dwarf nova. Magnetic fields may play a unifying role in determining the accretion flow in all of these systems (see Sect. 8.4).

8.2.1.5   *The mass of the white dwarf from optical emission line studies*

How reliable are estimates of the masses of the white dwarfs in CV systems? For the white dwarf, the measure often depends on getting an accurate value for K_1, the radial-velocity semi-amplitude of the white dwarf. R. Webbink (private communication, 1990) has derived white-dwarf mass estimates for 84 CVs from the literature (but see the discussion of selection effects in Ritter *et al.* 1991); the weighted logarithmic mean mass is $0.74 \pm 0.04 M_\odot$, but of novae alone it is $0.91 \pm 0.06 M_\odot$. Twenty-two of the 84 CVs contain white dwarfs more massive than $1 M_\odot$. In contrast, the mean mass of single white dwarfs is $0.56 \pm 0.1 M_\odot$, and high masses among them are rare (Bergeron, Liebert, and Saffer 1992). However, there is considerable doubt about the validity of using this estimator because, in some systems like the SW Sex subclass, the emission line radial velocities are not phased with respect to the eclipse light curve in such a way as to inspire confidence that the emission lines represent the motion of the white dwarf or inner disk (see Wade and Horne 1988; Thorstensen *et al.* 1991). Figure 8.5 provides a good illustration of the difficulty.

High-speed photometry of eclipses are potentially a very good way to derive, from the contact times that mark the white dwarf and bright spot ingress and egress, the mass ratio, inclination, and disk radius. These lead, with some assumptions, to the masses of the component stars. In the few CVs that are eclipsing, double-line spectroscopic binaries, redundant observational contraints, however, do not always lead to a consistent picture. For Z Cha, for example, there was a 3σ discrepancy between the masses calculated spectroscopically (Wade and Horne 1988) and the masses derived from eclipse data (Wood *et al.* 1986). One interpretation of this discrepancy was that the mass-accreting star is 50% larger than a Hamada–Salpeter white dwarf, and the velocities in the outer disk are sub-Keplerian by 25% (Wade and Horne 1988). More recently, Wood and Horne (1990) reduced the discrepancy to 2σ by better fitting the eclipse light curves.

8.2.2 *High-energy emission and the inner disk*

8.2.2.1 *Background*

The late 1970s represent a turning point in our knowledge about CVs. Payne-Gaposchkin's 1977 Russell Prize lecture on CVs was given on the eve of the launch of the *International Ultraviolet Explorer* satellite (January 1978) and the series of NASA's High-Energy Astrophysical Observatories. *HEAO*-1, launched in August of 1977, and *HEAO*-2 (also known as the *Einstein X-ray Observatory*), launched in November of 1978, would give astronomers their first view of the habitat of the white dwarf, the inner accretion disk. The expectation for discovery was high: a sounding rocket carrying low-energy X-ray proportional counters had detected SS Cyg (Rappaport *et al.* 1974) and the *SAS*-3 X-ray satellite had detected a little-known novalike star, AM Her (Hearn, Richardson, and Clark 1976). Soon thereafter, the all-important finding was made, from optical circular polarization studies, that AM Her is a highly magnetic binary (Tapia 1977). SS Cyg was detected also as a hard X-ray source during quiescence using the Dutch satellite *ANS* (Heise *et al.* 1978). The old nova GK Per was detected as a hard X-ray source during a dwarf novalike outburst using the *Ariel V* satellite (King, Ricketts, and Warwick 1979); it would later be discovered to be an intermediate polar. Thus, at the end of the 1970s, examples of all of the major subclasses of CVs were discovered to be X-ray emitters, with variable and complex spectra that were not related in a simple way to variations in the mass accretion rate.

Accretion disk theory predicted that up to half the accretion luminosity would be emitted in the boundary layer, and the high-energy experiments offered the opportunity to test that prediction. Ultraviolet observations promised a measure of the mass accretion rate, which was believed to determine the salient properties of CVs (Robinson 1976) and the opportunity to model the structure and behavior of disks and columns. Soft X-ray observations promised a way to search for the primal source of the low-amplitude optical oscillations observed in CVs during episodes of high mass accretion (Warner 1976).

HEAO-1 made several spectacular discoveries concerning CVs; among them was the detection from two dwarf novae of rapid, large-amplitude X-ray pulsations in their soft X-ray flux, and longer-term soft and hard X-ray variations corresponding to the rise and decline from optical outbursts (see Mason, Córdova, and Swank 1979, and Córdova and Mason 1983 for reviews). The observed variability and luminosity verified that the accreting 'blue star' to which Payne-Gaposchkin referred was, without doubt, a degenerate star. However, *HEAO*-1 surveyed only the tip of the iceberg. The much more sensitive *Einstein X-ray Observatory* was used to establish X-ray emission as a property of all CV subclasses, based on its detection of hard and soft X-ray emission from 70 or more CVs (Becker 1981; Córdova, Mason, and Nelson 1981; Córdova and Mason 1983, 1984). The *EXOSAT* satellite offered, for the first time, continuous viewing of CVs through their binary orbits because of its exceptionally high Earth-orbit. Launched in May of 1983, *EXOSAT* showed that these stars have many of the properties associated with neutron stars that are accreting from low-mass companions, namely, dips, bursts, quasi-periodic oscillations, and structured coronae, and offered a test for models of the dwarf nova outburst. The medium energy experiment on *EXOSAT* (2–20 keV) was particularly well suited to studying the pulsed X-ray emission of the intermediate polars, while its low-energy (CMA) detector (8–300 Å) observed CVs at the shortest EUV wavelengths.

During the 1980s and into the 1990s, a very large sample of CVs was studied by *IUE* during many luminosity states in the important band between 1150 Å and 3000 Å (see LaDous 1991 for a compendium of spectra). Venturing from the UV to the EUV were the *Voyager* Ultraviolet Spectrometers, which covered the range 1700 Å to 500 Å. *Voyager*, while not nearly as sensitive as *IUE*, provided coverage on several CVs in high states in an important, untapped wavelength passband. In 1990 the space shuttle *Columbia* carried the Hopkins Ultraviolet Telescope (HUT) aloft and observed, with unprecedented spectral resolution in the far UV, a handful of CVs from 850 Å to 1850 Å. The Japanese satellite *Ginga* observed with modest spectral resolution the hard X-ray emission (1.5–36 keV) of many CVs between February 1987 and November 1991. At the writing of this review, *ROSAT* is imaging X-rays from CVs with much greater sensitivity than *EXOSAT* or *Einstein*, using three types of detectors that cover the soft X-ray and EUV regimes. Between July 1990 and January 1991, *ROSAT* made a survey of the entire sky over these wavelengths, and a number of new sources have been identified with previously uncataloged CVs (e.g. Mason *et al.* 1992; Mittaz *et al.* 1992; Watson 1993). Also surveying the entire sky in the EUV as this article goes to press is the *Extreme Ultraviolet Explorer* satellite, which was launched in June of 1992. In this Section, I summarize the discoveries made using high-energy experiments in space, and discuss the implications in terms of models for accretion onto degenerate stars. The emphasis here is chiefly on the disk-accreting CVs; further discussion of the X-ray emission of magnetic CVs can be found in Sect. 8.4.1.

8.2.2.2   *High-energy emission and the boundary layer*

As noted in Sect. 8.2.1.1, the detection of high-energy emission from dwarf novae is to be expected under standard scenarios for disk accretion onto a white dwarf, and half the total accretion luminosity can be liberated in the BL if the white dwarf is rotating relatively slowly compared to breakup speed. Since it is small compared to the entire disk, and $\dot{m}_{accr}$ is high, the BL will be a source of high-energy radiation. The radiation will appear as EUV or soft X-rays for an optically thick boundary layer surrounding a white-dwarf star, and hard(er) X-rays for an optically thin BL. Ultrasoft X-rays, with a thermal temperature corresponding to 5–30 eV, have been detected during the outbursts of SS Cyg, U Gem, and VW Hyi (Mason *et al.* 1978; Córdova *et al.* 1984; Van der Woerd, Heise, and Bateson 1986). All of these dwarf novae are observed at relatively low inclination angles that afford a view of the inner disk and white dwarf. In all three dwarf novae, the soft X-ray flux increases 100-fold during the optical outburst. In VW Hyi, the spectral shape remains constant during this large change in intensity. This implies the unrealistic scenario of maintaining a constant temperature in spite of a large change in mass accretion rate or, alternatively, suggests that the spectrum is more complicated than a simple optically thick blackbody with one characteristic temperature (Van der Woerd and Heise 1987).

Some of the problems with the standard BL theory, discussed in this and subsequent sections, are the lack of sufficient EUV luminosity, especially notable in VW Hyi (because its measured low column density severely constrains models for its X-ray spectrum); the behavior of the rapid oscillations, which show that their source is not in a steady-state; the poorly understood nature of the soft X-ray spectrum (mentioned above for VW Hyi); the ionization of the helium lines, which show that the BL is cooler than expected; the location of the hard X-ray emission in an extended source the size of the disk (as observed in an *eclipsing* dwarf nova); and the unknown effects of the high-velocity wind in the vicinity of the inner disk.

The earliest surveys of X-ray emission from CVs (e.g. Córdova, Mason, and Nelson 1981 for quiescent dwarf novae, and Córdova *et al.* 1980a for dwarf novae in outburst) showed that the X-ray luminosity was far lower than expected in the thick, steady-disk picture. This result has since been confirmed for SS Cyg in outburst by Mauche, Raymond, and Córdova (1988), who used their measure of N_{HI} from *IUE* observations to constrain the observed soft X-ray luminosity to a value much less than the theoretical expectation. The discrepancy between observation and theory has been verified for VW Hyi during all phases of its light curve: Pringle *et al.* (1987) using *EXOSAT* data, and Belloni *et al.* (1991) using *ROSAT* data, show that the X-ray flux of this dwarf nova during quiescence is smaller than its optical and far-UV (FUV) flux by almost an order of magnitude; Polidan, Mauche, and Wade (1990), Mauche *et al.* (1991), and Van Teeseling, Verbunt, and Heise (1993) use multispectral data to verify an X-ray flux shortfall of a similar magnitude during VW Hyi's outburst (Sect. 8.2.2.5).

Hoare and Drew (1991) use the Zanstra method to derive limits on BL temperatures of highly mass-accreting CVs, employing the strengths of the He II $\lambda 1640$ and $\lambda 4686$ recombination lines. Their modeling assumes that the He II is produced in a wind, rather than in the disk. They find a range of ionization temperatures, i.e. 50 000 K to 100 000 K, that is substantially lower than the standard theoretical BL model, but not inconsistent with the soft X-ray measurements of dwarf novae in outburst. A cooler BL obviates some of the problems with a model for the CV winds in which the wind is radiatively driven from the degenerate star; too large a mass outflow rate is required for this model to work if the BL temperature is high, i.e. > 200 000 K. With this high a temperature, the wind will be more ionized than observed, and optically thin, also in contrast to observations and modeling. If instead the BL is cool, then lower mass loss rates are allowed and the problem with the energy source is diminished (the problem is discussed in Sect. 8.2.2.7).

The failure to detect the predicted BL luminosity has prompted the speculation that either much of this luminosity is being produced in the EUV (beyond the range of observability until recently; see below), or that soft X-rays produced at the source are absorbed, either by the ISM in the line of sight to the CV, or by a partially ionized wind from the inner disk. The apparently cooler-than-expected BL has been attributed to one of several possibilities: an emission area that covers the entire white dwarf, rather than a thin equatorial region; rotation of the white dwarf within 10% of breakup speed; or mass loss in a disk wind that inhibits mass accretion, dissipating the BL energy into kinetic energy in the wind rather than radiation.

8.2.2.3   *Rapid oscillations in the soft X-ray flux*

It was similarly expected that the optical oscillations, discovered in the early 1970s at an amplitude of 0.001–0.1 magnitudes, would find their original source in a soft X-ray component. This was satisfied with the detection of high-amplitude (~ 15–30%) soft X-ray pulsations from SS Cyg (with a period, p, ~ 7–11 s: Córdova *et al.* 1980b, 1984; Jones and Watson 1992), U Gem ($p \sim 25$–9 s: Córdova *et al.* 1984), and VW Hyi ($p \sim 14$ s: Van der Woerd *et al.* 1987) during optical outbursts. Their dominant property, which is their lack of phase and frequency stability, distinguishes them from the coherent pulsations of isolated (e.g. ZZ Ceti) white dwarfs and radio pulsars. The behavior of the dwarf nova oscillations is complex: variability in phase and rapid unpredictable changes in period are superimposed on a longer-term variation in the period and large-amplitude changes of both the periodic and the nonperiodic flux. These features have been detailed by Córdova *et al.* (1980b, 1984) and Jones and Watson (1992). We infer, then, that the instability of the dwarf nova oscillations is due to the accretion process. This idea is further strengthened by the presence of the oscillations only during states of high mass accretion, i.e. they are detected in optical and X-ray bands only in the dwarf novae outburst and in the optical flux of 'disk-accreting' novalike stars (the 2σ limit on 13–15 s X-ray pulsations of VW Hyi during quiescence is 5%; Belloni *et al.* 1991).

In SS Cyg, there is a strong negative correlation between the X-ray flux and the oscillation period: the X-ray oscillations, like the optical oscillations in this and other dwarf novae, show a systematic long-term change in period over the course of the outburst. The period decreases at the onset of the outburst from ~ 10 s to 7 s, with the period derivative $\sim -10^{-4}$ s s^{-1}, and on the decline the period derivative increases from a few times 10^{-6} s s^{-1} to a few times 10^{-5} s s^{-1} (Patterson 1981; Jones and Watson 1992). This argues for an origin that is not tied to the rigid body of the white dwarf (whose moment of inertia is too large for rapid spinup/spindown), but that moves with changes in the mass accretion rate. A differentially rotating surface layer (or layers) on the white dwarf is one possible origin. The detected X-ray and optical oscillations differ greatly in their coherence properties (the observed optical oscillations are much more coherent); it would be important to determine whether or not this is a selection effect. There are virtually no examples of simultaneous optical and X-ray detections of a dwarf nova oscillation (see discussion in Córdova and Mason 1983). This would seem to be a critical experiment for constraining the location and nature of the X-ray pulsation.

The oscillations have been variously attributed to Rossby-like waves in a thin layer on the white dwarf's surface or blobs circulating at Keplerian velocities in the inner disk (cf., review by Córdova and Mason 1983). Kley (1991) has done two-dimensional radiation hydrodynamic calculations of the structure of the BL and finds that, for a fairly realistic description of the turbulent viscosity, the flow is strongly time dependent and unstable. The luminosity of the BL in Kley's simulations oscillates with a large amplitude on time scales of the Keplerian period at the stellar surface, as shocks are generated and dissolved, thus giving an alternative possible explanation for the origin of the quasi-periodic oscillations (QPOs). Kley's calculations also reveal that above the disk the temperature increases dramatically, forming a hot corona. The heating is caused by the persistent dissipation and long cooling time of the very low-density gas. Another idea is that the pulsation could be reprocesssed somewhere in the system from a direct, possibly higher-energy pulsation at the white dwarf. The high amplitude of the soft X-ray pulsation might argue against this, but in truth the luminosity of this pulsation is not known so it is difficult to argue this point.

The morphologically similar horizontal-branch QPOs observed in close binary neutron star systems, albeit with a much lower amplitude and with a much higher frequency (see Ch. 6), have been interpreted as representing the beat frequency between the rotation period of the compact star and the period of rotation at the Alfven surface (Alpar and Shaham 1985). In the model of Lamb *et al.* (1985) the QPO arises due to a beat between the fixed period of the magnetic compact star and the Keplerian periods of blobs in the inner disk; the rate at which these blobs enter the magnetosphere of the compact star differs depending on the orientation of the blob with respect to the compact star's misaligned magnetic field at infall. The longer-term systematic variation in the period through a dwarf nova outburst would be due to the change in the Alfvén radius with changes in the mass accretion rate.

Of possible relevance are studies of two intermediate polars, AE Aqr and GK Per, both of which have stable X-ray and optical oscillations and optical QPOs (Patterson *et al.* 1980; Watson, King, and Osborne 1985; Norton, Watson, and King 1988). Unlike the usual dwarf novae, both of these systems have relatively long orbital periods and evolved companion stars. In AE Aqr, X-ray and optical pulse-timing studies, combined with knowledge of the orbit from optical spectral line studies, suggest that a direct X-ray pulse is not viewed in this system; instead, both the optical and X-ray pulsations represent the beat period of the rotational period of the white dwarf and something, perhaps the accretion stream from the companion star, which is fixed in the binary orbital frame (De Jager 1991). One inference drawn from this result is that the direct beam may emerge as gamma radiation due to particle acceleration in a collisionless shock in the accretion curtain of this intermediate polar (De Jager 1991). The reports of TeV emission from this and a few other CVs are suggestive (see Meintjes *et al.* 1992 and references therein); in addition, evidence for MeV emission from two magnetic CVs has been reported by Bhat, Richardson, and Wolfendale (1989). GK Per is of interest because it is an old nova that shows dwarf novalike outbursts. A 351 s hard X-ray ($kT \sim 20$ keV) pulsation is detected during both outburst and quiescence, when large-amplitude X-ray flickering is also detected. Like AE Aqr, GK Per evinces also an optical (blue) QPO whose origin is not understood (Patterson 1991).

8.2.2.4   *FUV and EUV emission from CVs*

Standard disk theory (Sect. 8.2.1.1; see also Ch. 10) predicts that for a luminous CV disk the spectrum will peak shortward of 1200 Å since the maximum temperature is 60 000 K or higher. Thus a major fraction of the disk luminosity, plus much of the BL luminosity if it is optically thick, should be emitted in the FUV (912–1200 Å) and EUV (200–912 Å).

SS Cyg was detected as a variable EUV source during an optical outburst by an EUV experiment aboard the *Apollo–Soyuz* spacecraft (Margon *et al.* 1978). It was more than a decade later that the *Voyager* Ultraviolet Spectrometers, the Hopkins Ultraviolet Telescope (HUT) aboard the *Astro-1* shuttle mission, the *ROSAT* Wide Field Camera and *EUVE* began observing CVs in this important waveband.

Voyager observations of several CVs with 18 Å resolution showed evidence of absorption features near 980 and 1030 Å and revealed a change in slope below about 1200 Å (Polidan and Carone 1987; Polidan and Holberg 1987). Polidan, Mauche, and Wade (1990) combine *Voyager* observations of five CVs with neutral hydrogen column densities derived by Mauche, Raymond, and Córdova (1988) from the curve-of-growth analysis of interstellar absorption lines in high-resolution *IUE* spectra to place upper limits on the emitted flux in the 600–700 Å band. No EUV emission was observed in any of the five CVs. For one of them, VW Hyi, the low neutral hydrogen column ($N_{HI} \sim 6 \times 10^{17}$ cm^{-2}) plus extensive *Voyager* data during a superoutburst place the observed upper limit to the 600–700 Å flux well below the expected EUV flux from model calculations. It is quite possible that the Kurucz model atmospheres used are wrong at the wavelengths and for the high temperatures and gravities being considered; for example, these models may underestimate the amount of line blocking at these wavelengths by a large factor (see discussion in Polidan, Mauche, and Wade 1990).

The HUT was used to observe with ~ 3 Å resolution the 830–1850 Å spectrum of a few CVs. The HUT also has EUV sensitivity in the 415–950 Å range using the second order of the grating. The dwarf nova Z Cam was viewed near the peak of a normal outburst (Long *et al.* 1991). In the sub-Lyα region, strong absorption lines that were detected include S VI $\lambda\lambda933,945$, C III $\lambda977$, N II $\lambda991$, O VI $\lambda\lambda1032,1038$, He II $\lambda1085$, P V $\lambda\lambda1118, 1128$, and C III $\lambda1176$ (see Figure 8.6). No strong FUV variability was detected in Z Cam over the 1870 s observation on time scales from 2 s to 20 min. The spectrum peaks at about 1050 Å and then falls; there is a sharp cutoff at the Lyman limit with no flux apparent shortward of 912 Å. The UV-plus-FUV spectrum fits a stellar atmosphere model better than a Planck model (which does not turn down at the shorter wavelengths), but the 1400 Å flux predicted by the best-fit stellar atmosphere model is a factor of 11 less than that observed. This is the usual problem discovered when trying to fit CV disk spectra over a broad wavelength range; see Sect. 8.2.1.3. Line blanketing may be responsible for the turnover of the continuum below 1050 Å. No lines except C IV $\lambda1549$ and He II $\lambda1640$ in Z Cam have wind-related features. The HUT data do not show obvious evidence for a boundary layer: the higher order Lyman lines are weak or absent.

U Gem was observed with the HUT ten days after a normal outburst had subsided. Its spectrum (see Figure 8.7) shows absorption features due to the Lyman series of hydrogen and the Balmer lines of He II, as well as weaker absorption lines due to

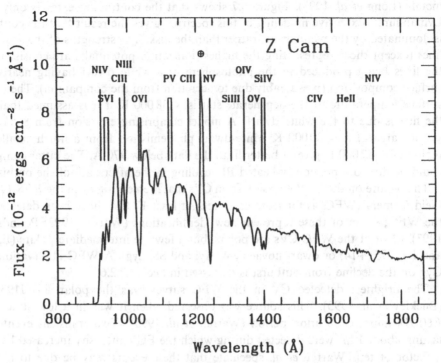

Fig. 8.6. HUT spectrum of Z Camelopardalis during a normal outburst. Lα is contaminated by airglow. From Long *et al.* (1991).

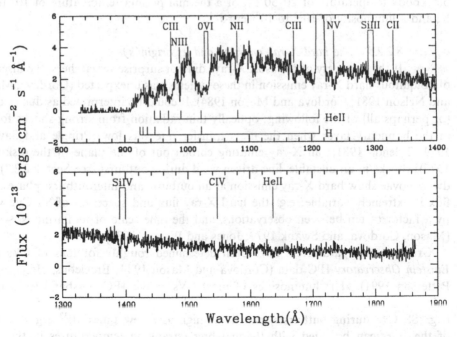

Fig. 8.7. HUT spectrum of U Geminorum during quiescence. From Long *et al.* (1993).

metals (Long *et al.* 1993). Figure 8.7 shows that the continuum extends only to the Lyman limit. Extensive modeling of this spectrum gives the result that the continuum is dominated by the white dwarf rather than the disk. The strengths of the absorption lines (except those representing the highest ionization potentials) are consistent with the lines being produced in the photosphere of a white dwarf having nearly solar surface composition (presumably due to accretion from the companion). The average surface temperature of the degenerate star is $\sim 38\,000$ K if it is assumed that all of the flux is due to the white dwarf. A model comprising emission from 85% of the surface area of a $\sim 30\,000$ K white dwarf plus emission from a much smaller, hot region of $\sim 57\,000$ K gives a better fit to the data below 970 Å. The latter component could be due to a previously heated BL, cooling in the aftermath of the outburst.

The picture on the EUV emission from CVs that is emerging from the *ROSAT* Wide Field Camera (WFC) is far from clear yet. At least 17 CVs have been detected with the WFC; seven of these represent new identifications (Watson 1992; Pounds *et al.* 1993). Most of the WFC CVs are polars, but a few are intermediate polars (EX Hya and RE 0751+14) or dwarf novae (VW Hyi and SS Cyg). A WFC observation of SS Cyg on the decline from outburst is discussed in Sect. 8.2.2.6.

The brightest detected CV in the WFC survey was the polar RE J1938−461 (Buckley *et al.* 1993). This source was observed when it was in a low state during *EUVE*'s initial calibration period (Warren *et al.* 1993). Two transient events, each lasting about 1 hr, were detected during which the EUV intensity increased by about a factor of ten. Warren *et al.* speculate that these events may be due to magnetic flaring activity on the companion star or coronal mass ejection from this star, causing episodic mass accretion onto the white dwarf. The transients are estimated to have a blackbody temperature of 20–50 eV, or a thermal plasma temperature of 10^6 to 10^8 K, and a luminosity of 10^{32} to 10^{33} erg s^{-1}.

8.2.2.5   *The hard X-ray emission and its origin(s)*

If the discovery of ultrasoft X-rays did not surprise researchers, the discovery of ubiquitous hard X-ray emission in these systems was unexpected (Córdova, Mason, and Nelson 1981; Córdova and Mason 1984). It could be interpreted as due to either (or perhaps all) of the following: optically thin radiation from strong shocks formed in the boundary layer when the mass transfer rate was low (Pringle and Savonije 1979; Tylenda 1981); an X-ray emitting corona out of the plane of the disk (Icke 1976); shocks in an unstable, line-driven wind during outburst (see Sect. 8.2.2.7). The dwarf novae show hard X-ray emission at all outburst and interoutburst phases. This flux is extremely variable; e.g. the hard X-ray flux and spectrum of SS Cyg varies by a factor of ten between observations, and the time scale of its flaring is ~ 100 s (Mason, Córdova, and Swank 1979; Jones and Watson 1992).

X-ray spectral parameters have been determined roughly for tens of CVs using *Einstein Observatory* IPC data (Córdova and Mason 1984; Eracleous, Halpern, and Patterson 1991). The luminosities of most CVs in the IPC passband (0.15 to 3.5 keV) are in the range $10^{30} - 10^{32}$ erg s^{-1}, although the luminosity of some CVs (e.g. SS Cyg during outbursts) can be as high as a few times 10^{33} erg s^{-1}. Most of the stars can be fitted with thermal bremsstrahlung temperatures in the range 1–5 keV, while a few (especially the magnetic CVs) have much higher temperatures (see, e.g., Ishida 1991). The X-ray emission from SS Cyg during quiescence is bright enough that detailed studies of its spectrum have been made using *OSO*-8, *HEAO*-1, *EXOSAT*, and *Ginga*. Both 6.7 keV and 7.9 keV emission lines are detected (Jones and Watson 1992; Yoshida, Inoue, and Osaki 1992); their energies are consistent with the K-alpha and K-beta lines of helium-like iron. Although the *Ginga* spectrum (1–37 keV) is well-fitted with a thermal bremsstrahlung model of temperature 16–18 keV, Yoshida, Inoue, and Osaki point out that in such a high-temperature plasma most of the iron should be fully ionized if the plasma is isothermal and in ionization equilibrium. These researchers propose, instead, a two-temperature stratified plasma, with $kT = 8 \pm 1$ keV and $kT \gtrsim 27$ keV, in which the higher-temperature plasma is twice as luminous as the lower-temperature plasma. A thermal plasma temperature of 7 keV is measured during an outburst of SS Cyg by Swank (1979). The hard X-ray emission of the intermediate polar EX Hya is discussed in Sect. 8.4.1.

Combined *EXOSAT* X-ray data and *Voyager* FUV data have been used to attempt to infer the high-energy spectrum of VW Hyi and the relative flux emitted by the accretion disk and BL (cf., Polidan, Mauche, and Wade 1990; Mauche *et al.* 1991; Van Teeseling, Verbunt, and Heise 1993). The latter group of authors find that a model which fits all of the data during superoutbursts in 1983, 1984, and 1985 consists of a single, optically thin component that varies between 10^5 K and $\geq 10^6$ K. They propose that an alternative model which satisfies the constraints of the data is a combination of a warm ($T > 10^6$ K) optically thin component with a cooler ($80\,000 < T < 100\,000$ K) optically thick component. Neither component contributes significantly to the flux at 1000 Å detected with *Voyager*. The cool component has an emitting surface 1/1000 of the surface of the white dwarf. Van Teeseling, Verbunt, and Heise note that the warm component may be the same as the 2.2 keV component measured during VW Hyi's quiescent interval by Belloni *et al.* (1991) using *ROSAT* data. Although the details of the spectral modeling differ among various groups of researchers, all studies point to the same conclusion: for VW Hyi, it appears that the ratio of the BL luminosity to the accretion disk luminosity is less than unity in all luminosity states.

The origin of the hard X-ray emission can be explored with observations of eclipsing CVs. An *EXOSAT* observation of the high-inclination ($i \sim 83°$) dwarf nova OY Car (Van der Woerd 1987; Naylor *et al.* 1988) during a superoutburst revealed no X-ray orbital modulation and no eclipse, unlike UV and optical observations, which do reveal an appreciable eclipse. This indicates that the inner disk and BL are hidden from view by extended vertical structure on the edge of the disk, a structure also suggested by the strong modulation of the UV continuum flux. The X-ray spectrum and count rate point to the presence of a hot, optically thin X-ray corona, with a temperature of 10^6–10^7 K, and comparable in size to the Roche lobe. This discovery further aggravates the problem of the lack of BL emission in disk-accreting CVs, as it assigns much of the detected high-energy emission to a location other than the BL. This result, it should be noted, is for an eclipsing system in which the white dwarf and any radiation produced near it are not viewed directly; the observed X-ray emission could be due to scattered BL emission. The character of the X-ray emission of OY Car (temperature, variability) differs substantially from that of SS Cyg, which is viewed closer to pole-on. X-ray spectra and light curves of samples of CVs spanning

354 *France Anne-Dominic Córdova*

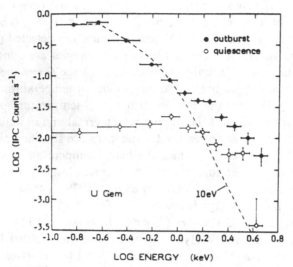

Fig. 8.8. *Einstein X-ray Observatory* IPC spectra of U Geminorum during outburst and quiescence. The dashed line is a 10 eV spectrum normalized to the low-energy part of the outburst spectrum. From Córdova and Mason (1984).

a range of inclinations will be needed in order to determine the existence and location of multiple medium/hard X-ray components.

8.2.2.6 *The morphology of the outburst in the UV and X-ray*

The observed X-ray behavior as a function of outburst phase is complex. Although U Gem has been observed as a bright soft X-ray source several times during outburst, no X-ray study has been performed through an entire outburst, except for the snapshot sampling of the original *HEAO*-1 discovery observation (Mason *et al.* 1978). Mason, Córdova and Swank (1979) show a composite picture of the optical, soft X-ray and hard X-ray light curves for the outburst covered by *HEAO*-1. These data show a possible delay in the soft X-ray outburst with respect to the optical outburst, as well as a possible early termination of the soft X-ray flux during the optical outburst, but the low fluxes at these epochs (and one other during the middle of the *HEAO*-1 sampling interval) could also be interpreted as due to phase-related 'dips' in the light curve, serendipitously sampled at phases at which dips were later seen with *EXOSAT* (see Sect. 8.3.5). The four-day *EXOSAT* observation occurred very late in a lengthy optical outburst and showed that the soft X-ray flux decrease was contemporaneous with the optical decrease. *HEAO*-1 observations of U Gem showed that there is a rise in the hard X-ray flux by a factor of three to five and a gradual decrease during outburst; *Einstein* spectra of U Gem during quiescent and outburst intervals confirm that the hard X-ray flux is greater during outburst (see Figure 8.8).

In SS Cyg, on the other hand, the hard X rays increase by a factor of two to three about one day after the rise to optical light, then drop to one-quarter of the quiescent value for most of the outburst, and then rise again by about a factor of two near the

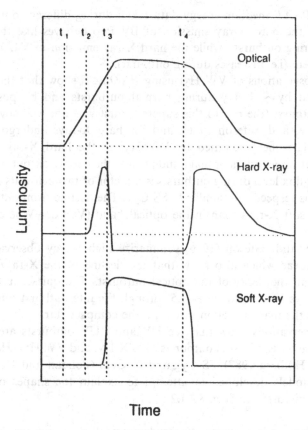

Fig. 8.9. A schematic of the behavior of the optical, soft X-ray (< 0.5 keV) and hard X-ray (> 1 keV) intensity of SS Cygni through an outburst. From Jones and Watson (1992).

end of the optical outburst. The decrease in SS Cyg's hard X-ray emission during quiescence is not accompanied by an increase in intrinsic absorption (Swank 1979; Jones and Watson 1992). The soft X-ray rise is delayed by 0.5–1.1 day with respect to the optical rise to outburst and the rise time is about 8 h, i.e. much shorter than the optical rise time of ~ 2 days (Jones and Watson 1992). This is similar to the delay in its far UV rise, as measured by Polidan and Holberg (1984) using *Voyager*. There is evidence that the decay time, too, is shorter at high energies than in the optical. A schematic of the behavior of the optical and X-ray components through outburst is shown in Figure 8.9. *ROSAT* EUV (WFC) data on SS Cyg late in the decline from an outburst were taken in conjunction with optical and hard X-ray (*Ginga*) data. Preliminary results show that the N_H to SS Cyg is higher than the interstellar value derived by Mauche, Raymond, and Córdova (1988) and that the (assumed blackbody) temperature probably declines with fading EUV light (Duck and Ponman 1992).

Thus far, comparative studies of the X-ray emission of dwarf novae during quiescence and outburst are still rare enough that no clear pattern of behavior has emerged. Szkody, Kii, and Osaki (1990) compare their *Ginga* observations of BV Pup and V426 Oph with *EXOSAT* medium-energy data taken during different outburst states and conclude that the hard X-ray emission of BV Pup behaves like that of SS Cyg (i.e. decreases during outburst), while the hard X-ray emission of V426 Oph behaves like that of U Gem (i.e. increases during outburst).

A number of X-ray observations of VW Hyi using *EXOSAT* show that the soft X-ray outburst is delayed by $\sim$ 1 day during normal outbursts and by possibly $\sim$ 2.5 days for a superoutburst (the rise to this superoutburst was not well covered; see Van der Woerd, Heise, and Bateson 1986), and the hard X-rays undergo only small changes in flux. The energy emitted by VW Hyi in the hard X-ray band during quiescence is about three orders of magnitude below the energy output during outburst. The hard X-ray flux level during outbursts is a factor of two below its mean quiescent flux level. In this respect, it is similar to SS Cyg. The decline from outburst in VW Hyi is steeper in soft X-rays than in the optical band (Van der Woerd and Heise 1987).

Van der Woerd, Heise, and Bateson (1986) summarize high-energy observations of several other dwarf novae which also show that the decline of the X-ray/EUV outburst starts earlier than the decay of the optical outburst. The quiescent X-ray emission of VW Hyi decays by a factor of $\sim$ 1.5 through the interoutburst interval, implying a diminution in the mass accretion rate onto the compact star.

IUE and *Voyager* observations show that the UV and FUV outbursts are also delayed with respect to the onset of visual outbursts of WX Hyi and VW Hyi (Hassall *et al.* 1983; Polidan and Holberg 1987), SS Cyg and U Gem (Polidan and Holberg 1984) by 0.5–1 day. Models developed to attempt to explain the shapes of the outburst light curves are discussed in Sect. 8.3.1.2.

8.2.2.7 *High-velocity winds, their nature and origin*

A discovery that resulted from the first UV spectral observations of CVs is that of high-velocity winds (Heap *et al.* 1978; see references in Córdova and Mason 1983). The winds are manifested by characteristic velocity-shifted profiles of the UV resonance lines of, chiefly, C IV λ1549, Si IV λ1398, and N V λ1240. These profiles are detected only during states of high mass accretion, i.e. in the dwarf novae during outbursts and some novalike stars and intermediate polars; wind line profiles are not detected during the quiescent states of dwarf novae. The shape of the line profile depends on the luminosity of the system, the inclination of the binary, and the binary orbital phase. The wind lines in dwarf novae reach maximum strength with maximum UV brightness during outburst (Woods, Drew, and Verbunt 1990). Deep absorption components are observed in low-inclination systems ($i \lesssim 65°$) during the peak and early decline of the UV outburst; these features have blue edge velocities as high as $\sim$ 5000 km s^{-1}. This is about the escape velocity from the surface of the white dwarf, and hence the origin of the wind has been inferred to be at or near the degenerate star (Córdova and Mason 1982). Figure 8.10(a) illustrates the change in the C IV line profile of the dwarf nova RX And from outburst, when strong P Cygni profiles are evident, to quiescence, when only emission lines are observed.

Several CVs show systematic variations in the line profiles of the UV resonance lines, including the intermediate polar TT Ari (Guinan and Sion 1981), the novalike objects IX Vel and LB 1800 (Mauche 1991; Mauche *et al.* 1994), and the dwarf novae

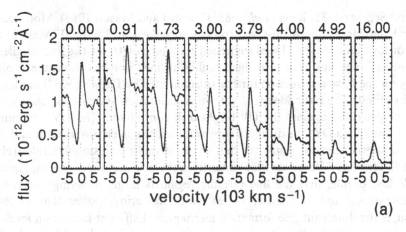

(a)

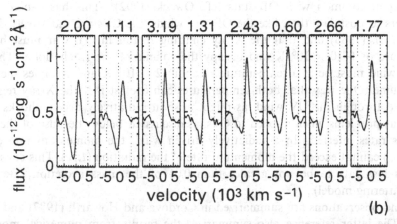

(b)

Fig. 8.10. Temporal evolution of the C IV line profile for two dwarf novae. (a) RX Andromedae from outburst maximum to quiescence, showing evolution from a P Cygni-like profile to a pure emission line profile. The numbers at the top give time in days since last optical maximum. (b) YZ Cancri during quiescence, sampled at different binary orbital phases. The figure shows repeated features in the line profile as a function of orbital phase (numbers at the top). From Verbunt (1991).

Z Cam, YZ Cnc, and SU UMa (Szkody and Mateo 1986; Drew and Verbunt 1988; Woods, Drew, and Verbunt 1990). Figure 8.10(b) shows the orbital related changes in the C IV line profile of YZ Cnc. On the basis of these observations, it is concluded that the wind has an asymmetry with respect to the disk's rotation axis. It is possible that a magnetic field on the white dwarf could control the wind loss geometry, as it does in the upper main sequence He-peculiar stars (Shore 1987).

Eclipsing systems can be used to probe the geometry of the wind. The first in depth observations of eclipsing novalike stars were of UX UMa and RW Tri, and these revealed broad, asymmetric emission line profiles for the UV resonance lines, which were only a little diminished in flux during deep eclipses of the UV continuum (Holm,

Panek, and Schiffer 1982; King *et al.* 1983; Córdova and Mason 1985). More recently, a similar behavior has been observed for the novalike star LB 1800 (Mauche *et al.* 1994). Córdova and Mason (1985) suggested that asymmetry of the line profiles and the absence of blue absorption in these highly inclined systems indicated a bipolar geometry for the flow, a deduction further substantiated by Drew's (1987) modeling of the wind outflow. The basic idea is that in high-inclination systems the mass outflow is not back-projected against the UV-bright disk continuum. The UV line emission from the dwarf nova OY Car during a superoutburst has been observed in and out of eclipse and shows behavior similar to that of the novalike eclipsing variables (Naylor *et al.* 1988). Drew and Verbunt (1985) infer from the absence of eclipses in, especially, the C IV line profile, that the line forming region is large, implying low densities and temperatures, and that therefore resonance scattering, rather than collisional excitation, is the dominant line formation mechanism. Different ionization levels may be present to account for the different ionization parameters deduced from the Si IV and N V lines.

In the presence of a hot corona, winds, and X-ray emission, the dwarf novae have something in common with OB stars (cf., Owocki 1992). This has caused many researchers to postulate that, like the OB star winds, the CVs are radiatively driven. Naylor *et al.* (1988) find that the X-ray emitting region in OY Car must have a temperature of 1–10×10^6 K and occupy the volume of the Roche lobe. The UV emitting wind must be at a temperature of about 2×10^4 K and occupies the same large volume. One way that both temperatures can coexist is if the X-ray region is distributed through the UV wind, as might arise in radiatively driven shocks. This picture has problems when applied to CVs, as discusssed below, because of the large mass loss rates inferred. In a radiatively driven wind, the momentum rate of the wind ($\dot{m}v_\infty$) cannot exceed the momentum rate of the radiation (L/c). This constrains the mass loss rate to be not greater than 0.01 of the mass accretion rate (in a single-scattering model).

CV wind observations are summarized in Córdova and Howarth (1987) and Drew (1990). The latter reference also summarizes the results from numerical modeling of the winds. One approach has been to determine mass loss rates by constructing a sensibly constrained ionizing radiation field that illuminates the disk. A second approach has been to constrain the wind geometry and velocity law by constructing UV line profiles that match the ones observed. The velocity law deduced from the shape of the absorption line profiles is a slow one, and this can be understood from the fact that continuum photons scattered in the wind can be produced in the disk out to radii ten times the radius of the degenerate star; this is in marked contrast to the spherical geometry of the radiation field in OB stars, in which winds can be accelerated only to about three times the stellar radius (Mauche and Raymond 1987). Estimates of the mass loss rate have been calculated from synthesized model profiles, which include the nonspherical nature of the disk's radiation field, but these are parameterized as $\dot{m}q$, where q is the fractional abundance of the scattering ion (Drew 1987; Mauche and Raymond 1987). Until an adequate model of the ionization structure of the wind can be made, a reliable mass loss rate cannot be derived.

The relatively low ionization level of the wind in the presence of what is assumed to be the strongly ionizing field of the BL is problematic (Drew and Verbunt 1985).

The ionization calculations of Mauche and Raymond (1987) indicate that the mass loss rate is greater than one-third of the mass accretion rate. Wind driven in a bipolar outflow relaxes somewhat Mauche and Raymond's constraint on $\dot{m}q$, but the wind must be shielded from the BL radiation in order to allow the observed ion densities to exist in a less massive outflow. For this reason, these authors conclude that a source of energy other than accretion may be required to drive the wind, or that the wind is produced in the inner disk, rather than at the white dwarf's surface. A similar conclusion was reached earlier by Drew and Verbunt (1985) and Kallman and Jensen (1985). Studies of the line profiles of systems at various inclinations show that it is not likely that the wind is produced in the outer disk (Drew 1990).

A model has been proposed that gets out of some of the difficulties with the high mass loss implied by a radiation-driven wind from the BL. In the three-dimensional kinematical model of Vitello and Shlosman (1993) the wind is bi-conical and driven from the disk. This model, in which the acceleration is slightly faster than linear and in which the wind acceleration scale height is much greater than the radius of the white dwarf, appears to produce most of the features observed, except the result that the wind lines vary with orbital phase. The model requires that the BL temperature be low, i.e. $\ll 100\,000$ K, in order to prevent the wind from becoming overionized and optically thin. There is support for a low BL temperature from the soft X-ray and EUV observations and the He II recombination line data (Sect. 8.2.2.2).

8.2.3 The long wavelength spectrum

8.2.3.1 Detecting the not-so-normal companion star

The companion star contributes relatively little to the total light of the system. The best way to try to isolate this star's emission from that of the accretion disk is to look at a CV at near infrared wavelengths during a state of low mass accretion. The previous paradigm was that the companion star was a 'normal' or main-sequence star. More recently, various researchers have used a variety of methods, including eclipse timings and observations of the TiO bands and the λ 8190 Å sodium doublet, as a diagnostic of this star. It has been found that many secondary stars have apparent spectral types that are cooler than predicted, and some are less massive than expected, if the companion obeys the empirical main-sequence mass–radius relation (Friend *et al.* 1990; see also Echevarría 1983). This affects the derivation of the inclination of the binary system, a calculation formerly made on assuming that the secondary was a main-sequence star. The derivation of the K_2 value is fraught with uncertainty, as the effects of irradiation of the companion star and contamination of the spectral lines by emission from the disk are not well known. Observations longward of 1 μm may find suitable lines for measuring K_2 without these difficulties. An alternative technique, using photometry of the eclipses to determine the parameters of the secondaries, reveals that the masses of these stars in several short-period eclipsing dwarf novae are extremely low ($< 0.1 M_\odot$) and that their radii are larger than those of main-sequence stars (Wood 1987). These stars may be out of thermal equilibrium and may be becoming degenerate. Mass transfer will force a CV secondary out of thermal equilibrium, but the magnitude of this effect is not well known. The longer-period CVs probably contain evolved secondaries.

8.2.3.2   The far-IR emission of dwarf novae

New observations in the infrared using *IRAS* archival data have yielded detections of more than 70% of the dwarf novae and novalike objects searched for (Harrison and Gehrz 1992). This is higher than the fraction of classical novae detected in the infrared, yet the latter are significantly higher in total luminosity than the dwarf novae (the known dwarf novae, however, are a closer population of sources than the novae). The radiation mechanism for the IR emission of dwarf novae is not understood. Harrison and Gehrz (1992) say that line emission from a circumstellar, highly ionized shell is the most plausible explanation (in analogy with the classical novae), but they note that there is no evidence for circumstellar shells around the dwarf novae that are massive enough to produce the line emission observed. The *IRAS* detections are not inconsistent with the observed optical–near-IR flux of these objects. Source confusion is a possible explanation for some of the *IRAS* detections.

8.2.3.3   Radio emission from CVs

Radio emission is apparently uncommon among CVs (Córdova, Mason, and Hjellming 1983; Dulk, Bastian, and Chanmugam 1983; Bookbinder and Lamb 1987), but positive detections have been made at 6 cm of the magnetic variable AM Her, the DQ Her star AE Aqr, and the dwarf nova EM Cyg. In AM Her, a 'quiescent' radio level of ~ 0.5 mJy has been repeatedly detected and a 10-min flare has also been observed at 20 times the quiescent level. Dulk, Bastian, and Chanmugam interpret the flare, which was 100% circularly polarized, as caused by electron-cyclotron maser action. AE Aqr was detected at a level of 8–16 mJy, but its emission was not circularly polarized (Bookbinder and Lamb 1987). EM Cyg, which was observed during an optical outburst, was detected with the VLA at 4.9 GHz at a level of about 0.2 mJy, but with considerable variability (Benz and Güdel 1991). If interpreted as due to synchrotron emission, a source size larger than the binary separation is implied; if, instead, the radio flux from EM Cyg is due to maser emission of nonthermal electrons in the vicinity of the white dwarf, as might be supported by the large radio circular polarization and variability detected, it implies a field strength of the order of 800–2000 G, depending on the maser mode. Radio emission may be one way to measure the field strengths of weakly magnetic CVs, but first the location of the emission (white dwarf, companion star, or interaction region between the binary components) must be determined.

8.3   The variability of CVs and its origins

One of the distinguishing features of CVs is their high-amplitude variability. The multiwavelength observations discussed in Sect. 8.2 revealed diverse types of variability among these systems. Some of these, like the regular broad humps in the quiescent orbital light curves of some dwarf novae, or the superhumps observed during the outbursts of the SU UMa subclass of dwarf novae, are observed in only one waveband, while others, like the dwarf nova outburst and, especially, the rapid intensity variability (flickering and oscillations), have a complex behavior across the spectrum.

8.3.1 The dwarf nova outburst

8.3.1.1 The nature of the debate

The debate over the origin of the dwarf nova outburst has brewed since outbursts were first detected in the 19th century. The answer is of consequence: determining the origin of the outburst can illuminate our understanding about disk physics and the physics of the companion star. Modeling the time-dependent behavior of the dwarf nova outburst is one method of deriving an estimate of the viscosity in these systems. This understanding is relevant to the behavior of the neutron star X-ray binaries that also have disks and low-mass companion stars and are characterized by many types of outburst behaviors. The knowledge of the outburst mechanism is important in understanding how disks behave in many other kinds of astrophysical settings, including that of soft X-ray transients, AGN and symbiotic stars. Knowledge of the nature of the outburst is also relevant to understanding mass loss and the evolution of CVs, as well as the relationship between the various CV subclasses.

In current models for the dwarf nova outburst, the transient brightening results from an increase in the accretion onto the white dwarf. Nuclear burning, which is believed to be the cause of the classical nova outburst, is not thought to apply to the dwarf nova outburst because of the short recurrence time of these outbursts. The low amplitude of the outburst, and the low amount of mass ejected in a dwarf nova outburst, are consistent with an origin in enhanced accretion. The delay in the rise of the UV and X-ray outburst with respect to the optical outburst (see Sect. 8.2.2.6) confirms that the white dwarf is not the site where the outburst begins.

Two scenarios have been proposed for the source of the enhanced accretion onto the degenerate star: an instability in the disk and a short-lived burst of matter from the companion star (see Sect. 10.3). The past decade has seen the development of detailed theoretical light curves based on these models. This activity has fortuitously coincided with the ability to test the models: the *EXOSAT* satellite, with its 4-day orbit, afforded the opportunity to observe objects uninterruptedly for days at X-ray wavelengths; the *IUE* satellite could be used to make extensive, coordinated observations in the UV wavelength region, from where much of the radiation from the disks was emitted; the *Voyager* Ultraviolet Spectrometer could be used to view the spectrum shortward of Lyman α; and ground-based telescopes geographically distributed and networked together could be used to provide fairly continuous coverage at optical wavelengths. A few large teams of CV researchers used this capability to monitor dwarf novae for days or months, covering, in part, the rise and fall from outbursts as well as the intervals between outbursts.

Below, I summarize the predictions of the competing models. Then I describe what recent extensive studies of dwarf novae through outbursts have revealed and their interpretation in the context of the proposed models. First, to orient the reader, I show in Figure 8.11 a collection of several example dwarf nova outburst light curves, including their amplitudes and time scales. Any model for the dwarf nova outburst must be able to explain the morphology of all the outbursts, wide and narrow. Figure 8.11(a) is for SS Aur and shows the 'classical' long and short outbursts of the 'SS Cyg' (also called the 'U Gem') subclass of dwarf nova. Figure 8.11(b) shows outbursts from RX And, a member of the 'Z Cam' subclass of dwarf novae, which have both ordinary

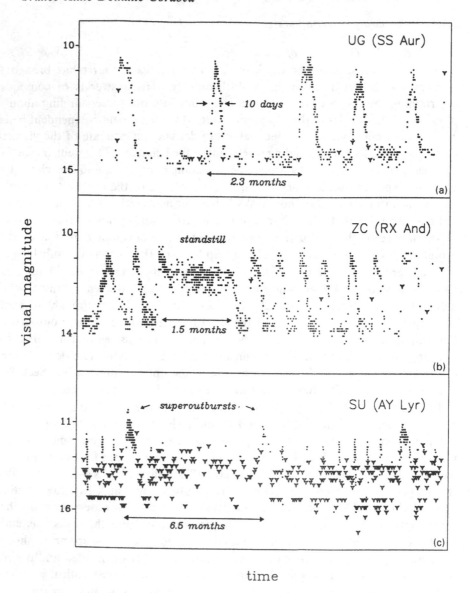

time

Fig. 8.11. Example light curves of the three subclasses of dwarf novae, using AAVSO data on SS Aurigae, RX Andromedae, and AY Lyrae. (a) The narrow and wide outbursts of the U Gem subtype. (b) A typical standstill of the Z Cam subtype. (c) The frequent narrow outbursts of the SU UMa subtype, interspersed with the more infrequent superoutbursts. Drawing by F.A. Córdova and H. Papathanassiou.

outbursts and long 'standstills' in brightness, part way down on the decline from an apparently ordinary outburst. Figure 8.11(c) illustrates the behavior of AY Lyr, one of the 'SU UMa' subclass of dwarf novae. In addition to their ordinary outbursts, these objects have 'superoutbursts' of longer duration and larger amplitude than ordinary outbursts; the superoutbursts, when they occur, always follow an ordinary outburst. Superoutbursts are accompanied by 'superhumps' in the orbital light curve; at outburst maximum, these occur with a period of a few per cent longer than the binary orbital period. An illustration of the superhump phenomenon is shown in Figure 8.14(c); see also Sect. 8.3.4. The AAVSO data on AY Lyr, which is replete with upper limits, also show well the limitations of amateur variable star observer data on CVs that are fainter than the reach of small telescopes, i.e. ~ magnitude 14.

8.3.1.2 Competing models

The mass-transfer instability model requires that there is a large increase in the mass loss from the companion on a semi-regular time scale and this enhanced mass is transferred inward to the white dwarf through the disk. Bath (1975) has proposed a particular mechanism for the instability in the companion star. Bath's instability is based on the recombination of plasma in the hydrogen and helium ionization zones, which can destabilize the envelope when the companion's photosphere is sufficently close to the inner Lagrange point. There is no agreement that this mechanism actually works, however, in CVs (see Gilliland 1985; Edwards 1988). In this model, the disk's brightness, which depends only on the mass transfer rate from the companion, decreases slowly or remains steady in the interval between outbursts. There has been no observation reported of a brightening of the companion star prior to an outburst. The value of the viscosity needed for this model to explain the rapid variations of the accretion disk luminosity is $\alpha \simeq 1$.

In the disk instability model (Sect. 10.3), the mass transfer rate from the companion star is fairly constant, but the transfer of this matter through the disk is unstable (Osaki 1974; Meyer and Meyer-Hofmeister 1981). The disk can only be in equilibrium if the mass transfer is very high or very low; at intermediate values, the disk becomes unstable and is forced to make rapid transitions between the two equilibrium states. This behavior is described by an S-shaped curve in a plot of surface density S (locally defined) versus temperature T (equivalent to local mass transfer rate). This model predicts that in the interoutburst interval the disk will slowly brighten as it ascends the bottom branch of the S-curve. When the upper end of this branch is reached, there will be a rapid transition to the top branch of the curve. The magnitude of this brightening has been calculated as 0.5 to 1 magnitudes, based on the diffusion of matter through the disk (Meyer and Meyer-Hofmeister 1984; Mineshige 1986). The disk instability model has provoked two schools of thought concerning the critical temperature at which an eruption is triggered. One school holds that the critical point is due to ionization of hydrogen and is determined by radiative processes; in this model, the critical temperature is ~ 6000 K (Smak 1982). The other school holds that convection is important and lowers the critical temperature to 2000–4000 K (Cannizzo, Ghosh, and Wheeler 1982). The disk instability models are very sensitive to the adopted viscosity prescription; in the time-dependent disk models constructed by Smak (1984a), alpha-disk models with $\alpha \simeq 0.2$, but with lower viscosity ($\alpha \simeq 0.05$) for lower temperatures, reproduce many, but not all, of the features of the outbursts. A disk instability model in which the outburst starts in the outer disk can reproduce many of the features of U Gem's outburst, like the brightening first in the outer disk and the expansion of the disk at the peak of the outburst, but does not produce the variable (i.e. long and short) outbursts observed in this star. A more detailed treatment of the disk instability model is found in Ch. 10.

Several workers have examined the evolution of the disk radius, since this will mirror changes in the angular momentum. Smak (1984b) was the first to do this for a disk instability, and Livio and Verbunt (1988) did this for a mass transfer instability. The latter found that, because the newly arrived matter has a lower specific angular momentum than the matter in the outer region of the disk, the disk will shrink to a smaller radius at the onset of a burst of mass transfer. As the density builds up, the disk starts expanding towards a new equilibrium radius. After the return of the mass transfer rate to the original rate, the disk radius relaxes to its initial radius before the burst. More recently, Ichikawa and Osaki (1992) used a numerical scheme to simulate the dwarf nova outburst. They modeled the time evolution of the accretion disk radius using both types of instability models. Their technique conserves the total mass and the total angular momentum during the evolution and takes into account the tidal torque and tidal dissipation exerted by the companion star on the disk. The resulting schema of the time evolution of the disk radius is shown in the paper by Ichikawa and Osaki (1992) for both the mass transfer burst model and the disk instability model. The mass transfer model is distinguished by the presence of a transient decrease in the disk radius at the onset of an outburst. In both models, the disk radius shrinks during interoutburst, but in the mass transfer model this shrinkage stops when the luminosity reaches a minimum.

8.3.1.3   Observations and their interpretations

The early attempts to decipher the nature of the outburst focused on the morphology of the outburst, its rise time, decay time, and the intervals between outbursts. One particularly remarkable new observation, made with high-energy satellite observatories, was that the UV and X-ray rise to outburst is delayed by 0.5 to 1 day with respect to the rise to optical outburst, and the rate of decline is faster at these higher energies (see discussion in Sect. 8.2.2.6). Observations of individual dwarf novae through the rise and peak of the outburst, however, revealed different time dependencies for different dwarf novae. An illustration of this is shown in Figure 8.12 for the UV/optical rise to maximum of two dwarf novae, VW Hyi and CN Ori. In VW Hyi, the rise to optical maximum is fast and there is a marked delay in the UV rise with respect to the optical rise. In CN Ori, in contrast, the optical rise is slow and the UV and optical rise together. Verbunt (1991) uses these observations as evidence against the (simplest, early version of the) disk instability model in which the disk is cool during quiescence and hot and ionized during outbursts. This model predicts the rise in temperature to spread rapidly through the disk in a fast outburst, and a delay between optical and UV rises in a slow outburst. This model, however, has undergone more recent accommodating variations so that observations through outburst are not, unfortunately, that constraining (Verbunt 1991).

It has gradually become apparent that one of the best ways to test the instability models is to monitor the *interoutburst* interval. It has turned out that the eclipsing systems provide especially useful tools for this study.

At a time when astronomers believed that the white dwarf was the origin of the outburst, Krezminski (1964) found that the width of the eclipse in U Geminorum

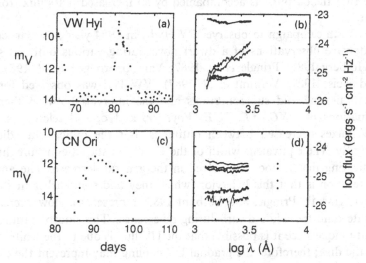

Fig. 8.12. Light curves, (a) and (c), and *IUE*-plus-*FES* data on the dwarf novae VW Hydri and CN Orionis during their rise to outburst maxima. In the fast-rising outburst of VW Hyi, the UV rise comes 0.5 day after the optical rise, while in the slow-rising outburst of CN Ori, the UV and optical rise together. From Pringle, Verbunt, and Wade (1986); see also Verbunt (1991).

decreased between outbursts. In this system, which is viewed at an inclination of 67°, only the bright spot is eclipsed. Smak (1971) tried to determine the geometrical elements of this system using the eclipses of the bright spot and deduced that the radius of the disk increased during an outburst by about 30% and then decreased, the decrease being fastest after outburst. Smak (1984b) confirmed this result using a new analysis that incorporated improved geometrical elements for the system.

O'Donoghue (1986) assembled all previously published light curves of another dwarf nova, Z Cha, and added new observations in white light photometry, to representatively sample the outburst cycle. The purpose of the analysis was to measure the contact times of the eclipses of the bright spot and white dwarf in this highly inclined ($i \sim 82°$) disk-accreting system as a function of outburst phase. O'Donoghue finds that the radius of the accretion disk expands during outbursts and declines between outbursts; this relationship is most significant when the radius is calculated as a function of the fractional time between outbursts. The decrease is most rapid immediately after outburst. Among the observations in this collection were those of Cook (1985), who found that the disk size was considerably smaller immediately prior to a normal outburst than any measurement previously reported. This observation is consistent with the work of Livio and Verbunt (1988), who predicted, in the context of the mass transfer model, a dip in the outburst light curve immediately prior to an outburst.

Hessman (1986) used spectroscopic observations of a noneclipsing system, SS Cyg, during the declining phase of a short outburst to deduce that a significant fraction of the increased λ4686 Å emission comes from the disk bright spot. Hessman interprets this to indicate that the outburst is accompanied by an increased mass flux from the companion star.

A multiwavelength campaign to observe VW Hydri in 1984 yielded a particularly good set of extensive observations of a dwarf nova during various outburst stages (Polidan and Holberg 1987; Pringle *et al.* 1987; Van Amerongen *et al.* 1987; Van der Woerd and Heise 1987; Verbunt *et al.* 1987). VW Hyi was observed for four months, through normal and superoutbursts and the intervals between them, by various combinations of *EXOSAT, IUE, Voyager,* and optical telescopes. The X-ray and UV fluxes decreased between outbursts, but the optical flux did not change significantly. The equivalent width of the C IV line stayed constant through quiescence, indicating that it, too, is involved in the general decrease. The authors' simplest interpretation is that this behavior (which they had seen earlier in the UV flux of WX Hyi; Hassall, Pringle, and Verbunt 1985) represents a slow decrease in the accretion rate onto the white dwarf during quiescence. This interpretation, they point out, is not unique since it is possible that the UV flux is due to the white dwarf, rather than to the disk; therefore, the gradual UV cooling may represent the cooling of either disk or white dwarf. The campaign yielded observations that are consistent with the accretion by the white dwarf of all the mass transferred on to the bright spot during the interoutburst (if the ultraviolet flux, F_{UV}, is due to accretion), or with the accretion of none of it (if F_{UV} is due to the white dwarf). The results of the campaign demonstrated the need for detailed UV observations through an interoutburst interval to determine the source of the UV flux.

Kiplinger, Sion, and Szkody (1991) made detailed UV interoutburst observations of U Gem, which is a system that has a much longer interoutburst interval than either VW Hyi or WX Hyi and is therefore potentially more interesting for monitoring secular changes during the interoutburst interval. They made seven sets of IUE observations over quiescent periods from 22 days to 103 days after outbursts (the last set ended two days before an outburst). They confirm a previous detection of a 30 000 K white dwarf, which dominates the UV light throughout the interoutburst. They find the presence of an additional, slowly decreasing UV component that they speculate is the inner disk or BL, and also a high temperature ($\sim 10^5$ K) region which gives rise to the N V and He II lines throughout quiescence. (See also Sect. 8.2.2.4 for an FUV snapshot spectrum of U Gem during quiescence.)

Long-term Walraven photometry of Z Cha during two 'quiescent' intervals (i.e. the periods between outbursts) by Van Amerongen, Kuulkers, and Van Paradijs (1990, hereafter AKP) made use of the multicolor eclipses of this highly inclined dwarf nova to probe the temperature changes in the white dwarf, bright spot, and disk as a function of the time between outbursts (normalized to the length of the outburst interval, since this varies quite a bit within the same system). These observations revealed that the flux of the bright spot increased by $\sim 40\%$ between phases 0.1 and 0.4 of the interoutburst interval, but the fluxes of the disk and white dwarf did not show a secular trend during interoutburst (although short-term variations were common). The increase of the bright spot's flux is expected with a secular decrease of the radius of the disk, as measured by O'Donoghue (1986). In its interoutburst behavior, Z Cha is seen as similar to VW Hyi. Both these observations, as well as that of U Gem by Kiplinger, Sion and Szkody, can be explained in the mass

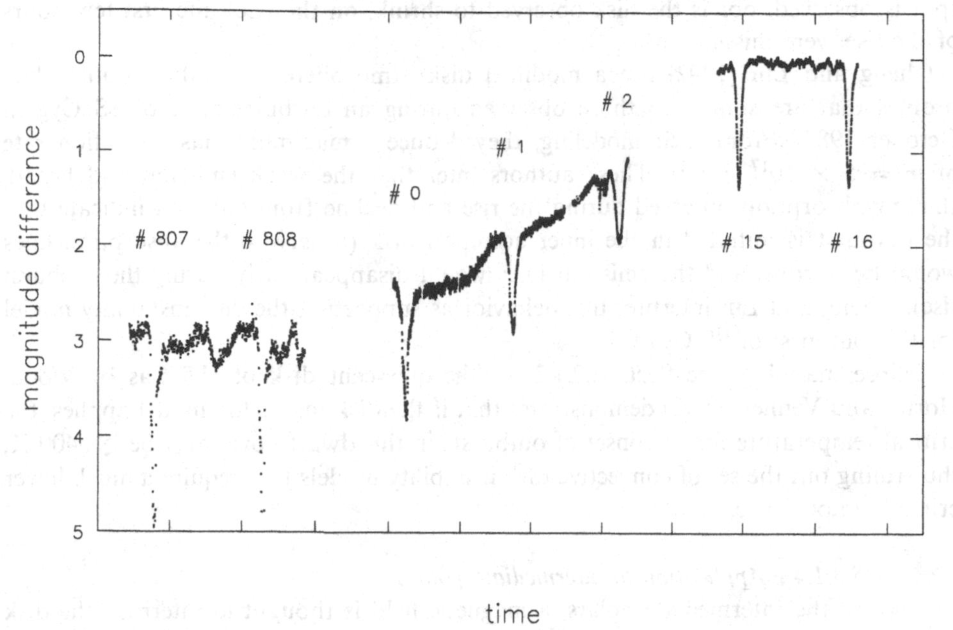

Fig. 8.13. Optical eclipse light curves of the dwarf nova OY Carinae, illustrating quiescence, the early rise to outburst, and outburst maximum. The ordinate represents the magnitude difference with a comparison field star. The tick marks on the abscissa mark hours of time. Numbers associated with each light curve indicate an eclipse cycle count. From Rutten *et al.* (1992).

transfer instability model but are more difficult to explain in the disk instability model. The latter half of the interoutburst interval of Z Cha was not well covered in the observations of AKP, and their conclusions, therefore, should be regarded as tentative.

Rutten *et al.* (1992) have employed the maximum-entropy eclipse-mapping technique (see Sect. 8.2.1.2) on eclipse photometry of the dwarf nova OY Car ($i = 83°$) during quiescence and the rise and maximum of a normal outburst. Figure 8.13 shows some of their eclipse data; the figure is one of the finest examples of eclipse light curves spanning the range from quiescence to rise to maximum. Rutten *et al.* use the resulting intensity maps to study the development of the disk radius, brightness structure, and the radial temperature dependence of the disk in the context of the two competing dwarf nova outburst models. They find that at outburst maximum, about 15% of the total light comes from an uneclipsed source that develops during the rise to outburst. This additional light source is much brighter than the entire system during quiescence. The speculation as to the origin of this extra light source ranges from a brightening of the entire companion star to disk light scattered into the line of sight by a wind high off the plane of the disk. Rutten *et al.* find that during the outburst rise the most dramatic changes occur in the outer disk; an interpretation is that a bright ring develops in the outer disk. While the observations of Rutten *et al.* do not show evidence for a mass transfer outburst (i.e. no enhancement of the bright spot is observed, nor is the disk observed to shrink, on the rise), the first few hours of the rise were missed.

Cheng and Lin (1992) use a modified disk atmosphere to produce optical line profiles that are similar to those observed during an outburst cycle of SS Cyg in October 1981. From their modeling, they deduce a maximum mass accretion rate of $\dot{m} = 5 \times 10^{17}$ g s^{-1}. These authors infer that the weak emission and broad, shallow absorption observed during the rise and decline from outburst indicate that the outburst is initiated in the inner accretion disk (otherwise the absorption lines would be narrow, and the emission lines would disappear early during the outburst rise). Cheng and Lin interpret this behavior as supporting the disk instability model for this outburst of SS Cyg.

Eclipse mapping (see Sect. 8.2.1.2) of the quiescent disk of HT Cas by Wood, Horne, and Vennes (1992) demonstrates that if the disk instability model applies, the critical temperature for the onset of outbursts in this dwarf nova must be $\gtrsim 5400$ K, thus ruling out the set of convective disk instability models that require a much lower critical temperature.

8.3.1.4   Application to intermediate polars

In the intermediate polars, a magnetic field is thought to interrupt the disk flow and there is no inner disk. A few of these systems have been observed to exhibit dwarf nova type outbursts, albeit of smaller amplitude than the usual dwarf nova outburst. Angelini and Verbunt (1989) discuss the intermediate polar outbursts in terms of an instability in a magnetically truncated disk. Kim, Wheeler, and Mineshige (1992) find, too, that they can explain the recent small outbursts in GK Per with a disk model in which the instability starts in the inner disk, and a modest increase in the mass transfer rate from the companion star that began about a decade ago. Photometry and red spectroscopy of EX Hya during one of its rare outbursts has been interpreted, on the other hand, by Hellier *et al.* (1989) as due to an increase in the rate of mass transfer from the companion, with the enhanced mass transfer stream skimming over the top of the initial impact region at the edge of the accretion disk to strike the magnetosphere directly. In this picture, the interaction of the stream with the magnetosphere gives rise to a broad, high-velocity base component in the wings of the Hα line, whose velocity is modulated with the orbital period. The interpretation of the outbursts of the intermediate polars is clearly in its initial stages.

8.3.1.5   The superoutburst

A subset of dwarf novae, called the SU UMa stars, show 'superoutbursts' in addition to the more ordinary outbursts of the average dwarf nova (see Sect. 10.4 for a discussion of theoretical models of the superoutburst phenomenon). There is some evidence to suggest that superoutbursts may be triggered by ordinary outbursts as they are sometimes observed to follow the latter by a few days. In the case of VW Hyi, the normal outburst and superoutburst are sometimes clearly separated by a drop in the light curve, which is more pronounced at higher energies (Bateson 1977; Marino and Walker 1979; Polidan and Holberg 1987). An interesting observation (Kuulkers *et al.* 1991) is that there are strong dips in the optical light curves during the initial stages of superoutbursts of Z Cha. The authors interpret this as due to an enhanced mass transfer rate from the companion star compared to the rate during normal outbursts, when no dips are observed. The dips are discussed in Sect. 8.3.5 and illustrated in Figure 8.14(a).

Many ideas for the nature of the superhumps have been proposed and discarded; a model that has survived is that they are caused by variations in tidal heating in the disk (Whitehurst 1988). In systems with small mass ratios, the disk can become non-axisymmetric and precess slowly in its inertial frame if it extends beyond its tidal instability radius. This results in the superhump phenomenon, which would then be a periodic tidal heating of the disk on a time scale within a few per cent of the binary orbital period. In this picture, the entire disk brightens; this is why the superhump phenomenon is not an observed function of inclination. Tidal instability has been explained invoking a disk instability (Osaki 1989) and, alternatively, a mass transfer instability (Whitehurst and King 1991). In the former model, the increased mass transfer rate may be induced in the disk by irradiation. The latter authors claim that the enhanced mass transfer is triggered by the ordinary, precursor outburst.

The issue – companion star instability or disk instability – is still not settled for the dwarf novae. The truth might lie with some combination of both kinds of instabilities, or in a new, presently unrecognized, instability.

8.3.2 Flickering

Flickering is characteristic of CVs at virtually every wavelength and must be regarded as a consequence of the mass transfer and accretion process. Little is understood, however, about the nature of the flickering. The usual time scale for optical flickering is one to several minutes and its amplitude is high, 0.1–1 mag. High-speed white-light photometry of the optical eclipses of several systems shows that the flickering may come from different locations in different systems: in U Gem the flickering appears to originate from its bright spot (Nather and Warner 1971); in RW Tri the entire disk participates in the flickering (Horne and Stiening 1985); and in HT Cas the flickering appears to come from the region near the white dwarf (Patterson 1981). Studies of the color of the flickering are rare: in SY Cnc the flickering is blue in color and has a Balmer jump in emission (Middleditch and Córdova 1982). AE Aqr can have long periods of no flickering which are punctuated by episodes of very high-amplitude flickering in the Balmer continuum or the emission lines, or both (Bruch 1991). SS Cyg also shows hydrogen emission line flares (Walker 1981).

X-ray flickering on time scales of minutes is also common in both the hard X-ray flux and soft X-ray flux of virtually all types of CVs, although the flickering in these bands is not correlated (see, e.g., the *ROSAT* observation of the polar EF Eri by Beuermann, Thomas, and Pietsch 1991). For a couple of systems, data have been taken that allow a cross-correlation of the optical and X-ray flickering. Simultaneous hard X-ray and optical photometry of the novalike variable (intermediate polar) TT Ari revealed correlated hard X-ray and optical variability over a broad range of time scales, from ~1000 s to the binary orbital time scale, as well as shorter time scale QPOs in both frequency bands (Jensen *et al.* 1983). The dwarf nova V426 Oph exhibited, during some portions of extended *Ginga* and overlapping optical observations, quasi-coherent hard X-ray and optical modulations with a time scale of about 28 min (Szkody, Kii, and Osaki 1990).

8.3.3   Oscillations

The regular, rapid optical and X-ray oscillations exhibited by dwarf novae in outburst and some novalike stars are discussed in Sect. 8.2.2.3. There are no published simultaneous studies of these optical and soft X-ray oscillations. Longer-period, less coherent oscillations are observed in the optical light of some dwarf novae at the same time that the higher-frequency oscillations are detected (Patterson 1981). These oscillations have the time scales corresponding to the range of Keplerian time scales expected in the disk. In X Leo, these oscillations were detected in the R, but not the B, band (Middleditch and Córdova 1982).

8.3.4   Humps

The binary orbital period in CVs often manifests itself as a broad hump in the light curve. For some of the apparently nonmagnetic, disk-accreting CVs, a broad enhancement occurs once per binary orbit at a phase of ~0.8, or just before the eclipse of the white dwarf. This is the expected location, where the mass stream from the companion impacts the disk. These humps, called bright spots, are observed in the quiescent, but not the outburst, state of dwarf novae. In some systems (e.g. U Gem and Z Cha), the bright spot is a prominent feature in the quiescent light curve, but in others, like HT Cas, it is weak or absent (Horne, Wood, and Stiening 1991). Periodic orbital humps are also associated with the polars; these are linked to the viewing angle of the magnetically controlled accretion column(s) and are used to derive the geometry of the accretion flow (see Cropper 1990). During the superoutbursts of the SU UMa dwarf novae, 'superhumps' appear in the optical light curve at the peak of the outburst (or during the final rise) and continue during the decline, with a period that is a few per cent longer than the orbital period (see Sect. 8.3.1.5). The amplitude of these humps (a few tenths of a magnitude) is insensitive to inclination. Patterson *et al.* (1993) have generalized the superhump phenomena beyond the subclass of dwarf novae, making a case that this behavior is common among CVs of high mass accretion rate and short orbital period. Figure 8.14 shows some of the variety of humps observed in CVs: the superhumps of Z Cha during a superoutburst, the orbital hump of U Gem during quiescence, and the hump of the polar QQ Vul.

8.3.5   Dips

Broad intensity dips, variable in nature but often related to a particular phase in the binary orbit, have been observed in the optical and X-ray light curves of several CVs. A series of *EXOSAT* observations of U Gem on the decline of a particularly long outburst revealed phase-related broad dips in the soft X-ray light (Mason *et al.* 1988). The morphology of the dips changed from cycle to cycle; two dips per orbit were observed to be centered at phases 0.7 and 0.15 with respect to the time of optical eclipse (phase 0.0). The relative amplitude of the dips changed throughout the four-day observation, as Figure 8.15 illustrates. The ingress to one deep dip on day 2 of the X-ray observations occurred in less than 15 s. The dips are remarkably similar in morphology and phasing to the dips that occur in some low-mass, accreting neutron star binaries (e.g. Parmar and White 1988). The dips in U Gem were not evident in simultaneous UV or medium-energy X-ray data.

Wide, transient dips have been observed in an optical superoutburst of the eclipsing

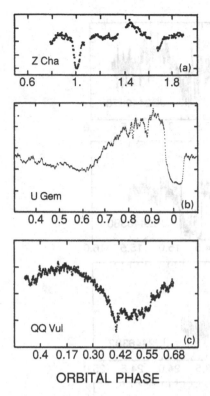

Fig. 8.14. Triptych of CV orbital light curves, illustrating a variety of humps and dips. (a) A superhump and a small dip (in addition to deep eclipse) during superoutburst of Z Chamaeleontis. The superhump phenomenon is thought to be associated with precession of the disk because of tidal forces. From Kuulkers *et al.* (1991). (b) The prominent hump in the quiescent light curve of U Geminorum is believed to be due to a bright spot on the outer accretion disk at the mass stream impact point. From Nather and Warner (1971). (c) The broad dips in polars, like QQ Vulpeculae, are associated with the geometry of the magnetically funneled, columnar flow. From Nousek *et al.* (1984).

Z Cha, beginning with the precursor outburst (Kuulkers *et al.* 1991). The dips, which can be seen in Figure 8.14(a), last from orbital phase 0.6 to 0.9 and are strongest at the shortest wavelengths (Walraven *U* and *W* bands). Less prominent dips appeared at other orbital phases. Other objects showing similar dips, OY Car and HT Cas, are also eclipsing systems.

The observation of broad dips at certain phases in the light curves of low-mass, nonmagnetic neutron star binaries has been used to develop simple models of the structure of the disk in which the rim of the disk is extended vertically, owing to the impact of the accretion stream, and occults the inner, emitting regions of the disk (see Ch. 1.2; also Livio 1993a). One example of modeling is the time-dependent, three dimensional particle simulations of an accretion disk interacting with an incoming stream in a close binary system performed by Hirose, Osaki, and Mineshige (1991). These researchers found that the half-thickness of the disk is much larger everywhere

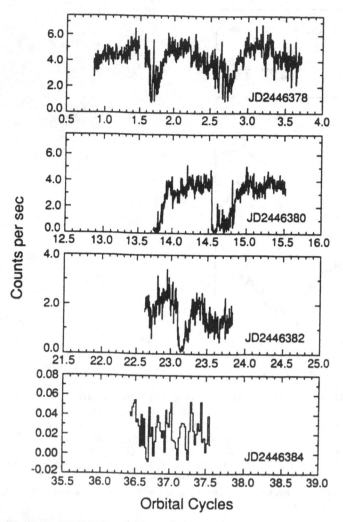

Fig. 8.15. *EXOSAT* soft X-ray (0.04–2 keV) light curves of U Geminorum over four consecutive days on the decline of a wide outburst, illustrating prominent dips that repeat at similar orbital phases, but with changes in relative amplitude. By the fourth day, the soft X-ray intensity has decreased by more than a factor of 200. The ingress to the deep dip in cycle 14 occurs in ~ 15 s. From Mason *et al.* (1988).

in the outer portions of the disk than the value expected from hydrostatic balance. They also found that, because of the violent collision of the incoming stream with the disk, the ratio of the vertical height of the disk, H, to the radius of the disk, r, is on average 10–15 %, and is especially enhanced at orbital phases 0.8, 0.2, and 0.5, where $H/r \sim 0.15$. These are the same phases at which the dips are observed. While this picture may satisfactorily explain the light curves of the neutron star binaries and also the optical dips of the eclipsing dwarf novae, it falls short of explaining similar behavior in U Gem because this object is not observed at a high enough inclination for the inner part of the disk to be eclipsed by structure on the rim of the disk.

Mason *et al.* (1988) argue that the dips in U Gem are caused by photoelectric absorption in material with a column density of 3×10^{20} atoms cm^{-2}, far from the X-ray source and high above the orbital plane. It may be that a scenario like the one suggested by Walter *et al.* (1982; see also Frank, King, and Lasota 1987), in which dips are caused by material in the accretion stream that skims over and above the disk, applies in the case of U Gem. A discussion of the possible cause of variable dips in low-mass X-ray binary light curves is given in Sect. 10.2.

8.3.6 *Bursts*

Rare instances of isolated, short time scale enhancements in the flux of CVs have been reported. An *Einstein X-ray Observatory* observation of U Gem during an optical outburst revealed an ~ 15 s burst of soft X-rays in which the count rate increased three-fold (Mason and Córdova 1983). This is the time scale of the ($\sim 15\%$ amplitude) soft X-ray QPOs detected in this object during an earlier (*HEAO-*1) observation. A large-amplitude flare lasting about 5 min was observed in the red optical light of HT Cas; the flare was undetectable in *U* light (Horne, Wood, and Stiening 1991). The flare's time scale is also similar to the flickering time scale in this star, albeit with a much larger magnitude.

8.4   The importance of magnetic fields in CVs

One of the most remarkable pieces of information we have learned about CVs since the late 1970s is the importance of magnetic fields and how their presence or absence influences the geometry of the accretion flow. The pre-1977 paradigm of CVs had no provision for magnetic fields. Yet today, largely because of the development of the coincident technologies of sensitive optical and infrared polarization studies and X-ray telescopes in space, we know that magnetic fields can significantly affect the mass flow between stars in these close binaries.

8.4.1 *X-ray emission*

Many magnetic CVs were discovered first in X-rays because of the copious hard and soft X-ray emission produced when matter falls radially onto a degenerate star. The radiation from this geometry, and under the mass accretion rates and magnetic field strengths of relevance to CVs, has been described by Lamb and Masters (1979) in the first, classic treatment of this problem for polars. Early X-ray experiments discovered that for some polars the flux of soft X-rays relative to the flux of hard X-rays exceeded the predictions of the standard model (see the review by Cropper 1990 and also the more recent EUV results on the polar UZ For by Watson 1992). The presently accepted cause for this is that the accretion flow onto the magnetic poles is not smooth, but blobby. Only the small, low-density blobs falling through the accretion column produce shocks above the white dwarf's surface that cool by bremsstrahlung (hard X-ray) and cyclotron (optical and IR) emission; denser, larger blobs penetrate all the way to the white dwarf's photosphere before cooling, releasing their energy as thermalized soft X-ray and EUV emission (Frank, King, and Lasota 1988). This results in a wide spectrum of observed ratios of soft to hard X-ray emission among the polars.

ROSAT observations of one polar, EF Eri, clearly distinguish between the hard and soft X-ray components and show that they vary independently and thus come from separate regions in the accretion flow; this object shows high-amplitude soft X-ray flaring that is not detected in hard X-rays (Beuermann, Thomas, and Pietsch 1991). Such observations can be used to constrain models for the detailed accretion geometry.

The emission mechanisms appropriate to the polars may hold also for the intermediate polars, except that in the latter the area over which the accretion occurs is believed to be greater (see Sect. 8.4.3). In addition, the intermediate polars (IPs) appear to have higher column densities than the polars, perhaps due to a higher accretion rate and a more complicated accretion geometry (Norton and Watson 1989). Polarized (cyclotron) radiation will be more difficult to detect in these systems because of dilution from disk light (if present) and the complex accretion flow.

EXOSAT medium energy (1–10 keV) data revealed that most IPs exhibit predominantly hard X-ray emission, in contrast to the polars, which during bright states usually exhibit a pronounced ultrasoft X-ray component and a harder X-ray emission component. Combining low- and medium-energy *EXOSAT* data showed that the X-ray spectra of IPs do not fit a single-temperature absorbed thermal-bremsstrahlung model. Norton and Watson (1989) model the X-ray spectrum as due to partial covering of the X-ray emission region ('patchy absorber' model), although they note that this model is not unique and a two-component model might also fit the data equally well.

A soft X-ray component from the white dwarf's surface could be present in IPs, but may have too low a temperature to detect if the accreting area is large, or the absorbing column is high (King and Lasota 1990). The intermediate polar RE 0751+14, discovered with the *ROSAT* WFC, is a nearly singular example of an IP evidencing a strong, distinctly soft X-ray/EUV component (Mason *et al.* 1992). The only other known IP from which EUV and soft X-ray emission has been detected is EX Hya (Córdova, Mason, and Kahn 1984; Watson 1992).

Data on EX Hya from the solid state spectrometer (SSS) on the *Einstein X-ray Observatory* affords a means of examining this object's soft X-ray emission with moderate spectral resolution ($\sim$ 160 eV FWHM) in the 0.5 to 4.5 keV band. Singh and Swank (1993) modeled the soft excess found with the SSS as a 0.74 keV optically thin plasma in collisional ionization equilibrium. A variation in the soft X-ray flux with the phase of the 67 min white-dwarf rotation period in EX Hya is modeled as arising because of absorbing material that covers 40% of the emission region at the minimum phase of this period. Singh and Swank also detected in the EX Hya SSS data a 1.72 keV Si fluorescence line, implying the existence of a 6.4 keV iron fluorescence line. There is insufficient spectral resolution in data from the host of X-ray observatories that have detected a 6.7 keV line to resolve whether this feature includes a 6.4 keV line. The analysis of Singh and Swank suggests that to produce the observed luminosity of the He II λ4686 lines there should be a soft X-ray/EUV component in addition to the extrapolation of the SSS data below 0.5 keV. The *ROSAT* WFC and PSPC data on EX Hya have not yet been subject to a rigorous analysis to determine whether or not the observed low-energy emission constitutes a component that is distinct from the SSS component.

The *Ginga* satellite, operating at higher X-ray energies, observed more than a dozen magnetic CVs, including the intermediate polar GK Per, and confirms in them the presence of substantial hard X-ray emission with an optically thin plasma temperature of tens of kilo-electron-volts. Providing a good model fit to the hard X-ray spectrum of a few of the systems requires a combination of two bremsstrahlung models having the same temperature but different column densities (i.e. the leaky absorber model). An interpretation of this is that the absorbing matter in the accretion column is inhomogeneous (see Ishida 1991 and Ishida *et al.* 1992 for a report of *Ginga* observations of many CVs). Rosen *et al.* (1991) made extensive *Ginga* observations of the intermediate polar EX Hya and found that the iron line energy of 6.7 ± 0.05 keV indicates that it is thermal in origin; this is in contrast to the proposed fluorescence origin for the iron line in many other IPs (Norton and Watson 1989).

8.4.2 *Magnetic field strength and evolution*

Before *ROSAT* discovered approximately 17 new magnetic CVs (Watson 1993; Beuermann and Thomas 1993), 19 CVs were known to exhibit optical circular polarization, indicating magnetic field strengths of 10 to 60 MG. All but three of these CVs are AM Her stars. Stockman *et al.* (1992) have carried out a decade-long optical and infrared polarimetric observing program on ~ 80 CVs, representing most of the known (pre-*ROSAT* epoch) AM Her stars and IPs, as well as many other subclasses of CV. They find a lack of AM Hers with field strengths higher than 60 MG and a lack of asynchronous CVs with $B > 5$ MG. They claim that this is not due to selection effects, but reflects a difference in the evolution of magnetic CVs with respect to nonmagnetic CVs. They postulate that diskless asynchronous CVs could evolve faster if they expel material and angular momentum in a wind. Wickramasinghe, Wu, and Ferrario (1991) also show, using detailed calculations of the polarization properties of extended accretion shocks on the surface of a magnetic white dwarf, that the null detection of circular polarization from most IPs implies fields of less than 5 MG. While polarization is not commonly found in IPs, there are at least two detections: West, Berriman, and Schmidt (1987) report optical and near-IR polarization (V(J) $\sim 1\%$) of BG CMi, and Piirola, Hakala, and Coyne (1993) report variable circular and linear polarization of the newly discovered RE 0751+14 in the IR (4% CP) and optical (2% CP).

Most polars have orbital periods less than 2 h, while most IPs have periods greater than 3 h (see discussion in Sect. 8.5.1 and Figure 8.16, based on pre-*ROSAT* data). The recent discovery of more magnetic CVs with the *ROSAT* WFC does not substantially change this picture (Watson 1993). There has been much discussion in the literature concerning the evolution of IPs into polars, for example whether IPs above the period gap evolve into polars below the gap, but this is not resolved. In known IPs above the gap, there is good observational evidence for the existence of an accretion disk (see Sect. 8.4.4 below). Wickramasinghe, Wu, and Ferrrario (1991) predict that IPs above the gap will evolve into asynchronous, diskless systems below the gap and will emit mainly in the EUV region because of the larger area over which accretion occurs. *ROSAT* and *EUVE* observations will be important in resolving the evolutionary picture of magnetic CVs. King and Lasota (1991) hold that the nonsynchronous precursors of the polars are not the known IPs, but are an unobservable (nonaccreting) class of diskless binary.

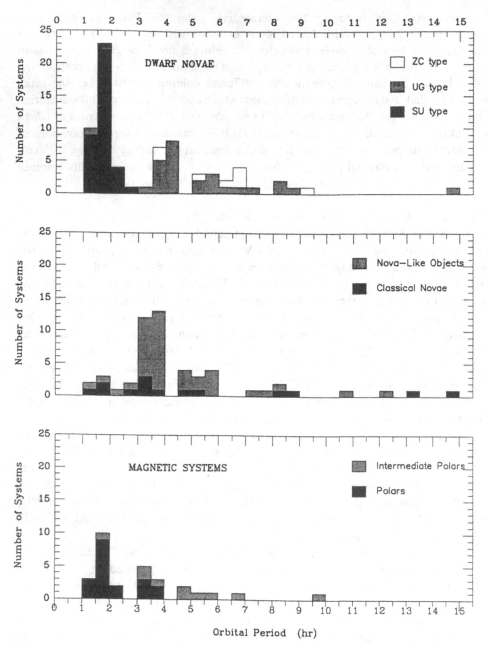

Fig. 8.16. Period distribution for various subclasses of CV (pre-*ROSAT*-discovered CVs). Drawing by F.A. Córdova and H. Papathanassiou.

8.4.3   *Accretion geometry*

The accretion geometry of the IPs is a much-debated topic which uses for its arguments the characteristics of the pulse modulation (see discussion and diagrams in Mason, Rosen, and Hellier 1988). The occultation model, proposed by King and Shaviv (1984), requires a rather large polecap in which the optical depth to absorption is least vertically and greatest horizontally. Thus the flux maximum occurs looking down into the pole. In order to produce sinusoidal light curves and to explain the magnitude of the absorption, this model requires each polecap to cover about one-third of the white dwarf surface. In order to keep the absorption in phase with the occultation, accretion must fill the polecap's area. This model requires that we see emission only from the upper pole. The absorption model, in contrast, requires a significantly smaller footprint. It proposes that the accretion regions are thin, azimuthally extended curtains whose footprints are arcs around the magnetic poles covering only 0.001–0.01 of the white dwarf surface. Flux maximum occurs when viewing the curtain side-on because this is where the absorption is least (see illustrations in Rosen, Mason, and Córdova 1988, and Hellier 1991).

Norton and Watson (1989) made a comparative study of all the IPs observed with *EXOSAT*. The modulation depths of the pulse profiles increase with decreasing energy. This is in contrast to the polars, whose strong modulation does not appear to vary as a function of energy. The interpretation is that in the latter, the modulation is due to self-occultation of the column, whereas in IPs the modulation is due to the combined effects of self-occultation of the footprint of the magnetic accretion column and photoelectric absorption in the same accretion column. The Norton and Watson conjecture requires that the emission region must occupy at least one-quarter of the white dwarf's surface. This is a much larger accretion region than that postulated in the accretion curtain model. The latter model can also incorporate occultation with absorption; the occultation arises by allowing a contribution from the second pole below the orbital plane. In the Norton and Watson model the occultation and absorption are constrained to operate in phase, while in the Rosen, Mason, and Córdova (1988) model they are not.

Ginga observations of EX Hya have proved particularly useful for constructing a model of the accretion geometry because in this IP the companion star occults part of the central X-ray emitting region every 98 min. The new information wrought by *Ginga* is that the pulse fraction of the 67 min white dwarf rotational modulation decreases with increasing X-ray energy between 1.5 and 6 keV, but is constant at ∼10% at higher energies up to at least 15 keV. In contrast, the fractional depth of the X-ray eclipse increases with increasing energy between 1.5 and 4 keV and at higher energies is level at 40–50%. Rosen *et al.* explain these and other complex behaviors of the light curve in terms of two emission regions, one located above and one below the plane of the disk; both occultation and photoelectric absorption are important in this model.

8.4.4   *Disks or no disks: the intermediate polars*

Although perhaps some theoretical considerations favor direct accretion onto the magnetosphere in an IP (e.g. Hameury, King, and Lasota 1986; King and Lasota 1991), several observations suggest that disks are present in many IPs (Hellier 1991). Eclipses have been used to give the spatial extent of the eclipsed material, and the S-wave shape of the emission line profiles give the location of this emission; both types of observations imply that mass from the companion star impacts far out in the Roche lobe of the white dwarf, suggesting that a disk is present. Hellier (1991) also argues that, in the case of FO Aqr, the X-ray spin pulse, with its sinusoidal profile, requires an azimuthally uniform accretion flow, which is naturally provided in the case of disk accretion. In FO Aqr, there is evidence that both accretion modes can occur simultaneously when the mass from the secondary overshoots the accretion disk and goes onto the magnetosphere directly (Hellier, Mason, and Cropper 1990).

8.4.5   Magnetic fields: common to all CVs?

The observational results discussed in Sect. 8.4.2 leave us with the conclusion that the magnetic fields of the majority of CVs are < 5 MG. Yet a field strength of 0.1–1 MG could exist in these objects and account for a number of unexplained behaviors. I give some examples below. The observational results are discussed earlier in this chapter; the hypothesis about the connection with magnetic fields is chiefly my own. These are offered in the spirit of stimulating work on the effects of weaker magnetic fields on the accretion process.

- The similarity of the coherence and time scales of the QPOs in the polar EF Eri (Beuermann, Thomas, and Pietsch 1991) and the dwarf nova U Gem (Córdova *et al.* 1984) suggests a common mechanism for the production of the QPOs.
- The phase-related asymmetries in the spectral wind lines of YZ Cnc (Drew and Verbunt 1988) and other CVs (Sect. 8.2.2.7) may arise owing to the presence of a magnetically controlled wind.
- Production of a wind-like spectral line profile during a flare in the IP TV Col (Szkody and Mateo 1984) illustrates that high-velocity winds do occur in magnetic species of CVs.
- Distorted optical and UV line profiles in some novalike objects (Sect. 8.2.1.4) suggest vertical flow of material.
- Dwarf novalike outbursts in IPs (e.g. GK Per, EX Hya) show that magnetism does not hinder the dwarf nova phenomenon. In fact, it may assist: Livio and Pringle (1992) find that a delay of the UV with respect to the optical maximum can be produced if the white dwarf has a magnetic field of $\sim 10^4$ G. In this case, during quiescence the inner disk is evacuated and there is no accretion. During outburst there is a delay before the hotter inner disk region is filled and radiating. (It should be noted that the disk instability model employed by Livio and Pringle, while producing the effect of a delay, does not produce the magnitude of the difference in optical and UV fluxes observed.)

A final word on magnetic fields, quoting Eggleton (1990): 'Just because a CV is not an AM Her or a DQ Her, it does not follow that it is non-magnetic.'

8.5   CV evolution

8.5.1   Angular momentum losses and secular changes in the orbital period

Ordering CVs by binary orbital period has led to insights about the angular momentum losses among these objects and their secular orbital evolution (Paczyński 1981; Rappaport, Joss, and Webbink 1982; Paczyński and Sienkiewicz 1983; King 1988; see also Sect. 11.3). Reliable orbital periods have been determined for more than 140 CVs (Shafter 1991; see also Watson 1993 for references to follow-up work on further *ROSAT* WFC discoveries). The period distribution has a minimum at 80 min and there is a dearth of CVs with periods between two and three hours. Figure 8.16 shows period distributions for the different subclasses and subtypes of CV.

It is thought that the progenitors of CVs may be short-period detached binaries with a degenerate or subdwarf component, like V471 Tau (Paczyński 1981). The loss of angular momentum will bring these systems into a semi-detached arrangement, whereupon mass transfer can ensue.

At periods greater than three hours (that is, above the period gap), the observed accretion rates are too large to be explained by gravitational radiation; instead it is thought that CVs lose angular momentum by magnetic braking through the stellar wind of the secondary, with the result that the orbital period decreases (Verbunt and Zwaan 1981; Mestel and Spruit 1987). During this time, the companion star is oversized for its mass compared with its main-sequence radius. In one evolutionary scenario, as the period decreases to about three hours, the secondary becomes fully convective and the efficiency of the magnetic braking is quenched (the theory to explain this is not well-developed; see, e.g., Livio 1993b). The secondary then shrinks inside its Roche lobe and mass transfer ceases. The system undergoes a long ($10^{8.5-9}$ yr) period of 'hibernation' during which it is unobservable because of its faintness. Eventually, through the contraction of the orbit brought about by gravitational radiation, or, perhaps, a residual magnetic braking, the Roche lobe shrinks until, at about two hours, it makes contact with the secondary, and mass transfer resumes. When the binary orbit, driven by gravitational radiation losses, reaches a period of about 80 min, the (hydrogen-rich) secondary becomes too small (M $\lesssim 0.1 M_\odot$) to support nuclear burning and it leaves the main sequence, gradually becoming degenerate (Paczyński and Sienkiewicz 1983).

The few CVs that have been found in the middle of the period gap may have been formed there, but the CVs found near the edges of the period gap may have a different evolutionary history (see, for example, Hameury, King, and Lasota 1991).

Paczyński (1981) briefly discusses the possibility that magnetic winds, perhaps driven by disk accretion, could provide a sink for angular momentum loss and therefore be important in the evolution of close binaries.

8.5.2   *Long-term changes in the mass transfer rate and orbital period*

Cyclical variations of outburst time intervals and cyclical variations of the interoutburst luminosity, both on the time scale of several years, have been reported for a number of dwarf novae. Bianchini (1992) gives a brief review of this phenomenon, examining especially the hypothesis that modulations in the mass transfer rate may result from solar cycles of the companion stars. The lengths of these apparent cycles in CVs are not correlated with the orbital periods of the systems. There is a large literature on observed *erratic* changes on time scales of 5–10 yr in the orbital periods of CVs (e.g. Africano *et al.* 1978; Rubenstein, Patterson, and Africano 1991). One idea, that these changes are caused by changes in the quadrupole moment of one star in the binary system, has been discounted by Marsh and Pringle (1990), who find that too much energy is required to effect this on the observed time scale.

8.5.3   The temperature of CV white dwarfs and implications for CV lifetimes

We know much about the details of the masses, temperatures, and composition of isolated white dwarfs (see contributed papers in *White Dwarfs*, edited by Vauclair and Sion 1991). Do we know anything about the white dwarfs in accreting binaries? Do we know how they cool, and hence how long they will live? Although it is *characterized* by the presence of the white dwarf, a cataclysmic variable's behavior is *dominated* by the accretion process. Only when the accretion rate is very low is the compact star unveiled. This happens for dwarf novae between outbursts and for magnetic variables and novalike objects during low states in luminosity. All such observations have been collected and analyzed by Sion (1991), and they reveal a distribution in effective temperature centered about 16 000 K. Sion notes that if repeated nova explosions do not appreciably erode the core mass of the white dwarf, thus accelerating its cooling, this distribution implies a mean lower limit to the total cooling lifetime of 5×10^8 yr for CVs. If CVs were born as detached binaries, this distribution implies that envelope heating is induced by the accretion process. If, on the other hand, CVs were born as semi-detached systems, or if the time scale to Roche lobe overflow is less than 10^9 yr, then 5×10^8 yr is a typical lower limit to the CV lifetime. This gives us an estimate of the lifetime of a CV binary, namely about 10^9 yr. The distribution of hot and cool CVs on either side of the period gap (chiefly the paucity of hot degenerates below the gap) has been used as weak evidence that CV systems evolve across the period gap. Studies are underway to try to determine whether magnetic CVs cool differently than nonmagnetic CVs. There is little understanding of how the long-term core cooling of the white dwarf is affected by heating due to continual accretion or cooled by the loss of core mass in repeated nova explosions.

8.5.4   With a whimper or a bang?

What becomes of a cataclysmic variable star? Do any of them end up as doubly degenerate binaries? While it may seem reasonable that some white dwarfs could, through accretion, grow to the Chandrasekhar limit and thereafter suffer a supernova explosion or accretion-induced collapse to a neutron star, in practice this is plausible for only a small subset of CVs (see Sect. 11.3). The existence among the classical novae of 'neon novae', that is, novae in which the ejected material is rich in intermediate mass elements, suggests that we are observing ONeMg white dwarfs in these systems. In order for the Ne rich core material to be dredged-up, the white dwarf has to be eroding away its helium and CO layers, and thus decreasing in mass as a result of nova explosions (Livio and Truran 1992).

This is not the situation for the known recurrent novae, for which observations may suggest that the white dwarf grows in mass. Livio and Truran (1992) model the outbursts of the recurrent novae, at least the ones with low-mass companion stars, as due to a thermonuclear runaway on a massive white dwarf ($M > 1.3 M_\odot$) that is accreting at a high rate ($\dot{m} > 10^{-8} M_\odot$ yr^{-1}). It is these two conditions, plus the fact that their recurrence periods between outbursts are short compared to the classical novae, that prohibit a great deal of mixing, and hence mass dredge-up, in the white dwarf.

The most likely white dwarfs to reach the Chandrasekhar limit are those which satisfy the conditions: $M_{WD} > 1.2M_\odot$ and $\dot{m} > 10^{-8}M_\odot$ yr^{-1} and in which the secondary is transferring helium-rich material. The CV systems that can grow to the Chandrasekhar mass can give way to an accretion-induced collapse, forming a low-mass neutron star binary, or a Type Ia supernova. Nomoto and Kondo (1991) favor the former for ONeMg white dwarfs, in which the collapse is triggered by electron capture on ^{24}Mg and ^{20}Ne, and C/O white dwarfs, in which the collapse results from carbon deflagration. The rate at which these recurrent novae can be expected to produce accretion-induced collapse is consistent with the birth rate of low-mass X-ray binaries. Wheeler (1991), on the other hand, makes the case for a Type Ia supernova as the endpoint of a CV, while Starrfield *et al.* (1991) argue specifically for a thermonuclear runaway on massive ONeMg white dwarfs in recurrent novae.

8.5.5 *Overlaps in CV subclasses*

In the late 1970s, there were four apparently mutually exclusive subclasses of CV. Today we know of several new subclasses (e.g. polars, intermediate polars, VY Scl stars, SW Sex stars), and we have seen that an object may be cataloged differently at different epochs. For example, the classical nova of 1901, GK Per, now exhibits dwarf novalike outbursts and has recently been discovered to be an IP. Angelini and Verbunt (1989) show theoretically how IPs might exhibit small, infrequent dwarf nova outbursts. Optical circular polarization has been discovered from the fast Nova Cygni 1975 (= V1500 Cyg) by Stockman, Schmidt, and Lamb (1988), revealing that this system was likely to have been a synchronized AM Her star before its classical nova outburst. O'Donoghue *et al.* (1990) have put forth the possibility, based on the slight difference between the photometric and spectroscopic periods of CP Pup, that this fast nova is, like V1500 Cyg which exhibits similar behavior, a magnetic CV. As discussed in Sect. 8.2.1.4, Williams (1989) argues on the basis of observational studies that the emission lines in many novalike stars and old novae may be formed in a magnetically confined accretion column. He cites the relative strength of He II in these stars as further evidence for a magnetic origin, since He II emission is a prominent feature in the spectra of polars. He offers the provocative idea that the short recurrence times of the recurrent novae could be aided by the presence of a high magnetic field for the white dwarf. A different idea that blends the CV subclasses has been put forward by Livio (1992a): the dwarf novalike outbursts of some old novae suggest that DNe and CNe are the same systems witnessed at different epochs in their cyclic evolution. This work does not take into account the effect of magnetic fields on the classical nova explosion.

8.6 Endpiece: areas for future studies

A review is both an end and a beginning, summarizing the current paradigms and revealing, in the ambiguities in interpretations of observations, some areas for new explorations. What follows is my personal view of some areas of CV research that have great potential for illuminating our understanding of the accretion process and its consequences in the environment of strong gravitational and magnetic fields.

(1) To model well the disk spectrum and calculate the mass transfer rate through the disk requires a unified treatment of the radiative transfer in the disk and its environs, without the arbitrary discrimination between optically thick and thin

components. The deficiencies in our present modeling are ably pointed out by Wade (1988). 'Unified model atmosphere' calculations are being developed (e.g. Gabler, Kudritzki, and Mendez 1991), in which a radiation-driven wind code is combined with a unified, non-LTE model atmosphere code for a spherical geometry to model stellar atmospheres more completely. As demonstrated by Shaviv and Wehrse (1991), making a model that solves the disk equations in a self-consistent way is essential. The currrently available X-ray spectral data are of particularly poor quality, limiting our understanding of the relative importance of BL and disk emission and our knowledge of any hot, optically thin region surrounding the accretion disk.

(2) It is not known whether or not the absence of velocity-shifted absorption components in the line profiles of quiescent CVs indicates the lack of a wind during low mass accreting states; at these times, the disk is not optically thick and would not provide the continuum light necessary to produce the absorption part of the line profile. High spectral resolution UV observations of the edge velocities of the emission lines as a function of the outburst state of an eclipsing dwarf nova like OY Car may be revealing.

(3) A model for the ionization structure of the CV wind is needed to infer mass loss rates. Presently there are many inconsistencies in our picture of the wind; its origin and nature, as well as its relation to the production of X-rays, hard or soft, and its effect on the stability of the rapid oscillations are not known. This is an area for productive new study and can draw on knowledge of winds in many other astrophysical systems.

(4) As studies of the line profiles of many novalike objects, especially the SW Sex stars, reveal, we do not yet have a picture of the location of the lines or understand how they are produced. In a two-dimensional approximation, the technique of Doppler tomography has some success, but this technique cannot be applied to mass flow out of the plane of the disk.

(5) Our knowledge of the BL temperature, size, and luminosity in disk-accreting CVs is especially poor. We have no good estimate of the energy budget of these systems, a situation exacerbated by the lack of EUV plus soft X-ray spectral data. There exists no simultaneous spectrum of a CV from hard X-ray to infrared wavelengths that would give us the relative flux contributions of the BL and disk. Such a spectrum is necessary in order to adequately model simultaneously the mass transfer and mass accretion rates, and to determine the characteristics of the BL. Broadband X-ray light curves of both eclipsing and pole-on systems are needed to resolve the location (and number) of X-ray components.

(6) The cause of the dwarf nova outburst eludes us still. Progress may come from spectroscopy in the infrared (to study the companion star) and the far-UV and EUV (to study the white dwarf and BL) throughout the entire outburst interval, including the interoutburst. What is needed is data of the quality of the *HUT* observation of U Gem (Long *et al.* 1993), but covering the extensive interoutburst interval to see the development of the instability, while simultaneously monitoring the 'health' of the white dwarf in the wake of an outburst.

(7) There is a need for high-time-resolution, simultaneous continuum measurements and spectroscopy of the flickering and rapid oscillations, at multiple wavelengths and over a range of accretion rates. This applies also to the study of the winds;

the relationship between these winds and the rapid soft X-ray/EUV oscillations is not understood, yet it is hard to imagine a scenario in which one is not affected by the other since both originate in the inner disk or BL of the white dwarf.

(8) Lack of knowledge of the vertical structure of the disk has hampered our ability to understand the CV emission line spectrum, the production of the hard X-ray emission and its relation to the optical flaring, and many of the dip and hump phenomena in the light curves. A variety of suitable viscosity mechanisms must continue to be explored. The effect of tides and resonances on the disk structure as a function of both the orbital elements of the system and the mass transfer rate is a new area that has yielded a new interpretation of the superhump phenomenon; it deserves further study for its possible application to other time-variable behavior in the disk.

(9) Long-wavelength studies (IR and radio) have great potential, especially for understanding the role of the companion star in the outbursts and evolution of CVs, and the importance of nonthermal processes in affecting the temporal behavior of these binaries.

(10) The effects of a small (10^4–10^5G) white dwarf magnetic field on the structure of the BL and its effect on the wind acceleration and structure should be examined. (A new paper relevant to this point came to my attention after a first draft of this manuscript was circulated: Livio and Pringle 1992a have produced a model for the outbursts of dwarf novae that depends on a small white dwarf magnetic field; see Sect. 8.4.5.)

(11) Young CV scholars can take heart. Amid the debates over the cause of the dwarf nova outburst, the presence of a disk in intermediate polars, the source of the emission lines in SW Sextantis stars, and the location of the wind in luminous CVs, there is one point upon which all the CV experts agree: 'The origin and evolution of cataclysmic variables is poorly understood' (Paczyński 1981), or recast with a 1990s outlook, 'The end-point of CV evolution seems to me to be as problematic as its starting-point' (Eggleton 1990). With the discovery of millisecond pulsars and unidentified soft X-ray sources in globular clusters, research on the evolution of CVs seems now more tantalizing than ever.

Acknowledgments

I thank my graduate students for supporting me while I wrote this review, especially Craig Robinson for providing a tower of references, and Haraklia Papathanassiou for executing two original figures for the review. Richard Wade provided a plethora of useful annotations on a first draft, and Mario Livio, Keith Mason, Chris Mauche, and Jim Pringle also made helpful comments. I thank my family, who often went fishing or ice-skating without me, for their patience during the protracted writing.

Note added in proof

This manuscript was completed early in 1993 and does not report therefore on the more recent spectacular spectral data taken on CVs with the *Extreme Ultraviolet Explorer Satellite* and Japan's X-ray satellite *ASCA*.

References

Africano, J. L., Nather, R. E., Paterson, J., Robinson, E. L. and Warner, B. 1978, *PASP*, 90, 568.

Alpar, A. and Shaham, J. 1985 *Nature*, 316, 239.

Angelini, L. and Verbunt, F. 1989, *MNRAS*, 238, 697.

Bailyn, C. and Grindlay, J. 1990, *ApJ*, 353, 159.

Bateson, F. M. 1977, *N.Z.J.Sci.*, 20, 73.

Bath, G. 1975, *MNRAS*, 171, 311.

Bath, G. T. and Pringle, J.E. 1985, in *Interesting Binary Stars*, eds. J.E. Pringle and R.A. Wade (Cambridge University Press), p. 177.

Becker, R. H. 1981, *ApJ*, 251, 626.

Belloni, T., Verbunt, F., Beuermann, K., Bunk, W., Izzo, C., Kley, W., Pietsch, W., Ritter, H., Thomas, H. C. and Voges, W. 1991, *A&A*, 246, L44.

Benz, A. O. and Güdel, M. 1991, *A&A*, 218, 137.

Bergeron, P., Liebert, J. and Saffer, R. A. 1992, *ApJ*, 394, 228.

Beuermann, K. and Thomas, H.-C. 1993, *Adv. Space Res.*, 13, (12), 115.

Beuermann, K., Thomas, H.-C. and Pietsch, W. 1991, *A&A*, 246, L36.

Bhat, C. L., Richardson, K. M. and Wolfendale, A. W. 1989, in Proc. of the GRO Science Workshop, ed. W. Neil Johnson (NRL), p. 4/57.

Bianchini, A. 1992, In Vina Del Mar Workshop on Cataclysmic Variable Stars, ed. N. Vogt, ASP Conf. Series 29, p. 284.

Bode, M. F. and Evans, B. 1989, eds., *Classical Novae* (Chichester: John Wiley & Sons).

Bookbinder, J. A. and Lamb, D. Q. 1987, *ApJ*, 323, L131.

Bruch, A. 1991, *A&A*, 251, 59.

Buckley, D. A. H., O'Donoghue, D., Hassall, B. J. M., Kellett, B. J., Mason, K. O., Sekiguchi, K., Watson, M. G., Wheatley, P. J. and Chen, A. 1993, *MNRAS*, 262, 93.

Cannizzo, J. K., Ghosh, P. and Wheeler, J. C. 1982, *ApJ*, 260, L83.

Casatella, A. and Viotti, R. 1990, eds., IAU Coloq. no. 122, *Physics of Classical Novae* (Springer-Verlag).

Cheng, F. H. and Lin, D. N. C. 1992, *ApJ*, 389, 714.

Cook, M. C. 1985, *MNRAS*, 216, 219.

Cool, A. M., Grindlay, J. E., Krockenberger, M. and Bailyn, C. D. 1993, *ApJ*, 410, L103.

Córdova, F. A. and Howarth, I. 1987, in *Exploring the Universe with the IUE Satellite*, eds. Y. Kondo *et al.* (Dordrecht; D. Reidel), p. 395.

Córdova, F. A. and Mason, K. O. 1982, *ApJ*, 260, 716.

Córdova, F. A. and Mason. K. O. 1983, in *Stellar-Driven Accreting X-ray Sources*, eds. W. H. G. Lewin and E. P. J. Van den Heuvel (Cambridge University Press), p. 147

Córdova, F. A. and Mason, K. O. 1984, *MNRAS*, 206, 879.

Córdova, F. A. and Mason, K. O. 1985, *ApJ*, 290, 671.

Córdova, F. A. and Mason, K. O. and Hjellming, R. 1983, *PASP*, 95, 69.

Córdova, F. A., Mason, K. O. and Kahn, S. 1984, *MNRAS*, 212, 447.

Córdova, F. A. and Mason K. O., and Nelson, J. E. 1981, *ApJ*, 245, 609.

Córdova, F. A., Nugent, J., Klein, S. and Garmire, G. 1980a, *MNRAS*, 190, 87.

Córdova, F. A., Chester, T., Tuohy, I. and Garmire, G. 1980b, *ApJ*, 235, 163.

Córdova, F. A., Chester, T., Mason, K. O., Kahn, S. and Garmire, G. 1984, *ApJ*, 278, 739.

Cropper, M. 1990, *Space Sci. Rev.*, 54, 195.

De Jager, O. C. 1991, *ApJ*, 378, 286.

Downes, R. A. 1986, *ApJ*, 307, 170.

Drew, J. 1987, *MNRAS*, 224, 595.

Drew, J. 1990, in IAU Coloq. no. 122, *Physics of Classical Novae*, eds. A. Casatella and R. Viotti (Springer-Verlag), p. 228.

Drew, J. and Verbunt, F. 1985, *MNRAS*, 213, 191.

Drew, J. and Verbunt, F. 1988, *MNRAS*, 234, 341.

Duck, S. R. and Ponman, T. J. 1992, in Vina Del Mar Workshop on Cataclysmic Variable Stars, ed. N. Vogt, ASP Conf. Series 29, p. 351.

Duerbeck, H. W. 1987, *Space Sci. Rev.*, 45, 1.

Dulk, F., Bastian, T. and Chanmugam, G. 1983, *ApJ*, 273, 249.

Echevarría, J. 1983, *Rev. Mex. Astr. Astrof.*, 8, 109.

Edwards, D. A. 1988, *MNRAS*, 231, 25.

Eggleton, P. P. 1990, in IAU Colloq. no. 122, *Physics of Classical Novae*, eds. A Casatella and R. Viotti (Springer-Verlag), p. 449.

Eracleous, M., Halpern, J. and Patterson, J. *ApJ*, 382, 290.

Frank, J., King, A. R. and Lasota, J.-P. 1987, *A&A*, 178, 137.

Frank, J., King, A. R. and Lasota, J.-P. 1988, *A&A*, 193, 113.

Friend, M. T., Martin, J. S., Smith, R. C. and Connon, R. 1990, *MNRAS*, 246, 654.

Gabler, R., Kudritzki, R. P. and Mendez, R. H. 1991, *A&A*, 245, 587.

Gilliland, R. L. 1985, *ApJ*, 292, 522.

Guinan, E. and Sion, E. 1981, in *The Universe at Ultraviolet Wavelengths: The Second Year of IUE*, ed. R. D. Chapman, NASA CP 2171, P. 477.

Hameury, J.-M., King, A. R. and Lasota, J.-P. 1986, *MNRAS*, 218, 695.

Hameury, J.-M., King, A. R. and Lasota, J.-P. 1991, *A&A*, 248, 525.

Harrison, T. and Gehrz, R. D. 1992, *AJ*, 103, 243.

Hassall, B. J. M. 1985, *MNRAS*, 216, 335.

Hassall, B., Pringle, J. and Verbunt, F. 1985, *MNRAS*, 216, 353.

Hassall, B. J. M., Pringle, J. E., Schwartzenberg-Czerny, A., Wade, R. A., Whelan, J. A. J. and Hill, P. W. 1983, *MNRAS*, 203, 865.

Heap, S. *et al.* 1978, *Nature*, 275, 385.

Hearn, D. R., Richardson, J. A. and Clark, G. W. 1976, *ApJ*, 210, L23.

Heise, J. *et al.* 1978, *A&A*, 63, L1.

Hellier, C. 1991, *MNRAS*, 251, 693.

Hellier, C., Mason, K. O. and Cropper, M. S. 1990, *MNRAS*, 242, 250.

Hellier, C., Mason, K. O., Smale, A.P., Corbet, R. H. D., O'Donoghue, D., Barrett, P. E. and Warner, B. 1989, *MNRAS*, 238, 1107.

Hertz, P. and Grindlay, J. E. 1983, *ApJ*, 275, 105.

Hertz, P. and Grindlay, J. E. 1984, *ApJ*, 278, 137.

Hertz, P. Bailyn, C. D., Grindlay, J. E., Garcia, M. R., Cohn, H. and Leyer, P. M. 1990, *ApJ*, 364, 251.

Hessman, F. V. 1986, *ApJ*, 300, 794.

Hind, J. R. 1856, *MNRAS*, 16, 56.

Hirose, M., Osaki, Y. and Mineshige, S. 1991, *PASJ*, 43, 809.

Hoare, M. G. and Drew, J. E. 1991, *MNRAS*, 249, 452.

Holm, A. V., Panek. R. J. and Schiffer, F. H., III. 1982, *ApJ*, 252, L35.

Horne, K. 1985, *MNRAS*, 213, 129.

Horne, K. 1991, in IAU Colloq. no. 129, *Structure and Emission Properties of Accretion Disks*, eds. C. Bertout, S. Collin, J. P. Lasota, and J. Tran Thanh Van (Gif sur Yvette Cedex, France: Editions Frontiéres), p. 3.

Horne, K. and Cook, M. 1985, *MNRAS*, 214, 307.

Horne, K. and Saar, S. H. 1991, *ApJ*, 374, L55.

Horne, K. and Stiening, R. 1985, *MNRAS*, 216, 933.

Horne, K., Wood, J. H. and Stiening, R. F. 1991, *ApJ*, 378, 271.

Hubeny, I. 1989, in *Theory of Accretion Disks*, ed. F. Meyer *et al.* (Dordrecht: Kluwer Academic Press), p. 445.

Ichikawa, S. and Osaki, Y. 1992, *PASJ*, 44, 27.

Icke, V. 1976, in IAU Symp. 73, *The Structure and Evolution of Close Binaries*, eds. P. Eggleton, S. Mitoon, and J. Whelan (Dordrecht: Reidel) p. 267.

Ishida, M. 1991, Ph.D. thesis, ISAS, Research Note 505.

Ishida, M. *et al.* 1992, *MNRAS*, 254, 647.

Jensen, K., Córdova, F. A., Middleditch, J., Mason, K. O., Grauer, A. D., Horne, K. and Gomer, R. 1983, *ApJ*, 270, 211.

Jones, M. and Watson, M. 1992, *MNRAS*, 257, 633.

Kallman, T. R. and Jensen, K. A. 1985, *ApJ*, 299, 277.

Kim, S.-W., Wheeler, J. C. and Mineshige, S. 1992, *ApJ*, 384, 269.

King, A. R. 1988, *QJR Astr. Soc.*, 29, 1.

King, A. R. and Lasota, J.-P. 1990, *MNRAS*, 247, 214.

King, A. R. and Lasota, J.-P. 1991, *ApJ*, 378, 674.

King, A. R. and Shaviv, G. 1984, *MNRAS*, 211, 883.

King, A. R., Ricketts, M. J. and Warwick, R. S. 1979, *MNRAS*, 197, 77.

King, A. R., Frank, J., Jameson, R. F. and Sherrington, M. 1983, *MNRAS*, 203, 677.

Kiplinger, A. L., Sion, E. M. and Szkody, P. 1991, *ApJ*, 366, 569.

Kley, W. 1991, *A&A*, 247, 95.

Koyama, K., Makishima, K., Tanaka. Y. and Tsunemi, H. 1986, *PASJ*, 38, 121.

Kraft, R. P. 1990, in IAU Colloq. no. 122, *Physics of Classical Novae*, eds. A. Casatella and R. Viotti (Springer-Verlag) p. 3.

Krezminski, W. 1964, *AJ*, 69, 549.

Kuulkers, E., Van Amerongen, S., Van Paradijs, J. and Röttgering, H. 1991, *A&A*, 252, 605.

LaDous, C. 1991, in *Cataclysmic Variables and Related Objects*, ed. M. Hack, NASA/CRNS Monograph Series on Non-Thermal Phenomena in Stellar Atmospheres, p. 14.

Lamb, D. Q. and Masters, A. R. 1979, *ApJ*, 234, L117.

Lamb, F. K., Shibazaki, N., Shaham, J. and Alpar, M. A. 1985, *Nature*, 317, 681.

Livio, M. 1992a, in Vina del Mar Workshop on Cataclysmic Variable Stars, ed. N. Vogt, ASP Conf. Series 29, p. 4

Livio, M. 1992b, *ApJ*, 393, 516.

Livio, M. 1993a, in *Accretion Disks in Compact Stellar Systems*, ed. J. C. Wheeler (World Scientific Publishing Co.).

Livio, M. 1993b, in *Interacting Binaries*, Proc. 22nd SAAS FEE Advanced Course, ed. H. Nussbaumer.

Livio, M. and Pringle, J. E. 1992, *MNRAS*, 259, 23P.

Livio, M. and Truran, J. W. 1992, *ApJ*, 389, 695.

Livio, M. and Verbunt, F. 1988, *MNRAS*, 232, 1P.

Long, K. *et al.* 1991, *ApJ*, 381, L25.

Long, K. S., Blair, W. P., Bowers, C. W., Sion, E. M. and Hubeny, I. 1993, *ApJ*, 405, 327.

Margon, B., Szkody, P., Bower, S., Lampton, M. and Paresce, F. 1978, *ApJ*, 224, 167.

Marino, B. F. and Walker, W. S. G. 1979, in *Changing Trends in Variable Star Research*, Proc. IAU Colloq. 46, ed. J. Smak, p. 29.

Marsh, T. R. and Pringle, J. E. 1990, *ApJ*, 365, 677.

Marsh, T., Horne, K., Schlegel, E. M., Honeycutt, R. K. and Kaitchuck, R. H. 1990, *ApJ*, 364, 637.

Mason, K. O. and Córdova, F. A. 1983, *Adv. Space, Res.*, 2, 109.

Mason, K. O., Córdova, F. and Swank, J. 1979, in *(COSPAR) X-ray Astronomy*, eds. W. A. Baity, and L. E. Peterson (Oxford: Pergamon Press), p. 121.

Mason, K. O., Rosen, S. and Hellier, C. 1988, *Adv. Space Res.*, 8, 293.

Mason, K. O., Lampton, M.L., Charles, P. A. and Bowyer, S. 1978, *ApJ*, 226, L129.

Mason, K., Córdova, F., Watson, M. and King, A. 1988, *MNRAS*, 232, 779.

Mason, K. *et al.* 1992, *MNRAS*, 258, 749.

Mauche, C. 1990, ed., *Accretion-Powered Compact Binaries* (Cambridge University Press).

Mauche, C. 1991, *ApJ*, 373, 624.

Mauche, C. W. and Raymond, J. C. 1987, *ApJ*, 323, 690.

Mauche, C. W., Raymond, J. C. and Córdova, F. A. 1988, *ApJ*, 335, 829.

Mauche, C. W., Wade, R. A., Polidan, R. S., Van der Woerd, H. and Paerels, F. B. S. 1991, *ApJ*, 372, 659.

Mauche, C. W., Raymond, J. C., Buckley, D. A. H., Mouchet, M., Bonnell, J., Sullivan, D. J. Bonnet-Bidaud, J.-M. and Bunk, W. H. 1994, *ApJ*, 424, 347.

Meintjes, P. J., Raubenheimer, B. C., De Jager, O. C., Brink, C., Nel, H. I., North, A. R., Van Urk, G. and Visser, B. 1992, *ApJ*, 401, 325.

Mestel, L. and Spruit, H. C. 1987, *MNRAS*, 226, 57.

Meyer, F. and Meyer-Hofmeister, E. 1981, *A&A*, 104, L10.

Meyer, F. and Meyer-Hofmeister, E. 1984, *A&A*, 132, 143.

Middleditch, J. and Córdova, F. A. 1982, *ApJ*, 255, 585.

Mineshige, S. 1986, in *Hydrodynamic and Magnetohydrodynamic Problems in the Sun and Stars*, ed. Y. Osaki (University of Tokyo), p. 275.

Mittaz, J. P. D., Rosen, S. R., Mason, K. O. and Howell, S. B. 1992, *MNRAS*, 258, 277.

Nather, E. and Warner, B. 1971. *MNRAS*, 152, 219.

Naylor, T., Bath, G. T., Charles, P. A., Hassall, B. J. M., Sonneborn, C., Van der Woerd, H. and Van Paradijs, J. 1988, *MNRAS*, 231, 237.

Nomoto. K. and Kondo, Y. 1991, *ApJ*, 367, L19.

Norton, A. J. and Watson, M. G. 1989, *MNRAS*, 237, 853.

Norton, A. J., Watson, M. G. and King, A. R. 1988, *MNRAS*, 231, 783.

Nousek, J. A. *et al.* 1984, *ApJ*, 277, 682.

O'Donoghue, D. 1986, *MNRAS*, 220, 23P.

O'Donoghue, D., Warner, B., Wargau, W. and Grauer, A.D. 1990, in IAU Colloq. no. 122, *Physics of Classical Novae*, eds. A. Casatella and R. Viotti (Springer-Verlag), p. 63.

Osaki, Y. 1974, *PASJ*, 26, 429.

Osaki, Y. 1989, *PASJ*, 41, 1005.

Osborne, J. 1988, in *X-ray Astronomy with EXOSAT*, eds. R. Pallavicini, and N. E. White, Mem. S.A. It., 59, p. 117.

Owocki, S. P. 1992, in *Atmospheres of Early-Type Stars*, eds. U. Heber and S. Jeffery (Berlin: Springer), p. 393.

Paczyński, B. 1981, *Acta, Astr.*, 31, 1.

Paczyński, B. and Sienkiewicz, R. 1983, *ApJ*, 268, 825.

Parmar, A. and White, N. 1988, in *X-ray Astronomy with EXOSAT*, eds. R. Pallavicini, and N. E. White, Mem. S.A. It., 59, p. 117.

Patterson, J. 1981, *ApJ*, 45, 517.

Patterson, J. 1984, *ApJ*, 54, 443.

Patterson, J. 1991, *PASP*, 103, 1149.

Patterson, J., Branch, D. Chincarini, G. and Robinson, E. L. 1980, *ApJ*, 240, L133.

Patterson, J., Thomas, G., Skillman, D. R. and Diaz, M. 1993, *ApJS*, 86, 235.

Payne-Gaposchkin, C. 1977, *AJ*, 82, 665.

Piirola, V., Hakala, P. and Coyne, G. V. 1993, *ApJ*, 410, L107.

Polidan, R. and Carone, T. 1987, *Astr. & Space Sci.*, 130, 235.

Polidan, R. and Holberg, J. 1984, *Nature*, 309, 528.

Polidan, R. and Holberg, J. 1987, *MNRAS*, 225, 131.

Polidan, R. S., Mauche, C. W. and Wade, R. A. 1990, *ApJ*, 356, 211.

Pounds, K. *et al.* 1993, *MNRAS*, 260, 77.

Pringle, J. E. 1977, *MNRAS*, 178, 195.

Pringle, J. E. 1981, *Ann. Rev. Astr. Astrophys.*, 19, 137.

Pringle, J. E. and Savonije, G. J. 1979, *MNRAS*, 187, 777.

Pringle, J. E., Verbunt, F. and Wade, R. A. 1986, *MNRAS*, 221, 169.

Pringle, J. *et al.* 1987, *MNRAS*, 225, 73.

Rappaport, S., Joss, P. C. and Webbink, R. 1982, *ApJ*, 254, 616.

Rappaport, S., Cash, W., Doxsey, R., McClintock, J. and Moore, G. 1974, *ApJ*, 187, L5.

Ritter, H. 1990, *A&A Suppl.*, 85, 1179.

Ritter, H., Politano, M., Livio, M. and Webbink, R. F. 1991, *ApJ*, 376, 177.

Robinson, E. L. 1976, *Ann. Rev. Astron. Ap.*, 14, 119.

Rosen, S. R., Mason, K. O. and Córdova, F. A. 1988, *MNRAS*, 231, 549.

Rosen, S. R., Mason, K. O., Mukai, K. and Williams, O. R. 1991, *MNRAS*, 249, 417.

Rubenstein, E. P., Patterson, J. and Africano, J. L. 1991, *PASP*, 103, 1258.

Rutten, R. G. M., Kuulkers, E., Vogt, N. and Van Paradijs, J. 1992, *A&A*, 265, 159.

Shafter, A. W. 1991, in *Fundamental Properties of Cataclysmic Variable Stars*, ed. A. W. Shafter (San Diego State University), p. 39.

Shafter, A., Szkody, P. and Thorstensen, J. R. 1986, *ApJ*, 308, 765.

Shakura, N.I. and Sunyaev, R. A. 1973, *A&A*, 24, 337.

Shara, M. M. 1989, *PASP*, 101, 5.

Shaviv, G. and Wehrse, R. 1991, *A&A*, 251, 117.

Shore, S. N. 1987, *Astr. J.*, 94, 731.

Singh, J. and Swank, J. 1993, *MNRAS*, 262, 1000

Sion, E. M. 1991, *AJ*, 102, 295.

Smak, J. 1971, Acta Astr., 21, 15.

Smak, J. 1982, Acta Astr., 32, 213.

Smak, J. 1984a, Acta Astr., 34, 161.

Smak, J. 1984b, Acta Astr., 34, 93.

Starrfield, S. 1990, in IAU Colloq. no. 122, *Physics of Classical Novae*, eds. A. Casatella and R. Viotti (Springer-Verlag), p. 127.

Starrfield, S., Sparks, W. M., Truran, J. W. and Shaviv, G. 1991, in *Supernovae*, ed. S. E. Woosley (New York: Springer-Verlag), p. 602.

Stockman, H. S., Schmidt, G. D. and Lamb, D. Q. 1988, *ApJ*, 332, 282.

Stockman, H. S., Schmidt, G. D., Berriman, G., Liebert, J., Moore, R. L. and Wickramasinghe, D. T. 1992, *ApJ*, 401, 628.

Swank, J. H. 1979 in IAU Colloq. no. 53, *White Dwarfs and Variable Degenerate Stars*, eds. H. M. Van Horn and V. Weidemann (NY: University of Rochester Press), p. 135.

Szkody, P. and Mateo, M. 1984, *ApJ*, 280, 729.

Szkody, P. and Mateo, M. 1986, *ApJ*, 301, 286.

Szkody, P., Kii, T. and Osaki, Y. 1990, *AJ*, 100, 546.

Tapia, S. 1977, *ApJ*, 193, L11.

Thorstensen, J. R., Ringwald, F. R., Wade, R. A., Schmidt, G. D. and Norsworthy, J. E. 1991, *AJ*, 102, 272.

Tylenda, R. 1981, *Acta. Astr.*, 31, 127.

Van Amerongen, S., Kuulkers, E. and Van Paradijs, J. 1990, *MNRAS*, 242, 522.

Van Amerongen, S., Damen, E., Groot, M., Kraakman, H. and Van Paradijs, J. 1987, *MNRAS*, 225, 93.

Van der Woerd, H. 1987, *Astra. Space. Sci.*, 130, 225.

Van der Woerd, H. and Heise, J. 1987, *MNRAS*, 225, 141.

Van der Woerd, H., Heise, J. and Bateson, F. 1986, *A&A*, 156, 252.

Van der Woerd, H., Heise, J., Paerels, F., Beuermann, K. Van der Klis, M., Motch, C. and Van Paradijs, J. 1987, *A&A*, 182, 219.

Van Teeseling, A., Verbunt, F. and Heise, J. 1993, *A&A*, 270, 159.

Vauclair, G. and Sion, E. 1991, eds. *White Dwarfs* (Dordrecht: Kluwer Academic Press).

Verbunt, F. 1991, Adv. Space Res., 11, (11), 57.

Verbunt. F. and Zwaan, C. 1981, *A&A*, 100, L7.

Verbunt, F., Hassall, B. J. M., Pringle, J. E., Warner, B. and Marang, F. 1987, *MNRAS*, 225, 113.

Vitello, P. and Shlosman, I. 1993, *ApJ*, 410, 815.

Wade, R. A. 1985, in *Interacting Binaries*, eds. P. P. Eggleton and J. E. Pringle (Dordrecht: D. Reidel), p. 289.

Wade, R. A. 1988, *ApJ*, 335, 394.

Wade, R. A. and Horne, K. 1988, *ApJ*, 324, 411.

Wade, R. and Ward, M. 1985, in *Interacting Binary Stars*, eds. J. E. Pringle and R. A. Wade (Cambridge University Press), p. 129.

Walker, M. 1981, *ApJ*, 248, 256.

Walter, F. *et al.* 1982, *ApJ*, 253, L67.

Warner, B. 1976, in *Structure and Evolution of Close Binary Systems*, eds. P. P. Eggleton, S. Mitton and J. A. J. Whelan (Dordrecht: D. Reidel), p. 85.

Warner, B. 1983, in IAU Colloq. no. 72: *Cataclysmic Variables and Related Objects*, eds. M. Livio and G. Shaviv (Dordrecht: D. Reidel), p. 269.

Warren, J. K., Vallerga, J. V., Mauche, C. W., Mukai, K. and Siegmund, O. H. W. 1993, *ApJ*, 414, L69.

Watson, M. G. 1993, *Adv. Space Res.*, 13, (12), 125.

Watson, M., King, A. and Osborne, J. 1985, *MNRAS*, 212, 917.

West, S. C., Berriman, G. and Schmidt, G. D. 1987, *ApJ*, 322, L35.

Wheeler, J. C. 1990, in *Frontiers of Stellar Evolution*, ed. D. L. Lambert (San Francisco: Astron. Soc. Pacific), p. 483.

Whitehurst, R. 1988, *MNRAS*, 232, 35

Whitehurst, R. and King, A. R. 1991, *MNRAS*, 249, 25.

Wickramasinghe, D. T., Wu, K. and Ferrario, L. 1991, *MNRAS*, 249, 460.

Williams, R. E. 1989, *AJ*, 97, 1752.

Wood, J. 1987, *Astrophys. Space Sci.*, 130, 81.

Wood, J. and Horne, K. 1990, *MNRAS*, 242, 606.

Wood, J., Horne, K. and Vennes, S. 1992, *ApJ*, 385, 294.

Wood, J., Horne, K., Berriman, G., Wade, R., O'Donoghue, D. and Warner, B. 1986, *MNRAS*, 219, 629.

Wood, J. H., Horne, K., Berriman, G. and Wade, R. A. 1989, *ApJ*, 341, 974.

Woods, J. A., Drew, J. E. and Verbunt, F. 1990, *MNRAS*, 245, 323.

Yoshida, K., Inoue, H. and Osaki, Y. 1992, *PASP*, 44, 537.

9

Normal galaxies and their X-ray binary populations

G. Fabbiano

Harvard-Smithsonian Center for Astrophysics, 60 Garden Street, Cambridge, MA 02138, USA

9.1 Introduction

X-ray binaries (XBs) are an important component of the X-ray emission of galaxies. Therefore the knowledge gathered from the study of Galactic X-ray sources can be used to interpret X-ray observations of external galaxies. Conversely, observations of external galaxies can provide us with uniform samples of XBs, in a variety of different environments.

Although detailed spectral/variability studies are at the moment only feasible for Galactic X-ray sources and perhaps for X-ray sources in Local Group galaxies, the study of sources in external galaxies presents unique advantages. These sources within a given galaxy do not suffer from the uncertainty in distances that affects Galactic binaries. Moreover, except for very edge-on galaxies, these sources are not affected by large line-of-sight absorption, as are, for example, Galactic bulge sources. Furthermore, in external galaxies it is easy to associate sources with different galactic components (e.g. bulge, disk, spiral arms). Finally, different galaxies provide different laboratories in which to test theories of Galactic source formation and evolution with a variety of boundary conditions: different metallicity/stellar population, star-formation activity, galaxy structure. All considered, the best place to study the overall properties of Galactic sources is in external galaxies!

The study of galaxies in X-rays began with the *Einstein Observatory* (Giacconi *et al.* 1979) and will be undoubtly greatly expanded with *ROSAT* (Trümper 1983). A review of the *Einstein* results can be found in Fabbiano (1989); see also Fabbiano (1990a, 1990b). Enlarged compilations of the *Einstein* data (images, fluxes and spectra), including almost 500 galaxies, have been published recently (Fabbiano, Kim and Trinchieri 1992; Kim, Fabbiano and Trinchieri 1992a). These observations have shown that most of the X-ray emission of normal spiral galaxies (i.e. galaxies whose X-ray emission is not dominated by a point-like non-thermal nuclear source) in the *Einstein* band ($\sim 0.2 - 4.0$ keV) is due to the integrated contribution of galactic X-ray sources. Three components of the X-ray emitting population have been identified: bulge, disk and spiral arm sources. In starburst galaxies, a hot gaseous component is also present. Elliptical galaxies can be much more luminous in X-rays than spirals because they may retain a hot gaseous component, which dominates the X-ray emission. However, a good fraction of the elliptical galaxies observed with *Einstein* have relatively low X-ray luminosities. In these galaxies the X-ray emission may be dominated by a population of low-mass XBs (LMXBs).

In this chapter, largely based on the *Einstein* observations, I will concentrate on the results relevant for the study of the binary X-ray source component in galaxies. First, I will discuss what we have learned about populations of X-ray sources both in the Galaxy and in Local Group galaxies; secondly, I will summarize the results of the observations of other relatively nearby galaxies, where a few discrete luminous sources were resolved with the *Einstein* instruments; thirdly, I will give a very brief summary of the major results of the observations of a large number of more distant galaxies; fourthly, I will discuss the implications of recent spectral results on the global emission properties of galaxies, with particular emphasis on the non-gaseous emission of elliptical galaxies. I will conclude by looking at the future: the very near future, accessible through *ROSAT* and *ASTRO-D*; the near future of *AXAF* and *XMM*; and the farther away future of an ideal observatory to study galaxies in X-rays.

9.2 X-ray sources in Local Group galaxies

It is not surprising that the most detailed observations of X-ray sources in galaxies are for Local Group galaxies. Besides the sources in the Galaxy, early non-imaging X-ray observations discovered sources in the Magellanic Clouds (see Clark *et al.* 1978 and references therein). The overall X-ray emission of M31 was also detected (e.g. Forman *et al.* 1978). With *Einstein*, galaxies in the Local Group have been surveyed in detail, achieving limiting sensitivities of 10^{34-35} erg s^{-1} in the Clouds, and of 10^{36-37} erg s^{-1} in M31 and M33 (see reviews of Helfand 1984a,b; Fabbiano 1989; Trinchieri and Fabbiano 1991; see also Wang and Helfand 1991a,b and Wang *et al.* 1991 for detailed studies of the diffuse emission component of the LMC).

9.2.1 *The Galaxy*

The properties and evolution of Galactic XBs are discussed in the other chapters of this book. For the purpose of this chapter, it is interesting to look at the global X-ray emission properties of the Galaxy and of the different X-ray emitting populations. Given our position in the Galactic plane, this is not straightforward. A fairly recent summary of the X-ray properties of the Galaxy can be found in Watson (1990), and we refer the reader to this paper. Table 9.1, adapted from Watson (1990), summarizes the different components of the X-ray emission. XBs account for most of this emission and are also the brightest individual sources ($L_X \sim 10^{37-38}$ erg s^{-1}). However stars, although much fainter individually ($L_X \sim 10^{28-33}$ erg s^{-1}), are also responsible for a significant amount of the emission below 1 keV. As shown in Table 9.1, there are also diffuse – or unresolved – emission components detected in the Galaxy.

9.2.2 *The Magellanic Clouds*

A comprehensive review of the *Einstein* X-ray observations of the Clouds can be found in Helfand (1984a,b); see also Fabbiano (1989). The X-ray sources of the LMC and SMC belong to the Population I component predominant in these galaxies. Among these sources are the brightest XBs known in the pre-*Einstein* era (Clark *et al.* 1978). The *Einstein* observations discovered a large number of supernova remnants, and a number of fainter (10^{34-36} erg s^{-1}) unidentified X-ray sources.

Table 9.1. *The Galaxy*[a]

Discrete Galactic sources		
Category	Spectrum	Total L_X(erg s^{-1})
Binaries Pop.I	hard	3.2×10^{38}
Binaries old disk and Pop.II	hard	1.6×10^{39}
SNR	soft	3.2×10^{37}
CV	hard	$\sim 1 \times 10^{37}$
Stars	soft	6.6×10^{38}

'Hard' diffuse emission components			
Category	Thickness	Radius	L_X(erg s^{-1})
Halo	$\sim$3 kpc	> 12 kpc	$\sim 5 \times 10^{38}$
Ridge	$\sim$300 pc	$\sim$16 kpc	$\sim 1.2 \times 10^{38}$
Inner ridge	$\sim$100 pc	$\sim$ 6.5 kpc	$\sim 1.5 \times 10^{38}$
Bulge	—	$\sim$ 3 kpc	$\sim 1 \times 10^{38}$

[a] Watson (1990 and references therein).

The differences between the X-ray source population of the Clouds and that of the Galaxy or M31 have been ascribed to the different stellar composition and metallicity of these galaxies. In particular, low metallicity has been sought as an explanation for the presence of very bright binary sources, emitting well in excess of the Eddington limit for a $1M_\odot$ accreting object (Clark *et al.* 1978).

9.2.3   *M33*

M33 is an Sc galaxy in the Local Group ($D \sim$720 kpc), which has been observed extensively with *Einstein* (Long *et al.* 1981; Markert and Rallis 1983) Figure 9.1 shows the *Einstein* IPC contour map superposed on a POSS image. The entire set of the *Einstein* data has been analyzed and discussed by Trinchieri, Fabbiano, and Peres (1988) and Peres *et al.* (1989). This work has resulted in the detection of 13 (perhaps 15) point-like sources associated with the disk and spiral arms, with luminosities between 1×10^{37} and 1×10^{38} erg s^{-1}. One of these sources (M33 X-7 of Trinchieri, Fabbiano, and Peres 1988) has a characteristic binary light curve with a 1.8 day period, reminiscent of an HMXB (Peres *et al.* 1989; Figure 9.2). A very bright source is found at the nucleus ($L_X = 2 \times 10^{39}$ erg s^{-1}). This source is variable (Peres *et al.* 1989), and has a relatively soft spectrum ($kT \sim$3–4 keV), different from

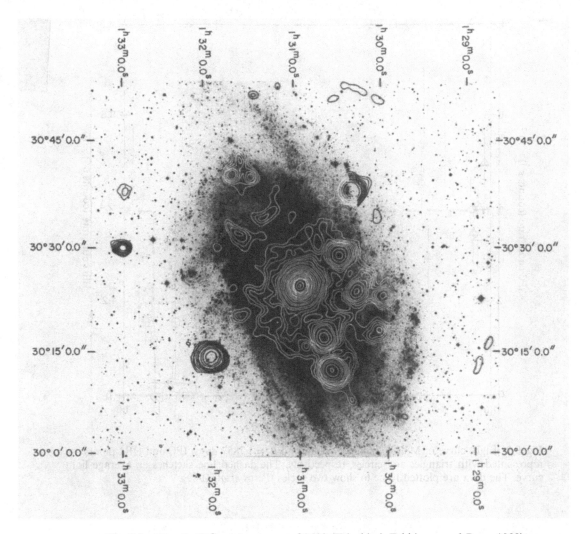

Fig. 9.1. *Einstein* IPC contour map of M33 (Trinchieri, Fabbiano, and Peres 1988).

typical spectra of XBs (Trinchieri, Fabbiano, and Peres 1988). It is possible that this source is a nearby example of an optically quiet, low-X-ray-luminosity active nucleus. M33 has been recently observed with *ROSAT*, but the data have not been fully analyzed. A preliminary look (K. Long, 1992, private communication) shows that all the *Einstein* sources are present in the *ROSAT* image, plus a similar number of new ones. Some of these new sources are bright enough to have been detected with *Einstein* unless their spectra are very soft. Source variability is also evident within the *ROSAT* observation.

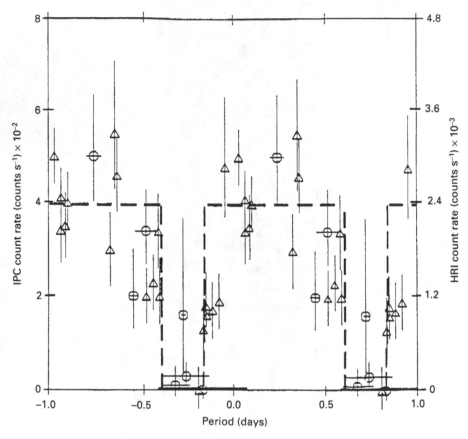

Fig. 9.2. Light curve of M33 X-7 folded with period $P = 1.7857$ days. IPC and HRI points are represented with triangles and circles, respectively. The dashed line sketches an average light curve. The data are plotted twice to show two cycles (Peres *et al.* 1989).

9.2.4   M31

The amount of data collected, and the detail of the work done on M31, grant a more in-depth discussion. At a distance of ~690 kpc, M31 is the nearest large spiral galaxy. This Sb galaxy may be similar to our own Galaxy, although it has been argued that the Galaxy may have a less prominent spheroidal bulge (Hodge 1983). Starting from the *Einstein* observations of Van Speybroeck *et al.* (1979), detailed comparisons have been made between the X-ray properties of M31 and of the Galaxy (see Fabbiano 1989). An analysis of the entire set of the *Einstein* observations of M31 has led to the detection of 108 X-ray sources, 101 of which are likely to belong to M31 (Figure 9.3; see Table 2 of Trinchieri and Fabbiano 1991). These sources have (0.2–4.0 keV) luminosities of 5×10^{36} to a few times 10^{38} erg s^{-1}, in the range of those of Galactic X-ray sources. A recent *ROSAT* HRI observation of the bulge of M31 has led to the discovery of an additional 37 sources (Primini, Forman, and Jones 1993), bringing the number of X-ray sources in M31 to 138.

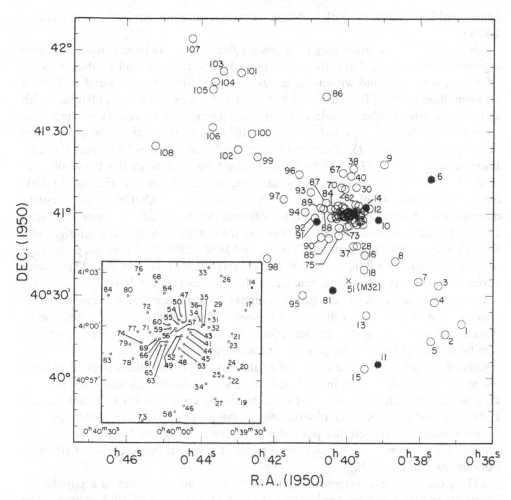

Fig. 9.3. Spatial distribution of the sources in M31. Filled dots identify foreground sources. The cross identifies a source at the position of M32. (Trinchieri and Fabbiano 1991.)

Recent *ROSAT* PSPC observations have detected 408 X-ray sources in the field of M31 (Supper *et al.* 1994).

Based on the associations with different optical components, X-ray sources have been separated into three groups: disk and spiral arm sources; globular cluster sources; and bulge sources. The latter have been separated further into inner and outer bulge sources (Van Speybroeck *et al.* 1979; Trinchieri and Fabbiano 1991). The *Einstein* spectrum of the bulge of M31 (which cannot be resolved spatially into individual components with the *Einstein* spectral imager, the IPC) suggests that the emission is dominated by a population of low mass binaries (Fabbiano, Trinchieri, and Van Speybroeck 1987). The *Ginga* observations of M31 confirm this result (Makishima *et al.* 1989). Spectra of individual bright sources outside of the bulge of M31 are reported by Trinchieri and Fabbiano (1991). Of notice are two sources, one of which is associated with a globular cluster, exhibiting very soft spectra ($kT \leq$ 1 keV).

Some of the M31 sources vary in between different *Einstein* observations or within a given observation (Collura, Reale and Peres 1990; Trinchieri and Fabbiano 1991). Of the 51 sources found within the central $\sim 20'$ of M31, 11 are found to vary in between observations (Trinchieri and Fabbiano 1991). For one source, identified with a globular cluster, Collura, Reale, and Peres (1990) report a light curve consistent with those of LMXBs, with a period of $\sim 3.25 \times 10^4$ s. More dramatic is the comparison between *Einstein* and *ROSAT* observations of the bulge, taken ~ 10 y apart (Primini, Forman, and Jones 1993). These authors detect 86 sources in the bulge of M31 above $1.3 - 2.2 \times 10^{36}$ erg s^{-1}. In the same region, Trinchieri and Fabbiano (1991) report 66 sources, 49 of which are detected with *ROSAT*. Of the 37 new sources detected with *ROSAT*, 16 are above the *Einstein* threshold. Of the *Einstein* sources, 17 are not detected with *ROSAT*. Based on the comparison between *Einstein* and *ROSAT* source detection, Primini, Forman, and Jones (1993) find that the percentage of variable sources within $\sim 7.5'$ of the nucleus of M31 is $\leq 42\%$. This source variability confirms that a large fraction of the M31 sources are XBs.

In M31 we can look at the global luminosity distributions of different types of sources in a way that it will never be possible for Galactic sources, because of distance uncertainties and extinction effects. Notwithstanding an earlier report of a significantly brighter inner bulge source population (Van Speybroeck *et al.* 1979), the results of the analysis of the entire set of the *Einstein* observations show that there are no significant differences in the luminosity distributions of disk and bulge sources, and, within the bulge, of inner and outer bulge sources (Figure 9.4; Trinchieri and Fabbiano 1991). The *ROSAT* observations of the bulge of M31 are deeper than the *Einstein* ones, and suggest the possibility of a flattening of the luminosity function around 2×10^{37} erg s^{-1}. A single power law fit gives an index of ~ 0.4 (Primini, Forman, and Jones 1993).

M31 gives us a unique opportunity to study X-ray bulge sources as a population. These observations can be used to test theories of formation of such sources. The results are quite puzzling. The coincidence of the crowded inner bulge X-ray region, with a reported 'hole' in the distribution of novae, led Vader *et al.* (1982) to suggest an evolutionary scheme by which the inner bulge X-ray sources would be the remnant of a dead cataclysmic variable population which had long since undergone its nova phase. However, Ciardullo *et al.* (1987) have since dispelled the notion of a nova hole. Trinchieri and Fabbiano (1991) find that the radial distribution of the M31 bulge sources could be consistent with that of the novae, but the azimuthal distributions differ. While the novae follow the general shape of the optical bulge, the X-ray sources are distributed in an east–west geometry (see Van Speybroeck and Bechtold 1981). This result is not likely to be due to source variability, since it is still present when the *ROSAT* 'new' sources are added to the *Einstein* sources (Figure 9.5). This global difference in the distribution of X-ray sources and novae in the bulge casts some doubts on the possibility that the X-ray sources are more evolved members of the same class of stellar system to which the novae belong. Moreover, Trinchieri and Fabbiano (1991) find an – unexplained – apparent similarity between the distribution of the inner bulge sources and that of the inner Hα arms

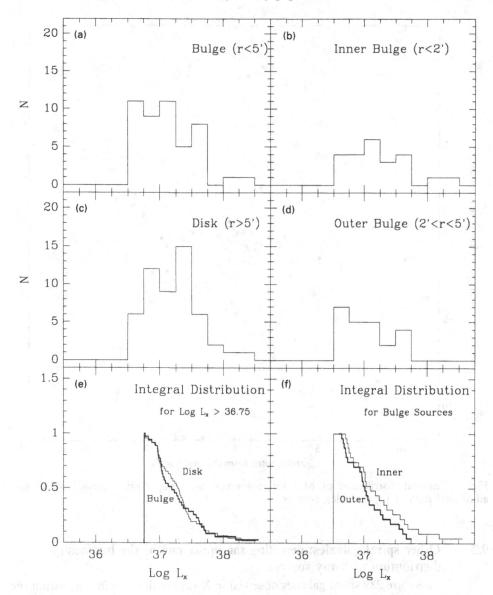

Fig. 9.4. Luminosity distributions of M31 sources within and without a 5' radius (a and c) and in the inner and outer bulge (b and d). Comparisons of the integral distributions of (e) disk and bulge sources, and (f) inner and outer bulge sources. In both comparisons, a KS test gives probabilities of $\sim 70\%$ of the two distributions being drawn from the same parent population. (Trinchieri and Fabbiano 1991.)

at the center of M31 (Ciardullo *et al.* 1988; Figure 9.6). One cannot help believing that these data may be giving us crucial information on the formation of bulge sources. However, one also feels that we are quite a long way from explaining these results.

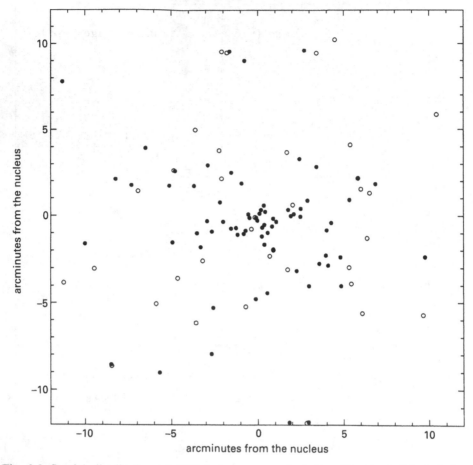

Fig. 9.5. Spatial distribution of M31 bulge sources, including *Einstein* (solid circles) and additional *ROSAT* (open circles) sources.

9.3   Other spiral galaxies: detecting the upper end of the luminosity distribution of X-ray sources

There are 289 spiral galaxies observed in X-rays with *Einstein* well within the instrumental fields. Of these, 157 have been detected (Fabbiano, Kim, and Trinchieri 1992). Only a few of these galaxies, however, are near enough to have been – even partially – resolved. Table 9.2 summarizes the results of the X-ray observations of nearby spirals where individual sources could be detected, listed in increasing order of distance. The Magellanic Clouds and other Local Group dwarfs are not included in this Table (see Fabbiano 1989). Since, with a few exceptions, the typical exposure times on each galaxy are comparable, it is not surprising that the detection threshold is higher for farther away galaxies. What is perhaps surprising is the number of very bright individual sources detected in these galaxies.

As discussed in Sect. 9.2., the X-ray sources of the Milky Way, M31 and M33 (with the exclusion of the nucleus of M33) typically have X-ray luminosities compatible

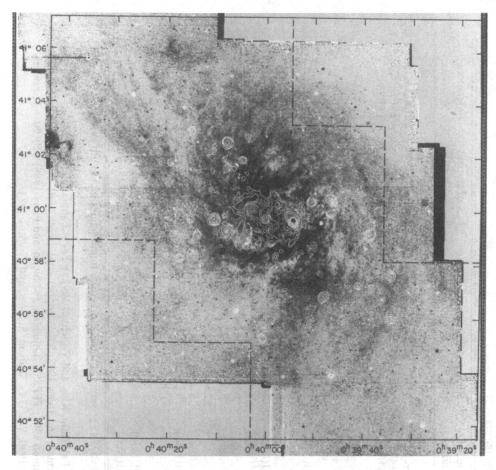

Fig. 9.6. Comparison of the X-ray source distribution and of the Hα emission in the bulge of M31. The Hα image is from Ciardullo *et al.* (1988). The X-ray contours are from the *Einstein* HRI data (Trinchieri and Fabbiano 1991).

with the Eddington limit for a $\sim 1 M_\odot$ object ($\sim 1.3 \times 10^{38}$ erg s^{-1}). Much brighter 'super-Eddington' sources have been detected in the Magellanic Clouds, and these have been ascribed to the low metallicity of these galaxies (Clark *et al.* 1978; see also Sect. 2.5). We now find that 'super-Eddington' sources are a fairly common occurrence. As discussed in Fabbiano (1989), some of these sources at least are likely to be single accretion binaries. These sources may represent the upper end of the luminosity distribution of binary X-ray sources in a given galaxy. Their presence suggests that very massive compact objects might be relatively common.

It is interesting to notice that the occurrence of very bright sources is not limited to galaxies of a definite morphological type, as was suggested on the basis of comparisons between the Galaxy, M31 and the Magellanic Clouds (Clark *et al.* 1978; Helfand 1984a). To illustrate differences in the X-ray emitting population of optically similar galaxies, it is interesting to compare two pairs of galaxies well studied in X-rays and

Table 9.2. *X-ray sources detected in nearby spirals*

Galaxy	Distance (Mpc)	Type	No. of sources	Range of L_X (erg s^{-1})	No. of sources $L_X \gtrsim 2 \times 10^{38}$	References
Milky Way	—	Sb-Sc	~200[a]	~10^{37}–10^{38}	—	W90,Ch.14
M31	0.69	Sb	~400[b]	~1×10^{36}–2.5×10^{38}	3	TF91,P92,S94
M33	0.72	Sc	12+1(Nuc.)	8×10^{36}–1×10^{38}, (Nuc.~ 9×10^{38})	1(Nuc.)	TFP88
NGC247	2.1	Sc	3	1–2×10^{38}	—	FKT92
M82	3.3	IO	1+7[c]	5–9×10^{38}	8	WSG84
NGC253	3.4	Sc	8	1–6×10^{38}	5	FT84
M81	3.5	Sb	9+1(Nuc.)	1.8×10^{38}–1.4×10^{39}, (Nuc.~ 1×10^{41})	~9+1(Nuc.)	F88
NGC4236	3.5	Sd	2	6×10^{37},1×10^{38}	2	FKT92
M83	3.7	Sc	4+1(Nuc.)	2–6×10^{38},(Nuc.~ 2×10^{39})	4+1(Nuc.)	TFP85
IC342	4.5	Scd	2+1(Nuc.)	2–4×10^{39}	2+1(Nuc.)	FT87
NGC4449	4.8	Sm	4	5×10^{38}–1×10^{39}	4	FKT92
NGC2403	6.8	Sc	3	8×10^{38}–5×10^{39}	3	FKT92
NGC6946	7.0	Sc	2	2×10^{39},5×10^{39}	2	FT87
M101	7.2	Sc	6	5×10^{38}–3×10^{39}	6	TFR90
M51	9.6	Sbc	3+1(Nuc.)	9.8×10^{38}–1.4×10^{39}, (Nuc.~ 8×10^{39})	3+1(Nuc.)	P85
NGC4258	10.9	Sb	2	1×10^{40},5×10^{40}	2	FKT92
NGC3628	12.4	Sbc	1+1(Nuc.)	8×10^{39},(Nuc.~ 2×10^{40})	1+1(Nuc.)	FKT92
NGC4631	13.0	Sc	1+1(Nuc.)	~1×10^{40},(Nuc.~ 6×10^{39})	1+1(Nuc.)	FT87

[a] Only X-ray binaries included; [b] *Einstein* and *ROSAT*; [c] seven of these sources are found with a maximum entropy deconvolution; [d] SN1980k.

References: F88 – Fabbiano 1988a; FKT92 – Fabbiano, Kim, and Trinchieri 1992; FT87 – Fabbiano and Trinchieri 1987; P85 – Palumbo *et al.* 1985; P92 – Primini, Forman, and Jones 1992; S94 – Supper *et al.* 1994; TF91 – Trinchieri and Fabbiano 1991; TFP85 – Trinchieri, Fabbiano, and Palumbo 1985; TFP88 – Trinchieri, Fabbiano, and Peres 1988; TFR90 – Trinchieri, Fabbiano, and Romaine 1990; W90 – Watson 1990; WSG84 – Watson, Stanger, and Griffiths 1984.

at many other wavelengths: the two Sb galaxies M31 and M81, and the two Sc galaxies M33 and M101.

9.3.1   Two Sb galaxies: M31 and M81

Table 9.2 shows that all the sources detected in M81 (Fabbiano 1988a) are at or above the luminosity of the brightest sources detected in M31. Unless the sources in M81 are remarkably clumped (e.g. $\sim$10 sources with $L_X \sim 10^{37}$ erg s^{-1} within $\sim$200 pc), these sources are likely to be individual X-ray emitting objects. The distance assumed for deriving these luminosities is near the low end of the distance estimates for M81 (3.3–5.75 Mpc; Bottinelli *et al.* 1984; Sandage 1984). Only if the distance were as low as 1 Mpc would the luminosities of the point sources in M81 be consistent with those of the bright sources of M31. M81 (nucleus excluded) is also four to six times more luminous than M31 in X-rays relative to their optical luminosities. Is this 'excess' X-ray emission due to an additional component of very bright sources, not present in M31, or are X-ray sources produced in M81 with the same luminosity function as in M31, but in larger number? In principle, these questions can be answered by comparing the luminosity distributions of X-ray sources in the two galaxies.

In M81, the bulge is not resolved with *Einstein*, and moreover it is totally obliterated by the bright nuclear source (Fabbiano 1988a). All the individual sources detected with *Einstein* in M81 (with the exception of the nucleus) are in the disk/spiral arms (Figure 9.7). These sources account for < 50% of the total – non-nuclear – luminosity of M81. Figure 9.8 shows a comparison of the luminosity distributions of the M81 sources and of the M31 disk sources. As it can be seen from Figure 9.8, with the *Einstein* data we cannot discriminate between the two options outlined above. *ROSAT* has observed M81 with the PSPC, and more observations, at higher resolution, may be achieved during the *ROSAT* lifetime. These observations will lower the detection threshold in M81 to the point where a good comparison with M31 will be possible.

Even if we cannot discriminate yet between different possibilities for the luminosity function of X-ray sources in M81, it is clear that M81 is a stronger X-ray emitter than M31. It is also a stronger emitter in radio continuum, and, to lesser amount, in the IR (Fabbiano 1988a and references therein). All of this points to a more active/recent star formation in M81. The reason for this activity is somewhat of a puzzle. Both galaxies have strong spiral shocks, and radio observations suggest stronger shocks in M31, where the magnetic fields are better aligned with the spiral arms (Beck, Klein, and Krause 1985). If star formation is triggered by these spiral wave shocks, we would expect the reverse of what is observed. Perhaps the gas supply in M31 has been depleted by past more active star formation, or the alignment of the magnetic fields might not be a good indicator of the strength of the shocks (Fabbiano 1988a). In any event, it is clear that the study of these galaxies in X-rays is uncovering fundamental questions on their structure and evolution. Moreover, these results demonstrate that the picture linking the 'anomalously bright' X-ray sources of the Magellanic Clouds with their lower metallicity (when compared to the Galaxy; Clark *et al.* 1978; Sect. 2.5) is not a generally valid explanation for the occurrence of very bright sources. It is possible that these sources are the result of accretion on a very massive compact

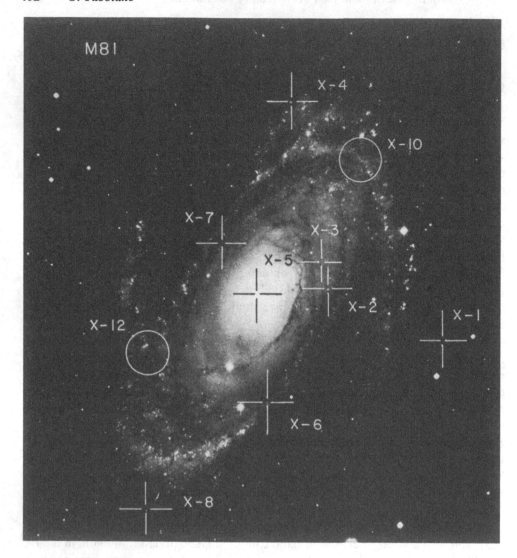

Fig. 9.7. Overlay of the X-ray source positions and error circle on the plate of M81 from Sandage (1961). The regions within the crosses represent the error regions of the HRI sources (90% positional uncertainties $\sim 4''$); the larger circles are the error regions for the IPC sources (90% positional uncertainties $\sim 45''$); Fabbiano (1988a).

object. Given our present knowledge (or lack thereof), they should be considered as black-hole candidates.

9.3.2   *Two Sc galaxies: M33 and M101*

A situation similar to that of M31 and M81 exists for the two Sc galaxies M33 and M101. With the exclusion of the nuclear source, the sources in M33 are all within the range of X-ray luminosities observed in Galactic sources. The six sources detected in M101 instead all have $L_X \geq 5 \times 10^{38}$ erg s^{-1} (Trinchieri, Fabbiano, and

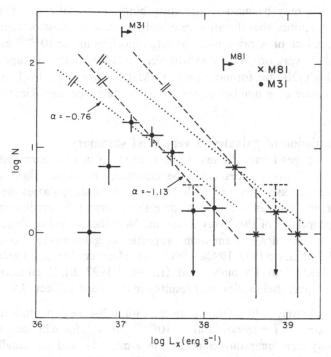

Fig. 9.8. Log–log plot of the number of individual X-ray sources of a certain X-ray luminosity vs. the luminosity for M31 (dots, data from Long and Van Speybroeck 1983) and M81 (crosses). The completeness limits for the two galaxies are indicated by the horizontal arrows. The error bars on the points are the 1σ level. The dashed upper limits are at the 2σ level. Dotted and dashed lines represent two power law approximations to the differential luminosity function of X-ray sources in M31. The lines crossing them at the low-luminosity end represent the points at which the luminosity function must turn off to avoid exceeding the total X-ray luminosity of M31. The same power laws are normalized to the M81 source luminosity distribution. (Fabbiano 1988a.)

Romaine 1990). To bring these luminosities within the range of those of M33, the distance of M101 would have to be ≤ 2 Mpc. This is highly unlikely. The distance of this Sc galaxy was first estimated at 3.5 Mpc (Sandage 1962), but has been revised to be of the order 7.1–7.6 Mpc, by using both the Cepheids method and a method based on the expansion of SN envelopes (7.2 Mpc, Sandage and Tamman 1974; 7.1 Mpc, Cook, Aaronson, and Illingworth 1986; 7.6 Mpc, Schmidt and Kirshner 1992).

Unless there is a remarkable amount of clumpiness in the X-ray sources of M101, the very bright sources detected with *Einstein* are single objects. Moreover, some of these sources may vary in time (Trinchieri, Fabbiano, and Romaine 1990). If one excludes the bright nucleus of M33, the overall X-ray emission of M101 is larger than that of M33, when compared with their optical emissions. However, the limit on the X-ray luminosity of the disk of M101 suggests that the X-ray source luminosity distribution of M101 may be flatter than that of M33 with a significantly larger component of very bright X-ray sources. Trinchieri, Fabbiano, and Romaine (1990) remark that the association of these bright sources with star formation regions in

M101 suggests that they may be massive accreting binaries, containing a massive collapsed star. They also notice that the absence of detected diffuse emission from the disk of M101 suggests a lack of X-ray sources with luminosities in the 10^{37-38} erg s^{-1} range. The associations of very bright sources with very massive binaries is supported by the exceptionally high OB star formation rate in M101 (Blitz *et al.* 1981), which could result in the presence of a number of very massive stars (see also Richter and Rosa 1984).

9.4   The X-ray emission of galaxies: a very brief summary

In the previous pages I have discussed observations of nearby spiral galaxies where individual bright X-ray sources could be detected. However, these are a minority among all the galaxies surveyed with *Einstein*. For most galaxies we only have a flux measurement, and, in the case of detection, some information on the spatial and spectral properties of the X-ray emission. Nonetheless, using these data we can learn about the general X-ray emission properties of galaxies (see Long and Van Speybroeck 1983; Fabbiano 1989, 1990a, 1990b and references therein; Fabbiano, Kim, and Trinchieri 1992; Kim, Fabbiano, and Trinchieri 1992a,b). A summary of some of these results is given below. Spectral results are discussed in Sect. 9.5.

- Galaxies are relatively faint X-ray sources, with 0.2–4 keV luminosities of $\sim 10^{39-41}$ erg s^{-1} for spirals, and $\sim 10^{39-43}$ erg s^{-1} for ellipticals (however, the very high luminosity ellipticals are a minority, and are usually the central galaxies in groups or clusters; e.g. M87 in Virgo).

- Several results suggest that the X-ray emission of spiral galaxies is dominated by sources belonging to the evolved stellar component. These include the observations of nearby galaxies where sources can be resolved; the fact that the X-ray surface brightness distribution generally follows the optical body of the galaxy; spectral results (see below); the linear correlation between X-ray and optical luminosities, which shows that the X-ray emission is a constant fraction of the optical (stellar) emission.

- The X-ray emission of spiral galaxies can be further attributed to three separate components of evolved sources: bulge sources (e.g. M31); disk sources; and spiral arm sources. This conclusion is based on the direct observations of nearby galaxies; on the radial profiles of the X-ray surface brightness of face-on spirals, which follows the exponential disk (e.g. Palumbo *et al.* 1985; Trinchieri, Fabbiano and Palumbo 1984; Fabbiano and Trinchieri 1987); and on differences in correlations between luminosities at different wavebands of bulge-dominated and disk-dominated spirals. These correlations show comparable excess in the optical and X-ray emission of bulge spirals, relative to their radio continuum and far-infrared emission, which can be attributed to the bulge (Fabbiano and Trinchieri 1985; Fabbiano, Gioia, and Trinchieri 1988). The presence of an exponential 'old disk' component of the X-ray emission has implications for our understanding of the formation of LMXBs, and favors the hypothesis that these systems evolve from 'native' low-mass binaries (Fabbiano 1985).

- Active star formation in galaxies is linked to an average increase of X-ray emission that may be, at least in part, attributed to an enhancement of the HMXBs and young SNR component (Fabbiano, Feigelson, and Zamorani

1982; Fabbiano, Trinchieri, and Macdonald 1984; Boller *et al.* 1992; David, Jones, and Forman 1992). Gaseous emission, associated with winds outflowing from starburst nuclei, has also been detected in nearby galaxies (e.g. Watson, Stanger, and Griffiths 1984; Fabbiano 1988b; Fabbiano, Heckman, and Keel 1990).

- For a given optical luminosity, the X-ray luminosity of elliptical galaxies can span a wide range of values, going from levels similar to those seen in spirals, to 10–100 times higher values. This excess X-ray luminosity has been attributed to the emission of the hot interstellar medium (ISM) of these galaxies, above a baseline emission due to a population of bulge-type sources. Additional evidence for the presence of a gaseous component in X-ray bright ellipticals is given by imaging and by spectral properties (see below). Considering this hot ISM as a test particle in the galaxy potential, X-ray observations can be used to measure the mass of elliptical galaxies (Fabricant and Gorenstein 1983; Forman, Jones, and Tucker 1985). It is therefore important to establish where the X-ray emission is dominated by the hot ISM. This proved, it is also important to be able to measure radial temperature profiles at high accuracy (see Trinchieri, Fabbiano, and Canizares 1986).

9.5   Recent spectral results

Since different types of X-ray sources have different spectral signatures, X-ray spectra are a useful tool for understanding the nature of the X-ray emission of galaxies and their X-ray source composition. Up until recently, spectral parameters had been derived only for a few galaxies, detected with fairly high signal to noise ratios (e.g. Trinchieri, Fabbiano, and Canizares 1986; Fabbiano and Trinchieri 1987; see Fabbiano 1989). As part of a systematic analysis of the galaxy sample observed with *Einstein*, we have now derived spectral parameters for 37 individual 'normal' galaxies (i.e. the emission is not dominated by an active nucleus), and we have also looked at the average spectral properties of different types of galaxies, using composite spectra and X-ray colors for a sample of 85 galaxies (Kim, Fabbiano, and Trinchieri 1992a,b). The following discussion is based mostly on these two papers. This work confirms that, on average, the X-ray emission temperature of spirals is higher than that of ellipticals. This is consistent with the understanding that accreting binaries are a major source of emission in spirals, while a hot ISM may be present in ellipticals. The X-ray spectra of Sa galaxies are intermediate between those of ellipticals and spirals, suggesting that these galaxies contain hot gaseous emission as well as emission from accreting binaries (Figures 9.9 and 9.10).

Two results of this work deserve a more in-depth presentation. The first is the discovery of spectral trends in elliptical galaxies, as a function of their relative X-ray brightness (for a given optical luminosity); the second is the discovery of a very soft emission component in some spiral galaxies.

9.5.1   *Spectral properties and L_X/L_B of elliptical (E and S0) galaxies*

Although everybody agrees that the X-ray emission of X-ray bright E and S0 galaxies is dominated by the hot ISM, there has been a certain amount of controversy on the nature of the emission of X-ray faint early-type galaxies (Canizares, Fabbiano

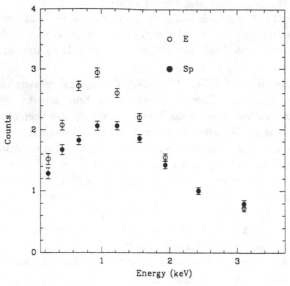

Fig. 9.9. Combined observed distributions of spectral counts of ellipticals (open circles) and spirals (solid circles), normalized at 2.4 keV. This comparison shows that the spectrum of ellipticals is softer than that of spirals. (Kim, Fabbiano, and Trinchieri 1992b.)

and Trinchieri 1987; Fabbiano 1989 and references therein; Fabbiano, Gioia, and Trinchieri 1989). These galaxies have X-ray to optical ratios in the range of those seen in spirals, where the X-ray emission is dominated by the discrete source component. Moreover, the bulge of M31, which could be considered a smaller example of an early-type galaxy, has an X-ray to optical ratio similar to those of X-ray faint ellipticals, and its emission is dominated by a population of binary X-ray sources (see Sect. 9.2.4 above). The X-ray faint ellipticals cannot be explained with cooling flow models for the gaseous ISM, under a variety of boundary conditions (Figure 9.11, from Fabbiano 1989). More recent evolutionary models for the ISM (D'Ercole *et al.* 1989; Ciotti *et al.* 1991; David, Forman, and Jones 1991) put them in a regime where galactic winds may prevail. An indication that these galaxies may have shallower potential wells, and may therefore be unable to retain their hot ISM, is also given by the fact that, on average, they tend to have smaller central velocity dispersions than X-ray bright ellipticals (Fabbiano, Gioia, and Trinchieri 1989). All of this circumstantial evidence and theoretical considerations suggest that the emission of the X-ray faint galaxies may be due to a collection of LMXBs, similar to those present in the bulge of M31 (Fabbiano, Trinchieri, and Van Speybroeck 1987).

A direct way to address this problem is given by the study of the X-ray spectra of elliptical galaxies, because the emission of a population of binary X-ray sources would have a significantly harder X-ray spectrum than the ∼ 1 keV emission seen in X-ray bright ellipticals. In Figure 9.9, we have shown that the composite spectrum of the elliptical sample of Kim, Fabbiano, and Trinchieri (1992b) is considerably softer than that of the spiral sample; but this composite spectrum is dominated by

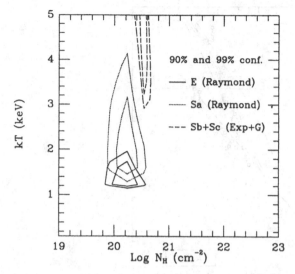

Fig. 9.10. 90% and 99% confidence regions for χ^2 fits to thermal emission models of the composite spectra of E, Sa, and Sb + Sc galaxies. The shift in N_H is due to the different models used for the fit. Simulations show that it would be extremely unlikely to obtain Sa-type confidence contours with E-type spectra. These contours are most easily obtained by fitting a composite spectrum (soft + hard). (Kim, Fabbiano, and Trinchieri 1992b.)

the spectra of the X-ray bright ellipticals. To explore a possible dependence of the emission temperature on the X-ray brightness, we have divided the elliptical sample into four subsamples according to the L_X/L_B ratio, and we have then repeated our analysis. A comparison of the composite distributions of spectral counts of the four subsamples with those of the composite spiral and elliptical spectra of Figure 9.9 is shown in Figure 9.12. It is clear from this Figure that the spectrum of elliptical galaxies hardens, to resemble that of spirals, with decreasing L_X/L_B. This confirms that L_X/L_B is a good indicator of the presence (or lack) of a hot ISM. This also implies that low X-ray luminosity ellipticals can be a good future hunting ground for samples of binary X-ray sources.

A totally unexpected result is given by the ellipticals with the lowest L_X/L_B ratio (see Figure 9.12). In these galaxies, the distribution of the spectral counts suggests the presence of a *very soft* component of the X-ray emission. We have postulated that the emission of these galaxies consists of two components, a hard – binary – component, plus a second component. Under these assumptions, we find a temperature for the very soft component of 0.16–0.22 keV (at 90% confidence). The luminosity of this component would be approximately one-third to one-half ($\sim 10^{40}$ erg s^{-1}) of the total X-ray emission.

What is the origin of this very soft emission? If it is cool gaseous emission, its X-ray luminosity is well in excess of that predicted by wind models (e.g. Ciotti *et al.* 1991). Therefore, in this case, it may suggest that these galaxies have a 'cooler' hot ISM than high L_x/L_B ellipticals. This possibility is rather unlikely (Pellegrini and Fabbiano 1994). More likely alternatives are that we may be detecting the integrated

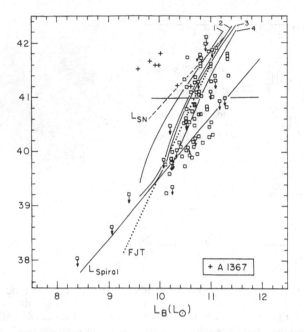

Fig. 9.11. X-ray and optical luminosities of E and S0 galaxies observed in X-rays (from Canizares, Fabbiano, and Trinchieri 1987). The solid line (L_{spiral}) is the best-fit line of early-type spirals from Fabbiano, Gioia, and Trinchieri (1988); the horizonthal solid line delimits the maximum X-ray luminosity of spiral galaxies. The dashed line (L_{SN}) is the estimate by Canizares *et al.* (1987) of supernova heating of the halos; the dotted line (FJT) is the estimate by Forman, Jones, and Tucker (1985); and the curves labeled 1, 2, 3, and 4 are the models of Sarazin and White (1988) for massive halos and supernova heating, supernova heating without massive halos, massive halos but no supernova heating, and no massive halos and no supernova heating, respectively. (From Fabbiano 1989.)

coronal emission of M stars in these galaxies, or perhaps ultra-soft binaries, similar to those detected with *ROSAT* in the LMC (Greiner, Hasinger, and Kahabka 1991). If this very soft emission is of stellar origin, then we would expect it to be present in all early-type galaxies, and to scale approximately with the optical luminosity. If this were the case, it would have been undetectable in more luminous ellipticals. The origin of this very soft emission is an open problem, which will require future spectral and spatially resolved spectral observations to unravel its nature.

9.5.2 *The very soft emission component of spiral galaxies*

As shown by Figure 9.9, the average spectrum of spirals is hard. However, there are galaxies in the *Einstein* sample that exhibit excess emission in the lowest spectral channels, above the level expected from a hard emission spectrum undergoing line of sight absorption. This is shown in Figure 9.13, where the spiral sample is plotted in a color–color plot. The two X-ray colors C21 and C32 are defined as follows (Kim, Fabbiano, and Trinchieri 1992a,b):

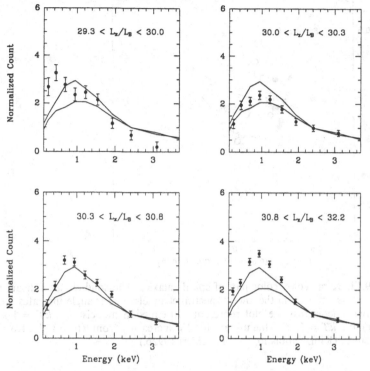

Fig. 9.12. Combined observed spectra of elliptical (E + S0) galaxies grouped by L_X/L_B. The two lines represent the composite spectra of ellipticals (upper) and spirals (lower), shown in Figure 9.9. The combined spectra of the four L_X/L_B groups have been arbitrarily renormalized to match these curves. (Kim, Fabbiano, and Trinchieri 1992b.)

- C21 = counts (0.8–1.4 keV) / counts (0.2–0.8 keV),
- C32 = counts (1.4–3.5 keV) / counts (0.8–1.4 keV).

C32 can be considered as an index of the emission temperature, while C21 is an index of the absorption experienced by the emitted radiation. The galaxies represented by filled dots in Figure 9.13 and the triangle (NGC 253) have values of C21 which – for a simple thermal model with low-energy cut-off – would imply a value of the absorbing column smaller than the Galactic N_H. This is clearly unphysical and suggests that the simple one component model is not valid, and that these galaxies have an additional very soft emission component. The distribution of the composite observed spectral counts of these galaxies is compared with the composite spiral spectrum in Figure 9.14, and shows clearly the excess counts at the low energies.

As in the case of the very soft component of X-ray faint ellipticals, it is not possible with the *Einstein* data to establish the origin of this emission. Possibilities include a warm phase of the ISM, some stellar emission, and very soft binaries (e.g. Greiner, Hasinger, and Kahabka 1991). *ROSAT* observations may help in discriminating between these possibilities.

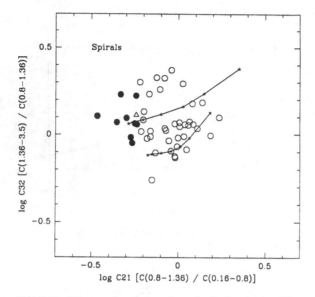

Fig. 9.13. X-ray color–color plot of spiral galaxies. The solid circles indicate galaxies with excess emission in the lowest spectral channels. The triangle indicates NGC 253. The two lines across the plot represent two emission models with $kT = 1$ keV (the lower) and $kT = 5$ keV (the upper), for N_H increasing from left to right, from 10^{19} to 10^{22} cm^{-2}. (Kim, Fabbiano, and Trinchieri 1992b.)

9.6   The future

The study of the X-ray emission of normal galaxies is a very recent part of astronomy. Before the *Einstein* satellite, only four 'normal' galaxies were known to emit X-rays: the Milky Way, M31 (Andromeda), and the Magellanic Clouds. Now, the sample of galaxies observed in X-rays is close to 500 (Fabbiano, Kim and Trinchieri 1992). In the preceding pages, I have summarized some of the *Einstein* results, with special reference to those relevant for the binary emission component of galaxies. I have also tried to emphasize some of the questions raised by these observations. Here I will talk of what can be done with present and future observatories to further this field.

Our knowledge of the X-ray properties of galaxies is presently evolving, thanks to the observations of the German/US/UK satellite *ROSAT*. With *ROSAT*, we can now try to explore some of the questions raised by the *Einstein* work. In particular, *ROSAT*'s improved spatial and spectral resolution (over *Einstein*), together with our present awareness of the X-ray properties of galaxies, is resulting in observations of galaxies that will push forward considerably the present state of knowledge. Preliminary reports of early *ROSAT* results show a very significant increase in the sample of X-ray sources detected in nearby spirals. *ROSAT*'s spectral capabilities and bandwidth should give us very exciting results on the softer emission component of galaxies of all morphological types. I expect that in a year from now this chapter could be entirely rewritten.

In early 1993 the Japanese/US satellite *ASCA* (Tanaka 1990) was launched. Al-

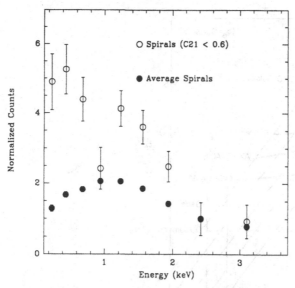

Fig. 9.14. Composite spectrum of spirals with soft excess (open circles) normalized at 2.4 keV, compared with average spectrum of spirals (solid circles). (Kim, Fabbiano, and Trinchieri 1992b.)

though the limited angular resolution of *ASCA* (2'.5 FWHM) will not allow the detailed spatial study of different components in galaxies, its spectral capabilities will let us explore a harder spectral band in a large number of galaxies. This will allow direct comparisons between our knowledge of Galactic sources and the properties of external galaxies. As demonstrated by a few *Ginga* results (e.g. Makishima *et al.* 1989), an X-ray population synthesis of nearby galaxies will be possible. By obtaining the hard X-ray spectra of nearby starburst galaxies, it will also be possible to explore directly the contribution of distant starburst galaxies to the hard X-ray background (e.g. Fabbiano 1988b, 1989; Griffiths and Padovani 1990).

This is the near future. In the following pages, I will talk of what we can expect from other more distant (in time) missions such as *AXAF* and *XMM*, and of what an ideal X-ray observatory for galaxies may be. The following pages are based entirely on a paper delivered in the workshop on 'High Energy Astrophysics in the 21st Century', held in Taos in December 1989 (Fabbiano 1990c).

9.6.1 Requirements for future observatories

9.6.1.1 Detecting galaxies

Normal galaxies are relatively faint X-ray sources ($\sim 10^{39-42}$ erg s^{-1}). Therefore, if we want to study them in any more depth than allowed by the exploratory *Einstein* survey, and if we want to be able to detect them in large numbers and out to large redshifts to pursue sensible statistical studies of their sample proportion and evolution, we need to have large collecting areas of imaging telescopes. Figure 9.15 shows a plot of the X-ray flux as a function of galaxy distance, for galaxies with X-ray luminosities in the range of those detected with *Einstein*.

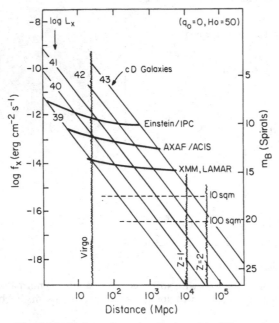

Fig. 9.15. Galaxy detection limits with different instruments (Fabbiano 1990c).

The corresponding B magnitudes are also given. These assume the average X-ray to optical ratio for spiral galaxies from the *Einstein* survey. For elliptical galaxies the corresponding X-ray fluxes could be larger by a factor of ~100 or more. Superimposed on this graph are curves that give the limiting sensitivity for a 5σ detection in a 5000 s observation for the *Einstein* IPC, the *AXAF* (see Tananbaum 1990) CCD experiment (ACIS), the planned European large throughput mission *XMM* (Peacock and Ellwood 1988), and the proposed large throughput mission *LAMAR* (see Gorenstein 1990). Although longer observing times are of course possible, I am not considering them here, since I assume that the observations of galaxies will only be one of the topics that will be pursued with future observatories, and that therefore there will be constraints on the available observing time. Since I wish to emphasize here the capability of detecting large samples of galaxies in a moderate observing time, I am not showing curves for the high resolution experiments on *Einstein* and *AXAF*.

Figure 9.16 gives the typical angular sizes for a typical galaxy diameter (~30 kpc) at different distances, and these range from something of the order of a degree for a few nearby objects to a few (2–3) arcseconds for galaxies at high z.

With *Einstein* we could detect the least luminous galaxies only in our immediate surroundings, and we could only study the higher luminosity galaxies in the Virgo cluster. *XMM* and *LAMAR* will allow the detection in reasonably short observing times of the entire luminosity range of galaxies in the Virgo cluster. This will produce excellent science. However, if we want to study galaxies at larger distances, and therefore sample a variety of cluster environments, we need to go deeper. This is especially true if we want to study in some detail galaxies at the epoch of formation.

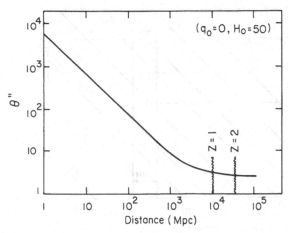

Fig. 9.16. Angular sizes corresponding to 30 kpc at different distances (Fabbiano 1990c).

In the absence of X-ray luminosity evolution in galaxies, larger collecting areas, 100–1000 times larger than those presently planned, and longer observing times will be needed to study galaxies at high z in X-rays. It is also important to remember that an angular resolution of $\sim 1''$ is needed to resolve galaxies at high z, and to distinguish them from point-like sources, such as QSOs (Figure 9.16). This angular resolution is ten times better than planned for *XMM* and *LAMAR*.

9.6.1.2 High resolution studies of individual galaxies

A different and no less important aspect of the study of galaxies is that based on detailed high resolution observations of individual objects. Here again the observing times should not be forbiddingly long, to be in the optimal situation of exploring the properties of the X-ray emission components as a function of a reasonably large range of galaxian properties.

Figure 9.17 shows a plot of the angular resolution needed to resolve structures of a given linear size at different distances from the Solar System, ranging from 100 pc to cosmological distances. This graph shows that the high throughput missions *XMM* and *LAMAR* will not be able, for example, to resolve supernova remnants in external galaxies, whereas the higher resolution *Einstein* HRI could resolve them in the LMC, and the *AXAF* HRC will resolve them as far as M31. However, to be able to resolve SNRs in Virgo galaxies, an angular resolution of $\sim 10^{-2}$ arcsec will be needed. This same angular resolution will allow the study of a few kiloparsec scale structures in high z galaxies, and of planetary scale structures within 100 pc from the Sun.

However, there is no point in studying galaxies at very high resolution without a very large collecting area. Figure 9.18 shows the limiting sensitivity for detecting individual sources of a given luminosity in galaxies from the LMC to the Virgo cluster and beyond. I assume 5σ detections in 10^4 s observations. The *AXAF* HRC will be able to detect down to the luminosity of the brightest stars in the LMC, but will

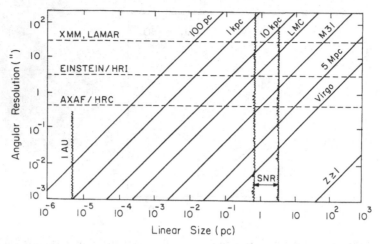

Fig. 9.17. Angular resolution needed to resolve structures of given linear sizes at different distances (Fabbiano 1990c).

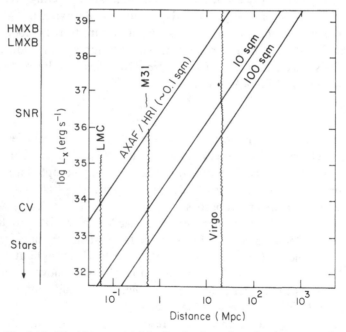

Fig. 9.18. Limiting sensitivities for the detection of point sources in galaxies (Fabbiano 1990c).

detect only the brightest binary X-ray sources in the Virgo cluster galaxies. A much larger collecting area will be needed to explore the luminosity function of the X-ray sources in Virgo galaxies down to the luminosities accessible with the *Einstein* data in Local Group galaxies.

9.6.1.3 Bandwidth

The X-ray band is extremely broad, covering easily two decades of the spectrum (equivalent to combining the mid-infrared, optical and ultraviolet bands). This breadth has great diagnostic power and the ideal X-ray telescope to study galaxies should cover most of this bandwidth (from $\sim$0.1 keV to 10–20 keV), because different galaxian components contribute to different energy bands. The stellar component, old SNRs and hot ISM contribute to energies ≤ 1 keV; young SNRs and XBs emit mostly at higher energies; and small active nuclei are likely to emit both in the soft and in the hard bands (e.g. Fabbiano 1988a).

9.6.1.4 Spectral studies

Spectral information is essential in the understanding of the physical mechanisms giving rise to the X-ray emission. We want to be able to gather spectral data on galaxies as a whole (e.g. to discriminate between gaseous emission and the integrated output of binary X-ray sources as the predominant source of X-ray emission in X-ray faint early-type galaxies); we want also to be able to study separately specific structures (e.g. bulges vs. spiral arms), or to obtain temperature profiles of elliptical galaxies to measure their masses (see Trinchieri, Fabbiano, and Canizares 1986); finally we want to be able to study individual sources in more nearby galaxies. Therefore, the first requirement for spectral studies of galaxies is the ability to obtain spatially resolved spectral information.

There are two approaches one can take to spectral studies. The first, similar to optical photometry, is to obtain low resolution spectra ($\lambda/\Delta\lambda \sim$a few) or colors over a larger bandwidth. In this case, the incoming photons will be divided in a relatively small number of spectral bins. The second approach, spectroscopy, is to obtain mid to high resolution spectra ($\lambda/\Delta\lambda \sim$100–1000) over different bandwidths. Low resolution spectra can be used to characterize the global spectral properties of different galaxies or populations of X-ray sources; high resolution spectra will be invaluable for exploring the physical state of gaseous components in galaxies (e.g. Canizares, Markert, and Donahue 1988).

Even without exploring in detail the characteristics of spectral detectors, it is clear that very high throughput missions (i.e. large collecting areas) will be needed for spectral studies if we do not want to limit ourselves to relatively nearby objects. This can be seen by looking at Figure 9.15 and considering that the number of incoming photons will have to be divided in a number of separate spectral channels. This is equivalent (in a first approximation) to moving the sensitivity threshold curves upwards by equivalent amounts.

9.6.1.5 Time variability studies

We know from X-ray observations of sources in the Milky Way that X-ray sources vary in time. This time variability gives us precious insights into the nature of sources, especially if joined with spectral analysis. As in the case of spectral studies, large collecting areas are needed to study source variability in external galaxies, even more so if one wishes to explore the spectral behavior in different intensity states. Moreover, it will be desirable to have the flexibility to devote the telescope, or even better a part of it, to long and repeated observations of nearby galaxies to obtain light curves of variable sources on the common characteristic time scales that are longer than a single observation.

9.6.2 *The 'ideal' X-ray telescope for the study of galaxies*

Based on the previous discussion, I will try to sketch here the characteristics of the ideal X-ray telescope for studying galaxies. I consider this a goal to which the high energy community and NASA should aim, but it is clear to me that the way to reach this goal is by intermediate incremental steps. These intermediate missions will allow us to keep studying galaxies in X-rays and to test new developments in X-ray optics and instrumentation that will be needed to achieve our final proposed goal.

The characteristics of this ideal X-ray telescope are given below:

- collecting area > 100 m^2 (similar to the optical Keck telescope);
- large bandwidth (0.1–20 keV);
- angular resolution of $\sim 1''$. High resolution studies of individual galaxies will require higher angular resolution (up to 10^{-3} arcsec, if there is enough collecting area);
- spectral capabilities for both X-ray photometry and X-ray spectroscopy;
- capability to devote part of the area to variability studies of nearby galaxies.

I think that the only possible way of achieving this X-ray facility is to put an array of X-ray telescopes on the Moon.

References

Beck, R., Klein, U., and Krause, M. 1985, *A&A*, 152, 237

Blitz, L., Israel, F. P., Neugebauer, G., Gatley, I., Lee, T. J., and Beattie, D. H. 1981, *Ap. J.*, 249, 76

Boller, Th., Meurs, E. J. A., Brinkman, W., Fink, H., Zimmerman, U., and Adorf, H.-M. 1992, *A&A*, 261, 57

Bottinelli, L., Gouguenheim, L., Paturel, G., and de Vaucouleurs, G. 1984, *A& A Suppl.*, 56, 381

Canizares, C. R., Fabbiano, G., and Trinchieri, G. 1987, *Ap. J.*, 312, 503

Canizares, C. R., Markert, T. H., and Donahue, M. E. 1988, in *Cooling Flows in Clusters and Galaxies*, ed. A. Fabian, p. 63. Dordrecht: Kluwer

Ciardullo, R., Ford, H. C., Jacoby, G. H., and Shafter, A. W. 1987, *Ap. J.*, 318, 520

Ciardullo, R., Ford, H. C., Shafter, A. W., Neill, J. D., and Jacoby, G. H. 1988, *A. J.*, 95, 438

Ciotti, L., D'Ercole, A., Pellegrini, S., and Renzini, A. 1991, *Ap. J.*, 376, 380

Clark, G., Doxsey, R., Li, F., Jernigan, J.G., and Van Paradjis, J. 1978, *Ap. J. Letters*, 221, L37

Collura, A., Reale, F., and Peres, G. 1990, *Ap. J.*, 356, 119

Cook, K. H., Aaronson, M., and Illingworth, G. 1986, *Ap. J. Letters*, 301, L45

David, L. P., Forman, W., and Jones, C. 1991, *Ap. J.*, 369, 121

David, L. P., Jones, C., and Forman, W. 1992, *Ap. J.*, 388, 82

D'Ercole, A., Renzini, A., Ciotti, L., and Pellegrini, S. 1989, *Ap. J. Letters*, 341, L9

Fabbiano, G. 1985, in *Japanese-U.S. Seminar on Galactic and Extragalactic Compact X-ray Sources*, eds. Y. Tanaka and W. H. G. Lewin, p. 233. Tokyo: ISAS

Fabbiano, G. 1988a, *Ap. J.*, 325, 544

Fabbiano, G. 1988b, *Ap. J.*, 330, 672

Fabbiano, G. 1989, *ARA& A*, 27, 87

Fabbiano, G. 1990a, in *Imaging X-ray Astronomy*, ed. M. Elvis, p. 155. Cambridge University Press

Fabbiano, G. 1990b, in *Windows on Galaxies*, eds. G. Fabbiano, J. S. Gallagher and A. Renzini, p. 231. Dordrecht: Kluwer

Fabbiano, G. 1990c, in *High Energy Astrophysics in the 21st Century*, ed. P. C. Joss, p. 74. New York: AIP

Fabbiano, G., Feigelson, E., and Zamorani, G. 1982, *Ap. J.*, 256, 397

Fabbiano, G., Gioia, I. M., and Trinchieri, G. 1988, *Ap. J.*, 324, 749

Fabbiano, G., Gioia, I. M., and Trinchieri, G. 1989, *Ap. J.*, 347, 127

Fabbiano, G., Heckman, T., and Keel, W. C. 1990, *Ap. J.*, 355, 442

Fabbiano, G., Kim, D.-W., and Trinchieri, G. 1992, *Ap. J. Suppl.*, 80, 531

Fabbiano, G., and Trinchieri, G. 1985, *Ap. J.*, 296, 430

Fabbiano, G., and Trinchieri, G. 1987, *Ap. J.*, 315, 46

Fabbiano, G., Trinchieri, G., and Macdonald, A. 1984, *Ap.J.*, 284, 65

Fabbiano, G., Trinchieri, G., and Van Speybroeck, L. S. 1987, *Ap. J.*, 316, 127

Fabricant, D., and Gorenstein, P. 1983, *Ap.J.*, 267, 535

Forman, W., Jones, C., and Tucker, W. 1985, *Ap. J.*, 293, 102

Forman, W., Jones, C., Cominsky, L., Julien, P., Murray, S., Peters, G., Tananbaum, H., and Giacconi, R. 1978, *Ap. J. Suppl.*, 38, 357

Giacconi, R., *et al.* 1979, *Ap. J.*, 230, 540

Gorenstein, P. 1990, in *High Energy Astrophysics in the 21st Century*, ed. P. C. Joss, p. 297. New York: AIP

Greiner, J., Hasinger, G., and Kahabka, P. 1991, *A&A*, 246, L17

Griffiths, R. E., and Padovani, P. 1990, *Ap. J*, 360, 483

Helfand, D. J. 1984a, *PASP*, 96, 913

Helfand, D. J. 1984b, in *Structure and Evolution of the Magellanic Clouds*, eds. S. Van den Bergh and K. deBoer, p. 293. Dordrecht: Reidel

Hodge, P. 1983, *PASP*, 95, 721

Kim, D.-W., Fabbiano, G., and Trinchieri, G. 1992a, *Ap. J. Suppl.*, 80, 645

Kim, D.-W., Fabbiano, G., and Trinchieri, G. 1992b, *Ap. J.*, 393, 134

Long, K. S., and Van Speybroeck, L. P. 1983, in *Accretion Driven X-ray Sources*, eds. W. Lewin and E. P. J. Van den Heuvel, p. 117. Cambridge University Press

Long, K. S., D'Odorico, S., Charles, P. A., and Dopita, M. A. 1981, *Ap. J. Letters*, 246, L61

Makishima, K., *et al.* 1989, *PASP*, 41, 697

Markert, T. H., and Rallis, A. D. 1983, *Ap. J.*, 275, 571

Palumbo, G. G. C., Fabbiano, G., Fransson, C., and Trinchieri, G. 1985, *Ap. J.*, 298, 259

Peacock, A., and Ellwood, J. 1988, *Space Sci. Rev.*, 48, 343

Peres, G., Reale, F., Collura, A., and Fabbiano, G. 1989, *Ap. J.*, 336, 140

Pellegrini, S., and Fabbiano, G. 1994, *Ap. J.*, in press

Primini, F., Forman, W., and Jones, C. 1993, *Ap. J.*, 410, 615

Richter, O.-G., and Rosa, M. 1984, *A&A*, 140, L1

Sandage, A. 1961, *The Hubble Atlas of Galaxies*, Carnegie Inst. Pub. No. 618

Sandage, A. 1962, *IAU Symp.* 15:359

Sandage, A. 1984, *A. J.*, 89, 621

Sandage, A., and Tamman, G. A. 1974, *Ap. J.*, 194, 223

Sarazin, C. L., and White, R. E. III 1988, *Ap. J.*, 331, 102

Schmidt, B. P., and Kirshner, R. 1992, preprint

Supper, R. *et al.* 1994, in preparation

Tanaka, Y. 1990, *Adv. Space Res.*, 10-2, 255

Tananbaum, H. 1990, in *Imaging X-ray Astronomy*, ed. M. Elvis, p. 15. Cambridge University Press

Trinchieri, G., and Fabbiano, G. 1991, *Ap. J.*, 382, 82

Trinchieri, G., Fabbiano, G., and Canizares, C. R. 1986, *Ap. J.*, 310, 637

Trinchieri, G., Fabbiano, G., and Palumbo, G. G. C. 1985, *Ap. J.*, 290, 96

Trinchieri, G., Fabbiano, G., and Peres, G. 1988, *Ap. J.*, 325, 531

Trinchieri, G., Fabbiano, G., and Romaine, S. 1990, *Ap. J.*, 356, 110

Trümper, J. 1983, *Adv. Space Res.*, 2, 241

Vader, J. P., Van den Heuvel, E. P. J., Lewin, W. H. G., and Takens, R. J. 1982, *A&A*, 113, 328

Van Speybroeck, L., and Bechtold, S. 1981, in *X-ray Astronomy with the Einstein Satellite*, ed. R. Giacconi, p. 153. Dordrecht: Reidel

Van Speybroeck, L., Epstein, A., Forman, W., Giacconi, R., Jones, C., Liller, W., and Smarr, L. 1979, *Ap. J. Letters*, 234, L45

Wang, Q., and Helfand, D. J. 1991a, *Ap. J.*, 370, 541

Wang, Q., and Helfand, D. J. 1991b, *Ap. J.*, 373, 497

Wang, Q., Hamilton, T., Helfand, D. J., and Wu, X. 1991, *Ap. J.*, 374, 475
Watson, M. G. 1990, in *Windows on Galaxies*, eds. G. Fabbiano, J. S. Gallagher and A. Renzini, p. 177. Dordrecht: Kluwer
Watson, M. G., Stanger, V., and Griffiths, R. E. 1984, *Ap. J.*, 286, 144

10

Accretion in close binaries

Andrew King

Astronomy Group, University of Leicester, Leicester LE1 7RH, UK

10.1 Introduction

The first suggestion that X-ray binaries were powered by accretion on to a compact object was made by Shklovsky[117] in 1967, five years after the discovery of Sco X-1[35], and three years after Salpeter's[111] original recognition of the possible importance of accretion as an astrophysical energy source (in quasars). The theory of accretion processes has advanced considerably since then, and has produced a vast literature; in particular, it is now clear that in many binaries the accreting material has significant angular momentum and is likely to form an accretion disc. This chapter aims to provide a concise summary of the field, particularly the theoretical ideas used in other chapters of this book.

10.2 Summary of accretion disc theory

This section summarizes the results of accretion disc theory used in the rest of the chapter. As there are now many reviews and books giving the derivations of these results (see, e.g., Frank *et al.*[33], Pringle[103]), these are for the most part omitted.

10.2.1 Disc formation

Accreting matter forms a disc when its specific angular momentum J is too large for it to hit the accreting object (mass M_1) directly. This typically requires that the circularization radius

$$R_{\text{circ}} = \frac{J^2}{GM_1} \qquad (10.1)$$

(where the matter would orbit if it lost energy but not angular momentum) should be larger than the effective size of the accretor. In X-ray binaries the accretor is a neutron star or black hole, and the condition always holds if accretion is via Roche lobe overflow. The outcome is much less clear if the accretion is from a wind, as J is much lower (see Sect. 10.8). Moreover, the neutron star in many wind-accreting systems has a strong magnetic field, increasing the size of the obstacle presented to the infalling matter. Cataclysmic variables, in which a low-mass companion star overflows its Roche lobe on to a white dwarf, provide (almost!) text-book examples of accretion discs if the white dwarf is 'non-magnetic' (surface fields $B \lesssim 10^5$ G) (see below). The marked outburst behaviour of a subset of these systems – the dwarf novae – also supplies the main observational tests of theories of disc instabilities (see Sect. 10.3).

White dwarfs can have magnetic moments $\mu = BR^3$ (R = stellar radius) amongst the largest of any astronomical object (up to several times 10^{34} G cm^3), making their magnetospheric radii r_μ sometimes comparable with the binary separation a. The cataclysmic variables (CVs) with the largest r_μ/a ratios are the AM Herculis systems (see Cropper[25] for a review), in which disc formation is suppressed. The most complex case is that of the intermediate polars, where r_μ/a is smaller, but still large enough to raise doubts that discs can form in the usual way (see Sect. 10.9).

Assuming that matter can orbit at R_{circ}, the fundamental assumption of disc theory, that energy is lost more rapidly than angular momentum, means that accretion can only take place through a sequence of circular orbits with gradually decreasing J. Without an external angular momentum sink such as a magnetic field, this can only be accomplished by transporting angular momentum outwards. The disc edge thus expands far beyond R_{circ}; most of the original angular momentum is carried out to this edge, where it is typically returned to the secondary star's orbit through tides.

The agency responsible for both energy dissipation and angular-momentum transport is called *viscosity*, assumed to provide a torque between the shearing Kepler orbits. The physical basis of the viscosity is almost entirely unclear (ordinary kinetic-theory viscosity is much too weak, for example). This is highly unsatisfactory, as viscosity drives the entire evolution of the disc: the lack of knowledge of its functional dependence means that disc theory is scarcely a theory at all in the usual sense. Needless to say, the freedom allowed by this ignorance has been eagerly seized on by theoreticians: for example, every analysis of disc instabilities makes implicit (or explicit) assumptions about the behaviour of the viscosity. Fortunately some properties of steady discs are independent of the viscosity, and in other cases one can put limits on its magnitude.

10.2.2   *The thin disc approximation*

Disc theory makes most headway if cooling is assumed always to be able to keep the local Kepler speed $v_{\text{K}} = (GM_1/R)^{1/2}$ supersonic. This is the *thin disc approximation*: the disc lies essentially in the orbital plane, with only a small 'vertical' extent

$$H \simeq \frac{c_S}{v_{\text{K}}} R << R, \tag{10.2}$$

where c_S is the local sound speed, and its azimuthal speed is close to the Kepler value v_{K}. Both the radial and vertical velocities are much smaller. It is important to realize that the three properties of being geometrically thin, Keplerian, and efficiently cooled are equivalent for any disc flow: if one of them breaks down, so do the other two.

If the thin disc approximation holds, the radial and vertical structures essentially decouple, and one can describe the disc largely by its surface density Σ. Mass and angular momentum conservation show that this obeys a nonlinear diffusion equation:

$$\frac{\partial \Sigma}{\partial t} = \frac{3}{R} \frac{\partial}{\partial R} \left(R^{1/2} \frac{\partial}{\partial R} [\nu \Sigma R^{1/2}] \right). \tag{10.3}$$

In a steady state this implies

$$\nu \Sigma = \frac{\dot{M}}{3\pi} \left[1 - \beta \left(\frac{R_{\text{in}}}{R} \right)^{1/2} \right], \tag{10.4}$$

where $\dot{M}$ is the accretion rate and the dimensionless quantity β depends on the boundary conditions at the inner disc edge R_{in}: for a disc around a non-rotating star we have $\beta = 1$ with R_{in} = stellar radius. The surface dissipation pattern of the disc is also fixed independently of viscosity in a steady state, and leads to a local effective temperature distribution

$$T_{eff} = \left(\frac{3GM_1\dot{M}}{8\pi\sigma R^3}\right)^{1/4}\left[1 - \beta\left(\frac{R_{in}}{R}\right)^{1/2}\right]^{1/4}. \qquad (10.5)$$

The thin disc picture also motivates the most common parametrization[115] of the unknown disc viscosity as

$$v = \alpha c_S H, \qquad (10.6)$$

for it is reasonable to assume that the sizes of any random motions contributing to v cannot greatly exceed the disk thickness, while supersonic motions are likely to suffer rapid dissipation in shocks. Thus one can hope that $\alpha \lesssim 1$, except possibly in some restricted regions of the disc where an external agency (such as the mass transfer stream) continually feeds energy into the random motions. Note that there is absolutely no justification for any particular functional form or dependence of α: in particular, there is no reason at all to assume that α is constant throughout a given disc. Through the diffusion equation (10.3), the value of α specifies the viscous time scale $\sim l^2/v$ on which the surface density can change in a region of spatial scale l. Observed time scales for luminosity changes thus imply estimates of the effective value of α. Dwarf nova outbursts (see Sect. 10.3) show that surface density changes involving most of the disc in a CV ($l \sim 3 \times 10^{10}$ cm) can occur in a few days, implying $\alpha \sim 0.1 - 1$ for them.

10.2.3 Accretion discs in CVs

Although in general the predictions of disc theory involve the unknown viscosity, this is not true of the effective temperature T_{eff}: indeed from Eq. (10.5) we see that for most of the disc ($R >> R_{in}$) we have the simple relation

$$T_{eff} \propto R^{-3/4}. \qquad (10.7)$$

Observational tests of this result are of prime importance: fortunately eclipse mapping of non-magnetic CVs makes them relatively tractable (see Sect. 8.2.1.2). The principle is straightforward: Eq. (10.7) predicts that the blue light of the disc will be more centrally concentrated than the red. Thus, an eclipse by a dark companion will produce a deep, narrow eclipse in the blue, and a broad, shallow eclipse in the red. In practice, one observes many eclipses of a given system at several wavelengths, and uses a maximum-entropy technique to find the most axisymmetric surface brightness distribution of the average disc. These techniques were pioneered by Horne and collaborators (see refs 45 and 46 for reviews). For an optically thick disc, it is reasonably uncomplicated to convert from colour to effective temperature, and results consistent with Eq. (10.7) are obtained in several cases (see Figure 10.1). Thus in some CVs at least it is difficult to deny that there exist accretion discs conforming to

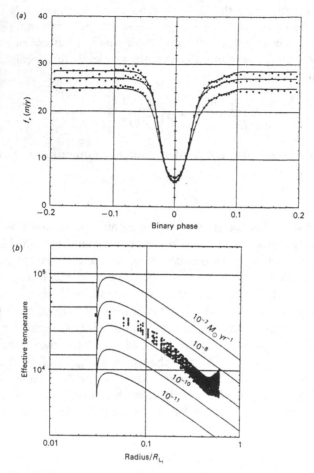

Fig. 10.1. The surface brightness of the accretion disc in the dwarf nova Z Cha (during outburst) as revealed by eclipse mapping. (a) Eclipse light curves (*B,U,V* from the top). (b) Effective temperature distribution resulting from maximum-entropy deconvolution of these data, compared with that predicted by Eq. (10.5). (Reproduced from ref. 45.)

simple theoretical ideas. This statement probably cannot be made of any other class of accreting compact binaries, as we shall see.

Even in the context of CVs, the simple disc picture is considerably complicated by the interaction of the accretion stream with the disc. Since the latter grows out well beyond R_{circ}, at least some of the gas stream from the secondary's inner Lagrange point must crash obliquely into it with a hypersonic relative velocity. (The stream thickness may well exceed the disc scale height, so some of the stream material can avoid the collision and skim over the disc faces – see Sect. 10.2.) The resultant dissipation makes part of the disc edge bright: the anisotropic radiation pattern of this region gives rise to a periodic hump in orbital light curves of some CVs (see Sect. 8.3.4). As the disc edge is much larger than the white dwarf radius, the total luminosity of this 'bright spot' is on average much smaller than the system's bolometric luminosity; it can nevertheless temporarily outshine the disc if the accretion rate at the disc edge greatly exceeds that at the centre. Generally, the spot is visible only in systems with low accretion rates, particularly quiescent dwarf novae (see Sect. 10.3). Spectroscopic studies of these systems[81,123] confirm that the stream follows essentially a ballistic trajectory and give estimates of the mass ratio M_2/M_1. In several cases, the stream continues within the disc radius, suggesting either that some of it penetrates the disc material, or skims over the disc faces as envisaged above.

Theoretical treatment of the bright spot is difficult as the problem is fully three-dimensional. Two-dimensional treatments confine the flow within a limited distance from the disc plane, and thus tend to exaggerate its azimuthal extent as the gas cannot expand or splash away from this plane.

10.3 Disc structure in LMXBs

Low-mass X-ray binaries (LMXBs) differ from CVs only in replacing the white dwarf by a neutron star or black hole. In particular, mass transfer is likely to be via Roche lobe overflow in most cases, so that the disc formation criterion (10.1) should hold. It is natural to hope then that some of the simple ideas that work quite successfully for CVs might find application here too. We can expect differences to arise where the presence of a more compact accretor is able to transmit itself. Virtually the only way this can happen is through irradiation: the same accretion rate falling on a neutron star or black hole produces a much greater luminosity than on a white dwarf, and the spectrum is likely to be much harder as well. The characteristic effective temperature dependence, Eq. (10.7), can easily be altered by this effect. Further, irradiation can lead to substantial mass loss (see Sect. 10.8 below), breaking the assumption of constant mass inflow made in deriving Eq. (10.7). But it is rather unlikely that irradiation can lead to substantial deviation from Kepler flow in the orbital plane (see below).

10.3.1 Difficulties with the standard picture

The above arguments suggest that one might expect a geometrically thin disc to provide a reasonable picture of accretion also in LMXBs. However, X-ray observations of these systems utterly confound this expectation (see Sect. 1.2.2). If the X-rays came from a point source (the immediate vicinity of the neutron star or black hole) surrounded by a thin Keplerian disc, one would expect to see either no orbital modulation at all, or a total eclipse with very sharp ingress and egress, depending on the system inclination. Yet *partial* eclipses are sometimes observed, while steep-sided total eclipses are rare. This strongly suggests that the X-ray source is effectively extended, presumably through the presence of some kind of scattering corona around it. Partial eclipses can result if the size of the corona is comparable with that of the companion star. This means that the corona subtends a larger angle at the central source than the companion, so at inclinations too low for an eclipse we may see continuous periodic modulation of the X-rays by persistent structure in the corona. Such behaviour is indeed observed, as well as 'dipping', irregular drops in X-ray intensity occurring in a stable phase interval (Figure 10.2). The corona will also tend to scatter X-rays away from the orbital plane, making edge-on systems rather faint in X-rays, and explaining the relative rarity of eclipsing systems. This

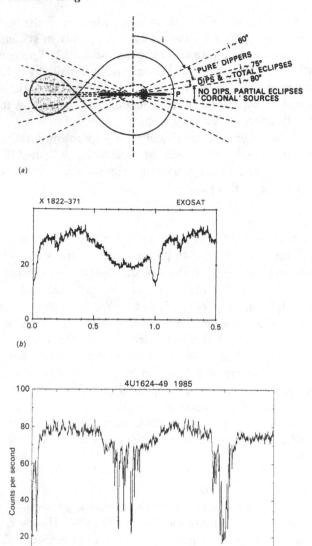

(a)

(b)

(c)

Fig. 10.2. (a) The light curves of LMXBs (reproduced from ref. 31). At high inclination, only X-rays scattered from an extended corona around the accreting neutron
star (or black hole) are seen, and eclipses by the companion star are partial, as in (b)
reproduced from ref. 97. At lower inclinations the central point X-ray source is seen,
and the X-rays may be modulated by dips, as in (c) from ref. 133.

idea is supported by the fact that continuously modulated systems, presumably seen
in scattered X-rays, have significantly lower X-ray to optical flux ratios than face-on
systems, where we see the X-rays directly (the optical flux comes mainly from the disc
and depends much less strongly on inclination).

10.3.2 *Vertical structure*

This coronal picture, put forward originally by White & Holt[135], is purely empirical: there is no physical model for the structures whose existence is posited. Theoretical attempts to provide them have met only partial success. There is good reason to expect a corona to exist, if some of the central X-rays irradiate the inner parts of the disc (see Sect. 10.6), and its scattering optical depth τ can be large enough ($\gtrsim 0.1$) to account for the scattered flux. But there is considerable difficulty in producing the vertical structure required to give periodic effects such as dips. White and Holt suggested that this structure should be near the disc edge. However this leads to several problems.

First, the vertical angular size of this structure implies a scale height h such that

$$h \gtrsim 0.15R \qquad (10.8)$$

at radial position R. If we identify h with H, the scale height of a thin disc, Eq. (10.2) shows that the local temperature T would have to satisfy

$$T \sim \frac{0.014 G M_1 m_H}{kR} \gtrsim 4.5 \times 10^5 \text{ K} \qquad (10.9)$$

at a typical outer disc radius of 3×10^{10} cm. This is more than irradiation can produce: the radiation temperature from even an Eddington-limited $1 M_\odot$ object at this radius is only

$$T_{\text{rad}} = \left(\frac{L_{\text{Edd}}}{4\pi R^2 \sigma} \right)^{1/4} = 1.6 \times 10^5 \text{ K.} \qquad (10.10)$$

Turbulence is sometimes invoked as a cause of the large scale height. However, this only shifts the problem elsewhere, as Eq. (10.9) shows that the turbulence would have to be highly supersonic. In fact, this idea, along with a number of others, are also open to the objection that they should work just as well for CVs, where, as we have seen, there is little obvious evidence of substantial deviation from the thin disc picture. The only way that most of the system 'knows' that the central object is a neutron star or black hole is through irradiation, suggesting that it must be involved here. Then Eq. (10.10) suggests that this will not radically alter the thin disc picture for LMXBs.

This brings us to the second problem with putting the structure at the disc edge. The only plausible way of getting material to the required heights above the disc plane is in ballistic orbits. Indeed, it is fairly clear that the disc rim will deviate somewhat from vertical hydrostatic balance, as the time scale to establish this is similar to the local Kepler period. Where we can determine it, the orbital phase of the dips suggests that matter might follow ballistic orbits as a result of the impact of the mass transfer stream from the secondary at the disc edge. But such orbits inevitably give a rather smooth azimuthal structure: inclined circular orbits intersect at two points separated by 0.5 in orbital phase. The disc edge picture requires a very structured 'Manhattan skyline' disc profile in order to reproduce the observed light curves. To retain these shapes would require matter to move in circular orbits in planes parallel to, but above or below, the orbital plane. These are not permissible test particle orbits, and it is hard to see how any gaseous material can be made to form itself into such shapes.

The third problem concerns the energy dependence of the dips. At higher photon energies, photoelectric absorption is negligible, and the dips must be caused by electron scattering; we infer equivalent hydrogen column densities $\gtrsim 10^{25}$ cm^{-2} along the line of sight from the observed dip depths. But this implies much greater photoelectric absorption depths at low X-ray energies than are observed (typically a few $\times 10^{23}$ cm^{-2}). There are two ways to avoid this problem: either the absorbing elements (CNO, He) are underabundant relative to H, or their K-shells, which do the absorbing, are ionized. But the latter possibility is ruled out if the matter causing the dips is near the disc edge, for the ionization parameter

$$\xi = \frac{L_X}{NR^2} \lesssim \frac{L_X}{10^{25}R} \lesssim \frac{10^{12}}{R} \sim 30 \qquad (10.11)$$

is then well below the values $\sim$ a few $\times 10^2$ required for this (Eq. (10.11) assumes $NR > 10^{25}$ cm^{-2} and an X-ray luminosity $L_X \sim 10^{37}$ erg s^{-1}). If the dip material is near the disc edge it must be very underabundant in metals (factors[136] up to $\sim 10^3$). These are not impossible, but would imply that LMXB secondaries belong to a rather extreme stellar population. Without independent supporting evidence, this seems a rather contrived expedient.

10.3.3   Gas streams

It thus seems worthwhile to look for ways of making the required vertical structure at locations nearer to the accreting star. The main problem is how to transmit a knowledge of the orbital phase to this structure so as to produce the observed periodic behaviour. The two possible ways of doing this involve either the tides produced on the disc by the secondary, or effects of the mass transfer stream penetrating or skimming over the disc. Tides probably only produce marked effects at the edge of the disc (but see Sect. 10.5), so models have so far concentrated on the second idea. Frank et al. [31] showed that it is quite reasonable for much of the stream to skim over the disc faces. The self-intersection of the gas stream near the circularization radius $R_{\text{circ}} \sim 10^{10}$ cm will lead to matter being thrown to large heights ($\sim R_{\text{circ}}$) above the disc. If the accreting star is a neutron star or black hole, this material is subject to a two-phase instability, quite independently of the accretion rate. This means that much of it condenses into cool filaments, which absorb X-rays photoelectrically. This therefore provides a physical reason for the prime observational feature of dips sources, cool material at large scale heights, at a fixed binary phase. This picture also predicts roughly the number, size and density of the filaments, in quite good agreement with the observations. Lubow[78] pointed out that the stream intersects the disc plane near the point of closest approach to the accreting star; he suggested that this would inhibit the formation of the structure proposed by Frank et al. Lubow's structure is close enough to the central star that a two-phase instability is likely here too; this would give dips with different phasing. In both models, all of the structure should occur in CVs too, but there will be no two-phase instability because of the lower ionizing luminosity for a given accretion rate. The disc regions obscured by this structure probably radiate in the ultraviolet or soft X-rays; it is possible that this might be revealed by Hubble Space Telescope observations.

This type of model provides physical reasons for the vertical structure and dip behaviour: its main defect is that it does not straightforwardly explain all of the

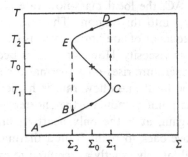

Fig. 10.3. Effective temperature vs. surface-density curve giving a thermal-viscous disc instability (see text for detailed explanation). The vertical axis (arbitrary scale) also describes the local mass flow rate $\dot{M}$, or the combination $\nu\Sigma$.

phase dependence observed in some LMXBs. It is important to realize that the disc-edge picture only achieves this by *fiat*, as it is physically unconstrained. A relatively unexplored possibility is that resonances (either horizontal or vertical) between disc orbits and the binary motion (cf. Sect. 10.4) might somehow produce vertical structure. A difficulty here is that the occurrence of resonances is a strong function of binary mass ratio (Sect. 10.4), whereas the vertical structure does not seem to be.

10.4 Disc instabilities

We have seen that some steady-state CVs provide a near-perfect observational paradigm for disc accretion. This makes it natural to try and interpret time-dependent behaviour in terms of discs also, particularly in systems where we have good indications that discs are present at certain epochs.

10.4.1 Thermal–viscous instability in dwarf novae

The most striking type of variability is that of the dwarf novae: outbursts (factors ~ 10 in luminosity) lasting for a few days, at intervals of weeks to months. There have been strenuous efforts to model these as disc instabilities, following the original suggestion by Smak[118,119] and later Osaki[93]. (For reviews, see refs 88, 120, 127.) The most successful approach uses the fact that the unknown dependence of viscosity means that the diffusion equation (10.3) does not necessarily imply smooth adjustment to a steady equilibrium.

Let us consider a single disc radius R, and assume that the equilibrium $\nu\Sigma$–Σ curve has the S-shape shown in Figure 10.3. From Eqs. (10.4) and (10.5) we have

$$\nu\Sigma \propto \dot{M} \propto T_{\text{eff}}^4, \tag{10.12}$$

where $\dot{M}$ is the local mass-flow rate. One can show that to the left of this equilibrium curve local cooling dominates local viscous heating, whereas the reverse holds to the right of the curve. This means that parts of the curve with $\partial(\nu\Sigma)/\partial\Sigma > 0$ define equilibria which are stable against thermal perturbations, while parts such as CE in Figure 10.3, with $\partial T_{\text{eff}}/\partial\Sigma > 0$, are thermally unstable. Thus we cannot have a locally steady structure with mass flow in the range $\dot{M}_C - \dot{M}_E$. If conditions outside the disc annulus at R impose such a flow, the structure will undergo limit cycles:

for example, if the structure is at a point on AC, the local mass-flow rate is lower than the externally determined value $\dot{M}_0 \propto T_0^4$ into this region. The local surface density increases, and the disc moves along a sequence of equilibria towards C, on the long viscous time scale corresponding to the low viscosity. However, at C it becomes unstable to thermal perturbations, and the temperature rises on a thermal time scale until the system reaches point D. But here the local mass-flow rate is higher than $\dot{M}_0$, so Σ decreases, taking the system towards point E, now on the shorter viscous time scale given by the larger value of v. Again, at E the only nearby equilibria are thermally unstable, so the system must move back to point B on a thermal time scale. It is clear that this local limit-cycle is qualitatively exactly as required to explain dwarf nova outbursts: long states of low accretion rate alternating with short, high accretion-rate states, with rapid (thermal time scale) transitions between them. This basic limit-cycle behaviour was first pointed out by Bath & Pringle[9].

Despite this initial success, we evidently require much more in order to claim this as the correct explanation of dwarf nova outbursts. Here we run into to the basic problem of all disc physics, the lack of a predictive theory of disc viscosity. We might hope that the details of the instability would be sufficiently robust that the α-prescription (10.6) would be enough to characterize the disc equilibria. Indeed, $T_{\text{eff}} - \Sigma$ equilibrium curves with constant α do give the desired S-shape[47,87], provided that the local temperature is close to 10^4 K: the kinks in the curve correspond to the onset of hydrogen ionization, and the instability corresponds to jumps between a cool, low-viscosity, low-opacity structure and a hot, high-viscosity, high-opacity structure.

Two further points have to be clarified: first, the instability described above is purely local, i.e. at one value of R. To get a global disc instability giving a large luminosity increase requires that a substantial part of the disc should become unstable; we therefore need the local instability at one radius to trigger instabilities at other radii. Secondly, all of the above is *qualitative*: we have to check that the mechanism produces the correct time scales, luminosities and spectra. In principle, we can use the unconstrained functional form of α to arrange this. For example, it is impossible to get long enough outbursts[120] with a single value of α: the S-curves for fixed α are too narrow. To get outbursts as wide as observed (see Sect. 8.3.1), α has to be manipulated, with, e.g., $\alpha = 0.05$ in the low state and $\alpha = 0.2$ in the high state. This is because the time spent by a given disc annulus on either the hot or the cool branch of the S-curve is rather short at constant α; the local outbursts are then out of phase, leading to little overall luminosity change, unless α is changed in the way indicated. Indeed, if we reverse the direction of change and take $\alpha = 0.2$ in the cool state and $\alpha = 0.05$ in the hot state even the local instability disappears.

As α is a purely *ad hoc* parameter, this is hardly surprising, but it emphasizes the lack of predictive power of disc theory. One should regard the requirements on α as constraints that the real viscosity mechanism will have to satisfy when it is discovered, if dwarf nova outbursts are to be understandable as disc instabilities of this type.

Another unsatisfactory feature of current disc instability calculations is that they are essentially one-dimensional, whereas the propagation of the thermal fronts is clearly a three-dimensional problem. The one-dimensional numerical calculations show that the instability can start in different regions of the disc depending on the external accretion rate $\dot{M}_0$. At low values of this rate, much of the mass accreted at the outer edge has time to diffuse inwards in the quiescent state, building up the density throughout the disc. The critical density Σ_C for the instability is actually lowest at the inner edge, and is consequently achieved there first, triggering an 'inside-out' outburst. If $\dot{M}_0$ is somewhat larger, the critical density is reached more quickly and the outbursts repeat at shorter intervals. At higher values of $\dot{M}_0$, mass arrives at the outer disc edge more rapidly than it can be transported inwards by the quiescent viscosity, and the density there reaches the critical value first, triggering an 'outside-in' outburst. In order to conserve angular momentum as most of the matter moves inwards, some of the matter near the disc edge must move outwards, increasing the disc size during the outburst. After the outburst the material accreting at the disc edge has lower specific angular momentum and causes the disc to shrink again. There is some evidence that this theoretical prediction is born out in outbursts of a few systems (see Sect. 8.3.1.3).

Many of the potential tests of the disc instability picture depend on the unknown functional form of the viscosity, so it is important to look for tests which are generic, i.e. independent of this detailed form. The most promising of these uses the fact that the instability is assumed to result from the change of disc structure caused by hydrogen ionization. If the disc is too hot, i.e. has $T_{\rm eff} \gtrsim$ a critical temperature $T_{\rm ion} \simeq 6000$ K everywhere, it should be steady. Thus from Eq. (10.5) we see that the condition for instability is that the accretion rate $\dot{M}_0$ should be less than a critical value $\dot{M}_{\rm inst} \propto R_{\rm out}^3 / M_1$, where $R_{\rm out}$ is the outer radius of the disc. This is typically a fairly constant fraction f of the binary separation a, which in turn can be related to the binary period P and total mass M through Kepler's law, giving

$$\dot{M}_{\rm inst} \simeq 1.0 \times 10^{-9} \left(\frac{f}{0.5}\right)^3 \frac{M}{M_1} \left(\frac{T_{\rm ion}}{6000 \text{ K}}\right)^4 \left(\frac{P}{3 \text{ hr}}\right)^2 M_\odot \text{ yr}^{-1}. \qquad (10.13)$$

Comparison of the predictions of Eq. (10.13) with observations of CVs is mildly encouraging. CVs with periods below the famous 2–3 hr period gap are expected to have mass transfer rates $\lesssim 5 \times 10^{-11} M_\odot$ yr^{-1} (see Sect. 11.3.2), about one-half of the estimated $\dot{M}_{\rm inst}$ for such systems ($M \simeq M_1, P \simeq 1.5$ hr, $f \gtrsim 0.5$). This agrees with the observation that all nonmagnetic CVs below the period gap are observed to be dwarf novae[106]. Above the period gap, things are less clear. The estimate $\dot{M}_{\rm inst} \gtrsim 10^{-9} M_\odot$ yr^{-1} is probably above the transfer rates expected near the period gap at $P = 3$ hr (ref. 53); but the $\lesssim P^2$ dependence of $\dot{M}_{\rm inst}$ is rather flatter than that of the expected average mass transfer rates near this period (see Sect. 11.3.2). Taking this at face value, one would expect dwarf novae near 3 hr; but some nonmagnetic CVs should be dwarf novae at periods $P \gtrsim 4$ hr. There is some indication[113,114] that the first of these predictions fails, but at longer periods CVs do display both dwarf nova and steady (novalike) behaviour (see Figure 8.16). These predictions are probably affected by long-term (say 10^5 yr) fluctuations around the transfer rates predicted from secular evolution, which can take a given system across the instability boundary represented by $\dot{M}_{\rm inst}$. Indeed, it is an interesting question why such fluctuations do not upset the apparent agreement between theory and observation at shorter periods. A more direct challenge to theory is provided by the VY Sculptoris systems, sometimes called anti-dwarf-novae. These apparently nonmagnetic CVs have intermittent low states lasting for weeks. The cause of these is unclear, but the interesting point here is that the mass transfer rate must be below the relevant value of $\dot{M}_{\rm inst}$. Yet there are no reports of outbursts in these low states. This may reflect incomplete monitoring, so that outbursts have been missed, or that the low states may be shorter than the viscous time scale for the instability to occur.

This section has highlighted some of the theoretical difficulties in current versions of the thermal–viscous disc instability. Detailed comparison with observations (see Ch. 8) certainly does not justify any complacency in believing that the connection between theory and observation is proven. However, it seems likely that some form of disc global instability picture will survive theoretical advances.

10.4.2 Disc instabilities in other systems

The application of the disc instability theory outlined here to systems other than dwarf novae remains problematical. The obvious candidates are LMXBs: unless there are systematic differences in mass transfer rates (which is conceivable – see Sect. 11.3.5), one might expect dwarf nova behaviour in a large subclass of this group. But the discs themselves may be rather different (see Sect. 10.2), and in particular irradiation by the central X-ray source may keep the whole disc above the hydrogen ionization temperature (cf. Eq. (10.10)). A more promising application is supplied by the soft X-ray transients[20] (to be distinguished from the hard X-ray transients, which are often periodic and occur in long-period binaries with eccentric orbits: see Sect. 10.8, 2.3.6, and Ch. 3). These are LMXBs with orbital periods $\sim 4 - 20$ hr which undergo outbursts lasting $\sim$ months at intervals of $\sim 1 - 10$ yr or even more. Encouragingly for a disc instability picture, their long-term average mass transfer rates do appear to be rather low ($\sim 10^{-10} M_\odot$ yr^{-1}), as their luminosities are extremely low ($\lesssim 10^{34}$ erg s^{-1}) in quiescence. However, the major obstacle to a straightforward transplanting of the dwarf nova-type instability is the vast difference in the time scales, which are a factor of $\sim 10^2$ longer for soft X-ray transients. Detailed fitting[49,89] produces tolerable (although not perfect) agreement with the observed light curves provided that the values of α are reduced by a factor of this order compared with dwarf novae. But there is no physical justification for this step.

Of course, the dwarf-nova thermal–viscous instability described above is not the only type of disc instability which has been discussed. In particular, the inner regions of discs around neutron stars and black holes are dominated by radiation pressure, and are formally unstable in at least some versions of the α-prescription[100]. There have been attempts to use this in models of the Rapid Burster, but the theoretical situation is too unclear[77] to have much confidence in such models (see Sect. 4.6 and 4.7).

10.5 Tides, resonances and superhumps

In superoutbursts of SU UMa stars (see Sect. 8.3.1.5 and Figure 10.4) a photometric modulation of the optical light appears with a period longer, by a small percentage, than the orbital period. Numerical studies of discs by means of a particle code[137] do reproduce these 'superhumps' provided that (a) the disc is large enough, and (b) the ratio $q = M_2/M_1$ of secondary star to white-dwarf mass is small enough ($\lesssim 0.25 - 3$). Under these conditions, the disc gradually develops an eccentric shape which slowly precesses in a prograde fashion in inertial space (Figure 10.5). The dissipation pattern (and hence the luminosity) of the disc is affected by the relative

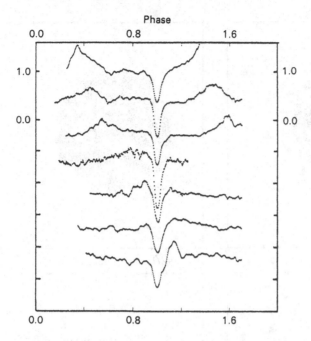

Fig. 10.4. Superhumps observed during superoutbursts of the dwarf nova Z Cha. The figure shows orbital light curves, aligned by the deep central eclipse. The superhump is the large maximum seen near phase 0.35 in the upper curve, migrating to later phases in the following curves, appearing near phase 0.1 in the lowest curve. This shows that the superhump period is slightly longer than the orbital period. The curves are assembled from different superoutbursts to give sufficient phase coverage. (Figure by B. Warner and D. O'Donoghue.)

position of the secondary and the disc's major axis, and therefore varies on the beat period between these two, which is slightly longer than the orbit. Superhumps thus appear as a spectacular manifestation of tidal effects on accretion discs.

10.5.1 *Tides*

An accretion disc which spreads out enough within the primary's Roche lobe will be subject to tidal torques from the secondary star. This has long been recognized as the main effect limiting the size of a disc in a binary system[98]: the tides take angular momentum from orbits at the disc edge, and return it to the binary orbit. A steady state is reached when the tides remove the angular momentum at the rate it is transported outwards in the disc by viscosity. Note that the tidal torque would vanish identically if there were no viscosity, as the disc would be symmetrical about the line of centres: a full calculation[138] shows that the sign of the torque varies around the disc edge, successive quadrants losing and gaining angular momentum, the net effect being always to cut the disc off at about 90% of the primary's Roche lobe radius.

10.5.2 *Resonances and superhumps*

Small perturbations of the gas elements in a disc oscillate around the mean Kepler motion with the epicyclic frequency κ, which is almost equal to the angular

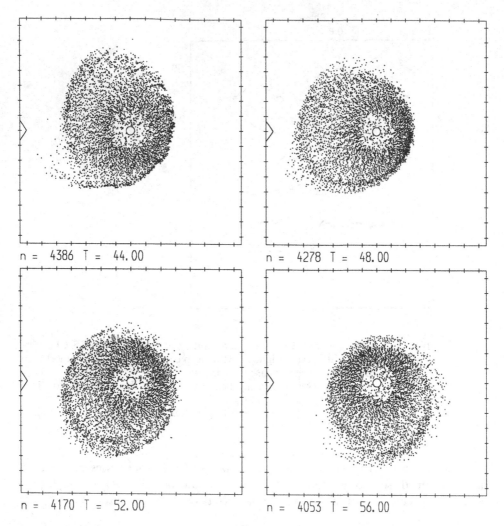

n = 4386 T = 44.00   n = 4278 T = 48.00

n = 4170 T = 52.00   n = 4053 T = 56.00

Fig. 10.5. Numerical simulation of an accretion disc in an SU UMa system. The mass ratio is $q = 0.12$: the white dwarf is the small circle, and part of the secondary's Roche lobe is visible at the left of each frame. n is the number of particles present in each frame, and T is the number of orbital cycles since the start of the simulation. The disc precesses on a period slightly longer than the orbital period. (Figure by R. Whitehurst.)

frequency Ω. If κ is such that the particle and secondary star are periodically in the same relative configuration, we have a resonance; the resonance condition is thus

$$k\kappa = j(\Omega - \Omega_{\mathrm{orb}}), \qquad (10.14)$$

where j and k are positive integers, since the particle sees the secondary move with angular frequency $(\Omega - \Omega_{\mathrm{orb}})$. As κ is close to Ω, resonances appear near commensurabilities of the form $(j : j-k)$ between particle and binary frequencies. At resonance, perturbations by the secondary can build up cumulatively, and can have a profound effect on the disc orbits. A simple argument shows that it is the $j = 3, k = 2$ (3:1 commensurability) which probably causes superhumps.

Even quite extreme aperiodic orbits do not differ greatly from Keplerian circles, so we may measure the typical radii of orbits near the general (j,k) resonance as

$$R_{jk} = (GM_1/\Omega_{jk}^2)^{1/3},$$

where Ω_{jk} is the value of Ω in Eq. (10.14). Using Kepler's law, the binary separation is

$$a = [G(M_1 + M_2)/\Omega_{orb}^2]^{1/3} \tag{10.15}$$

so that

$$R_{jk}/a = [(j-k)/k]^{2/3}(1+q)^{-1/3}, \tag{10.16}$$

where $q = M_2/M_1$ is the mass ratio. To see what resonances are allowed for a given q, we can compare this result with the primary's Roche lobe size[30]

$$R_{Roche}/a = 0.49/[0.6 + q^{2/3}\ln(1 + q^{-1/3})] \tag{10.17}$$

and ask when $R_{jk} \lesssim 0.9R_{Roche}$, corresponding to the radius at which tides cut the disc off.

From Eqs. (10.16) and (10.17), we see that $k = 1$ resonances can only occur for very small mass ratios: $j = 2, k = 1$ requires $q \lesssim 0.025$, with even smaller ratios required for larger j. The 2:1 resonance thus cannot occur in most CVs or LMXBs, which generally have $q \gtrsim 0.1$. But $k = 2$ resonances are allowed: the $j = 3, k = 2$ resonance (near the 3:1 commensurability) lies inside the tidal radius for $q \lesssim 0.3$. Following Whitehurst's numerical simulations, Hirose & Osaki[44] and Whitehurst & King[140] independently suggested this resonance as the basic cause of superhump behaviour. Once the disc orbits become eccentric they are likely to precess, as the Roche potential is not spherically symmetric, so one expects a superhump period exceeding P_{orb} by an amount ΔP, with $\Delta P/P_{orb} \propto q$. The SU UMa systems do appear to follow such a relation[90]. As pointed out by Lubow[79,80], arguments in terms of single-particle orbits are inapplicable near resonances, where collective effects dominate: he showed analytically that an eccentric perturbation of the disc stationary in inertial space grows under the influence of the $m = 3$ part of the Roche potential if the resonance is allowed. This would produce superhumps with exactly the orbital period. However, the disc will precess in a slightly prograde fashion under the full potential, consistent with observation.

The tidal-instability picture explains the period distribution of SU UMa systems. The mass ratio requirement $q \lesssim 0.3$ can be combined with the usual period–secondary mass relation[129] for Roche-lobe filling lower-main-sequence stars

$$M_2 \simeq 0.11P_{hr}M_\odot, \tag{10.18}$$

where P_{hr} is P_{orb} measured in hours, to give the requirement

$$M_1 \gtrsim 0.36P_{hr}M_\odot. \tag{10.19}$$

All CVs below the period gap (i.e. with $P_{hr} < 2$) are SU UMa systems unless they contain a magnetic white dwarf, suggesting that $M_1 \gtrsim 0.7M_\odot$ for this group, which seems very likely to be true. The longest-period SU UMa system (TU Men) is the first non-magnetic system above the gap ($P_{hr} = 2.8$) and requires $M_1 \gtrsim 1.0 M_\odot$.

10.5.3   *The nature of superoutbursts*

What causes superoutbursts themselves is still not understood. There are a number of properties to explain: (a) superhumps occur only during superoutbursts, and not in normal outbursts or quiescence; (b) at least some superoutbursts appear to evolve from outbursts[125]; (c) superoutbursts seem to occur at fairly regular intervals; (d) the disc radius expands during the outburst and then slowly contracts during the following cycle[92,143] in Z Cha; and (e) the intervals between outbursts, and the mass transfer rate, both correlate with the superoutburst phase in VW Hyi[125], the latter increasing in the last few outbursts before a superoutburst. In addition, the very definition of a superoutburst is disputed: some authors restrict the term to those systems showing superhumps, while others argue that the wide outbursts seen in longer-period systems such as SS Cygni and U Geminorum should be included[126]. The latter view has some force in that the mass-ratio condition *alone* rules out superhumps in such systems (see Sect. 10.4.2 above), and the known correlation[6] between orbital period and *normal* outburst width blurs the distinction between normal and superoutbursts for long-period systems. We can certainly ask if a given model could explain (f) the existence of long outbursts in long-period dwarf novae.

Two basic types of models for superoutbursts have been suggested. One (EMT: enhanced mass transfer) assumes that for some reason the mass transfer rate from the secondary star ($\dot{M}_0$ in the notation of Sect. 10.4.2 and 10.4.3) is suddenly enhanced. This may be because of irradiation of this star, perhaps during a normal outburst[54] (but note that one version[94] of this mechanism suffers from crucial defects[39]).

The other type of model (TTI: thermal and tidal instability[95]) combines the usual thermal–viscous model for normal outbursts with the tidal effects which can give superhumps. During normal outbursts, the disc is assumed to be small enough that the tides have little effect, so the disc radius gradually increases at each successive outburst. Eventually the radius reaches a size at which the tides are able to clear the outer disc, giving the superoutburst, and also to produce superhumps. The higher accretion from the outer disc makes the usual thermal instability last longer and hence reduce the disc to its original size.

We can ask how these models account for the list of properties (a)–(f). Property (a) (superhumps only in superoutbursts) is satisfied by construction in the TTI model, and is reasonable in the EMT model: numerical simulations[139] suggest that the superhumps appear as promptly as observed only if stimulated by the impact of the enhanced gas stream on the disc. Property (b) (outbursts stimulate superoutbursts) is again natural in the TTI model, and also in the EMT model using irradiation as the trigger. The regularity of the superhumps (c) is not clearly explained by the EMT model, but in the TTI model is supposed to result from the constancy of the external mass transfer rate $\dot{M}_0$. One could object that ordinary outbursts show no such periodicity, even in long-period systems where there are no superhumps and $\dot{M}_0$ is presumably constant.

Smak[121] has recently performed calculations checking the radius behaviour (d) and cycle length variations (e) predicted by the models. These reveal precisely opposite predictions of the EMT and TTI models for (d), making the radius variation a valuable test. In the EMT model, the disc radius expands after a superoutburst and then slowly contracts during the supercycle, while the reverse occurs in the TTI model. These features follow straightforwardly from the basic physics of the two models: in the EMT model, the large burst of low angular momentum material from the secondary makes the disc first shrink, then rapidly expand to conserve angular momentum as the central accretion rate rises and the superoutburst becomes visible. In the subsequent supercycle the disc expands at outbursts and contracts in quiescence, but the overall effect is a gradual shrinking towards the quasi-steady size. In the TTI model, the superoutburst results from a sudden loss of angular momentum through tides, so the disc shrinks after the superoutburst, gradually expanding back to the tidally limited size during the subsequent supercycle. Observations of the disc size in Z Cha[92,143] show that it is largest just after a superoutburst, and gradually decreases during the rest of the supercycle. This is consistent with the predictions of the EMT model and contradicts those of the TTI model. The length-of-cycle variations (e) also discriminate sharply in the predictions of the two models. In the EMT model, the disc returning to quiescence after the superoutburst is more massive and has a higher surface density. This makes subsequent normal outbursts more frequent, because the thermal instability is more easily triggered. As the supercycle proceeds, the surface density drops and the outbursts become less frequent. However, we noted under (e) that there is an observed tendency for the mass transfer rate to increase slightly just before a superoutburst. This effect can be included 'by hand' in the EMT model: it causes no change in the radius variations, which remain consistent with observations of Z Cha, but it does cause a dramatic change in the cycle length: the last few outbursts now become *more* frequent. The enhancement of the mass transfer rate is excluded, by definition, from the TTI model, so here the outbursts always become *less* frequent as the disc density decreases. Observations of VW Hyi clearly show that normal outbursts become less frequent as the supercycle proceeds (which would support both models), but that in supercycles showing a large number of outbursts (Type S) the last few outbursts become *more* frequent, just as predicted by the modified EMT model. Thus both (d) and (e) tend to support the (modified) EMT picture. Finally, (f) (the extension of superoutbursts to longer period systems not showing superhumps) is actually expected in the EMT picture: the superoutbursts can still be triggered by normal outbursts even if the mass ratio condition $q \lesssim 0.3$ for superhumps fails. In the TTI picture, a separate mechanism is needed for these long outbursts (at least once of order 50 days in U Gem[82]); the superoutbursts are assumed to result from the disc expanding beyond the resonant radius, so superoutbursts and superhumps must occur together. As can be seen, the evidence tends to favour the enhanced mass transfer picture, which copes with (d)–(f) rather better than the thermal tidal instability model. But the case is not yet overwhelming.

10.5.4 *Tidal instabilities in other systems*[56]

The discussion of Sect. 10.4.2 suggests that three conditions are required for the appearance of superhumps: (i) the mass ratio must allow the 3:1 resonance in the accretion disc, so that $q \lesssim 0.3$; (ii) the disc must be large enough that there is a significant density near this radius; and (iii) the superhumps must have time to grow.

In the simulations by Whitehurst[137,139], the growth time was much shorter in the EMT picture, where the disc is perturbed by an enhanced mass transfer stream, than when a dense disc was simply left to evolve with a steady external accretion rate. If this is true of the real superhumps, it would tend to favour the EMT picture; without the enhanced stream the growth time might be longer than the superoutburst (so that condition (ii) would fail before the superhumps appeared). We have already seen that the mass ratio condition (i) alone explains the period distribution of the SU UMa systems: conditions (ii) and (iii) essentially define superoutbursts (Sect. 10.4.3). In CVs with a *steady* accretion rate, it would appear that (ii) and (iii) are automatically satisfied. Thus, any steady systems satisfying $q \lesssim 0.3$ should be good candidates for *permanent* superhumps. We saw in Sect. 10.4.1 (Eq. (10.12)) that steady CVs must have quite high accretion rates, and thus tend to be found at longer orbital periods, where q (Eq. (10.18)) is larger. The overlap between the two conditions lies *within* the 2–3 hr period gap: CVs born sufficiently close to 3 hr do evolve into the gap with appreciable mass transfer rates[41], which are above the limit, Eq. (10.12), for $P \gtrsim 2.5$ hr (such systems are of course rare: the gap is a statistical phenomenon). There are currently three known CVs within the gap which are neither dwarf novae nor AM Herculis (strongly magnetic) systems: V795 Her (2.60 hr), V Per (2.57 hr) and 1H 0709–36 (2.44 hr). Two of these (V795 Her and 1H 0709–36) show photometric periods which are longer than the spectroscopic orbital periods. (Because of this, they were tentatively classified as intermediate polars, but would be unprecedented in that the white dwarf would have to spin more slowly than the orbit.) These are strong candidates for permanent superhumps[107]. One objection is that the eclipse of 1H 0709–36 is rather long, and if interpreted in terms of a steady-state disc eclipse implies a mass ratio $q \gtrsim 0.6$. However, if the photometric modulation really is a permanent superhump, a standard disc will not be a good model: we would expect indeed that the disc would be brighter at the edge, making the eclipses longer.

We should ask what the tidal model predicts for other systems. As usual, we consider first LMXBs, as they are the systems most similar to CVs. There are no confirmed LMXBs in the period range 80 min–2 hr occupied by all of the SU UMa systems except TU Men, but from Eq. (10.19) the mass ratio condition $q \lesssim 0.3$ is satisfied for any period $\lesssim 3.9$ hr, for neutron star masses $M_1 = 1.4 M_\odot$. This range includes about seven LMXBs, three of them ultrashort-period systems with $P_{hr} < 1$. The mass ratio condition is certainly satisfied in the soft X-ray transient black-hole candidate A0620−00, where[85] $M_2 \lesssim 0.8 M_\odot$, $M_1 \gtrsim 3.2 M_\odot$. There have been four suggestions of superhump-like behaviour in LMXBs (see below). This may be a consequence of the relatively sparse observational coverage of these optically faint systems, where the superhump modulation could get lost in the complications of the light curves (see Sect. 10.2). If the absence of superhumps in steady LMXBs is real, we would need to investigate the effect of irradiation on a tidally distorted disc. For the transient systems, the lack of superhumps in quiescence may again reflect observational difficulties, or conceivably mean that the disc viscosity is so low that the disc never spreads to the resonant radius even during the ~ 1–10 yr quiescent intervals. Note that this difficulty is independent of whether the outbursts are modelled as disc instabilities[20,49,89] or mass transfer events[39].

Four suggestions of superhump behaviour in LMXBs have appeared in the literature. White[134] proposed this to explain the discrepancy between the optical (50.46 min) and X-ray (50.0 min) periods in XB1916−05 (see Sect. 2.6.3). The secondary here must be degenerate and have $M_2 < 0.1M_\odot$, so the mass ratio condition is certainly satisfied for any reasonable neutron star mass. Bailyn[7] suggested that the optical quasi-period of ~ 10.5 hr in the recent outburst of the soft X-ray transient Nova Muscae 1991 may be a superhump modulation, completing the analogy with SU UMa systems. Radial-velocity measurements[84] support this idea, giving a period of 10.42 hr and a primary mass function of $3.1M_\odot$. The system is therefore a black-hole candidate (see Sect. Ch.3.3), and satisfies the mass-ratio condition for superhumps. The transient sources GS2000+25 and GRO J0442+32 have also shown similarly discrepant periods.

A totally unexplored field is the possibility of finding superhumps in other semidetached binaries. In fact, there are no obvious candidates: in many cases (e.g. symbiotic stars, β Lyrae systems), the mass ratios are too large, and the best candidates seem to be some Algols. The mass ratios are suitable in some cases, but there may be problems in that discs either cannot form (the accreting star is too large, Eq. (10.1)) or are too faint compared with the component stars. It is worth remarking finally that there are some unexplained periodicities in certain magnetic CVs which it is tempting to interpret in terms of resonances, but probably not of the type discussed here (see Sect. 10.9).

10.6 Spiral shocks

The fundamental failing of disc theory is an understanding of the viscosity, or how angular momentum is transported. Spiral shock theory is a very interesting attempt to provide an explicit model of this process.

Any non-axisymmetric perturbation in a disc, caused, for example, by tidal interaction with the companion star, is rapidly sheared into a trailing spiral wave. These waves can interact gravitationally and transport angular momentum. If the waves remain small in amplitude, they do not interact with the disc and their angular momentum is conserved separately. But as the spiral waves propagate inwards they steepen into shocks. The resulting dissipation mixes some of the wave angular momentum with the gas in the inner regions of the disc. This angular momentum is negative, as a disc with trailing spiral waves has lower angular momentum than one without. Thus the shocks have the effect of transporting disc angular momentum outwards, just as required to drive accretion. The only viscosity involved here is the usual atomic viscosity causing dissipation in the shocks.

Numerical simulations of this process show that a two-armed spiral shock pattern is set up by the tides of the companion star. As the shocks propagate inwards, a balance is reached between amplification and dissipation, giving a rate of angular momentum transport which is independent of the perturbation at the disc edge. The effective α-value given by current simulations is $\sim 10^{-4}$–10^{-2}, much too small, e.g., for dwarf nova outbursts. Indeed, if α were much larger than this the thin disc approximation might be in danger, as the shocks would dissipate an appreciable fraction of the local Kepler velocity of the gas. The simulations have $H/R \sim 0.25$–0.33, already too large to be realistic for CVs. However, more sophisticated calculations including cooling are needed before definite conclusions can be drawn. (For a recent review, see ref. 91.)

10.7   Coronae and winds in discs

There is abundant observational evidence of copious mass loss from CVs. Quite apart from the question of exactly what drives the winds at the observed velocities (see Sect. 8.2.2.7), their existence is likely to be a universal feature of disc accretion. For there is no good reason to assume that the dissipation we observe as the disc luminosity is somehow confined to regions that are optically thick to their own radiation, and therefore efficiently radiate away the heat generated. Even without invoking external heating by irradiation, the optically thin layers of the disc are unlikely to be able to cope with this dissipation without reaching temperatures beyond the local escape value, as simple analytic arguments show[26]. In LMXBs the possibility of irradiation by the centrally-produced X-rays enhances the mass loss to rates comparable with the accretion rate: this was first demonstrated by Begelman *et al.*[10], who considered only Compton heating and cooling of the optically thin layers. This confines the winds to disc radii r where the Compton temperature ($\sim 10^7$ K, set by the X-ray spectrum) exceeds the local escape temperature $\sim GM_1 \mu m_H / kr$, requiring $r \gtrsim 10^{11}$ cm for typical parameters. This in turn restricts the effect to longer period ($\gtrsim 15$ hr) systems. However, if atomic heating and cooling processes are included[27], most of the mass loss takes place at small disc radii ($\sim 10^7$–10^8 cm). All LMXBs should therefore have substantial winds with up to half the mass transfer rate being blown away. (Note that it is energetically possible to blow away an arbitrary fraction of the transferred mass from a sufficiently large radius r_{wind}: one requires an accretion luminosity $GM_1 \dot{M}_{\text{acc}} / r_{\text{acc}} >$ the wind mechanical luminosity $\gtrsim GM_1 \dot{M}_{\text{wind}} / r_{\text{wind}}$. However, the wind will make Roche lobe overflow of the companion dynamically unstable if it carries off specific angular momentum greater than about two-thirds that of this star[57].) The scattering optical depth through these winds is of order 0.1–1, and not very sensitive to the accretion rate. This is in good agreement with observations of the 'coronal' (ADC) LMXBs (see Sect. 10.2 and 1.2). Detailed modelling of CV winds is discussed in Sect. 8.2.2.7.

10.8   Boundary layers

We have seen that a geometrically thin disc provides a good model of the accretion flow in CVs (probably) and LMXBs (possibly). But the required Kepler angular velocity $\Omega_K = (GM_1 R^{-3})^{1/2}$ cannot be maintained at the inner edge of the disc if this is to join smoothly on to a non-magnetic accreting star spinning at below the breakup rate $\Omega_K(R_*)$.

10.8.1   *Boundary layer around a slowly rotating star*

If the star rotates slowly, i.e. with angular velocity $\omega < \Omega_K(R)$ for disc radii $R >> R_*$, the simplest possibility is that the disc remains thin and Keplerian until very close to the star's surface, with a rapid decrease to the stellar angular velocity ω in a narrow boundary layer (Figure 10.6). Then $\Omega(R)$ must reach a maximum close to the star's surface, so that there exists a point $R = R_* + b$, $b << R_*$, with $d\Omega/dR = 0$ and $\Omega \simeq \Omega_K$. Taking this as the inner boundary condition for a steady disc leads to the value $\beta = 1 + O(b/R_*)$ in the surface density equation (10.4) (see, e.g., ref. 33). The existence of a maximum of $\Omega(R)$ means that the shear vanishes at this point. Thus viscous shear stresses ($\propto \nu d\Omega/dR$ for an outer annulus acting on an

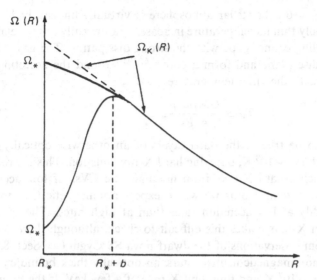

Fig. 10.6. Angular velocity distribution $\Omega(R)$ near the inner edge of an accretion disc. $\Omega_K(R)$ and Ω_* are the Kepler and stellar angular velocities, respectively. If $\Omega_* \ll \Omega_K(R_*)$, $\Omega(R)$ curve reaches a maximum at $R = R_* + b$ before decreasing towards the stellar surface (lower branch of curve). The region $R_* \leq R \leq R_* + b$ is a boundary layer: angular momentum cannot be transported outwards as $\Omega' = 0$ at $R_* + b$, so the star is spun up by the advected Kepler angular momentum. If, instead, Ω_* is comparable with $\Omega_K(R_*)$ (upper branch of curve) there is no maximum of $\Omega(R)$: angular momentum is transported away from the star by viscosity ($\Omega' < 0$), and this effect can balance advection.

inner one) cannot transport angular momentum into or out of the star, and the star simply gains the angular momentum of the accreting matter, which has essentially the Kepler value. The star is therefore subject to an accretion torque

$$G_{acc} \simeq \dot{M}(GM_1R_*)^{1/2}. \tag{10.20}$$

If the star spins more slowly than the breakup value, the boundary layer must release a large amount of energy as the accreting matter comes to rest at the stellar surface. Some of this is used to spin up the star, but there remains an amount[72]

$$L_{BL} = \frac{GM_1\dot{M}}{2R_*}\left(1 - \frac{\omega}{\Omega_K}\right)^2 \tag{10.21}$$

to be dissipated. For $\omega \ll \Omega_K$ this is one-half of the total accretion luminosity (this is obvious as a particle in Kepler rotation has half the binding energy of one at rest at the same radius). Not all of L_{BL} has to appear as radiation, but it is clear that for accretion on to a slowly rotating star the boundary layer can emit a luminosity comparable with the whole of the rest of the disc. As its area is small, the radiation emerges at high photon energies (see Sect. 8.2.2). For accretion on to a nonrotating white dwarf an optically thick boundary layer has a temperature[102] $\gtrsim 10^5$ K, and so produces soft X-rays. Detailed numerical treatments of the coupled radiation–hydrodynamical equations[69–71] confirm this picture. The accreting matter remains fairly close to the equator rather than spreading over most of the star; at latitudes greater than $\sim$ 4–6°, the stellar atmosphere is virtually undisturbed. If the boundary layer is optically thin its temperature increases dramatically as the relatively inefficient radiative cooling cannot cope with the strong dissipation. The gas is likely to expand out of the disc plane and form a corona[65,105] around the accreting star with a temperature close to the virial temperature

$$T_V \simeq \frac{GM_1 \mu m_H}{4kR_*}.$$ (10.22)

(This will of course also be true of the outer layers of an otherwise optically thick flow.) For a white dwarf $T_V \sim 10^8$ K, implying hard X-ray emission. This is probably the only way of producing hard X-rays from nonmagnetic CVs. If low accretion rates correspond to low optical depths we would expect nonmagnetic CVs to emit hard X-rays more readily at low accretion rates than at high rates. The faintness of most such systems in X-rays makes this difficult to check, although there is some favourable evidence from observations of the dwarf nova SS Cygni (see Sect. 8.2.2).

For accretion on to non-magnetic neutron stars, an optically thick boundary layer will have a temperature $\sim 10^7$ K and thus emit X-rays of a few keV. If the boundary layer is optically thin, a corona would again form, but only the ions will reach the virial temperature (Eq. (10.22)), which is now of order 10^{11} K. The electrons, characterizing the radiation spectrum, will be much cooler ($\sim 10^9$ K) as Coulomb heating by the ions is otherwise too slow to compensate their radiative losses[59]. An interesting alternative possibility has been pointed out by Kluźniak & Wagoner[73]: they note that some neutron star models predict stellar radii R_* smaller than the radius R_{MS} of the last stable general-relativistic test particle orbit around the star. Unless there are strong radiation pressure effects, the disc must terminate here, as it would around a black hole. Within this radius the matter falls freely inwards and accretes at the neutron star equator. The flow is optically thin and very hot, producing a $\sim$ 100 keV X-ray spectrum.

Black-hole accretors are automatically smaller than the last stable test-particle orbit, so matter always falls freely into them from this radius. However, they have no hard surface to produce an accretion shock or reprocess some of the radiation produced further out in the flow. This has motivated suggestions of X-ray spectral differences as potential black-hole indicators (see Sect. 1.5.2 and 3.4).

10.8.2 *Accretion on to a rapidly rotating star*

The size of the accretion torque, Eq. (10.20), means that the central star (moment of inertia $k^2 M_1 R_*^2$) will be spun up near to the breakup value $\omega = \Omega_K(R_*)$ once it has accreted a mass[64] $\Delta M \sim k^2 M_1$. For the white dwarfs and neutron stars in CVs and LMXBs, this is typically $\lesssim 0.1 M_\odot$, whereas they are likely to accrete several times this amount in their active lifetimes (cf. Eq. (10.17)). This severely reduces L_{BL} (see Eq. (10.21)), and probably accounts for the relatively feeble X-ray luminosities of nonmagnetic CVs and the low Zanstra temperatures deduced from ultraviolet emission lines (see Sect. 8.2.2.2): the loss of angular momentum in nova outbursts is unlikely[64] to stabilize ω at a value significantly below $\Omega_K(R_*)$.

Once the star reaches $\sim \Omega_K$, it would appear at first sight from Eq. (10.20) that further accretion would cause it to break up. The problem is even more acute for a massive accreting white dwarf, where adding mass reduces the moment of inertia and thus requires angular momentum to be *removed* from the star rather than accreted. What happens in this limit has recently been clarified by Paczyński[96] and Popham & Narayan[101] (see also Pringle[102–104]). The purely advective accretion torque Eq. (10.20) results from assuming that the angular velocity profile $\Omega(R)$ has a maximum value outside R_*. But for values of ω near $\Omega_K(R_*)$ a second type of profile is possible, in which $\Omega(R)$ has no maximum but increases monotonically all the way in to the star's surface, i.e. $d\Omega/dR < 0$ throughout. The viscous torque therefore transports angular momentum outwards at all radii. This reduces G_{acc} below the value given in Eq. (10.20), and for ω sufficiently close to breakup can make it negative. Hence the spin of an accreting star will first increase towards breakup, and then stabilize at an 'equilibrium' value analogous to those for magnetic accretors – see Sect. 10.9. The star can now continue to accrete without spinning up any further, or even spin down if this is required. In this case the distinction between the disc and boundary layer becomes largely semantic: $\Omega(R)$ remains close to the Kepler value throughout, the deviation in the inner disc being slight enough that the disc remains geometrically thin in the vertical direction. The distinction between the disc flow and the star is also somewhat blurred, and a full discussion should treat the circulations in the latter simultaneously. Some analytic results for rigidly rotating polytropic stars are given by Bisnovatyi–Kogan[11]. In the general case the mixing of the arriving material with that of the star is an important element of nova outburst models.

10.9 Accretion from a wind

In many X-ray binaries the mass-losing star does not fill its Roche lobe and the compact component must accrete from a wind from this star. This is an inherently less efficient process than Roche lobe overflow: mass leaves the star in all directions, not just towards the accretor, and has enough kinetic energy to escape the system except where it passes close to this star. In general only a small fraction of the wind is accreted, in contrast to the Roche lobe case, where almost all the mass is captured by the accreting component. Further, the captured wind material has rather low specific angular momentum compared with mass overflowing the Roche lobe: it is difficult to decide whether the disc formation condition, Eq. (10.1), is satisfied in wind accretion, whereas it clearly holds for most cases of Roche lobe overflow. An added complication is that the accreting neutron star in many X-ray binaries has a strong magnetic field, increasing the effective size of the accretor. The spin of this star is readily detectable as an X-ray pulse period, and gives information on the angular momentum accreted by the neutron star (see Sect. 10.9), potentially distinguishing between cases where a disc does or does not form. A hybrid situation can occur in binaries with very eccentric orbits, where accretion near periastron may result in the formation of a disc.

10.9.1 *Mass accretion from a wind*

Analytic[13,48] and numerical[12,34,83] treatments of the bow shock of a compact star moving with orbital velocity v in a stellar wind of velocity v_w show that most of the wind material passing within a cylindrical radius

$$r_{\text{acc}} = \frac{2GM_1}{v_{\text{rel}}^2} \qquad (10.23)$$

of the star is captured and ultimately accreted, where $v_{\text{rel}}^2 = v_w^2 + v^2$. If the wind is emitted over solid angle Ω and has total mass-loss rate $\dot{M}_w$ this implies an accretion rate

$$\dot{M} \sim \frac{\pi r_{\text{acc}}^2}{\Omega a^2} \dot{M}_w. \qquad (10.24)$$

In X-ray binaries containing OB supergiants, the winds are isotropic ($\Omega \sim 4\pi$) and have high velocities ($v_w \sim 1\text{–}2 \times 10^3$ km s^{-1}). Thus, $r_{\text{acc}} \sim 10^{10}$ cm is much smaller than the binary separation $a \sim 10^{12}$ cm, and only a fraction, $\lesssim 10^{-4}$, of the wind is accreted. The mass-loss rates are so high ($\dot{M}_w \sim 10^{-4} M_\odot$ yr^{-1}) that this is quite adequate to power Eddington-limited X-ray sources. In these binaries at least the theoretical estimate, Eq. (10.22), appears to be fairly secure (see Sect. 2.2.3 and 2.2.4).

In Be-star X-ray binaries the high-velocity spherical winds ($\dot{M}_w \sim 10^{-8} M_\odot$ yr^{-1}) would not support the observed X-ray emission, and the compact component (a magnetic neutron star) accretes from a dense low-velocity ($v_w \sim 100$ km s^{-1}) wind confined close to the equatorial plane $\Omega << 4\pi$). Too little is known[131] about these winds to check the estimate, Eq. (10.24), for these systems (see Sect. 2.2.1.1).

10.9.2   *Angular momentum accretion from a wind*

To see if an accretion disc can form in a wind-fed binary we need an estimate of the mean specific angular momentum J of the captured matter (cf. Eq. (10.1)). If the wind were symmetrical around the accreting star, J would vanish identically, severely reducing the chances of disc formation (although this might nevertheless occur if the matter picked up angular momentum from the rotating magnetosphere of a neutron star accretor – see Sect. 10.9.2). But since the wind density and velocity both vary radially outwards from the mass-losing star, while the accretor moves azimuthally through the wind, we might argue that the symmetry around the accretor is broken and $J \neq 0$. Arguments of this type[50,116] lead to the estimate

$$J = \eta \frac{r_{\text{acc}}^2}{4} \frac{2\pi}{P_{\text{orb}}}, \qquad (10.25)$$

where $\eta \sim 1$ is an efficiency factor: this is the specific angular momentum of a flat circular plate with axis along the line of centres and radius r_{acc}, rotating about the accretor once per binary period. From Eq. (10.1), this value of J would imply a circularization radius somewhat bigger than the expected size of the neutron star magnetosphere in a pulsing X-ray binary (see Sect. 10.9), although perilously sensitive to the wind velocity ($\propto v_w^8$). However, it is inconsistent to use the recipe, Eq. (10.24), for r_{acc}, as this holds only for a symmetrical flow. Eq. (10.24) should be a reasonable estimate if the asymmetries are small, but then J is the difference between two large and almost equal angular momenta from the two sides of the accretor, so Eq. (10.25) is probably unreliable. A further objection[29] is the existence of analytic examples showing that not all velocity or density gradients produce a nonzero angular momentum J. Numerical treatments are unavoidable in deciding the order of magnitude of the efficiency η in Eq. (10.25). The problem is three-dimensional, but computational limitations have confined most calculations to two-dimensional study of the flow in the orbital plane. While the largest angular momentum contributions are in this plane, a two-dimensional calculation automatically suppresses any effects of focussing or spreading out in the other dimension, so the results[12,34,83] are not yet definitive. However, almost all agree that the efficiency η in Eq. (10.22) is rather small (10^{-1}–10^{-2}). An interesting feature is found if the numerical resolution is high enough[12,34,83]: the accretion flow changes the net sign of J in an erratic manner, with the long-term average giving a small positive η (see also Sect. 1.3.2 and 1.6.2). This 'flip-flop' behaviour seems to be independent of density or velocity gradients, and is reminiscent of a von Kármán vortex street; there are indications of similar behaviour in three-dimensional calculations[83]. At each reversal the flow thus tries to form a disc rotating in the opposite sense to that already present. The sudden decrease in the net angular momentum can cause a burst of accretion on to the central star. The time scales of these variations do not explain observed variations in the neutron-star spin periods of wind-fed X-ray binaries, however, being[83] of the order of the flow (sound-crossing) time $r_{\rm acc}/c_S$, which is typically ~ 1 hr, compared with observed variations on time scales $\gtrsim 10$ days. Variations on shorter time scales may exist, but current observational data are too sparse to give much information here.

Thus, calculations tend to suggest that the simplest criterion for disc formation is not met in most wind-fed binaries, as $R_{\rm circ}$ (cf. Eq. (10.1)) is probably smaller than the magnetospheric radius $\sim r_\mu$ (see Sect. 10.9.2 below). This is supported by considerations of the angular momentum flow in such systems, as revealed by the spin periods of magnetic neutron stars in such systems (see Sect. 10.9).

10.9.3 Accretion in eccentric binaries

A number of the Be X-ray binaries have quite large orbital eccentricities[132] $e(\gtrsim 0.3)$. These probably arise because tides have not yet circularized the binary after the supernova explosion giving rise to the neutron star (see Sect. 11.2.4 and 11.2.7). If the supernova was sufficiently asymmetric, the orbital plane need not coincide with the Be star equator; however, observational evidence[131] suggests that it does, in line with the theoretical expectation[99] that spin-orbit misalignments are tidally damped on shorter time scales than are required for circularizing the orbit. Thus the neutron star moves in and out within the equatorial wind by a factor $(1+e)/(1-e)$; depending on the uncertain wind density and velocity structure, a variety of orbital modulations are possible[131]. Often the system is an X-ray source only for a short fraction of each orbit, and is sometimes called a hard X-ray transient.

If the separation at periastron is short enough, and the Be star is spinning rapidly enough, it may be impossible for the outer layers of this star to remain bound to it as the neutron star passes. This 'tidal lobe overflow' resembles Roche lobe overflow at periastron, and disc formation is likely. If the interaction between the stars is confined to periastron the binary orbit evolves with constant periastron distance[58]: the combination $P(1-e^2)^{3/2}$ remains fixed and approximately equal to the orbital period before the supernova creating the neutron star. Evolution to both smaller or larger P, e appears possible, depending on the precise tidal and accretion interaction. The first case leads to a circular orbit, which may be tidally unstable, as the Be star probably has spin angular momentum exceeding one-third of that of the orbit[24]; the neutron star spirals down into this star through a sequence of circular orbits. The second case leads to a small periastron distance and a rapidly spinning Be star in a highly eccentric binary, favouring tidal lobe overflow. Two very eccentric Be-star binaries with rapidly spinning neutron stars may be examples of this type of evolution[58]: the X-ray binary A 0538−66 ($e \gtrsim 0.4$, $P = 16.7$ days, $P_{\text{spin}} = 69$ ms) and the Be star radio pulsar binary PSR 1259−63 ($e = 0.87$, $P = 1237$ days, $P_{\text{spin}} = 47$ ms).

10.10   Accretion on to a magnetic star

Both neutron stars and white dwarfs can possess quite substantial magnetic moments $\mu = BR_*^3$ ($B =$ polar field). If the magnetic stresses are comparable with material ones, the accretion flows in X-ray binaries and CVs can be channelled on to restricted regions of the accreting star. The spin of the accreting star, and sometimes its rate of change, are often readily observable through the resulting rotational modulations, offering insight into the angular momentum flow in the system (see Sect. 1.3). The beat-frequency model of horizontal-branch quasi-periodic oscillations (QPOs) in LMXBs invokes flow on to a weakly magnetized neutron star (see Ch. 6). Magnetic accretors also produce different radiation spectra from nonmagnetic ones: inflow on to the stellar surface is quasi-radial rather than tangential, favouring the emission of higher energy radiation.

A measure of the relative strength of the magnetic field and the accretion flow is given by the spherical Alfvén radius, defined as the radius r_μ at which magnetic stresses would balance the ram pressure of a spherically symmetrical accretion flow. Assuming a dipole field at large distance from the star, the magnetic stresses fall off as r^{-6}, and one finds

$$r_\mu \simeq 2.7 \times 10^{10} \mu_{33}^{4/7} \dot{M}_{16}^{-2/7} \text{ cm} \text{ (CVs)} \tag{10.26}$$

and

$$r_\mu \simeq 2.7 \times 10^{8} \mu_{30}^{4/7} \dot{M}_{17}^{-2/7} \text{ cm} \text{ (X-ray binaries),} \tag{10.27}$$

where the units of magnetic moment and accretion rate are chosen to represent conditions for white-dwarf and neutron-star accretion, respectively. Despite their potentially much larger surface fields ($\gtrsim 10^{12}$ G), neutron star accretors have much smaller magnetic moments than the corresponding white dwarfs, where the Alfvén radius can be comparable with the binary separation and produce qualitatively different behaviour. It is important to realise that Eqs. (10.26) and (10.27) are extremely rough measures: the actual dynamics of the accretion flow depends on many factors such as the degree to which the matter is threaded by the field, plasma instabilities, and the field geometry, and we will see (e.g. Sect. 10.9.3) that there are cases where criteria involving Eqs. (10.26) and (10.27) are very misleading. Nevertheless, the very steep radial dependence of the magnetic stresses means that Eqs. (10.26) and (10.27) are useful indicators of where magnetic effects are potentially significant, even if the accretion flow is far from spherical.

We can thus distinguish several likely physical regimes: for a binary to show significant magnetic behaviour at all (e.g. pulsing at the accretor's spin period) requires

$$r_\mu \gtrsim R_*. \tag{10.28}$$

Disc formation may be affected if

$$r_\mu \gtrsim R_{\text{circ}}, \tag{10.29}$$

while there will be strong spin-orbit torques if

$$r_\mu \gtrsim a, \tag{10.30}$$

where a is the binary separation. The minimum condition, Eq. (10.28), requires typical surface fields $\gtrsim 10^4$ G and $\gtrsim 10^8$ G for white dwarfs and neutron stars, respectively.

10.10.1 Disc accretion on to magnetic neutron stars

In some pulsing X-ray binaries (particularly where the companion overflows the Roche lobe) the inequality (10.29) does not hold, and a disc can form in the normal way. We expect the disc to be disrupted by the magnetic field at some radius of order r_μ. Beyond this bare description, there is no consensus on how or where the matter attaches to the field lines and flows to the neutron star surface, or on the interaction of the field with the disc outside the disruption radius, despite a large number of papers on the subject. There are two main approaches to the problem. In one[36,37,52,128] the field is assumed to thread a large fraction of the disc because of Kelvin–Helmholtz instabilities, field line reconnection with tangled fields within the disc, and turbulent diffusion. The other approach[1–3,5,112] assumes that the disc is a perfect conductor, completely excluding the field. In both approaches the matter is often assumed to leave the disc in a narrow transition zone at the inner edge (near r_μ), thereafter flowing along field lines to the neutron star.

A principal aim of such models is to predict the torques acting on the neutron star and thus its spin behaviour. Accreting matter gives up its Kepler angular momentum to this star at the point where it begins to interact strongly with the field, producing an accretion torque $G_{\text{acc}} \simeq \dot{M}(GM_1 r_\mu)^{1/2}$. Equally, any matter attaching to the field lines at cylindrical radii R too far from the spin axis will be flung out by centrifugal forces and carry off specific angular momentum $\simeq (GM_1 R)^{1/2}$: this occurs if R exceeds the 'corotation radius'

$$R_{\text{co}} = \left(\frac{GM_1 P_{\text{spin}}^2}{4\pi^2} \right)^{1/3}, \tag{10.31}$$

where the field lines move with the local Kepler speed. Evidently a detailed model of the disc–field interaction is required to specify how much material is expelled in this way, and from which radii. The inner edge of the disc will probably contribute some such material; as pointed out by Ghosh & Lamb[36,37] and Scharlemann[112] the currents induced in the disc will pinch the field lines towards the neutron star, particularly near the inner edge. Hence, matter flowing along field lines from the disc plane has first to move outwards (e.g. under thermal pressure) against gravity. Spruit and Taam[122] show that there are lobes formed by these field lines in which the field can support a mass of gas against gravity and centrifugal forces. Much of this corotating gas probably drifts inwards across field lines through interchange instabilities, but some opens field lines to form a magnetically accelerated wind. Finally there is another torque $-G_{int}$ resulting from the interaction (if any) of the field lines with parts of the disc outside R_{co}: again a detailed model is needed to specify this, but it must act to spin the star down as the field lines move faster than the matter. We can thus write the angular momentum equation for the neutron star symbolically as

$$I\dot{\omega} \simeq \dot{M}(GM_1 r_\mu)^{1/2} - \dot{M}_{ex}(GM_1 R_{ex})^{1/2} - G_{int}, \qquad (10.32)$$

where I is the neutron star's moment of inertia, $\omega = 2\pi/P_{spin}$ is its angular velocity, while $\dot{M}_{ex}$ is the rate at which mass is expelled and R_{ex} is the average radius at which it breaks free of the field.

A model of the disc–field interaction specifies the right-hand side of Eq. (10.32) in terms of $\dot{M}$ and μ: given an estimate of $\dot{M}$ from the X-ray luminosity, we can in principle compare the predictions of this equation with the observations of pulsing X-ray sources that have accumulated over the past two decades and hope to fit a single value of μ for each one. In practice, the physics has so far proved too complex for a generally accepted theory to emerge. Fits to observations, even of systems where there is good evidence for a disc, are not uniformly successful for any current theory; the fact that entirely unreasonable models seem to fit at least as well[3] suggests that we would not learn much even if they were. We are limited to drawing some general conclusions from the basic form of Eq. (10.32). First, it is clear that the neutron star cannot accrete if it spins more rapidly than the rate at which

$$r_\mu \simeq R_{co}, \qquad (10.33)$$

for then centrifugal force expels all the matter approaching the magnetosphere: the torques on the right-hand side of Eq. (10.32) must balance at some 'equilibrium' value ω_{eq} smaller than the ω satisfying Eq. (10.33). The ratio of these quantities is sometimes called the fastness parameter. Models typically find equilibrium fastness parameters close to unity. For $\omega \ll \omega_{eq}$ we expect the accretion torque to dominate and spin the neutron star up towards ω_{eq} on a time scale $t_{spinup} \sim I\omega/\dot{M}(GM_1 r_\mu)^{1/2}$. If this time scale is shorter than the binary lifetime, we would thus expect $R_{co} \sim r_\mu$ and $\omega \sim \omega_{eq}$, and we can compare the observed spinup time scale $\omega/\dot{\omega}$ with the unresisted spinup time scale

$$t_{spinup} \simeq \frac{I}{\dot{M}R_{co}^2} \simeq \frac{I}{\dot{M}}\left(\frac{4\pi^2}{GM_1 P_{spin}^2}\right)^{2/3}. \qquad (10.34)$$

Observations of pulsing X-ray binaries which probably have discs (see Sect. 1.3) generally reveal spinup time scales $\sim 10^2$ longer than implied by Eq. (10.34), suggesting that they are close to equilibrium. By contrast we would expect Eq. (10.34) to give a rough estimate of the spinup time scale of systems not too close to equilibrium. Note that these requirements are common to all detailed models of the disc–field interaction, and thus do not allow us to discriminate among them.

10.10.2 Wind Accretion on to Neutron Stars: P_{spin}–P_{orb} relations

We saw in Sect. 10.8 that theoretical calculations suggest that wind accretion is unlikely to supply matter with enough specific angular momentum J to satisfy the disc formation condition. However, a significant correlation $P_{spin} \propto P_{orb}^{2.3}$ between spin and orbital periods in the Be-star X-ray binaries was noticed by Corbet[21-23], and it is tempting to try and interpret this as nevertheless reflecting some kind of disc accretion equilibrium spin rate[132] (averaged over time scales $\sim t_{\text{spinup}}$: the small *observed* spin fluctuations on time scales much less than this do not themselves imply that the spin is not close to an equilibrium value). However, a simple argument[55] suggests that the neutron star accretes too little angular momentum for this to be the case, and that the spin-orbit correlation is probably established at an earlier evolutionary stage. For if the spin rate is near equilibrium, we see from Eqs. (10.31) and (10.33) that the neutron star must accrete angular momentum at the rate

$$G_{\text{acc}} \simeq \dot{M}(GM_1)^{2/3}\left(\frac{P_{\text{spin}}}{2\pi}\right)^{1/3}. \tag{10.35}$$

Of course, the other torques on the right-hand side of Eq. (10.33) balance this term, but we can show that there is too little angular momentum in the accreting matter to supply this equilibrium value of G_{acc}. For in disc accretion the star accretes a fraction $\psi < 1$ of the captured specific angular momentum J, the remainder being lost from the disc in some way (e.g. tides at its edge). Using the estimate, Eq. (10.25), and Kepler's law for the binary we can write

$$J = \eta\left(\frac{M_1}{M}\right)^2 (GM)^{2/3}\left(\frac{P_{\text{orb}}}{2\pi}\right)^{1/3}\left(\frac{v}{v_{\text{rel}}}\right)^4, \tag{10.36}$$

where M is the total binary mass. Equating G_{acc} and $\dot{M}\psi J$ gives the equilibrium requirement

$$P_{\text{spin}} = 51(\eta\psi)^3\left(\frac{v}{v_{\text{rel}}}\right)^{12}\left(\frac{15M_1}{M}\right)^4\left(\frac{P_{\text{orb}}}{30\,\text{d}}\right)^4 \text{ s}. \tag{10.37}$$

At first sight this equation looks quite promising, as the observed spin-orbit relation has $P_{\text{spin}} \sim 10$ s for $P_{\text{orb}} \sim 30$ days. However, the first two factors on the right-hand side are probably $\ll 1$: quite apart from the theoretical expectation $\eta \ll 1$ and the requirement $\psi < 1$ for disc accretion, the ratio v/v_{rel} is likely to be small for systems with long orbital periods: when raised to the power 12 in Eq. (10.37) this makes the predicted P_{spin} much too short compared with observation. The same argument applied to OB-star X-ray binaries also predicts equilibrium spin periods far shorter than the observed periods: this is no surprise, as the lifetime of these systems is probably shorter than t_{spinup} (see Sect. 1.3.2.1 and Ch. 11).

The argument above supports the theoretical expectation that wind-fed binaries do not in general form accretion discs in the usual way. An idea of how they accrete has been put forward by Anzer[4] *et al.* Since J is very small, matter is assumed to accrete in rough spherical symmetry on to the neutron star magnetosphere and is stopped by a shock near r_μ. Behind this shock the matter interacts with the rotating magnetic field through Kelvin–Helmholtz instabilities, gaining some angular momentum from the neutron star. It thus tends to form a torus around the rotational equator: as matter accretes from this torus its angular momentum is returned to the neutron star. Hence there is no net gain of angular momentum by this star, other than the small amount $\dot{M}J$ carried by the captured matter. (The argument of the last paragraph does not apply in this case, as we formally have $\eta \ll 1$ and $\psi \gg 1$.) Fluctuations in the mass and angular momentum of the torus resulting from changes in accretion rate make the spin rate fluctuate somewhat rather than changing systematically. As before, we conclude that the spin rates of neutron stars in wind-fed systems are established at an earlier stage of their evolution.

10.10.3 Accretion on to magnetic white dwarfs

Accretion on to magnetic white dwarfs has a rather different character from the neutron-star case: typically, the magnetic moments are larger and the binary separations smaller, so the more extreme inequalities $r_\mu \gtrsim R_{circ}$ and $r_\mu \gtrsim a$ occur commonly. This leads to a much richer variety of possible behaviour.

Almost all known white-dwarf accretors are in CVs, where mass transfer is by Roche lobe overflow. The closest analogues to neutron-star binaries are the following.

DQ Herculis systems. We follow Warner[130] in restricting this name to systems with $P_{spin} \lesssim 100$ s. This class is small, as most nonsynchronous systems have P_{spin} of order 10 min: two indisputable members are DQ Her itself ($P_{spin} = 71$ s) and AE Aqr ($P_{spin} = 33$ s). The star cannot spin faster than the rate specified by $R_{co} \simeq r_\mu$, so Eq. (10.27) implies

$$P_{spin} \gtrsim 2700 \mu_{33}^{6/7} \dot{M}_{16}^{-3/7} \text{ s}, \qquad (10.38)$$

and it is apparent that these systems must have rather weak magnetic moments $\mu \sim 10^{31}$, corresponding to surface fields $\sim 10^4$ G. Clearly, $r_\mu \sim 10^9$ cm is well inside $R_{circ} \sim 10^{10}$ cm, and an accretion disc will presumably form in the usual way. Unlike other magnetic CVs these systems are weak X-ray sources if at all; as they do not generally have outbursts either there is little to pick them out, and their space density could be much higher than suggested by the current sample.

Intermediate polars. Most nonsynchronous systems have spin periods of order 10–20 min, and there is a strong tendency[8] to have $P_{spin} \simeq 0.1 P_{orb}$. This relation is exactly equivalent[61] to the equality $R_{circ} \simeq R_{co}$, implying that in spin equilibrium the accreting matter has just the orbital specific angular momentum $(GM_1 R_{circ})^{1/2}$. This is incompatible with long-term accretion (i.e. for times comparable with t_{spinup}) from a well-developed disc coupling back part of the captured angular momentum to the companion star via viscosity and tides (cf. Sect. 10.4 above): this would reduce the accreted specific angular momentum below $(GM_1 R_{circ})^{1/2}$ and imply a *shorter* spin period. One requires instead that the time scale t_{mag} for the accreting gas to give up its angular momentum to the magnetic field (and hence the white dwarf) should be shorter than the viscous time scale t_{visc}, leading to surface fields $\gtrsim 10^4$ G (see below), in agreement with the upper limit on the fields in the DQ Her systems.

Recent work[57] gives a possible picture of how accretion may occur in this case. It is assumed that the matter overflowing the companion's Roche lobe is blobby rather than a continuous stream, as appears to be the case in the AM Her systems (see below and Sect. 10.9.4). The gas blobs are diamagnetic and not easily penetrated by the magnetic field, but interact with it via surface currents, producing a drag force proportional to the relative velocity. Thus the blobs exchange angular momentum and energy with the rotating magnetosphere. Blobs of orbital energy less than a certain critical value lose energy in the interaction: their orbits circularize and they are ultimately accreted, giving up all their original angular momentum (specified by R_{circ}) to the white dwarf. More energetic blobs gain energy and are ultimately flung out (usually to be mopped up by the companion star), removing some angular momentum from the white dwarf. The critical blob energy is $E_E \simeq -\omega J/2$, with ω, as before, the white dwarf spin rate (the precise value depends on the eccentricity of the blob orbit and the radial dependence of the drag term). For low spin rates this is considerably greater than the typical orbital energy $E < 0$ of blobs leaving the companion, so all the blobs are accreted and ω increases. Ultimately, $-\omega J/2$ reduces to E, and all the blobs would be ejected if it increased further. There is thus an equilibrium with $-\omega J/2 \simeq E$: given any distribution of blob energies about E, the lower energy ones are accreted, while a comparable number of higher energy blobs are returned to the companion. The accreting blobs each have specific angular momentum $(GM_1 R_{circ})^{1/2}$; equating their orbital energies E to E_E now reproduces the $P_{spin} \simeq 0.1 P_{orb}$ relation. This picture offers a potential explanation for periodicities unrelated to P_{spin} or P_{orb} (or combinations of them) observed in some intermediate polars: these might arise from blobs forced by the field into resonant orbits which would remain unpopulated in nonmagnetic systems.

In view of the complexity of the matter–field interaction in intermediate polars, it is hardly surprising that there is no consensus as to their accretion geometry. In the picture presented in the last paragraph, the accreting blob orbits become circular through interaction with the magnetosphere. The time scale t_{mag} for this is longer than the orbital time scale $t_{dyn} \lesssim P_{orb}$ but much shorter than the viscous time scale t_{visc} to form a disc, so the blobs form only a ring near R_{circ} (the spin behaviour above is equivalent to saying that the blobs spinup the white dwarf until $R_{co} = R_{circ}$). But low-density blobs, which probably produce the observed medium-energy X-rays (see Sect. 10.9.4 below) might couple to the field lines almost immediately on impact (i.e. $t_{mag} \ll t_{dyn}$). The dissipation pattern of these blobs and their accretion geometry at the white dwarf would both show the beat period $(P_{spin}^{-1} - P_{orb}^{-1})^{-1}$, as the field lines see the secondary star rotating with this period. In fact, most intermediate polars do show this periodicity in both the optical and X-rays[43]; in some systems the X-ray modulation is fairly weak (a few per cent) while in others it dominates, suggesting that diffuse material is present to some degree in most intermediate polars. On the other hand, very dense blobs might ignore the field almost completely and form a disc in the usual way (i.e. $t_{mag} \gg t_{visc}$). This geometry cannot give X-ray beat-period modulations as the matter loses all memory of the orbital phase before accreting. (*Optical* beat-period modulations can result from reprocessing of harder radiation in some structure fixed in the binary frame, which is implausible in X-rays.) The long-term presence of a disc would destroy the $P_{spin} \simeq 0.1 P_{orb}$ relation, but discs lasting for times $\ll t_{spinup} \sim 10^5$ yr would not affect it. However, if discs dominated the accretion process in most observed intermediate polars the relation would require a selection effect strongly favouring discovery of systems when they have discs.

In both the disc and diamagnetic-blob pictures, an essentially plane circular flow is eventually disrupted (at some inner disc radius, or in the ring at R_{circ}) and flows in towards the white dwarf magnetic poles. In this disc picture, this happens because the matter is assumed to follow field lines inside $\sim r_\mu$, while in the blob picture the diamagnetic blobs oscillate vertically about the orbital plane at frequency ω as the field lines sweep past: at large radii the net excursion from the plane is small, but it becomes comparable with the azimuthal motion near $R_{co} \simeq R_{circ}$. In either case this implies two fairly broad accretion funnels spinning with the white dwarf: matter in these funnels may absorb and scatter radiation from the accreting polecaps and also radiate in its own right. The geometry and kinematics of this 'accretion curtain' have been considered extensively for accretion from a disc[109,110]. For example, photoelectric absorption in this curtain could explain the relative phasing of the medium-energy X-ray and radial-velocity variations at the spin period. However, dynamical calculations[109] show that a homogeneous curtain flow has far too little optical depth to cause any such modulation. On the other hand, the same geometry and kinematics would also apply to the blob picture, making the curtain a 'leaky absorber': significant optical depths are possible in this case. At photon energies $\gtrsim 10$ keV, photoelectric absorption is negligible, and the observed modulations at the spin period must result from occultations by the white dwarf body[66] or scattering in the accretion curtain. The roughly sinusoidal modulations mean in the first case that at least one dimension of the accretion region must be comparable with the white dwarf radius. Scattering in the curtain requires larger column densities than are available in a homogeneous flow, as we have seen. If an accretion curtain is present, its large area ($\gtrsim 10^{20}$ cm^2) will make it a significant optical emitter, even with a small filling factor for optically thick blobs, and could well dominate the optical spin-modulated light. As most of the area is in regions of fairly low magnetic field well away from the white dwarf, this light would not show any polarization. This is also true of spin-modulated dissipation from blobs whose orbits have circularized. Arguments that the fields in intermediate polars are low, as they do not show optical polarization, should therefore be treated with some caution.

AM Herculis systems: $r_\mu \gtrsim a$. These unique systems are the most strongly affected by magnetic fields (see Cropper[25] for a comprehensive review). Observationally they are distinguished by the presence of strong linear and circular optical polarization varying on the orbital period. Modelling of this, and the detection of Zeeman features in states of low accretion rate (see below), show that the white dwarfs have fields ~ 1–3×10^7 G and thus magnetic moments $\mu \gtrsim 10^{34}$ G cm^3. In eclipsing systems, the spin and orbital periods differ by less than a few parts per million, and indeed all radiations, from infrared to X-rays, show the same period. This locking of spin and orbital rotations is probably a direct consequence of the inequality $r_\mu \gtrsim a$, which is unique to these systems: from Eq. (10.26) with $\mu \gtrsim 10^{34}$ G cm^3 we find $r_\mu \gtrsim 10^{11}$ cm, while the short orbital periods (all less than 4 hr, and most below 2 hr) imply $a \gtrsim 10^{11}$ cm. There are several candidates for the torque holding the white dwarf in synchronous rotation against the spinup effect of the accretion torque[14–19,51,75]. The limits on asynchronous rotation probably eliminate[140] dissipative torques[14] as a means of locking the rotations (although they are vital in damping perturbations away from the synchronous state[140]), and the simplest candidate is the interaction of the white dwarf dipole with an intrinsic field in the secondary star[15].

AM Her systems necessarily satisfy $r_\mu > R_{circ}$, so it is not surprising that there is no evidence of accretion discs in them. For gas flow concentrated in a stream (rather than spread out over a sphere as assumed in calculating r_μ), the ram pressure exceeds magnetic stresses over much of the binary separation. The accretion stream thus moves freely under gravity until a distance $\sim 10^{10}$ cm from the white dwarf, where it is thought to be threaded rather rapidly by the field, thereafter following field lines down to the white dwarf surface. The fact that some of the accretion evidently manages to reach the other magnetic pole in most systems, even when this would be difficult along field lines, suggests that t_{mag} is comparable with, rather than $\ll t_{\text{orb}}$, for at least some of the accreting matter.

The relation between intermediate polars and AM Her systems. The distinctive features of AM Her systems result from the inequality $r_\mu \gtrsim a$. But we saw above that these quantities are comparable at binary periods ~ 4 hr: at longer periods, a increases as $P_{\text{orb}}^{3/2}$, while r_μ is likely to be smaller, since average accretion rates are higher at longer periods. Since CVs are believed to evolve towards shorter orbital periods (see, e.g., ref. 53, Sect. 8.5 and 11.3.2), and the period distribution of the AM Her systems requires that their progenitors were formed as CVs at periods > 4 hr[42,108], this suggests that the progenitors of AM Her systems cannot satisfy the condition $r_\mu \gtrsim a$, and are presumably asynchronous. Indeed, if there are medium-term ($\sim 10^5$ yr) fluctuations in $\dot{M}$ around the secular mean transfer rate it is conceivable that AM Her systems at periods 3–4 hr could unlock themselves in this way[40]. It is tempting to conclude that some of the intermediate polars are progenitors of AM Herculis systems[62], particularly as there is relatively little overlap in orbital period beween the two classes (most intermediate polars have $P_{\text{orb}} \gtrsim 4$ hr; see Figure 8.16). But this requires that at least some of these systems have magnetic fields $\gtrsim 10^7$ G, which is in *prima facie* conflict with their lack of optical polarization[141]. There may be ways around this constraint (see e.g. the end of the subsection on intermediate polars), but if not, we are apparently the victims of selection effects making strong-field systems difficult to observe when they are asynchronous[61]. Very recently, the ROSAT-discovered asynchronous system RE0751+14 has been found to exhibit optical polarization (S. Rosen, J. Mittaz & P. Hakala, 1993, private communication; V. Piirola, 1993, private communication), making it a prime candidate for an AM Her progenitor.

10.10.4 Accretion columns

Matter channelled by a magnetic field will fall roughly radially on to the stellar surface at a speed comparable with the free-fall velocity ($\sim 10^8$–10^9 cm s^{-1} for white dwarfs and $\sim 10^{10}$ cm s^{-1} for neutron stars). These velocities are highly supersonic, suggesting that the accretion energy can be liberated in a strong shock; we might thus expect emission characterized by the shock temperature $T_s \simeq T_V$ (cf. Eq. (10.22) above) to dominate the accretion luminosity. This expectation is largely unfulfilled, as we shall see: in many cases the emission has the characteristic effective temperature

$$T_{\text{eff}} = \left(\frac{L_{\text{acc}}}{4\pi R_*^2 f \sigma} \right)^{1/4}, \tag{10.39}$$

where L_{acc} is the accretion luminosity and f is the emitting surface fraction. This is hardly surprising, as in any conceivable geometry a large part of the emitted radiation is likely to be thermalized by interacting with the stellar surface.

Neutron stars.   This case involves the most extreme physics: for example, a simple collisional shock is probably not allowed, as the collisional mean free path in the accretion flow is much longer than the neutron star radius[86]. At high accretion rates, radiation pressure becomes dominant, and the accretion flow can be decelerated by prompt conversion of the infall energy into radiation. The result is a 'radiative shock', in which matter makes a rapid transition from near free-fall to almost stagnant flow[28]: this type of structure has large optical depth, and the emergent radiation is roughly blackbody at the temperature in Eq. (10.39). At lower accretion rates, two other structures are possible: there may be a collisionless shock, in which the infall energy of the ions is thermalized. Electron–ion coupling is inefficient compared with the radiative losses from the former, so the electron temperature is much lower than the virial temperature. If the neutron star has a strong magnetic field, most of the emission is cyclotron radiation[76]. In the absence of a collisionless shock, the infalling ions are stopped below the neutron star photosphere[67,68,142] by Coulomb collisions. The energy released is carried off as thermal bremsstrahlung and Compton cooling.

White dwarfs.   Here $T_s \sim 10^8$ K and $T_{eff} \sim 3 \times 10^5$ K, so the expected radiation components are medium (10 keV) and soft (0.03–0.1 keV) X-rays, respectively. Clearly the latter is very vulnerable to interstellar absorption if T_{eff} is closer to 10^5 K than 3×10^5 K. In the synchronously rotating magnetic CVs (AM Her systems: see below and Ch. 8), the soft component is almost always detected, but the medium-energy component is usually weak or absent. In the nonsynchronous magnetic CVs, the medium-energy component is detected, but its luminosity is far smaller than the expected accretion luminosity ($\lesssim 1\%$). Indirect arguments[60] suggest that the bulk of L_{acc} is again radiated as the soft component, which is too soft ($T_{eff} \lesssim 10^5$ K) to evade interstellar extinction. In both types of system the most likely cause of this 'soft X-ray excess' is that the accretion occurs in narrow filaments or blobs[32,74] rather than a single uniform-density stream, possibly because of the way gas threads on to the magnetic field lines. The higher density ($\rho \gtrsim 10^{-7}$ g cm^{-3}) filaments shock below the white dwarf photosphere so that their primary radiation at $\sim T_s$ is degraded by the white dwarf atmosphere before emerging at $\sim T_{eff}$. These dense filaments evidently account for most of the accretion flow; there is an upper limit of $\sim 10^{15}$ g s^{-1} on the instantaneous accretion rate carried by any single such filament[32,33], so that, except at very low luminosities, around ten (and possibly many more) of them must accrete simultaneously. Filaments with lower densities shock above the photosphere, producing mainly medium-energy X-rays or self-absorbed thermal cyclotron emission, depending on the density and magnetic field strength. For the $\sim 10^7$ G fields found in AM Her systems, X-ray filaments must have 10^{-6} g cm^{-3} $\gtrsim \rho \gtrsim 10^{-8}$ g cm^{-3}: the cyclotron emission from filaments with $\rho \lesssim 10^{-8}$ g cm^{-3} is responsible for the strong optical polarization characteristic of these systems. The filament or blob picture makes it fairly easy to understand the transitions of AM Her soft X-ray light curves from roughly sinusoidal in usual states to intrinsically noisy, roughly square-wave in 'anomalous' states[38]. In normal states many filaments accrete simultaneously, and the accreting region appears simply as a corrugated radiating surface, implying a sinusoidal modulation as the white dwarf rotates. In anomalous states the accretion is apparently concentrated into rather few (~ 15) filaments. This small number makes the light curve intrinsically noisy: further, the filaments produce individual splashes that do not shadow each other, and so are seen to appear over the limb as the star rotates and give sharp square-wave light curves. Evidently, future progress in understanding accretion in AM Her systems will require an explanation of what produces the filamentary structure in the first place: this is completely unknown at present. In the nonsynchronous systems, the problem is further complicated by the fact that the basic accretion geometry is not understood.

Acknowledgements
I am very grateful to Drs J.P. Lasota, M. Livio, S. Lubow and J.E. Pringle for carefully reading various drafts of this chapter and offering detailed suggestions for improvement, and to the many colleagues who alerted me to suitable references.

References
[1] Aly, J.J., 1980, *Astr. Ap.*, **86**, 192.
[2] Anzer, U. & Börner, G., 1980, *Astr. Ap.*, **83**, 133.
[3] Anzer, U. & Börner, G., 1983, *Astr. Ap.*, **122**, 73.
[4] Anzer, U., Börner, G. & Monaghan, J.J., 1987, *Astr. Ap.*, **176**, 235.
[5] Arons, J., Burnard, D.J., Klein, R.I., McKee, C., Pudritz, R.E. & Lea, S.M., 1984, in *High Energy Transients in Astrophysics*, AIP Conf. Proc. 115, ed. S.E. Woosley, AIP, New York, p. 215.
[6] Bailey, J.A., 1975, *J. Br. Astr. Assoc.*, **86**, 30.
[7] Bailyn, C.D., 1992, *Ap. J.*, **391**, 298.
[8] Barrett, P., O'Donoghue, D. & Warner, B., 1988, *M.N.R.A.S.*, **233**, 759.
[9] Bath, G.T. & Pringle, J.E., 1982, *M.N.R.A.S.*, **199**, 267.
[10] Begelman, M.C., McKee, C.F. & Shields, G.A., 1983, *Ap. J.*, **271**, 70.
[11] Bisnovatyi–Kogan, G.S., 1992, MPA preprint 641.
[12] Blondin, J.M., Kallman, T.R., Fryxell, B.A. & Taam, R.E., 1990, *Ap. J.*, **335**, 862.
[13] Bondi, H. & Hoyle, F., 1944, *M.N.R.A.S.*, **104**, 273.
[14] Campbell, G.C., 1983, *M.N.R.A.S.*, **205**, 1031.
[15] Campbell, G.C., 1984, *M.N.R.A.S.*, **211**, 83.
[16] Campbell, G.C., 1985, *M.N.R.A.S.*, **215**, 509.
[17] Campbell, G.C., 1986, *M.N.R.A.S.*, **219**, 589.
[18] Campbell, G.C., 1986, *M.N.R.A.S.*, **221**, 599.
[19] Campbell, G.C., 1989, *M.N.R.A.S.*, **236**, 475.
[20] Cannizzo, J.K., Wheeler, J.C. & Ghosh, P., 1985, in *Proceedings of the Cambridge Workshop on Cataclysmic Variables and Low-Mass X-Ray Binaries*, eds. D.Q. Lamb & J. Patterson, Reidel, Dordrecht.
[21] Corbet, R.D.H., 1984, *Astr. Ap.*, **141**, 91.
[22] Corbet, R.D.H., 1985, *Space Sci. Rev.*, **40**, 409.
[23] Corbet, R.D.H., 1986, *M.N.R.A.S.*, **220**, 1047.
[24] Counselman, C.C., 1973, *Ap. J.*, **180**, 307.
[25] Cropper, M.S., 1990, *Space Sci. Rev.*, **54**, 195.
[26] Czerny, M. & King, A.R., 1989, *M.N.R.A.S.*, **236**, 843.
[27] Czerny, M. & King, A.R., 1989, *M.N.R.A.S.*, **241**, 839.
[28] Davidson, K., 1973, *Nature Phys. Sci.*, **246**, 1.
[29] Davies, R.E. & Pringle, J.E., 1980, *M.N.R.A.S.*, **191**, 599.
[30] Eggleton, P.P., 1983, *Ap. J.*, **268**, 368.

[31] Frank, J., King, A.R. & Lasota, J.P., 1987, *Astr. Ap.*, **178**, 137.
[32] Frank, J., King, A.R. & Lasota, J.P., 1988, *Astr. Ap.*, **193**, 113.
[33] Frank, J., King, A.R. & Raine, D.J., 1992, *Accretion Power in Astrophysics*, 2nd Edn, Cambridge University Press.
[34] Fryxell, B.A. & Taam, R.E., 1988, *Ap. J.*, **335**, 862.
[35] Giacconi, R., Paolini, F.R., Gursky, H. & Rossi, B.B., 1962 *Phys. Rev. Lett.* **9**, 439.
[36] Ghosh, P. & Lamb, F.K., 1979, *Ap. J.*, **232**, 259.
[37] Ghosh, P. & Lamb, F.K., 1979, *Ap. J.*, **234**, 296.
[38] Hameury, J.M. & King, A.R., 1988, *M.N.R.A.S.*, **235**, 433.
[39] Hameury, J.M., King, A.R. & Lasota, J.P., 1986, *Astr. Ap.*, **162**, 71.
[40] Hameury, J.M., King, A.R. & Lasota, J.P., 1989, *M.N.R.A.S.*, **237**, 39.
[41] Hameury, J.M., King, A.R. & Lasota, J.P., 1991, *Astr. Ap.*, **248**, 525.
[42] Hameury, J.M., King, A.R., Lasota, J.P. & Ritter, H. 1988, *M.N.R.A.S.*, **231**, 535.
[43] Hellier, C., 1992, *M.N.R.A.S.*, **258**, 578.
[44] Hirose, M. & Osaki, Y., 1990, *Pub. Astr. Soc. Japan*, **42**, 135.
[45] Horne, K. & Marsh, T.M., 1987, in *The Physics of Accretion on to Compact Objects*, eds. K.O. Mason, M.G. Watson & N.E. White, Springer-Verlag, Berlin, p. 1.
[46] Horne, K., 1991, in *Accretion Discs in Compact Stellar Objects*, ed. J.C. Wheeler, World Scientific, in press.
[47] Hoshi, R., 1979, *Prog. Theor. Phys.*, **61**, 1307.
[48] Hoyle, F. & Lyttelton, R.A., 1939 *Proc. Camb. Phil. Soc.*, **35**, 405.
[49] Huang, M. & Wheeler, J.C., 1989, *Ap. J.*, **343**, 229.
[50] Illarionov, A.F. & Sunyaev, R.A., 1975, *Astr. Ap.*, **39**, 185.
[51] Joss, P.C., Katz, J.I. & Rappaport, S.A., 1979, *Ap. J.*, **230**, 176.
[52] Kaburaki, O., 1986, *M.N.R.A.S.*, **220**, 321.
[53] King, A.R., 1988, *Q.J.R.A.S.*, **29**, 1.
[54] King, A.R., 1989, *M.N.R.A.S.*, **241**, 365.
[55] King, A.R., 1991, *M.N.R.A.S.*, **250**, 3P.
[56] King, A.R., 1992, in *Proceedings of IAU Symposium 151*, Cordoba, Argentina, eds. Y. Kondo, R. Sistero & R. Polidan, Kluwer, Dordrecht, p. 195.
[57] King, A.R., 1993, *M.N.R.A.S.*, **261**, 144.
[58] King, A.R., 1993, *Ap. J.*, **405**, 727.
[59] King, A.R. & Lasota, J.P, 1987, *Astr. Ap.*, **185**, 155.
[60] King, A.R. & Lasota, J.P. 1990, *M.N.R.A.S.*, **247**, 214.
[61] King, A.R. & Lasota, J.P, 1991, *Ap. J.*, **378**, 674.
[62] King, A.R., Frank, J. & Ritter, H., 1985, *M.N.R.A.S.*, **213**, 181.
[63] King, A.R., Frank, J. & Whitehurst, R., 1990, *M.N.R.A.S.*, **244**, 731.
[64] King, A.R., Regev, O. & Wynn, G.A., 1991, *M.N.R.A.S.*, **251**, 30P.
[65] King, A.R. & Shaviv, G., 1984, *Nature*, **308**, 519.
[66] King, A.R. & Shaviv, G., 1984, *M.N.R.A.S.*, **211**, 883.
[67] Kirk, J.G. & Galloway, D.J., 1981, *M.N.R.A.S.*, **195**, 45P.
[68] Kirk, J. G. & Galloway, D.J., 1982 *Plasma Phys.*, **24**, 339 and 1025 (erratum).
[69] Kley, W., 1989, *Astr. Ap.*, **208**, 98.
[70] Kley, W., 1989, *Astr. Ap.*, **222**, 141.
[71] Kley, W., 1991, *Astr. Ap.*, **247**, 95.
[72] Kluźniak, W., 1987, Ph. D. Thesis, Stanford University.
[73] Kluźniak, W. & Wagoner, R.V., 1985, *Ap. J.*, **297**, 548.
[74] Kuijpers, J. & Pringle, J.E., 1982, *Astr. Ap.*, **114**, L4.
[75] Lamb, F.K., Aly, J.J., Cook, M.C. & Lamb, D.Q., 1983, *Ap. J.*, **274**, L71.
[76] Langer, S.H. & Rappaport, S., 1982, *Ap. J.*, **257**, 733.
[77] Lasota, J.P. & Pelat, D., *Astr. Ap.*, **249**, 574.
[78] Lubow, S.H., 1989, *Ap. J.*, **340**, 1064.
[79] Lubow, S.H., 1991, *Ap. J.*, **381**, 259.
[80] Lubow, S.H., 1991, *Ap. J.*, **381**, 268.
[81] Marsh, T.R., Horne, K., Schlegel, E.M., Honeycutt, K. & Kaitchuck, R.H., 190, *Ap. J.*, **364**, 637.
[82] Mason, K.O., Córdova, F.A., Watson, M.G. & King, A.R., 1988, *M.N.R.A.S.*, **232**, 779.

[83] Matsuda, T., Sekino, N., Sawada, K., Shima, E., Livio, M., Anzer, U. & Börner, G., 1991, *Astr. Ap.*, **248**, 301.

[84] McClintock, J.E., Bailyn, C.D. & Remillard, R.A., 1992, *IAU Circ.* 5499.

[85] McClintock, J.E. & Remillard, R.A., 1986, *Ap. J.*, **308**, 110.

[86] Mészáros, P., 1992, *High-Energy Radiation from Magnetized Neutron Stars*, University of Chicago Press.

[87] Meyer, F. & Meyer–Hofmeister, E., 1981, *Astr. Ap.*, **104**, L10.

[88] Meyer–Hofmeister, E. & Ritter, H., 1991, in *The Realm of Interacting Binary Stars*, eds. J. Sahade, Y. Kondo & G. McCluskey, Kluwer, Dordrecht, to appear.

[89] Mineshige, S. & Wheeler, J.C., 1989, *Ap. J.*, **343**, 241.

[90] Molnar, L.A. & Kobulnicky, H.A., 1992, *Ap. J.*, **392**, 678.

[91] Morfill, G., Spruit, H.C. & Levy, E.H., 1991, in *Protostars and Planets III*, eds. E.H. Levy, J. Lunine & M.S. Matthews, University of Arizona Press, Tucson.

[92] O'Donoghue, D., 1986, *M.N.R.A.S.*, **220**, 23P.

[93] Osaki, Y., 1974, *Pub. Astr. Soc. Japan*, **26**, 429.

[94] Osaki, Y., 1985, *Astr. Ap.*, **144**, 369.

[95] Osaki, Y., 1989, *Pub. Astr. Soc. Japan*, **41**, 1005.

[96] Paczńyski, B., 1991, *Ap. J.*, **370**, 597.

[97] Parmar, A.N., White, N.E., Giommi, P. & Gottwald, M., 1986, *Ap. J.*, **308**, 199.

[98] Papaloizou, J.C.B. & Pringle, J.E., 1977, *M.N.R.A.S.*, **181**, 441.

[99] Papaloizou, J.C.B. & Pringle, J.E., 1982, *M.N.R.A.S.*, **200**, 49.

[100] Piran, T., 1978, *Ap. J.*, **221**, 652.

[101] Popham, R. & Naryan, R., 1991, *Ap. J.*, **370**, 604.

[102] Pringle, J.E., 1977, *M.N.R.A.S.*, **178**, 195.

[103] Pringle, J.E., 1981, *Ann. Rev. Astr. Ap.*, **19**, 137.

[104] Pringle, J.E., 1989, *M.N.R.A.S.*, **236**, 107.

[105] Pringle, J.E. & Savonije, G.J., 1979, *M.N.R.A.S.*, **198**, 177.

[106] Ritter, H., 1990 *Astr. Ap. Supp.* **85**, 1179.

[107] Ritter, H., King, A.R. & Lasota, J.P., 1994, in preparation.

[108] Ritter, H. & Kolb, U., 1992, *Astr. Ap.*, **259**, 159.

[109] Rosen, S.R., 1992, *M.N.R.A.S.*, **254**, 493.

[110] Rosen, S.R., Mason, K.O. & Córdova, F.A., 1988, *M.N.R.A.S.*, **231**, 549.

[111] Salpeter, E.E., 1964, *Ap. J.*, **140**, 796.

[112] Scharlemann, E.T., 1978, *Ap. J.*, **219**, 617.

[113] Shafter, A.W., Wheeler, J.C. & Cannizzo, 1986, *Ap. J.*, **305**, 261.

[114] Shafter, A.W., 1992, in *Fundamental Properties of Cataclysmic Variable Stars*, ed. A.W. Shafter, Mount Laguna Observatory.

[115] Shakura, N.I. & Sunyaev, R.A., 1973, *Astr. Ap.*, **24**, 337.

[116] Shapiro, S.L. & Lightman, A.P., 1976, *Ap. J.*, **204**, 555.

[117] Shklovsky, I.S., 1967, *Ap. J.*, **148**, L1.

[118] Smak, J.I., 1971, in *New Directions and New Frontiers in Variable Stars Research*, Veröff. Remeis–Sternwarte Bamberg, p. 248.

[119] Smak, J.I., 1971, *Acta Astr.*, **21**, 15.

[120] Smak, J.I., 1984, *Pub. Astr. Soc. Pacific*, **96**, 5.

[121] Smak, J.I., 1991, *Acta Astr.*, **41**, 269.

[122] Spruit H.C. & Taam, R.E., 1990, *Astr. Ap.*, **229**, 475.

[123] Stover, R.J., 1981, *Ap. J.*, **249**, 673.

[124] Taam, R.E. & Fryxell, B.A., 1988, *Ap. J.*, **327**, L73.

[125] Van der Woerd, H. & Van Paradijs, J, 1987, *M.N.R.A.S.*, **224**, 271.

[126] Van Paradijs, J., 1983, *Astr. Ap.*, **125**, L16.

[127] Verbunt, F., 1987, in *The Physics of Accretion on to Compact Objects*, eds. K.O. Mason, M.G. Watson & N.E. White, Springer-Verlag, Berlin, p. 59.

[128] Wang, Y.M., 1987, *Astr. Ap.*, **183**, 257.

[129] Warner, B., 1976, in *Structure and Evolution of Close Binary Systems*, eds. P. Eggleton, S. Mitton & J. Whelan, Reidel, Dordrecht, p. 85.

[130] Warner, B., 1983, in *Cataclysmic Variables and Related Objects*, eds. M. Livio & G. Shaviv, Reidel, Dordrecht, p.155.

456   *Andrew King*

[131] Waters, L.B.F.M., de Martino, D., Habets, G.M.H.J. & Taylor, A.R., 1989, *Astr. Ap.,* **223**, 218.
[132] Waters, L.B.F.M. & van Kerkwijk, M.H., 1989, *Astr. Ap.,* **223**, 196.
[133] Watson, M.G., Willingale, R., King, A.R., Grindlay, J.E. & Halpern, J., 1985, unpublished.
[134] White, N.E., 1989, *Astr. Ap. Rev.,* **1**, 85.
[135] White, N.E. & Holt, S.S., 1982, *Ap. J.,* **257**, 318.
[136] White, N.E., Parmar, A.N., Sztajno, M., Zimmermann, H.U., Mason, K.O. & Kahn, S.M., 1984, *Ap. J.,* **283**, L9.
[137] Whitehurst, R., 1988, *M.N.R.A.S.,* **232**, 35.
[138] Whitehurst, R., 1989, in *Theory of Accretion Disks,* eds. F. Meyer, W.J. Duschl, J. Frank & E. Meyer–Hofmeister, Kluwer, Dordrecht, p. 213.
[139] Whitehurst, R. & King, A.R., 1991, *M.N.R.A.S.,* **249**, 25.
[140] Whitehurst, R. & King, A.R., 1991, *M.N.R.A.S.,* **250**, 152.
[141] Wickramasinghe, D.T., Wu, K. & Ferrario, L., 1991, *M.N.R.A.S.,* **249**, 460.
[142] Zel'dovich, Ya. B. & Shakura, S.S., 1969, *Sov. Astron.,* **13**, 175.
[143] Zola, S., 1989, *Acta Astr.* **39**, 45.

11

Formation and evolution of neutron stars and black holes in binaries

F. Verbunt

Astronomical Institute, University of Utrecht, Postbox 80.000,
NL-3508 TA Utrecht, The Netherlands
Center for High-Energy Astrophysics, Kruislaan 403, 1098 SJ,
Amsterdam, The Netherlands

E. P. J. van den Heuvel

Astronomical Institute 'Anton Pannekoek', University of Amsterdam, and
Center for High-Energy Astrophysics,
Kruislaan 403, NL-1098 SJ Amsterdam, The Netherlands

11.1 Introduction and brief observational overview

11.1.1 *Introduction*

The existence of X-ray binaries and binary pulsars shows that binary systems can, under certain circumstances, survive the supernova explosion of one – or in some cases: both – of the components. In the case of the high-mass X-ray binaries, this survival is a clear consequence of the large-scale *mass transfer* which precedes the supernova explosion of the initially more massive component of the system, such that at the moment of the explosion this star has become the less massive of the two. Explosive mass ejection from the less massive component will in general not lead to disruption (Blaauw 1961) unless the effects of impact and ablation are very large – which for high-mass systems is not expected to be the case (Fryxell and Arnett 1981).

In the case of low-mass X-ray binaries, it is much more difficult to see why the systems were not disrupted. Here, however, one may show semi-empirically that – in contrast to the case of high-mass X-ray binaries – the formation of such a system is an extremely rare event and therefore must require a very exceptional evolutionary history.

In this chapter, we explore how the various types of binaries that contain compact objects may have formed. We first summarize in Sect. 11.1.2 the observed characteristics of the various types of these binaries. In Sect. 11.2 and 11.3 we review the evolution of close binaries – with mass transfer and mass loss – and discuss the various possible final states of the systems: either as radio pulsar binaries or as single pulsars resulting from disrupted binaries. We discuss the X-ray sources and pulsars in globular clusters in Sect. 11.4. Some remarks on the statistics of X-ray and radio pulsar binaries conclude the chapter.

11.1.2 *Brief observational overview of compact objects in binaries*

The observations of X-ray binaries and of radio pulsar binaries are reviewed extensively elsewhere in this book (see Ch. 2–5). We limit ourselves here to a brief discussion of those observational facts that are necessary in understanding the evolution of these systems.

11.1.2.1 X-ray binaries

The binary X-ray sources that contain compact objects can be roughly divided into two groups: the high-mass ones, in which the mass of the companion to the compact star is $M_s \gtrsim 10 M_\odot$; and the low-mass ones, in which $M_s \lesssim 1 M_\odot$. Each can be subdivided further into several subclasses, as follows (see Table 11.1 and Figures 11.1 and 11.2).

High-mass X-ray binaries.   Two broad groups can be distinguished (see Table 11.1 for some examples) which differ in a number of physical characteristics. The 'standard' systems such as Cen X-3 and SMC X-1 are strong sources, characterized by the occurrence of regular X-ray eclipses and double-wave ellipsoidal light variations produced by tidally deformed giant and/or supergiant companion stars that (nearly) fill their critical equipotential lobes (see Sect. 2.2.2). With one exception (GX301-2), their binary periods are between 1.4 and 10 days. The optical luminosities and spectral types of the companions indicate orginal main-sequence masses $\gtrsim 20 M_\odot$, corresponding to O-type progenitors. Among the standard high-mass X-ray binaries, there are two that are thought to harbor black holes: Cyg X-1 and LMC X-3 (see Ch. 3). On the other hand, in the Be X-ray binaries, first recognized as a group by Maraschi, Treves and Van den Heuvel (1976), the companions are rapidly rotating B-emission stars belonging to the main sequence (luminosity class III–V). At present, some 25 such systems are known, making them the most numerous class of high-mass X-ray binaries (see, for example, Van den Heuvel and Rappaport 1987 for a review; see also Ch. 14). The Be stars are deep within their Roche lobes, as is indicated by the generally long binary periods ($\gtrsim 15$ days) and by the absence of X-ray eclipses and of ellipsoidal light variations. According to the luminosities and spectral types, the companion stars have masses in the range about 8–20 $M_\odot$ (spectral types O9–B3, III–V). The X-ray emission from the Be X-ray systems tends to be extremely variable, ranging from complete absence to large transient outbursts. These X-ray outbursts are most probably related to the irregular optical outbursts generally observed in Be stars, which indicate sudden outbursts of mass ejection, presumably generated by rotation-driven instability in the equatorial regions of these stars (see, for example, Slettebak 1988).

Low-mass X-ray binaries.   For some 30 of these systems, the orbital period has been measured (see Ch. 14). These orbital periods range from 11 min to 17 days, similar to the binary period distribution of cataclysmic variables, as illustrated in Figure 11.2. A notable difference is the small number of low-mass X-ray binaries with $80 \text{ min} < P_b < 2 \text{ hr}$, compared with the preponderance of cataclysmic variables in this period range.

A few of the low-mass X-ray binaries contain X-ray pulsars (4U1626 − 67 and, if it is a low-mass system (see Verbunt, Wijers and Burm 1990), GX1 + 4); these have hard X-ray spectra, like the pulsars in the high-mass systems (see Ch. 1). All others have softer X-ray spectra. Only in systems with longer orbital periods (Cyg

Table 11.1. *Some representative X-ray binaries (A) and radio pulsar binaries and their parameters (B)*

For references see Ch. 1, 2, 3, 5 and 14. For the X-ray binaries, Table 11.1A lists name, position, pulse period, X-ray luminosity, orbital period and eccentricity, and (for high-mass systems) the spectral type of the donor. Binaries containing a black hole and transients are indicated with B and T, respectively. For transients, the luminosity is the luminosity at outburst maximum. It should be noted that luminosities are uncertain due to uncertain distances for many sources. For the radio pulsar binaries, Table 11.1B lists position, pulse period, characteristic age ($\tau_c \equiv P/(2\dot{P})$), magnetic field strength, orbital period and eccentricity, and companion mass. The companion masses marked * were calculated for an assumed inclination of 60°.

A Selected X-ray binaries

Name	Position	P (s)	log L_x (erg s^{-1})	P_b (d)	e	Spectral type
High-mass X-ray binaries						
LMC X-4	0532 − 66	13.5	38.6	1.4	0.011	O7III
LMC X-3	0538 − 64	-	B38.5	1.7	~0	BIII–IV
Cen X-3	1119 − 60	4.8	37.9	2.1	0.0007	O6.5II
SMC X-1	0115 − 74	0.7	38.8	3.9	<0.0008	B0I
Cyg X-1	1956 + 35	-	B37.3	5.6	~0	O9.7I
Vela X-1	0900 − 40	283	36.8	9.0	0.092	B0.5I
LMC tran	0535 − 67	0.069	T39.0	16.7	~0.7	B2IV
V635 Cas	0115 + 63	3.6	T36.9	24.3	0.34	Be
BQ Cam	0331 + 53	4.4	T35.8	34.3	0.31	Be
GX301-2	1223 − 62	696	37.0	41.5	0.47	B1–1.5
V725 Tau	0535 + 26	104	T37.3	111.0	0.3–0.4	Be
Low-mass X-ray binaries						
KZ TrA	1627 − 67	7.7	36.8	0.029		
V1405 Aql	1916 − 05		36.9	0.035		
UY Vol	0748 − 68		T37.0	0.159		
V4134 Sgr	1755 − 34		36.8	0.186		
V616 Mon	0620 − 00		BT38.3	0.323		
N Mus 1991	1124 − 68		BT37.6	0.427		
Cen X-4	1455 − 31		T38.0	0.629		
Sco X-1	1617 − 16		37.5	0.787		
V404 Cyg	2023 + 33		BT38.4	6.500		
Cyg X-2	2142 + 38		38.0	9.843		
Peculiar systems						
Her X-1	1656 + 35	1.2	36.8	1.7	< 0.0003	A9–B
Cyg X-3	2030 + 41		38.0	0.2		WN
Cir X-1	1516 − 57		T38.9	16.6		
SS433	1909 + 05		35.8	13.2		

Table 11.1. *cont.*

B Selected binary radio pulsars

Position	P (ms)	$\log \tau_c$ (yr)	$\log B$ (G)	P_b (d)	e	M_c ($M_\odot$)
High-mass binary radio pulsars						
1534 + 12	37.9	8.4	10.0	0.42	0.2737	1.36
1913 + 16	59.0	8.0	10.4	0.32	0.6171	1.39
0655 + 64	195.6	9.7	10.1	1.03	<0.00005	>0.7
2303 + 46	1066.4	7.5	11.9	12.34	0.6584	1.5
Low-mass binary radio pulsars						
1957 + 20	1.6	9.2	8.2	0.38	<0.001	0.02
1831 − 00	521.0	8.8	10.9	1.81	<0.005	0.07*
J0437 − 47[a]	5.8	8.9	8.9	5.74	0.000018	0.17*
1855 + 09	5.4	9.7	8.5	12.33	0.000021	0.23
1953 + 29	6.1	9.5	8.6	117.35	0.00033	0.22*
0820 + 02	864.9	8.1	11.5	1232.40	0.0119	0.23*
Antediluvian radio pulsars						
1259 − 63	47.8			1236.8	0.870	Be
1820 − 11[b]	279.8	6.5	11.8	357.8	0.795	0.8*
Single recycled radio pulsars						
1937 + 21	1.6	8.4	8.6			
1257 + 12[c]	6.2					

[a] Johnston *et al.* (1993).
[b] This pulsar is tentatively listed as antediluvian, i.e. in an evolutionary stage preceding mass transfer (Phinney and Verbunt 1991); alternatively, this system may be a high-mass binary radio pulsar.
[c] This pulsar has two planets (Wolszczan and Frail 1992).

X-2, GX1 + 4) can one observe the spectrum of the optical companion. In all other systems, the optical spectrum is that of the hot accretion disk (see Ch. 2). The X-ray bursts observed in many of these systems leave no doubt that the compact objects in them are neutron stars (see Ch. 4).

There are three low-mass X-ray binaries for which there is strong evidence that the compact object is a black hole (see Ch. 3). All three are X-ray transients: they appeared as bright X-ray novae, with luminosities $L_x \sim 10^{38}$ erg s^{-1} for at least several weeks. At the same time, the optical luminosity of these systems brightened by 6 to 10 magnitudes, making them optical novae as well. After the decay of the X-ray and optical emission, the spectrum of a K or G star became visible, and

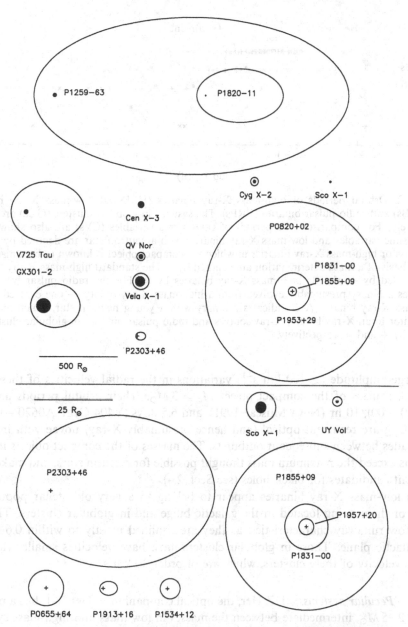

Fig. 11.1. Relative orbits of selected X-ray binaries and radio pulsar binaries, plotted to scale. High-mass binaries are shown on the left, low-mass binaries on the right. A system like the Be-star + PSR1259 − 63 binary, shown at the top, evolves into a high-mass X-ray binary like those shown middle left, which in turn can evolve into a high-mass radio pulsar binary like PSR2303 + 46. Some high-mass radio pulsar binaries are shown on an enlarged scale at bottom left. PSR1820 − 11 may have a low mass main sequence companion. If so, it evolves into a low-mass X-ray binary, like Cyg X-2 or Sco X-1, shown middle right, which in turn may evolve into a low-mass radio binary pulsar, like those also shown middle right. Two low-mass X-ray and three radio pulsar binaries are shown on an enlarged scale at bottom right.

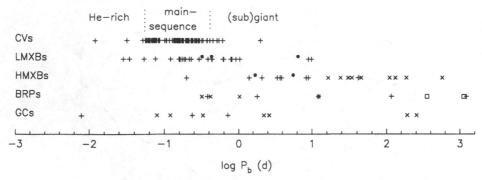

Fig. 11.2. Orbital periods of high-mass X-ray binaries (HMXBs), low-mass X-ray binaries (LMXBs) and radio pulsar binaries (BRPs). The systems in globular clusters (GCs) are shown separately. For comparison, the periods of cataclysmic variables (CVs) are also shown. The cataclysmic variables and low-mass X-ray binaries with a neutron star are denoted by +. The low-mass or high-mass X-ray binaries in which the compact object is known to be a black hole on the basis of a mass determination are denoted by •. The standard high-mass X-ray binaries are denoted by +, the Be high-mass X-ray binaries by ×. For the radio pulsar binaries, × indicates a binary presumably evolved from a high-mass X-ray binary, + one evolved from a low-mass X-ray binary. A □ indicates a binary with a young neutron star, which may be a progenitor to an X-ray binary. X-ray sources and radio pulsar binaries in globular clusters are shown by + and ×, respectively.

the large-amplitude ($\gtrsim 100$ km s^{-1}) variations in the radial velocities of these stars indicate a mass of the compact object $M_c \geq 3 M_\odot$. Their orbital periods are 8 hr (A0620 − 00), 10 hr (Nova Muscae 1991), and 6.5 days (V404 Cyg). A0620 − 00 and V404 Cyg are recurrent optical (and hence presumably X-ray) novae with intervals of decades between subsequent outbursts. The masses of the compact objects in these systems exceed the maximum mass thought possible for neutron stars, and make them excellent candidates for black holes (see Sect. 2.4).

The low-mass X-ray binaries appear to belong to a very old stellar population, many of them being located in the galactic bulge and in globular clusters. They do not show runaway characteristics, as they are confined mostly to within 0.6 kpc of the galactic plane. Those in globular clusters must have velocities smaller than the escape velocity of these clusters, which are of order 30 km s^{-1}.

Peculiar systems.   HZ Her, the optical companion to Her X-1, has a mass of about $2.35 M_\odot$, intermediate between the masses in low-mass and high-mass systems. The system is therefore a population I object, as the lifetime of a $2.35 M_\odot$ star is $\simeq 5 \times 10^8$ yr. The location of Her X-1, at 3 kpc from the galactic plane, is very unusual for a population I object, and shows that the binary was shot out of the plane with a velocity $\gtrsim 75$ km s^{-1} (Sutantyo 1975a). This velocity may well be the result of the supernova explosion that created the neutron star.

Three strongly radio-emitting, peculiar X-ray binaries, Cyg X-3, Cir X-1 and SS 433, might form a separate category (see Table 11.1). They are characterized by occasional strong radio outbursts with a synchrotron spectrum and with large IR luminosities (see Ch. 7). Cyg X-3 is unique in showing a nearly sinusoidal X-ray light curve, with a period of 4.8 hr. At a distance of 10 kpc, it is one of the brightest X-ray sources in the Galaxy (see Table 11.1). VLA and VLBI observations show that it has small radiojets. The companion of Cyg X-3 was recently shown to be a nitrogen-type Wolf–Rayet (i.e. helium) star (Van Kerkwijk *et al.* 1992). The infrared luminosity suggests a mass of the helium star between 5 and $10 M_\odot$. The period increases on a time scale of 6×10^5 yr, due most probably to the strong wind mass loss of the Wolf–Rayet star.

SS433 is an emission-line object, coinciding with the X-ray source A1909 + 04, and at the center of a radio synchrotron shell, W50. Optical spectroscopy and photometry indicate an orbital period of 13.2 days (Crampton, Cowley and Hutchings 1980). The optical luminosity is $\geq 10^4 L_\odot$. The mass transfer rate in SS433 is apparently highly super-Eddington, as witnessed by the disk brightness. This may be related to the formation of the jets (Shklovskii 1979). The emission-line spectrum shows the strong Balmer emission lines that are associated with Wolf–Rayet stars, and in addition satellite lines to the Balmer lines that have been interpreted as due to two jets directed in opposite directions from the compact object, at velocities $v \simeq 0.26c$. The jets have also been seen in the radio and in X-rays, and are sufficiently powerful to have produced W50, if they have existed for some 10^4 yr (Begelman *et al.* 1980, Van den Heuvel, Ostriker and Petterson 1980). The jets precess with a period of 164 days. Optical photometry indicates that the disk, which emits 80% of the optical flux, precesses also at this period (Cherepashchuk 1981). Nodding in the jet precession indicates a mass for the donor of $\sim 14 M_\odot$. For reviews of this system, see Margon (1984) and Vermeulen (1989).

The 3.49 s X-ray pulsar 1E2259 + 59 may belong to the same category (Fahlman and Gregory 1983; Iwasawa, Koyama and Halpern 1992). SS 433 and 1E2259 + 59 are both surrounded by large radio and X-ray shells, similar in appearance to supernova remnants, and both show beamlike structures extending from the source at the center of the shell.

11.1.2.2 Radio pulsar binaries

Like the binary X-ray sources, the radio pulsar binaries also fall into two broad groups with distinctly different characteristics, as follows (see Figure 11.1 and Ch. 5).

*The PSR*1913 + 16 *group.* These systems have, in general, narrow orbits, and companion masses in the range 0.8–1.45 $M_\odot$. Five of the six known systems of this type have very eccentric orbits. In these five systems, the companions to the radio pulsars are most probably neutron stars. In two systems, study of the general relativistic effects provides direct evidence for this (Taylor and Weisberg 1989; Wolszczan 1991). In the system with the circular orbit, PSR0655 + 64, the white-dwarf companion to the radio pulsar has been optically identified (Kulkarni 1986).

*The PSR*1953 + 29 *group.* These systems have, on average, much wider, circular orbits, and their companion masses tend to be low, between 0.02 and 0.45 $M_\odot$. All evidence, from optical identifications or the absence thereof (the distances can be estimated from the dispersion measure of the radio pulsars), indicates that the companions are white dwarfs. Possible exceptions are the eclipsing millisecond pulsars that appear to be evaporating their companions: PSR1957+20 and PSR1744−24A; in these binaries the companions are red dwarfs, with masses of only $0.02 M_\odot$ and about $0.1 M_\odot$, respectively.

11.2   Origin and evolution of high-mass X-ray binaries

The compact objects in X-ray binaries are remnants of relatively massive stars that evolved in a binary. Our understanding of the origin and evolution of X-ray binaries is thus based in good part on our understanding of the evolution of single massive stars. It is adversely affected by the appreciable uncertainty in our knowledge of the later evolutionary stages of such stars. Aspects specifically related to binaries include mass transfer between the stars, mass loss from the binary, and evolution of stars that have lost or gained an appreciable fraction of their mass.

In this section we briefly reiterate some aspects of single-star evolution, and apply them to the evolution of a high-mass binary that evolves into a high-mass X-ray binary. We describe the continued evolution of such a binary into a binary radio pulsar. Most of our discussion is kept relatively simple; some of the more intricate details are touched upon at the end.

In Sect. 11.3 we discuss the evolution of binaries with stars of initially very unequal mass that may form low-mass X-ray binaries. Such binaries in turn may become low-mass radio binary pulsars. In Sect. 11.4, we note that X-ray binaries in globular clusters most probably have followed very different evolutionary routes, and describe our at present rather rudimentary ideas about these.

11.2.1   *The evolution of a massive single star*

The evolution of a star is driven by a rather curious property of a self-gravitating gas in hydrostatic equilibrium, described by the virial theorem, namely that the loss of energy of such a gas causes it to contract and therewith to increase its temperature. The contraction of the primordial gas cloud of mainly hydrogen and helium causes an increase in pressure and temperature in the center until hydrogen starts fusing into helium, landing the star on the main sequence. Upon exhaustion of the hydrogen, the core contracts further, until pressure and temperature have risen enough for the fusion of helium into carbon and oxygen. Thus, cycles of nuclear burning alternate with stages of exhaustion of nuclear fuel in the stellar core. A massive star continues this cycle until its core is made of iron, at which point further fusion costs rather than produces energy. The core of such a star implodes to form a neutron star or a black hole. The gravitational energy released in this implosion is enough to explode the outer layers of the star.

In a less massive star, degenerate pressure in the core at the end of one nuclear burning stage may be high enough to prevent contraction to the density required for the ignition of the next fusion process. Extensive mass loss will erode such a star, until the degenerate core remains only. This core cools into a white dwarf. For the lowest initial stellar mass, this core is a pure helium white dwarf; somewhat more massive stars leave a carbon–oxygen white dwarf, and even more massive stars may leave white dwarfs consisting of neon–oxygen–magnesium.

Table 11.2. *End products of stellar evolution as a function of initial mass*

Initial mass	He core mass	Final product	
		Single star	Binary member
$< 3\,M_\odot$	$< 0.45\,M_\odot{}^a$	CO white dwarf	He white dwarf
3–$8\,M_\odot$	0.5–$1.9\,M_\odot$	CO white dwarf	CO white dwarf
8–$11\,M_\odot$	1.9–$2.9\,M_\odot$	neutron star	O–Ne–Mg white dwarf or neutron star^b
11–$40\,M_\odot$	2.9–$17.5\,M_\odot$	neutron star	neutron star
$> 40\,M_\odot$	$> 17.5\,M_\odot$	black hole	black hole

^{a} Before the helium flash.
^{b} Depending on the separation.

Each end-product of the evolution of a single star thus corresponds to a range of initial stellar masses. The possible end-products and the corresponding initial mass ranges are listed in Table 11.2. It should be noted that the actual values of the different mass ranges are only known approximately due to considerable uncertainty in our knowledge of the evolution of massive stars. Causes of this uncertainty include limited understanding of the mass loss undergone by stars in their various evolutionary stages, as well as fundamental problems in understanding convection, in particular in stars that consist of layers with very different chemical composition.

Another unsolved question is whether the velocity of convective gas cells may carry them beyond the boundary of the region in the star which is convective according to the Schwarzschild criterion. Different treatment of this so-called overshooting leads to different predictions for the evolution of a massive star. For example, inclusion of significant overshooting in the evolutionary calculations decreases the lower mass limit for neutron star progenitors to about $6\,M_\odot$.

A feature of stellar evolution that is of particular importance for the evolution of stars in a binary is the variation of the size of the star. During the phases of nuclear burning in the core, the star expands slowly and mildly. Each time that the nuclear fuel is exhausted, however, the outer layers of the star expand rapidly and strongly, only to shrink again at the onset of the next stage of nuclear burning. Two such cycles are illustrated in Figure 11.3, which shows the evolution of a star of (initially) $5\,M_\odot$ during hydrogen burning, exhaustion of hydrogen in the core, helium burning, and exhaustion of helium in the core. To illustrate the uncertainty in the details, we show results of different calculations. It is seen that the sizes of the stars and their lifetimes are markedly different for the two different calculations.

Several time scales of single-star evolution are also important for binary evolution (Morton 1960). The time spent by the star on the main sequence is given by the time scale on which the star burns its nuclear fuel during its stay on the main sequence. This nuclear time scale is set by the ratio of the total amount of nuclear energy available to the rate at which energy is lost. The energy available is the amount of mass available for fusion, typically of the order of one-tenth of the total mass of the star times the amount of energy ϵ released per unit of mass by the fusion process.

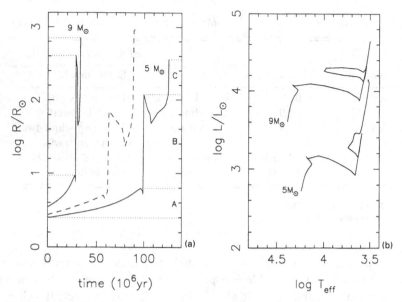

Fig. 11.3. (a) Evolution of the radius of stars of $5\,M_\odot$ and $9\,M_\odot$ until the onset of carbon burning. The solid lines give the results of calculations by Maeder and Meynet (1988), which include overshooting. For comparison, the dashed line gives radii from calculations by Paczyński (1970) for a $5\,M_\odot$ star that do not include effects of overshooting. The horizontal dotted lines delineate the radius ranges on the main-sequence and on first and second ascent of the giant branch. Cases of mass transfer while a stellar radius is in these ranges are referred to as case A, case B and case C, respectively, as indicated for the $5\,M_\odot$ track. (b) The evolutionary tracks in the Hertzsprung–Russell diagram.

The rate of energy loss is given by the stellar luminosity L. For hydrogen burning we have $\epsilon_{pp} \simeq 0.007c^2$, and thus

$$\tau_n \simeq \frac{0.1 M \epsilon_{pp}}{L} \simeq 10^{10} \frac{M}{M_\odot} \frac{L_\odot}{L} \text{ yr.} \qquad (11.1)$$

The expansion of the star upon exhaustion of the nuclear fuel in the core occurs on the thermal time scale. It is given by the ratio of thermal energy content of the star to the energy loss, i.e to the luminosity:

$$\tau_t \simeq \frac{GM^2}{RL} \simeq 3 \times 10^7 \left(\frac{M}{M_\odot}\right)^2 \frac{R_\odot}{R} \frac{L_\odot}{L} \text{ yr.} \qquad (11.2)$$

Rough estimates of these time scales for a star on the hydrogen-burning main sequence can be obtained with use of the empirical mass–radius and mass–luminosity relations for such stars, listed in Table 11.3.

For a star of $9\,M_\odot$, Eqs. (11.1), (11.2) and Table 11.3 give an estimated life-time on the hydrogen main sequence of 2×10^7 yr, and a thermal time scale on which the giant expands upon hydrogen exhaustion of 10^5 yr. Because of the uncertainty in the details of the evolution of a massive star, illustrated in Figure 11.3 by the difference between the results of two calculations for a $5\,M_\odot$ star, we may use the

Table 11.3. *Empirical mass–radius and mass–luminosity relations of stars on the hydrogen main sequence*

For data see Andersen (1991).

	$M \gtrsim M_\odot$	$M \lesssim M_\odot$
Mass–radius relation	$R/R_\odot \simeq (M/M_\odot)^{0.75}$	$R/R_\odot \simeq (M/M_\odot)$
Mass–luminosity relation	$L/L_\odot \simeq (M/M_\odot)^{3.8}$	$L/L_\odot \simeq (M/M_\odot)^3$

rough estimates provided by Eqs. (11.1) and (11.2) in our application of the evolution of single stars to the description of binary evolution.

11.2.2 Roche lobes and the onset of mass transfer in a binary

The potential surfaces in a binary are determined by the gravitational attraction of both stars and by the motion of the two stars around one another. In a frame co-rotating with the binary revolution, one may write the potential as

$$\Phi = -\frac{GM_1}{r_1} - \frac{GM_2}{r_2} - \frac{\omega^2 r_3^2}{2}, \tag{11.3}$$

where r_1 and r_2 are the distances to the center of the stars with mass M_1 and M_2, respectively; ω is the orbital angular velocity; and r_3 is the distance to the axis of rotation of the binary (see Figure 11.4). Consider first the case of stars that are small with respect to the distance between them, and revolving in circular orbits, i.e. $\omega = [G(M_1 + M_2)/a^3]^{1/2}$. Close to each star, the potential is dominated by the gravity of that star, and the equipotential surface is a sphere around the center of that star. Further out, the equipotential surfaces are deformed, and for a critical value of the potential the equipotential surfaces of the two stars touch, in the inner Lagrange point. Along the line connecting the two stars, the potential therefore reaches an extremum in the inner Lagrange point, i.e. the net force on a particle there is zero. The Roche lobe is surrounded by equipotential surfaces closed around both stars. At other critical values of the potential, first one of the two lobes, and then the other one, opens up, at the outer Lagrange points (see Figure 11.4).

Consider a star that is expanding. If the stellar surface reaches the Roche lobe, mass will start flowing out via the inner Lagrange point in the direction of the companion star. If the star expands so rapidly that its surface reaches an outer Lagrange point, mass may be lost from the binary. Some care should be taken in applying this latter conclusion to a real binary, because the Roche potentials lose their validity once mass stops co-rotating, as it is wont to do once it is removed from the stellar surface. Mass lost from the outer Lagrange point may well end up in a disk around the binary, rather than being lost to infinity (see, however, Flannery 1977).

By rewriting Eq. (11.3) in dimensionless units (mass in units of the total mass, and distances in units of the separation a between the stars), one easily shows that the form of the equipotential surfaces depends only on the mass ratio $q \equiv M_1/M_2$. The volume of the Roche lobe can be calculated numerically (for efficient algorithms, see

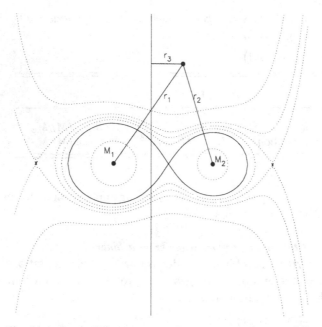

Fig. 11.4. Roche-lobe geometry for a mass ratio $q = 5/3$. The graph shows cuts through equipotential surfaces of the co-rotating frame, parallel to the rotation axis, which is indicated with the vertical solid line.

Mochnacki 1984). The average radius R_{L1} of the Roche lobe of the star with mass M_1 can be approximated to within 1% by

$$R_{L1} = \frac{0.49a}{0.6 + q^{-2/3}\ln(1 + q^{1/3})} \qquad (11.4)$$

for all values of q (Eggleton 1983). For $q \lesssim 0.8$, the following useful approximation is valid to within 2% (Paczyński 1967b)

$$R_{L1} = 0.46a \left(\frac{M_1}{M_1 + M_2}\right)^{1/3}. \qquad (11.5)$$

From the masses of the two binary stars and the orbital period, we can calculate a from Kepler's law and R_L for each star from Eq. (11.4) or Eq. (11.5). This gives us the size that the star must reach before mass transfer sets in. In a very close binary, mass transfer can be started while the donor is still on the main sequence. More often, however, mass transfer begins during the first or second ascent of the giant branch, which occurs upon central exhaustion of hydrogen and helium, respectively. For example, in a binary consisting of stars of $9 M_\odot$ and $5 M_\odot$, we learn from Figure 11.3 and Eq. (11.4) that the more massive star fills its Roche lobe already on the main sequence if the binary orbital period P_b is less than 3 days, on the first ascent of the giant branch for $3 \text{ days} \lesssim P_b \lesssim 900 \text{ days}$, and on the second ascent if $900 \text{ days} \lesssim P_b \lesssim 2000 \text{ days}$. These three cases are referred to as case A, B and C, respectively.

11.2.3   *Conservative mass transfer and non-conservative mass transfer*

Because massive stars evolve faster, according to Eq. (11.1), the more massive star in a binary expands first to fill its Roche lobe – provided that P_b is sufficiently short. To illustrate the effect of the ensuing mass transfer, we first assume that mass and angular momentum are conserved in the process, i.e. we assume *conservative* evolution of the binary. We assume for simplicity that the orbit is circular. By combining the assumptions of conservation of mass and of angular momentum,

$$M_1 + M_2 = \text{const.} \quad \text{and} \quad J_b = M_1 M_2 \left(\frac{Ga}{M_1 + M_2} \right)^{1/2} = \text{const.,} \quad (11.6)$$

with Kepler's law, one readily shows that the distance between the stars a and the binary period P_b change according to

$$\frac{a_f}{a_i} = \left(\frac{M_{1i} M_{2i}}{M_{1f} M_{2f}} \right)^2 \quad \text{and} \quad \frac{P_{bf}}{P_{bi}} = \left(\frac{M_{1i} M_{2i}}{M_{1f} M_{2f}} \right)^3, \quad (11.7)$$

where i and f index the initial and final values, respectively.

Because it is the more massive star that transfers mass, the orbital period decreases. Once the mass transfer has reversed the mass ratio, the orbital period may increase again. The final period is longer or shorter than the initial period if the final mass ratio is closer to unity or further away from it, respectively.

The radius of the mass donor is affected by its mass loss. Consider the case of most interest for the formation of X-ray binaries, in which the donor is a moderately massive star, transferring mass in case B. Upon losing some mass from its surface, the star first adjusts itself on a dynamical time scale, given by the ratio of the stellar radius and the average sound speed c_s in the star:

$$\tau_d \simeq \frac{R}{c_s} \simeq 0.04 \left(\frac{M_\odot}{M} \right)^{1/2} \left(\frac{R}{R_\odot} \right)^{3/2} \text{ days.} \quad (11.8)$$

This adjustment causes the star to shrink. As explained by Webbink (1985), this shrinking is the consequence of the rapid rise of the stellar entropy near the surface of a radiative star. The star subsequently tries to readjust to a new thermal equilibrium. The mass loss causes the ratio of core-mass to envelope-mass to increase, mimicking the situation in a more evolved star. Accordingly, the thermal equilibrium radius gradually increases with the mass loss. The consequence of this is that, once started, the mass transfer will continue on a thermal time scale until the mass ratio of the binary stars is reversed, so that the Roche radius of the donor increases again. In practice, mass transfer will continue until the donor has been almost denuded of its envelope (Paczyński 1967a). It will stop, once helium fusion in the core ignites, causing the radius of the star to shrink dramatically.

For a binary consisting of stars of $9\,M_\odot$ and $5\,M_\odot$ and with $P_b = 200\,\text{days}$, this first stage of the binary evolution is depicted in Figure 11.5, for the conservative scenario.

In reality, we know that the conservative scenario cannot hold. Stars, in particular massive ones, lose much mass in the form of a stellar wind. Thus, some mass will be lost from the system, taking angular momentum with it; if the specific angular momentum of the mass leaving the binary is that of the mass-losing star, the orbit

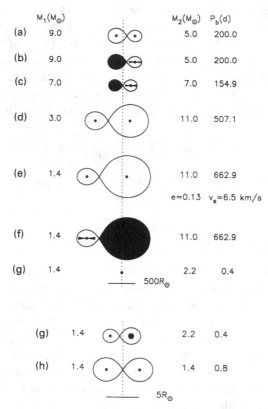

Fig. 11.5. The evolution of a wide high-mass binary into a high-mass X-ray binary, according to the conservative scenario, in which mass and angular momentum of the binary are conserved. The more massive star evolves through its main-sequence stage on a nuclear time scale (a), after which it expands (b). It then transfers mass on a thermal time scale to its companion, during which the orbit shrinks at first (b–c), and expands once the mass ratio is reversed (c–d). Upon completion of the mass transfer only the helium core of the initial primary is left (d). When this core explodes to leave a neutron star, the orbit becomes eccentric, and the system obtains a finite velocity (e). The neutron star can become an X-ray source by capturing mass from the stellar wind of its companion (e) or receiving mass via Roche overflow, once the companion has evolved into a giant (f). This latter stage lasts only a fraction of a thermal time scale, upon which the neutron star spirals into the envelope of its companion. At this point, evolution becomes highly non-conservative! If the envelope of the companion is expelled, the helium core forms a binary with the neutron star (g), which may evolve into a binary neutron star, with a highly eccentric orbit, if it survives the second supernova explosion (h). The helium core may also cool into a white dwarf, which orbits its neutron-star companion in a circular orbit.

widens somewhat. However, as long as the amount of mass lost is not too large, the binary evolution will qualitatively still be fairly similar to that depicted in Figure 11.5.

For a binary whose initial mass ratio is of order unity, conservative evolution will lead to wide final binaries, with long orbital periods, due to the more extreme mass ratio at the end of the mass transfer. Conservative, or almost conservative, evolution is therefore applicable to evolved binaries with fairly wide orbits. Evolved binaries with orbits of the order of a few days, however, cannot have evolved via conservative mass transfer. Such binaries must have lost a large amount of mass and of angular momentum in the course of their evolution. The mechanism that provides these losses is the spiral-in mechanism (Paczyński 1976).

We describe this by considering a binary with an initially fairly large mass ratio. The onset of mass transfer in this binary causes the distance between the two stars to decrease so rapidly (see Eq. (11.7)) that the expanding envelope of the mass donor engulfs the companion star. The orbital motion of the companion star is braked by its interaction with the expanding envelope, and as a result the orbit shrinks. The details of this highly complicated process, the correct description of which requires three-dimensional hydrodynamics beyond the capacity of current computers, are not understood (e.g. Soker and Livio 1989; Taam and Bodenheimer 1989). It is thought, however, that the process may end by the ejection of the envelope as soon as the binding energy released by the shrinking binary is large enough to do this. By equating the binding energy of the stellar envelope to the difference in binding energies of the binary at the onset and at the end of the spiral-in process, one may estimate the size of the final orbit. For a donor of mass M_1, with a core of mass M_c and an envelope of mass M_e and radius R_1 at the moment of first mass transfer, we thus find

$$\frac{GM_1 M_e}{\lambda R_1} = \alpha \left(\frac{GM_c M_2}{2a_f} - \frac{GM_1 M_2}{2a_i} \right), \tag{11.9}$$

where λ is a weighting factor for the gravitational binding of the envelope to the core, and a_i and a_f are the distances between the binary stars before and after the spiral-in, respectively. The factor α takes into account the efficiency with which the released energy is used in the expulsion of the envelope. This efficiency is, of necessity, smaller than unity, but by how much is not known. If the efficiency is too small, or if the binding energy of the envelope is too large (i.e. if the donor has not expanded much), spiral-in may continue all the way to a merger of both stars. Otherwise, a close binary may emerge from the spiral-in, in which the core of the donor revolves around its companion. The spiral-in phase is thought to proceed very rapidly, which prevents the accretion of much mass onto the companion of the donor. This companion is therefore expected to emerge virtually unchanged.

An illustration of a first stage of the evolution of a high-mass binary that involves moderate mass loss by stellar winds and a violent mass loss via the spiral-in process is given in Figure 11.6.

11.2.4 Supernova explosion in a binary

Both the conservative and the non-conservative scenarios described above lead to a binary with the core of an initially fairly massive star in orbit around a relatively massive companion. This core continues its evolution, not much affected by the absence of its erstwhile envelope. In this evolution, the radius of the star may increase again (see Figure 11.3), so that renewed mass transfer is possible. In a straightforward extension of the nomenclature discussed above, such renewed mass transfer is called case AB, BB, etc.

Provided the core is sufficiently massive, it will evolve into a supernova that leaves a neutron star. When it does so, mass is lost from the binary, causing the orbit to

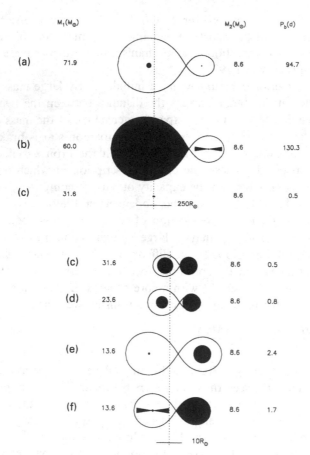

Fig. 11.6. The evolution of a high-mass binary into a high-mass X-ray binary, according to a non-conservative scenario, in which a large amount of mass and angular momentum are lost from the binary. The initial binary widens as mass is lost (a–b). Once the primary expands, it engulfs its companion, and a spiral-in follows (b–c). If a binary survives, the core of the initial primary evolves until its supernova explosion (d–e). The remaining compact object – here taken to be a black hole – becomes an X-ray source as it captures mass from a wind of its companion (e) or via Roche overflow, once the companion expands in its turn (f). (After Van den Heuvel and Habets 1984.) Continued evolution of this system probably leads to a binary consisting of a black hole and a neutron star.

change. In a simple estimate of the effect of this mass loss, one assumes the only change wrought by the instantaneous explosion is in the mass of the exploding star. For an orbit that is circular before the explosion, this implies that the periastron distance between the two stars in the post-explosion orbit is equal to the distance between the stars before the explosion, and that the relative velocity at that point is the velocity of the circular pre-explosion orbit. From this, one readily derives the eccentricity e of the new orbit, as well as the velocity v_s imparted by the explosion to the center of gravity (Boersma 1961; Dewey and Cordes 1987):

$$e = \frac{\Delta M}{M_1 + M_2 - \Delta M} \quad \text{and} \quad v_s = \frac{M_2 v_2 - (M_1 - \Delta M) v_1}{M_1 + M_2 - \Delta M} = e v_1, \quad (11.10)$$

where v_i is the orbital velocity of the star with mass M_i before the explosion.

Whether these equations really hold must be considered rather doubtful, as observations of radio pulsars show that young neutron stars tend to have large velocities. This suggests that a kick velocity of the order of 100 km s^{-1} or more is imparted to a neutron star as it is born, presumably due to a small asymmetry in the collapse of the iron core.

Nonetheless, Eqs. (11.10) show that a binary may be disrupted by a supernova explosion (i.e. $e > 1$), if more than half of the binary mass is lost. This is the reason why most scenarios for the formation of a neutron star in a binary involve mass transfer between the binary stars before the supernova event. After all, it is the more massive star that evolves and explodes first, and if it still has its initial mass at the moment of explosion, the mass lost in the supernova event exceeds half of the binary mass unless the initial mass ratio is very close to unity. If, however, the initially more massive star has lost most of its envelope before it explodes, the binary may remain bound, even in the event of a finite kick velocity (Van den Heuvel and Heise 1972).

In the conservative scenario, the initially more massive star has transferred its envelope to the companion. In the spiral-in scenario, the envelope mass has been lost from the binary. A third possibility would be the loss of most of the stellar envelope in the form of a stellar wind; in this case, no interaction between the binary stars is required to reverse the mass ratio. For example, in the calculations with overshooting by Maeder and Meynet (1988), a single star with initial mass of $9\,M_\odot$ loses all but $3.5\,M_\odot$ before exploding. It would therefore remain bound even with a rather low-mass companion.

The observations show that massive companions to neutron stars in relatively wide orbits are virtually always Be stars. Be stars are known to rotate rapidly. The rapid rotation of the companion to the neutron star is readily explained in a conservative or modestly non-conservative scenario, by the large amount of matter added with a high specific angular momentum to the initially least massive star. If the progenitor of the neutron star had lost all its mass to infinity rather than donated it to its companion, the companion would not be expected to rotate more rapidly than an ordinary single B star. Thus, the conservative scenario is a more promising description of the history of a wide binary with a massive star and a neutron star. The radio pulsar PSR1259 $-$ 63 is in a very eccentric orbit around a Be star companion (see Fig. 11.1). The high eccentricity of its orbit indicates that the neutron star was born with an appreciable kick velocity. We will consider this in more detail in Sect. 11.2.7.

The close binaries in which a massive star accompanies a neutron star must have undergone extensive loss of mass and angular momentum in their past, presumably in a spiral-in event.

11.2.5 *The high-mass X-ray binary*

The neutron star may show up as a bright X-ray source if it captures mass from its companion. As we will argue below, most high-mass X-ray binaries are wide systems, in which the neutron star captures mass from the stellar wind of its companion. Be stars have a bi-modal wind. In addition to spherical mass loss, these stars occasionally lose much matter in an equatorial wind, which possibly flows out in a disk geometry. Many X-ray binaries consist of a neutron star that accretes at a low level from the stellar wind of its Be star companion, and occasionally becomes very bright as an X-ray transient, when it captures mass from the transient wind disk. Examples of such systems are A0535 + 26 and V0332 + 53. A possible origin of such systems is depicted in Figure 11.5, where they would correspond to phase (e). The lifetime of these X-ray binaries is set by the main-sequence lifetime of the Be star, and can therefore be estimated for a given donor mass with Eq. (11.1).

The permanently bright X-ray binaries, which were amongst those first discovered, are mostly close binaries, in which the donor has evolved away from the main sequence. The supergiant emits a strong stellar wind, part of which is caught by the neutron star. In other systems, the massive star has expanded enough to fill its Roche lobe and transfers mass to the neutron star via the inner Lagrange point, in a phase of beginning Roche-lobe overflow (Savonije 1979, 1983). The latter is possible only if the orbital period is shorter than 3–4 days, such that the massive star is still in the core-hydrogen-burning phase when it begins to overflow its Roche lobe: such stars expand so slowly that during $\sim 10^5$ yr mass can be transferred at rates below the Eddington limit. Once the mass transfer rises much above this limit, the X-rays will probably be absorbed by the surplus of accreting matter, and the binary will not be a strong X-ray source. A possible origin of such systems is shown in Figure 11.6, where we have chosen an initially most massive star of sufficient mass to form a black hole. An example of such a system in which the donor is still smaller than the Roche lobe is Vela X-1; an example of a system in which the donor fills the Roche lobe is Cyg X-1. Because the donors in these bright systems have already left the main sequence, the lifetime of the X-ray phase is set roughly by the thermal time scale, given by Eq. (11.2).

11.2.6 *Continued evolution of a high-mass X-ray binary*

The massive donor in a high-mass X-ray binary will evolve to a sufficiently large radius to fill its Roche lobe, and in some cases has already done so. If the compact star is a neutron star, the mass ratio is extreme, and a spiral-in phase must be expected to follow soon after the onset of mass transfer. This time it is the neutron star that spirals into the envelope of its companion. The outcome of the spiral-in process may well be a merger, in particular in the short-period binaries, in which the donor will fill its Roche lobe already at small radius; see Eq. (11.9). In wider systems the donor may have lifted its envelope mass to a larger radius, allowing the neutron star to expel it. In that case, the core of the massive star will emerge in a binary with the neutron star (phase (g) in Figure 11.5). Recently it has be shown that Cyg X-3 may very well be an example of such a binary (Van Kerkwijk *et al.* 1992).

The helium core proceeds with its evolution, and may end as a relatively massive white dwarf. Due to the frictional processes during the spiral-in phase, this massive white dwarf is expected to revolve in a circular orbit around its neutron-star companion. An example of such a system is the binary that contains PSR0655 + 64. Alternatively, the evolved core may explode as a supernova in its turn and produce another neutron star. If the explosion unbinds the binary, it causes the addition of two neutron stars, one newly born and one slightly older, to the population of single neutron stars. If the binary survives, it will consist of two neutron stars. PSR1913+16 and PSR2303 + 46 are examples of such binaries (Smarr and Blandford 1976).

In LMC X-3, the black hole is more massive than the donor star. No spiral-in is expected in the continued evolution of this system. The binary is very likely to survive the second supernova explosion, after which it will consist of a black hole and a neutron star.

11.2.7 Kick velocities and other complications

In the global evolutionary picture as descibed above there are a number of loose ends that we address in this section, in particular: (i) the occurrence of more than one phase of mass transfer in the origin of high-mass X-ray binaries; (ii) the need of kick velocities imparted to the newly born neutron stars to explain the high eccentricities of Be X-ray binaries; and (iii) the nature of some peculiar systems.

11.2.7.1 More than one phase of mass transfer in the origin of Be X-ray binaries

Habets (1985) has computed the evolution of helium stars with masses between 2 and $4\,M_\odot$ (corresponding to initial main-sequence masses between 8.5 and $14\,M_\odot$) in the Hertzsprung–Russell diagram. He finds that helium stars with masses below $\sim 3.5\,M_\odot$ still undergo considerable radius expansion before they explode as a supernova. In a close binary this leads to a second phase of mass transfer, called case BB of mass transfer (Delgado and Thomas 1981; Law and Ritter 1983). The mass of the helium star is thus further reduced before the supernova explosion.

11.2.7.2 Evidence for kicks from Be X-ray binaries

The orbits of several Be X-ray binaries have appreciable eccentricities (see Table 11.1). In the absence of kick velocities, this would imply appreciable mass loss, according to Eq. (11.10). For example, a current donor star with a mass between 10 and $15\,M_\odot$, in an orbit with eccentricity $e = 0.3$, implies a mass loss in the explosion of between 3.4 and $5\,M_\odot$, which, added to a neutron star mass of $1.4\,M_\odot$, leads to helium stars between 4.8 and $6.4\,M_\odot$. Helium stars of this mass are Wolf–Rayet stars, i.e. they have prodigious mass-loss rates via stellar winds of $10^{-5}\,M_\odot\,\mathrm{yr}^{-1}$. There is, however, only one known Wolf–Rayet star with a companion of spectral type B. About half of all Wolf–Rayet stars known have O-star companions (Van der Hucht 1992). The absence of Wolf–Rayet + Be star binaries as suitable progenitor systems for the Be X-ray binaries indicates that we must look for other suitable progenitor systems. Such systems would be helium-burning stars with masses less than $\simeq 4\,M_\odot$ with Be star companions. Helium stars of such low mass have much smaller stellar winds, and do not become Wolf–Rayet stars. For such low masses the observed eccentricities $e \geq 0.3$ in the current orbit imply that the neutron star must have had a kick velocity at birth of order ≥ 50 km s^{-1}. In the case of PSR1259 − 63, the eccentricity $e = 0.87$ implies a kick velocity in excess of 200 km s^{-1}.

11.2.7.3 Origin of peculiar systems

The system Her X-1 has been the subject of some debate because it combines a neutron star with a strong magnetic field with a relatively old donor star (Lamb

1981). It has therefore been argued that the neutron star in this old binary was formed only recently, via accretion-induced collapse of a white dwarf (Taam and Van den Heuvel 1986). The problem with this scenario is that accretion-induced collapse does not provide a system velocity that is required to explain the current distance of the binary to the galactic plane; see Eq. (11.10). A kick velocity imparted to the neutron star does not solve this, as sufficiently large kick velocities will disrupt the binary. One must conclude therefore that the neutron star was formed from the initially most massive star, and that it has kept its field in the $\sim 5 \times 10^8$ yr that the companion spent as a main-sequence star, before it expanded into a giant (Verbunt, Wijers and Burm 1990).

The system Cyg X-3 is precisely the post-common-envelope system that one expects to result after the spiral-in evolution of a high-mass X-ray binary (Van den Heuvel and De Loore 1973), i.e. phase (g) in Figure 11.5. It thus is a perfect progenitor system for a binary like PSR1913 + 16 (Flannery and Van den Heuvel 1975).

The system SS433 is similar in also having a massive evolved donor for the compact star, but different in having a much wider orbit. It is thus also a high-mass X-ray binary which has evolved beyond the initial stage. Somehow, it has hitherto avoided the onset of the spiral-in stage. To see whether we can understand how, we first note that the mass transfer rate is much higher than the Eddington limit, which suggests that virtually all the mass that is transferred subsequently leaves the system. In terms of Eq. (11.22) below, this means $\beta \simeq 0$. If we assume that the mass leaves the binary with the specific angular momentum of the compact star, we may further write $\alpha = (M_2/M_1)^2$. The discussion following Eq. (11.22) shows that the orbit widens if $\alpha < 1 + M_2/(2M_1)$. In the present case, this implies $M_2/M_1 \lesssim 1.3$. If the accreting star is a $1.4 M_\odot$ neutron star, a donor mass of $14 M_\odot$ therefore implies that spiral-in is unavoidable. If, however, the accreting star is a relatively massive black hole, spiral-in could be avoided.

11.3   Origin and evolution of low-mass X-ray binaries

As noted in Sect. 11.1, the nature of the low-mass X-ray binaries is rather less secure than that of the high-mass X-ray binaries. The donor of the low-mass X-ray binary only rarely shows a clear signature in the observation, and, as a result, a system like Cyg X-3, now thought to belong in the category of high-mass systems (see Sect. 11.1.2.1 and 11.2.7.3 above), has been long classified as a low-mass X-ray binary.

The assumption that the companion to many low-mass X-ray binaries is a low-mass star on or close to the main sequence derives mainly from the comparison with cataclysmic variables with similar orbital periods (see Figure 11.2). This comparison further tells us that low-mass X-ray binaries with $P_b \lesssim 80$ min probably contain helium-rich donors, and those with $P_b \gtrsim 0.5$ days contain subgiants or giants.

In view of our discussion of the supernova explosion in a binary, it will be clear that the main problem in understanding the origin of a binary with a low-mass companion to a neutron star is to explain how the binary could remain bound in the explosion. Three basic solutions have been suggested. The first one postulates a spiral-in scenario, in which the binary lost most of its initial mass and angular momentum. The second one postulates a relatively quiet supernova explosion, resulting from a collapse of a white dwarf induced by accretion of mass; this scenario also requires a spiral-in phase in the previous evolution. The third scenario postulates that the neutron star was born outside the binary in which it now resides. The discussion of this latter scenario, thought to apply to the X-ray sources in globular clusters, is deferred to Sect. 11.4.

11.3.1 Origin of low-mass X-ray binaries

As the progenitor of a neutron star or black hole must have been a fairly massive star, the progenitor binary of a binary with a neutron star and a low-mass companion must have had an extreme mass ratio. We therefore consider the evolution of such a binary, first suggested by Van den Heuvel (1983). The massive star evolves first, and because of the extreme mass ratio will engulf its companion soon after the onset of mass transfer Eq. (11.7). Because the mass of this companion is low, the energy released as its orbit shrinks is not enough to expel the envelope of the massive star, unless this envelope is very extended; see Eq. (11.9). This will be the case only in extremely wide binaries. Consider, for example, a star of $9\,M_\odot$, with a core of $2.5\,M_\odot$ and a low-mass companion of $1\,M_\odot$. If mass transfer is case C, Figure 11.3 shows that $R_1 \simeq 600\,R_\odot$, which for $M_1/M_2 = 9$ implies $a_i \simeq 1000\,R_\odot$. According to Eq. (11.9) the final orbit will have $a_f = 6\,R_\odot$ if $\alpha = 1$.

The core continues its evolution, and may turn into a supernova. In order for the binary to survive the supernova explosion, this core should not be too massive; see Eq. (11.10). To some extent, the mass loss during the supernova explosion can be offset by a kick velocity of the neutron star that is fortuitously in the right direction.

With minor variations, we can now envisage scenarios for the formation of a variety of low-mass X-ray binaries (Verbunt 1988a). A 12 hr orbital period is short enough for loss of angular momentum to bring a $1\,M_\odot$ companion to the neutron star into contact with its Roche lobe within the Hubble time. As mass transfer starts, the binary turns into a low-mass X-ray binary with a main-sequence donor. If the orbit is somewhat wider, loss of angular momentum may not suffice to bring the prospective donor into contact, and mass transfer will start only when the donor evolves away from the main sequence, and expands into a giant. At which value of the orbital period loss of angular momentum is no longer effective in shrinking it is not well known because the loss processes are not known in any detail. Another interesting possibility arises if the companion to the neutron star progenitor was slightly more massive, as in Her X-1. When such a star starts filling its Roche lobe, a second spiral-in may ensue, leaving the neutron star in orbit around the core of an evolved star. If the emerging binary is very close, this core may come into contact, and start transferring mass to, the neutron star. For an inert core, only loss of angular momentum may produce this contact, and the donor star will be a white dwarf. A slightly more massive core may burn helium as it comes into contact.

One of the more crucial phases in these scenarios is the supernova explosion, which may easily dissolve the binary. One may take the view that low-mass X-ray binaries are rare, and therefore accept a low probability of survival of the binary. However, some researchers have proposed alternative solutions to avoid this crucial phase.

One of these uses the known fact that many stars are in multiple systems. Suppose, for example, that a system like Cen X-3 would have a low-mass companion in a

Table 11.4. *Mass–radius relations and derived mass–orbital-period relations for low-mass X-ray binaries*

Main sequence	$R_2/R_\odot = M_2/M_\odot$	$P_b = 8.9\,\mathrm{hr}\,M_2/M_\odot$
He main sequence	$R_2/R_\odot = 0.2\,M_2/M_\odot$	$P_b = 0.89\,\mathrm{hr}\,M_2/M_\odot$
White dwarf	$R_2/R_\odot = 0.0115\,(M_2/M_\odot)^{-1/3}$	$P_b = 40\,\mathrm{s}\,M_\odot/M_2$

wide orbit around the high-mass binary. The neutron star is bound to be engulfed by its massive O-star companion, and the short orbital period ensures that a merger will result. This merger might become a quasi-stable star, whose nucleus produces energy by accretion onto the neutron star, rather than by nuclear fusion, but whose outer layers expand into a big giant. Such a giant is called a Thorne–Zytkow object (Thorne and Zytkow 1977; Biehle 1991; Cannon *et al.* 1992). If this giant in its turn engulfs the low-mass star, the latter may spiral in towards the vicinity of the neutron star. In this ingenious scenario, the low-mass star is shielded from the impact of the supernova by the massive companion to the exploding star (Eggleton and Verbunt 1986).

Another suggestion starts again from a high-mass binary of extreme mass ratio, but now with an initially most massive star that evolves into a relatively massive white dwarf. The spiral-in process brings the low-mass star close enough that at some point, be it via loss of angular momentum or via evolutionary expansion of the donor, mass transfer to the white dwarf starts. Under some conditions, the white dwarf may accrete enough matter for its mass to exceed the Chandrasekhar limit to the maximum possible mass for a white dwarf. At that point the white dwarf implodes, presumably without losing much mass, and is therefore possibly less disruptive to the binary system (Whelan and Iben 1973; Canal and Schatzman 1976; Canal, Isern and Labay 1990).

How can we describe the evolution of the X-ray binaries after they have been formed? Once a low-mass companion to a neutron star fills its Roche lobe, the binary period is uniquely related to its average density, as can be seen when one combines Eq. (11.5) with Kepler's third law, to find

$$P_b \simeq 8.9\,\mathrm{hr} \left(\frac{R_2}{R_\odot}\right)^{3/2} \left(\frac{M_\odot}{M_2}\right)^{1/2}. \qquad (11.11)$$

Table 11.4 summarizes how the mass–radius relations may be used to derive the relation between donor mass and binary period for stars on the hydrogen or helium main sequences, and for white dwarfs. It is seen that a main-sequence star less massive than the sun fills its Roche lobe at binary periods less than about 9 hr. A low-mass binary with a period longer than this must have a star whose radius is larger than the main-sequence radius, for example a subgiant or giant. Ultra-short periods, less than 80 min, suggest a helium-rich donor star, i.e. a helium-burning star or a white dwarf.

The structure of low-mass X-ray binaries is similar to that of cataclysmic variables, in which a white dwarf accretes matter from a low-mass companion (see Ch. 8). Our description of the evolution therefore starts by following ideas originally developed for cataclysmic variables. We assume for the moment that all mass lost by the donor is accreted by the compact star, i.e. $\dot{M}_1 = -\dot{M}_2$. We may then write the change in orbital angular momentum as (see Eq. (11.6))

$$\frac{\dot{J}_b}{J_b} = \frac{1}{2}\frac{\dot{a}}{a} + \left(1 - \frac{M_2}{M_1}\right)\frac{\dot{M}_2}{M_2}. \qquad (11.12)$$

Conservation of angular momentum ($\dot{J}_b = 0$) therefore drives the two stars apart when mass is transferred from the least massive to the more massive star. With the expansion of the orbit, the Roche lobe of the donor expands also; see Eq. (11.5). If mass transfer is to continue, some mechanism therefore has to remove angular momentum from the orbit, or the donor star must expand. We discuss these possibilities in turn.

11.3.2 Evolution of a low-mass X-ray binary via loss of angular momentum

Stable mass transfer via the Roche lobe implies that the donor star fills its Roche lobe, $R_2 = R_{L2}$, and continues to do so, $\dot{R}_2 = \dot{R}_{L2}$. With Eq. (11.5) we write this condition as

$$\frac{\dot{R}_{L2}}{R_{L2}} = \frac{\dot{a}}{a} + \frac{1}{3}\frac{\dot{M}_2}{M_2} = \frac{\dot{R}_2}{R_2} = n\frac{\dot{M}_2}{M_2}, \qquad (11.13)$$

where we use a generic mass–radius relation $R_2 \propto M_2^n$. We use this equation to eliminate a from Eq. (11.12):

$$\frac{\dot{J}}{J} = \frac{\dot{M}_2}{M_2}\left(\frac{5}{6} + \frac{n}{2} - \frac{M_2}{M_1}\right). \qquad (11.14)$$

This shows that loss of angular momentum may drive stable mass transfer at $\dot{M}_2/M_2 \sim \dot{J}_b/J_b$, provided that $M_2 < (n/2 + 5/6)M_1$. A first suggestion for such a mechanism was suggested by Kraft, Mathews and Greenstein (1962) who noted that two stars revolving around one another lose angular momentum via gravitational waves according to

$$-\left(\frac{\dot{J}}{J}\right)_{GR} = \frac{32G^3}{5c^5}\frac{M_1 M_2(M_1 + M_2)}{a^4}. \qquad (11.15)$$

Eqs. (11.14) and (11.15) can be solved together to calculate the evolution of a binary. In Figure 11.7 we show the orbital period, according to Table 11.4, and the rate of mass transfer, according to Eqs. (11.14) and (11.15), for different assumed donor stars. For a main-sequence donor, the mass transfer rate is $\sim 10^{-10}\,M_\odot\,\mathrm{yr}^{-1}$, almost independent of the donor mass. As mass is transferred, the mass of the donor decreases, and with it the radius of the donor and the binary period (Faulkner 1971). For a white-dwarf donor, on the other hand, the mass transfer rate is a very strong function of the donor mass, and the stellar radius and orbital period increase as mass is transferred (Paczyński 1967b).

Quite a few low-mass X-ray binaries have X-ray luminosities that imply accretion rates in excess of $10^{-10}\,M_\odot\,\mathrm{yr}^{-1}$. This has led to the suggestion of additional mechanisms for loss of angular momentum from the binary, to boost the mass transfer. One such mechanism is magnetic braking. This process is thought to be responsible

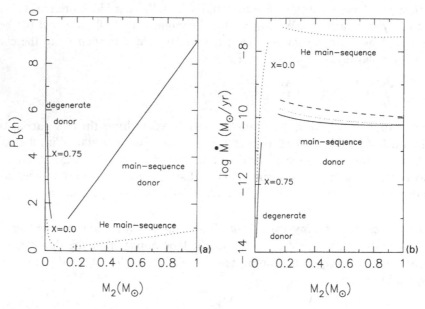

Fig. 11.7. (a) Orbital periods of low-mass X-ray binaries and (b) predicted mass-transfer rates for loss of angular momentum driven by gravitational radiation only, as a function of donor mass. Orbital periods according to Table 11.4 for degenerate donors with hydrogen fraction $X = 0.75$ (solid line) and $X = 0.0$ (dotted line) at donor masses $M_2 < 0.1\,M_\odot$, and for main-sequence (solid line) and helium-main-sequence (dotted line) donor stars at masses $M_2 > 0.15\,M_\odot$. (b) Mass transfer rates are shown with corresponding curves, i.e. solid for $X = 0.75$ degenerate donors and for main-sequence donors, dashed for $X = 0.0$ degenerate donors and helium-main-sequence donors. These curves assume that no mass is lost from the binary, and a mass of the accreting object $M_1 = 1.4\,M_\odot$, appropriate for a neutron star. For a main-sequence donor, two additional curves are shown: the dashed curve assumes no mass loss, and an accreting object of $M_1 = 7\,M_\odot$, appropriate for a black hole; the dotted curve assumes $M_1 = 1.4\,M_\odot$ and that half of the mass lost by the donor leaves the system with the angular momentum of the donor, i.e. $\beta = 0.5, \alpha = 1$.

for the loss of angular momentum of late-type single main-sequence stars, which is observed as a decrease of the rotational velocity with age. Late-type stars have strong magnetic fields, especially if they rotate rapidly, which force the stellar wind to co-rotate out to several stellar radii. As a result, even a modest amount of mass loss can correspond to an appreciable loss of angular momentum. Rough estimates suggest that this mechanism can indeed boost the transfer of mass in a low-mass binary (Verbunt and Zwaan 1981): in such a binary, tidal forces keep the donor star in co-rotation with the orbital revolution, thereby causing the stellar loss of angular momentum to become a drain of the orbital angular momentum. Unfortunately, the stellar wind as well as the structure of the magnetic field of late-type stars are not known observationally. Scaling laws of the loss of angular momentum with mass and rotational velocity of the star may be devised (e.g. Verbunt 1984; Mestel and Spruit 1987), but cannot be directly compared with observation. As a result, the actual importance of magnetic braking as a driving mechanism for the mass tranfer in a low-mass binary remains highly uncertain.

The observed distribution of orbital periods of cataclysmic variables (see Figure 11.2) shows two remarkable features: there is a dearth of systems at binary periods between 2 and 3 hr, and few systems have periods shorter than 80 min. These features are usually referred to as the period gap and the period minimum. Although mass transfer rates in cataclysmic variables are notoriously difficult to determine (e.g. Wade 1988), it has been suggested that the rates below the period gap are lower than above it (e.g. Patterson 1984). This suggests that there may be a mechanism for the loss of angular momentum that operates above, but not below, the period gap. Sudden switch-off of this mechanism at an orbital period of 3 hr can explain the period gap as follows (Rappaport, Verbunt and Joss 1983; Spruit and Ritter 1983). As the donor mass decreases in the course of the binary evolution, its thermal time scale increases according to Eq. (11.2). Once the thermal time scale exceeds the time scale for mass transfer, $\tau_{\dot{M}} \equiv M_2/\dot{M}$, the donor star is driven out of thermal equilibrium, and becomes larger than a main-sequence star of the same mass. Suppose now that, at some point, the mass transfer decreases, due to the fact that one mechanism for the loss of angular momentum switches off. This causes $\tau_{\dot{M}}$ to increase, and allows the donor to shrink somewhat closer to its main-sequence radius. As the donor shrinks within its Roche lobe, mass transfer stops altogether, allowing the star to shrink onto its main-sequence radius. The remaining mechanism for loss of angular momentum, gravitational radiation, continues to drive the two stars together, until at a shorter binary period the donor fills its Roche lobe once more, and mass transfer resumes. This explanation for the period gap implies that the stars just above the period gap have radii significantly in excess of the main-sequence stars of the same mass. Observational evidence for this prediction is ambiguous (Verbunt 1984).

As the main-sequence star loses more and more mass, it reaches a point where its central temperature and pressure drop enough for the stellar core to become degenerate. Further loss of mass now causes the star to expand again, leading to an increase in orbital period. As first pointed out by Faulkner (1971), the low-mass binary evolution thus passes through a minimum binary period. This minimum period may be identified with the observed minimum period of 80 min (Paczyński and Sienkewicz 1981).

As the nature of the mass-receiving star did not enter into the above considerations, the above arguments may be applied to low-mass X-ray binaries as well as to cataclysmic variables. What about the systems with binary periods less than 80 min? Three possible explanations for these systems have been put forward. It has been noted in Sect. 11.3.1 that a second spiral-in phase may occur, after which a neutron star is left with the core of a giant as its companion. If this core is still burning helium once loss of angular momentum brings it into contact with its Roche lobe, the orbital period may be small. Similar considerations as in the case of a main-sequence donor lead to a predicted period of about 10 min, at which the core of the donor becomes degenerate, and the predicted minimum period for helium burning donor stars is accordingly about 10 min (Savonije, De Kool and Van den Heuvel 1986). If the core of the giant has cooled into a white dwarf before coming into contact, orbital periods as short as minutes are possible (Table 11.4, Figure 11.7). A third, intermediate possibility has been suggested by Pylyser and Savonije (1988), who study a binary that comes into contact just after the initial expansion onto the subgiant branch of the donor star. The ensuing mass transfer stalls the stellar evolution of the donor, and further evolution of the binary is driven by loss of angular momentum. The donor star has a helium-enriched core, and as a result may be whittled down to lower mass before it becomes degenerate. The minimum period for such a system may be on the order of 40–50 min.

11.3.3 *Evolution of a low-mass X-ray binary via donor expansion*

A wide binary has a large angular momentum, and will be little affected by loss of some angular momentum. In such a binary, mass transfer is more easily driven by expansion of the donor star. Using once more the condition that the donor fills its Roche lobe and continues to do so, we may use the first three members of Eq. (11.13) to eliminate a from Eq. (11.12), so as to obtain

$$\frac{\dot{R}_2}{R_2} = -2\frac{\dot{M}_2}{M_2}\left(\frac{5}{6} - \frac{M_2}{M_1}\right). \tag{11.16}$$

The transfer of mass from the less massive to the more massive star causes the orbit to widen; mass is transferred until the orbit has expanded enough to make the expanded donor fit its Roche lobe. Hence, the rate of mass transfer is set by the rate of expansion of the donor star. This rate may be obtained from a full stellar evolution model, and used to calculate the binary evolution.

As noted by Webbink, Rappaport and Savonije (1983), a simplification is possible, thanks to the fact that the properties of the (sub)giant depend mainly on the mass of the core, and much less on the mass of the envelope. With $y \equiv \ln M_c/0.25\, M_\odot$, the radius and luminosity of a low-mass giant can be fitted to polynomials

$$\ln(R_2/R_\odot) = a_0 + a_1 y + a_2 y^2 + a_3 y^3 \tag{11.17}$$

$$\ln(L_2/L_\odot) = b_0 + b_1 y + b_2 y^2 + b_3 y^3 \tag{11.18}$$

with constants as given in Table 11.5, whereas the change in core mass is related to the luminosity as

$$\dot{M}_c \simeq 1.37 \times 10^{-11}\left(\frac{L}{L_\odot}\right) M_\odot\, \mathrm{yr}^{-1}. \tag{11.19}$$

This relation arises because the luminosity on the giant branch is almost completely due to hydrogen shell burning, and because the newly produced helium adds to the core mass. If we now take the time derivative of Eq. (11.17):

$$\frac{\dot{R}_2}{R_2} = [a_1 + 2a_2 y + 3a_3 y^2]\frac{\dot{M}_c}{M_c}, \tag{11.20}$$

we have all ingredients required to calculate the evolution.

In a binary consisting of a compact star and a companion, the binary period and component masses determine at which radius the expanding (sub)giant fills its Roche lobe. This radius sets the core mass, via Eq. (11.17), which in turn sets the rate of radius expansion, via Eqs. (11.18)–(11.20), and thus the mass transfer rate, via Eq. (11.16). Figure 11.8 shows the results of calculations made with this simplified

Table 11.5. *Constants for the fits to the core-mass–radius and core-mass–luminosity relations for low-mass giants, according to Webbink, Rappaport and Savonije (1983)*

For $Z = 0.02$, the constants can be used for core masses $0.16 < M_c/M_\odot < 0.45$; for $z = 0.0001$, the constants can be used for core masses $0.20 < M_c/M_\odot < 0.37$.

	a_0	a_1	a_2	a_3	b_0	b_1	b_2	b_3
$Z = 0.02$	2.53	5.10	−0.05	−1.71	3.50	8.11	−0.61	−2.13
$Z = 0.0001$	2.02	2.94	2.39	−3.89	3.27	5.15	4.03	−7.06

prescription. The main results, which tally well with those obtained with more complete calculations, are as follows. The mass transfer rate is highest in the systems with the longest binary periods. The reason for this is that large giants expand faster than small giants. Most of the envelope mass of the donor is transferred to the compact star; the remainder is added to the degenerate core of the giant. This core increases its mass by $\sim 0.05\,M_\odot$ during the mass transfer stage. The mass transfer leads to a dramatic increase in the binary period, which may be calculated from Eq. (11.7). Even if the binary was originally eccentric, we expect the tidal forces exerted by the Roche-lobe filling donor to be large enough to circularize the orbit very rapidly.

11.3.4 *Origin of binary and single radio pulsars from low-mass X-ray binaries*

The mass transfer stops when the mass of the envelope drops below $\sim 0.01\,M_\odot$, at which point the envelope collapses. The binary then consists of the degenerate core of the giant, which cools into a white dwarf, and a compact star. If the compact star is a neutron star, the accretion will have spun it up, and after the disruption of the mass transfer it may show up as a radio pulsar. A number of radio pulsars with low-mass white-dwarf companions have indeed been found, and their orbits are nearly circular, as expected. We thus see that the evolution of wide low-mass X-ray binaries results in a straightforward fashion in the production of radio pulsar binaries of the PSR1953 + 29 class, described in Sect. 11.1.2.2 (see Ch. 5). In view of the long duration of the mass transfer stage, enough mass can be transferred to bring a neutron star with weak magnetic field, $B \sim 10^9$ G, say, into rapid rotation. Transfer of only $\sim 0.01\,M_\odot$ already may lead to periods of the order of a millisecond (see Sect. 10.9).

Since the final radius of the giant is related to its final core mass, and the latter is close to the initial core mass, we expect that there is a correlation between binary period and mass of the white dwarf in the binary that contains the radio pulsar. Joss, Rappaport and Lewis (1987) write this as

$$P_b \simeq 8.4 \times 10^4 \, \text{days} \left(\frac{M_{wd}}{M_\odot} \right)^{11/2}. \tag{11.21}$$

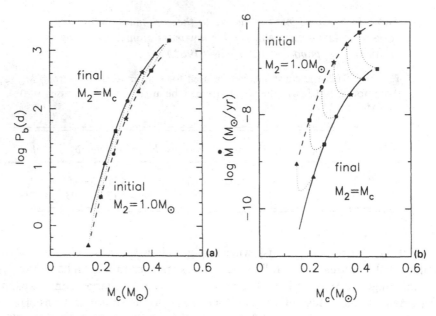

Fig. 11.8. (a) Orbital period and (b) mass-transfer rate as a function of the mass M_c of the core of the expanding donor star. A neutron star of mass $M_1 = 1.4\,M_\odot$ and an initial donor mass of $1\,M_\odot$ are assumed. Dashed and solid curves show the values at the onset and end of mass transfer, respectively. For six different initial core masses, evolutionary tracks are shown as dotted lines.

This equation is valid for $P_b \gtrsim 2\,\text{days}$, and for circular orbits, for stars with normal abundances. In globular clusters, the stars may have much lower metallicities; according to stellar-evolution calculations, they are smaller for the same core mass, and therefore have larger core mass while filling the Roche lobe at the same binary period.

The first discovered radio pulsar with a millisecond period, PSR1937+21, is single. It has nonetheless been suggested that this pulsar also found its origin in a binary, and that the mass donor has been completely destroyed. Such complete destruction can follow highly unstable initial mass transfer, which will arise if a relatively massive white dwarf starts transferring mass to a neutron star (Bonsema and Van den Heuvel 1985; see also Verbunt and Rappaport 1988). If the mass of the destroyed star forms a high-mass accretion disk around the neutron star, and if enough matter of this disk is accreted by the neutron star, a rapidly rotating pulsar may be left. This scenario has attracted attention once more following the discovery of two planets around a radio pulsar (Wolszczan and Frail 1992), which may have condensed out of such a high-mass disk (Van den Heuvel 1992, 1993; Phinney and Hansen 1993).

Another scenario for the formation of a single millisecond pulsar is attrition of the mass donor, either by X-rays during the mass transfer phase, or by particle and electromagnetic radiation from the radio pulsar that switches on when mass transfer stops. This scenario gained in popularity following the discovery of a millisecond pulsar, PSR1957+20, that appears to be evaporating its low-mass companion (Fruchter,

Stinebring and Taylor 1988). The binary period of 9.17 hr is decreasing at an extremely rapid rate, on a time scale of $\sim 10^7$ yr. It is not clear how this time scale is connected to the evaporation time scale, which appears to be much longer; an interesting possibility is that the evaporated matter forms a disk surrounding the binary, the disk then draining angular momentum from the binary (Ryba and Taylor 1991; Banit and Shaham 1992; Tavani and Brookshaw 1992).

11.3.5 *Effects of irradiation of the donor*

A number of models have investigated acceleration of the binary evolution through irradiation of the donor star by the compact object. All of these models start from the notion that the compact star is a source of power which is huge compared to the puny luminosity of the donor star in low-mass X-ray binaries. If the donor therefore intercepts even a small fraction of this power, its structure, and with it the binary evolution, may be dramatically altered. During accretion, the neutron star or black hole is a source of X-rays. When accretion has switched off, the rotational energy of a neutron star can be emitted in the form of electromagnetic or particle radiation. In more exotic models, an accreting neutron star is also a source of high-energy gamma rays.

Interception of radiation by the donor may cause it to expand and accelerate mass transfer. It may also evaporate mass from the donor surface. If this matter leaves the binary with sufficient angular momentum, this will also accelerate mass transfer.

To calculate these effects, we have to know what type of radiation is emitted by the compact object, and how the donor absorbs it. We know that much of the accretion energy is emitted in the form of X-rays, and from the optical light curves of a number of low-mass X-ray binaries, which show that the side of the donor facing the compact object is heated, we also know that some fraction of the X-rays is intercepted (see Sect. 2.3.4). Early studies of the irradiation and its effect on the mass transfer were made in relation to Her X-1, the first clear observational example of heating by X-rays (Alme and Wilson 1974; Basko, Sunyaev and Titarchuk 1974). More recently, Podsiadlowski (1991), Ergma and Fedorova (1991), and Harpaz and Rappaport (1991) have calculated the effect of irradiation of a low-mass companion to a neutron star. They find that irradiation of a low-mass main-sequence star suppresses convection in it, and brings the originally convective zones into radiative equilibrium. During the change, the star expands, and mass transfer is enhanced. Once the change is effected, however, the star settles in a new equilibrium and does not expand anymore. Mass transfer is then driven only by whatever other causes are present, such as loss of angular momentum. If the irradiated donor is a giant, no such equilibrium is reached, and mass transfer is accelerated until the whole giant's envelope is transferred to the compact star (Podsiadlowski 1991).

Ruderman *et al.* (1989a,b) have calculated the effect of irradiation by various forms of radiation, such as very-high-energy gamma rays and electron–positron pairs. A problem that arises here is that the fluxes of various forms of radiation emitted by neutron stars are not known from observation, and have to be predicted from theoretical models. Ruderman *et al.* argue that interception of these fluxes may boost the mass transfer to the Eddington limit, and keep it there, in particular for very-low-mass donors. This limitation to very-low-mass donors limits the importance of these effects, as the majority of low-mass X-ray binaries have more massive donors.

Models for the interception of pulsar radiation have of course been guided by the observations of PSR1957 + 20, the pulsar that evaporates its companion. In contrast to the expectations raised by more sanguine model-makers, no detection of gamma rays from this pulsar has been announced by the GRO observers so far; the X-ray flux – which may find its origin in reprocessed high-energy gamma rays – is also rather low (Kulkarni *et al.* 1992). A more fruitful approach may therefore be to determine the amount of energy intercepted from the properties of the matter surrounding the pulsar in the form of an Hα emission cloud. The pulse-period derivative, and thus the loss of rotational energy, of PSR1957 + 20 is low, and would lead to an evaporation time scale in excess of 10^8 yr (Phinney *et al.* 1988). The decrease of the orbital period on a time scale of 10^7 yr is much more rapid, and must have its cause in other effects, as discussed in Sect. 11.3.4.

If mass is lost from the binary, the binary orbit may expand or shrink, depending on the amount of angular momentum carried by the lost matter. To see this, we assume that a fraction β of the mass lost by the donor is accreted by the primary, i.e. $\dot{M}_1 = -\beta\dot{M}_2$, whereas the rest leaves the system, carrying a specific angular momentum of α times the specific angular momentum of the donor star. If the loss of angular momentum not related to this mass loss is denoted $\dot{J}$, we may replace Eq. (11.12) by

$$\frac{\dot{a}}{a} = 2\frac{\dot{J}}{J} - 2\frac{\dot{M}_2}{M_2}\left(1 - \frac{\beta M_2}{M_1} - \frac{(1-\beta)M_2}{2(M_1 + M_2)} - \alpha(1-\beta)\frac{M_1}{M_1 + M_2}\right). \quad (11.22)$$

As an example, consider the case where all mass leaves the system, i.e. $\beta = 0$ and with no other loss of angular momentum, i.e. $\dot{J} = 0$. The mass loss then acts to widen the orbit if $\alpha < 1 + M_2/(2M_1)$. If α is too large, the mass transfer becomes unstable. To see this, we take the derivative of Eq. (11.5), allowing for mass loss, and combine the result with Eq. (11.22) to find

$$-\frac{\dot{J}}{J} = -\frac{\dot{M}_2}{M_2}\left(\frac{5}{6} + \frac{n}{2} - \frac{\beta M_2}{M_1} - \frac{(1-\beta)M_2}{3(M_1 + M_2)} - \alpha(1-\beta)\frac{M_1}{M_1 + M_2}\right). \quad (11.23)$$

If we again consider the case where all mass lost by the donor is also lost from the system, we see that stable mass transfer is possible only if $\alpha < 5/6 + n/2 + (1/2 + n/2)M_2/M_1$.

11.4   X-ray sources in globular clusters

While one can assume in the galactic disk that the two stars in a binary evolved from their progenitors in the same binary, this need not be true in the dense core of a globular cluster. In the core, the stars may be so closely packed that encounters between the binary and other cluster stars become an important factor in its development. Routes of binary evolution are thus opened that are not available to binaries in the galactic disk.

That something special is happening in globular clusters is obvious from the census of X-ray sources in our Galaxy. Some 10% of the X-ray binaries that we know are located in globular clusters (see Ch. 14), even though the clusters only contain about

10^{-4} of the number of stars of our Galaxy. A similar situation holds for the nearby galaxy M31, in which about 20 out of several hundred X-ray sources are located in globular clusters (Long and van Speybroeck 1983; Crampton *et al.* 1984). Recycled radio pulsars are also present in globular clusters in larger numbers than expected from simply scaling with total numbers of stars (see Ch. 5). In the cluster 47 Tuc alone, ten such pulsars have been discovered (Manchester *et al.* 1991; these authors list 11 pulsars; one of them, K, may be a harmonic of another one, D – see Phinney 1992).

The observation that the X-ray sources are located especially in the clusters with the densest cores points to close encounters between stars as the physical mechanism for the formation of binaries with neutron stars (Clark 1975).

Fabian, Pringle and Rees (1975) showed that tidal capture of neutron stars by main-sequence stars may occur with sufficient frequency in dense cluster cores to explain the observed number of X-ray sources. Tidal capture occurs because the tidal wave caused on the main-sequence star by a closely passing neutron star drains enough energy out of the orbital motion to bind the neutron star. A simple estimate shows that this happens if the closest passage is within approximately three stellar radii. Giants can similarly capture a passing neutron star. They may also do so in a direct collision (Sutantyo 1975b). Tidal capture or direct collisions therefore may lead both to short-period and long-period binaries, as reviewed by Verbunt (1988b).

In the last years, however, it has become clear that tidal capture may be less efficient in forming binaries than previously thought. One reason for this is that the energy dissipated as the orbit of the neutron star is circularized is comparable to the total binding energy of the main-sequence star, or of the giant's envelope (McMillan, McDermott and Taam 1987; Ray, Kembhavi and Antia 1987). A simple estimate may serve to illustrate this. Consider a star of mass M and radius R that captures a compact star of mass m. Immediately after capture, the orbit has a large semi-major axis a_e, as its eccentricity is close to unity. As tidal interaction circularizes the orbit, angular momentum is conserved, and this allows us to derive that the radius a_c of the circularized orbit is twice the distance of the initial close passage, i.e. $a_c \ll a_e$. Thus, the energy difference ΔE between the initial, highly eccentric, and the final circularized orbit is comparable to the binding energy E_* of the main-sequence star:

$$\frac{\Delta E}{E_*} = \left(\frac{-GMm}{2a_e} - \frac{-GMm}{2a_c} \right) \bigg/ \left(\frac{3GM^2}{5R} \right) \simeq \frac{5}{6} \frac{m}{M} \frac{R}{a_c}. \qquad (11.24)$$

The circularization process destroys the donor.

One might try to alleviate the problem by assuming that the energy dissipation is slow enough for the dissipated energy to be radiated away. This solution evokes a new problem, as the prolonged existence of a strong tidal wave makes it likely that the energy in the wave is fed back into the orbital motion, allowing the captured star to escape again (Kochanek 1992).

Tidal capture may therefore not be very efficient in the production of binaries with neutron stars, but it could well be effective in the production of neutron stars with a high-mass disk around them, formed from the remnants of the disrupted star. If the neutron star manages to accrete enough mass from this disk, it may be spun up to become a rapidly rotating, recycled pulsar. Tidal capture therefore could be efficient in producing single recycled pulsars.

A process that is less detrimental to the prospective companion to the neutron star is the exchange of the neutron star into a binary. In this process, the neutron star encounters a binary and forms a temporary triple with it. The triple is not stable, and tends to eject the lightest star. Thus, neutron stars may take the place of the less massive main-sequence star(s) in binaries in the core of a globular cluster. As the newly formed binary may undergo more encounters, a whole Pandora's box of new possibilities is opened.

The efficiency of this process is directly proportional to the number of suitable binaries present in globular clusters. After many years of fruitless searches, a number of ordinary binaries have recently been discovered in clusters (as reviewed by Hut *et al.* 1992). This suggests that the number of binaries in globular clusters is sufficient for the frequent occurrence of exchange encounters. A first investigation into the formation of binaries containing neutron stars has been made by Phinney and Sigurdsson (1991). If mass is transferred in the binary newly containing a neutron star, an X-ray source becomes visible. Once the mass transfer stops, the neutron star may switch on as a pulsar. Phinney and Sigurdsson (1991) note that it is possible that both original members of a binary are exchanged for a neutron star, and form a binary of neutron stars. The recoil velocity of the three-body interaction may be large enough to explain the distance to the cluster core of the neutron-star binary in M15 (Prince *et al.* 1991).

11.5   Statistical considerations

By combining the birth rate of binaries containing massive stars with evolutionary scenarios, one may try to predict the numbers of X-ray binaries and radio pulsar binaries of various types. In practice, such predictions are highly uncertain, due to our limited knowledge of the binary birth rate and of binary evolution. For a comparison with observation, we need to correct the observed numbers of X-ray and radio pulsar binaries for selection effects and for finite lifetimes. These corrections again are highly uncertain, due to our very limited understanding of them. This section illustrates how various estimates may be made; it cannot claim to provide accurate numbers. We first discuss the high-mass and low-mass binaries in the galactic disk, and then the sources in globular clusters.

11.5.1   High-mass binaries

To calculate the evolution of the galactic population of high-mass binaries, Meurs and Van den Heuvel (1989) assume (1) a distribution of massive primaries according to the initial mass function for single stars as determined by Scalo (1986); (2) that 25% of the massive primaries will interact with their companion; and (3) that during this interaction on average one-third of the envelope mass of the primary is lost from the system. From this they derive a theoretical prediction that there are about 10^4 binaries in our Galaxy in which a neutron star or black hole has an early B star companion with a mass between 8 and $15\,M_\odot$, and about half this number with an O star companion more massive than $15\,M_\odot$.

Indeed, the most numerous observed type of high-mass X-ray binary has a Be star mass donor. As these systems tend to be transients, their absolute numbers are highly uncertain. Meurs and Van den Heuvel (1989) estimate an observed number, after correction for selection effects, in the range of 10^3–10^4 of such systems in our Galaxy (see also Sect. 2.2.1.4). The high end of this range is compatible with the theoretical prediction. As the lifetime of these systems is essentially the lifetime of a B star, several million years, the birth rate of such binaries is of the order of 10^{-3} yr^{-1}.

The standard high-mass X-ray binaries, in which a massive star transfers mass steadily to a compact companion, are only visible as X-ray sources during the short-lived onset of mass transfer (see Sect. 11.2.5), which is only a few per cent of the lifetime of the O star. Thus, Meurs and Van den Heuvel predict that the observed number of such standard systems active as X-ray sources is two orders of magnitude below that of the Be transient X-ray binaries. This is compatible with the small observed number of such systems (see Sect. 2.2.1.4).

Due to the extensive mass transfer, the first supernova explosion in the interacting high-mass binaries is expected to be the explosion of a bare helium core. Such supernovae most probably can be identified with the type Ib supernovae (Nomoto, Filippenko and Shigeyama 1990; Wheeler and Harkness 1990). Blaauw (1993) has recently traced seven runaway OB stars back to their parent OB-association, and finds that six of them (1) are blue stragglers in the Hertzsprung–Russell diagram of their parent association; (2) rotate rapidly; and (3) have high helium abundances. These properties are exactly those expected for the initial secondaries in binaries that have accreted the envelope of the initially more massive companion. The runaway velocity can be explained as the consequence of the explosion of the initial primary; see Eq. (11.10).

Taylor and Stinebring (1986) estimate that the total number of single radio pulsars in our Galaxy is of order 3×10^5. Since out of the ~ 450 observed pulsars, three are in high-mass radio pulsar binaries, one may estimate that a few thousand such binaries are present in the Galaxy. If the lifetimes of the high-mass radio pulsar binaries listed in Table 11.1, namely $\sim 10^8$ yr, are typical, the estimated birth rate of these systems is of the order of 10^{-5} yr^{-1}. This number is two orders of magnitude below our estimate above of the birth rate of Be X-ray binaries. Although one may argue that at least part of this difference is due to the appreciable uncertainties in the estimated numbers, this does seem to suggest that only a small fraction of the Be X-ray binaries evolve into radio pulsar binaries.

This is not wholly unexpected: close Be X-ray binaries are prone to become mergers, whereas wide Be X-ray binaries are unlikely to form. To illustrate this, we consider a Be X-ray binary in which a Be star of $12\,M_\odot$ has a core of $2.5\,M_\odot$, which evolves to the point where the B star starts Roche-lobe overflow. We rewrite Eq. (11.9) for the case $a_f \ll a_i$ as

$$\frac{a_f}{a_i} \simeq \lambda \alpha \frac{M_c M_2}{M_e M_1} \frac{R_1}{a_i}, \tag{11.25}$$

where R_1 is the radius of the Roche lobe of the B star, and a_i is the separation between the stars at onset of mass transfer. According to Eq. (11.4) $R_1/a_i \approx 0.566$. Taking $\lambda = 0.5$, and substituting the other numbers, we find $a_f/a_i \simeq 0.004\alpha$, so that our assumption $a_f \ll a_i$ is justified. The constraint that the helium core must fit within its Roche lobe after spiral-in requires $a_f \gtrsim 1.2\,R_\odot$. Thus, the Be X-ray binary must have $a_i \gtrsim 300 R_\odot/\alpha$ if a merger is to be avoided. However, the formation of a Be X-ray binary requires that the binary survives the first supernova. In the presence of kick velocities imparted to the neutron star at birth, this favors short-period binaries. In our example, the helium-star progenitor to the neutron star had an orbital velocity before the first supernova of less than roughly $80\alpha^{1/2}$ km s^{-1}. A kick velocity larger than this is likely to disrupt the binary and prevent the formation of the Be X-ray binary.

Thus, the period range of primordial binaries sufficiently close to survive the first supernova, but sufficiently wide to avoid a merger once the first-born neutron star spirals into its companion, may be rather small, in particular if the spiral-in processs is not fully efficient in lifting the stellar envelope, i.e. when $\alpha < 1$. Because the binary emerging from the spiral-in may still be disrupted in the second supernova explosion, it appears quite possible that indeed only a small fraction of the Be X-ray binaries will evolve into high-mass binary radio pulsars.

The standard high-mass X-ray binary is in all cases too close to survive the spiral-in.

11.5.2 *Low-mass binaries*

With about 100 persistently bright low-mass X-ray binaries in the Galaxy (see Ch. 14), an estimated lifetime of 10^8–10^9 yr gives a birth rate of order 10^{-7}–10^{-6} yr^{-1}. This number is uncertain by at least an order of magnitude, because of the appreciable uncertainty in our knowledge of the lifetime of low-mass X-ray binaries. Nonetheless, it appears that the formation of a low-mass X-ray binary is an extremely rare event. The number of transiently bright low-mass X-ray binaries, including those which contain a black hole, and the birth rate of these systems may be similar to the number and birth rate of the permanently bright systems. Due to the uncertainty in correcting the observed number of transient low-mass X-ray binaries to the total number actually present in our Galaxy, the birth rate of these systems is even more uncertain than that of the permanently bright systems (see Sect. 3.6).

With birth rates for neutron stars of order 10^{-2} yr^{-1} and for black holes a few times lower, these low birth rates imply that only a tiny fraction of the neutron stars and black holes formed in our Galaxy are formed as members of low-mass X-ray binaries.

As discussed in Sect. 11.3.1, there are several scenarios for the formation of low-mass X-ray binaries outside globular clusters. In these scenarios, the progenitor systems are either rather wide binaries with extreme mass ratios, or triple systems in which a high-mass binary has a low-mass companion at large distance. All scenarios involve one or more spiral-in phases. Low-mass companions to a high-mass primary or to a high-mass binary are difficult to detect, especially if their orbit has a long period. It should be no surprise, then, that we do not have reliable statistics for any of the proposed progenitor systems to low-mass X-ray binaries that would enable us to determine their numbers and birth rates. Because the low-mass star must be able to remove the envelope of its companion during the spiral-in phase, its original distance must be large; because interaction must occur, it cannot be too large. Thus, the orbital period of the progenitor system must be extremely fine-tuned (Webbink 1992). It does not appear impossible that the number of suitable progenitor systems is sufficiently low to explain the extremely low birth rates of low-mass X-ray binaries.

Of the currently known (~ 600) radio pulsars, around 20 (i.e. a small percentage) are members of a low-mass binary; if this fraction also holds for all pulsars in the Galaxy, there should be of order 10^4 low-mass binary pulsars in the Galaxy. The characteristic age $\tau_c \equiv P/(2\dot{P})$ gives an estimate for the age of a radio pulsar. For lifetimes of about 10^9 yr, as suggested by Table 11.1, this gives an estimated birth rate of 10^{-5} yr^{-1}. The birth rate of low-mass radio pulsar binaries should be lower than that of the low-mass X-ray binaries, if all such pulsar binaries are formed from the low-mass X-ray binaries. Our estimate for the birth rate of low-mass radio pulsar binaries is higher than the estimated range of birth rates for the low-mass X-ray binaries. This suggests that our estimate for the lifetime of low-mass X-ray binaries is too high. Kulkarni and Narayan (1988) also argued that there are too many low-mass radio pulsar binaries, and that other channels for the formation of such binaries are required. These authors used underestimated lifetimes for several radio pulsar binaries, however. In addition, Johnston and Bailes (1991) use revised distances and more accurate fluxes for the known millisecond pulsars to argue that the total number of recycled pulsars in the Galaxy may well be low enough to remove any discrepancy between their birth rate and that of low-mass X-ray binaries.

In the above argument it is assumed that the low-mass X-ray binaries are the sole progenitors to the low-mass binary radio pulsars. An alternative route could be the direct production of a millisecond pulsar via accretion-induced collapse of a massive white dwarf, bypassing the X-ray phase. The adherents of this scenario pose *ad hoc* that accretion-induced collapse leads to a neutron star with a weak magnetic field. There is no reason for such an assumption (see, e.g., Bhattacharya and Van den Heuvel 1991). More seriously, the collapse would cause appreciable eccentricity in the binary orbit, according to Eq. (11.10), as the mass loss ΔM is at least equal to the difference in gravitational binding energy of a white dwarf and a neutron star, i.e. $\sim 0.15\,M_\odot$. The collapse follows mass transfer to the white dwarf; continued mass transfer would be required to circularize the orbit. During such continued mass transfer, however, the binary would be a bright X-ray source, and the original paradox of discrepant numbers of radio pulsar descendants and X-ray source progenitors arises anew.

Once more low-mass radio pulsar binaries are discovered, we can set our estimates for their numbers and birth rates on a statistically more sound footing.

11.5.3 Globular clusters

The number of radio pulsars in globular clusters is high, as witnessed by their presence in clusters with relatively low density, such as M 4 (Sect. 5.7). Triple-star interaction may be sufficiently common in clusters with lower central densities to explain this. The absence of X-ray sources in low-density clusters (with one exception, the source in NGC6712) suggests that the cluster environment acts to shorten the lifetime of the X-ray sources. A similar argument may be derived from the presence of ten radio pulsars in 47 Tuc, in the absence of a single known bright X-ray source. However, Verbunt *et al.* (1993) have discovered four low-luminosity X-ray sources in the core of 47 Tuc, and argue that these may be low-mass X-ray binaries in a low state, suitable progenitors for radio pulsars. Alternatively, we may argue that the current absence of bright X-ray sources does not necessarily imply an absence in the past; in other words, that the binary population in the cluster has changed over the last 10^9 yr.

Acknowledgements

F.V. gratefully acknowledges support from the Netherlands Organization for Scientific Research (NWO) under grant PGS 78-277. We thank Sterl Phinney and Shri Kulkarni for many stimulating discussions.

References

Alme, M.L. and Wilson, J.R. 1974, *Ap. J.* 194, 147

Andersen, J. 1991, *Astron. Astrophys. Rev.* 3, 91

Banit, M. and Shaham, J. 1992, *Ap. J. Lett.* 388, L19

Basko, M.M., Sunyaev, R.A. and Titarchuk, L.G. 1974, *Astron. Astrophys.* 31, 249

Begelman, M.C., Sarazin, C.L., Hatchett, S.P., McKee, C.F. and Arons, J. 1980, *Ap. J.* 238, 722

Bhattacharya, D. and Van den Heuvel, E.P.J. 1991, *Phys. Reports* 203, 1

Biehle, G.T. 1991, *Ap. J.* 380, 167

Blaauw, A. 1961, *Bull. Astron. Inst. Neth.* XV, 265

Blaauw, A. 1993, in: *Massive Stars, Their Lifes in the Insterstellar Medium*, eds. J.P. Cassinelli and E.B. Churchwell, ASP Conf. Ser. 35, p. 207

Boersma, J. 1961, *Bull. Astron. Inst. Neth.* XV, 291

Bonsema, P.F.J. and Van den Heuvel, E.P.J. 1985, *Astron. Astrophys.* 146, L3

Canal, R., Isern, J. and Labay, J. 1990, *Ann. Rev. Astron. Astrophys.* 28, 183

Canal, R. and Schatzman, E. 1976. *Astron. Astrophys.* 46, 229

Cannon, R., Eggleton, P.P., Zytkow, A.N. and Podsiadlowski, P. 1992, *Ap. J.* 386, 206

Cherepashchuk, A.M. 1981, *MNRAS* 194, 761

Clark, G.W. 1975, *Ap. J. Lett.* 199, L143

Crampton, D., Cowley, A.P. and Hutchings, J.B. 1980, *Ap. J. Lett.* 235, L131

Crampton, D., Cowley, A.P., Hutchings, J.B., Schade, D.J. and van Speybroeck, L.P. 1984, *Ap. J.* 284, 663

Delgado, A.J. and Thomas, H.C. 1981, *Astron. Astrophys.* 96, 142

Dewey, R.J. and Cordes, J.M. 1987, *Ap. J.* 321, 780

Eggleton, P.P. 1983, *Ap. J.* 268, 368

Eggleton, P.P. and Verbunt, F. 1986, *MNRAS* 220, 13P

Ergma, E.V. and Fedorova, A.V. 1991, *Astron. Astrophys.* 242, 125

Fabian, A.C., Pringle, J.E. and Rees, M.J. 1975, *MNRAS* 172, 15P

Fahlman, G.G. and Gregory, P.C. 1983, in: *Supernova Remnants and their X-ray Emission*, eds. J. Dantziger and P. Gorenstein, IAU Symp. 101, Reidel, Dordrecht, p. 445

Faulkner, J. 1971, *Ap. J. Lett.* 170, L99

Flannery, B.P. 1977, *Ann. N.Y. Acad. Sci.* 362: 36

Flannery, B.P. and Van den Heuvel, E.P.J. 1975, *Astron. Astrophys.* 39, 61

Fruchter, A.S., Stinebring, D.R. and Taylor, J.H. 1988, *Nature* 333, 237

Fryxell, B.A. and Arnett, W.D. 1981, *Ap. J.* 243, 994

Habets, G.M.H.J. 1985, Ph.D. Thesis, Amsterdam

Harpaz, A. and Rappaport, S. 1991, *Ap. J.* 383, 739

Hut, P., McMillan, S., Goodman, J., Mateo, M., Phinney, E.S., Pryor, C., Richer, H.B., Verbunt, F. and Weinberg, S. 1992, *Publ. Astr. Soc. Pac.* 104, 981

Iwasawa, K., Koyama, K. and Halpern, J.P. 1992, in: *Frontiers of X-ray Astronomy*, eds. Y. Tanaka and K. Koyama, Universal Academy Press, Tokyo, p. 49

Johnston, S. and Bailes, M. 1991, *MNRAS* 252, 277

Johnston, S., Lorimer, D.R., Harrison, P.A., Bailes, M., Lyne, A.G., Bell, J.F., Kaspi, V.M., Manchester, R.N., D'Amico, N., Nicastrol, L. and Shengzhen, J. 1993, *Nature* 361, 613

Joss, P.C., Rappaport, S. and Lewis, W. 1987, *Ap. J.* 319, 180

Kochanek, C.S. 1992, *Ap. J.* 385, 604

Kraft, R.P., Mathews, J. and Greenstein, J.L. 1962, *Ap. J.* 136, 312

Kulkarni, S.R. 1986, *Ap. J. Lett.* 306, L85

Kulkarni, S.R. and Narayan, R. 1988, *Ap. J.* 335, 755

Kulkarni, S.R., Phinney, E.S., Evans, C.R. and Hasinger, G. 1992, *Nature* 359, 300

Lamb, F.K. 1981, in: *Pulsars*, eds. W. Sieber and R. Wielebinski, IAU Symp. 95, p. 309

Law, W.Y. and Ritter, H. 1983, *Astron. Astrophys.* 123, 33

Long, K.S. and van Speybroeck, L.P. 1983, in: *Accretion-driven Stellar X-ray Sources*, eds. W.H.G. Lewin and E.P.J. van den Heuvel, Cambridge University Press, p. 118

McMillan, S.L.W., McDermott, P.N. and Taam, R.E. 1987, *Ap. J.* 318, 261

Maeder, A. and Meynet, G. 1988, *Astron. Astrophys. Suppl.* 76, 411

Manchester, R.N., Lyne, A.G., Robinson, C., D'Amico, N., Bailes, M. and Lim, J. 1991, *Nature* 352, 219

Maraschi, L., Treves, A. and Van den Heuvel, E.P.J. 1976, *Nature* 259, 292

Margon, B. 1984, *Ann. Rev. Astron. Astrophys.* 22, 507

Mestel, L. and Spruit, H.C. 1987, *MNRAS* 226, 57

Meurs, E.J.A. and Van den Heuvel, E.P.J. 1989, *Astron. Astrophys.* 226, 88

Mochnacki, S.W. 1984, *Ap. J. Suppl.* 55, 551

Morton, D.C. 1960, *Ap. J.* 132, 146

Nomoto, K., Filippenko, A.V. and Shigeyama, T. 1990, *Astron. Astrophys.* 240, L1

Paczyński, B. 1967a, *Acta Astron.* 17, 193 & 355

Paczyński, B. 1967b, *Acta Astron.* 17, 287

Paczyński, B. 1970, in: *Mass Loss and Evolution in Close Binaries*, eds. K. Gyldenkerne and R.M. West, Copenhagen University Observatory, p. 139

Paczyński, B. 1976, in: *Structure and Evolution of Close Binary Systems*, IAU Symp. 73, eds. P. Eggleton, S. Mitton and J. Whelan, Reidel, Dordrecht, p. 75

Paczyński, B. and Sienkewicz, R. 1981, *Ap. J. Lett.* 248, L27

Patterson, J. 1984, *Ap. J. Suppl.* 54, 443

Phinney, E.S. 1992, *Phil. Trans. R. Soc. Lond. A* 341: 39

Phinney, E.S. and Hansen, B.M.S. 1993, in: Planets Around Pulsars, eds. J.A. Philips, S.E. Thorsett and S.R. Kulkarni, ASP, San Francisco, p. 371

Phinney, E.S. and Sigurdsson, S. 1991, *Nature* 349, 220

Phinney, E.S. and Verbunt, F. 1991, *MNRAS* 248, 21P

Phinney, E.S., Evans, C.R., Blandford, R.D. and Kulkarni, S.R. 1988, *Nature* 333, 832

Podsiadlowski, Ph. 1991, *Nature* 350, 136

Prince, T.A., Anderson, S.B., Kulkarni, S.R. and Wolszczan, A. 1991, *Ap. J. Lett.* 374, L41

Pylyzer, E. and Savonije, G.J. 1988, *Astron. Astrophys.* 191, 57

Rappaport, S., Verbunt, F. and Joss, P.C. 1983, *Ap. J.* 275, 713

Ray, A., Kembhavi, A.K. and Antia, H.M. 1987, *Astron. Astrophys.* 184, 164

Ruderman, M., Shaham, J. and Tavani, M. 1989a, *Ap. J.* 336, 507

Ruderman, M., Shaham, J., Tavani, M. and Eichler, D. 1989b, *Ap. J.* 343, 292

Ryba, M.F. and Taylor, J.H. 1991, *Ap. J.* 380, 557

Savonije, G.J. 1979, *Astron. Astrophys.* 71, 352

Savonije, G.J. 1983, in: *Accretion-driven Stellar X-ray Sources*, eds. W.H.G. Lewin and E.P.J. van den Heuvel, Cambridge University Press, p. 343

Savonije, G.J., de Kool, M. and Van den Heuvel, E.P.J. 1986, *Astron. Astrophys.* 155, 51

Scalo, J.M. 1986, *Fundamentals Cosmic Phys.* 11, 1

Shklovskii, I.S. 1979, *Pis'ma Astron. Zh.* 5, 644

Slettebak, A. 1988, *Publ. Astr. Soc. Pac.* 100, 770

Smarr, L.L. and Blandford, R. 1976, *Ap. J.* 207, 574

Soker, N. and Livio, M. 1989, *Ap. J.* 339, 268

Spruit, H.C. and Ritter, H. 1983, *Astron. Astrophys.* 124, 267

Sutantyo, W. 1975a, *Astron. Astrophys.* 41, 47

Sutantyo, W. 1975b, *Astron. Astrophys.* 44, 227

Taam, R.E. and Bodenheimer, P. 1989, *Ap. J.* 337, 849

Taam, R.E. and Van den Heuvel, E.P.J. 1986, *Ap. J.* 305, 235

Tavani, M. and Brookshaw, L. 1992, *Nature* 356, 320

Taylor, J.H. and Stinebring, D.R. 1986, *Astron. Astrophys. Rev.* 24, 285

Taylor, J.H. and Weisberg, J.M. 1989, *Ap. J.* 345, 434

Thorne, K.S. and Zytkow, A.N. 1977, *Ap. J.* 212, 832

Van den Heuvel, E.P.J. 1983, in: *Accretion-driven Stellar X-ray Sources*, eds. W.H.G. Lewin and E.P.J. van den Heuvel, Cambridge University Press, p. 303

Van den Heuvel, E.P.J. 1992, *Nature* 356, 668

Van den Heuvel, E.P.J. 1993, in: Planets Around Pulsars, eds. J.A. Philips, S.E. Thorsett and S.R. Kulkarni, ASP, San Francisco, p. 123

Van den Heuvel, E.P.J. and De Loore, C. 1973, *Astron. Astrophys.* 25, 387

Van den Heuvel, E.P.J. and Habets, G.M.H.J. 1984, *Nature* 309, 598

Van den Heuvel, E.P.J. and Heise, J. 1972, *Nature Phys. Sci.* 239, 67

Van den Heuvel, E.P.J., Ostriker, J.P. and Petterson, J.A. 1980, *Astron. Astrophys.* 81, L7

van den Heuvel, E.P.J. and Rappaport, S. 1987, in *Physics of Be Stars*, eds. A. Slettebak and T.P. Snow, Cambridge University Press, p. 291

Van der Hucht, K.A. 1992, *Astron. Astrophys. Rev.* 4, 123

Van Kerkwijk, M.H., Charles, P.A., Geballe, T.R., King, D.L., Miley, G.K., Molnar, L.A., Van den Heuvel, E.P.J., Van der Klis, M. and Van Paradijs, J. 1992, *Nature* 355, 703

Verbunt, F. 1984, *MNRAS* 209, 227

Verbunt, F. 1988a, in: *The Physics of Neutron Stars and Black Holes*, ed. Y. Tanaka, ISAS, Tokyo, p. 159

Verbunt, F. 1988b, in: *The Physics of Compact Objects, Theory vs. Observation*, eds. N.E. White and L. Fillipov, Pergamon Press, Oxford, p. 529

Verbunt, F. and Rappaport, S.A. 1988, *Ap. J.* 332, 193

Verbunt, F., Wijers, R.A.M.J. and Burm, H.H.M.G. 1990, *Astron. Astrophys.* 234, 195

Verbunt, F. and Zwaan 1981. *Astron. Astrophys.* 100, L7

Verbunt, F., Hasinger, G., Johnston, H.M. and Bunk, W. 1993, *Adv. Space Res.* 13, (12), 151

Vermeulen, R. 1989, Ph.D. Thesis, Leiden

Wade, R.A. 1988, *Ap. J.* 335, 394

Webbink, R.F. 1985, in: *Interacting Binary Stars*, eds. J.E. Pringle and R.A. Wade, Cambridge University Press, p. 39

Webbink, R.F. 1992, in *X-ray Binaries and Recycled Pulsars*, eds. E.P.J. van den Heuvel and S.R. Rappaport, NATO ASI, Kluwer Academic Publishers, Dordrecht, p. 269

Webbink, R.F., Rappaport, S. and Savonije, G.J. 1983, *Ap. J.* 270, 678

Wheeler, J.C. and Harkness, R.P. 1990, *Reports Progr. Phys.* 53: 467

Whelan, J. and Iben, I. 1973, *Ap. J.* 186, 1007

Wolszczan, A. 1991, *Nature* 350, 688

Wolszczan, A. and Frail, D.A. 1992, *Nature* 355, 145

12

The magnetic fields of neutron stars and their evolution

D. Bhattacharya and G. Srinivasan
Raman Research Institute, Bangalore 560080, India

12.1   Introduction

The magnetic field plays an important role in the life history of a neutron star. The vast majority of neutron stars discovered so far are radio pulsars; the strong magnetic fields of neutron stars are crucial to this pulsar activity. The derived field strengths of more than 500 known pulsars cluster around $\sim 10^{12}$ G. The polarization of the observed radio radiation indicates a predominantly dipolar structure of the field at distances ranging from a few stellar radii to a few hundred stellar radii. In neutron stars accreting matter from binary companions, the magnetic field serves to collimate the incoming matter onto the polar caps, causing the observed X-ray intensity to be modulated at the spin period. About 30 such pulsating X-ray sources are known presently (see Ch. 1 and 14), and they account for about half the population of known neutron stars in binary systems.

Since pulsars are strongly magnetized and rapidly rotating, they spin down with time as the energy radiated in the form of electromagnetic radiation and relativistic plasma comes at the expense of the stored rotational energy. Similarly, the interaction between the magnetosphere of accreting neutron stars with the inflowing matter causes angular momentum exchange between the neutron star and the infalling matter. Usually, when the mass transfer from the companion is weak (stellar wind phase), the neutron star will be spun down, whereas it will be rapidly spun up during the phase of heavy mass transfer and accretion. Such a spin-up can give a new lease of life to a 'dead' pulsar. The degree to which a neutron star can be spun up is restricted by the maximum possible accretion rate and the magnetic field strength of the neutron star. In an equilibrium situation, the angular velocity of the neutron star will match the Keplerian angular velocity at the magnetospheric boundary, which corresponds to a spin period $P_{eq} = 1.9$ ms $(B/10^9 \text{ G})^{6/7}(\dot{M}/\dot{M}_{Edd})$, where B is the dipole field strength at the surface of the neutron star and $\dot{M}_{Edd}$ is the Eddington accretion rate (see also Sect. 5.3).

The characteristic times of the various evolutionary phases mentioned above are rather short by astronomical standards. The typical lifetime of a neutron star as an active radio pulsar is $\sim 10^7$ yr, while spin-up via accretion torques could occur on time scales as short as a few hundred years. The observed population therefore represents a mixture of neutron stars which have undergone different evolutionary phases. Connection between these phases can be understood only with the help of a proper evolutionary model.

As mentioned above, the rotational history of a neutron star is controlled to a large extent by its magnetic field. However, the magnetic field itself may change with time – both in strength and in structure. Thus, the evolution of the magnetic field should be an integral part of evolutionary models for neutron stars.

Does the magnetic field of a neutron star evolve, and if so why? Soon after the discovery of neutron stars, Ostriker and Gunn [54] argued that their magnetic fields would undergo spontaneous ohmic decay with a time-constant of a few million years. Although it was soon pointed out that this conclusion was based on an erroneous theoretical modelling [7], this simple hypothesis became generally accepted. The apparent discrepancy between the large spin-down ages of some pulsars compared with their kinematic ages, the lower field strengths observed in old recycled radio pulsars, etc., seemed to fit in nicely if one invoked an exponential field decay. During the last several years the situation has changed considerably. New observational evidence has emerged which has forced one to either modify this hypothesis or to abandon it. A host of new theoretical ideas have also been put forward. In this chapter we review the classical arguments and the new developments both from an observational and a theoretical point of view. In Sect. 12.2 we discuss how the magnetic fields of neutron stars are inferred. Sect. 12.3 is devoted to the origin of the magnetic fields and the structure of the field. In Sect. 12.4 observational evidence for the evolution of the magnetic field is presented, and in Sect. 12.5 possible mechanisms for the field evolution are described. Sect. 12.6 discusses some physical consequences of the field evolution.

12.2   Estimation of the magnetic field

12.2.1   *Radio pulsars*

Perhaps the most reliable estimates of the magnetic field strengths of neutron stars are from radio pulsars. It is customary to assume that the spin-down torque on a pulsar is similar to that on a spinning magnet emitting magnetic dipole radiation. Although it is far from clear whether pulsars, in fact, emit low frequency dipole radiation, this model is generally used in estimating their field strengths. In the dipole radiation model, the spin-down torque is given by

$$N = I\dot{\Omega} = -\frac{2}{3c^3}B^2R^6\Omega^3\sin^2\alpha, \qquad (12.1)$$

where B is the polar field strength at the stellar surface, R is the radius of the star, I its moment of inertia, Ω its spin rate, and α is the angle between the spin axis and the magnetic axis. Thus, in this simple model the torque will be maximum when the magnetic axis is orthogonal to the spin axis, and will vanish when the two axes are aligned. That this is a gross oversimplification was appreciated quite early on. As shown in the classical paper of Goldreich and Julian [32], an aligned rotator will also slow down, the energy in this case being carried away by a relativistic plasma. In fact, it turns out that the energy loss rate of such an aligned rotator is of the same order as that of an orthogonal rotator in the dipole radiation model. Although there have been many attempts, there has not been much progress in understanding the electrodynamics of an inclined rotator. To illustrate the state of confusion, in some models an aligned rotator will hardly slow down, whereas in other models the spin-down torque vanishes in an orthogonal rotator! (See, e.g., ref. [8].)

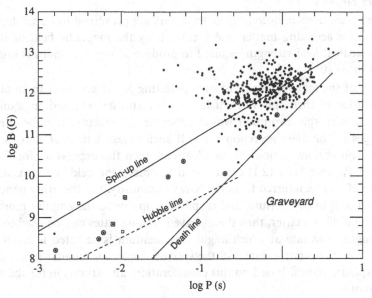

Fig. 12.1. The derived magnetic fields of known radio pulsars plotted against their spin periods. Pulsars in binaries are shown encircled. Squares denote pulsars in globular clusters; binaries among them are shown with dots inside the squares. The 'death line' corresponds to a polar cap voltage below which the pulsar activity is likely to switch off. The 'Hubble line' represents a spin-down age $\tau \equiv P/2\dot{P}$ of 10^{10} yr for a pulsar. The 'spin-up line' represents the minimum period to which a neutron star can be spun up in an Eddington-limited accretion.

In the absence of a clear theoretical prediction of the dependence of the spin-down torque on the angle α, one must resort to observational evidence to shed some light on this important question. The angle of inclination between the magnetic axis and the rotation axis has now been estimated for about 100 pulsars from their observed pulse width and polarization characteristics [47, 58]. Quite remarkably, the distribution of the spin-down torques of these pulsars shows no evidence for a dependence on the angle of inclination [10, 47, 66].

In view of the above discussion, Eq. (12.1) can be used to estimate the magnetic field of a neutron star by dropping the angle-dependent factor. Rewriting the expression in terms of the observed quantities, namely the spin period P and the spin-down rate $\dot{P}$, one obtains

$$B = \left(\frac{3c^3 I P \dot{P}}{8\pi^2 R^6} \right)^{1/2} = 3.2 \times 10^{19} (P\dot{P})^{1/2} \text{ G}. \tag{12.2}$$

In making the numerical estimate, typical values have been used for the radius and moment of inertia of the neutron star. The magnetic fields of pulsars are derived from this expression, and, as one sees from the distribution shown in Figure 12.1, most neutron stars have fields in the range $\sim 10^8$–$10^{13.5}$ G.

12.2.2   X-ray binaries

X-ray pulsations in accreting neutron stars are presumed to occur due to the collimation of the accreting matter into a column by the magnetic field of the star. Although the critical field strength required to produce X-ray pulsations is uncertain, it is likely to be $\sim 10^{12}$ G [2].

An estimate of the magnetic fields of such pulsating X-ray sources can be obtained from the evolution of their spin rates. If the neutron star has reached an equilibrium period, then secular spin evolution would either be imperceptible or be driven by secular changes in the mass accretion rates. If such cases of neutron stars spinning close to their equilibrium period can be identified, then the expression for the equilibrium period P_{eq} (see Sect. 12.1) can be used to obtain the field strength, since the accretion rate $\dot{M}$ can be inferred from the X-ray luminosity. On the other hand, if the object is far from spin equilibrium, and is spinning up due to the angular momentum carried by the infalling matter, then the observed spin-up rates can be used to obtain the field strength. The rate at which angular momentum is accreted is given by the product of the mass accretion rate and the specific Keplerian angular momentum at the magnetospheric radius. Based on this consideration, an expression for the spin-up rate can be written:

$$-\dot{P} \propto \mu^{2/7} P^2 L^{6/7}, \tag{12.3}$$

where μ is the magnetic dipole moment of the neutron star and L is the luminosity generated due to accretion. The proportionality constant depends on the detailed model of interaction of the incoming matter with the magnetic field. Unfortunately, there is as yet no 'consensus model' for the interaction of the accretion disc with the magnetic field of the star, and therefore the magnetic moment inferred from Eq. (12.3) is quite model-dependent. Further, it is not straightforward to distinguish between sources which are close to spin equilibrium and those far from it. Given all these uncertainties, all one can say is that the observed spin-up rates are consistent with surface dipole fields of these accreting neutron stars being $\sim 10^{12}$ G (see, e.g., ref. [51] and references therein).

Yet another observational handle on the magnetic field strength of an accreting neutron star comes from the presence of cyclotron absorption lines in the observed X-ray spectrum since the energy of the cyclotron resonance is directly proportional to the field strength. Cyclotron absorption lines (Sect. 1.5.1.4) have been observed in nine X-ray binaries and possibly two γ-ray bursters (Ch. 13), and these suggest field strengths $\sim (0.5\text{–}2) \times 10^{12}$ G [49, 51].

It should be kept in mind, however, that whereas the cyclotron line gives information about the *total* field strength close to the stellar surface, the field strength derived from the spin evolution of neutron stars or the degree of collimation of the accretion flow refers mainly to the *dipole component* of the field. If strong multipole components are present, then the field strength derived from cyclotron line frequencies would not in general agree with those inferred using other means.

12.3 The origin and structure of the magnetic field

12.3.1 Introduction

In this section we turn to a discussion of the origin of the observed magnetic fields in neutron stars. There are two main hypotheses in this regard: (a) that the magnetic field is inherited from its progenitor, the magnetic flux being conserved during the process of collapse; and (b) that the magnetic field is built up after the formation of the neutron star through a battery mechanism or a dynamo process. We shall discuss these two alternatives below.

12.3.2 Fossil field

Even before pulsars were discovered, Woltjer [78] and Ginzburg [30] had conjectured that since magnetic flux in stellar cores will be conserved during the collapse of their cores, the collapsed remnants could have field strengths as large as 10^{14}–10^{16} G (when the flux is conserved, the magnetic field $B \propto R^{-2}$; this will result in an enormous field amplication). Since the derived magnetic fields of radio pulsars turned out to be $\sim 10^{12}$ G, the fossil field hypothesis became generally accepted. The situation remains so even today, despite some alternative suggestions that have been made in recent times.

The fossil field hypothesis cannot, however, be conclusively proved since one knows very little about the field strengths in the cores of massive stars which are the progenitors of neutron stars. No measurements of magnetic fields exist for stars massive enough (main sequence mass $\gtrsim 8M_\odot$) to leave neutron stars as end products. Among the somewhat lower mass stars in the upper main sequence, only the chemically peculiar ones, which account for $\sim 10\%$ to 15% of stars in spectral classes between B5 and F0, seem to possess surface field strength measurable by current techniques [20]. These stars exhibit field strengths in the range of several hundred to several thousand gauss. The field geometry in most cases is predominantly dipolar, with the dipole axis inclined at some angle to the spin axis of the star [18, 27, 70]. The angle of inclination appears to be randomly distributed [19, 35, 44, 55]. The structure of the field in the interior is not known, but the existence of strong poloidal components with scale length of the order of the stellar radius suggests that a substantial amount of flux may reside in the core.

To what extent the processes generating the observed fields in gaseous stars may be relevant for neutron stars is difficult to judge. But one common characteristic of the upper main sequence stars is the presence of a convective core which is thought to be capable of generating magnetic fields through dynamo processes (see, e.g., ref. [20]). If this is, indeed, the case, then the cores of most upper main sequence stars are likely to possess strong magnetic fields. The more massive the star, the larger is the size of its convective core–and it is of some interest to note that white dwarfs descending from more massive progenitors tend to have stronger magnetic fields than those from less massive progenitors [39, 45]. The magnetic flux threading the core of a massive star need not, however, be generated in the main sequence phase itself. It has been argued by Ruderman and Sutherland [63] that the highly convective carbon-burning phase prior to core collapse is the most likely stage during which flux is generated by a dynamo process. If this is correct, then the field strengths of neutron stars bear no direct relation to those observed in magnetic stars which are in or near the main sequence, or those in white dwarfs, which (except for the very massive and rare O–Ne–Mg dwarfs) do not experience carbon burning. The equipartition field (for which the magnetic energy density equals the mechanical energy density in turbulence) in the convective carbon-burning phase would naturally lead to a field strength of $\sim 10^{12}$ G in the post-collapse neutron star [63]. This may explain the uniformity of the field strength among the young neutron stars.

If the magnetic field of a neutron star is the fossil field amplified during the core collapse, then the magnetic flux is expected to permeate the whole neutron star more or less uniformly. The interior of the neutron star consists of a fluid core, which is essentially a neutron fluid, with an admixture of $\sim 1\%$ protons and electrons (which stabilize the neutrons against beta decay). If the proton fluid in the interior is in a 'normal' state, then the exterior magnetic field could be understood in terms of *currents* in the interior; the predominantly poloidal magnetic field arising due to toroidal currents. However, both the neutrons and the protons in the interior are expected to be in a superfluid state, although the electrons will remain as a normal relativistic degenerate plasma.

The possibility of superfluid states in the interior of neutron stars was first pointed out by Migdal [48] nearly a decade before the discovery of pulsars. The attractive nuclear force between the nucleons provides a natural pairing mechanism, and the newly discovered microscopic theory of superconductivity by Bardeen, Cooper and Schrieffer (BCS) provided the theoretical framework for estimating the transition temperature T_c. Following this suggestion, Ginzburg and Kirzhnits [31] estimated the transition temperature using the BCS relation: $\Delta \equiv kT_c \sim E_0 \exp(-1/NV)$, where $E_0 \sim p_F \hbar/ma$ is the energy-width of the interaction region, p_F is the Fermi momentum, m is the nucleon mass, and a is the range of nuclear force. N is the density of states at the Fermi surface, and V is the strength of the attractive potential between the nucleons. For densities in the range $\sim 10^{13}$–10^{15} g cm^{-3}, the estimated transition temperature was $\sim$1–20 MeV. Thus, when the interior temperature of the neutron star falls below $\sim 10^{10}$ K, the neutrons and protons will undergo a transition to a superfluid state. The more recent and refined estimates [23, 36, 77] have somewhat lowered the transition temperature to ~ 0.5 MeV, but the above conclusion remains unaltered because the interior temperatures are expected to be $\lesssim 10^8$ K.

If one accepts the conclusion that the protons in the interior will be in a superconducting state, then one has to ask whether this will have any consequences for the magnetic properties of the neutron star. This depends upon two important lengths: the *coherence length* ξ and the *London penetration depth* λ. The *coherence length* is the measure of the distance within which the properties of the superconductor cannot change appreciably in the presence of a magnetic field. The *London penetration depth* is the *screening length*: the strength of an externally applied mangetic field would fall to $1/e$ of its surface value at a depth λ inside the superconductor. If $\xi > \sqrt{2}\,\lambda$, the superconductor will be in a Type I state exhibiting perfect diamagnetism. At applied field strengths below a critical value B_c, the field will be completely expelled (Meissner effect); for fields exceeding the critical value, superconductivity will be destroyed.

If $\xi < \sqrt{2}\,\lambda$, the superconductor will exhibit Type II behaviour: below a lower critical field B_{c1} there will be a complete expulsion of the field; above an upper critical

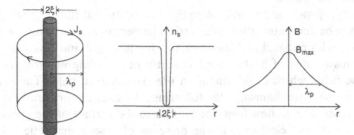

Fig. 12.2. The structure of an Abrikosov fluxoid. The fluxoid consists of a normal core of radius equal to the coherence length ξ surrounded by superconducting material. The maximum magnetic field strength B_{max} resides in the core and decreases exponentially with a scale length λ, the London penetration depth. Supercurrents J_s responsible for maintaining the magnetic field are also confined within this characteristic distance from the core. When the spacing between fluxoids is of order λ or more, the value of B_{max} nearly equals the lower critical field B_{c1}. In the diagram, n_s stands for the number density of superconducting carriers and B denotes the magnetic field strength.

field B_{c2} superconductivity will be destroyed; in the intermediate range $B_{c1} < B < B_{c2}$ the superconductor will allow the magnetic field to penetrate, not homogeneously but confined to quantized flux tubes (known as Abrikosov fluxoids). Each fluxoid carries a quantum of magnetic flux $\phi_0 = hc/2e = 2 \times 10^{-7}$ G cm^2. Within the core of a fluxoid (of radius $\sim \xi$), the matter is in its 'normal' state. The field strength rises to $\sim B_{c1}$ within this normal core. Around this the matter is in the superconducting state, and the field strength decreases exponentially away from the core with a scale length λ (see Figure 12.2). In terms of ξ, λ and ϕ_0 the critical fields can be expressed as

$$B_{c1} = \frac{\phi_0}{\pi \lambda^2} \ln\left(\frac{\lambda}{\xi}\right) \quad \text{and} \quad B_{c2} = \frac{\phi_0}{\pi \xi^2}. \tag{12.4}$$

For protons in the interior of a neutron star, the coherence length is $\xi_p = 0.6 \times 10^{-12} \rho_{p,13}^{1/3} \Delta_p^{-1}$ cm and the London penetration depth is $\lambda_p = 0.9 \times 10^{-11} \rho_{p,13}^{-1/2}$ cm, where $\rho_{p,13}$ is the mass density of protons in units of 10^{13} g cm^{-3}, and Δ_p is the energy gap of the proton superconductor in MeV. For current estimates of Δ_p, ξ_p is $\ll \lambda_p$ in the neutron star interior and hence the proton superconductor is expected to exhibit Type II behaviour.

There is a subtle point here worth noting. Estimates suggest that the lower critical field for the proton superconductor would be $B_{c1} \sim 10^{15}$ G. Since the observed field strength of a neutron star is $\sim 10^{10}$–10^{12} G, much smaller than B_{c1}, one would expect a complete expulsion of the magnetic field from the superconductor. That this does not happen is due to the enormously high electrical conductivity of the interior in the normal state. We defer the discussion of this important point till Sect. 12.4. Here we merely remark that, since the flux cannot be expelled from the interior, superconductivity will nucleate in the presence of the magnetic field, i.e. the superconductor will be in a metastable state. The total number of fluxoids in the neutron star interior will be $N_f \simeq 10^{31} B_{12}$ ($B_{12} = B/10^{12}$ G).

12.3.3   *Field generation after birth*

Woodward [79] was the first to explore the possibility that the magnetic fields of neutron stars may be generated after their birth. The next major step was taken by Blandford *et al.* [16] who pursued an idea originally due to Urpin and Yakovlev [74], namely that the outward flux of heat through the crust of a cooling neutron star can give rise to electric fields which could amplify a pre-existing seed field. The essence of the underlying physical mechanism is the following: the temperature gradient ∇T_0 radially inwards causes heat to flow from the interior to the surface of the crust. The heat transport is due to the electrons; in the presence of a seed magnetic field $\vec{B}_0$ these electrons are deflected in a direction perpendicular to $\vec{B}_0$. The deflection of the outward moving electrons (hotter and hence faster) is larger than and in a direction opposite to that of the deflection suffered by the inward flowing electrons (cooler and hence slower). This creates a heat flux Q_1 and an associated temperature gradient ∇T_1 perpendicular to both ∇T_0 and $\vec{B}_0$. The additional pressure gradient due to ∇T_1 is balanced by a thermoelectric field which has a non-zero 'curl'. This generates a magnetic field $\vec{B}_1$: $\nabla \times \vec{E}_{\text{th}} = \partial \vec{B}_1 / c \partial t$. If $\vec{B}_1 \cdot \vec{B}_0 > 0$, this process will amplify the seed field (thermomagnetic instability). To obtain a net field growth, the amplification rate must exceed the ohmic dissipation rate.

Linearized computations by Blandford *et al.* [16] showed that this process in the solid crust could generate the observed fields in neutron stars on astrophysically interesting time scales only if the crustal conductivities are at least a factor of three higher than the existing estimates, or if the neutron star surface is covered by a large quantity of helium. These authors suggest that a faster process may operate if there is a liquid layer on top of the solid crust. The thermoelectric effect mentioned above would cause circulatory motions to occur in the liquid, which could then amplify the seed field by a dynamo process. In principle, if the field so generated could be transported to the solid crust and anchored there quickly enough, one would expect the magnetic field to grow. Urpin *et al.* [73] considered the thermomagnetic instability in the liquid layer in the presence of hydrodynamic motions, as well as stellar rotation, and concluded that large magnetic fields $\sim 10^{12}$ G could be generated in $\lesssim 100$ days.

It appears from these computations that irrespective of whether the field generation takes place in the solid or the liquid phase, the scale length of the generated field will be of the order of the melting depth, ~ 100 m, since this characterizes the region of the highest temperature gradient ∇T_0. Whether this small scale field (or 'spots') will evolve to produce a large scale field remains a matter of conjecture.

If the abovementioned process is, indeed, operative and effective, then the structure of the magnetic field in the stellar interior will be very different from that expected for a fossil field. This generated flux will be anchored in the solid crust by means of either the inward transport of the flux from the liquid layer due to ohmic diffusion and Hall convection, or the gradual solidification of the liquid layer itself. The latter is relevant if the magnetic field is generated very quickly (say within a few days) after the birth of the neutron star. In any case, these processes will confine the magnetic flux to the outer layers of the crust; since the conductivity will rise very rapidly with depth, only a small amount of further inward diffusion will occur in a Hubble time [64]. As a result, most of the magnetic flux would never enter the fluid interior, and the amount of flux passing through the core would not exceed that responsible for the original seed field, $\lesssim 10^8$ G, in the crust.

A serious objection to the idea that the observed fields are generated by thermomagnetic instabilities has been recently raised by Yabe *et al.* [80]. These authors show that to generate magnetic fields in excess of $\sim 10^3$ G with a scale size of ~ 100 m by the dynamo process the fluid velocities must be highly relativistic. They argue that if the condition that fluid velocities must not become supersonic is imposed, then the maximum field strength generated will be only ~ 10 G. Based on this consideration, Yabe *et al.* conclude that it is very unlikely that the observed fields are generated by thermomagnetic processes.

Given the various theoretical uncertainties surrounding this mechanism, one may ask whether there is any observational evidence for the magnetic field of a neutron star being generated after its birth. If the field generation takes place within, say, a few years, then, for all practical purposes, one will never be able to conclude anything from observations. Obviously, the neutron star which triggered the supernova SN 1987A holds a very important clue, and its direct detection is eagerly awaited. On the other hand, if the field is generated over a time scale $\sim 10^4$–10^5 yr, as proposed by Blandford *et al.* [16], then one can expect to find interesting pulsar–supernova remnant associations. To be specific, one could expect to find young pulsars in old supernova remnants. In fact, there may already be such a case. The pulsar PSR1509−58 in the supernova remnant (SNR) MSH15−52 has a spin-down age of ~ 1700 yr, while, according to standard estimates, the age of the SNR is $\sim 10^4$ yr. Blandford *et al.* [16] have invoked field growth to explain this discrepancy in the ages. However, it has also been pointed out [69, 9] that the conventional estimate of the age of the SNR may be grossly incorrect, and the observed properties of MSH 15−52 are well fit by assuming an age of the same order as the pulsar's spin-down age and an expansion in an ambient medium of relatively low density $\sim 10^{-2}$ atoms cm^{-3}. Recently, Thorsett [72] has argued that MSH15−52 may, in fact, be the remnant of the supernova AD185. If this is the case, then the discrepancy between the age of the remnant and the age of the pulsar would be removed.

We would like to conclude this section by remarking that at present there is no compelling observational evidence to suggest that the magnetic fields of neutron stars are generated after their birth.

12.4 Evolution of the magnetic field

12.4.1 *Early evidence from radio pulsars*

As mentioned earlier, Ostriker and Gunn [54] were the first to suggest that the magnetic fields of pulsars may decay; their estimate yielded a characteristic decay time scale $\sim 10^6$ yr. There was apparent evidence for such a decay from the dozen or so pulsars which were known at that time [34]. Although this conclusion should have been regarded as very tentative, particularly in view of the large scatter in the period derivatives in the observed sample compared to the spread in the periods, the inference that there is observational evidence for the decay of the magnetic fields of pulsars became quickly accepted and became an essential ingredient in most evolutionary scenarios for radio pulsars. The case for field decay seemed to be

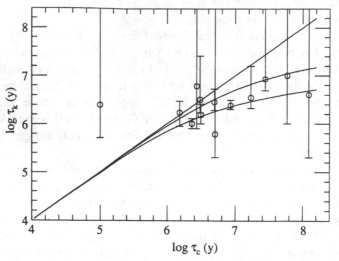

Fig. 12.3. Observed relationship between the kinetic ages and the spin-down ages of 13 pulsars, from Lyne, Anderson and Salter [46]. The three curves correspond to (from top) the expected relationship in case of no field decay, and exponential field decay with time scales of 8 Myr and 2 Myr, respectively.

strengthened when the measurement of proper motions of pulsars became available. The velocity transverse to the line of sight has now been measured for over 75 pulsars [4, 5, 24, 46]. These measurements confirm the early conjecture that pulsars are high velocity objects, and that the observed large scale height of pulsars compared to that of their progenitors must be understood in terms of their migration from their birth place. Thus, the distance z of a pulsar from the mean galactic plane can be used as an indicator of its 'kinetic age' t_k defined as the ratio of the z-distance and the z-velocity. From a plot of the spin-down ages versus the kinetic ages (Figure 12.3), Lyne, Anderson and Salter [46] came to the conclusion that there is evidence for field decay, with a preferred decay constant ~ 5 Myr. Unfortunately, there is a large inherent error in the estimates of the 'kinetic ages' of the pulsars in question. This arises due to the fact that the radial component of the velocities are unknown, and also the fact that there is considerable uncertainty in the birth place of the pulsars in question. In view of this, no strong conclusion should be drawn from the apparent discrepancy between the kinetic ages and the spin-down ages. Nevertheless, field decay in radio pulsars became an accepted paradigm. For example, field decay was an essential ingredient in trying to understand the distribution of the derived magnetic fields and the periods of the observed pulsars (Figure 12.4). As can be seen, although the majority of pulsars have fields in excess of 10^{12} G there are a substantial number with fields in the range $10^{10.5}$–10^{12} G. The most logical way to understand the low fields of these pulsars seemed to be in terms of field decay, as illustrated in Figure 12.4. The alternative explanation in terms of these pulsars being born with the presently observed fields seemed to be ruled out in view of the fact that there are no corresponding low field pulsars with short periods in the observed sample. If the field does not decay, then the trajectories of the pulsars must be horizontal in this

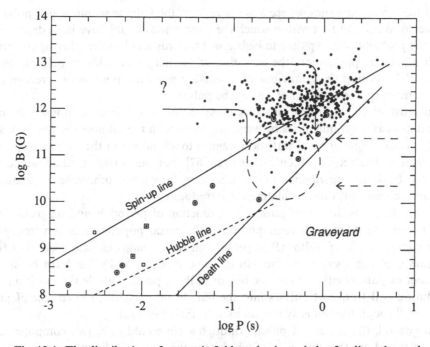

Fig. 12.4. The distribution of magnetic fields and spin periods of radio pulsars, shown with evolutionary tracks corresponding to different models. The pulsars within the dashed circle have traditionally been believed to have come from higher field pulsars through a spontaneous decay of their magnetic fields (curved lines). It now seems more likely that these pulsars have been processed in binaries which were disrupted in a second supernova explosion. Mass-transfer evolution in binaries could have reduced the magnetic fields of these neutron stars and spun them up from much longer periods (dashed arrows).

diagram, and therefore one would expect to find pulsars with low fields and short periods. The absence of such pulsars in the distribution, in spite of the fact that these should spend much longer times as short period objects than the high field pulsars, seemed to lend credence to the field decay hypothesis [56].

To summarize, the early evidence from radio pulsars seemed to indicate that their magnetic fields may be decaying on time scales less than their lifetimes as radio pulsars.

12.4.2   The present situation

However, several recent and more careful analyses of the present sample of pulsars (now more than 500) seem to suggest exactly the opposite, namely that *there is no evidence for field decay in the population of radio pulsars*.

This result has been arrived at independently from two entirely different approaches to pulsar statistics. We briefly describe them below.

12.4.2.1   Current analysis

The first approach attempts to construct a 'galactic' population of pulsars by assuming that for each known pulsar, the galaxy contains S/f undetected ones with the same properties, where S is the ratio of the volume within which pulsars are expected to exist to that within which the given pulsar could have been detected, and f is the probability of a pulsar to be beamed towards an observer. Having constructed such a galactic population, the evolution of the magnetic fields of pulsars could be inferred by noting the behaviour of a suitably weighted population average of the magnetic field as a function of age of the pulsars.

In most of the statistical analyses done so far, it has been customary to derive population average quantities by weighting them with the number distribution, which amounts to assigning equal statistical weights to all pulsars in the galactic population. However, it has recently been realized [26, 67] that the correct method would be to use the *birth rate distribution* as the weighting function – otherwise an undue bias towards objects with longer lifetimes is introduced.

The galactic birth rate of pulsars as a function of period P and magnetic field B can be estimated from the assumption that the pulsar population is in a steady state. One chooses a bin of width ΔP in period and ΔB in magnetic field centred at (B, P). Pulsars evolving away from this bin must be replenished by the birth of an equal number of pulsars either inside the bin, or in those places outside the bin from which evolution will transport pulsars into the bin. In other words, the *current* of pulsars flowing through the bin is synonymous with their birth rate.

In general, the current of pulsars through a bin would have two components, one due to $\dot{P}$ and another due to $\dot{B}$, of which only the former is directly measurable. This component of the current is defined as

$$J(B,P) = \sum_i J_i = \frac{1}{\Delta P} \sum_i \frac{S(P_i, \dot{P}_i)\dot{P}_i}{f(P_i)}, \tag{12.5}$$

where $S(P, \dot{P})$ is the ratio of the volumes mentioned above, and $f(P)$ is the period-dependent beaming probability (for a detailed discussion, see Narayan [52]). The sum extends over all *observed* pulsars in the bin. While the $\dot{B}$-component of the current is not directly measurable, an estimate of the importance of this component can be obtained by comparing the value of $J(B, P)$ in adjacent bins with the same B. If the unknown $\dot{B}$-component is very significant, it will show up as a rapid drop in $J(B, P)$ with increasing P at constant B. An analysis of the observed pulsar population shows that this *does not* happen – the quantity $J(B, P)$ defined above remains the dominant component of the current during the entire active lifetime of a pulsar [11]. Figure 12.5 shows a plot of $J(B, P)$ in the B–τ plane, where $\tau \equiv P/2\dot{P}$ is the spin-down age.

One can now use this approach to estimate the time scale of the decay of the magnetic field. Assuming that the decay of the field is exponential with a time-constant t_d, and that the initial period of the pulsar is zero, the initial field B_0 is related to the present value B through the relation

$$B_0 = B \left[1 + \left(\frac{2\tau}{t_d} \right) \right]^{1/2}. \tag{12.6}$$

It is reasonable to assume that the above equation is also satisfied by the *average value of the field* (averaged over all pulsars with the same spin-down age). Then the average initial field $\bar{B}_0$ is given by

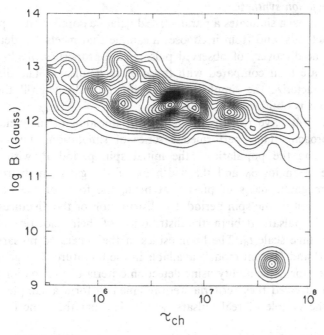

Fig. 12.5. Pulsar current as a function of the magnetic field and the spin-down age in years. A current-weighted average of the magnetic field as a function of the spin-down age suggests a field decay time scale $\gtrsim 20$ Myr. From Srinivasan [67].

$$\bar{B}_0 = \bar{B} \left[1 + \left(\frac{2\tau}{t_d} \right) \right]^{1/2}. \qquad (12.7)$$

The quantity $\bar{B}(\tau)$ is obtained by taking a current-weighted average of $B(\tau)$:

$$\bar{B}(\tau) = \frac{\sum_i B_i(\tau) J_i(\tau, B_i)}{\sum_i J_i(\tau, B_i)}. \qquad (12.8)$$

The summation is carried out over all observed pulsars with spin-down age within a narrow bin around τ. J_i is the current of the ith pulsar in the bin, as defined in Eq. (12.5). These values in different τ bins are used to define a χ^2 as

$$\chi^2 = \sum_\tau \left[\bar{B}_0 \left/ \left(1 + \frac{2\tau}{t_d} \right)^{1/2} - \bar{B}(\tau) \right. \right]^2, \qquad (12.9)$$

which is minimized with respect to t_d and B_0. The result of such a procedure yields $t_d > 20$ Myr [26].

One of the methods followed earlier for the estimation of t_d was to minimize the *width of the distribution of the initial field* B_0 with respect to t_d [41, 56]. This can also be repeated by weighting each pulsar by its corresponding *current*. Such an exercise again yields $t_d > 20$ Myr, consistent with the estimate by Krishnamohan [41].

12.4.2.2   Population synthesis

The second approach simulates a parametrized galactic population of pulsars using Monte Carlo methods, and from it chooses a sample that meets the detection criteria which led to the discovery of observed pulsars; the properties of objects in the simulated sample are then compared with those of real pulsars. The different parameters which characterize the simulated population are adjusted till the best match with observations is obtained.

Bhattacharya *et al.* [14] have recently addressed the question of magnetic field decay using this approach. They performed a series of simulations with several parameters characterizing the population: the initial spin period P_i with which pulsars are born, the mean $\log B_0$ and the width σ_B of the gaussian distribution in logarithm of the magnetic fields of pulsars at birth, the form of the beaming probability f as a function of the spin period, the distribution of the distances from the galactic plane of the pulsars at birth, the distribution of their velocities, and, of course, the field decay time scale t_d. The luminosities of the generated pulsars were computed from their P and $\dot{P}$ from models available in the literature. The generated pulsars were then tested for detectability using detection criteria of four major pulsar surveys. Those which survived these criteria were retained to form a sample, which was compared with the sample of real pulsars, which also met the same detection criteria.

For each assumed value of t_d, the values of B_0 and σ_B were varied to find the best match with the sample of real pulsars. The properties compared were the distributions of $\log P$, $\log B$, DMsin$b$ (where DM is the dispersion measure and b is the galactic latitude), $\log L$ (where L is the luminosity of a pulsar) and the logarithm of the spin-down age.

Having obtained the best-fit population for each t_d, the quality of fit was compared between these different 'best fits'. This entire exercise was repeated for a few different assumptions about the other simulation parameters.

The result of this exercise clearly showed that simulations with longer values of t_d produce significantly better fits with observations than those with shorter decay times (see Figure 12.6 for a comparison of best fits for t_d of 10 and 100 Myr). The overall conclusion reached was that the field decay time scale t_d must exceed ~ 100 Myr. This is tantamount to saying that there is no appreciable decay of the magnetic field during the active lifetime of a normal radio pulsar.

These results, therefore, indicate that most radio pulsars are too shortlived to undergo any field evolution during the time they are visible to us. This conclusion, however, does not preclude the possibility of field decay on much longer time scales. To test this, one has to look for evidence of field decay among 'dead pulsars'. During the last couple of years, the existence of strong magnetic fields $\gtrsim 10^{12}$ G has been inferred in some of the γ-ray burst sources (see Sect. 13.4) through the discovery of cyclotron absorption lines in their X-ray spectrum [49]. If these γ-ray bursters are very old solitary neutron stars, as argued by some authors (e.g. [60]), then one may be forced to conclude that the magnetic fields of solitary neutron stars do not decay significantly even on time scales comparable to the age of the galaxy.

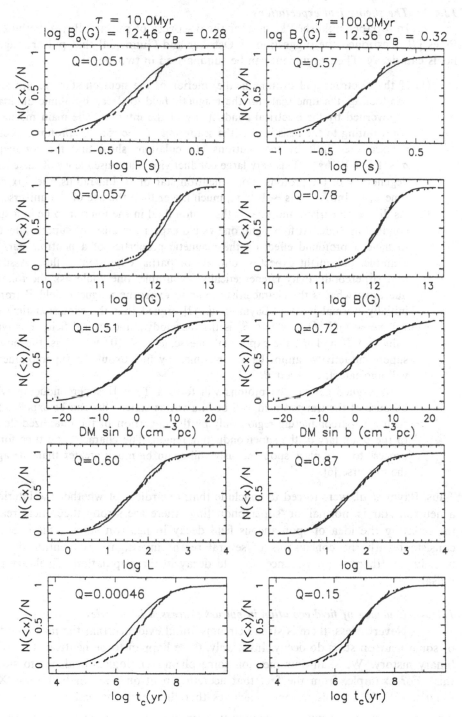

Fig. 12.6. Kolmogorov–Smirnov tests on the similarity between the distributions of (from top to bottom) the pulse period, magnetic field strength, 'vertical component' of the dispersion measure, radio luminosity and spin-down age for simulated and real pulsars, for the best simulations with field decay time scales of 10 Myr (left) and 100 Myr (right). The dots show individual real pulsars, the solid lines the simulated pulsars. Each graph is headed with the probability Q that the two distributions shown in the graph are drawn from the same population. (From Bhattacharya *et al.* [14].)

12.4.3   The theoretical expectations

In a paper published soon after the discovery of pulsars, Baym, Pethick and Pines [7] questioned the suggestion by Ostriker and Gunn [54] that pulsar magnetic fields can decay. Their argument can be summarized in two steps:

(1) If the electrons and protons in the interior of the neutron star are not super-conducting, the time scale for the magnetic field to decay by ohmic dissipation is governed by the electrical conductivity of the interior. The main mechanism contributing to the resistivity is the scattering of electrons by protons; electrons are scattered little by the neutrons. Calculations show that the conductivity $\sigma \sim 1.5 \times 10^{29}$ s^{-1}. This very large conductivity is a consequence of the extreme degeneracy of the protons. For a neutron star of 10 km radius, the flux decay time $\tau_d \sim 4\pi\sigma R^2/c^2$ is $\sim 10^{13}$ yr, much larger than the age of the universe.

(2) As discussed earlier, one expects the proton fluid in the interior to be in a super-conducting state. At first sight, one would expect the onset of superconductivity to have a profound effect on the magnetic properties of a neutron star. For example, one might expect a complete or partial expulsion of flux. Assuming that superconductivity is energetically favourable, one could ask the following question: what is the characteristic time to expel the magnetic field B from an initially normal region? Baym *et al.* [7] argued that the expulsion time scale will be $\sim \tau_d(B^2/B_c^2)$, where B_c is the thermodynamic critical field. For typical values of B and B_c, the expulsion time scale is $\sim 10^8$ yr. Thus, the onset of superconductivity cannot be accompanied by flux expulsion–superconductivity will nucleate at constant B.

As argued earlier, the protons will form a Type II superconductor, which, because of the trapped field, will be in a vortex state. Flux can be expelled from such a superconducting region only by the migration of the quantized fluxoids to the boundary of the superconducting region. The characteristic time for this process to take place spontaneously will again be much greater than the age of the universe [6].

Thus, Baym *et al.* were forced to conclude that, regardless of whether the interior of a neutron star is 'normal' or 'superconducting', there are strong theoretical reasons for rejecting the idea of spontaneous field decay in neutron stars. This is entirely consistent with the conclusions of several recent investigations mentioned above, namely that there is no evidence for field decay in the population of solitary radio pulsars.

12.4.4   Evidence of field evolution in pulsars processed in binaries

Nevertheless, there is strong circumstantial evidence that the magnetic fields of some neutron stars do decay. Invariably, these happen to be neutron stars with a binary history. We shall now mention some phenomenological evidence to support this. For example, from the fact that accreting neutron stars in high-mass X-ray binaries (HMXBs) pulsate, one concludes that they must be endowed with strong magnetic fields: accretion is collimated towards the polar cap by the magnetic field, and the modulation of the X-ray intensity is due to the rotation of the neutron star. In contrast, neutron stars in low-mass X-ray binaries (LMXBs) with very few exceptions (see Ch. 1) *do not* pulsate, and this is understood in terms of their magnetic fields being several orders of magnitude smaller than their counterparts with massive companions. Since the neutron stars in HMXBs are relatively young compared to those in LMXBs, it is natural to conclude that the magnetic fields of neutron stars in binaries do decay.

This conclusion has been strengthened by the observed characteristics of binary radio pulsars. The first clue to this came with the discovery of the Hulse–Taylor pulsar PSR 1913+16. The most satisfactory and the most widely accepted explanation for the anomalous combination of its short period of rotation and low magnetic field has been in terms of the 'recycling' or 'spin-up' scenario: this pulsar is the first born in the binary system, and it died a natural death when it crossed the *death line* due to period lengthening; it has subsequently been *resurrected* from its *graveyard* as a consequence of its having been spun up during the mass accretion phase; 'somehow' its magnetic field had decayed in the interval between its formation and the onset of mass transfer from its companion. Its rapid spin rate is a direct consequence of its low magnetic field during accretion (for a detailed discussion, as well as references, see refs. [13, 66]; see also Ch. 11). This recycling hypothesis has been substantially strengthened by the discovery of several millisecond pulsars with very low magnetic fields. It should be mentioned, however, that this explanation for the anomalous combination of short periods and low fields of several pulsars is not unanimously accepted.

An alternative point of view would be that these pulsars were just born with their presently observed fields. There are at least two main difficulties with this scenario:

(1) *The fraction of binaries among low-field pulsars:* whereas about 3% of all known pulsars in the galactic disk are in binary systems, pulsars in binaries account for $\sim 70\%$ of known disk pulsars with $B \lesssim 10^{10.5}$ G. The majority of these have white dwarf companions in highly circular orbits. In these cases the neutron stars have without doubt experienced mass transfer resulting in spin-up, while this has most probably happened also in the rest of these binaries, namely the double neutron star systems. This huge preponderance of 'binary-processed' neutron stars among the low-field pulsars points to an intimate connection between the reduction in the field strength and their binary history. If these pulsars were merely 'born' with such low fields, it is hard to see why hardly any solitary pulsar makes its appearance with a similar field strength.

(2) All the low-field pulsars in the galactic disk are located exclusively to the right of the spin-up line in Figure 12.1. This is precisely what one would expect if they were spun up. If pulsars were born with such low magnetic field strengths, one would have expected to see a fraction of them to the left of the spin-up line, i.e. with shorter periods.

In citing evidence for field evolution in neutron stars in binaries, we have so far discussed only the pulsars which are *still* in binary systems. Let us now return to the subset of solitary pulsars in the main population with relatively low fields (the pulsars with $B < 10^{11.5}$ G inside the dashed-circle in Figure 12.4). The paucity of such low-field pulsars with periods < 100 ms was earlier regarded as evidence for rapid field decay [56]. As we mentioned, recent analyses show that this conclusion is no longer tenable; but it remains difficult to accommodate these pulsars into a single-population, long-t_d model in which all pulsars are born with $P < 100$ ms and evolve by spin-down [14, 26]. If one rejects the possibility that these low field pulsars evolved from the *left* of the diagram (with or without field decay), then one is forced to conclude that these must have been *injected* into the population very close to their present location near the spin-up line at a spin-down age $\sim 10^{7.5}$ yr. This coincidence of these pulsars being injected close to the spin-up line suggests that they may be recycled pulsars from binary systems which disrupted during the second explosion [11, 26]. Further evidence for this comes from the fact that an analysis of the 'current' of these pulsars suggests that they are born at varying distances from the galactic plane all the way up to 600 to 700 parsec. The most reasonable explanation for this seems to be that these are the first-born pulsars in binary systems that had migrated away from the galactic plane during the time interval between the first and second supernova explosions, and released from them during the second explosion.

There is an additional property of these pulsars that strengthens this scenario. As remarked earlier, pulsars are high-velocity objects with typical velocities ~ 200 km s^{-1}. However, there is a population of low-velocity pulsars ($v \sim 50$ km s^{-1}) which predominantly have low fields ($B < 10^{11.5}$ G) and spin-down ages $\gtrsim 10^{7.5}$ yr [1, 3, 57]. The most attractive explanation for this correlation between low velocities and low fields is due to Bailes [3]. He has argued that this combination can be best understood in terms of these pulsars being the first-born in wide binaries.

While these conclusions need to be examined more carefully, they do lend credence to the hypothesis that whereas the magnetic fields of solitary pulsars *do not* decay, the fields of neutron stars born and processed in binaries *do* decay. Further, there is a wide range in the magnitude of the magnetic fields after the mass transfer and spin-up phase is over.

12.4.5 Asymptotic fields

While the explanation for the observed low magnetic fields in binary pulsars as due to decay may be contentious, there is strong observational evidence that there is no decay after the mass transfer phase is over:

(1) The estimated cooling age of the rather low-surface-temperature white dwarf companion of PSR0655+64 is in excess of $\sim 10^9$ yr. Since one expects the pulsar to be the first-born member of the binary, it must be even older than the white dwarf. Nevertheless, the observed magnetic field of this pulsar is $\sim 10^{10}$ G. This suggests that the presently observed field must be a long-lived *residual* or *asymptotic* field [42]. The upper limit on the surface temperature of the white dwarf companion of PSR1855+09 forces one to the same conclusion [21]. A recent determination of the surface temperature of the white dwarf companion to the pulsar PSR0820+02, which has a magnetic field strength $\sim 10^{11.5}$ G, suggests that it could be $\gtrsim 2 \times 10^8$ yr old, again indicating a long-term stability of the pulsar's magnetic field [40].

(2) The second piece of evidence comes from the statistics of millisecond pulsars [12, 75]. If one accepts the scenario that millisecond pulsars have been spun up to their presently observed periods, then their progenitors must be the LMXBs; only then is the required accretion of mass and angular momentum possible. The small volume of the galaxy in which the known millisecond pulsars are located,

and various 'selection effects' against the detection of rapidly spinning pulsars, suggest that the total number of active millisecond pulsars in the galaxy may exceed that of their progenitors by a factor of at least 100 [12, 25, 43]. This is only possible if the active lifetimes of millisecond pulsars are larger than those of their progenitors by a similar factor. Since the duration of the X-ray phase of LMXBs is estimated to be in the range 10^7–10^8 yr, millisecond pulsars must function for much longer than 10^9 yr. This would not be possible if their fields continue to decay after they have been spun up to ultrashort periods.

(3) The neutron star in the X-ray binary Her X-1 seems to be at least $\sim 6 \times 10^8$ yr old [76]. However, the X-ray spectrum of this object shows the presence of a cyclotron line, indicating a magnetic field strength $\sim 10^{12}$ G.

To summarize, it seems very likely that the magnetic fields of the first-born pulsars in interacting binary systems may decay due to as yet unidentified mechanisms. However, there are convincing reasons to believe that there is no further decay after the mass transfer phase is over.

12.5 Possible mechanisms for field decay

Before outlining some of the suggestions that have been made for the underlying physical mechanism that may be responsible for field decay, it may be worth recalling the various things that must find an explanation in any theoretical scenario:

(1) There is no evidence for field decay in solitary neutron stars.
(2) Very low field pulsars ($B \lesssim 10^{10.5}$ G) occur almost exclusively in binaries. This suggests a causal connection between the field decay mechanism and the evolution of a neutron star in a binary.
(3) Whereas the magnetic fields of neutron stars in binaries may decay, this does not seem to continue indefinitely. There appears to be an asymptotic or residual value of the field beyond which no further decay occurs.
(4) The strength of the residual field spans a wide range from 10^8 to $\gtrsim 10^{10}$ G, and could even be as high as 10^{12} G.

12.5.1 Field decay due to instabilities

We shall first discuss a possible mechanism for the evolution of pulsar magnetic fields suggested by Flowers and Ruderman [29]. In this model, the interior of the neutron star is assumed to be in a *normal* state. As is well known, a uniform conducting fluid with an axisymmetric polar magnetic field is unstable against fluid motions which will diminish the exterior dipole field. The characteristic time scale τ_{MHD} for the magnetic field to readjust itself through internal motions in a rotating conducting fluid is given by

$$\tau_{\mathrm{MHD}} \sim \frac{2\Omega}{k v_{\mathrm{A}}}, \qquad (12.10)$$

where v_{A} is the Alfvén velocity in the absence of rotation, Ω is the angular velocity of rotation, and k is the characteristic wavevector. Thus, rapid rotation will inhibit this instability.

Such an unstable poloidal field of a liquid star can in principle be stabilized by a comparable internal toroidal field. The currents that are responsible for such a stabilizing toroidal field will, of course, be poloidal. In a neutron star, these stabilizing currents are expected to flow along the poloidal field lines which close within the star. Thus, stabilizing currents flow through the crust of the star. As these stabilizing crustal currents decay due to ohmic dissipation, the dipole field can decrease in magnitude due to internal fluid motions. While this mechanism is quite attractive, it is hard to understand why field decay is observed only in neutron stars that have undergone a binary history.

An alternative suggestion is due to Chanmugam and is an elaboration of earlier suggestions along similar lines (ref. [22] and references therein). The suggestion is that, provided a neutron star is spinning sufficiently slowly ($P > 1$ s) and is strongly magnetized with the surface field in excess of 10^{12} G, its fluid interior can become *convectively* unstable. Such a convective instability could, in principle, transport the field radially upwards towards the crust of the star, where it can decay due to ohmic dissipation. If one takes the estimates mentioned above seriously, then the mechanism just mentioned will result in field decay in all neutron stars – solitary or binary. But it is conceivable that such a convective instability can set in only at much longer periods than suggested by simple estimates. In that case, one would expect field decay only in neutron stars that have been braked to rather long periods, such as the wind-driven pulsating X-ray sources.

12.5.2   *Possible field decay due to mass accretion*

Even before the discovery of the first binary pulsar, Bisnovatyi-Kogan and Komberg [15] had anticipated that pulsars in binaries may have low fields. The particular suggestion they made is that the accreted matter may 'screen' or 'bury' the field. This scenario has been revisited by several authors in recent times. From evolutionary models, Taam and Van den Heuvel [71] have argued that there is a correlation between the amount of matter accreted from the companion and the magnitude of field decay. Romani [59] has argued that accretion may actually be able to destroy the crustal magnetic field. For example, heating of the crust due to accretion will hasten ohmic decay in the crust due to a reduction in the electrical conductivity. A compression of the current carrying layers may also result in the reduction of the dipole moment. Blondin and Freese [17] have invoked an inverse thermoelectric battery effect to destroy the crustal flux. Such mechanisms of field decay in the crust would naturally lead to a correlation between the amount of matter accreted and the amount of flux destroyed. Although the abovementioned suggestions appear attractive, there are difficulties. One counter-example is the neutron star in the X-ray binary 4U1626−67. This may have accreted a considerable amount of mass and yet has a magnetic field $\sim 10^{12}$ G [76]. A more basic difficulty is that the various mechanisms just mentioned, while they can destroy or modify the crustal field, are unlikely to have any significant influence on the magnetic field that resides in the core of the neutron star. Even if the accreted matter is able to temporarily screen the field, reducing the exterior dipole moment, the field would resurface if the flux resides in the core. As mentioned at the beginning, there are sufficient reasons to believe that the fossil field of the pre-supernova core that collapsed to form the neutron star will be frozen in the core of the neutron star.

12.5.3 Field decay due to spin-down

As discussed in Sect. 12.3, the fossil field is most likely trapped in the superconducting core of the neutron star as an array of quantized fluxoids. The only way this field can be destroyed is by 'transporting' the fluxoids from the core to the crust where the field can decay due to ohmic dissipation. Although spontaneous expulsion of the flux trapped in a superconductor is not possible, it may happen under certain circumstances. In the suggestion described below, *the mechanism of field expulsion from the interior is related to the spinning down of the neutron star.*

Before elaborating on this, it is necessary to recall the nature of the neutron fluid in the core of the star. Following Migdal's original suggestion, Ginzburg and Kirzhnits showed that the neutrons in the core will be in a *superfluid* state, and because of the rotation of the star this superfluid will be in a *vortex state*, with each vortex having a quantum of circulation $h/2m$ (h = Planck's constant, m = nucleon mass). The necessity for a superfluid in a rotating bucket to perforate itself with a number of vortices was first pointed out by Onsager [53] and Feynman [28] (for a discussion of the underlying physics we refer to Sauls [65]). The number of these Onsager–Feynman vortices N_v will be related to the angular velocity of the star as follows:

$$N_v = \frac{2\pi R(\Omega R)}{h/2m} = 2 \times 10^{16} P^{-1},$$ (12.11)

where R is the radius of the neutron star and P is the rotation period, in seconds.

Thus, there are two families of 'vortices' in the neutron star – the Abrikosov fluxoids parallel to the magnetic axis and the Onsager–Feynman vortices parallel to the rotation axis (Figure 12.7). Although the superfluid vortices and the fluxoids have been invoked in the past to understand some aspects of the rotational dynamics and the decay of the magnetic field, respectively [50], the *interaction* between these two families of vortices and the consequences of such an interaction had not been explored till recently. There are several reasons to believe that the fluxoids and vortices may be strongly 'interpinned' (see the Appendix). If such an interpinning is important, then the expulsion of magnetic flux from the core will be a natural consequence of the spinning down of a neutron star.

It is now well established that the solid crust and the core superfluid are strongly coupled [65]. This arises because of the strong magnetism of the neutron vortices; electron scattering off the magnetic core of the vortices provides the necessary coupling between the crust and the superfluid. Therefore, as the crust spins down the superfluid will respond by destroying the required number of vortices. This will happen by a radial outward movement of the vortices and their annihilation at the interface between the inner crust and the superfluid core. If the fluxoids and vortices are strongly interpinned, then as the vortices move radially outward the fluxoids will also be *dragged* to the crust, and the field will eventually decay there due to ohmic dissipation.

It has been argued [68] that this simple mechanism of flux expulsion from the superconducting interior due to the spinning down of the neutron star is able to explain, at least at a qualitative level, the observational features that one set out to explain.

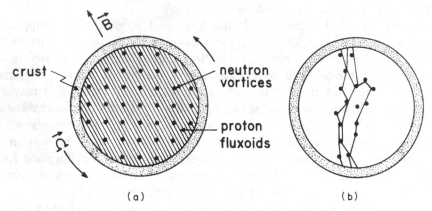

Fig. 12.7. (a) An idealized geometry showing the quantized fluxoids in the proton superconductor and the Onsager–Feynman vortices in the neutron superfluid. The former are parallel to the magnetic axis, and the latter to the rotation axis. (b) A more realistic geometry taking into account the strong interaction between the two families of 'vortices'. (From Srinivasan *et al.* [68].)

12.5.3.1 Solitary pulsars

In principle, the mechanism outlined above will operate even in isolated neutron stars. But, as we shall see, it would be rather ineffective. Since the energy of the electromagnetic radiation and the relativistic wind emitted by a pulsar comes at the expense of its stored rotational energy, the equation governing the spinning down of the neutron star is given by Eq. (12.1). If the evolution of the magnetic field is directly related to the secular spinning down, as assumed in the model, then one can substitute $B(t) \propto \Omega(t)$ in Eq. (12.1). Integrating this, one finds that, at late times,

$$B(t) \propto (t/\tau)^{-1/4}, \tag{12.12}$$

where τ is the characteristic spin-down time scale of the pulsar at late times (typically $\sim 10^6$–10^7 yr). Thus, the magnetic fields of isolated neutron stars will not decay significantly even over a time scale comparable to the age of the galaxy. The underlying reason is that an isolated neutron star will not spin down significantly for a reasonable fraction of the flux to be dragged out from its interior.

12.5.3.2 Neutron stars from binaries

This difficulty will not be encountered if the neutron star is in an interacting binary system. The main evidence for this comes from the 20 or so binary X-ray pulsars whose rotation periods range from a few seconds to 835 s [51] (Ch. 1). It is generally accepted that such a dramatic braking of the neutron star is a consequence of the electromagnetic torque experienced by it during the stellar wind phase of its companion [37]. During this spin-down phase, a substantial fraction of the field trapped in the interior can be expelled to the crust. Whether or not the expelled field will decay in the crust *while* the star spins down will depend upon whether the spin-down time scale is short or long compared to the ohmic dissipation time scale. In either case, given sufficient time the field will decay in the crust.

12.5.3.3 The residual field

When the companion fills its Roche lobe and heavy mass transfer sets in, the neutron star will be spun up (see Sect. 10.9). Consequently, the degree to which the flux is expelled from the interior will be determined by the *maximum* period to which the neutron star was spun down prior to spin-up – in other words, the magnetic field of the neutron star would have attained its asymptotic or residual value.

Regarding why there is no further decay, there are two reasons. First, during the spin-up the superfluid core will respond by creating new vortices, which will now move radially inward, thus pushing the core field even deeper into the superconducting core. Secondly, after the spin-up phase is over, the spin-down rate of the recycled pulsar will be determined by its electromagnetic luminosity appropriate to the reduced value of the field. Hence, further field decay due to spinning down will be even less important than for the high-field solitary pulsars.

We thus find that this simple idea of dragging out the flux trapped in the superconducting interior by exploiting the interpinning between the fluxoids and the superfluid vortices appears attractive.

12.5.4 Discussion

As we see, the three different mechanisms for field decay discussed in this section make specific assumptions about the nature of the magnetic field of neutron stars. Field decay due to mass accretion in a binary system may be relevant only if the field is predominantly confined to the crust. This is likely to be the case only if the field is built up after the birth of the neutron star, possibly due to mechanisms such as those discussed in Sect. 12.3. While discussing these mechanisms, we concluded that, given the various uncertainties in the model, and the fact that there is as yet no observational evidence to suggest that the observed fields are built up after birth, the fossil field hypothesis appears more reasonable. In any case, the detailed mechanism for accretion-induced field decay needs to be worked out more carefully. At present it suffers from the large uncertainities in the solid state properties of the crustal matter that make the field growth mechanisms themselves suspect. But the attractive feature of this model is that it is built upon the apparent correlation between the amount of accreted mass and the magnitude of the field decay.

The other two mechanisms discussed in this section assume that the observed field is the fossil field. The magnetohydrodynamic-instability model due to Flowers and Ruderman [29] further assumes that the interior is not superconducting, and this may be a serious limitation for two reasons: (i) the argument for proton superconductivity in the interior is quite compelling, and (ii) it is not obvious whether such fluid motions can occur in a superconductor. Further uncertainty arises from the fact that the magnetic flux threads the solid crust. Even if the electrical conductivity of the outer crust is small enough for currents passing through it to decay, the conductivity of the inner crust will be so large that the flux will be frozen into it, thus inhibiting rearrangement of the current loops in the fluid interior. In any case, it is not clear why this mechanism should preferentially occur in neutron stars in binaries.

The third mechanism, which relates the flux expulsion from the superconducting interior to the rotational history of the neutron star, hinges on the interpinning between the fluxoids in the proton superconductor and the superfluid vortices. As discussed in

the Appendix, although such an entanglement had been overlooked before, there are
several reasons why the fluxoids and the vortices should be interpinned. The main
conceptual difficulty with this picture is, however, the following: when the fluxoids
are transported from the superconducting interior to the interface between the fluid
core and the crust, how is the field 'transferred' from the superconductor to the
crust? Another major uncertainty – which is common to all models – is the electrical
conductivity of the crust. It has been assumed that once the field is deposited in
the crust it will decay due to ohmic dissipation. This may indeed be possible provided the field can be transported from the inner crust to the outer crust where the
conductivity is expected to be smaller. Jones [38] has suggested that Hall drift may
aid such an outward transport of the field. In this context the recent speculation by
Goldreich and Reisenegger [33] that the magnetic field in the crust may undergo a
turbulent cascade terminated by ohmic dissipation at small scales is of considerable
interest.

12.6   Some consequences of field evolution

So far we have discussed the observational evidence for field evolution and
some of the suggested mechanisms for field decay. In this section we wish to mention
some possible astrophysical consequences of field evolution. Although these ideas,
largely due to Ruderman [60–62], are still at a speculative stage, they are sufficiently
interesting to warrant inclusion in an article devoted to field evolution.

In what follows we shall assume that the field permeates the superconducting core
as an array of quantized fluxoids. This flux threads the crust and opens out as a
normal field. As already mentioned, because of the very high conductivity of the base
of the crust the flux will be frozen into it. We have argued in the previous section
that as the neutron star slows down the fluxoids will be dragged outwards by the
superfluid vortices. Since the magnetic flux is frozen into the crustal matter, this will
result in elastic stresses building up. Each quantized flux tube which terminates at the
base of the crust contains an average magnetic field $B_{c1} \sim 10^{15}$ G. Therefore, if the
crust were to remain rigid and immobile, the shear stress at the bottom of the crust
would grow to

$$S(B) \sim \frac{BB_{c1}}{8\pi} \sim \left(\frac{B}{3 \times 10^{12} \text{ G}} \right) 10^{26} \text{ dyne cm}^{-2}, \tag{12.13}$$

where B is the *average* field through the crust. One of the basic ideas underlying the
recent suggestions by Ruderman is that the magnetic stress may exceed the maximum
possible shear stress the crust could bear before yielding ($S_{max} \sim 10^{23}$–10^{25} dyne cm^{-2}).
When the maximum shear stress of a solid is exceeded, it will *yield* either plastically
or through a series of 'cracks'. Laboratory experiments show that plastic flow is more
likely close to the melting temperature or if the density of dislocations is sufficiently
low for the dislocations to be mobile. On the other hand, far from the melting
temperature or when the density of dislocations is rather large (as is expected to be
the case in the crust of the neutron star), brittle response is more likely. Whether
or not the thick crust of a neutron star will yield by numerous microscopic cracks
or through *faults* developing across the entire thickness of the crust is a matter of
speculation. Let us assume that major cracks and faults do appear in the crust in

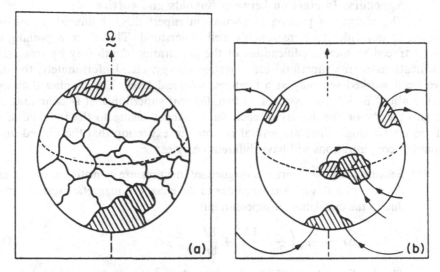

Fig. 12.8. A schematic illustration of the change in the configuration of the magnetic field of a neutron star due to the motion of crustal plates in a spinning down neutron star (from Ruderman [61]).

response to the shear stress arising out of the evolution of the core field. Thus, the crust will eventually break up into *plates*, and these plates will move like a viscous fluid towards the equatorial zone in a spinning-down neutron star. The equatorial region will be a 'subduction zone', where the plates will be pushed into the core and will dissolve into the superfluid sea of neutrons and protons. As the plates develop and move, new matter from the core will flow up into the cracks. Since the magnetic flux was frozen into the original crustal solid, plate motions can obviously alter the magnetic field structure. Such a rearrangement of the stellar field due to plate motions induced by the core field evolution is schematically sketched in Figure 12.8. The example shown is that of a high-field neutron star which has been spun down. It is interesting to note that, in the final configuration, the far field deduced from spin-down torques or accretion torques would yield rather small values representing the average field, while *local fields*, such as those deduced from cyclotron lines, may yield much larger values. It is quite obvious that if these ideas are correct then the magnetic fields of neutron stars in binaries which are first spun down and then spun up could have rather complex structures.

The cracking of the crustal solid and the movement of the plates towards the equatorial region may also have other interesting consequences. For example, Ruderman has suggested that the elastic energy released in a crack would be of the same order as the energy observed in a giant *glitch* of a pulsar. Such a mechanism for pulsar glitch has the added attractive feature that one can roughly predict the interval between the subsequent glitches (this will be the time needed for the elastic stresses to once again build up to the critical value). It is conceivable that such a cracking of the crustal solid may give rise to some γ-ray bursts.

Appendix. Interaction between fluxoids and vortices

The physics of pinning of vortices in superfluids, or fluxoids in supercon-
ductors, to point defects is reasonably well understood. That there is pinning at all
can be traced to local modifications of the penetration depth (say by strain), local
modifications of the superfluid condensation energy, etc. Unfortunately, the details
have been worked out only at a phenomenological level. The pinning that we are
talking about is between 'vortices' in two *different* superfluids. It is clear that unless
the gap energy or the density of each fluid changes owing to the interaction there
will be no pinning. There are several reasons for expecting that the pinned and the
unpinned configurations will have different energies:

(1) Because the proton vortex is magnetized, the pressure of matter in the vicinity of
the core of a fluxoid is reduced due to the enhanced magnetic pressure, and also
due to the circulation of supercurrents:

$$\Delta P \sim \frac{1}{2}\rho_{\mathrm{p}}\left(\frac{h}{2m}\cdot\frac{1}{r}\right)^2 + \frac{B^2}{8\pi}. \tag{12.14}$$

This implies a density fluctuation $\Delta\rho \sim \Delta P/c_s^2$, where c_s is the sound velocity.

(2) It turns out that the neutron superfluid in the core will be in a $^3P_2$ state [65].
The qualitatively new feature of vortices in the $^3P_2$ phase is that the condensate
in the core of the vortex is *spin-polarized*:

$$\langle S_z \rangle = |\psi_{\uparrow\uparrow}(r)|^2 - |\psi_{\downarrow\downarrow}(r)|^2. \tag{12.15}$$

This spin-polarization will lead to a density fluctuation in the core of the neutron
vortex. Since the neutrons have a magnetic moment, the vortex itself carries a
magnetization of order

$$M_{\mathrm{vortex}} \simeq (\gamma_{\mathrm{n}}\hbar)n_{\mathrm{n}}\left(\frac{\Delta_{\mathrm{n}}}{E_{\mathrm{F}}}\right)^2 \simeq 10^{11}\ \mathrm{G}. \tag{12.16}$$

There will be an additional density fluctuation due to the magnetization of the
vortex analogous to the discussion above.

(3) The neutron and proton fluids are strongly coupled due to strong interactions.
An important consequence of this is that the superfluid mass current of the
neutron and the proton are modified due to strong interactions. As a result of
such a superfluid drag, the neutron vortex acquires a very strong magnetization,
$B_{\mathrm{vortex}} \sim 10^{15}\ \mathrm{G}$.

Thus, the fluxoids will not only see the 'normal' cores of vortices as possible pinning
sites due to density fluctuations, but there will also be a strong electromagnetic
interaction between them due to the huge magnetic fields associated with both the
fluxoids and the vortices. As a result of this, strong pinning between the fluxoids and
the neutron vortices seems likely. According to simple estimates by Sauls [65] the
pinning energy per intersection may be as large as ~ 0.1–1 MeV. The rough estimate
of the *pinning force* per connection F_{pin} may be obtained by dividing the pinning
energy by the neutron coherence length ξ_{n}, which is a measure of the size of the
interaction region

$$F_{\mathrm{pin}} \approx \frac{\epsilon_{\mathrm{pin}}}{\xi_{\mathrm{n}}} \sim (0.1\text{–}1) \times 10^6\ \mathrm{dyne/connection}. \tag{12.17}$$

Admittedly, the arguments given above for the interpinning between the fluxoids and vortices are rather qualitative. Given the numerous astrophysical consequences of such a pinning this subtle question deserves careful study.

References

[1] Anderson, B. and Lyne, A. G., 1983, *Nature*, **303**, 597.
[2] Arons, J. and Lea, S. M., 1980, *Astrophys. J.*, **235**, 1016.
[3] Bailes, M., 1989, *Astrophys. J.*, **342**, 917.
[4] Bailes, M., Manchester, R. N., Kesteven, M. J., Norris, R. P., and Reynolds, J. E., 1989, *Astrophys. J.*, **343**, L53.
[5] Bailes, M., Manchester, R. N., Kesteven, M. J., Norris, R. P., and Reynolds, J. E., 1990, *Nature*, **343**, 240.
[6] Bardeen, J. and Stephen, M. J., 1965, *Phys. Rev.*, **140**, A1197.
[7] Baym, G., Pethick, C., and Pines, D., 1969, *Nature*, **224**, 673.
[8] Beskin, V. S., Gurevich, A. B., and Istomin, Ya. N., 1983, *Sov. Phys. JETP*, **58**, 235.
[9] Bhattacharya, D., 1990, *J. Astrophys. Astron.*, **11**, 125.
[10] Bhattacharya, D., 1989, In Hunt, J. and Battrick, B., editors, *X-ray Binaries, Proc. 23rd ESLAB Symp.*, p. 179, ESA, Noordwijk, The Netherlands.
[11] Bhattacharya, D. and Srinivasan, G., 1991, In Ventura, J. and Pines, D., editors, *Neutron Stars: Theory and Observation*, p. 219, Kluwer Academic Publishers, Dordrecht.
[12] Bhattacharya, D. and Srinivasan, G., 1986, *Curr. Sci.*, **55**, 327.
[13] Bhattacharya, D. and Van den Heuvel, E. P. J., 1991, *Phys. Reports*, **203**, 1.
[14] Bhattacharya, D., Wijers, R. A. M. J., Hartman, J. W., and Verbunt, F., 1992, *Astron. Astrophys.*, **254**, 198.
[15] Bisnovatyi-Kogan, G. S. and Komberg, B. V., 1974, *Sov. Astr.*, **18**, 217.
[16] Blandford, R. D., Applegate, J. H., and Hernquist, L., 1983, *Mon. Not. R. Astr. Soc.*, **204**, 1025.
[17] Blondin, J. M. and Freese, K., 1986, *Nature*, **323**, 786.
[18] Borra, E. F., 1980, *Astrophys. J.*, **235**, 915.
[19] Borra, E. F. and Landstreet, J. D., 1980, *Astrophys. J. Suppl.*, **42**, 421.
[20] Borra, E. F., Landstreet, J. D., and Mestel, L., 1982, *Ann. Rev. Astr. Astrophys.*, **20**, 191.
[21] Callanan, P. J., Charles, P. A., Hassal, B. M. J., Machin, G., Mason, K. O., Naylor, T., Smale, A. P., and Van Paradijs, J., 1989, *Mon. Not. R. Astr. Soc.*, **238**, 25P.
[22] Chanmugam, G., 1984, In Reynolds, S. P. and Stinebring, D. R., editors, *Millisecond Pulsars*, p. 213, NRAO, Green Bank.
[23] Chao, N. C., Clark, J. W., and Yang, C. H., 1972, *Nucl. Phys.*, **A179**, 320.
[24] Cordes, J. M., 1986, *Astrophys. J.*, **311**, 183.
[25] Coté, J. and Pylyser, E. H. P., 1989, *Astron. Astrophys.*, **218**, 131.
[26] Deshpande, A. A., Ramachandran, R. and Srinivasan, G., 1994, *J. Astrophys. Astr.*, submitted.
[27] Deutsch, A. J., 1956, *Publ. Astr. Soc. Pacific*, **68**, 92.
[28] Feynman, R. P., 1955, *Prog. Low Temp. Phys.*, **1**, 17.
[29] Flowers, E. and Ruderman, M. A., 1977, *Astrophys. J.*, **215**, 302.
[30] Ginzburg, V. L., 1964, *Sov. Phys. Doklady*, **9**, 329.
[31] Ginzburg, V. L. and Kirzhnits, J., 1965, *Sov. Phys. JETP*, **20**, 1346.
[32] Goldreich, P. and Julian, W. H., 1969, *Astrophys. J.*, **157**, 869.
[33] Goldreich, P. and Reisenegger, A., 1992, *Astrophys. J.*, **395**, 250.
[34] Gunn, J. E. and Ostriker, J. P., 1970, *Astrophys. J.*, **160**, 979.
[35] Hensberge, H., Van Rensbergen, W., Goossens, M., and Deridder, G., 1979, *Astron. Astrophys.*, **75**, 83.
[36] Hoffberg, M., Glassgold, A. E., Richardson, R. W., and Ruderman, M., 1970, *Phys. Rev. Lett.*, **24**, 175.
[37] Illarionov, A. F. and Sunyaev, R. A., 1975, *Astron. Astrophys.*, **39**, 185.
[38] Jones, P. D., 1987, *Mon. Not. R. Astr. Soc.*, **228**, 513.
[39] Koester, D. and Chanmugam, G., 1990, *Rep. Prog. Phys.*, **53**, 837.
[40] Koester, D., Chanmugam, G., and Reimers, D., 1992, *Astrophys. J.*, **395**, L107.
[41] Krishnamohan, S., 1987, In Helfand, D. J. and Huang, J. H., editors, *IAU Symp 125: Origin and Evolution of Neutron Stars*, p. 377, D. Reidel, Dordrecht.

[42] Kulkarni, S. R., 1986, *Astrophys. J.*, **306**, L85.

[43] Kulkarni, S. R. and Narayan, R., 1988, *Astrophys. J.*, **335**, 755.

[44] Landstreet, J. D., 1970, *Astrophys. J.*, **159**, 1001.

[45] Liebert, J., 1988, *Publ. Astr. Soc. Pacific*, **100**, 1302.

[46] Lyne, A. G., Anderson, B., and Salter, M. J., 1982, *Mon. Not. R. Astr. Soc.*, **201**, 503.

[47] Lyne, A. G. and Manchester, R. N., 1988, *Mon. Not. R. Astr. Soc.*, **234**, 477.

[48] Migdal, A. B., 1959, *Zh. Eksp. Teor. Phys.*, **37**, 249.

[49] Murakami, T., Fujii, M., Hayashida, K., Itoh, M., Nishimura, J., Yamagami, T., Conner, J. P., Evans, W. D., Fenimore, E. E., Klebesadel, R. W., Yoshida, A., Kondo, I., and Kawai, N., 1988, *Nature*, **335**, 234.

[50] Muslimov, A. G. and Tsygan, A. I., 1985, *Sov. Astron. Lett.*, **11**, 80.

[51] Nagase, F., 1989, *Publ. Astr. Soc. Japan*, **41**, 1.

[52] Narayan, R., 1987, *Astrophys. J.*, **319**, 162.

[53] Onsager, L., 1949, *Nuovo Cimento Suppl.*, **6**, 249.

[54] Ostriker, J. P. and Gunn, J. E., 1969, *Astrophys. J.*, **157**, 1395.

[55] Preston, G. W., 1971, *Publ. Astr. Soc. Pacific*, **83**, 571.

[56] Radhakrishnan, V., 1982, *Contemp. Phys.*, **23**, 207.

[57] Radhakrishnan, V. and Shukre, C. S., 1985, In Srinivasan, G. and Radhakrishnan, V., editors, *Supernovae, Their Progenitors and Remnants*, p. 155, Indian Academy of Sciences, Bangalore.

[58] Rankin, J. M., 1990, *Astrophys. J.*, **352**, 247.

[59] Romani, R. W., 1990, *Nature*, **347**, 741.

[60] Ruderman, M., 1991, *Astrophys. J.*, **382**, 587.

[61] Ruderman, M., 1991, *Astrophys. J.*, **366**, 261.

[62] Ruderman, M., 1991, *Astrophys. J.*, **382**, 576.

[63] Ruderman, M. A. and Sutherland, P. G., 1973, *Nature Phys. Sci.*, **246**, 93.

[64] Sang, Y. and Chanmugam, G., 1987, *Astrophys. J.*, **323**, L61.

[65] Sauls, J., 1989, In Ögelman, H. and Van den Heuvel, E. P. J., editors, *Timing Neutron Stars*, p. 457, Kluwer, Dordrecht.

[66] Srinivasan, G., 1989, *Astron. Astrophys. Rev.*, **1**, 209.

[67] Srinivasan, G., 1991, *Ann. N. Y. Acad. Sci.*, **647**, 538.

[68] Srinivasan, G., Bhattacharya, D., Muslimov, A. G., and Tsygan, A. I., 1990, *Curr. Sci.*, **59**, 31.

[69] Srinivasan, G., Dwarakanath, K. S., and Radhakrishnan, V., 1982, *Curr. Sci.*, **51**, 596.

[70] Stibbs, D. W. N., 1950, *Mon. Not. R. Astr. Soc.*, **110**, 395.

[71] Taam, R. E. and Van den Heuvel, E. P. J., 1986, *Astrophys. J.*, **305**, 235.

[72] Thorsett, S. E., 1992, *Nature*, **356**, 690.

[73] Urpin, V. A., Levshakov, S. A., and Yakovlev, D. G., 1986, *Mon. Not. R. Astr. Soc.*, **219**, 703.

[74] Urpin, V. A. and Yakovlev, D. G., 1980, *Sov. Astr.*, **24**, 425.

[75] Van den Heuvel, E. P. J., Van Paradijs, J. A., and Taam, R. E., 1986, *Nature*, **322**, 153.

[76] Verbunt, F., Wijers, R. A. M. J., and Burm, H., 1990, *Astron. Astrophys.*, **234**, 195.

[77] Wambach, J., Ainsworth, T. L., and Pines, D., 1991, In Ventura, J. and Pines, D., editors, *Neutron Stars: Theory and Observations*, p. 37, Kluwer, Dordrecht.

[78] Woltjer, L., 1964, *Astrophys. J.*, **140**, 1309.

[79] Woodward, J. F., 1978, *Astrophys. J.*, **225**, 574.

[80] Yabe, T., Shibazaki, N., and Hanami, H., 1991, *Publ. Astr. Soc. Japan*, **43**, L51.

13

Cosmic gamma-ray bursts

K. Hurley

University of California, Space Sciences Laboratory, Berkeley, CA 94720, USA

13.1 Foreword

A few years ago, the appearance of an article on cosmic gamma-ray bursts in a book on X-ray binaries would not have seemed too inappropriate; today it requires justification. Although the mystery of gamma-ray bursts was far from solved, several pieces of the puzzle appeared to fall into place and strongly suggest a compact object origin. One was the presence of rapid fluctuations in burster light curves, indicative of a small emitting region. Another was the detection of features in the energy spectra, which could be interpreted as emission and absorption lines, giving magnetic field strengths and redshifts compatible with those attributed to neutron stars. Yet another was the evolution of gamma-burst spectra with time, consistent with blackbody X-ray emission following the gamma-ray emission, and suggesting radiation from a region the size of a neutron star polar cap. But several pieces of the puzzle stubbornly refused to fit into place. One was the absence of burster quiescent counterparts in any wavelength range where deep searches were performed. And equally troubling, the burster spatial distribution was isotropic, consistent with a spherically symmetric distribution, but giving no indication of gamma-ray burster distances.

It was generally assumed that with the advent of large-area, long-lived, sensitive satellite experiments, weaker bursts would be detected whose spatial distribution would reveal their origin, and that this might well turn out to be in the galactic disk. Two years after the launch of the Compton Observatory, it is apparent that this is not the case. Not only is the burster spatial distribution still isotropic, but there is also evidence that current instrumentation is sensitive enough to detect bursts to their distance limit (whatever it may be) and beyond. None of the previous evidence in favor of a compact object origin has yet been contradicted; yet none of the newer data support this idea directly either.

Not surprisingly, new ideas about the origin of cosmic gamma-ray bursts have been proposed, and some old ideas resurrected. They often involve an extragalactic origin. Some are consistent with a compact object, binary system scenario, while others are cosmological. And still others attempt to reconcile the new data with the old galactic neutron star hypothesis. The final chapter in this persistent mystery cannot yet be written.

13.2 Introduction

When the discovery of cosmic gamma-ray bursts (GRBs) was announced in 1973 [33], it was not long before a compact object origin was suggested for them based on rather general considerations [37]. When X-ray bursts were discovered several years later, compact objects were again proposed rapidly, and today the accepted explanation for them involves galactic neutron stars. But, whereas the evidence is quite good in the case of X-ray bursters, it remains circumstantial and even problematic for the sources of gamma-bursts. Furthermore, the two phenomena are unrelated in the sense that gamma-ray bursts (GRBs) are not observed to originate from X-ray burst sources, and vice-versa. GRBs occupy a parameter space by themselves, typically with short durations (10 s on average), high-energy spectra (extending to 1 GeV in one case) and an apparently isotropic source distribution. Of the order of 1000 GRBs have now been observed, and with only three noteworthy exceptions, none have been observed to repeat. There is no model-independent estimate of the source distances.

As a class, the gamma-burst sources are extremely diverse. Whether one attempts to categorize them by duration, intensity, or energy spectrum, their characteristic parameters span at least several, and sometimes many, orders of magnitude. This may be an indication that the causes of the phenomenon are themselves quite diverse; although this is pure speculation, it *is* fair to say that there is certainly at present no single model or even class of models which satisfactorily accounts for all the observational properties.

In this review, the observational characteristics (time histories, energy spectra, counterpart searches, and statistical properties) of cosmic gamma-ray bursts will be described and, where appropriate, indications will be given of their possible interpretation. A short comparison of galactic and extragalactic models is provided, and prospects for future progress are speculated upon. Two other reviews have appeared recently; one treats the theoretical aspects of bursts [19], and the other, observations and theories [23]. The reader is referred to both for a second opinion.

13.3   Time histories

A gamma-ray burst may last as short as 8 ms [7] or as long as 1000 s [34]. Within that range, the distribution appears to peak at around 10 s, with evidence for a smaller, second peak around 0.1 s [27]. The time history, or light curve, of each burst appears to be unique. However, there are enough general similarities to encourage a morphological classification based on the time histories alone. One method uses the burst durations to establish classes. Short events, tens of milliseconds long, 1–2 s long events with a single peak, and complex, multipeaked events lasting several tens of seconds appear to define distinct categories [3]. An example of the third type appears in Figure 13.1. A variation on this theme was suggested by Klebesadel *et al.* [32], who defined the classes as brief, double-peaked and quasi-periodic, and long and irregular. They also pointed out that some time histories could be cross-correlated, shrinking the time scale of one to match the other, with the result indicating a degree of similarity consistent with a single origin, even though the events in question were known to originate from different locations. Similar analyses appear in Wood *et al.* [76].

Structure within the time history of an event can be characterized in various ways. When the time history is simple, consisting of a single peak, the e-folding rise and decay times may be calculated. They generally fall into the several milliseconds to one second range [2], with one exceptional event, the March 5, 1979, burst, having

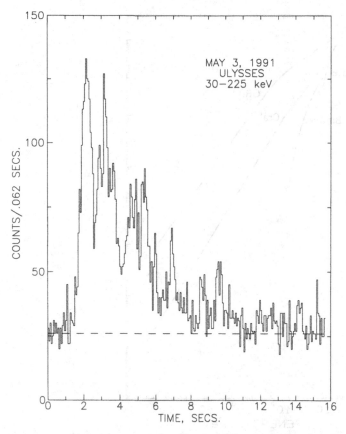

Fig. 13.1. Time history of the May 3, 1991, gamma-ray burst, as observed with the Ulysses detector [28]. Dashed line indicates background level.

a rise time less than 0.25 ms [10]. When the event is more complex, the widths of individual peaks may be calculated. The shortest structures observed have durations < 5 ms [39] and around 0.2 ms [7]. Periodic structure is generally not observed in burst time histories, although there has been one significant exception, the March 5, 1979, event, in which pulsations with an 8 s period were clearly detected [4, 45]. Possible periodicities in other bursts have been reported, e.g. at 4.2 s [77] and 2.2 s [35].

The interpretation of structure in light curves which immediately comes to mind is that the emission arises from an object with characteristic dimension < 60 km (from light travel time arguments), and density $> 5 \times 10^6$ g cm^{-3} (to survive breakup under rotation). As periods of the order of seconds are familiar in X-ray pulsator systems known to contain neutron stars, a neutron star origin for bursters would be consistent with the light curve structure. However, other interpretations are possible. Fine time structure could be caused by gravitational lensing [49], or possibly be inherent to burst generation by superconducting cosmic strings [60]. Furthermore, fine time

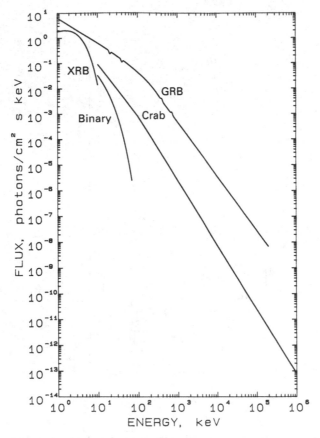

Fig. 13.2. Typical energy spectra of various astrophysical objects. GRB: a gamma-ray burst spectrum; this is a composite of many different observations; absorption and emission features are shown. XRB: an X-ray burster spectrum (1 keV blackbody). Binary: binary X-ray source (8 keV exponential). Crab: Crab nebula and pulsar.

structure is not a characteristic of all bursts, and periodicities appear to be quite rare, indicating that the time histories do not require a single explanation.

13.4   X- and gamma-ray energy spectra

In general, the spectra of bursts have been measured from around 1 keV [31, 78] to over 100 MeV [69, 70], with the current record now standing at 1 GeV [71]. Figure 13.2 shows a generic GRB photon spectrum, and compares it with the spectra of other astrophysical objects. Thermal bremsstrahlung [47], power law [44], and thermal synchrotron [41] are among the functions which provide acceptable fits to the continua. The actual energy emitted in the 3–10 keV range is quite small, amounting to only a few per cent of the energy above 30 keV [38]. Indeed, the power per logarithmic bandwidth in the spectra rises; this X-ray paucity constraint can be used to set limits on GRB emission regions [30], since the gamma-ray emission cannot be significantly degraded into X-rays before escaping.

Just as the time histories are variable, so are the energy spectra. The best fitting

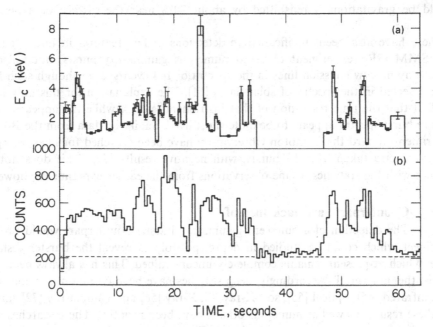

Fig. 13.3. (a) Time evolution of the critical energy E_c for a thermal synchrotron fit. (b) Time history of the March 13, 1982A, GRB. In the first part of (a), E_c displays a quasi-random evolution, while in the second portion, it appears to be well correlated with the time history [29].

parameters for a given function may change in correlation with the time history, or independently of it [29]; an example is shown in Figure 13.3. In addition, X-ray precursors [57] and tails [78] have been noted for several events. Their energy spectra are consistent with those of a 1–2 keV blackbody whose radius is about 1 km × D (kpc), where D is the source distance in kiloparsec. This is clearly suggestive of a galactic neutron star origin.

Absorption features at 20 and 40 keV have been reported in the spectra of some 50 bursts. Their typical full widths at half maximum are 25%, and up to 90% of the continuum flux may be absorbed. Their existence is still controversial, even though they have been found in the data of five different experiments [5, 16, 26, 47]. The classical interpretation is that they are caused by transitions between Landau levels in teragauss fields – again, strongly indicative of a neutron star origin. However, a more exotic interpretation has been proposed recently, namely 'femtolensing' [17]. A femtolens is a hypothetical small, dark matter object, which would gravitationally lens bursters at cosmological distances, producing interference fringes in the energy spectrum.

In a slightly smaller number of bursts, emission features around 400 and 700 keV have been detected. Like the absorption features, they have been detected with five different experiments [26, 47, 53, 73], yet they remain controversial. Here the classical interpretation is that the 400 keV feature is due to positron–electron annihilation, while the higher energy line features may be due to nuclear de-excitations. Both would be gravitationally redshifted by about 20% near the surface of a neutron star.

There have also been significant non-detections of line features in burst spectra. The SMM GRS experiment observed numerous gamma-ray bursts, but failed to detect any narrow emission lines in the spectra of 144 events, even though such lines were detected in the spectra of solar flares [51]. The explanation for this may lie in the fact that the time resolution of that instrument is 16 s, while the appearance of lines in burst spectra appears to be quite short lived. Similarly, data from the BATSE experiment aboard the Compton Observatory have been searched for lines, using 48 energy spectra taken from 52 bursts, with negative results [74]. This does not yet conflict with the statistics of line observations from the earlier experiments, however.

13.5   Counterparts and lack thereof

The detection of a quiescent gamma-ray burster counterpart would provide an object which could be studied in depth, possibly to reveal the burster distance scale, which at present remains completely unconstrained. This has always been, and still is, the 'holy grail'. Accordingly, deep searches have been carried out in the radio [66], infrared [65], optical [55], soft X-ray [8], X-ray [56] and gamma-ray [72] ranges. All these results, as well as numerous others, have been negative. These searches were generally carried out years after the burst. Figure 13.4 compares these upper limits to the intensity of a typical gamma burst. The only exception to this apparent lack of counterparts remains the March 5, 1979, burst [11], whose position is consistent with that of the N49 supernova remnant in the LMC. Because this burst reached a record intensity of 2×10^{-3} erg cm^{-2} s^{-1}, a source in the LMC would have had a luminosity of around 10^{44} erg s^{-1}. It is this fact, more than any other, which has prevented this possible counterpart identification from being widely accepted, although, with cosmological models now gaining acceptance, this may change. No point X-ray source has been detected in N49, which is known, however, as a diffuse X-ray source; a ROSAT observation has taken place with the objective of finding a point source, but no results are available as of this writing.

Sensitive searches for emission simultaneous with a burst have also taken place. Upper limits in the radio [12], optical [25], GeV gamma-ray [6] (but see also [71] for a positive detection), VHE [58], and UHE [9] ranges have been established for numerous events.

Another type of search which has been carried out extensively is for optical transients whose positions are consistent with those of gamma-ray bursts, but which are not necessarily time-correlated. Most of the searches have taken place using archival photographic plates. Numerous detections of transient optical sources have now been reported [24, 54, 63]. The general characteristics of these flashes are the following.

(1) Only upper limits to the durations can be established; they fall in the minutes to 1 hr range.
(2) If a flash duration of 1 s is assumed (consistent with a typical GRB time scale), the flashes reach 3rd magnitude.
(3) The ratio of the gamma-ray burst to archival optical intensity is of the order

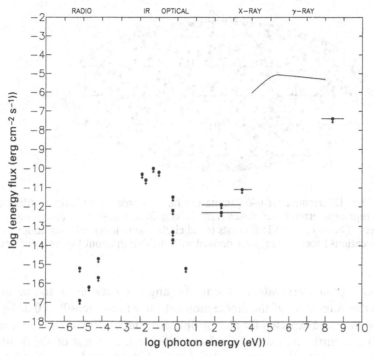

Fig. 13.4. Upper limits to quiescent emission from gamma-ray burst sources in various energy ranges. A typical burst spectrum is shown for comparison (solid line).

of 10^4, indicating that the optical emission is unlikely to represent the tail of a gamma-ray spectrum; reprocessing of the gamma rays is required.

(4) Deep optical searches of the transient positions have so far failed to reveal any likely quiescent candidates.

The reality of these optical transients has been debated extensively [18, 79]. The argument centers on whether it is possible to distinguish the transient images from subtle plate defects. No consensus has been reached, but searches for optical flashes occurring simultaneously with bursts, which are now taking place with CCD cameras, should eventually settle the question of whether intense optical emission accompanies bursts. The Explosive Transient Camera at Kitt Peak reaches a magnitude limit of 10 for a 1 s exposure, and is sensitive to gamma-ray to optical luminosity ratios of 10^5 [75].

13.6 Statistical properties of gamma-ray bursters

If the sources of gamma-ray bursts are galactic, their distribution might be expected to cluster towards the galactic center or plane; if their origin is extragalactic, but nearby, the distribution might display excesses at the positions of bright galaxies or clusters of galaxies. These ideas and others may be tested by calculating the dipole and quadrupole moments of the burster distribution [20]. The distribution of 646 gamma bursts is shown in Figure 13.5. Meegan *et al.* [50] have tested the subset of 447 bursts detected with the Burst and Transient Source Experiment (BATSE)

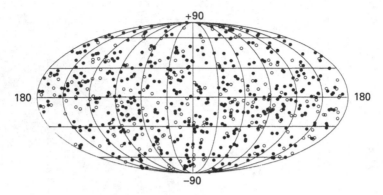

Fig. 13.5. Distribution of 646 gamma-ray burst sources in galactic coordinates. The dots represent error boxes which vary in size from roughly an arcminute to many degrees. The 199 pre-BATSE bursts (solid circles) were localized over about a decade; 447 locations (open circles) were derived with BATSE in about two years of operation.

aboard the Compton Observatory. Using the angle θ between a source and the galactic center as a measure of the dipole moment, they found $\langle \cos\theta \rangle = 0.034 \pm 0.027$, where -0.014 would be expected for an isotropic distribution (taking non-uniform sky coverage into account), and using the galactic latitude β as a test of the quadrupole moment with respect to the galactic plane, they derived $\langle \sin^2\beta \rangle = 0.316 \pm 0.014$, where 0.329 is expected for isotropy (again taking non-uniform sky coverage into account). No clustering of bursts around nearby galaxies was found. These results are quite consistent with previous studies which, however, utilized the approximately 100 brighter bursts which had been detected up to 1989. Those studies also examined the angular correlation function for bursts, to test the idea of distant extragalactic sources, and concluded that bursts had been sampled to a depth of at least 100 Mpc [21]; a similar test has not yet been carried out on the BATSE data. The earlier results did not contradict a galactic neutron star origin. Detailed models of the motion and distribution of galactic neutron stars were constructed [22, 59], which showed that a limited sampling depth could produce an apparently isotropic distribution. The fact that BATSE samples bursts which are at least an order of magnitude weaker than previous instruments, however, is indeed difficult to reconcile with a purely galactic disk origin.

Conclusions about the true spatial distribution of bursters obviously rely on estimates of the sampling depth and the related question of instrument sensitivity. Similar questions arose when quasars were first detected, and a statistical test, $V/V_{\max}$, was devised to answer them [67]. As applied to gamma-ray bursts, the prescription is the following [68]. For each burst detected by a given instrument, calculate the maximum observed count rate, $C_{\max}$, and the minimum count rate, $C_{\min}$, which would have resulted in detection of that burst under identical observing conditions. In Euclidean space, the peak count rate is proportional to $(\text{distance})^{-2}$, so the volume V which contains the source and the volume $V_{\max}$ within which it could have been detected are related by $(V/V_{\max}) = (C_{\max}/C_{\min})^{-1.5}$. If the source distribution is uniform, the average value $\langle V/V_{\max} \rangle = 0.5$, whereas a distribution sampled beyond its characteristic

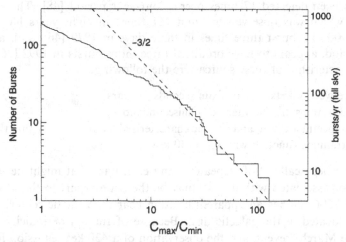

Fig. 13.6. The BATSE cumulative C_{max}/C_{min} distribution [52].

distance would give lower values. The cumulative C_{max}/C_{min} distribution for BATSE is shown in Figure 13.6 [52]. It appears to be consistent with a slope of -1.5 for the bright bursts, but is clearly deficient of weaker bursts, indicating that the full volume has been sampled.

The implication of these results – that we are at or near the center of a distribution which does not depart significantly from spherical symmetry – is difficult, although not impossible, to reconcile with known distributions of galactic mass. For example, Mao and Paczynski [43] conclude that all galactic disk models are excluded, whatever their thickness, while halo models with a minimum core radius of 18 kpc are acceptable (this minimum radius increases approximately with the square root of time as new bursts are added to the distribution). Lingenfelter and Higdon [42], on the other hand, have considered a two-population model, in which nearby low luminosity galactic neutron stars are not detected beyond their characteristic distance, while distant high luminosity halo neutron stars, sampled beyond their limit, provide a backdrop.

A final example of the statistical properties of bursters is the time between two bursts from a source. No classical gamma-ray burster has yet been observed to repeat. Since the intrinsic burster luminosity function is unknown, the interpretation of this result is uncertain. However, model-dependent lower limits to the recurrence time may be derived, and have typical values of around 10 yr [1, 64]. The galactic neutron star hypothesis requires that bursters repeat, since with some 10^8 neutron stars in the galaxy, and a full-sky burst rate of around 1000 events yr^{-1} (Figure 13.6), if each neutron star burst only once, there would only be enough to last $100\,000$ yr. On the other hand, if all galactic neutron stars are bursters, the required recurrence rate is only one burst every 10^5 yr on average, and recurrence would not necessarily be observed.

There are three high-energy transient sources which have been observed to recur, although their relation to the gamma bursters is unclear. One is the March 5, 1979, source, which has been detected 17 times over a four-year period [48]. The second is SGR 1806-20, which was observed to burst 111 times in eight years [40]. The third source, B 1900+14, burst three times in three days in 1979 [46], and, after a long quiescent period, appears to have produced three more bursts in 1992 [36]. The observational characteristics of these sources are the following:

- the time between bursts may be from seconds to years;
- the burst durations are from tens of milliseconds to 10 s;
- the energy spectra are soft, and may be characterized by an optically thin thermal bremsstrahlung function with $kT = 40$ keV.

These sources have been called 'soft repeaters', and each has what might be termed a distance indicator associated with it. N49 may be the counterpart to the March 5, 1979, source. SGR1806−20 has an apparent location near the galactic center; and B 1900+14 may be located in the galactic disk. Because of the 8 s periodicity in the time history of the March 5 event, and the observation of a 430 keV emission feature in its energy spectrum, a neutron star origin seems assured for it, whatever the source distance. Through 'guilt by association', the other soft repeaters might be assumed to be neutron stars as well. However, Paczynski [60] has shown that repeating sources could actually be at cosmological distances, with gravitational microlensing producing the apparent recurrence.

13.7   The great debate redux

In a sense, little has changed since the discovery of cosmic gamma-ray bursters, when the burning issue was their origin, galactic or extragalactic. Today there is considerably more data, but no single class of models can explain all the observational aspects. In favor of a neutron star origin, one can cite the periodicities and rapid fluctuations in the time histories, the emission and absorption features in the gamma-ray energy spectra, and the spectral evolution consistent with blackbody emission. But, apparently contradicting this evidence, there is the isotropic spatial distribution and the C_{max}/C_{min} statistic. Before abandoning the Galaxy as the source of bursts, several possibilities should be considered. One is a nearby, heliospheric origin. A now-defunct theory proposed that gamma-ray bursts could be generated by the evaporation of relativistic dust grains in the solar neighborhood; this proposal had the unusual and laudable distinction of making an unambiguous, testable prediction, namely that narrow beams of gamma rays would be produced. It met its fate with the advent of burst detectors on interplanetary spacecraft, whose data were inconsistent with locally generated, well-focused beams. The heliosphere has not been seriously reconsidered since, but if a mechanism could be found, it seems likely that such a distribution could be made to fit the statistical data. A two-component model, with bursting neutron stars both very nearby and in the distant halo, has recently been proposed [42]. This, as well as the halo-only possibility [43], could also salvage the galactic neutron star hypothesis, but the question then becomes how the required number of neutron stars ended up in the distant halo. (One possible answer is that they were born there [15].) This model, too, may eventually lead to a testable prediction, for, as the minimum acceptable halo radius increases, bursts from the halo of M31 could become observable with BATSE [61]. Some cosmological models, such as binary neutron star mergers [14, 62], do not attempt to explain all gamma-ray bursts – they exclude, for example, the ones with spectral features – but instead provide a framework for understanding the statistical properties. In general, it seems unlikely that ideas which place the sources at redshifts around 1 or more will be unambiguously testable in the near future. Optical counterparts will be difficult or impossible to identify, even in error boxes with dimensions of around an arcminute, and observation of the bursts of gravitational radiation which are predicted to accompany such events require the next generation of detectors.

To return to the subject of this book, X-ray binaries, a test for the origin of gamma-ray bursts in binary systems has been proposed [13]. It involves searching for an echo of reflected radiation from the binary companion of a neutron star. Because of the nature of Compton scattering in the binary's atmosphere, the echo has a softer spectrum than the original radiation. The time delay between the original and reflected components is a measure of the separation between the neutron star and its companion. This method is currently being applied to several events.

Despite the difficulty of the problem, there are many reasons to be optimistic for at least a partial solution in the near future. There is more powerful ground – and space – based instrumentation at work today than ever in the past: the Compton Observatory, the third interplanetary network, the Explosive Transient Camera, and large optical telescopes. If some of the sources are nearby, it seems likely that they can be identified.

Acknowledgments

This work was partially supported under NASA Grant NASA NAG5-1560. The author is grateful to the BATSE team for allowing the use of their spatial distribution data prior to publication.

Note added in proof

In the years which have intervened since this article was written, numerous new data have been accumulated, but the GRB mystery remains intact. BATSE has now observed over 1000 GRBs, but isotropy and inhomogeneity still characterize the distribution. One GRB has been observed to an energy of 18 GeV, with the high energy emission trailing the low energy emission by over 1 hour. New observations of the soft repeaters, however, have partially resolved the mystery of their origin: it is now quite certain that they are associated with the neutron stars in galactic supernova remnants, and that they may emit bursts which exceed the Eddington luminosity by up to six orders of magnitude. Their identification holds out the hope that the GRB puzzle is ultimately solvable by counterpart searches.

References

[1] Atteia, J.-L., *et al.*, *Ap. J. Suppl.* 64, 305, 1987
[2] Barat, C., *et al.*, *Ap. J.* 285, 791, 1984
[3] Barat, C., *et al.*, *Astrophys. Space Sci.* 75, 83, 1981
[4] Barat, C., *et al.*, *Astron. Astrophys. Lett.* 79, L24, 1979
[5] Barat, C., *Astron. Astrophys. Suppl. Ser.* 97, 43, 1993
[6] Bhat, P., *et al.*, *Phil. Trans. Roy. Soc. London A*, 301, 659, 1981
[7] Bhat, P., *et al.*, *Nature* 359, 217, 1992

[8] Boer, M., *et al.*, *Astron. Astrophys.* 202, 117, 1988
[9] Clay, R., *et al.*, *Astrophys. Space Sci.* 83, 279, 1982
[10] Cline, T., *et al.*, *Ap. J. Lett.* 237, L1, 1980
[11] Cline, T., *et al.*, *Ap. J. Lett.* 255, L45, 1982
[12] Cortiglioni, S., *et al.*, *Astrophys. Space Sci.* 75, 153, 1981
[13] Dermer, C., *et al.*, *Ap. J.* 370, 341, 1991
[14] Eichler, D., *et al.*, *Nature* 340, 126, 1989
[15] Eichler, D., and Silk, J., *Science* 257, 937, 1992
[16] Fenimore, E., *et al.*, *Ap. J. Lett.* 335, L71, 1988
[17] Gould, A., *Ap. J. Lett.* 386, L5, 1992
[18] Greiner, J., *et al.*, *Astron. Astrophys.* 234, 251, 1990
[19] Harding, A., *Phys. Rep.* 206(6), 327, 1991
[20] Hartmann, D., and Epstein, R., *Ap. J.* 346, 960, 1989
[21] Hartmann, D., and Blumenthal, G., *Ap. J.* 342, 521, 1989
[22] Hartmann, D., *et al.*, *Ap. J.* 348, 625, 1990
[23] Higdon, J., and Lingenfelter, R., *Ann. Rev. Astron. Astrophys.* 28, 401, 1990
[24] Hudec, R., *et al.*, *Astron. Astrophys.* 235, 174, 1990
[25] Hudec, R., *et al.*, *Astron. Astrophys.* 175, 71, 1987
[26] Hueter, G., Ph.D. Thesis, University of California, San Diego, Department of Physics (1987)
[27] Hurley, K., in *Proc. Huntsville Gamma-Ray Burst Workshop*, eds. W. Paciesas and G. Fishman, AIP Conf. Proc. 265 (AIP Press, New York), p. 3, 1992
[28] Hurley, K., *et al.*, *Astron. Astrophys. Suppl.* 92, 401, 1992
[29] Hurley, K., *et al.*, in *Proc. Huntsville Gamma-Ray Burst Workshop*, eds. W. Paciesas and G. Fishman, AIP Conf. Proc. 265 (AIP Press, New York), p. 195, 1992
[30] Imamura, J., and Epstein, R., *Ap. J.* 313, 711, 1987
[31] Katoh, M., *et al.*, in *High Energy Transients In Astrophysics*, ed. S. Woosley, AIP Conf. Proc. 115, 1984, p. 390 (AIP Press, New York).
[32] Klebesadel, R., *et al.*, in *Gamma-Ray Transients and Related Astrophysical Phenomena*, eds. R. Lingenfelter, H. Hudson, and D. Worrall, AIP Conf. Proc. 77 (AIP Press, New York), p. 1, 1982
[33] Klebesadel, R., *et al.*, *Ap. J. Lett.* 182, L85, 1973
[34] Klebesadel, R., *et al.*, *B.A.A.S.* 16, no. 4, 1016, 1984
[35] Kouveliotou, C., *et al.*, *Ap. J. Lett.* 330, L101, 1988
[36] Kouveliotou, C. *et al.*, *Nature*, 362, 728, 1993
[37] Lamb, D., *et al.*, *Nature Phys. Sci.* 246, 52, 1973
[38] Laros, J., *et al.*, *Ap. J.* 286, 681, 1984
[39] Laros, J., *et al.*, *Nature* 318, 448, 1985
[40] Laros, J., *et al.*, *Ap. J. Lett.* 320, L111, 1987
[41] Liang, E., *et al.*, *Ap. J.* 271, 766, 1983
[42] Lingenfelter, R., and Higdon, J., *Nature* 356, 132, 1992
[43] Mao, S., and Paczynski, B., *Ap. J. Lett.* 389, L13, 1992
[44] Matz, S., Ph. D. Thesis, University of New Hampshire, Physics Department, 1986
[45] Mazets, E., *et al.*, *Nature*, 282, 587, 1979
[46] Mazets, E., *et al.*, *Astrophys. Space Sci.* 84, 173, 1982
[47] Mazets, E., *et al.*, in *e+ e- Pairs In Astrophysics*, AIP Conf. Proc. 101 (AIP Press, New York), p. 36, 1983
[48] Mazets, E., and Golenetskii, S., *Sov. Sci. Rev. E. Astrophys. Space Phys.* 6, 281, 1988
[49] McBreen, B., and Metcalfe, L., *Nature* 332, 234, 1987
[50] Meegan, C., *et al.*, *I.A.U.C.* 5641, October 21, 1992
[51] Messina, D., and Share, G., in *Proc. Huntsville Gamma-Ray Burst Workshop*, eds. W. Paciesas and G. Fishman, AIP Conf. Proc. 265 (AIP Press, New York), p. 206, 1992
[52] Meegan, C., *et al.*, in *Proc. Huntsville Gamma-Ray Burst Workshop*, eds. W. Paciesas and G. Fishman, AIP Conf. Proc. 265 (AIP Press, New York), p. 61, 1992
[53] Mitrofanov, I., *et al.*, *Planet. Space Sci.* 39(1/2), 23, 1991
[54] Moskalenko, E., *et al.*, *Astron. Astrophys.* 223, 141, 1989
[55] Motch, C., *et al.*, *Astron. Astrophys.* 145, 201, 1985
[56] Murakami, T., *et al.*, *Astron. Astrophys.* 227, 451, 1990

[57] Murakami, T., *et al.*, *Nature* 350, 592, 1991
[58] O'Brien, S., and Porter, N., *Astrophys. Space Sci.* 42, 73, 1976
[59] Paczynski, B., *Ap. J.* 348, 485, 1990
[60] Paczynski, B., *Ap. J. Lett.* 317, L51, 1987
[61] Paczynski, B., private communication, 1992
[62] Piran, T., *Ap. J. Lett.* 389, L45, 1992
[63] Schaefer, B., *Nature*, 294, 722, 1981
[64] Schaefer, B., and Cline, T., *Ap. J.* 289, 490, 1985
[65] Schaefer, B., *et al.*, *Ap. J.* 313, 226, 1987
[66] Schaefer, B., *et al.*, *Ap. J.* 340, 455, 1989
[67] Schmidt, M., *Ap. J.* 151, 393, 1968
[68] Schmidt, M., *et al.*, *Ap. J. Lett.* 329, L85, 1988
[69] Schneid, E., *et al.*, *Astron. Astrophys. Lett.* 255, L13, 1992
[70] Share, G., *et al.*, *Adv. Space Res.*, 6(4), 15, 1986
[71] Sommer, M., *et al.*, *Ap. J.*, 422, L63, 1994
[72] Sumner, T., *et al.*, *Astron. Astrophys. Suppl. Ser.* 71, 557, 1987
[73] Teegarden, B. and Cline, T., *Ap. J. Lett.* 236, L67, 1980
[74] Teegarden, B., *et al.*, in *Compton Gamma Ray Observatory*, eds. M. Friedlander, N. Gehrels, and D. Macomb, AIP Conf. Proc. 280 (AIP Press, New York), p. 860, 1993
[75] Vanderspek, R., *et al.*, in *Proc. Huntsville Gamma-Ray Burst Workshop*, eds. W. Paciesas and G. Fishman, AIP Conf. Proc. 265 (AIP Press, New York), p. 404, 1992
[76] Wood, K., *et al.*, in *Gamma Ray Bursts*, eds. E. Liang and V. Petrosian, AIP Conf. Proc. 141 (AIP Press, New York), p. 4, 1986
[77] Wood, K., *et al.*, *Ap. J.* 247, 632, 1981
[78] Yoshida, A., *et al.*, *Publ. Astron. Soc. Japan* 41, 509, 1989
[79] Zytkow, A., *Ap. J.* 359, 138, 1990

14

A catalogue of X-ray binaries

Jan van Paradijs

Astronomical Institute 'Anton Pannekoek', University of Amsterdam, and
Center for High-Energy Astrophysics,
Kruislaan 403, NL-1098 SJ Amsterdam, The Netherlands

In this chapter, I present a catalogue with information on X-ray binaries, i.e. semi-detached binary stars in which matter is transferred from a usually more or less normal star to a neutron star or a black hole. Thus, cataclysmic variables are not included. The aim of this catalogue is to provide the reader with some basic information on the X-ray sources and their counterparts in other wavelength ranges (UV, optical, IR, radio). It has not been my intention to compile complete reference lists, but to help the reader gain easy access to the recent literature (up to about Christmas 1992) on individual sources. The information is tabulated separately for low-mass X-ray binaries (LMXBs; Table 14.1) and high-mass X-ray binaries (HMXBs; Table 14.2). These two tables contain 124 and 69 sources, respectively. In case there is some doubt about the nature of the X-ray source this is mentioned.

The format of the tables is similar to that of the well-known work of Bradt & McClintock [85], of which the present catalogue is meant to be an update. In each of the two tables the sources are ordered according to right ascension; part of the (mainly numerical) information on a source is arranged in seven columns, below which for each source additional information is provided in the form of key words with reference numbers [in square brackets]. The columns have been arranged as follows.

In column 1, the first line contains the source name, with rough information on its sky location according to the convention hhmm±ddd. Here hh and mm indicate the hours and minutes of right ascension, ddd the declination in units of 0.1 degree (in a small number of cases, the coordinates shown in the name are given with more, or less, digits). The prefix J indicates a name based on J2000 coordinates. Otherwise, 1950 coordinates were used in the name. Alternative source names are given in the second line. The third line of column 1 lists survey catalogues and experiments in which the source was listed and detected, respectively. The following abbreviations have been used.

- A: Ariel V sky survey [795,1255];
- C: Compton γ-ray Observatory;
- E: Einstein Observatory;
- Exo: Exosat;
- G: Ginga;
- Gr: Granat;
- H: HEAO A-1 sky survey [1305];
- Ha: Hakucho;

- K: Kvant;
- M: MIT OSO-7 sky survey [754];
- OAO: Orbiting Astronomical Observatory;
- R: ROSAT;
- S: SAS 3;
- SL: Space Lab;
- T: Tenma;
- U: Uhuru sky survey [333];
- V: Vela-5 and -6 satellites.

(References quoted in the above list refer to published sky survey catalogues.)

In the first line of column 2, the source types are indicated with a letter code, as follows:

- A: atoll source (11 LMXBs) [437];
- B: X-ray burst source (42 LMXBs);
- D: 'dipping' LMXBs (9) (see Ch. 1);
- G: globular-cluster X-ray source (12 LMXBs);
- P: X-ray pulsar (3 LMXBs, 29 HMXBs);
- T: transient X-ray source (41 LMXBs, 30 HMXBs);
- U: ultra-soft X-ray spectrum (17 LMXBs, 3 HMXBs). These sources include black-hole candidates (see Ch. 1 and 4); some 'extreme ultra-soft' (EUS) source may be white dwarfs on whose surface steady nuclear burning takes place [1184];
- Z: Z-type (6 LMXBs)

In the third line of column 2, we provide some information on the type of observation from which the source position has been derived. The following abbreviations have been used: o, optical; x, X-ray; r, radio; IR, infrared. A reference on the source position is given below the columnar information under '*Pos.*'. In addition, I give an indication of the accuracy of this position, in the form of equivalent (90% confidence level) error radii, but in several cases this can only be considered an approximation (e.g. when the error box is not circular). When no accuracy is quoted, it is about one arcsecond or better.

Column 3 contains in the first two lines the right ascension (RA) and declination (DEC) of the source (for epoch 1950). RA is given as hhmmss.s to an accuracy of 0.1 s, DEC is given in ° ' ", to an acuracy of 1". The third line gives the galactic longitude and latitude to an accuracy of 0.1° (except for sources close to the galactic center, where these coordinates are given to 0.01°).

The first and second lines of column 4 give names of an optical counterpart. The third line contains a reference to a finding chart. An asterisk followed by a number or letter refers to star numbers used in the finding chart; "star" refers to a star in the finding chart that has not been assigned a number or letter. Many optical counterparts have been indicated with a variable-star name, as given in the *General Catalogue of Variable Stars* and in recent name lists of variable stars as published regularly in the *IAU Information Bulletin on Variable Stars*, or a number in a well-known catalogue (e.g., HD, SAO). For X-ray sources in globular clusters, the cluster name is here given, in addition to the name of a stellar optical counterpart.

The fifth column contains some photometric information on the optical counterpart. In the first line, the apparent visual magnitude, V, and the color indices $B - V$, and $U - B$, are listed. The second line contains an estimate of the interstellar reddening, E_{B-V}; for HMXBs, in addition, the spectral type of the optical counterpart is given.

In column 6, the average X-ray flux, or the range of observed X-ray fluxes (2–10 keV, unless otherwise indicated), is given, in units of

$$1 \, \mu\text{Jy} = 10^{-29} \text{erg cm}^{-2} \, \text{s}^{-1} \, \text{Hz}^{-1} = 2.4 \times 10^{-12} \, \text{erg cm}^{-2} \, \text{s}^{-1} \text{keV}^{-1}.$$

The first line in column 7 gives the orbital period, which for LMXBs is given in hours, and for HMXBs in days. The second line contains for X-ray pulsars the pulse period, in seconds.

In Table 14.3, cross references are given from often-used nomenclature to the names based on sky coordinates.

Acknowledgements

I am grateful to T. Augusteijn, C. Chevalier, J. Grindlay, S. Ilovaisky, E. Kuulkers, W. Lewin, J. McClintock, T. Oosterbroek, A. Parmar, S. Prins, and M. van der Klis for their helpful remarks.

Table 14.1. *Low-mass X-ray binaries*

Source	Type Pos.	RA (1950) DEC (1950) ℓ^{II}, b^{II}	Opt. cpt [FC]	V, B–V, U–B E_{B-V}	F_X (μJy)	P_{orb}(hr) P_{puls}(s)
0042+327	T	00 42 08.5 +32 44 53	*3	19.3, 0.6, - - 0.2	<0.5-55	
U,M,A	x 60"	121.3, −29.8	[124]	[124]	[85]	

Pos.: [980]; *transient*: [645,980,1261]; *optical counterpart uncertain*: [741].

- -

0142+614		01 42 53.9 +61 30 08		0.8	4	
U,M,A,H,S	x 10"	129.4, −0.4	[1292]	[878]	[1292]	

Pos.: [1292]; *very soft X-ray spectrum*: [1292]; *X-ray period 25 min.* [1292] *suggests this is an intermediate polar,
but X-ray to optical flux ratio does not fit that idea* [1292].

- -

J0422+32	T	04 18 29.9 +32 47 24		13.2, 0.3, −0.5 0.4	3000: (20-300 keV)	
C,Gr	o,r	197.3, −11.9		[1039,1043]	[368,897]	

Pos.: [794]; *outburst*: [121,897]; *hard X-rays*: [111]; *QPO*: [619]; *opt. sp.*: [1249]; *UV sp.*: [1043]; *radio obs.*: [431].

- -

0512−401	GB	05 12 27.9 −40 06 00	NGC1851	0.1:	6 0.1	
U,M,A,H	x 2"	244.5, −35.0	[412]	[8]	[920,1228]	

Pos: [412,454]; *opt. studies glob. cl.*: [8,34,235,457,1008]; *cluster center*: [454,1041]; *cluster moderately metal
poor*: [22]; *X rays*: [300]; *X-ray bursts*: [161,173,329]; *radio obs.*: [714].

- -

0521−720 LMC X-2		05 21 18.0 −72 00 26	*22	18.0-19.0, 0.0, −0.8 0.1	9-44	
U,M,A,H	o 3"	283.1, −32.7	[561]	[110,224,701]	[85]	

Pos.: [85]; *X-ray obs.*: [1093]; *opt. id.*: [697,900,904]; *orb. period uncertain (8.15 h and 12.5 d proposed)*: [110,224];
coordinated X-ray/opt. obs.: [75]; *opt. spectrum*: [75,841]; *near-IR spectrum*: [210].

- -

J0527.8−6954		05 28 15.9 −69 56 21		0.1	0.1: (0.2-0.5 keV)	
R	x 30"	280.5, −32.5		[396,701]	[396]	

Pos.: [1160]; *very soft X-ray source*: [396]; *in LMC*: [396]; *HV 2554 possible optical counterpart*: [655,1160].

- -

0543−682 CAL 83	U	05 43 48.0 −68 23 34	*V	16.2-17.3, 0.0, −1.0 0.1	1 (0.1-4.5 keV)	24.96
E	o	278.6, −31.3	[205,223,1064]	[223,701,906,1064]	[697]	[1064]

Pos.: [1064]; *ultra-soft X-ray spectrum*: [396]; *UV*: [223]; *near IR spectrum*: [210]; *optical spectroscopy and
photometry*: [221,223,906]; *optical light curve*: [1064]; *compact star a white dwarf(?)*: [1184].

- -

0547−711 CAL 87	U	05 47 26.8 −71 09 50	*X	18.8-20, −0.15, −0.7 0.1	0.2 (0.1-4.5 keV)	10.62
E	x 10"	281.8, −30.7	[908]	[209,701]	[697]	[109]

Pos.: [205]; *opt. eclipses*: [109,908]; *ultra-soft X ray sp*: [205,908]; *opt. sp.&phot.*: [209]; *near IR sp.*: [210];
compact star a white dwarf(?): [1184].

- -

Table 14.1. *(continued)*   LMXB

Source	Type Pos.	RA (1950) DEC (1950) ℓ^{II}, b^{II}	Opt. cpt [FC]	V, B–V, U–B E_{B-V}	F_X (μJy)	P_{orb}(hr) P_{puls}(s)
0614+091	B?	06 14 22.8 +09 09 22	V1055 Ori	18.5, 0.3, –0.5 0.3	50	
U,M,A,H,S	o	200.9, –3.4	[280,383]	[240]	[1255,1305]	

Pos.: [85]; *X-ray bursts (?)*: [93,705,1110]; *opt. cpt*: [240,860]; *optical spectrum*: [715]; *coordinated X-ray/opt. obs.*: [715]; *long-term variability*: [715,759]; *radio obs.*: [293].

0620–003	TU	06 20 11.1	V616 Mon	11.2, 0.2, –0.8	<0.02-50000	7.75
N Mon 1975		–00 19 11		0.4		
A,S	o	210.0, –6.5	[290,383]	[1273]	[697,1273]	[783]

Pos.: [85]; *bright transient 1975*: [299,1273]; *previous outburst 1917*: [296]; *hard X-rays.*: [186]; *optical outburst*: [893,1273]; *7.8 day modulation in outburst*: [291,693,776,965,998,1163,1260]; *quiescent X rays*: [258,697]; *quiescent opt. spectra (K5) & mass function (black hole)*: [442,563,783,784,863]; *UV obs.*: [1309]; *near-IR sp.*: [210]; *quiescent opt. phm.*, V_{quiesc} = 18.2: [783,784]; *opt. polarimetry*: [273]; *radio obs.*: [248,293,349,896].

0656–072	T	06 56 01			20-80	
		–07 11.7				
A	x 3'	220.2, –1.7			[116,159,576]	

Pos.: [116]; *relatively soft X-ray spectrum*: [179].

0748–676	TBD	07 48 25.0	UY Vol	16.9-17.5, 0.1, –0.9	0.1-60	3.82
		–67 37 32		0.42		
Exo	o	280.0, –19.3	[1244]	[1020,1023]	[378]	[919,923]

Pos.: [1244]; *orbital decay*: [24,923,1066]; *X-ray dips*: [919,1066,1067]; *X-ray spectrum*: [1293]; *X-ray bursts*: [237,378,380]; *optical spectra & photometry*: [207,222,848,1020,1223]; *near-IR sp.*: [210]; V_{quiesc}>23: [1244].

0836–429	TB	08 35 37			1-55	
		–42 42.6				
U,M	x 1'	261.9, –1.1			[179,441,720]	

Pos.: [441,1101]; *northern of two X-ray sources*: [386,721]; *X-ray bursts*: [18,720,721]; *likely MX 0836-42*: [179,753].

0918–549		09 18 54.7	*X	21.0, 0.3, –0.9	10	
		–54 59 37		0.3		
U,M,A,H,S	x 5"	275.9, –3.8	[140]	[140,1218]	[140]	

Pos.: [140]; *optically underluminous (?)* : [140].

0921–630	D	09 21 25.1	V395 Car	15.3, 0.6, –0.5	3	216.2
		–63 04 48		0.2		
A,S	o	281.8, –9.3	[684]	[96,139,773,1218]	[684,773]	[771,773]

Pos.: [85]; *partial X-ray eclipse (ADC source)*: [773]; *optical sp., secondary star F-G giant*: [203]; *halo object*: [203]; *opt. obs.*: [138,139]; *opt/X-ray obs.*: [96]; *near-IR sp.*: [210]; *radio*: [1313]; *orbital period 13.25 d (?)*: [632].

Table 14.1. *(continued)* LMXB

| Source | Type | RA (1950) | Opt. cpt | V, B–V, U–B | F_X | P_{orb}(hr) |
| | | DEC (1950) | | E_{B-V} | (μJy) | P_{puls}(s) |
	Pos.	ℓ^{II}, b^{II}	[FC]			
1124–684	TU	11 24 18.5		13.6, 0.3, - -	<4-3000	10.4
N Mus '91		–68 24 02		0.25		
Gr,G	o	295.0,–6.1	[259]	[259,370]	[1097,1103]	[33,792,988]

Pos.: [259]; *bright X-ray transient Jan. 1991*: [92,388,610,704,1097]; *hard X-ray spectrum, strong soft component*: [1097]; *annihilation line near 500 keV*: [359,367,1105]; *QPO*: [1118]; *similar to 0620–000*: [260]; *optical outburst*: [33]; *mass function suggests black-hole accretor*: [792,988]; *UV obs.*: [132,370]; *radio obs.*: [596]; *quiescent ellipsoidal light curve*: [792,1024]; *in quiescence B=20.9, V=20.4, R=19.4*: [259,792]; *secondary K0-4V*: [988].

1254–690	BD	12 54 21.0	GR Mus	19.1, 0.3, - -	25	3.93
		–69 01 08		0.35		
U,M,A,H,S	o	303.5, –6.4	[398]	[845]	[198]	[198,845]

Pos. (source D): [85]; *X-ray bursts, X-ray dips*: [198]; *X-ray spectrum*: [466,768,1293]; *opt. burst*: [768]; *opt. spectrum*: [207]; *opt. lt crv. & spectrum*: [845]; *near-IR sp.*: [210].

1323–619	BD	13 23 16.8			7	2.93
		–61 52 36		~5:		
U,M,A	x 3"	307.1, +0.2	[922]	[922]	[922,1255]	[922]

Pos. (source D): [922]; *X-ray bursts & dips*: [922,1194].

1354–645	TU	13 54 27.5	BW Cir	16.9, 1.1, –0.1	5-120	
Cen X-2?		–64 29 29		~1:		
M,G	o	310.0, –2.8	[607]	[607]	[607,1034]	

Pos.: [607]; *1987 outburst*: [607]; *position consistent with transient Cen X-2 (1967)*: [154,336,436,753]; *opt. id.* [607]; *opt. spectrum*: [207]; *possible 50.0 hour period during decline*: [146]; R_{quiesc} ~ 22: [607].

1455–314	TB	14 55 19.6	V822 Cen	12.8, 0.05, –0.9	0.1-20000	15.10
Cen X-4		–31 28 09		0.1		[149,208,
V,A,Ha	o	332.2, +23.9	[114]	[66,114]	[180,1220]	784]

Pos.: [85]; *transient*: [81,180,301,578,778]; *quiescent X rays*: [948,1220]; *opt. outburst sp. & phm.*: [114]; *K5-7 secondary*: [149,784,948,1214]; *near-IR sp.*: [210]; V_{quiesc} ~ 18.3: [784]; *optical quiescent lt. crv.*: [149,784]; *UV*: [66]; *X-ray bursts*: [53,778]; *no X-ray eclipses*: [1298]; *radio obs.*: [469,476].

1516–569	TB	15 16 48.4	BR Cir	21.4, - -, - -	5-3000	398.4
Cir X-1		–56 59 12		>1.0		
U,M,A,H,S	r	322.1, +0.0	[833]	[833,1085]	[85]	[575]

Pos.: [21]; *orbital variations*: [575,880,1137]; *X ray obs.*: [281]; *X-ray spectrum*: [703]; *distance*: [374]; *QPO*: [1137,1138]; *X-ray bursts*: [1139,1140]; *opt. cpt*: [21,833]; *not the opt. cpt in [1274]*: *opt. lt. crv.*: [833]; *IR obs.*: [364]; *long-term weakening in K band*: [833]; *radio obs.*: [293,443,1274]; *correlated radio/X-ray obs.*: [1085]; *connection SNR (?)*: [158]; *VLBI obs.*: [963]; *model*: [865].

1524–617	TU	15 24 05.8	KY TrA	17.5B, - -, - -	<5-950	
TrA X-1		–61 42 35	*N	0.7		
A,S	o 3"	320.3, –4.4	[861]	[861]	[85]	

Pos.: [85]; *X-ray outburst in 1977*: [574]; *ultra-soft X-ray spectrum*: [1278]; *hard component in quiescent X-ray spectrum*: [45,1090]; *optical outburst*: [861]; V_{quiesc} > 21: [861].

Table 14.1. *(continued)*   LMXB

Source	Type Pos.	RA (1950) DEC (1950) ℓ^{II}, b^{II}	Opt. cpt [FC]	V, B–V, U–B E_{B-V}	F_X (μJy)	P_{orb}(hr) P_{puls}(s)
1543–624		15 43 34.1 –62 24 51	*6	≥20(B), - -, - - 0.5	35	
U,M,A,H,S	o	321.8, –6.3	[19]	[785]	[1255]	

Pos.: [85]; *opt. cpt*: [785].

- -

| 1543–475 | TU | 15 43 33.9 –47 30 54 | star | 14.9, 0.6, - - 0.7 | <1-15000 | |
| U,M,T | o | 330.9, +5.4 | [134] | [934,1201] | [85,602] | |

Pos.: [934]; *opt. id.*: [930,934]; *ultra-soft X-ray spectrum, Fe Kα emission*: [152,1077,1201]; *X-ray outbursts*: [434,602,682,775]; *quiescent opt. phm.*: [134]; *quiescent (V,B) = (16.7, 17.5)*: [134]; *triple star(?)*: [134].

- -

| 1556–605 | | 15 56 45.8 –60 35 52 | LU TrA *X | 18.6-19.2, 0.45, –0.7 0.6 | 16 | 9.1: |
| U,M,A,H,S | o | 324.1, –5.9 | [847] | [125,847,1022] | [847,1255] | [1055] |

Pos.: [85]; *opt/X-ray obs.*: [847]; *opt. phm.*: [1022]; *opt. spectrum*: [207]; *proposed orbital period of 9.1 h. requires confirmation*: [1055].

- -

| 1603.6 +2600 | | 16 03 40.5 +25 59 48 | star | 19.7, - -. - - <0.1 | 0.15 (0.3-3.2 keV) | 1.85 |
| E | o | 42.8, +46.8 | [711] | [839] | [839] | [839] |

Pos.: [839]; *period of a CV but opt. spectrum of an LMXB*: [839].

- -

| 1608–522 | TBA | 16 08 52.2 –52 17 43 | star | I ~18.2, - -, - - 1.5: | <1-110 | |
| U,M,H,S | o | 330.9, –0.9 | [405] | [405,937] | [85] | |

Pos.: [405]; *outbursts*: [306,573,694,753,1125,1290]; *off-state X-ray emission*: [754,1125]; *X-ray bursts*: [54,381,402,853,854,859,872,873,937]; *X-ray sp.*: [466,818,1106,1293]; *atoll source*: [437]; $I_{quiesc} > 20$: [405].

- -

| 1617–155 Sco X-1 | Z | 16 17 04.5 –15 31 15 | V818 Sco | 12.2, 0.2, –0.8 0.3 | 14000 | 18.90 |
| U,M,A,H | o | 359.1, +23.8 | [383,1011] | [464,1239] | [1255] | [377,646] |

Pos.: [85]; *X-ray spectrum*: [466,548,569,571,703,1106,1240,1291]; *QPO*: [439,677,804,973,1185,1196]; *Z source*: [437,1028]; *no orb. X-ray variation*: [966]; *no X-ray pulsations*: [460,1306]; *hard X rays*: [548,1006,1172]; *opt. obs.*: [464]; *X-ray/opt./radio obs.*: [86,112]; *X-ray/opt. obs.*: [27,530,836,949]; *X-ray/radio obs*: [479]; *radio obs.*: [7,475,1243]; *opt. sp.*: [1017]; *near-IR sp.*: [210]; *UV obs.*: [572,1239,1300]; *no radio lobes*: [328]; *review (1977)*: [822].

- -

| 1624–490 | D | 16 24 17.8 –49 04 46 | | ~7: | 55 | 21 |
| U,M,A,H,S | x 12" | 334.9, –0.3 | [986] | [1266] | [1255] | [567] |

Pos.: [85]; *X-ray dips*: [567,1266].

- -

Table 14.1. *(continued)*   LMXB

Source	Type / Pos.	RA (1950) DEC (1950) ℓ^{II}, b^{II}	Opt. cpt [FC]	V, B–V, U–B E_{B-V}	F_X (μJy)	P_{orb}(hr) P_{puls}(s)
1627–673	P	16 27 14.7	KZ TrA	18.5, 0.1, –1.2	25	0.69
		–67 21 18	*4	0.1		7.7
U,M,A,H,S	o	321.8, –13.1	[88,555]	[786,1218]	[795]	[805,866]

Pos.: [85]; *X-ray pulsations*: [601,866]; *X-ray spectrum*: [1240]; *X-ray pulse-phase spectr.*: [960,1288]; *1000 sec flaring*: [687]; *QPO*: [1042]; *aperiodic variability*: [56]; *orbital parameters*: [663]; *opt. pulsations*: [528,805]; *opt. spectrum*: [207]; *evolution*: [1229].

1630–472	TU	16 30 19.4			<2-1400	
		–47 17 24		4.5:		
U,M,A	x 10"	336.9, +0.3	[918]	[918]	[85]	

Pos.: [918]; *recurrent transient (~600 d interval)*: [566,965]; *very soft X-ray sp. with hard X-ray tail*: [918].

1632–477		16 32 46			13	
		–47 43 32				
K	x 1.1'	336.9, –0.4			[541]	

Pos.: [1102].

1636–536	BA	16 36 56.4	V801 Ara	17.5, 0.7, –0.7	220	3.80
		–53 39 18	*3	0.8		
U,M,A,H,S	o	332.9, –4.8	[555]	[648,786]	[1255]	[1225]

Pos.: [85]; *X-ray spectrum*: [466,703,1181,1240,1293]; *X-ray bursts*: [97,237,485,539,676,890,1089,1107,1113, 1167,1198,1219,1251]; *atoll source*: [437,1028]; *opt. cpt.*: [786]; *opt. lt crv.*: [931,1020,1058,1225]; *opt. bursts*: [648,779,932,933,1020,1158,1168]; *opt. spectrum*: [113,207]; *reddening*: [648,1218].

1642–455	Z	16 42 09.5			500	
GX 340+0		–45 31 13		12:		
U,M,A,H,S	r	339.6, –0.1	[19]	[456]	[1255]	

Pos.: [942]; *Z source*: [437,1028]; *QPO*: [677,941,1185,1224]; *X-ray spectrum*: [1240]; *no X-ray pulsations*: [1306]; *radio obs.*: [406,942].

1656+354	P	16 56 01.7	HZ Her	13.0-14.6, –0.2, –0.9	15-50	40.80
Her X-1		+35 25 05		<0.05		1.24
U,M,A,H	o	58.2, +37.5	[282,383]	[82,417]	[795,1305]	[253,866]

Pos.: [85]; *orb. period decrease*: [257]; *X-ray pulse profile, neutron star precession (?)*: [1071,1072,1159]; *X-ray spectrum*: [703,810,1240,1288]; *cyclotron line*: [808,1157,1165]; *X-ray pulse-phase spectr.*: [1073]; *aperiodic variability*: [56,57]; *X-ray behaviour chaotic*: [883,1234]; *35-day cycle*: [888]; *X-ray eclipse*: [251]; *X-ray orbit*: [253]; *opt. lt crv.*: [17,31,32,68,709]; *35 day effect opt. lt crv.*: [82,135,252,352]; *extended low state*: [916]; *low-state opt. lt crv.*: [261,815]; *off-state opt. lt crv.*: [502,565]; *opt. spectr.*: [213,214,215,616]; *optical radial-velocity curve*: [523]; *near-IR sp.*: [210]; *opt. pulsations*: [803,806,807]; *35-day effects opt. spectrum*: [523]; *optical polarimetry*: [298]; *UV*: [294,417,496,497]; *disk precession(?)*: [615,950,951,978]; *evolution*: [1229].

1658–298	TBD	16 58 55.4	V2134 Oph	18.3, 0.45, –0.4	<5-80	7.11
		–29 52 28	*T	0.3		
A,H	o	353.8, +7.3	[286]	[286,878]	[85]	[177]

Pos. [85]; *X-ray dips*: [174,176]; *X-ray bursts*: [173,670,673,1040]; *opt. spectrum*: [113,207].

Table 14.1. *(continued)*   LMXB

Source	Type Pos.	RA (1950) DEC (1950) ℓ^{II}, b^{II}	Opt. cpt [FC]	V, B–V, U–B E_{B-V}	F_X (µJy)	P_{orb}(hr) P_{puls}(s)
1659–487 GX339-4 U,M,A,H	TU o	16 59 02.0 −48 43 07 338.9, −4.3	V821 Ara *V [107,286]	15.5, 0.8, −0.1 1.1 [206,727,1218]	1.5-900 [533,824]	14.83 [107,491]

Pos.: [85]; *high-low-off Xray states/ optical state correlated*: [533,727,750,824,843,844]; *off state B>21*: [519,527]; *X-ray QPO*: [387,824]; *rapid X-ray var.*: [56,716,825]; *hard X rays*: [274,325]; *X-ray halo*: [962]; *X-ray spectrum*: [1240]; *very soft X-ray sp. component*: [727,824]; *opt. QPO*: [537,842,843,1076]; *opt. spectrum*: [195,206,207,286, 399]; *opt. millisec. pulsations (?)*: [536]; *near-IR sp.*: [210].

| 1702–429 U,M,S | BA x 5" | 17 02 41.0 −42 58 09 343.9, −1.3 | [456] | 2.5: [456] | 45 [556,892] | |

Pos.: [456]; *atoll source*: [892]; *X-ray bursts*: [724,892,925,1108]; *radio obs.*: [406].

| 1702–363 GX349+2 U,M,A,H,S | Z r | 17 02 22.9 −36 21 20 349.1, +2.7 | *6 [187,935] | 18.6, 1.5, - - 1.3: [187,935] | 825 [1255] | |

Pos.: [187]; *Z source*: [437]; *X ray obs.*: [466,1028]; *X-ray spectrum*: [703,1240,1293]; *QPO*: [677,957,1185]; *radio obs.*: [187,349,406].

| 1704+240 A,H | o | 17 04 29.7 +24 02 14 45.2, +33.0 | HD 154791 [342] | 7.8, 1.3, 2.1 0.3 [342] | <0.5-11 [85] | |

Pos.: [85]; *sp. type opt. cpt M3 II*: [342]; *UV obs.*: [342]; *high-mass X-ray binary (?)*: [236].

| 1705–440 U,M,A,S | BA x 13" | 17 05 17.9 −44 02 13 343.8, −2.3 | [986] | 2: [640] | 10-280 [640,1255] | |

Pos.: [85]; *atoll source*: [437]; *X-ray bursts*: [382,640]; *X-ray power spectra*: [641]; *X rays*: [1028]; *long-term on-off (223 d?)*: [964,965]; *X-ray spectrum*: [1240]; *radio obs.*: [293].

| 1705–250 N Oph 1977 A,H | T o 2" | 17 05 10.4 −25 01 38 358.6, +9.1 | V2107 Oph [397] | 15.9, 0.6, - - 0.5 [397] | <2-3600 [85] | |

Pos.: [85]; *X-ray outburst 1977*: [1260]; *very soft X-ray spectrum with hard tail*: [186,1302]; $B_{quiesc} \sim 21$: [397].

| 1708-408 U,M,A,H | x 1' | 17 08 53 −40 47 02 346.3, −0.9 | | | 32 [541] | |

Pos.: [541].

| 1711–339 H,S | x 40" | 17 11 01.3 −33 59 32 352.1, +2.7 | [389] | | 16-130 [85] | |

Pos.: [85]; *radio obs.*: [389].

Table 14.1. (*continued*)   LMXB

Source	Type Pos.	RA (1950) DEC (1950) ℓ^{II}, b^{II}	Opt. cpt [FC]	V, B–V, U–B E_{B-V}	F_X (µJy)	P_{orb}(hr) P_{puls}(s)
1715–321	B	17 15 32.3 –32 07 34			28	
M,A,H,S	x 23"	354.1, +3.1	[986]		[752,754,1255]	

Pos.: [85]; *X-ray bursts*: [487,722,1129].

- -

| 1724–356 | | 17 24 18 –35 41 36 | | | 32 | |
| K | x 1.3' | 352.2,–0.5 | | | [541] | |

Pos.: [541].

- -

| 1724–307 | GB | 17 24 20.1 –30 45 39 | Ter 2 | 1.4 | 15 | |
| E | x 3" | 356.3, +2.3 | [412] | [22,733] | [920,1228] | |

Pos.: [412,454]; *opt. study globular cluster*: [155]; *cluster metal rich*: [22]; *structure gl. cl.*: [457]; *X-ray bursts*: [409,1109]; *hard X-ray obs.*: [44]; *IR obs.*: [155]; *radio obs.*: [406].

- -

1728–337	BA	17 28 39.2			150	
GX 354–0		–33 47 55		3.5:		
U,M,A,H,S	x 5"	354.3, –0.2	[404]	[456]	[1228,1255]	

Pos.: [454]; *atoll source*: [437]; *not in globular cl.*: [404,1209]; *X-ray bursts*: [47,334,483,484,488,489,581,925]; *X rays*: [1028]; *X-ray spectrum*: [1240]; *radio obs.*: [406].

- -

1728–169	A	17 28 50.2	V2216 Oph	16.8, 0.3, –0.7	300	4.20
GX 9+9		–16 55 32		0.3		
U,M,A,H,S	o	8.5, +9.0	[284,383]	[878,1018]	[1255]	[458,1018]

Pos.: [85]; *X-ray obs.*: [466,571,1028]; *X-ray spectrum*: [703,1236,1240]; *opt. cpt*: [207,241]; *near-IR sp.*: [210]; *radio*: [293].

- -

1728–247	P	17 28 57.9	V2116 Oph	19.0, - -, - -	100	304d?
GX 1+4		–24 42 35	*GF	1.7P:		114s
U,M,A,H,S	o	1.9, +4.8	[284]	[241,242]	[1255]	[229,866]

Pos.: [85,1049]; *X-ray pulsations and period history*: [276,390,668,729,834,866,1009]; *X-ray spectrum*: [994]; *X-ray pulse phase spectroscopy*: [1288]; *opt. cpt M6 giant*: [207,242]; *GX 1+4 unlikely to be source of 511 keV emission* [782,926]; *IR obs.*: [365,366].

- -

1730–335	GTB	17 30 06.6	Lil 1		<0.1-200	
Rapid Burster		–33 21 13		3.0:		
S	x 3"	354.8, –0.2	[412,689]	[22,733]	[400,1083]	

Pos.: [412,454]; *opt./IR obs. glob. cluster*: [457,611]; *very metal rich gl. cluster*: [22]; *recurrent transient (0.5 year?)*: [403,665]; *type II bursts*: [486,669]; *type II bursts behave as relaxation oscillator*: [669,758,1083]; *type II burst profiles*: [699,1120,1131]; *pre- & post-burst dips*: [758,1083]; *type II burst patterns*: [43,635,758,868,1083]; *type II burst spectra*: [586,634,678,700,758,1120]; *type I bursts*: [43,486,635]; *radio/IR/X-ray obs.*: [647]; *QPO*: [278,678, 698,1083,1128]; *radio obs.*: [406].

- -

Table 14.1. *(continued)*   LMXB

Source	Type Pos.	RA (1950) DEC (1950) ℓ^{II}, b^{II}	Opt. cpt [FC]	V, B–V, U–B $E_{B–V}$	F_X (µJy)	P_{orb}(hr) P_{puls}(s)
1730–220	T	17 30 56 −22 00 07			<10-130	
U	x 1.7'	4.5, +5.9			[179]	

Pos.: [333]; *X-ray outburst in 1972, relatively soft X-ray spectrum*: [179].

| 1731–260 | TB | 17 31 06.8 −26 03 10 | | 3.2: | <10-110 | |
| K | x 1' | 1.1, +3.6 | | [1092] | [1092] | |

Pos.: [1092]; *X-ray bursts*: [1092,1094]; *hard X-ray spectrum*: [46].

| 1732–304 | GB | 17 32 34.8 −30 27 03 | Ter 1 | 1.5 | 10 [920, | |
| Ha | x 8" | 357.6, +0.9 | [723] | [22,1268] | 1047,1228] | |

Pos.: [920]; *moderately rich globular cluster*: [22]; *X-ray bursts*: [538,723]; *X-ray obs.*: [1047]

| 1732–273 | TU | 17 32 54 −27 23 42 | | | <5-50 | |
| K,G | x 1' | 0.16, +2.59 | | | [542,718] | |

Pos.: [542,1102]; *position consistent with that of GS1734-275*: [718,1311]; *very soft X-ray spectrum*: [718].

| 1734–292 | | 17 34 14 −29 09 02 | | | 3.4 | |
| Gr | x 1' | 358.84, +1.39 | | | [928] | |

Pos.: [1102]; *X-ray observation*: [928]; *outburst of hard X rays*: [156].

| 1735–269 | | 17 35 08 −26 58 34 | | | 10 | |
| SL | x 2' | 0.78, +2.40 | | | [1047] | |

pos.: [1047].

| 1735–444 | BA | 17 35 19.3 −44 25 20 | V926 Sco *5 | 17.5, 0.2, −0.8 0.15 | 160 | 4.65 |
| U,M,A,H,S | o | 346.1, −7.0 | [555] | [428,555,1218] | [1255] | [1183] |

Pos.: [85]; *atoll source*: [437]; *X-ray obs.*: [466,1028]; *X-ray spectrum*: [1240]; *QPO*: [938]; *opt. cpt*: [786]; *opt. lt crv.*: [193,1061,1183]; *opt. spectrum*: [113,207,787,1057,1059]; *near-IR sp.*: [210]; *opt/X-ray variability*: [196,1061]; *UV*: [428]; *X-ray bursts*: [674,1222]; *optical bursts*: [408,788]; *radio obs.*: [293,406].

| 1735–28 | T | 17 35.4 −28.45 | | | <0.4-565 | |
| U,M | x 7' | 359.57, +1.56 | | | [179,1047] | |

Pos.: [333]; *transient*: [591]; *soft X-ray spectrum*: [179].

Table 14.1. *(continued)*   LMXB

Source	Type	RA (1950) DEC (1950)	Opt. cpt	V, B–V, U–B	F_X	P_{orb}(hr)
	Pos.	ℓ^{II}, b^{II}	[FC]	E_{B-V}	(μJy)	P_{puls}(s)
1736–297		17 36 21 −29 41 50			2	
Gr	x 1'	358.63, +0.71			[644]	

Pos.: [1102]; *X-ray observation*: [644,928].

| 1737–282 | | 17 37 47 −28 17 06 | | | 3 | |
| SL | x 3' | 0.01, +1.17 | | | [1047] | |

Pos.: [1047]

| 1739–304 | | 17 39 31 −30 29 29 | | | 9 | |
| K | x 1.6' | 358.33, −0.29 | | | [541] | |

Pos.: [1102].

| GC X-4 | T(?) | 17 40.6 −29 25 | | | 30 | |
| | x 1.2' | 359.36, +0.08 | | | [226] | |

Pos.: [226].

| 1740.7–2942 | | 17 40 42.9 −29 43 26 | | | 4-30 [1047, | |
| E | x 12" | 359.12, −0.11 | [654,801,1051] | | 1050,1095] | |

Pos: [1051]; *hard X-ray spectrum (black-hole candidate)*: [584,734,1051,1095,1098,1100]; *source of 511 keV annihilation line from galactic-center region*: [926]; *hard X-ray obs.*: [48]; *radio obs.*: [813,975,985]; *double-sided radio jet*: [814]; *relation with filamentary radio emission (?)*: [385,649]; *accretion from ISM (?)*: [35,813].

| 1741.2–2859 | | 17 41 14.7 −28 59 30 | | | <1.5-300 | |
| G | x 1' | 359.80, +0.18 | | | [819] | |

Pos.: [819]; *alternative position RA = 17 41 39.8, DEC = −28 49 24* [819].

1741–293	TB	17 41 38 −29 19 53			<5-26	
MXB1743–29?						
K	x 1'	359.55, −0.07			[542]	

Pos.: [542]; *X-ray bursts (MXB1743–29?)*: [542,671].

| 1741–322 | TU | 17 41 46 −32 12 25 | | | <2-770 | |
| H | x 20" | 357.1, −1.6 | | | [85,1050,1305] | |

Pos.: [1305]; *transient*: [285,416,573,1304]; *ultra-soft X-ray spectrum*: [1279]; *hard X-ray excess*: [186]; *also known as 1743-32*: [1305].

Table 14.1. *(continued)*   LMXB

Source	Type Pos.	RA (1950) DEC (1950) ℓ^{II}, b^{II}	Opt. cpt [FC]	V, B–V, U–B E_{B-V}	F_X (µJy)	P_{orb}(hr) P_{puls}(s)
1741.9–2853	T	17 41 52 −28 52 55			7	
Gr	x 1.6'	359.96, +0.13			[1095]	

Pos.: [1095]; *X-ray obs.*: [1099,1102].

| 1742–326 | | 17 42 12 −32 40 28 | | | <2-3 | |
| Exo | x 9' | 356.8, −1.9 | | | [1050,1256] | |

Pos.: [1256]; *X-ray spectrum*: [1311].

| 1742.2–2857 | | 17 42 15.9 −28 57 49 | | | 0.1 (0.9-4.0 keV) | |
| E | x 1' | 359.94, +0.01 | [1262] | | [1262] | |

Pos.: [1262].

| GC X-2 | T(?) | 17 42 26 −29 26 | | | 45 | |
| | x 3' | 359.56, −0.26 | | | [226] | |

Pos.: [226]; *in error boxes of MXB1743–29 and MXB 1742–29*: [671].

| 1742–289 | TB(?) | 17 42 26.3 −28 59 57 | | | <9-2000 | |
| A | r 3" | 359.93, −0.00 | [1262] | | [85] | |

Pos.: [85]; *transient*: [94,226,304,976,1262]; *X-ray bursts (MXB1743-29?)*: [671]; *radio obs.*: [243]; *possible optical counterpart (K dwarf)*: [863].

| 1742.5–2859 | | 17 42 30.0 −28 59 01 | | | 1-5 [1047, 1095,1262] | |
| E | x 1' | 359.95, −0.05 | [1262] | | | |

Pos.: [1262]; *Sgr A West (?)*: [1262]; *in error box MXB 1743–29*: [671].

| 1742.5–2845 | | 17 42 32.5 −28 45 44 | | | 0.1 (0.9-4.0 keV) | |
| E | x 1' | 0.14, +0.06 | [1262] | | [1262] | |

Pos.: [1262].

| 1742.7–2902 | | 17 42 42.1 −29 02 13 | | | 0.2 (0.9-4.0 keV) | |
| E | x 1' | 359.93, −0.11 | [1262] | | [1262] | |

Pos.: [1262]; *in error box MXB 1743–29*: [671].

Table 14.1. (*continued*)   LMXB

Source	Type	RA (1950)	Opt. cpt	V, B–V, U–B	F_X	P_{orb}(hr)
		DEC (1950)		$E_{B–V}$	(μJy)	P_{puls}(s)
	Pos.	ℓ^{II}, b^{II}	[FC]			

1742.8–2853		17 42 49.4			0.2
		–28 53 41			(0.9-4.0 keV)
E	x 1'	0.06, –0.06	[1262]		[1262]

Pos.: [1262]; *in error box MXB 1743–29*: [671].

1742.9–2852		17 42 54.3			0.2
		–28 52 05			(0.9-4.0 keV)
E	x 1'	0.09, –0.06	[1262]		[1262]

Pos.: [1262]; *in error box MXB 1743–29*: [671].

1742–294	B(?)	17 42 54.7			60-180
GC X-1		–29 29 58			[1047,
A	x 1'	359.56, –0.39	[556]		1095,1255]

Pos.: [455]; *transient, X-ray bright since 1975*: [94,226,304,584,976,1047,1050,1099,1262]; *X-ray bursts (MXB 1742-29?)*: [671,925,1095,1102].

1742.9–2849		17 42 59.2			0.2
		–28 49 57			(0.9-4.0 keV)
E	x 1'	0.13, –0.06	[1262]		[1262]

Pos.: [1262]

1743.1–2843		17 43 08.9			0.5-12
		–28 43 00		45:	[1047,
E	x 1'	0.25, –0.03	[1262]	[1095]	1095,1262]

Pos.: [1262]; *X-ray obs.*: [584,1099]; *in error box MXB 1743–28*: [671].

1743.1–2852		17 43 08.9			0.2
		–28 52 36			(0.9-4.0 keV)
E	x 1'	0.11, –0.11	[1262]		[1262]

Pos.: [1262]; *in error box MXB 1743–29*: [671].

1743–288	T	17 43.9			<1-40
GX +0.2,–0.2		–28 52.6			[226,976,
A	x 2'	0.21, –0.25			1047,1262]

Pos.: [976]; *X-ray history*: [976]; *probably same source as GC X-3*: [226].

1744–299		17 44 13.4			6
		–29 58 41		>7.5:	
SL	x 1'	359.30, –0.89		[1050]	[1050]

Pos.: [1050]; *northern component of two X-ray sources separated by 3'*: [1050].

Table 14.1. *(continued)*   LMXB

Source	Type	RA (1950) DEC (1950) ℓ^{II}, b^{II}	Opt. cpt	V, B–V, U–B E_{B-V}	F_X (μJy)	P_{orb}(hr)
	Pos.		[FC]			P_{puls}(s)
1744–300	B	17 44 13.6 −30 01 29		>7.5:	4	
SL	x 1'	359.26, −0.91		[1050]	[1050]	

Pos.: [1050]; *southern component of two X-ray sources separated by 3'*: [1050]; *X-ray bursts*: [925,1050,1102].

| 1744–265 GX3+1 | BA | 17 44 48.9 −26 32 49 | | 3.7 | 400 | |
| U,M,A,H,S | x 3" | 2.3, +0.8 | [986] | [456] | [1255] | |

Pos.: [456]; *atoll source*: [437]; *X-ray bursts*: [725]; *QPO*: [675,677,730]; *X-ray spectrum*: [1240]; *X-ray obs.*: [1028]; *IR obs.*: [876]; *radio obs.*: [406].

| 1744–361 | T | 17 44 50.9 −36 06 54 | | | <25-275 | |
| A | x 40" | 354.1, −4.2 | | | [85] | |

Pos.: [85]; *X-ray transient 1976*: [117,179,249]; *uncertain if LMXB or HMXB.*

| 1745–248 | GB | 17 45 51 −24 52 45 | Ter 5 | 2.1 | <0.1-110 | |
| Ha,Exo | x 9' | 3.8, +1.5 | | [1268] | [454,1256] | |

Pos.: [1256]; *X-ray bursts*: [540,723]; *radio obs.*: [340]; *glob. cl. properties*: [1268].

| 1745–203 | GT | 17 45 55.0 −20 21 07 | NGC 6440 | 1.1 | <0.1-180 | |
| U,M | x 1' | 7.7, +3.8 | [332,383,751] | [878,1268] | [300,332] | |

Pos.: [454]; *transient (1971) probably in NGC 6440*: [332,751]; *metal rich glob. cluster*: [22]; *optical study glob. cluster*: [762,1041]; *X-rays*: [300]; *radio obs.*: [340].

| 1746.7–3224 | | 17 46 47.3 −32 24 52 | | | 0.1 | |
| E | x 1' | 357.5, −2.6 | | | [455,1050] | |

Pos.: [455].

| 1746–331 | U | 17 46 33.2 −33 11 03 | | | 27 | |
| SL | x 35" | 356.9, −3.1 | | | [1050] | |

Pos.: [1050].

| 1746–370 | GB | 17 46 48.5 −37 02 18 | NGC 6441 | 0.50 | 32 | 5.7 |
| U,M,A,H,S | x 2" | 353.5, −5.0 | [412] | [1115] | [920,1228] | [1014] |

Pos.: [412,454]; *metal rich glob. cluster*: [22]; *optical properties glob. cluster*: [34,457,461]; *cluster center*: [454,1041]; *X-ray bursts*: [681,1115]; *X-ray obs.*: [1028]; *X-ray dips(?)*: [920]; *long-term X-ray behaviour*: [964]; *radio obs.*: [406].

Table 14.1. *(continued)*

Source	Type Pos.	RA (1950) DEC (1950) ℓ^{II}, b^{II}	Opt. cpt [FC]	V, B–V, U–B E_{B-V}	F_X (μJy)	P_{orb}(hr) P_{puls}(s)
1747–214	TB	17 47 25.7 –21 24 33			70	
Exo	x 7"	19.8, +22.7			[917,1256]	

Pos.: [383]; *X-ray bursts*: [717].

1747-313	G	17 47 31.2 –31 16 45	Ter 6	1.5	1.5-20	
R,Gr	x 1'	358.6, –2.2		[1268]	[928,961]	

Pos.: [961]; *ROSAT and Granat positions within 2', likely same source*: [961,1102]; *X-ray observations*: [928]; *globular cluster properties*: [1268].

1749–285	T	17 49 06 –28 29 41			60	
GX +1.1,–1.0	x 5'	+1.1, –1.0			[976]	

Pos.: [976].

1755–338	DU	17 55 21.5 –33 48 14	V4134 Sgr	18.5, 0.7, - - 0.5:	100	4.46
U,M,A,H,S	o	357.2, –4.9	[770]	[770,878]	[1255]	[770,1289]

Pos.: [85]; *ultra-soft X-ray spectrum*: [1279,1293]; *X-ray dips*: [157]; *opt. spectrum*: [207]; *opt. study*: [770].

1758–250	Z	17 58 03.1 –25 04 43		7.5:	1250	
GX 5–1						
U,M,A,H,S	x 3"	5.1, –1.0	[876,986]	[456]	[1255]	

Pos.: [456]; *Z source*: [437]; *X-ray sp.*: [703,1240]; *QPO*: [677,679,820,885,1185,1193,1195,1197]; *X-ray obs.*: [466,1028]; *long- term X-ray record*: [1199]; *no X-ray pulsations*: [1306]; *radio obs.*: [91,406]; *X-ray/radio obs.*: [1122]; *IR obs.*: [876].

1758–258		17 58 06.7 –25 44 25		3:	20	
Gr	r 2"	4.5, –1.4	[801,1045]	[1045]	[1100]	

Pos.: [998a]; *black-hole candidate*: [1100]; *hard X-ray spectrum*: [926,1100]; *radio obs.*: [998a].

1758–205	A	17 58 33.5 –20 31 44		3.7:	700	
GX9+1						
U,M,A,H,S	x 3"	9.1, +1.2	[383,876,986]	[456]	[1255]	

Pos.: [456]; *atoll source*: [437,1185]; *no X-ray pulsations*: [1306]; *QPO*: [459]; *X-ray spectrum*: [1240,1293]; *X-ray obs.* : [639,1028]; *IR obs.*: [876]; *radio obs.*: [406].

1803–245	T	18 03 45.8 –24 35 38			<2-1000	
S	x 30"	6.1, –1.9	[556]		[179]	

Pos.: [556]; *transient 1976*: [554,556]; *uncertain if LMXB or HMXB*

Table 14.1. *(continued)*   LMXB

Source	Type	RA (1950) DEC (1950) ℓ^{II}, b^{II}	Opt. cpt [FC]	V, B–V, U–B E$_{B-V}$	F$_X$ (μJy)	P$_{orb}$(hr) P$_{puls}$(s)
1811–171	BA	18 11 36.7	IR star	K=12	350	
GX13+1		–17 10 23		5.7		
U,M,A,H,S	r,IR	13.5, + 0.1	[284,383,876]	[122]		[1255,1305]

Pos.: [345,406]; *atoll source*: [437,1185]; *X-ray spectrum*: [1240]; *X-ray obs.*: [1028]; *X-ray bursts(?)*: [327,680]; *IR counterpart*: [122]; *IR obs.*: [345,876]; *X-ray/radio obs.*: [343]; *radio obs.*: [406].

1812–12	B	18 12.4			15	
		–12 06.0				
M,A,H	x 12'	18.1, +2.3				[1255,1305]

Pos.: [856]; *X-ray bursts*: [327,856]; *two Einstein IPC sources in error box*: [302].

1813–140	ZB	18 13 10.9	NP Ser	17.5, 1.3,1.0	700	
GX 17+2		–14 03 15		2.5:		
U,M,A,H,S	o	16.4, +1.3	[284,383,876]	[737,1028,1114]		[1255]

Pos.: [85]; *Z source*: [437,1185]; *QPO*: [642,940,1085]; *X-ray spectrum*: [1240,1293]; *X-ray obs.*: [1028]; *no X-ray pulsations*: [1306]; *X-ray bursts*: [568,1114,1130]; *opt. cpt (G star?)*: [241,468,481,667,1126]; *opt. spectrum*: [207]; *IR obs.*: [876]; *radio obs.*: [406,474]; *radio/X-ray obs.*: [936,1283]; *no radio lobes*: [939].

1820–303	GBA	18 20 27.8	NGC 6624		250	0.19
		–30 23 17		0.3		
U,M,A,H,S	x 3"	2.8, –7.9	[412]	[1180]	[1228,1255]	[1080]

Pos.: [412,454]; *metal rich glob. cluster*: [22]; *structure glob. cluster*: [457,702]; *cluster dynamics*: [977]; *cluster HR diagram*: [34]; *atoll source*: [437,1185]; *no X-ray pulsations*: [1306]; *X-ray spectrum*: [703,1240,1293]; *176 day X-ray high-low period*: [970,1068]; *QPO*: [277,677,1081,1185]; *X-ray obs.*: [466,571,1028,1078,1236]; *orbital period*: [1062,1121]; *orbital period decrease*: [1013,1121,1200]; *X-ray bursts*: [162,238,407,420,1180,1211]; *radio obs.*: [323,349,406].

1822–371	D	18 22 22.7	V691 CrA	15.3–16.3, 0.1, –0.9	10–25	5.57
		–37 08 04		0.15		
U,M,A,H	o	356.9, –11.3	[398]	[763,764,767]	[795,1305]	[767,1286]

Pos.: [85]; *orbital period change*: [448]; *X-rays*: [447]; *opt. spectrum*: [127,202,769]; *near-IR sp.*: [210]; *UV obs.*: [763,764]; *accretion disk corona*: [447,764,1277].

1822–000		18 22 48.3	star	22, 1:, - -	25–62	
		–00 02 29		1.3		
U,M,A,H,S	x 3"	29.9, +5.8	[148,383]	[144,456]	[85]	

Pos.: [456]; *X-ray obs.*: [462]; *long-term X-ray obs.*: [969]; *radio obs.*: [406].

1826–238	T(B?)	18 26 24			30	
		–23 49 31				
G	x 0.9'	9.3, –6.0			[541,719]	

Pos.: [541]; *X-ray obs.*: [1123]; *transient*: [719]; *X-ray bursts (?)*: [49].

Table 14.1. *(continued)*   LMXB

Source	Type	RA (1950) DEC (1950) Pos. ℓ^{II}, b^{II}	Opt. cpt [FC]	V, B–V, U–B E_{B-V}	F_X (µJy)	P_{orb}(hr) P_{puls}(s)
1832-330	G	18 32 27.3 −33 01 24	NGC 6652	0.1	8 (0.5-2 keV)	
R	x 1'	1.5, −11.4	[1268]		[961]	

Pos.: [961]; *moderately metal rich glob. cluster:* [22]; *opt. study gl. cl. :* [444].

1837+049	B	18 37 29.5 +04 59 20	MM Ser *DS	19.2B, - -, −0.5 ≥0.4	225	
Ser X-1						
U,M,A,H,S	o	36.1, +4.8	[383,1150]	[1150]		[969,1255]

Pos.: [85]; *X-ray obs.:* [466,571,1028,1236]; *X-ray spectrum:* [703,1240]; *X-ray bursts:* [685,1112]; *opt./X-ray bursts:* [422]; *opt. spectrum:* [207]; *radio obs.:* [406]; *X-ray/radio obs.:* [1176].

| 1846−031 | TU | 18 46 39.8 −03 07 12 | | | 300 | |
| EXO | x 15" | 29.9, −0.9 | [923a] | | [923a] | |

Pos.: [923a]; *alternative position (18 45 26.8, −03 25 39) also possible:* [923a]; *ultra-soft X-ray spectrum with hard tail:* [923a]; *archival optical plate search:* [1270a].

| 1850−087 | GB | 18 50 21.1 −08 46 04 | NGC 6712 star S | 21.0, 0.2, −0.9 0.42 | 7 [920, | |
| U,A,H,S | o | 25.4, −4.3 | [28,227,383] | [227,881] | 1228,1255] | |

Pos.: [227]; *optical studies glob. cluster:* [34,227,454,457,1041]; *X-ray obs.:* [178,609]; *radio source:* [406,656,714]; *long-term X-ray obs.:* [969]; *X-ray bursts:* [489,1108].

| 1905+000 | B | 19 05 53.4 +00 05 18 | star | 20.5, 0.5, −0.5 0.5 | 10 | |
| U,A,H,S | x 5" | 35.0, −3.7 | [147,383] | [147] | [141,1255] | |

Pos.: [147]; *X-rays:* [141]; *X-ray bursts:* [141,672,777]; *radio obs.:* [406].

1908+005	TB	19 08 42.8 +00 30 05	V1333 Aql	14.8, 0.6, −0.4 0.4	<0.1-1300	19.0
Aql X-1						
U,M,A,S	o	35.7, −4.1	[383,1148]	[126,144,1148]	[85,1220]	[143]

Pos.: [85]; *X-ray outbursts:* [103,126,577,969]; *opt. outbursts:* [126,708,744,1148]; *long-term X-ray behaviour:* [969]; *X-ray bursts:* [230,621]; *quiescent X-ray flux:* [230,1220]; *7.6 Hz pulsations in X-ray burst:* [1019]; *secondary G8-K0 star:* [144,1148]; $V_{quiesc} = 19.2$: [144]; *radio obs.:* [480].

| 1915+105 | T | 19 12 55 +10 52.8 | | | 300 | |
| Gr | x 3' | 45.3, −0.9 | | | [197] | |

Pos.: [197]; *outburst history:* [435]; *uncertain if this is a LMXB.*

| 1916−053 | BD | 19 16 08.4 −05 19 41 | V1405 Aql | 21.0, 0.4, −0.5 0.7 | 25 | 0.83 [413,1021, |
| U,M,A,H,S | o | 31.4, −8.5 | [383,1021,1253] | [413,1021] | [1063,1255] | 1063,1280] |

Pos.: [383]; *X-ray dips:* [1063,1067,1167,1253,1280]; *opt. lt crv.:* [413,1021]; *X-ray and optical periods differ:* [401,1066]; *X-ray bursts:* [51,1063,1108,1111]; *long-term X-ray obs.:* [969,1068]; *radio obs.:* [406].

Table 14.1. *(continued)*   LMXB

Source	Type Pos.	RA (1950) DEC (1950) ℓ^{II}, b^{II}	Opt. cpt [FC]	V, B–V, U–B E_{B-V}	F_X (μJy)	P_{orb}(hr) P_{puls}(s)
1918+146	T	19 18.0 +14 36.0			<5-45	
U,A	x 40'	49.3, +0.4			[179]	

Pos: [1033]; *soft X-ray spectrum*: [179].

1940–04	B	19 40 –4.0			<50	
Ha	x 1°	35.3, –13.1			[856]	

Pos.: [856]; *X-ray bursts*: [856].

1957+115	U	19 57 02.2 +11 34 16	V1408 Aql	18.7, 0.3, –0.6 0.4	30	9.33
U,M,A,H,S	o	51.3, –9.3	[283,383]	[745]	[1255]	[1146]

Pos.: [745]; *ultrasoft X-ray spectrum*: [1279]; *X-ray obs*.: [1028]; *long-term X-ray obs*.: [969]; *opt. spectr*: [207,745].

2000+251	TU	20 00 42.9 +25 05 44	QZ Vul *B	16.9, 1.3, 0.0 1.5	<0.5-11000	8.26 [108,131,
G	o	63.4, –3.1	[108,383]	[79,1164]	[812,1123]	142,144]

Pos.: [894]; *X-ray outburst*: [1164]; *soft X-ray component*: [1164]; *hard X-ray tail*: [1091]; *opt. obs. during outburst*: [131,142,927]; *quiescence opt. phot*.: [108,145]; $R_{quiesc} = 21.2$: [108]; *radio outburst*: [476].

2023+338	TU	20 22 06.3 +33 42 18	V404 Cyg	12.7, 1.5, 0.3 1.0:	0.4-20000 [812,1096,	155.4
G	o	73.2, –2.2	[290,1246]	[104,130,1248]	1123]	[120]

Pos.: [1246]; *X-ray outburst*: [604]; *hard X-ray tail*: [1096,1123]; *rapid X-ray variability*: [825,1123]; *opt. outburst*: [119,661,1173,1248]; *previous outbursts 1938, 1956*: [992,1241]; *opt. photometry*: [1173]; *opt. spectrum*: [118,119,376]; *orbital period & mass function (black hole candidate)*: [120]; *triple (?)*: [120,1250,1252]; *high Li abundance in secondary*: [761,1252]; *high-speed opt. photometry*: [375]; *radio obs*.: [431a,472,477]; *quiescent ellipsoidal variations*: [1250]; $V_{quiesc} \sim 19$: [120]; *UV obs*.: [1250].

2127+119	GB	21 27 33.3 +11 56 51	M15 AC 211	15.8-16.4, –0.1, –1.2 <0.06	6	17.1 [453,534,
U,M,A,H	o	65.0, –27.3	[29]	[30,65,877]	[1228,1255]	535,875]

Pos.: [348]; *cluster center*: [454,1041]; *metal poor globular cluster*: [22]; *cluster dynamics*: [946]; *opt. studies glob. cl.*: [235,312,457,702,1041]; *opt. id.*: [30,129]; *opt. spectrum*: [874]; *X-ray bursts*: [279,1227]; *UV obs*.: [877]; *radio cpt*: [714]; *radio obs*.: [406]; *orbital period twice 8.5 hr*: [146].

2129+470	B	21 29 36.2 +47 04 08	V1727 Cyg	16.4-17.5, 0.65, –0.3 0.5:	9	5.24
U,M,A,H	o	91.6, –3.0	[383,1149]	[789,1149]	[1255]	[789,790]

Pos.: [85]; *X-ray orbital variations*: [1177,1277]; *opt. phot. & spectr*.: [493,1147,1149]; *X-ray bursts*: [341]; *low state*: [346,503,828,954]; *no low-state ellipsoidal var*.: [150,199,344,579,1151]; *triple(?)*: [344]; *accretion disk corona*: [1277].

Table 14.1. *(continued)* LMXB

Source	Type	RA (1950)	Opt. cpt	V, B–V, U–B	F_X	P_{orb}(hr)
		DEC (1950)		E_{B-V}	(μJy)	P_{puls}(s)
	Pos.	ℓ^{II}, b^{II}	[FC]			
2142+380	ZB(?)	21 42 36.9	V1341 Cyg	14.7, 0.5, –0.2	450	236.2
Cyg X-2		+38 05 28		0.45		
U,M,A,H	o	87.3, –11.3	[353,383]	[151,708,791]	[1255]	[201]

Pos.: [85]; *Z source*: [437,440]; *X-ray spectrum*: [153,466,571,703,1235,1240,1293]; *long-term X-ray record*: [490,1068,1237]; *no X-ray pulsations*: [1306]; *QPO*: [438,817,884,1197]; *X-ray obs.*: [465,1028]; *X-ray bursts(?)*: [569]; *multi-frequency campaign*: [440,478,1226,1238]; *opt. spectrum*: [629]; *companion F giant*: [201]; *near-IR sp.*: [210]; *opt. lt crv.*: [201,371]; *UV obs.*: [151,735,791,1238]; *radio source*: [471].

Source	Type	RA (1950)	Opt. cpt	V, B–V, U–B	F_X	P_{orb}(hr)
2259+587	P	22 59 02.6			1.0	
		+58 36 38		0.5:		6.98
E	x 6"	109.1, –1.0	[244]	[391]	[85]	[246]

Pos.: [244]; *associated with SNR G109-1(?)*: [391,504]; *no optical/IR counterpart (V665 Cas = star D is not the counterpart)*: [244,597]; *pulse period history*: [246,622,623,866]; *proposed orbital period 2300 s* [310,806] *not confirmed*: [245,838]; *merged white dwarfs(?)*: [899]; *X-ray cyclotron line*: [545].

Source	Type	RA (1950)	Opt. cpt	V, B–V, U–B	F_X	P_{orb}(hr)
2318+620		23 18 22.6			2.4	
		+62 01 07		0.3		
U,A	r	112.6, +1.3		[1135]	[333]	

Pos.: [1135]; *radio source with jet, identification with 4U2316+61 and 20th mag. star not certain*: [1135].

Table 14.2. *High-mass X-ray binaries*

Name(s) Type Pos.	RA (1950) DEC (1950) ℓ^{II}, b^{II}	Opt. cpt [FC]	V, B–V, U–B Sp. type, E_{B-V}	F_X (µJy)	P_{orb} (d) P_{pulse}(s)
0050–727 T SMC X-3 A,H,S o 3"	00 50 19.5 –72 42 24 302.9, –44.7	*4 [163]	~14, –0.3, –1.0 O9 III-Ve, 0.03 [9,218]	<1-5 [163]	

Pos.: [85]; *transient*: [163,164]; $v_r sin\ i \sim 200\ km/s$: [218].

0053–739 T SMC X-2 A,S o	00 52 53.1 –73 57 19 302.6, –43.4	*5 [163]	16.0, –0.3, –0.5 B1.5 Ve, 0.03 [862]	<1-7 [163]	

Pos.: [862]; *transient*: [163,164]; *southern component of a close pair*: [862]; $v_r sin\ i \sim 200\ km/s$: [218].

0053+604 U,A,H,S o	00 53 40.3 +60 26 47 123.6, –2.1	γ Cas [280]	1.6-3.0, –0.15, –1.08 B0.5 III-Ve, 0.05 [826]	5-11 [85]	

Pos.: [85]; *X-ray spectrum*: [339,857]; *UV observations*: [69,268,427,449,755]; *variable Be star*: [373,756,1242, 1295]; *radial-velocity variations*: [553]; $v_r sin\ i \sim 300$-500 km/s: [506,512,826]; *system velocity*: [1206]; *comparison with X Per*: [826,1287]; *radio/mm obs*: [287,1259]; *polarimetry*: [166]; *optical interferometry Hα envelope*: [852, 1144]; *non-radial pulsations*: [1312]; *wind structure*: [1258].

0103–762 T H o	01 07 45.1 –75 00 38 301.9, –41.1	Be star [1166]	17, - -, - - Be, 0.03 [1166]	2.3 [1305]	

Pos.: [1166]; *variable X-ray source*: [1166]; *member SMC*: [1166].

0114+650 A,S o 3"	01 14 41.8 +65 01 32 125.7, +2.6	V662 Cas LS I+65°010 [280]	11.0, 1.2, 0.1 B0.5 Ib, 1.4 [3,220]	4 [85]	11.6 850 (?) [220,1310]

Pos.: [85]; *X rays*: [322,614,1310]; *opt. spectra*: [3,220,738]; $v_r sin\ i \sim 45\ km/s$: [3]; *opt. polarimetry*: [61]; *review multi-wavelength behaviour*: [361].

0115+634 TP U,M,H,S o	01 15 13.8 +63 28 38 125.9, +1.0	V635 Cas [559]	14.5-16.3, 1.4, 0.3 OBe, 1.7 [557,798]	<2-350 [85]	24.3 3.61 [866,981]

Pos.: [559]; *X-ray outbursts*: [965,996,1000,1119,1162]; *recurrence time 3 yr (?)*: [1296]; *X-ray spectrum*: [996, 1288]; *X-ray cyclotron line*: [870,1119]; *optical pre-outburst behaviour*: [630,798]; *long-term opt. phot.*: [798]; *opt. spectra*: [511]; $v_r sin\ i \sim 365\ km/s$: [511].

0115–737 P SMC X-1 U,M,A o 3"	01 15 45.6 –73 42 22 300.4, –43.6	Sk 160 [1012]	13.3, –0.14, –0.98 B0 Ib, 0.03 [514,1270]	0.5-57 [85]	3.89 0.71 [866,974,1026]

Pos.: [163]; *highly variable X-ray source*: [1032]; *pulse timing & X-ray orbit*: [974]; *aperiodic X-ray variability*: [14]; *long-term X-ray obs.*: [72,415]; *X-ray sp.*: [757,1288]; *opt. lt crv. & system parameters*: [600,688,1152,1210]; *opt. spectrum & orbit*: [514,990]; $v_r sin\ i \sim 200\ km/s$: [514]; *UV obs*: [74,429,1189].

Table 14.2. (continued)   HMXB

Name(s)	Type Pos.	RA (1950) DEC (1950) ℓ^{II}, b^{II}	Opt. cpt [FC]	V, B–V, U–B Sp. type, E_{B-V}	F_X (µJy)	P_{orb} (d) P_{pulse}(s)
J0146.9 +6121 R	o 2"	01 43 32.6 +61 06 26 129.9, –0.5	LS I +61°235 [433]	11.33, 0.82, –0.39 B5 IIIe, 0.87 [212,662,1053]	1-3 (0.1-2.4 keV) [850]	

Pos.: [433]; in open cluster NGC 663: [713]; $v_r sin\, i \sim 250$ km/s: [1053].

| 0236+610 E | o,r | 02 36 40.6 +61 00 54 135.7, +1.1 | LS I +61°303 V615 Cas [100,433] | 10.7, 0.8, –0.3 B0e:, 0.75: [288,510,1003] | 0.2 (0.2-5 keV) [64] | 26.45 [1133] |

Pos.: [394]; X-ray source: [64]; radio outbursts: [1133,1134]; four-year modulation radio outbursts: [395,911]; VLBI radio jet: [774,1136]; optical light curve: [797,909,910]; opt. spectra (luminosity class uncertain): [5,510]; $v_r sin\, i \sim$ 200 km/s: [216]; system velocity: [1206]; long-term optical obs.: [394]; IR obs.: [239]; distance: [335]; UV obs.: [495,509,510,736]; wind structure: [1258]; related to γ source CG 135+1(?): [392,452].

| 0331+530 V,T,Exo | TP o | 03 31 14.9 +53 00 24 146.1, –2.2 | BQ Cam [612] | 15.1-15.4, 1.6-2.3, - - Be, 1.9 [194,492] | <0.5-1250 [1079,1141] | 34.25 4.4 [866,1079,1297] |

Pos.: [612]; X-ray outbursts: [1079,1141,1297]; X-ray obs.: [732]; rapid X-ray variability: [56]; X-ray cyclotron line: [731]; X-ray pulse phase sp.: [1178]; optical id.: [59,612]; optical spectrum: [1087]; $v_r sin\, i \sim 150$ km/s: [194]; system velocity: [1206]; Hα emission: [547]; IR obs.: [170]; wind structure: [1258].

| 0352+309 U,M,A,H,S | P o | 03 52 15.1 +30 54 01 163.1, –17.1 | X Per [87] | 6.0-6.6, 0.29, –0.82 O9 III-Ve, 0.4 [513,826,1052] | <9-37 [85] | 835 [858,866] |

Pos.: [85]; X rays: [997,1271]; X-ray sp.: [1288]; optical id.: [101]; long-term optical record: [501,835]; reported 580 day period unlikely to be orbital: [508,513,944]; disappearance emission lines: [886]; system velocity: [1206]; $v_r sin\, i \sim 250-400$ km/s: [508,826,1282]; UV obs.: [60,427]; stellar parameters: [308]; comparison with γ Cas: [826,1287]; wind structure: [1258].

| 0521+373 U,H | o | 05 19 10.7 +37 37 44 170.0, +0.7 | HD34921 SAO57950 [1069] | 7.51, 0.14, –0.86 B0 IVpe, 0.42 [106,956,1002] | 1 [333,1305] | |

Pos.: [1069]; hard X-ray spectrum: [956]; UV/opt./IR obs.: [956,1002].

| 053109 –6609.2 Exo,SL | T x 10" | 05 31 09 –66 09 12 276.2, –32.7 | [432] | Be?, 0.1 [432] | 1 [432] | |

Pos.: [907]; hard X-ray spectrum: [432]; not detected with Einstein: [432]; optical cpt probably Be star: [432].

| 0532–664 LMC X-4 U,M,A,H | P o 2" | 05 32 47.3 –66 24 13 276.3, –32.5 | Sk-Ph [137] | 14.0, –0.1, –1.1 O7 III-V, 0.1 [463,588,701] | <3-60 [85] | 1.40 13.5 [588,866] |

Pos.: [85]; X-ray orbital parameters: [588,664,953]; X-ray eclipses: [683,1275]; 30 day period X rays: [638]; X-ray spectrum: [1093,1288]; rapid X-ray variability: [56]; X-ray flare: [953]; opt. light curve: [136,446,532]; opt.spectrum & radial velocity curve: [516,947]; $v_r sin\, i \sim 170$ km/s: [516]; opt. 30 day period: [446,532]; precessing-disk model: [446]; UV obs.: [74,1189].

Table 14.2. *(continued)*   HMXB

Name(s)	Type	RA (1950) DEC (1950)	Opt. cpt	V, B–V, U–B	F_X	P_{orb} (d)
	Pos.	ℓ^{II}, b^{II}	[FC]	Sp. type, E_{B-V}	(μJy)	P_{pulse}(s)
0535–668	TP	05 35 42.4	*Q	12.3-14.9, 0.1, –0.9	<0.01-180	16.7
		–66 53 39		B2 III-IVe, 0.1		0.069
A	o 2"	276.9, –32.2	[128,562]	[128,265,701,903]	[85]	[866,1044]

Pos.: [562]; *in LMC*: [560,903]; *X-ray outbursts*: [560,1276]; *X-ray pulsations*: [1046]; *optical outbursts*: [128,191, 265,1044,1217]; *long-term off states*: [697,903]; *quiescent opt. phot. & spectroscopy*: [191,524,1056,1060,1182, 1217]; *very strong emission lines during outbursts*: [128]; *IR obs.*: [10]; *opt. polarimetry*: [167]; *UV obs.*: [128,498,984]; *wind structure*: [1258].

- -

0535+262	TP	05 35 48.0	V725 Tau	8.9-9.6, 0.45-0.62, –0.54	<3-2800	111
		+26 17 18	HD 245770	O9.7 IIe, 0.8		104
U,A,H	o	181.4, –2.6	[686]	[355,743,1086]	[85]	[866,967]

Pos.: [85]; *transient*: [867,1001]; *long-term X-ray obs.*: [967]; *hard X-ray obs.*: [172,595,991,1031]; *55.7 d orbit?* [749]; *long-term optical record*: [58,1086]; *opt. phot.*: [1038]; *opt. spectr.*: [4,515,522]; $v_r sin\, i \sim 300\, km/s$: [515]; *system velocity*: [1206]; *multi-wavelength obs.*: [263,551]; *X-ray/opt. relation*: [710]; *UV obs.*: [263,929,1309]; *accretion model*: [849]; *wind structure*: [1258]; *1992 review*: [360].

- -

0538–641	U	05 38 39.7	*1	16.7-17.5, –0.2, –0.6	<1.7-44	1.70
LMC X-3		–64 06 34		B3 Ve, 0.1		
U,M,A,H	o 3"	273.6, –32.1	[1192]	[204,701,1180]	[85]	[204]

Pos.: [561]; *X-ray obs.*: [561]; *rapid X-ray variability*: [56]; *X-ray sp.*: [1093,1154,1279,1293]; *199 d X-ray period*: [211,1093]; *opt. lt crv.*: [67,1221]; *mass function*: [204,633,898]; $v_r sin\, i \sim 130\, km/s$: [204]; *UV obs.*: [1155,1156]; *X-ray/UV/opt. obs.*: [1156]; *near IR sp.*: [210].

- -

0540–697	U	05 40 05.5	*32	14.5, 0.29, –0.70	3-25	4.22
LMC X-1		–69 46 04		O7-9 III, 0.37		
U,M,A,H	o	280.2, –31.5	[200,292]	[63,521]	[85]	[526]

Pos.: [85]; *ultra-soft X-ray spectrum*: [1279]; *X-ray spectrum & QPO*: [297]; *rapid X-ray variability*: [56]; *X-ray obs.*: [561,1093]; *mass function*: [526]; $v_r sin\, i \sim 150\, km/s$: [521]; *UV obs.*: [63,526]; *in He III region*: [902].

- -

0544–665		05 44 15.6	*1	15.4, –0.20, –0.96	1.8	
		–66 34 59		B1 Ve, 0.1		
H	o 3"	276.5, –31.4	[561]	[701,1191]	[561]	

Pos.: [561]; *optical id. not completely certain*: [1191]; **1 member of LMC*: [1191].

- -

0556+286		05 52 44.3	HD249179	9.2, - -, - -	1.1	
		+28 46 41		B5ne, - -		
H	o	181.3, +1.9		[958,1166]	[1305]	

Pos.: [1166]; *Be star*: [1242].

- -

0726–260		07 26 50.0	star	11.6, 0.3, –0.6	1.2-4.7	
		–26 00 13		B0-1e, 0.6		
U,A,H	o	240.3, –4.1	[1075]	[1075]	[1255,1305]	

Pos.: [1075]; *X-ray flare*: [1075].

- -

Table 14.2. *(continued)*   HMXB

Name(s)	Type	RA (1950) DEC (1950)	Opt. cpt	V, B–V, U–B	F_X	P_{orb} (d)
	Pos.	ℓ^{II}, b^{II}	[FC]	Sp. type, E_{B-V}	(μJy)	P_{pulse}(s)
0739–529		07 46 09.8	HD63666	7.62, 0.02, –0.24	0.7	
		–53 12 28	SAO235515	B7 IV-Ve, - -		
H	o	266.4, –13.7	[1069]	[494,712,879,1166]	[1305]	

Pos.: [1166].

0749–600		07 55 27.7	HD65663	6.73, 0.05, –0.25	0.7	
		–60 57 54	SAO250018	B8 IIIe, 0.09		
H	o	274.0, –16.2	[1069]	[231,232,1166]	[1305]	

Pos.: [1166]; *Be star*: [1242]; *in open cluster NGC 2516*: [231]

0834–430	TP	08 34 10			30-300	110
		–43 00.6				12.3
G,Gr,R	x 1'	262.0, –1.5			[386,1101]	[386,1301]

Pos.: [441,1101]; *southern of two neighboring sources*: [386,721]; *recurrent transient*: [644,1104]; *not the same transient as MX 0836-42*: [179,753]; *X-ray obs.*: [18].

0900–403	P	09 00 13.2	HD77581	6.9, 0.47, –0.51	2-1100	8.96
Vela X-1		–40 21 25	GP Vel	B0.5 Ib, 0.7		283
U,M,A,H	o	263.1, +3.9	[330]	[295,837,879,1202]	[85]	[256,866,1186]

Pos.: [85]; *X-ray sp.*: [869,891,1016,1288]; *cyclotron line*: [595]; *pulse timing & X-ray orbit*: [83,84,255,256,983, 1161,1186]; *X-ray eclipses*: [330,1174]; *limits on apsidal motion*: [254]; *rapid X-ray variability*: [56]; *orbital variation X-ray absorption*: [419]; *energy dependence pulse profile*: [982]; *opt. light curve*: [599,1152]; *long-term cycles (?)*: [965,1152]; *opt. radial-velocity curve*: [1205,1212]; *system velocity*: [1206]; $v_r \sin i \sim 90\text{-}130$ km/s: [811,1152,1299]; *UV obs.*: [295,570,793,929,1007]; *opt. polarimetry*: [271]; *IR obs.*: [365].

1024.0	P	10 24 05.4	Wack 2134	12.7, 1.5, - -	1-10	
–5732		–57 33 24	THα 35-42	O5:, 1.8	(0.2-4.5 keV)	0.061
E	o 6"	284.5, –0.2	[1143]	[115]	[115]	[115]

Pos.: [115]; *no optical pulsations*: [267]; *emission line star*: [1242].

1036–565		10 28 28.3	HD91188	6.64, –0.10, –0.56	3.3	
		–56 49 14	SAO238130	B4 IIIe, - -		
A	o	284.6, +0.7	[1069]	[347,879]	[1255]	

Pos.: [1166]; *Be star*: [1242]; *X-ray flare 1974 Nov.*: [1255]; *periodic (2.924d) optical brightness variations reflect rotation of Be star*: [37,38].

1048.1	P	10 48 09.0			<0.1-1	
–5937		–59 37 21				6.4
E	x 9"	288.2, –0.5	[802]		[1036]	[190,1036]

Pos.: [802]; *discovery*: [1036]; *Be/X or LMXB(?)*: [190]; *opt. phm. and sp. of possible counterparts*: [802].

1118–615	PT	11 18 45.2	He3-640	12.1, 0.96, –0.30	0.1-70	
		–61 38 31		O9.5 III-Ve, 1.2		405
A	o	292.5, –0.9	[549]	[549]	[85,846]	[866]

Pos.: [85]; *X-ray transient*: [303,544,706]; *opt. phot. & spectr.*: [549]; $v_r \sin i \sim 300$ km/s: [549]; *system velocity*: [1206]; *X-ray/opt. obs.*: [846]; *UV obs.*: [169]; *radio obs.*: [293].

Table 14.2. *(continued)*   HMXB

Name(s)	Type Pos.	RA (1950) DEC (1950) ℓ^{II}, b^{II}	Opt. cpt [FC]	V, B–V, U–B Sp. type, E_{B-V}	F_X (μJy)	P_{orb} (d) P_{pulse}(s)
1119–603	P	11 19 01.9	V779 Cen	13.3, 1.07, –0.04	10-312	2.09
Cen X-3		–60 20 57		O6.5 II-III, 1.4		4.84
U,M,A,H	o 3"	292.1, +0.3	[102,631]	[517,631,993]	[85]	[589,866]

Pos.: [85]; *X-ray pulsations*: [354,1161]; *X-ray orbit*: [589]; *period decay*: [305,589]; *X-ray spectrum*: [703, 1288]; *long-term X-ray history*: [414,543,959,968,1027]; *X-ray eclipses*: [1025]; *X-ray obs.*: [499,855,871]; *aperiodic X-ray variability*: [56]; *QPO*: [1116]; *opt. light curve*: [598,1152,1215]; *opt. radial velocity curve*: [517,851,895]; $v_r \sin i \sim$ 250 km/s: [517]; *system velocity*: [1206]; *IR obs.*: [365]; *structure companion star*: [165]; *distance*: [505].

1145.1	P	11 45 02.3	V830 Cen	13.1, 1.5, 0.15	4-40	
–6141		–61 40 33		B2 Iae, 1.6		298
A,E	o 2"	295.5, –0.0	[518]	[264,531]	[518]	[866]

Pos.: [518]; *X-ray pulsations*: [1284]; *proposed orbital periods 10.76d (opt. sp.) and 5.65 d (opt. phot.)*: [525,531]; *system velocity*: [1206]; *15' away from 1145-619*: [636,1285].

1145–619	PT	11 45 33.6	Hen 715	9.3, 0.18, –0.81	4-1000	187.5
		–61 55 44	HD102567	B1 Vne, 0.35		292
U,M,A,H,S	o	295.6, –0.2	[87]	[314,879,1258]	[85]	[866,1263]

Pos.: [85]; *X-ray spectrum*: [1288]; *X-ray obs.*: [799]; *long-term X-ray history*: [968,1263]; *opt. spectrum*: [550, 1284]; $v_r \sin i \sim 270$ km/s: [427,550]; *system velocity*: [1206]; *coordinated X-ray/opt. obs.*: [185]; *UV obs.*: [62,262, 427,929]; *wind structure*: [1258].

1223–624	PT	12 23 49.7	BP Cru	10.8, 1.76, 0.42	9-1000	41.5
GX 301–2		–62 29 37	Wra 977	B1-1.5 Ia, 1.8		696
U,M,A,H,S	o 3"	300.1, –0.0	[87]	[78,426,1145,1233]	[85]	[866,1015]

Pos.: [85]; *X-ray spectrum*: [1288]; *long-term X-ray record (recurrent outbursts near periastrion)*: [966,1004,1264]; *X-ray obs.*: [418,653,994,995,1281]; *pulse-phase spectr.*: [651]; *pulse profile*: [816]; *X-ray orbit*: [1015]; *X-ray dips*: [652]; *aperiodic X-ray variability*: [56,1127]; *optical spectrum*: [77,426,520,914,1145]; *system velocity*: [1206]; *optical photometry*: [901,1208]; *IR obs.*: [365].

1239-599	P	12 39 07.5			3-16	
		–59 55 39			[85,1255,	191
A,H,S	x 30"	301.8, +2.6	[280]		1305]	[500]

Pos.: [280]; *hard X-ray spectrum*: [500]; *see also*: [117].

1244-604	T	12 44 38			<24-100	
		–60 22.2				
A	x 6.2'	302.5, +2.2			[117]	

Pos.: [117].

1246-588	T	12 46 39			<24-300	
		–58 51.0				
U,A,H	x 4.5'	302.7, +3.8			[117]	

Pos.: [117].

Table 14.2. *(continued)*   HMXB

Name(s)	Type Pos.	RA (1950) DEC (1950) ℓ^{II}, b^{II}	Opt. cpt [FC]	V, B–V, U–B Sp. type, E_{B-V}	F_X (μJy)	P_{orb} (d) P_{pulse}(s)
1249–637		12 39 53.2 –62 47 06	HD110432 SAO252002	5.31, 0.27, –0.79 B0 IIIe, 0.40	2.2	
H	o	302.0, –0.2	[1069]	[168,879]		[1305]

Pos.: [1166]; *Be star*: [1242]; *UV obs.*: [168]; *interstellar abs. lines*: [225]; *optical obs.*: [234,315,326]; *variable radial velocity*: [105,1142]; *white-dwarf accretor(?)*: [1257].

| 1253–761 | | 12 35 59.8
–75 05 43 | HD109857
SAO256967 | 6.49, 0.08, –0.24
B7 Vne, 0.20 | 0.6 | |
| H | o | 302.1, –12.5 | [1069] | [188,879] | | [1305] |

Pos.: [1166]; *Be star*: [1242]; *visual double (sep. 2.2")*: [318]; *no optical brightness variations*: [38]; *white-dwarf accretor(?)*: [1257].

| 1255–567 | | 12 51 39.6
–56 53 50 | μ^2 Cru
HD112091 | 5.17, –0.12, –0.51
B5 Ve, 0.04 | 0.8 | |
| H | o | 303.4, +5.7 | | [879,1052] | | [1305] |

Pos.: [1166]; *Be star*: [1242]; $v_r \sin i \sim 220$ km/s: [1052]; *visual double with μ^1 Cru*: [482]; *slow & small optical variability*: [945].

1258–613	PT?	12 58 11.8	V 850 Cen	13.5-14.2, 1.7, 0.8	0.3-200	133?
GX 304–1		–61 19 58	*2 (MMV)	B2 Vne, 2.0		272
U,M,A,S	o 2"	302.1, +1.2	[766]	[192,423,913]	[85,955]	[866,968]

Pos.: [766]; *long-term X-rays*: [965,968]; *X-ray off state*: [955]; *X-ray spectrum*: [994,1288]; *optical spectrum*: [189, 766,1145]; *long-term opt. variations*: [192,423]; $v_r \sin i \sim 600$ km/s: [913]; *system velocity*: [1206]; *IR obs.*: [365].

| 1417–624 | PT | 14 17 25.5
–62 28 11 | *7 | 17, 2:, 0.7:
OBe, 2: | 2-43 | 17.6 |
| U,M,A,S | o | 313.0, –1.6 | [20] | [410] | [85] | [866] |

Pos.: [85]; *X-ray obs.*: [20,587]; *opt. id.*: [410]; *Centaurus 1971-2 transient (?)*: [179,1272].

| 1538–522 | P | 15 38 38.6
–52 13 37 | QV Nor
*12 | 14.4, 1.9, 0.6
B0 Iab, 2.1 | <3-30 | 3.73
529 |
| U,M,A,H,S | o | 327.4, +2.1 | [19] | [219,529,912] | [85] | [175,866] |

Pos.: [85]; *X-ray obs.*: [175,228,728,997a]; *X-ray eclipses*: [50,250]; *X-ray orbit*: [728]; *X-ray spectrum*: [728,1288]; *X-ray cyclotron line*: [160]; *opt. light curve*: [905]; *radial-velocity curve*: [219,989]; $v_r \sin i \sim 200$ km/s: [219]; *system velocity*: [1206]; *opt. obs.*: [912].

| 1553–542 | P | 15 53 55.6
–54 16 15 | | | 27 | 30.6
9.3 |
| S | x 35" | 327.9, –0.9 | [19] | | [85] | [590,866] |

Pos.: [85]; *opt. counterpart likely a Be star*: [590]; *X-ray orbit*: [590].

| 1555–552 | | 15 50 26.4
–55 10 54 | HD141926
SAO243098 | 8.60, 0.56, –0.43
B2nne, - - | 1.7 | |
| H | o | 327.0, 1.2 | [1069] | [312,191] | | [1305] |

Pos.: [1166]; *Be star*: [1242].

Table 14.2. *(continued)*   HMXB

| Name(s) | Type | RA (1950)
DEC (1950) | Opt. cpt | V, B–V, U–B | F_X | P_{orb} (d) |
	Pos.	ℓ^{II}, b^{II}	[FC]	Sp. type, E_{B-V}	(μJy)	P_{pulse}(s)
1657–415	P	16 57 16.8			4-42	10.4
		–41 35 59				38
OAO	x 15"	344.4, +0.3	[23]		[85]	[321,866]

Pos.: [23]; *error box #2*: [23,800,915]; *V861 Sco not the optical counterpart*: [23]; *X-ray obs.*: [580]; *X-ray spectrum*: [1288]; *hard X-ray obs.*: [800].

1700–377		17 00 32.7	HD153919	6.6, 0.27, –0.72	<11-110	3.41
		–37 46 29	V884 Sco	O6.5f, 0.52		
U,M,S	o	347.8, +2.2	[564]	[425,430,943,1303]	[85]	[95,419,1216]

Pos.: [85]; *X-ray variability*: [56,275,421]; *hard X-ray spectrum*: [324,952]; *no X-ray pulsations*: [275,379]; *opt. light curve*: [36,133,1216]; *opt. spectrum*: [182,233,311,450]; *radial-velocity curve*: [275,424]; $v_r sin\, i \sim 140\text{-}300$ *km/s*: [182,233,507,1303]; *system velocity*: [1206]; *UV obs.*: [294,425,430,582]; *soft X-ray Raman scattering*: [582]; *opt. polarimetry*: [270,271]; *IR obs.*: [365,943]; *system parameters*: [445].

1722–363	P	17 22 33			0.2-5	
		–36 22 05				413
Exo	x 9'	351.5, –0.6			[1117,1132]	[866,1132]

Pos.: [1256]; *hard X-ray spectrum*: [1132]; *limits on orbital period*: [1117].

1807–10	T	18 07.9			<2-10	
		–10 53				
U	x 1.3°	18.6, +3.9			[333]	

Pos.: *(large error box)*: [333]; *uncertain if HMXB or LMXB*.

1833–076	PT	18 33 46.3			1.6-200	
Sct X-1		–07 38 54				111
A,H	x 30"	24.5, –0.2	[383]		[624,969,1255]	[628,866]

Pos.: [986]; *alternative position (18 34 49.5, –07 38 05)*: [986]; *X-ray obs.*: [123,462]; *X-ray pulsations*: [628]; *hard X-ray spectrum*: [186,986]; *long-term X-ray record*: [969]; *transient*: [186].

1839–06	T	18 39.0			1	
		–05.9				
G	x 30'	26.6, –0.5			[624]	

Pos.: [624].

1839–04	PT	18 39.2			2.5	
		–04.5				81.1
G	x 24'	27.9, +0.1			[624]	[624]

Pos.: [624].

1843+009	PT	18 43.0			<0.4-33	
		+0.9				29.5
G	x 10'	33.1, +1.7			[624]	[624]

Pos.: [625]; *X-ray obs.*: [624,625].

Table 14.2. *(continued)*

Name(s)	Type	RA (1950) DEC (1950)	Opt. cpt	V, B–V, U–B	F_X	P_{orb} (d)
				Sp. type, E_{B-V}	(μJy)	P_{pulse} (s)
	Pos.	ℓ^{II}, b^{II}	[FC]			
1845–03	T	18 44.7			1	
		–03.2				
G	x 24'	29.7, –0.5			[624]	

Pos.: [624].

1845–024	PT	18 45 41.1			1-44	
		–02 28 37				94.8
A,H,S,G	x 30"	30.4, –0.4	[383]		[85,624]	[624]

Pos.: [85]; *X-ray observations*: [283,624,626,1033].

1855–02	T	18 55.4			2	
		–02.8				
G	x 24'	31.3, –2.7			[624]	

Pos.: [624].

1901+03	T	19 01.7			<2-87	
		+03 06.0				
	x 10'	37.2, –1.4			[179]	

Pos.: [333]; *transient 1971*: [179,331]; *hard X-ray spectrum*: [179].

1907+097	PT	19 07 15.1	star	16.4, 3.2, - -	4-275	8.38
		+09 44 54		B I, 3.3		438
U,M,A,H	o	43.7, +0.5	[1029]	[1029,1203]	[85]	[726,866]

Pos.: [1029]; *long-term X-ray record*: [969]; *1980 outburst*: [760]; *orbital parameters*: [184,726]; *orbital modulation X-ray flux*: [760]; *opt. spectr.*: [546,1203]; $v_r\sin i \sim 85\ km/s$: [1203].

1909+048		19 09 21.3	SS433	14.2, 2.1, 0.6	2-10	13.1
		+04 53 54	V1343 Aql	pec, 2.6		
U,A,H	o,r	39.7, –2.2	[658]	[746,864,1245]	[85]	[217]

Pos.: [85]; *X-ray obs.*: [39,56,98,99,411,585]; *Doppler shift X-ray Fe line*: [780,1267]; *extended X-ray lobes*: [1035, 1265]; *γ-ray obs.*: [351,637]; *Doppler shifted opt. emission lines*: [742,747]; *stationary emision lines from accretion disk*: [320]; *kinematic model precessing (164 d) high-speed jets*: [6]; *opt. spectra*: [26,309,313,618,748]; *opt. spectrophot.*: [11,617,1245]; *opt. light curve*: [15,16,25,362,451,657,659,660]; *long-term optical phot. record*: [363,451,593,781]; *X-ray/opt. eclipse*: [1084]; *IR obs.*: [613,1254]; *precession clock*: [12]; *radio structure*: [40,316, 317,999,1074,1231]; *radio variability*: [76,319]; *distance 5.5 kpc*: [473]; *nature of the compact star*: [269]; *reviews*: [739,740,1230,1314].

1936+541		19 31 42.6	DM+53°2262	9.8, - -, - -	0.7	
		+53 46 12		Be, - -		
H	o	85.9, +15.9	[1166]	[1305]		

Pos.: [1166]; *Be star*: [1242].

1942+274	T	19 42 58			<1-25	
		+27 29 24				
A	x 10'	63.4, +1.7			[1255]	

Pos.: [1255]; *transient Nov-Dec 1976*: [1255]; *uncertain if HMXB or LMXB*.

Table 14.2. *(continued)*   HMXB

Name(s)	Type Pos.	RA (1950) DEC (1950) ℓ^{II}, b^{II}	Opt. cpt [FC]	V, B–V, U–B Sp. type, E_{B-V}	F_X (μJy)	P_{orb} (d) P_{pulse}(s)
1947+300	T	19 47 36.3 +30 04 54	*3	14.2, 0.9, –0.3 1.1	<10-84	
K	o	66.1, +2.1	[384]	[372,384]	[80]	

Pos.: [384]; *Hα emission*: [372]; *likely, but not certain this is a HMXB*: [372,384]; *see also* [1048].

| 1954+319 | | 19 53 46.2 +31 58 02 | | | <1.5-80 | |
| U,A | x 15" | 68.4, +1.9 | [183] | | [183,1255] | |

Pos.: [302]; *heavily reddened supergiant system (?)*: [1169].

| 1956+350 Cyg X-1 | U | 19 56 28.9 +35 03 55 | HD226868 V1357 Cyg | 8.9, 0.84, –0.26 O9.7 Iab, 1.06 | 235-1320 | 5.60 |
| U,M,A,H | o,r | 71.3, +3.1 | [979] | [181,247,879] | [85] | [356] |

Pos.: [85]; *hard X- and γ-ray obs.*: [691,1010,1030,1170,1171]; *long-term X-ray obs.*: [490,972]; *X-ray low/high states*: [690,1124]; *X-ray dips*: [605,987]; *rapid X-ray variability*: [55,56,696,796,821,823,825,1005]; *QPO*: [338, 620,1232]; *chaotic(?)*: [695,1179]; *300 d period*: [357,592,594,692,972]; *X-ray spectrum*: [41,42,307,606,703,1240, 1279]; *radio obs.*: [89,90,467,474]; *opt. spectra*: [1,2,181,358,882,1070]; *radial-velocity curve*: [356,357,882]; v_r*sin i ~ 100 km/s*: [358]; *system velocity*: [1206]; *nature of the compact star*: [70,1037,1269]; *opt. light curve*: [594, 707]; *opt. polarimetry*: [272,337]; *UV obs.*: [247,294,1153,1308]; *IR obs.*: [650]; *review (1977)*: [887].

| 2030+375 | PT | 20 30 22.1 +37 28 00 | * 2 | 19.7, 3.3, - - Be, 3.8 | <0.5-1400 | 46.0 41.8 |
| Exo | o 2" | 77.2, –1.3 | [171,840] | [171,840] | [921] | [866,921,1087a] |

Pos.: [840]; *outburst*: [921]; *rapid X-ray variability*: [56]; *0.2 Hz QPO*: [13]; *opt. spectrum*: [171,552].

| 2030+407 Cyg X-3 | | 20 30 37.6 +40 47 13 | V1521 Cyg | I=20.0 Wolf-Rayet, 6.3 | 90-430 | 0.20 |
| U,M,A,H | r | 79.9, +0.7 | [1247] | [831,1204,1247] | [85] | [924] |

Pos.: [85]; *distance*: [266]; *orb. period change*: [608,1187]; *long-term X-ray obs.*: [490,971]; *no X-ray pulsations*: [608, 1306]; *orbital X-ray curve*: [73,603,832]; *aperiodic variability*: [56]; *X-ray halo*: [827]; *transient QPO*: [1188]; *radio outbursts*: [393,470,558,830]; *quiescent radio flares (period 4.95 h?)*: [829]; *radio jet & lobes*: [73,350,832, 1088]; *IR phot.*: [52,765,772]; *IR spectra (secondary helium star)*: [583,1204]; *no X-ray pulsations*: [1306]; *X-ray spectrum*: [603,1277]; *1989 review (incl. TeV/PeV γrays)*: [71].

| 2138+568 Cep X-4? | PT | 21 38.0 +56 50.0 | | | <6-100 | 66.2 |
| M,H,G | x 7' | 99.0, +3.3 | | | [627] | [627,866] |

Pos.: [627]; *likely Be/X-ray system*: [627]; *X-ray obs.*: [627]; *X-ray cyclotron line*: [809]; *same as Cep X-4(?)*: [754,1175].

| 2202+501 | | 21 59 44.1 +49 55 35 | DM+49°3718 SAO51568 | 8.8, - -, - - Be, - - | 0.7 | |
| H | o | 97.3, –4.0 | [1069] | [1166] | [1305] | |

Pos.: [1166]; *Be star*: [1242].

Table 14.2. *(continued)* HMXB

Name(s) Pos.	Type	RA (1950) DEC (1950) ℓ^{II}, b^{II}	Opt. cpt [FC]	V, B–V, U–B Sp. type, E_{B-V}	F_X (μJy)	P_{orb} (d) P_{pulse}(s)
2206+543		22 06 07.4 +54 16 23	star	9.9, 0.2, –0.6 B1e, 0.5	0.6-5.5	392 (?)
U,M,A,H	o 2"	100.6, –1.1	[1075]	[1075]	[1255,1305]	[1014a]

Pos.: [1075]; *X-ray observations*: [1014a]; *Hα emission*: [189].

| 2214+589 | | 22 24 47.8 +60 58 59 | GG3 71 | 11, - -, - - Be, - - | 0.5 | |
| H | o | 106.4, +3.1 | [369] | [1166] | [1305] | |

Pos.: [1166]; *Be star*: [1242].

Table 14.3. *Source names*

Variable Star	Source	Table	Other Names	Source	Table
V1333 Aql	1908+005	14.1	AC 211	2127+119	14.1
V1343 Aql	1909+048	14.2	CAL 83	0543−682	14.1
V1405 Aql	1916−053	14.1	CAL 87	0547−711	14.1
V1408 Aql	1957+115	14.1	γ Cas	0053+604	14.2
V801 Ara	1636−536	14.1	μ^2 Cru	1255−567	14.2
V821 Ara	1659−487	14.1	DM +49°3718	2202+501	14.2
BQ Cam	0331+530	14.2	DM +53°2262	1936+541	14.2
V395 Car	0921−630	14.1	GG3 71	2214+589	14.2
V615 Cas	0236+610	14.2	HD 34921	0521+373	14.2
V635 Cas	0115+634	14.2	HD 63666	0739−529	14.2
V662 Cas	0114+650	14.2	HD 65663	0749−600	14.2
V779 Cen	1119−630	14.2	HD 77581	0900−403	14.2
V822 Cen	1455−314	14.1	HD 91188	1036−565	14.2
V830 Cen	1145.1−6141	14.2	HD 102567	1145−619	14.2
V850 Cen	1258−613	14.2	HD 109857	1253−761	14.2
BR Cir	1516−569	14.1	HD 110432	1249−637	14.2
BW Cir	1354−645	14.1	HD 112091	1255−567	14.2
V691 CrA	1822−371	14.1	HD 141926	1555−552	14.2
BP Cru	1223−624	14.2	HD 153919	1700−377	14.2
V404 Cyg	2023+338	14.1	HD 154791	1702+240	14.1
V1341 Cyg	2142+380	14.1	HD 226868	1956+350	14.2
V1357 Cyg	1956+350	14.2	HD 245770	0535+262	14.2
V1521 Cyg	2030+407	14.2	HD 249179	0556+286	14.2
V1727 Cyg	2129+470	14.1	He3-640	1118−615	14.2
HZ Her	1656+354	14.1	Hen 715	1145−619	14.2
V616 Mon	0620−003	14.1	LS I +61°235	J0146.9+6121	14.2
GR Mus	1254−690	14.1	LS I +61°303	0236+610	14.2
QV Nor	1538−522	14.2	LS I +65°010	0114+650	14.2
V2107 Oph	1705−250	14.1	MMV	1258−613	14.2
V2116 Oph	1728−247	14.1	N Mon 1975	0620−003	14.1
V2134 Oph	1658−298	14.1	N Mus 1991	1124−684	14.1
V2216 Oph	1728−169	14.1	N Oph 1977	1705−250	14.1
V1055 Ori	0614+091	14.1	Rapid Burster	1730−335	14.1
X Per	0352+309	14.2	SAO 51568	2202+501	14.2
V818 Sco	1617−155	14.1	SAO 57950	0521+373	14.2
V884 Sco	1700−377	14.2	SAO 235515	0739−529	14.2
V926 Sco	1735−444	14.1	SAO 238130	1036−565	14.2
MM Ser	1837+049	14.1	SAO 243098	1555−552	14.2
NP Ser	1813−140	14.1	SAO 250018	0749−600	14.2
V4134 Sgr	1755−338	14.1	SAO 252002	1249−637	14.2
V725 Tau	0535+262	14.2	SAO 256967	1253−761	14.2
KY TrA	1524−617	14.1	Sk160	0115−737	14.2
KZ TrA	1627−673	14.1	Sk-Ph	0532−664	14.2
LU TrA	1556−605	14.1	SS 433	1909+048	14.2
GP Vel	0900−403	14.2	THα 35-42	1024.0−5732	14.2
UY Vol	0748−676	14.1	Wack 2134	1024.0−5732	14.2
QZ Vul	2000+251	14.1	Wra 977	1223−624	14.2

Table 14.3. *(continued)*

GX number	Source	Table
GX 1+4	1728–247	14.1
GX 3+1	1744–265	14.1
GX 5–1	1758–250	14.1
GX 9+1	1758–205	14.1
GX 9+9	1728–169	14.1
GX 13+1	1811–171	14.1
GX 17+2	1813–140	14.1
GX 301–2	1223–624	14.2
GX 304–1	1258–613	14.2
GX 339–4	1659–487	14.1
GX 340+0	1642–455	14.1
GX 349+2	1702–363	14.1
GX 354–0	1728–337	14.1

Glob. Cluster	Source	Table
Lil 1	1730–335	14.1
M15	2127+119	14.1
NGC 1851	0512–401	14.1
NGC 6440	1745–203	14.1
NGC 6441	1746–370	14.1
NGC 6624	1820–303	14.1
NGC 6652	1832–330	14.1
NGC 6712	1850–087	14.1
Ter 1	1732–304	14.1
Ter 2	1724–307	14.1
Ter 5	1745–248	14.1
Ter 6	1747–313	14.1

Constellation	Source	Table
Aql X-1	1908+005	14.1
Cen X-2	1354–645	14.1
Cen X-3	1119–603	14.2
Cen X-4	1455–314	14.1
Cep X-4	2138+568	14.2
Cir X-1	1516–569	14.1
Cyg X-1	1956+350	14.2
Cyg X-2	2142+380	14.1
Cyg X-3	2030+407	14.2
Her X-1	1656+354	14.1
LMC X-1	0540–697	14.2
LMC X-2	0521–720	14.1
LMC X-3	0538–641	14.2
LMC X-4	0532–664	14.2
Sco X-1	1617–155	14.1
Sct X-1	1833–076	14.2
Ser X-1	1837+049	14.1
SMC X-1	0115–737	14.2
SMC X-2	0053–739	14.2
SMC X-3	0050–727	14.2
TrA X-1	1524–617	14.1
Vela X-1	0900–403	14.2

References

[1]   Aab, O.E. 1987, Bull. Spec. Aph. Obs. North Caucasus 25, 26.
[2]   Aab, O.E. et al. 1983, SvA 27, 603.
[3]   Aab, O.E. et al. 1984, SvA Lett. 9, 285.
[4]   Aab, O.E. et al. 1984, SvA Lett. 10, 386.
[5]   Aab, O.E. et al. 1984, Bull. Spec. Ap. Obs. North Caucasus 17, 1.
[6]   Abell, G.O. & Margon, B. 1979, Nat 279, 701.
[7]   Ables, J.G. 1969, Proc. Astron. Soc. Aust. 1, 237.
[8]   Alcaino, G. et al. 1990, AJ 99, 817.
[9]   Allen, D. 1977, IAU Circular 3143.
[10] Allen, D.A. 1984, MNRAS 207, 45P.
[11] Anderson, S.F. et al. 1983, ApJ 269, 605.
[12] Anderson, S.F. et al. 1983, ApJ 273, 697.
[13] Angelini, L. et al. 1989, ApJ 346, 906.
[14] Angelini, L. et al. 1991, ApJ 371, 332.
[15] Antokhina, E.A. & Cherepashchuk, A.M. 1988, SvA 31, 295.
[16] Antokhina, E.A. & Cherepashchuk, A.M. 1988, SvA Lett. 11, 4.
[17] Antokhina, E.A. & Cherepashchuk, A.M. 1990, SvA Lett. 16, 182.
[18] Aoki, T. et al. 1993, PAS Japan (in press).
[19] Apparao, K.M.V. et al. 1978, Nat 271, 225.
[20] Apparao, K.M.V. et al. 1980, A&A 89, 249.
[21] Argue, A.N. et al. 1984, MNRAS 209, 11P.
[22] Armandroff, T.E. & Zinn, R. 1988, AJ 96, 92.
[23] Armstrong, J.T. et al. 1980, ApJ 236, L131.
[24] Asai, K. et al. 1993, PAS Japan (in press).
[25] Aslanov, A.A. et al. 1987, SvA Lett. 13, 369.
[26] Assadulaev, S.S. & Cherepashchuk, A.M. 1986, SvA 30, 57.
[27] Augusteijn, T. et al. 1992, A&A 265, 177.
[28] Auriere, M. & Koch-Miramond, L. 1992, A&A 263, 82.
[29] Auriere, M. et al. 1984, A&A 138, 415
[30] Auriere, M. et al. 1986, A&A 158, 158.
[31] Bahcall, J.N. & Bahcall, N.A. 1972, ApJ 178, L1.
[32] Bahcall, J.N. & Bahcall, N.A. 1975, PASP 87, 141.
[33] Bailyn, C. 1992, ApJ 391, 298.
[34] Bailyn, C.D. et al. 1988, ApJ 331, 303.
[35] Bally, J. & Leventhal. M. 1991, Nat 353, 234.
[36] Balog, N.I. et al. 1983, SvA 27, 310.
[37] Balona, L.A. 1990, MNRAS 245, 92.
[38] Balona, L.A. et al. 1992, A&AS 92, 533.
[39] Band, D.L. 1989, ApJ 336, 937.
[40] Band, D.L. & Gordon, M.A. 1989, ApJ 338, 945.
[41] Barr, P. & Van der Woerd, H. 1990, ApJ 352, L41.
[42] Barr, P. et al. 1985, MNRAS 216, 65P.
[43] Barr, P. et al. 1987, A&A 176, 69.
[44] Barret, D. et al. 1991, ApJ 379, L21.
[45] Barret, D. et al. 1992, ApJ 392, L19.
[46] Barret, D. et al. 1992, ApJ 394, 615.
[47] Basinska, E.M. et al. 1984, ApJ 281, 337.
[48] Bazzano, A. et al. 1992, ApJ 385, L17.
[49] Becker, R.H. et al. 1976, IAU Circular 2953.
[50] Becker, R.H. et al. 1977, ApJ 216, L11.
[51] Becker, R.H. et al. 1977, ApJ 216, L101.
[52] Becklin, E.E. et al. 1972, Nat 245, 302.
[53] Belian, R.D. et al. 1972, ApJ 171, L87.
[54] Belian, R.D. et al. 1976, ApJ 206, L135.
[55] Belloni, T. & Hasinger, G. 1990, A&A 227, L33.
[56] Belloni, T. & Hasinger, G. 1990, A&A 230, 103.
[57] Belloni, T. et al. 1991, A&A 245, L29.
[58] Berdnik, E.V. *et al.* 1990, SvA Lett. 16, 472.
[59] Bernacca, P.L. 1984, A&A 132, L8.
[60] Bernacca, P.L. & Bianchi, L. 1981, A&A 94, 345.
[61] Beskrovnaya, N.G. 1989, SvA Lett. 14, 314.
[62] Bianchi, L. & Bernacca, P.L. 1980, A&A 89, 214.
[63] Bianchi, L. & Pakull, M. 1985, A&A 146, 242.
[64] Bignami, G.F. et al. 1981, ApJ 247, L85.
[65] Bingham, E.A. et al. 1984, MNRAS 209, 765.
[66] Blair, W.P. et al. 1984, ApJ 278, 270.

[67] Bochkarev, N.G. et al. 1988, SvA 32, 405.
[68] Bochkarev, N.G. et al. 1988, SvA Lett. 14, 421.
[69] Bohlin, R.C. 1970, ApJ 162, 571.
[70] Bolton, C.T. 1972, Nat 235, 271.
[71] Bonnet-Bidaud, J.M. & Chardin, G. 1989, Phys. Rept. 170, 325.
[72] Bonnet-Bidaud, J.M. & Van der Klis, M. 1981, A&A 97, 134.
[73] Bonnet-Bidaud, J.M. & Van der Klis, M. 1981, A&A 101, 299.
[74] Bonnet-Bidaud, J.M. et al. 1981, A&A 101, 184.
[75] Bonnet-Bidaud, J.M. et al. 1989, A&A 213, 97.
[76] Bonsignori-Facondi, S.R. et al. 1986, A&A 166, 157.
[77] Bord, D.J. 1979, A&A 77, 309.
[78] Bord, D.J. et al. 1976, ApJ 203, 689.
[79] Borisov, N.V. et al. 1989, in ref. [1294], p. 305.
[80] Borozdin, K. et al. 1990, SvA Lett. 16, 345.
[81] Bouchacourt, P. et al. 1984, ApJ 285, L67.
[82] Boynton, P.E. et al. 1973, ApJ 186, 617.
[83] Boynton, P.E. et al. 1984, ApJ 283, L53.
[84] Boynton, P.E. et al. 1986, ApJ 307, 545.
[85] Bradt, H.V. & McClintock, J.E. 1983, ARA&A 21, 63.
[86] Bradt, H.V. et al. 1975, ApJ 197, 443.
[87] Bradt, H.V. et al. 1977, Nat 269, 21.
[88] Bradt, H.V. et al. 1977, Nat 269, 496.
[89] Braes, L.L.E. & Miley, G.K. 1971, Nat 232, 246.
[90] Braes, L.L.E. & Miley, G.K. 1976, Nat 264, 731.
[91] Braes, L.L.E. et al. 1972, Nat 236, 392.
[92] Brandt, S. et al. 1992, A&A 254, L39.
[93] Brandt, S. et al. 1992, A&A 262, L15.
[94] Branduardi, G. et al. 1976, MNRAS 175, 47P.
[95] Branduardi, G. et al. 1978, MNRAS 185, 137.
[96] Branduardi-Raymont, G. et al. 1983, MNRAS 205, 403.
[97] Breedon, L.M. et al. 1986, MNRAS 218, 487.
[98] Brinkmann, W. et al. 1989, A&A 218, L13.
[99] Brinkmann, W. et al. 1991, A&A 241, 112.
[100] Brodskaya, E.J. & Shajn, P.F. 1958, Izw. Krim. Ap. Obs. 20, 299.
[101] Brucato, R.J. et al. 1972, ApJ 173, L105.
[102] Brucato, R.J. et al. 1972, ApJ 175, L137.
[103] Buff, J. et al. 1977, ApJ 212, 768.
[104] Buie, M.W. & Bond, H.E. 1989, IAU Circular 4786.
[105] Buscombe, W. 1962, MNRAS 124, 189.
[106] Buscombe, W. 1980, MK Classification, 4th General Catalogue (Evanston).
[107] Callanan, P.J. et al. 1992, MNRAS 259, 395.
[108] Callanan, P. & Charles, P.A. 1991, MNRAS 249, 573.
[109] Callanan, P.J. et al. 1989, MNRAS 241, 37P.
[110] Callanan, P.J. et al. 1990, A&A 240, 346.
[111] Cameron, R.A. et al. 1992, IAU Circular 5587.
[112] Canizares, C.R. et al. 1975, ApJ 197, 457.
[113] Canizares, C.R. et al. 1979, ApJ 234, 556.
[114] Canizares, C. et al. 1980, ApJ 236, L55.
[115] Caraveo, P.A. et al. 1989, ApJ 338, 338.
[116] Carpenter, G.F. et al. 1975, IAU Circular 2852.
[117] Carpenter, G.F. et al. 1977, MNRAS 179, 27P.
[118] Casares, J. & Charles, P.A. 1992, MNRAS 255, 7.
[119] Casares, J. et al. 1991, MNRAS 250, 712.
[120] Casares, J. et al. 1992, Nat 355, 614.
[121] Castro-Tirado, R.A. et al. 1992, IAU Circular 5587.
[122] Charles, P.A. & Naylor, T. 1992, MNRAS 255, 6P.
[123] Charles, P.A. et al. 1975, Ap. Lett. 16, 145.
[124] Charles, P.A. et al. 1978, MNRAS 183, 29P.
[125] Charles, P.A. et al. 1979, BAAS 11, 720.
[126] Charles, P.A. et al. 1980, ApJ 237, 154.
[127] Charles, P.A. et al. 1980, ApJ 241, 1148.
[128] Charles, P.A. et al. 1983, MNRAS 202, 657.
[129] Charles, P.A. et al. 1986, Nat 323, 417.
[130] Charles, P.A. et al. 1989, in ref. [1294], p. 103.
[131] Charles, P.A. et al. 1991, MNRAS 249, 567.
[132] Cheng, F.H. et al. 1992, ApJ (in press).
[133] Cherepashchuk, A.M. & Khruzina, T.S. 1981, SvA 25, 697.

[134] Chevalier, C. 1989, in ref. [1294], p. 341.
[135] Chevalier, C. & Ilovaisky, S.A. 1974, A&A 35, 407.
[136] Chevalier, C. & Ilovaisky, S.A. 1977, A&A 59, L9.
[137] Chevalier, C. & Ilovaisky, S.A. 1978, ESO Messenger 9, 4.
[138] Chevalier, C. & Ilovaisky, S.A. 1981, A&A 94, L3.
[139] Chevalier, C. & Ilovaisky, S.A. 1982, A&A 112, 68.
[140] Chevalier, C. & Ilovaisky, S.A. 1987, A&A 172, 167.
[141] Chevalier, C. & Ilovaisky, S.A. 1990, A&A 228, 115.
[142] Chevalier, C. & Ilovaisky, S.A. 1990, A&A 238, 163.
[143] Chevalier, C. & Ilovaisky, S.A. 1991, A&A 251, L11.
[144] Chevalier, C. & Ilovaisky, S.A. 1992, private communication.
[145] Chevalier, C. & Ilovaisky, S.A. 1992, A&A (in press).
[146] Chevalier, C. & Ilovaisky, S.A. 1992, A&A (in preparation).
[147] Chevalier, C. et al. 1985, A&A 147, L3.
[148] Chevalier, C. et al. 1985, Sp. Sci. Rev. 40, 443.
[149] Chevalier, C. et al. 1989, A&A 210, 114.
[150] Chevalier, C. et al. 1989, A&A 217, 108.
[151] Chiappetti, L. et al. 1983, ApJ 265, 354.
[152] Chiappetti, L. et al. 1985, Sp. Sci. Rev. 40, 207.
[153] Chiappetti, L. et al. 1990, ApJ 361, 596.
[154] Chodil, G. et al. 1968, ApJ 152, L45.
[155] Christian, C.A. & Friel, E.D. 1992, AJ 103, 142.
[156] Churazov, E. et al. 1992, IAU Circular 5623.
[157] Church, M.J. & Balucinska-Church, M. 1993, MNRAS 260, 59.
[158] Clark, D.H. et al. 1975, Nat 254, 674.
[159] Clark, G.W. 1975, IAU Circular 2843.
[160] Clark, G.W. 1990, ApJ 353, 274.
[161] Clark, G. & Li, F. 1977, IAU Circular 3092.
[162] Clark, G.W. et al. 1977, MNRAS 179, 651.
[163] Clark, G.W. et al. 1978, ApJ 221, L37.
[164] Clark, G.W. et al. 1979, ApJ 227, 54.
[165] Clark, G.W. et al. 1988, ApJ 324, 974.
[166] Clarke, D. 1990, A&A 227, 151.
[167] Clayton, G.C. et al. 1989, MNRAS 236, 901.
[168] Codina, S.J. et al. 1984, A&AS 57, 239.
[169] Coe, M.J. & Payne, B.J. 1985, ApSS 109, 175.
[170] Coe, M.J. et al. 1987, MNRAS 226, 455.
[171] Coe, M.J. et al. 1988, MNRAS 232, 865.
[172] Coe, M.J. et al. 1990, MNRAS 243, 475.
[173] Cominsky, L.R. 1980, Ph.D. Thesis M.I.T.
[174] Cominsky, L.R. et al. 1983, ApJ 270, 226.
[175] Cominsky, L. & Moraes, F. 1991, ApJ 370, 670.
[176] Cominsky, L. & Wood, K.S. 1984, ApJ 283, 765.
[177] Cominsky, L. & Wood, K.S. 1989, ApJ 337, 485.
[178] Cominsky, L. et al. 1977, ApJ 211, L9.
[179] Cominsky, L. et al. 1978, ApJ 224, 46.
[180] Conner, J.P. et al. 1969, ApJ 157, L157.
[181] Conti, P.S. 1978, A&A 63, 225.
[182] Conti, P.S. & Cowley, A.P. 1975, ApJ 200, 133.
[183] Cook, M.C. et al. 1984, in "X-ray Astronomy '84", eds M. Oda & R. Giacconi, p. 225.
[184] Cook, M.C. & Page, C.G. 1987, MNRAS 225, 381.
[185] Cook, M.C. & Warwick, R.S. 1987, MNRAS 225, 369; 227, 661.
[186] Cooke, B.A. et al. 1984, ApJ 285, 258.
[187] Cooke, B.A. & Ponman, T.J. 1991, A&A 244, 358.
[188] Corbally, C.J. et al. 1984, ApJS 55, 657.
[189] Corbet, R.H.D. 1987, in ref. [1054], p. 311.
[190] Corbet, R.H.D. & Day, C.S.R. 1990, MNRAS 243, 553.
[191] Corbet, R.H.D. et al. 1985, MNRAS 212, 565.
[192] Corbet, R.H.D. et al. 1986, MNRAS 221, 961.
[193] Corbet, R.H.D. et al. 1986, MNRAS 222, 15P.
[194] Corbet, R.H.D. et al. 1986, A&A 162, 117.
[195] Corbet, R.H.D. et al. 1987, MNRAS 227, 1055.
[196] Corbet, R.H.D. et al. 1989, MNRAS 239, 533.
[197] Cordier, B. et al. 1992, Proc. Compton Symp. (St Louis).
[198] Courvoisier, T. J.-L. et al. 1986, ApJ 309, 265.
[199] Cowley, A.P. & Schmidtke, P.C. 1990, AJ 99, 678.
[200] Cowley, A.P. et al. 1978, AJ 83, 1619.
[201] Cowley, A.P. et al. 1979, ApJ 231, 539.
[202] Cowley, A.P. et al. 1982, ApJ 255, 596.
[203] Cowley, A.P. et al. 1982, ApJ 256, 605.
[204] Cowley, A.P. et al. 1983, ApJ 272, 118.
[205] Cowley, A.P. et al. 1984, ApJ 286, 196.
[206] Cowley, A.P. et al. 1987, AJ 93, 195.
[207] Cowley, A.P. et al. 1988, ApJ 333, 906.
[208] Cowley, A.P. et al. 1988, AJ 95, 1231.
[209] Cowley, A.P. et al. 1990, ApJ 350, 288.
[210] Cowley, A.P. et al. 1991, ApJ 373, 228.
[211] Cowley, A.P. et al. 1991, ApJ 381, 526.
[212] Coyne, G.V. & Otten, L.B. 1978, Vatican Obs. Publ. 1 (12), 257.
[213] Crampton, D. 1974, ApJ 187, 345.
[214] Crampton, D. & Hutchings, J.B. 1972, ApJ 178, L65.
[215] Crampton, D. & Hutchings, J.B. 1974, ApJ 191, 483.
[216] Crampton, D. & Hutchings, J.B. 1978, IAU Circular 3180.
[217] Crampton, D. & Hutchings, J.B. 1981, ApJ 251, 604.
[218] Crampton, D. et al. 1978, ApJ 223, L79.
[219] Crampton, D. et al. 1978, ApJ 225, L63.
[220] Crampton, D. et al. 1985, ApJ 299, 839.
[221] Crampton, D. et al. 1985, AJ 90, 43.
[222] Crampton, D. et al. 1986, ApJ 306, 599.
[223] Crampton, D. et al. 1987, ApJ 321, 745.
[224] Crampton, D. et al. 1990, ApJ 355, 496.
[225] Crawford, I.A. 1991, A&A 246, 210.
[226] Cruddace, R.G. et al. 1978, ApJ 222, L95.
[227] Cudworth, K.M. 1988, AJ 96, 105.
[228] Cusumano, G. et al. 1990, in ref. [1294], p. 369.
[229] Cutler, E.P. et al. 1986, ApJ 300, 551.
[230] Czerny, M. et al. 1987, ApJ 312, 122.
[231] Dachs, J. 1970, A&A 5, 312.
[232] Dachs, J. 1972, A&A 21, 373.
[233] Dachs, J. 1976, A&A 47, 19.
[234] Dachs, J. et al. 1989, A&AS 78, 487.
[235] Da Costa, G.S. & Armandroff, T.E. 1990, AJ 100, 162.
[236] Dal Fiume, D. et al. 1990, Nuovo Cimento C 13, 481.
[237] Damen, E. et al. 1989, MNRAS 237, 523.
[238] Damen, E. et al. 1990, A&A 237, 103.
[239] D'Amico, N. et al. 1987, A&A 180, 114.
[240] Davidsen, A. et al. 1974, ApJ 193, L25.
[241] Davidsen, A. et al. 1976, ApJ 203, 448.
[242] Davidsen, A. et al. 1977, ApJ 211, 866.
[243] Davies, R.D. et al. 1976, Nat 261, 476.
[244] Davies, S.R. & Coe, M.J. 1991, MNRAS 249, 313.
[245] Davies, S.R. et al. 1989, MNRAS 237, 973.
[246] Davies, S.R. et al. 1990, MNRAS 245, 268.
[247] Davis, R. & Hartmann, L. 1983, ApJ 270, 671.
[248] Davis, R.J. et al. 1975, Nat 257, 659.
[249] Davison, P.J.N. et al. 1976, IAU Circular 2925.
[250] Davison, P.J.N. et al. 1977, MNRAS 181, 73P.
[251] Day, C.S.R. et al. 1988, MNRAS 231, 69.
[252] Deeter, J. et al. 1976, ApJ 206, 861.
[253] Deeter, J.E. et al. 1981, ApJ 247, 1003.
[254] Deeter, J.E. et al. 1987, ApJ 314, 634.
[255] Deeter, J.E. et al. 1987, AJ 93, 877.
[256] Deeter, J.E. et al. 1989, ApJ 336, 376.
[257] Deeter, J.E. et al. 1991, ApJ 383, 324.
[258] De Kool, M. 1988, ApJ 334, 336.
[259] Della Valle, M. et al. 1991, A&A 247, L33.
[260] Della Valle, M. et al. 1991, Nat 353, 50.
[261] Delgado, A.J. et al. 1983, A&A 127, L15.
[262] De Loore, C. et al. 1981, A&A 104, 150.
[263] De Loore, C. et al. 1984, A&A 141, 279.
[264] Densham, R.H. & Charles, P.A. 1982, MNRAS 201, 171.
[265] Densham, R.H. et al. 1983, MNRAS 205, 1117.
[266] Dickey, J.M. et al. 1983, ApJ 273, L71.
[267] Dieters, S.W. et al. 1990, IBVS 3500.
[268] Doazan, V. et al. 1987, A&A 182, L25.

[269] D'Odorico, S. et al. 1991, Nat 353, 329.
[270] Dolan, J.F. & Tapia, S. 1984, A&A 139, 249.
[271] Dolan, J.F. & Tapia, S. 1988, A&A 202, 124.
[272] Dolan, J.F. & Tapia, S. 1989, ApJ 344, 830.
[273] Dolan, J.F. & Tapia, S. 1989, PASP 101, 1135.
[274] Dolan, J.F. et al. 1987, ApJ 322, 324.
[275] Doll, H. & Brinkmann, W. 1987, A&A 173, 86.
[276] Dotani, T. et al. 1989, PASJ 41, 427.
[277] Dotani, T. et al. 1989, PASJ 41, 577.
[278] Dotani, T. et al. 1990, ApJ 350, 395.
[279] Dotani, T. et al. 1990, Nat 347, 534.
[280] Dower, R.G. et al. 1978, Nat 273, 364.
[281] Dower, R.G. et al. 1982, ApJ 261, 228.
[282] Doxsey, R. et al. 1973, ApJ 182, L25.
[283] Doxsey, R.E. et al. 1977, Nat 269, 112.
[284] Doxsey, R.E. et al. 1977, Nat 270, 586.
[285] Doxsey, R. et al. 1977, IAU Circular 3113.
[286] Doxsey, R. et al. 1979, ApJ 228, L67.
[287] Drake, S.A. 1990, AJ 100, 572.
[288] Drilling, J.S. 1973, AJ 80, 128.
[289] Duchouroux, P. & Prantzos, N. (eds), 1991, Gamma Ray Line
      Astrophysics (AIP Proc. 232).
[290] Duerbeck, H.W. 1987, Sp. Sci. Rev. 45, 1.
[291] Duerbeck, H.W. & Walter, K. 1976, A&A 48, 141.
[292] Dufour, R.J. & Duval, J.E. 1975, PASP 87, 769.
[293] Duldig, M.L. et al. 1979, MNRAS 187, 567.
[294] Dupree, A. et al. 1978, Nat 275, 400.
[295] Dupree, A. et al. 1980, ApJ 238, 969.
[296] Eachus, L.J. et al. 1976, ApJ 203, L17.
[297] Ebisawa, K. et al. 1989, PASJ 41, 519.
[298] Egonsson, J. & Hakala, P. 1991, A&A 244, L41.
[299] Elvis, M. et al. 1975, Nat 257, 656.
[300] Ercan, N. et al. 1988, Ap. Lett. Comm. 26, 349.
[301] Evans, W.D. et al. 1970, ApJ 159, L57.
[302] Exosat & Einstein Data Bases.
[303] Eyles, C.J. et al. 1975, Nat 254, 577.
[304] Eyles, C.J. et al. 1975, Nat 257, 291.
[305] Fabbiano, G. & Schreier, E.J. 1977, ApJ 214, 235.
[306] Fabbiano, G. et al. 1978, ApJ 221, L49.
[307] Fabian, A.C. et al. 1989, MNRAS 238, 729.
[308] Fabregat, J. et al.. 1992, A&A 259, 522.
[309] Fabrika, S.N. & Bychkova, L.V. 1990, A&A 240, L5.
[310] Fahlmann, G.G. & Gregory, P.C. 1983, IAU Symp. 101, 445.
[311] Fahlmann, G.G. & Walker, G.A.H. 1980, ApJ 240, 169.
[312] Fahlmann, G.G. et al. 1985, ApJS 58, 225.
[313] Falomo, R. et al. 1987, MNRAS 224, 323.
[314] Feast, M.W. et al. 1961, MNRAS 122, 239.
[315] Feinstein, A. 1968, ZfA 68, 29.
[316] Fejes, I. 1986, A&A 168, 69.
[317] Fejes, I. et al. 1988, A&A 189, 124.
[318] Ferrer, O.E. 1980, A&A 84, 108.
[319] Fiedler, R.L. et al. 1987, AJ 94, 1244.
[320] Filippenko, A.V. et al. 1988, AJ 96, 242.
[321] Finger, M.H. et al. 1992, IAU Circular 5430.
[322] Finley, J.P. et al. 1992, A&A 262, L25.
[323] Fisher, M.L. & Gibson, D.M. 1990, BAAS 22, 804.
[324] Fishman, G.J. et al. 1991, IAU Circular 5394.
[325] Fishman, G.J. et al. 1991, IAU Circular 5395.
[326] Fitzgerald, M.P. 1973, A&AS 9, 297.
[327] Fleischman, J.R. 1985, A&A 153, 106.
[328] Fomalont, E.B. & Geldzahler, B.J. 1991, ApJ 383, 289.
[329] Forman, W. & Jones, C. 1976, ApJ 207, L177.
[330] Forman, W. et al. 1973, ApJ 182, L103.
[331] Forman, W. et al. 1976, ApJ 206, L29.
[332] Forman, W. et al. 1976, ApJ 207, L25.
[333] Forman, W. et al. 1978, ApJS 38, 357.
[334] Foster, A.J. et al. 1986, MNRAS 221, 409.
[335] Frail, D.A. & Hjellming, R.M. 1991, AJ 101, 2126.

[336] Francey, R.J. 1971, Nat. Phys. Sci. 229, 229.
[337] Friend, D.B. & Cassinelli, J.P. 1986, ApJ 303, 292.
[338] Frontera, F. & Fuligni, F. 1975, ApJ 198, L105.
[339] Frontera, F. et al. 1987, ApJ 320, L127.
[340] Fruchter, A.S. & Goss, W.M. 1990, ApJ 365, L63.
[341] Garcia, M.R. & Grindlay, J.E. 1987, ApJ 313, L59.
[342] Garcia, M. et al. 1983, ApJ 267, 291.
[343] Garcia, M.R. et al. 1988, ApJ 328, 552.
[344] Garcia, M.R. et al. 1989, ApJ 341, L75.
[345] Garcia, M.R. et al. 1992, AJ 103, 1325.
[346] Garcia, M. et al. 1992, IAU Circular 5578.
[347] Garrison, R.F. et al. 1977, ApJS 35, 111.
[348] Geffert, M. et al. 1989, A&A 209, 423.
[349] Geldzahler, B.J. 1983, ApJ 264, L49.
[350] Geldzahler, B.J. et al. 1983, ApJ 273, L65.
[351] Geldzahler, B.J. et al. 1989, ApJ 342, 1123.
[352] Gerend, D. & Boynton, P. 1976, ApJ 209, 562.
[353] Giacconi, R. et al. 1967, ApJ 148, L129.
[354] Giacconi, R. et al. 1971, ApJ 167, L67.
[355] Giangrande, A. et al. 1980, A&AS 40, 289.
[356] Gies, D.R. & Bolton, C.T. 1982, ApJ 260, 240.
[357] Gies, D.R. & Bolton, C.T. 1984, ApJ 276, L17.
[358] Gies, D.R. & Bolton, C.T. 1986, ApJ 304, 371 & 389.
[359] Gilfanov, M. et al. 1991, SvA Lett. 17, 437.
[360] Giovannelli, F. & Graziali, L.S.: 1992, Sp. Sci. Rev. 59, 1.
[361] Giovannelli, F. et al. 1985, in: Multifrequency Behaviour of
      Galactic Accreting Sources, ed. F. Giovannelli, p. 284.
[362] Gladyshev, S.A. et al. 1983, SvA Lett. 9, 1.
[363] Gladyshev, S.A. et al. 1987, SvA 31, 541.
[364] Glass, I.S. 1978, MNRAS 183, 335.
[365] Glass, I.S. 1979, MNRAS 187, 807.
[366] Glass, I.S. & Feast, M.W. 1973, Nat. Phys. Sci. 245, 39.
[367] Goldwurm, A. et al. 1992, ApJ 389, L79.
[368] Goldwurm, A. et al. 1992, IAU Circular 5589.
[369] Gonzales, G. & Gonzales, G. 1956, Bol. Obs. Tonantzintla y
      Tacubaya 2, No. 15, p. 16.
[370] Gonzalez-Riestra, R. et al. 1991, IAU Circular 5174.
[371] Goranskij, V.P. & Lyutyj, V.M. 1988, SvA 32, 193.
[372] Goranskij, V.P. et al. 1991, SvA Lett. 17, 399.
[373] Gorayo, P.S. & Tur, N.S. 1988, Ap&SS 145, 263.
[374] Goss, W.M. & Mebold, U. 1977, MNRAS 181, 255.
[375] Gotthelf, E. et al. 1991, ApJ 374, 340.
[376] Gotthelf, E. et al. 1992, AJ 103, 219.
[377] Gottlieb, E.W. et al. 1975, ApJ 195, L33.
[378] Gottwald, M. et al. 1986, ApJ 308, 213.
[379] Gottwald, M. et al. 1986, MNRAS 222, 21P.
[380] Gottwald, M. et al. 1987, ApJ 323, 575.
[381] Gottwald, M. et al. 1987, MNRAS 229, 395.
[382] Gottwald, M. et al. 1989, ApJ 339, 1044.
[383] Gottwald, M. et al. 1991, A&AS 89, 367.
[384] Grankin, K.N. et al. 1991, SvA Lett. 17, 415.
[385] Gray, A.D. et al. 1991, Nat 353, 237.
[386] Grebenev, S. & Sunyaev, R. 1991, IAU Circular 5294.
[387] Grebenev, S.A. et al. 1991, SvA Lett. 17, 413.
[388] Grebenev, S.A. et al. 1992, SvA Lett. 18, 5.
[389] Greenhill, J.G. et al. 1979, Nat 279, 620.
[390] Greenhill, J.G. et al. 1989, A&A 208. L1.
[391] Gregory, P.C. & Fahlmann, G.G. 1980, Nat 287, 805.
[392] Gregory, P.C. & Taylor, A.R. 1978, Nat 272, 704.
[393] Gregory, P.C. et al. 1972, Nat 239, 440.
[394] Gregory, P.C. et al. 1979, AJ 84, 1030.
[395] Gregory, P.C. et al. 1989, ApJ 339, 1054.
[396] Greiner, J. et al. 1991, A&A 246, L17.
[397] Griffiths, R.E. et al. 1978, ApJ 221, L63.
[398] Griffiths, R.E. et al. 1978, Nat 276, 247.
[399] Grindlay, J.E. 1979, ApJ 232, L33.
[400] Grindlay, J.E. 1981, in "X-ray Astronomy with the Einstein
      Satellite", ed. R. Giacconi (Reidel), p. 79.

[401] Grindlay, J.E. 1989, in ref. [1294], p. 121.
[402] Grindlay, J. & Gursky, H. 1976, ApJ 209, L61
[403] Grindlay, J.E. & Gursky, H. 1977, ApJ 218, L117.
[404] Grindlay, J.E. & Hertz, P. 1981, ApJ 247, L17.
[405] Grindlay, J.E. & Liller, W. 1978, ApJ 220, L127.
[406] Grindlay, J.E. & Seaquist, E.R. 1986, ApJ 310, 172.
[407] Grindlay, J. et al. 1976, ApJ 205, L127.
[408] Grindlay, J.E. et al. 1978, Nat 274, 567.
[409] Grindlay, J.E. et al. 1980, ApJ 240, L121.
[410] Grindlay, J.E. et al. 1984, ApJ 276, 621.
[411] Grindlay, J.E. et al. 1984, ApJ 277, 286.
[412] Grindlay, J.E. et al. 1984, ApJ 282, L13.
[413] Grindlay, J.E. et al. 1988, ApJ 334, L25.
[414] Gruber, D.E. 1988, ApJ 328, 265.
[415] Gruber, D.E. & Rothschild, R.E. 1984, ApJ 283, 546.
[416] Gursky, H. et al. 1978, ApJ 223, 973.
[417] Gursky, H. et al. 1980, ApJ 237, 163.
[418] Haberl, F. 1991, ApJ 376, 245.
[419] Haberl, F. & White, N.E. 1990, ApJ 361, 225.
[420] Haberl, F. et al. 1987, ApJ 314, 266.
[421] Haberl, F. et al. 1989, ApJ 343, 409.
[422] Hackwell, J.A. et al. 1979, ApJ 233, L115.
[423] Haefner, R. 1988, IBVS 3260.
[424] Hammerschlag-Hensberge, G. 1978, A&A 64, 399.
[425] Hammerschlag-Hensberge, G. & Wu, C.C. 1977, A&A 56, 433.
[426] Hammerschlag-Hensberge, G. et al. 1979, A&A 76, 245.
[427] Hammerschlag-Hensberge, G. et al. 1980, A&A 85, 119.
[428] Hammerschlag-Hensberge, G. et al. 1982, ApJ 254, L1.
[429] Hammerschlag-Hensberge, G. et al. 1984, ApJ 283, 249.
[430] Hammerschlag-Hensberge, G. et al. 1990, ApJ 352, 698.
[431] Han, X. & Hjellming, R.M. 1992, IAU Circular 5593.
[431a] Han, X. & Hjellming, R.M. 1992, ApJ 400, 304.
[432] Hanson, C.G. et al. 1989, MNRAS 240, 1P.
[433] Hardorp, J. et al. 1959, Luminous Stars in the Northern Milky
      Way I (Hamburger Sternwarte - Warner and Swasey Observatory).
[434] Harmon, B.A. et al. 1992, IAU Circular 5504.
[435] Harmon, B.A. et al. 1992, IAU Circular 5619.
[436] Harries, J.R. et al. 1967, Nat 215, 38.
[437] Hasinger, G. & Van der Klis, M. 1989, A&A 225, 79.
[438] Hasinger, G. et al. 1986, Nat 319, 469.
[439] Hasinger, G. et al. 1989, ApJ 337, 843.
[440] Hasinger, G. et al. 1990, A&A 235, 131.
[441] Hasinger, G. et al. 1990, IAU Circular 5142.
[442] Haswell, C.A. & Shafter, A.W. 1990, ApJ 359, L47.
[443] Haynes, R.F. et al. 1978, MNRAS 185, 661.
[444] Hazen, M.L. 1989, AJ 97, 771.
[445] Heap, S. & Corcoran, M.F. 1992, ApJ 387, 340.
[446] Heemskerk, M.H.M. & Van Paradijs, J. 1989, A&A 223, 154.
[447] Hellier, C. & Mason, K.O. 1989, MNRAS 239, 715.
[448] Hellier, C. et al. 1990, MNRAS 244, 39P.
[449] Henrichs, H. et al. 1982, ApJ 268, 807.
[450] Hensberge, G. et al. 1973, A&A 29, 69.
[451] Henson, G.D. et al. 1983, ApJ 275, 247.
[452] Hermsen, W. et al. 1977, Nat 269, 494.
[453] Hertz, P. 1987, ApJ 315, L119.
[454] Hertz, P. & Grindlay, J.E. 1983, ApJ 275, 105.
[455] Hertz, P. & Grindlay, J.E. 1984, ApJ 278, 137.
[456] Hertz, P. & Grindlay, J.E. 1984, ApJ 282, 118.
[457] Hertz, P. & Grindlay, J.E. 1985, ApJ 298, 95.
[458] Hertz, P. & Wood, K.S. 1988, ApJ 331, 764.
[459] Hertz, P. et al. 1990, ApJ 354, 267.
[460] Hertz, P. et al. 1992, ApJ 396, 201.
[461] Hesser, J. et al. 1976, ApJ 203, 97.
[462] Hill, R. et al. 1974, ApJ 189, L69.
[463] Hiltner, W.A. 1977, IAU Circular 3039.
[464] Hiltner, W. A. & Mook, D. E. 1970, ARA&A 8, 139.
[465] Hirano, T. et al. 1984, PASJ 36, 769.
[466] Hirano, T. et al. 1987, PASJ 39, 619.

[467] Hjellming, R.M. 1973, ApJ 182, L29.
[468] Hjellming, R.M. 1978, ApJ 221, 225.
[469] Hjellming, R.M. 1979, IAU Circular 3369.
[470] Hjellming, R.M. 1988, in: 'Galactic and Extragalactic Radio Astro-
      nomy', eds. G.L. Verschuur & K.I. Kellerman (Springer), p. 381 .
[471] Hjellming, R.M. & Blankenship, L.C. 1973, Nat Ph. Sci. 243, 81.
[472] Hjellming, R.M. & Han, X.H. 1989, IAU Circular 4879.
[473] Hjellming, R.M. & Johnston, K.J. 1981, ApJ 246, L141.
[474] Hjellming, R.M. & Wade, C.M. 1971, ApJ 168, L21.
[475] Hjellming, R.M. & Wade, C.M. 1971, ApJ 164, L1.
[476] Hjellming, R.M. et al. 1988, ApJ 335, L75.
[477] Hjellming, R.M. et al. 1989, IAU Circular 4790.
[478] Hjellming, R.M. et al. 1990, A&A 235, 147.
[479] Hjellming, R.M. et al. 1990, ApJ 365, 681.
[480] Hjellming, R.M. et al. 1990, IAU Circular 5112.
[481] Hoag, A. & Weisberg, J.M. 1976, ApJ 209, 908.
[482] Hoffleit, D. & Jaschek, C. 1982, Bright Star Catalogue (Yale).
[483] Hoffman, J.A. et al. 1976, ApJ 210, L13.
[484] Hoffman, J.A. et al. 1977, MNRAS 179, 57P.
[485] Hoffman, J.A. et al. 1977, ApJ 217, L23.
[486] Hoffman, J.A. et al. 1978, Nat 271, 630.
[487] Hoffman, J.A. et al. 1978, ApJ 221, L57.
[488] Hoffman, J.A. *et al.* 1979, ApJ 233, L51.
[489] Hoffman, J.A. et al. 1980, ApJ 240, L27.
[490] Holt, S.S. et al. 1979, ApJ 233, 344.
[491] Honey, W.B. et al. 1988, IAU Circular 4532.
[492] Honeycutt, R.K. & Schlegel, E.M. 1985, PASP 97, 300.
[493] Horne, K. et al. 1986, MNRAS 218, 63.
[494] Houk, N. & Cowley, A.P. 1975, Michigan Catalogue of Two-
      dimensional Spectral Types for the HD stars I (Ann Arbor).
[495] Howarth, I.D. 1983, MNRAS 203, 801.
[496] Howarth, I.D. & Wilson, B. 1983, MNRAS 202, 347.
[497] Howarth, I.D. & Wilson, B. 1983, MNRAS 204, 1091.
[498] Howarth, I.D. et al. 1984, MNRAS 207, 287.
[499] Howe, S.K. et al. 1983, ApJ 272, 678.
[500] Huckle, H.E. et al. 1977, MNRAS 180, 21P.
[501] Hudec, R. & Rätz, K. 1989, in ref. [1294], p. 427.
[502] Hudec, R. & Wenzel, W. 1986, A&A 158, 396.
[503] Hudec, R. & Wenzel, W. 1990, Bull. Astr. Inst. Chechosl. 41, 355.
[504] Hughes, V.A. et al. 1981, ApJ 246, L127.
[505] Humphreys, R.M. & Whelan, J. 1975, Observatory 95, 171.
[506] Hutchings, J.B. 1970, MNRAS 150, 55.
[507] Hutchings, J.B. 1974, ApJ 192, 677.
[508] Hutchings, J.B. 1977, MNRAS 181, 619.
[509] Hutchings, J.B. 1979, PASP 91, 657.
[510] Hutchings, J.B. & Crampton, D. 1981, PASP 93, 486.
[511] Hutchings, J.B. & Crampton, D. 1981, ApJ 247, 222.
[512] Hutchings, J.B. & Stoeckley, T.R. 1977, PASP 89, 19.
[513] Hutchings, J.B. et al. 1975, MNRAS 170, 313.
[514] Hutchings, J.B. et al. 1977, ApJ 217, 186.
[515] Hutchings, J.B. et al. 1978, ApJ 223, 530.
[516] Hutchings, J.B. et al. 1978, ApJ 225, 548.
[517] Hutchings, J.B. et al. 1979, ApJ 229, 1079.
[518] Hutchings, J.B. et al. 1981, AJ 86, 871.
[519] Hutchings, J.B. et al. 1981, IAU Circular 3585.
[520] Hutchings, J.B. et al. 1982, PASP 94, 541.
[521] Hutchings, J.B. et al. 1983, ApJ 275, L43.
[522] Hutchings, J.B. et al. 1984, PASP 96, 312.
[523] Hutchings, J.B. et al. 1985, ApJ 292, 670.
[524] Hutchings, J.B. et al. 1985, PASP 97, 418.
[525] Hutchings, J.B. et al. 1987, PASP 99, 420.
[526] Hutchings, J.B. et al. 1987, AJ 94, 340.
[527] Ilovaisky, S.A. & Chevalier, C. 1981, IAU Circular 3586.
[528] Ilovaisky, S.A. et al. 1978, A&A 70, L19.
[529] Ilovaisky, S.A. et al. 1979, A&A 71, L17 [erratum A&A 75, 258].
[530] Ilovaisky, S.A. et al. 1980, MNRAS 191, 81.
[531] Ilovaisky, S.A. et al. 1982, A&A 114, L7.
[532] Ilovaisky, S.A. et al. 1984, A&A 140, 251.

[533] Ilovaisky, S.A. et al. 1986, A&A 164, 67.
[534] Ilovaisky, S.A. et al. 1987, A&A 179, L1.
[535] Ilovaisky, S.A. et al. 1992, A&A (in press).
[536] Imamura, J.N. et al. 1987, ApJ 314, L11.
[537] Imamura, J.N. et al. 1990, ApJ 365, 312.
[538] Inoue, H. et al. 1981, ApJ 250, L71.
[539] Inoue, H. et al. 1984, PASJ 36, 831.
[540] Inoue, H. et al. 1984, PASJ 36, 855.
[541] In 't Zand, J.J.M. et al. 1989, in ref. [1294], p. 693.
[542] In 't Zand, J.J.M. et al. 1990, Adv. Space Res. 11 (8), 187.
[543] Iping, R.C. & Petterson, J.A. 1990, A&A 239, 221.
[544] Ives, J.C. et al. 1975, Nat 254, 578.
[545] Iwasawa, K. et al. 1992, PAS Japan 44, 9.
[546] Iye, M. 1986, PASJ 38, 463.
[547] Iye, M. & Kodaira, K. 1986, PASP 97, 1186.
[548] Jain, A. et al. 1984, A&A 140, 179.
[549] Janot-Pacheco, E. et al. 1981, A&A 99, 274.
[550] Janot-Pacheco, E. et al. 1982, IAU Symp. 98, 151.
[551] Janot-Pacheco, E. et al. 1987, A&A 177, 91.
[552] Janot-Pacheco, E. et al. 1988, A&A 202, 81.
[553] Jarad, M.M. et al. 1989, MNRAS 238, 1085.
[554] Jernigan, G. 1976, IAU Circular 2957.
[555] Jernigan, J.G. et al. 1977, Nat 270, 321.
[556] Jernigan, J.G. et al. 1978, Nat 272, 701.
[557] Johns, M. et al. 1978, IAU Circular 3171.
[558] Johnston, K.J. et al. 1986, ApJ 309, 707.
[559] Johnston, M. et al. 1978, ApJ 223, L71.
[560] Johnston, M.D. et al. 1979, ApJ 230, L11.
[561] Johnston, M.D. et al. 1979, ApJ 233, 514.
[562] Johnston, M.D. et al. 1980, Nat 285, 26.
[563] Johnston, H.M. et al. 1989, ApJ 345, 492.
[564] Jones, C. et al. 1973, ApJ 181, L43.
[565] Jones, C. et al. 1973, ApJ 182, L109.
[566] Jones, C. et al. 1976, ApJ 210, L9.
[567] Jones, M.H. & Watson, M.G. 1989, in ref. [1294], p. 439.
[568] Kahn, S.M. & Grindlay, J.E. 1984, ApJ 281, 826.
[569] Kahn, S.M. & Grindlay, J.E. 1984, ApJ 283, 286.
[570] Kallman, T.R. et al. 1987, ApJ 317, 746.
[571] Kallman, T.R. et al. 1989, ApJ 345, 498.
[572] Kallman, T.R. et al. 1991, ApJ 370, 717.
[573] Kaluzienski, L.J. & Holt, S.S. 1977, IAU Circular 3099 & 3106.
[574] Kaluzienski, L.J. et al. 1975, ApJ 201, L121.
[575] Kaluzienski, L.J. et al. 1976, ApJ 208, L71.
[576] Kaluzienski, L.J. et al. 1976, IAU Circular 2935.
[577] Kaluzienski, L.J. et al. 1977, Nat 265, 606.
[578] Kaluzienski, L.J. et al. 1980, ApJ 241, 779.
[579] Kaluzny, J. 1988, Acta Astron. 38, 207.
[580] Kamata, Y. et al. 1990, PASJ 42, 785.
[581] Kaminker, A.D. et al. 1989, A&A 220, 117.
[582] Kaper, L. et al. 1990, Nat 347, 652.
[583] Katz, J.I. et al. 1984, AJ 89, 1604.
[584] Kawai, N. et al. 1988, ApJ 330, 130.
[585] Kawai, N. et al. 1989, PASJ 41, 491.
[586] Kawai, N. et al. 1990, PASJ 42, 115.
[587] Kelley, R.L. et al. 1981, ApJ 243, 251.
[588] Kelley, R.L. et al. 1983, ApJ 264, 568.
[589] Kelley, R.L. et al. 1983, ApJ 268, 790.
[590] Kelley, R.L. et al. 1983, ApJ 274, 765.
[591] Kellogg, E. et al. 1971, ApJ 169, L99.
[592] Kemp, J.C. et al. 1983, ApJ 271, L65.
[593] Kemp, J.C. et al. 1986, ApJ 305, 805.
[594] Kemp, J.C. et al. 1987, SvA31, 170.
[595] Kendziorra, E. et al. 1992, Proc. 28th Yamada Conf. "Frontiers in X-ray Astronomy", p. 51.
[596] Kesteven, M.J. & Turtle, A.J. 1991, IAU Circular 5181.
[597] Kholopov, P.N. et al. 1989, IBVS 3323.
[598] Khruzina, T.S. & Cherepashchuk, A.M. 1986, SvA 30, 295.
[599] Khruzina, T.S. & Cherepashchuk, A.M. 1986, SvA 30, 422.

[600] Khruzina, T.S. & Cherepashchuk, A.M. 1987, SvA 31, 180.
[601] Kii, S. et al. 1986, PASJ 38, 751.
[602] Kitamoto, S. et al. 1984, PASJ 36, 799.
[603] Kitamoto, S. et al. 1987, PASJ 39, 259.
[604] Kitamoto, S. et al. 1989, Nat 342, 518.
[605] Kitamoto, S. et al. 1989, PASJ 41, 81.
[606] Kitamoto, S. et al. 1990, PASJ 42, 85.
[607] Kitamoto, S. et al. 1990, ApJ 361, 590.
[608] Kitamoto, S. et al. 1992, ApJ 384, 263.
[609] Kitamoto, S. et al. 1992, ApJ 391, 220.
[610] Kitamoto, S. et al 1992, ApJ 394, 609.
[611] Kleinmann, D.E. et al. 1978, ApJ 210, L83.
[612] Kodaira, K. et al. 1985, PASJ 37, 97.
[613] Kodaira, K. et al. 1985, ApJ 296, 232.
[614] Koenigsberger, G. et al. 1983, ApJ 268, 782.
[615] Kondo, Y. et al. 1983, ApJ 273, 716.
[616] Koo, D.C. & Kron, R.G. 1977, PASP 89, 285.
[617] Kopylov, I.M. et al. 1986, SvA 30, 408.
[618] Kopylov, I.M. et al. 1987, SvA 31, 410.
[619] Kouveliotou, C. et al. 1992, IAU Circular 5576.
[620] Kouveliotou, C. et al. 1992, IAU Circular 5592.
[621] Koyama, K. et al. 1981, ApJ 247, L27.
[622] Koyama, K. et al. 1987, PASJ 39, 801.
[623] Koyama, K. et al. 1989, PASJ 41, 461.
[624] Koyama, K. et al. 1990, Nat 343, 148.
[625] Koyama, K. et al. 1990, ApJ 356, L47.
[626] Koyama, K. et al. 1990, PASJ 42, L59.
[627] Koyama, K. et al. 1991, ApJ 366, L19.
[628] Koyama, K. et al. 1991, ApJ 370, L77.
[629] Kraft, R.P. & Miller, J.S. 1969, ApJ 155, L159.
[630] Kriss, G.A. et al. 1983, ApJ 266, 806.
[631] Krzeminski, W. 1974, ApJ 192, L135.
[632] Krzeminski, W. & Kubiak, M. 1991, Acta Astr. 41, 117.
[633] Kuiper, L. et al. 1988, A&A 203, 79.
[634] Kunieda, H. et al. 1984, PASJ 36, 215.
[635] Kunieda, H. et al. 1984, PASJ 36, 807.
[636] Lamb, R.C. et al. 1980, ApJ 239, 651.
[637] Lamb, R.C. et al. 1983, Nat 305, 37.
[638] Lang, F.L. et al. 1981, ApJ 246, L21.
[639] Langmeier, A. et al. 1985, Sp. Sci. Rev. 40, 367.
[640] Langmeier, A. et al. 1987, ApJ 323, 288.
[641] Langmeier, A. et al. 1989, ApJ 340, L21.
[642] Langmeier, A. et al. 1990, A&A 228, 89.
[643] Lapshov, I. et al. 1992, SvA Lett. 18, 1.
[644] Lapshov, I. et al. 1992, SvA Lett. 18, 12.
[645] Laros, J. & Wheaton, W.A. 1980, Nat 284, 324.
[646] LaSala, J. & Thorstensen, J.R. 1985, AJ 90, 2077.
[647] Lawrence, A. et al. 1983, ApJ 267, 301.
[648] Lawrence, A. et al. 1983, ApJ 271, 793.
[649] Leahy, D.A. 1991, MNRAS 251, 22P.
[650] Leahy, D.A. & Ananth, A.G. 1992, MNRAS 256, 39P.
[651] Leahy, D.A. & Matsuoka, M. 1990, ApJ 355, 627.
[652] Leahy, D.A. et al. 1988, PASJ 40, 197.
[653] Leahy, D.A. et al. 1989, MNRAS 237, 269.
[654] Leahy, D.A. et al. 1992, A&A 259, 209.
[655] Leavitt, H.S. 1908, Harvard Ann. 60, 87.
[656] Lehto, H. et al. 1990, Nat 347, 49.
[657] Leibowitz, E.M. 1984, MNRAS 210, 279.
[658] Leibowitz, E.M. & Mendelson, H. 1982, PASP 94, 977.
[659] Leibowitz, E.M. et al. 1984, MNRAS 206, 751.
[660] Leibowitz, E.M. et al. 1984, Nat 307, 341.
[661] Leibowitz, E.M. et al. 1991, MNRAS 250, 385.
[662] Leisawitz, D. 1988, NASA Ref. Publ. 1202.
[663] Levine, A. et al. 1988, ApJ 327, 732.
[664] Levine, A. et al. 1991, ApJ 381, 101.
[665] Lewin, W.H.G. 1977, American Scientist 65, 605.
[666] Lewin, W.H.G. & Van den Heuvel, E.P.J. (eds), 1983, Accretion Driven Stellar X-ray Sources (Cambridge University Press).

[667] Lewin, W.H.G. & Van Paradijs, J. 1985, A&A 142, 361.
[668] Lewin, W.H.G. et al. 1971, ApJ 169, L17.
[669] Lewin, W.H.G. et al. 1976, ApJ 207, L95.
[670] Lewin, W.H.G. et al. 1976, IAU Circular 2994.
[671] Lewin, W.H.G. et al. 1976, MNRAS 177, 83P.
[672] Lewin, W.H.G. et al. 1976, MNRAS 177, 93P.
[673] Lewin, W.H.G. et al. 1978, IAU Circular 3193.
[674] Lewin, W.H.G. et al. 1980, MNRAS 193, 15.
[675] Lewin, W.H.G. et al. 1987, MNRAS 226, 383.
[676] Lewin, W.H.G. et al. 1987, ApJ 319, 893.
[677] Lewin, W.H.G. et al. 1988, Sp. Sci. Rev. 46, 273.
[678] Lewin, W.H.G. et al. 1991, A&A 248, 538.
[679] Lewin, W.H.G. et al. 1992, MNRAS 256, 545.
[680] Lewin, W.H.G. et al. 1992, private communication.
[681] Li, F.K. & Clark, G.W. 1977, IAU Circular 3095.
[682] Li, F.K. et al. 1976, ApJ 203, 187.
[683] Li, F. et al. 1978, Nat 271, 37.
[684] Li, F.K. et al. 1978, Nat 276, 799.
[685] Li, F.K. et al. 1979, MNRAS 179, 21P.
[686] Li, F. et al. 1979, ApJ 228, 893.
[687] Li, F.K. et al. 1980, ApJ 240, 628.
[688] Liller, W. 1973, ApJ 184, L37.
[689] Liller, W. 1977, ApJ 213, L21.
[690] Ling, J.C. et al. 1983, ApJ 275, 307.
[691] Ling, J.C. et al. 1987, ApJ 321, L117.
[692] Lloyd, C. & Walker, E.N. 1990, in ref. [1294], p. 511.
[693] Lloyd, C. et al. 1979, MNRAS 179, 675.
[694] Lochner, J.C. & Roussel-Dupre, D. 1990, BAAS 22, 804.
[695] Lochner, J.C. et al. 1989, ApJ 337, 823.
[696] Lochner, J.C. et al. 1991, ApJ 376, 295.
[697] Long, K.S. et al. 1981, ApJ 248, 925.
[698] Lubin, L.M. et al. 1991, MNRAS 249, 300.
[699] Lubin, L.M. et al. 1991, MNRAS 252, 190.
[700] Lubin, L.M. et al. 1992, MNRAS 256, 624.
[701] Lucke, P. 1974, ApJS 28, 73.
[702] Lugger, P.M. et al. 1987, ApJ 320, 482.
[703] Lum, K.S.K. et al. 1992, ApJS 78, 423.
[704] Lund, N. & Brandt, S.; Makino, F. & Ginga Team, 1991, IAU Circular 5161.
[705] Lund, N. et al. 1991, Adv. Space Res. 11 (8), 17.
[706] Lund, N. et al. 1991, IAU Circular 5448.
[707] Lyutyi, V.M. 1985, SvA 29, 429.
[708] Lyutyi, V.M. & Shugarov, S. Yu. 1979, SvA Lett. 5, 206.
[709] Lyutyi, V.M. & Voloshina, I.B. 1989, SvA Lett. 15, 347.
[710] Lyutyi, V.M. et al. 1989, SvA Lett. 15, 182.
[711] Maccacaro, T. et al. 1992 (in preparation; EMSS III).
[712] MacConnell, D.J. 1982, A&AS 48, 355.
[713] MacConnell, D.J. & Coyne, G.V. 1983, Vatican Obs. Pub. 2 (5), 63
[714] Machin, G. et al. 1990, MNRAS 246, 237.
[715] Machin, G. et al. 1990, MNRAS 247, 205.
[716] Maejima, Y. et al. 1984, ApJ 285, 712.
[717] Magnier, E. et al. 1989, MNRAS 237, 729.
[718] Makino, F. 1988, IAU Circular 4571.
[719] Makino, F. & the Ginga Team, 1988, IAU Circular 4653.
[720] Makino, F. et al. 1990, IAU Circular 5142.
[721] Makino, F. et al. 1990, IAU Circular 5148.
[722] Makishima, K. et al. 1981, ApJ 244, L79.
[723] Makishima, K. et al. 1981, ApJ 247, L23.
[724] Makishima, K. et al. 1982, ApJ 255, L49.
[725] Makishima, K. et al. 1983, ApJ 267, 310.
[726] Makishima, K. et al. 1984, PASJ 36, 679.
[727] Makishima, K. et al. 1986, ApJ 308, 635.
[728] Makishima, K. et al. 1987, ApJ 314, 619.
[729] Makishima, K. et al. 1988, Nat 333, 746.
[730] Makishima, K. et al. 1989, PASJ 41, 531.
[731] Makishima, K. et al. 1990, ApJ 365, L59.
[732] Makishima, K. et al. 1990, PASJ 42, 295.
[733] Malkan, M. et al. 1980, ApJ 237, 432.
[734] Mandrou, P. et al. 1990, IAU Circular 5140.
[735] Maraschi, L. et al. 1980, ApJ 241, L23.
[736] Maraschi, L. et al. 1981, ApJ 248, 1010.
[737] Margon, B. 1978, ApJ 219, 613.
[738] Margon, B. 1980, Ann. NY Ac. Sci. 336, 550.
[739] Margon, B. 1983, in ref. [666], p. 287.
[740] Margon, B. 1984, ARA&A 22, 507.
[741] Margon, B. 1992, private communication.
[742] Margon, B. & Anderson, S.F. 1989, ApJ 347, 448.
[743] Margon, B. et al. 1977, ApJ 216, 811.
[744] Margon, B. et al. 1978, Nat 271, 633.
[745] Margon, B. et al. 1978, ApJ 221, 907.
[746] Margon, B. et al. 1979, ApJ 230, L41.
[747] Margon, B. et al. 1979, ApJ 233, L63.
[748] Margon, B. et al. 1984, ApJ 281, 313.
[749] Margoni, R. et al. 1988, A&A 195, 148.
[750] Markert, T.H. et al. 1973, ApJ 184, L67.
[751] Markert, T.H. et al. 1975, Nat 257, 32.
[752] Markert, T.H. et al. 1976, ApJ 208, L115.
[753] Markert, T.H. et al. 1977, ApJ 218, 801.
[754] Markert, T.H. et al. 1979, ApJS 39, 573.
[755] Marlborough, J.M. et al. 1977, ApJ 216, 446.
[756] Marlborough, J.M. et al. 1978, ApJ 224, 157.
[757] Marshall, F.E. et al. 1983, ApJ 266, 814.
[758] Marshall, H.L. et al. 1979, ApJ 227, 555.
[759] Marshall, N. & Millit, J.M. 1981, Nat 293, 379.
[760] Marshall, N. & Ricketts, M.J. 1980, MNRAS 193, 7P.
[761] Martin, E.L. et al. 1992, Nat 358, 129.
[762] Martins, D.H. et al. 1980, AJ 85, 521.
[763] Mason, K.O. & Cordova, F.A. 1982, ApJ 255, 603.
[764] Mason, K.O. & Cordova, F.A. 1982, ApJ 262, 253.
[765] Mason, K.O. et al. 1976, ApJ 207, 78.
[766] Mason, K.O. et al. 1978, MNRAS 184, 45P.
[767] Mason, K.O. et al. 1980, ApJ 242, L109.
[768] Mason, K.O. et al. 1980, Nat 287, 516.
[769] Mason, K.O. et al. 1982, MNRAS 200, 793.
[770] Mason, K.O. et al. 1985, MNRAS 216, 1033.
[771] Mason, K.O. et al. 1985, Sp. Sci. Rev. 40, 225.
[772] Mason, K.O. et al. 1986, ApJ 309, 700.
[773] Mason, K.O. et al. 1987, MNRAS 226, 423.
[774] Massi, M. et al. 1992, A&A (in press).
[775] Matilsky, T.A. et al. 1972, ApJ 174, L53.
[776] Matilsky, T. et al. 1976, ApJ 210, L127.
[777] Matsuoka, M. 1980, Symp. Space Astrophysics, (ISAS), p. 88.
[778] Matsuoka, M. et al. 1980, ApJ 240, L137.
[779] Matsuoka, M. et al. 1984, ApJ 283, 774.
[780] Matsuoka, M. et al. 1986, MNRAS 222, 605.
[781] Mazeh, T. et al. 1987, ApJ 317, 824.
[782] McClintock, J.E. & Leventhal, M. 1989, ApJ 346, 143.
[783] McClintock, J.E. & Remillard, R.A. 1986, ApJ 308, 110.
[784] McClintock, J.E. & Remillard, R.A. 1990, ApJ 350, 386.
[785] McClintock, J. et al. 1978, IAU Circular 3251.
[786] McClintock, J.E. et al. 1978, Nat 270, 320.
[787] McClintock, J.E. et al. 1978, ApJ 223, L75.
[788] McClintock, J.E. et al. 1979, Nat 279, 47.
[789] McClintock, J.E. et al. 1981, ApJ 243, 900.
[790] McClintock, J.E. et al. 1982, ApJ 258, 245.
[791] McClintock, J.E. et al. 1984, ApJ 283, 794.
[792] McClintock, J.E. et al. 1992, IAU Circular 5499.
[793] McCray, R. et al. 1984, ApJ 282, 245.
[794] McCroskey, R.E. 1992, IAU Circular 5597
[795] McHardy, I.M., et al. 1981, MNRAS 197, 893.
[796] Meekins, J.F. et al. 1984, ApJ 278, 288.
[797] Mendelson, H. & Mazeh, T. 1989, MNRAS 239, 733.
[798] Mendelson, H. & Mazeh, T. 1991, MNRAS 250, 373.
[799] Mereghetti, S. et al. 1987, ApJ 312, 755.
[800] Mereghetti, S. et al. 1991, ApJ 366, L23.
[801] Mereghetti, S. et al. 1992, A&A 259, 205.

[802] Mereghetti, S. et al. 1992, A&A 263, 172.
[803] Middleditch, J. 1983, ApJ 275, 278.
[804] Middleditch, J. & Priedhorsky, W.C. 1986, ApJ 306, 230.
[805] Middleditch, J. et al. 1981, ApJ 244, 1001.
[806] Middleditch, J. et al. 1983, ApJ 274, 313.
[807] Middleditch, J. et al. 1984, ApJ 292, 267.
[808] Mihara, T. et al. 1990, Nat 346, 250.
[809] Mihara, T. et al. 1991, ApJ 379, L61.
[810] Mihara, T. et al. 1991, PASJ 43, 501.
[811] Mikkelsen, D.R. & Wallerstein, G. 1974, ApJ 194, 459.
[812] Mineshige, S. et al. 1992, PAS Japan 44, 117
[813] Mirabel, I.F. et al. 1991, A&A 251, L43.
[814] Mirabel, I.F. et al. 1992, Nat 358, 215; IAU Circular 5655.
[815] Mironov, A.V. et al. 1986, SvA 30, 68.
[816] Mitani, K. et al. 1984, Ap&SS 103, 345.
[817] Mitsuda, K. & Dotani, T. 1989, PASJ 41, 557.
[818] Mitsuda, K. et al. 1989, PASJ 41, 97.
[819] Mitsuda, K. et al. 1990, ApJ 353, 480.
[820] Mitsuda, K. et al. 1991, PASJ 43, 113.
[821] Miyamoto, S. & Kitamoto, S. 1989, Nat 342, 773.
[822] Miyamoto, S. & Matsuoka, M. 1977, Sp. Sci. Rev. 20, 687.
[823] Miyamoto, S. et al. 1989, Nat 336, 450.
[824] Miyamoto, S. et al. 1991, ApJ 383, 784.
[825] Miyamoto, S. et al. 1992, ApJ 391, L21.
[826] Moffat, A.F.J. et al. 1973, A&A 23, 433.
[827] Molnar, L.A. & Mauche, C.W. 1986, ApJ 310, 343.
[828] Molnar, L. & Neely, M. 1992, IAU Circular 5595.
[829] Molnar, L.A. et al. 1984, Nat 310, 662.
[830] Molnar, L.A. et al. 1985, in: 'Radio Stars', eds. R.M. Hjellming & D. Gibson (Reidel), p. 329.
[831] Molnar, L.A. et al. 1988, BAAS 20, 736.
[832] Molnar, L.A. et al. 1988, ApJ 331, 494.
[833] Moneti, A. 1992, A&A 260, L7.
[834] Mony, B. et al. 1991, A&A 247, 405.
[835] Mook, D.E. et al. 1974, PASP 86, 894.
[836] Mook, D.E. et al. 1975, ApJ 197, 425.
[837] Morgan, W.W. et al. 1955, ApJS 2, 41.
[838] Morini, M. et al. 1988, ApJ 333, 777.
[839] Morris, S.L. et al. 1990, ApJ 365, 686.
[840] Motch, C. & Janot-Pacheco, E. 1987, A&A 182, L55.
[841] Motch, C. & Pakull, M.W. 1989, A&A 214, L1.
[842] Motch, C. et al. 1982, A&A 109, L1.
[843] Motch, C. et al. 1983, A&A 119, 171.
[844] Motch, C. et al. 1985, Sp. Sci. Rev. 40, 219.
[845] Motch, C. et al. 1987, ApJ 313, 792.
[846] Motch, C. et al. 1988, A&A 201, 63.
[847] Motch, C. et al. 1989, A&A 219, 158.
[848] Motch, C. et al. 1989, in ref. [1294], p. 545.
[849] Motch, C. et al. 1991, ApJ 369, 490.
[850] Motch, C. et al. 1991, A&A 246, L24.
[851] Mouchet, M. et al. 1980, A&A 90, 113.
[852] Mourard, D. et al. 1989, Nat 342, 520.
[853] Murakami, T. et al. 1980, ApJ 240, L143.
[854] Murakami, T. et al. 1980, PASJ 32, 543.
[855] Murakami, T. et al. 1983, ApJ 264, 563.
[856] Murakami, T. et al. 1983, PASJ 35, 531.
[857] Murakami, T. et al. 1986, ApJ 310, L31.
[858] Murakami, T. et al. 1987, PASJ 39, 253.
[859] Murakami, T. et al. 1987, PASJ 39, 879.
[860] Murdin, P. et al. 1974, MNRAS 169, 25.
[861] Murdin, P. et al. 1977, MNRAS 178, 27P.
[862] Murdin, P. et al. 1979, MNRAS 186, 43P.
[863] Murdin, P. et al. 1980, MNRAS 192, 709.
[864] Murdin, P. et al. 1980, MNRAS 193, 135.
[865] Murdin, P. et al. 1980, A&A 87, 292.
[866] Nagase, F. 1989, PASJ 41, 1.
[867] Nagase, F. et al. 1982, ApJ 263, 814.
[868] Nagase, F. et al. 1984, PASJ 36, 215.

[869] Nagase, F. et al. 1986, PASJ 38, 547.
[870] Nagase, F. et al. 1991, ApJ 375, L49.
[871] Nagase, F. et al. 1992, ApJ 396, 147.
[872] Nakamura, N. et al. 1988, PASJ 40, 209.
[873] Nakamura, N. et al. 1989, PASJ 41, 617.
[874] Naylor, T. & Charles, P.A. 1989, MNRAS 236, 1P.
[875] Naylor, T. et al. 1988, MNRAS 233, 285.
[876] Naylor, T. et al. 1991, MNRAS 252, 203.
[877] Naylor, T. et al. 1992, MNRAS 255, 1.
[878] Neckel, T. & Klare, G. 1980, A&AS 42, 251.
[879] Nicolet, B. 1978, A&AS 34, 1.
[880] Nicolson, G.D. et al. 1980, MNRAS 191, 293.
[881] Nieto, J.L. et al. 1990, A&A 239, 155.
[882] Ninkov, Z. et al. 1987, ApJ 321, 425 & 438.
[883] Norris, J.P. & Matilsky, T.A. 1989, ApJ 346, 912.
[884] Norris, J.P. & Wood, K.S. 1987, ApJ 312, 732.
[885] Norris, J.P. et al. 1990, ApJ 361, 514.
[886] Norton, A.J. et al. 1991, MNRAS 253, 579.
[887] Oda, M. 1977, Sp. Sci. Rev. 20, 757.
[888] Ögelman, H. 1987, A&A 172, 79.
[889] Ögelman, H. & Van den Heuvel, E.P.J. (eds), 1989, Timing Neutron Stars, (Kluwer Acad. Publ.).
[890] Ohashi, T. et al. 1982, ApJ 258, 254.
[891] Ohashi, T. et al. 1984, PASJ 36, 699.
[892] Oosterbroek, T. et al. 1991, A&A 250, 389.
[893] Oke, J.B. 1977, ApJ 217, 181.
[894] Okumura, S. & Noguchi, T. 1989, IAU Circular 4589.
[895] Osmer, P.S. et al. 1975, ApJ 195, 705.
[896] Owen, F.N. et al. 1976, ApJ 203, L15.
[897] Paciesas, W.S. et al. 1992, IAU Circular 5580.
[898] Paczynski, B. 1983, ApJ 273, L81.
[899] Paczynski, B. 1990, ApJ 365, L9.
[900] Pakull, M. 1978, IAU Circular 3313.
[901] Pakull, M.W. 1982, Proc. Workshop on Accreting Neutron Stars, MPE Report 177, p. 53.
[902] Pakull, M.W. & Angebault, L.P. 1986, Nat 322, 511.
[903] Pakull, M. & Parmar, A. 1981, A&A 102, L1.
[904] Pakull, M. & Swings, J.P. 1979, IAU Circular 3318.
[905] Pakull, M. et al. 1983, A&A 122, 79.
[906] Pakull, M. et al. 1985, Sp. Sci. Rev. 40, 229.
[907] Pakull, M. et al. 1985, Sp. Sci. Rev. 40, 379.
[908] Pakull, M.W. et al. 1988, A&A 203, L27.
[909] Paredes, J.-M. 1987, Rev. Mex. Astron. Astrofis. 14, 395.
[910] Paredes, J.-M. & Figueras, F. 1986, A&A 154, L30.
[911] Paredes, J.-M. et al. 1990, A&A 232, 377.
[912] Parkes, G.E. et al. 1978, MNRAS 184, 73P.
[913] Parkes, G.E. et al. 1980, MNRAS 190, 537.
[914] Parkes, G.E. et al. 1980, MNRAS 191, 547.
[915] Parmar, A.N. et al. 1980, MNRAS 193, 49P.
[916] Parmar, A.N. et al. 1984, Nat 313, 119.
[917] Parmar, A.N. et al. 1985, IAU Circular 4058.
[918] Parmar, A.N. et al. 1986, ApJ 304, 664.
[919] Parmar, A.N. et al. 1986, ApJ 308, 199.
[920] Parmar, A.N. et al. 1989, A&A 222, 96.
[921] Parmar, A.N. et al. 1989, ApJ 338, 359 & 373.
[922] Parmar, A.N. et al. 1989, ApJ 338, 1024.
[923] Parmar, A.N. et al. 1991, ApJ 366, 253.
[923a] Parmar, A.N. et al. 1993, A&A (in press).
[924] Parsignault, D.R. et al. 1972, Nat. Phys. Sci. 239, 123.
[925] Patterson, T.G. et al. 1989, in ref. [1294], p. 567.
[926] Paul, J. 1990, in ref. [289], p. 17.
[927] Pavlenko, E.P. et al. 1989, SvA Lett. 15, 262.
[928] Pavlinsky, M.N. et al. 1992, SvA Lett. 18, 88.
[929] Payne, B.J. & Coe, M.J. 1987, MNRAS 225, 985.
[930] Pedersen, H. 1983, ESO Messenger 34, 21.
[931] Pedersen, H. et al. 1981, Nat 294, 725.
[932] Pedersen, H. et al. 1982, ApJ 263, 325.
[933] Pedersen, H. et al. 1982, ApJ 263, 340.

[934] Pedersen, H. et al. 1983, IAU Circular 3858.
[935] Penninx, W. & Augusteijn, Th. 1991, A&A 246, L81.
[936] Penninx, W. et al. 1988, Nat 336, 146.
[937] Penninx, W. et al. 1989, A&A 208, 146.
[938] Penninx, W. et al. 1989, MNRAS 238, 851.
[939] Penninx, W. et al. 1990, A&A 240, 317.
[940] Penninx, W. et al. 1990, MNRAS 243, 114.
[941] Penninx, W. et al. 1991, MNRAS 249, 113.
[942] Penninx, W. et al. 1992, A&A 267, 92.
[943] Penny, A. J. et al. 1973, MNRAS 163, 7p.
[944] Penrod, D.G. & Vogt, S.S. 1985, ApJ 299, 653.
[945] Percy, J. et al. 1981, AJ 86, 53.
[946] Peterson, R.C. et al. 1989, ApJ 347, 251.
[947] Petro, L. & Hiltner, W. unpublished (see ref. [1207]).
[948] Petro, L.D. et al. 1981, BAAS 13, 900.
[949] Petro, L.D. et al. 1981, ApJ 251, L7.
[950] Petterson, J.A. 1975, ApJ 201, L61.
[951] Petterson, J.A. 1977, ApJ 216, 827.
[952] Pietsch, W. et al. 1980, ApJ 237, 964.
[953] Pietsch, W. et al. 1985, Sp. Sci. Rev. 40, 371.
[954] Pietsch, W. et al. 1986, A&A 157, 23.
[955] Pietsch, W. et al. 1986, A&A 163, 93.
[956] Polcaro, V.F. et al. 1990, A&A 231, 354.
[957] Ponman, T.J. et al. 1988, MNRAS 231, 999.
[958] Popper, D. 1950, ApJ 111, 495.
[959] Pounds, K.A. et al. 1975, MNRAS 172, 473.
[960] Pravdo, S.H. et al. 1979, ApJ 231, 912.
[961] Predehl, P. et al. 1991, A&A 246, L21.
[962] Predehl, P. et al. 1991, A&A 246, L40.
[963] Preston, R.A. et al. 1983, ApJ 268, L23.
[964] Priedhorsky, W. 1986, Astroph. Sp. Sci. 126, 89.
[965] Priedhorsky, W.C. & Holt, S.S. 1987, Sp. Sci. Rev. 45, 291.
[966] Priedhorsky, W.C. & Holt, S.S. 1987, ApJ 312, 743.
[967] Priedhorsky, W.C. & Terrell, J. 1983, Nat 303, 681.
[968] Priedhorsky, W.C. & Terrell, J. 1983, ApJ 273, 709.
[969] Priedhorsky, W. & Terrell, J. 1984, ApJ 280, 661.
[970] Priedhorsky, W. & Terrell, J. 1984, ApJ 284, L17.
[971] Priedhorsky, W. & Terrell, J. 1986, ApJ 301, 886.
[972] Priedhorsky, W.C. et al. 1983, ApJ 270, 233.
[973] Priedhorsky, W. et al. 1986, ApJ 306, L91.
[974] Primini, F. et al. 1977, ApJ 217, 543.
[975] Prince, T. et al. 1991, IAU Circular 5252.
[976] Proctor, R.J. et al. 1978, MNRAS 185, 745.
[977] Pryor, C. et al. 1990, AJ 98, 596.
[978] Qiao, G. & Cheng, J. 1989, ApJ 340, 503.
[979] Rappaport, S.A. et al. 1971, ApJ 168, L17.
[980] Rappaport, S.A. et al. 1977, Nat 268, 705.
[981] Rappaport, S.A. et al. 1978, ApJ 224, L1.
[982] Raubenheimer, B.C. 1990, A&A 234, 172.
[983] Raubenheimer, B.C. & Ögelman, H. 1990, A&A 230, 73.
[984] Raymond, J.C. 1982, ApJ 258, 240.
[985] Reich,W. & Schlickeiser, R. 1992, A&A 256, 408.
[986] Reid, C.A. et al. 1980, AJ 85, 1062.
[987] Remillard, R.A. & Canizares, C.R. 1984, ApJ 278, 761.
[988] Remillard, R.E. et al. 1992, ApJ 399, L145.
[989] Reynolds, A.P. et al. 1992, MNRAS 256, 631.
[990] Reynolds, A.P. et al. 1992, MNRAS (in press).
[991] Richer, G. et al. 1976, ApJ 204, L73.
[992] Richter, G.A. 1987, IBVS 3362.
[993] Rickard, J.J. 1974, ApJ 189, L113.
[994] Ricker, G.R. et al. 1973, ApJ 184, 237.
[995] Ricker, G.R. et al. 1976, ApJ 207, 333.
[996] Ricketts, M.J. et al. 1981, Space Sci. Rev. 30, 399.
[997] Robba, N.R. & Warwick, R.S. 1989, ApJ 346, 469.
[997a] Robba, N.R. et al. 1992, ApJ 401, 685.
[998] Robertson, B.S.C. et al. 1976, IBVS 1173.
[998a] Rodriguez, L.F. et al. 1992, ApJ 401, L15.
[999] Romney, J.D. et al. 1987, ApJ 321, 822.
[1000] Rose, L.A. et al. 1979, ApJ 231, 919.
[1001] Rosenberg, F.D. et al. 1975, Nat 256, 628.
[1002] Rossi, C. et al. 1991, A&A 249, L19.
[1003] Rössiger, S. 1978, IAU Circular 3210.
[1004] Rothschild, R.E. & Soong, Y. 1987, ApJ 315, 154.
[1005] Rothschild, R.E. et al. 1977, ApJ 213, 818.
[1006] Rothschild, R.E. et al. 1980, Nat 286, 786.
[1007] Sadakane, K. et al. 1984, ApJ 288, 284.
[1008] Sagar, R. et al. 1988, MNRAS 232, 131.
[1009] Sakao, T. et al. 1990, MNRAS 246, 11P.
[1010] Salotti, L. et al. 1992, A&A 253, 145.
[1011] Sandage, A.R. et al. 1966, ApJ 146, 316.
[1012] Sanduleak, N. 1968, AJ 73, 246.
[1013] Sansom, A.E. et al. 1989, PASJ 41, 591.
[1014] Sansom, A.E. et al. 1992, COSPAR Meeting Washington D.C.
[1014a] Saraswat, P. & Apparao, K.M.V. 1992, ApJ 401, 678.
[1015] Sato, N. et al. 1986, ApJ 304, 241.
[1016] Sato, N. et al. 1986, PASJ 38, 731.
[1017] Schachter, J. et al. 1989, ApJ 340, 1049 [erratum: ApJ 362, 379].
[1018] Schaefer, B.E. 1990, ApJ 354, 720.
[1019] Schoelkopf, R.J. & Kelley, R.L. 1991, ApJ 375, 696.
[1020] Schoembs, R. & Zoeschinger, G. 1990, A&A 227, 105.
[1021] Schmidtke, P. 1988, AJ 95, 1528.
[1022] Schmidtke, P. 1990, PASP 102, 144.
[1023] Schmidtke, P.& Cowley, A.P. 1987, AJ 93, 374.
[1024] Schmidtke, P.& Cowley, A.P. 1992, IAU Circular 5451.
[1025] Schreier, E. et al. 1972, ApJ 172, L79.
[1026] Schreier, E. et al. 1972, ApJ 178, L71.
[1027] Schreier, E.J. et al. 1976, ApJ 204, 539.
[1028] Schulz, N.S. et al. 1989, A&A 225, 48.
[1029] Schwartz, D.A. et al. 1980, AJ 85, 549.
[1030] Schwartz, R.A. et al. 1991, ApJ 376, 312.
[1031] Sembay, S. et al. 1990, ApJ 351, 675.
[1032] Seward, F.D. & Mitchell, M. 1981, ApJ 243, 736.
[1033] Seward, F.D. et al. 1976, MNRAS 175, 39P.
[1034] Seward, F.D. et al. 1976, MNRAS 177, 13P.
[1035] Seward, F. et al. 1980, Nat 287, 806.
[1036] Seward, F.D. et al. 1986, ApJ 305, 814.
[1037] Shafter, A.W. et al. 1980, ApJ 240, 612.
[1038] Shakhovskaya, N.I. et al. 1987, Bull. Crimean Aph. Obs. 75, 110.
[1039] Shakhovskoj, N. 1992, IAU Circular 5590.
[1040] Share, G. et al. 1978, IAU Circular 3190.
[1041] Shawl, S.J. & White, R.E. 1980, ApJ 239, L61.
[1042] Shinoda, K. et al. 1990, PASJ 42, L27.
[1043] Shrader, C.R. et al. 1992, IAU Circular 5591.
[1044] Skinner, G.K. 1980, Nat 288, 141.
[1045] Skinner, G.K. 1991, in ref. [289], p. 358.
[1046] Skinner, G.K. et al. 1982, Nat 297, 568.
[1047] Skinner, G.K. et al. 1987, Nat 330, 544.
[1048] Skinner, G.K. et al. 1989, IAU Circular No. 4850.
[1049] Skinner, G.K. et al. 1989, IAU Circular No. 4879.
[1050] Skinner, G.K. et al. 1990, MNRAS 243, 72.
[1051] Skinner, G.K. et al. 1991, A&A 252, 172.
[1052] Slettebak, A. 1982, ApJS 50, 55.
[1053] Slettebak, A. 1985, ApJS 59, 769.
[1054] Slettebak, A. & Snow, T. P. (eds) 1987: Physics of Be Stars, (Cambridge Univ. Press).
[1055] Smale, A.P. 1991, PASP 103, 636.
[1056] Smale, A.P. & Charles, P.A. 1989, MNRAS 238, 595.
[1057] Smale, A.P. & Corbet, R.H.D. 1991, ApJ 383, 853.
[1058] Smale, A.P. & Mukai, K. 1988, MNRAS 231, 663.
[1059] Smale, A.P. et al. 1984, MNRAS 207, 29P.
[1060] Smale, A.P. et al. 1984, MNRAS 210, 855.
[1061] Smale, A.P. et al. 1986, MNRAS 223, 207.
[1062] Smale, A.P. et al. 1987, MNRAS 225, 7P.
[1063] Smale, A.P. et al. 1988, MNRAS 232, 647.
[1064] Smale, A.P. et al. 1988, MNRAS 233, 51.
[1065] Smale, A.P. et al. 1989, PASJ 41, 607.

[1066] Smale, A.P. et al. 1989, in ref. [1294], p. 607.
[1067] Smale, A.P. et al. 1992, ApJ 400, 330.
[1068] Smale, A.P. et al. 1992, ApJ 395, 582.
[1069] Smithsonian Astrophysical Observatory Star Catalog (Washington D.C. 1966).
[1070] Sokolov, V.V. 1987, SvA 31, 419.
[1071] Soong, Y. et al. 1987, ApJ 319, L77.
[1072] Soong, Y. et al. 1990, ApJ 348, 634.
[1073] Soong, Y. et al. 1987, ApJ 348, 641.
[1074] Spencer, R.E. 1984, MNRAS 209, 869.
[1075] Steiner, J.E. et al. 1984, ApJ 280, 688.
[1076] Steinman-Cameron, T. et al. 1990, ApJ 359, 197.
[1077] Stella, L. 1990, Nat 344, 747.
[1078] Stella, L. et al. 1984, ApJ 282, 713.
[1079] Stella, L. et al. 1985, ApJ 288, L45.
[1080] Stella, L. et al. 1987, ApJ 312, L17.
[1081] Stella, L. et al. 1987, ApJ 315, L49.
[1082] Stella, L. et al. 1987, ApJ 321, 418.
[1083] Stella, L. et al. 1988, ApJ 324, 379.
[1084] Stewart, G.C. et al. 1987, MNRAS 228, 293.
[1085] Stewart, R.T. et al. 1991, MNRAS 253, 212.
[1086] Stier, M. & Liller, W. 1976, ApJ 206, 257.
[1087] Stocke, J. et al. 1985, PASP 97, 126.
[1087a] Stollberg, M.T. et al. 1992, Proc. Compton Symp. (St Louis).
[1088] Strom, R.G. et al. 1989, Nat 337, 234.
[1089] Sugimoto, D. et al. 1984, PASJ 36, 839.
[1090] Sunyaev, R. 1990, IAU Circular 5104.
[1091] Sunyaev, R.A. et al. 1988, SvA Lett.14, 327.
[1092] Sunyaev, R. et al. 1989, in ref. [1294], p. 641.
[1093] Sunyaev, R.A. et al. 1990, SvA Lett. 16, 55.
[1094] Sunyaev, R.A. et al. 1990, SvA Lett. 16, 59.
[1095] Sunyaev, R.A. et al. 1991, SvA Lett. 17, 42.
[1096] Sunyaev, R.A. et al. 1991, SvA Lett. 17, 123.
[1097] Sunyaev, R.A. et al. 1991, IAU Circular 5176.
[1098] Sunyaev, R.A. et al. 1991, ApJ 383, L49.
[1099] Sunyaev, R.A. et al. 1991, A&A 247, L29.
[1100] Sunyaev, R.A. et al. 1991, SvA Lett.17, 50 & 54
[1101] Sunyaev, R. et al. 1991, IAU Circular 5180.
[1102] Sunyaev, R.A. et al. 1991, Adv. Space Res. 11(8), 177.
[1103] Sunyaev, R.A. et al. 1991, IAU Circular 5398.
[1104] Sunyaev, R.A. et al. 1992, IAU Circular 5437.
[1105] Sunyaev, R. et al. 1992, ApJ 389, L75.
[1106] Suzuki, K. et al. 1984, PASJ 36, 761.
[1107] Swank, J.H. et al. 1976, IAU Circular 3000.
[1108] Swank, J.H. et al. 1976, IAU Circular 3010.
[1109] Swank, J.H. et al. 1977, ApJ 212, L73.
[1110] Swank, J.H. et al. 1978, MNRAS 182, 349.
[1111] Swank, J.H. et al. 1984, ApJ 277, 274.
[1112] Sztajno, M. et al. 1983, ApJ 267, 713.
[1113] Sztajno, M. et al. 1985, ApJ 299, 487.
[1114] Sztajno, M. et al. 1986, MNRAS 222, 499.
[1115] Sztajno, M. et al. 1987, MNRAS 226, 39.
[1116] Takeshima, T. et al. 1991, PASJ 43, L43.
[1117] Takeuchi, Y. et al. 1990, PASJ 42, 287.
[1118] Takisawa, M. et al. 1992, preprint.
[1119] Tamura, K. et al. 1992, ApJ 389, 676.
[1120] Tan, J. et al. 1991, MNRAS 251, 1.
[1121] Tan, J. et al. 1991, ApJ 374, 291.
[1122] Tan, J. et al. 1992, ApJ 385, 314.
[1123] Tanaka, Y. 1989, in ref. [1294], p. 3.
[1124] Tananbaum, H. et al. 1972, ApJ 177, L5.
[1125] Tananbaum, H. et al. 1976, ApJ 209, L125.
[1126] Tarenghi, M. & Reina, C. 1972, Nat Phys. Sci. 240, 53.
[1127] Tashiro, M. et al. 1991, MNRAS 252, 156.
[1128] Tawara, Y. et al. 1982, Nat 299, 38.
[1129] Tawara, Y. et al. 1984, ApJ 276, L41.
[1130] Tawara, Y. et al. 1984, PASJ 36, 861.
[1131] Tawara, Y. et al. 1985, Nat 318, 545.

[1132] Tawara, Y. et al. 1989, PASJ 41, 473.
[1133] Taylor, A.R. & Gregory, P.C. 1982, ApJ 255, 210.
[1134] Taylor, A.R. & Gregory, P.C. 1984, ApJ 283, 273.
[1135] Taylor, A.R. et al. 1991, Nat 351, 547.
[1136] Taylor, A.R. et al. 1992, ApJ 395, 268.
[1137] Tennant, A.F. 1987, MNRAS 226, 971.
[1138] Tennant, A.F. 1988, MNRAS 230, 403.
[1139] Tennant, A.F. et al. 1986, MNRAS 219, 871.
[1140] Tennant, A.F. et al. 1986, MNRAS 221, 27P.
[1141] Terrell, J. & Priedhorsky, W.C. 1984, ApJ 285, L15.
[1142] Thackeray, A.D. et al. 1972, Mem. RAS 77, 199.
[1143] The, P.S. 1966, Contr. Bosscha Obs. No. 35.
[1144] Thom, C. et al. 1986, A&A 165, L13.
[1145] Thomas, R.M. et al. 1979, MNRAS 188, 19.
[1146] Thorstensen, J.R. 1987, ApJ 312, 739.
[1147] Thorstensen, J.R. & Charles, P.A. 1982, ApJ 253, 756.
[1148] Thorstensen, J.R. et al. 1978, ApJ 220, L131.
[1149] Thorstensen, J.R. et al. 1979, ApJ 233, L57 [erratum 237, L25].
[1150] Thorstensen, J.R. et al. 1980, ApJ 238, 964.
[1151] Thorstensen, J.R. et al. 1988, ApJ 334, 430.
[1152] Tjemkes, S.A. et al. 1986, A&A 154, 77.
[1153] Treves, A. et al. 1980, ApJ 242, 1114.
[1154] Treves, A. et al. 1988, ApJ 325, 119.
[1155] Treves, A. et al. 1988, ApJ 335, 142.
[1156] Treves, A. et al. 1990, ApJ 364, 266.
[1157] Trümper, J. et al. 1978, ApJ 219, L105.
[1158] Trümper, J. et al. 1985, Sp. Sci. Rev. 40, 255.
[1159] Trümper, J. et al. 1986, ApJ 300, L63.
[1160] Trümper, J. et al. 1991, Nat 349, 579.
[1161] Tsunemi, H. 1989, PASJ 41, 453.
[1162] Tsunemi, H. & Kitamoto, S. 1988, ApJ 334, L21.
[1163] Tsunemi, H. et al. 1977, ApJ 211, L15.
[1164] Tsunemi, H. et al. 1989, ApJ 337, L81.
[1165] Tueller, J. et al. 1984, ApJ 279, 177.
[1166] Tuohy, I.R. et al. 1988, in: "Physics of Neutron Stars and Black Holes", ed. Y. Tanaka, (Universal Academic Press, Tokyo), p. 93.
[1167] Turner, M.J.L. & Breedon, L.M. 1984, MNRAS 208, 29P.
[1168] Turner, M.J.L. et al. 1985, Sp. Sci. Rev. 40, 249.
[1169] Tweedy, R.W. et al. 1989, in ref. [1294], p. 661.
[1170] Ubertini, P. et al. 1991, ApJ 366, 544.
[1171] Ubertini, P. et al. 1991, ApJ 383, 263.
[1172] Ubertini, P. et al. 1992, ApJ 386, 710 [erratum ApJ 396, 378].
[1173] Udalsky, A. & Kaluzny, J. 1991, PASP 103, 198.
[1174] Ulmer, M.P. et al. 1972, ApJ 178, L121.
[1175] Ulmer, M.P. et al. 1973, ApJ 184, L117.
[1176] Ulmer, M.P. et al. 1978, Nat 276, 799.
[1177] Ulmer, M.P. et al. 1980, ApJ 235, L159.
[1178] Unger, S.J. et al. 1992, MNRAS 256, 725.
[1179] Unno, W. et al. 1990, PASJ 42, 269.
[1180] Vacca, W.D. et al. 1986, MNRAS 220, 339.
[1181] Vacca, W.D. et al. 1987, A&A 172, 143.
[1182] Van Amerongen, S.F. et al. 1986, IBVS 2901
[1183] Van Amerongen, S. et al. 1987, A&A 185, 147.
[1184] Van den Heuvel, E.P.J. et al. 1992, A&A 262, 97.
[1185] Van der Klis, M. 1989, ARA&A 27, 517.
[1186] Van der Klis, M. & Bonnet-Bidaud, J.M. 1984, A&A 135, 155.
[1187] Van der Klis, M. & Bonnet-Bidaud, J.M. 1989, A&A 214, 203.
[1188] Van der Klis, M. & Jansen, F. 1984, Nat 313, 768.
[1189] Van der Klis, M. et al. 1982, A&A 106, 339.
[1190] Van der Klis, M. et al. 1983, A&A 126, 265.
[1191] Van der Klis, M. et al. 1983, MNRAS 203, 279.
[1192] Van der Klis, M. et al. 1985, A&A 151, 322.
[1193] Van der Klis, M. et al. 1985, Nat 316, 225.
[1194] Van der Klis, M. et al. 1985, Sp. Sci. Rev. 40, 287.
[1195] Van der Klis, M. et al. 1987, ApJ 313, L19.
[1196] Van der Klis, M. et al. 1987, ApJ 316, 411.
[1197] Van der Klis, M. et al. 1987, ApJ 319, L13.
[1198] Van der Klis, M. et al. 1990, ApJ 360, L19.

[1199]   Van der Klis, M. et al. 1991, MNRAS 248, 751.
[1200]   Van der Klis, M. et al. 1992, A&A (submitted).
[1201]   Van der Woerd, H. et al. 1989, ApJ 344, 320.
[1202]   Van Genderen, A.M. 1981, A&A 96, 82.
[1203]   Van Kerkwijk, M. et al. 1989, A&A 209, 173.
[1204]   Van Kerkwijk, M. et al. 1992, Nat 355, 703.
[1205]   Van Kerkwijk, M. et al. 1992, A&A (submitted).
[1206]   Van Oijen, J.G.J. 1989, A&A 217, 115.
[1207]   Van Paradijs, J. 1983, in ref. [666], p. 189.
[1208]   Van Paradijs, J. 1991, in 'Neutron Stars, Theory and Observation', eds J. Ventura & D. Pines (Kluwer), p. 289.
[1209]   Van Paradijs, J. & Isaacman, R. 1989, A&A 222, 129.
[1210]   Van Paradijs, J. & Kuiper, L. 1984, A&A 138, 71.
[1211]   Van Paradijs, J. & Lewin, W.H.G. 1987, A&A 172, L20.
[1212]   Van Paradijs, J. et al. 1977, A&AS 30, 195.
[1213]   Van Paradijs, J. et al. 1979, MNRAS 189, 387.
[1214]   Van Paradijs, J. et al. 1980, ApJ 241, L161.
[1215]   Van Paradijs, J. et al. 1983, A&A 124, 294.
[1216]   Van Paradijs, J. et al. 1984, A&AS 55, 7.
[1217]   Van Paradijs, J. et al. 1984, MNRAS 210, 863.
[1218]   Van Paradijs, J. et al. 1986, A&AS 63, 71.
[1219]   Van Paradijs, J. et al. 1986, MNRAS 221, 617.
[1220]   Van Paradijs, J. et al. 1987, A&A 182, 47.
[1221]   Van Paradijs, J. et al. 1987, A&A 184, 201.
[1222]   Van Paradijs, J. et al. 1988, A&A 192, 147.
[1223]   Van Paradijs, J. et al. 1988, A&AS 76, 185.
[1224]   Van Paradijs, J. et al. 1988, MNRAS 231, 379.
[1225]   Van Paradijs, J. et al. 1990, A&A 234, 181.
[1226]   Van Paradijs, J. et al. 1990, A&A 235, 156.
[1227]   Van Paradijs, J. et al. 1990, PASJ 42, 633.
[1228]   Verbunt, F. et al. 1984, MNRAS 210, 899.
[1229]   Verbunt, F. et al. 1990, A&A 234, 195.
[1230]   Vermeulen, R. 1989, Ph. D. Thesis, Univ. Leiden.
[1231]   Vermeulen, R.C. et al. 1987, Nat 328, 309.
[1232]   Vikhlinin, A. et al. 1992, IAU Circular 5576.
[1233]   Vidal, N.V. 1973, ApJ 186, L81.
[1234]   Voges, W. et al. 1987, ApJ 320, 794.
[1235]   Vrtilek, S.D. et al. 1986, ApJ 307, 698.
[1236]   Vrtilek, S.D. et al. 1986, ApJ 308, 644.
[1237]   Vrtilek, S.D. et al. 1988, ApJ 329, 276.
[1238]   Vrtilek, S.D. et al. 1990, A&A 235, 162.
[1239]   Vrtilek, S.D. et al. 1991, ApJ 376, 278.
[1240]   Vrtilek, S.D. et al. 1991, ApJS 76, 1127.
[1241]   Wachmann, A.A. 1948, Erg. Astron. Nachrichten 11, No. 5.
[1242]   Wackerling, L.R. 1970, Memoirs RAS 73, 153.
[1243]   Wade, C.M. & Hjellming, R.M. 1971, ApJ 170, 523.
[1244]   Wade, R.A. et al. 1985, PASP 97, 1092.
[1245]   Wagner, R.M. et al. 1986, ApJ 308, 152.
[1246]   Wagner, R.M. et al. 1989, IAU Circular 4783.
[1247]   Wagner, R.M. et al. 1989, ApJ 346, 971.
[1248]   Wagner, R.M. et al. 1991, ApJ 378, 293.
[1249]   Wagner, R.M. et al. 1992, IAU Circular 5589.
[1250]   Wagner, R.M. et al. 1992, ApJ 401, L97.
[1251]   Waki, I. et al. 1984, PASJ 36, 819.
[1252]   Wallerstein, G. 1992, Nat 356, 569.
[1253]   Walter, F. et al. 1982, ApJ 253, L67.
[1254]   Wang, Z.-R. et al. 1990, A&A 240, 98.
[1255]   Warwick, R.S. et al. 1981, MNRAS 197, 865.
[1256]   Warwick, R.S. et al. 1988, MNRAS 232, 551.
[1257]   Waters, L.B.F.M. 1989, in ref. [1294], p. 25.
[1258]   Waters, L.B.F.M. et al. 1988, A&A 198, 200.
[1259]   Waters, L.B.F.M. et al. 1989, A&A 213, L19.
[1260]   Watson, M.G. et al. 1978, ApJ 221, L69.
[1261]   Watson, M.G. et al. 1978, MNRAS 183, 35P.
[1262]   Watson, M.G. et al. 1981, ApJ 250, 142.
[1263]   Watson, M.G. et al. 1981, MNRAS 195, 197.
[1264]   Watson, M.G. et al. 1982, MNRAS 199, 915.
[1265]   Watson, M.G. et al. 1983, ApJ 273, 688.
[1266]   Watson, M.G. et al. 1985, Sp. Sci. Rev. 40, 195.
[1267]   Watson, M.G. et al. 1986, MNRAS 222, 261.
[1268]   Webbink, R.F. 1985, Proc. IAU Symposium 113, 541.
[1269]   Webster, B.L. & Murdin, P. 1971, Nat 235, 37.
[1270]   Webster, B.L. et al. 1972, Nat PS 240 183.
[1270a] Wenzel, W. 1985, IAU Circular 4059.
[1271]   Weisskopf, M.C. et al. 1984, ApJ 278, 711.
[1272]   Wheaton, W.A. et al. 1975, IAU Circular 2761.
[1273]   Whelan, J.A.J. et al. 1976, MNRAS 180, 657.
[1274]   Whelan, J.A.J. et al. 1977, MNRAS 181, 259.
[1275]   White, N.E. 1978, Nat 271, 38.
[1276]   White, N.E. & Carpenter, G.F. 1978, MNRAS 183, 11P.
[1277]   White, N.E. & Holt, S.S. 1982, ApJ 257, 318.
[1278]   White, N.E. & Marshall, F.E. 1983, IAU Circular 3806.
[1279]   White, N.E. & Marshall, F. E. 1984, ApJ 281, 354.
[1280]   White, N.E. & Swank, J.H. 1982, ApJ 253, L61.
[1281]   White, N.E. & Swank, J.H. 1984, ApJ 287, 856.
[1282]   White, N.E. et al. 1976, MNRAS 176, 201.
[1283]   White, N.E. et al. 1978, ApJ 220, 600.
[1284]   White, N.E. et al. 1978, Nat 274, 664.
[1285]   White, N.E. et al. 1980, ApJ 239, 655.
[1286]   White, N.E. et al. 1981, ApJ 247, 994.
[1287]   White, N.E. et al. 1982, ApJ 263, 277.
[1288]   White, N.E. et al. 1984, ApJ 270, 711.
[1289]   White, N.E. et al. 1984, ApJ 283, L9.
[1290]   White, N.E. et al. 1984, in ref. [1307], p. 31.
[1291]   White, N.E. et al. 1985, ApJ 296, 475.
[1292]   White, N.E. et al. 1987, MNRAS 226, 645.
[1293]   White, N.E. et al. 1988, ApJ 324, 363.
[1294]   White, N.E. et al. (eds) 1989, Proc. 23rd ESLAB Symp. (ESA SP-296).
[1295]   Whitehorne, M.L. 1989, JRAS Canada 83, 277.
[1296]   Whitlock, L. et al. 1989, ApJ 338, 381.
[1297]   Whitlock, L. et al. 1989, ApJ 344, 371.
[1298]   Whitlock, L. et al. 1990, A&A 238, 140.
[1299]   Wickramasinghe, D.T. et al. 1974, ApJ 188, 167.
[1300]   Willis, A.J. et al. 1980, ApJ 237, 596.
[1301]   Wilson, C.A. et al. 1992, Proc. Compton Symp. (St Louis).
[1302]   Wilson, C.K. & Rothschild, R.E. 1983, ApJ 274, 717.
[1303]   Wolff, S.C. & Morrisson, N.D. 1974, ApJ 187, 69.
[1304]   Wood, K.S. et al. 1978, IAU Circular 3203.
[1305]   Wood, K.S. et al. 1984, ApJS 56, 507.
[1306]   Wood, K.S. et al. 1991, ApJ 379, 295.
[1307]   Woosley, S.E. (ed.) High Energy Transients in Astrophysics, AIP Proceedings 115.
[1308]   Wu, C.C. et al. 1982, PASP 94, 149.
[1309]   Wu, C.C. et al. 1983, PASP 95, 391.
[1310]   Yamauchi, S. 1990, PASJ 42, L53.
[1311]   Yamauchi, S. & Koyama, K. 1990, PASJ 42, L83.
[1312]   Yang, S. et al. 1988, PASP 100, 233.
[1313]   Zwarthoed, G.A.A. et al. 1992, A&A 267, 101.
[1314]   Zwitter, T. et al. 1989, Fund. Cosm. Physics 13, 309.

15

A compilation of cataclysmic binaries with known or suspected orbital periods

Hans Ritter and Ulrich Kolb
Max-Plank-Institut für Astrophysik, Karl-Schwarzschild-Strasse 1,
D-85748 Garching, Germany

In this chapter, we present a compilation of cataclysmic binaries with known or suspected orbital periods. The justification for restricting ourselves to this particular subset of the known cataclysmic variables is that (1) in general, these objects are well investigated, and (2) it limits the number of objects to be dealt with. Furthermore, a recent and comprehensive compilation of known and suspected cataclysmic variables, which provides among other things accurate coordinates and finding charts, has been given by Downes and Shara (1993), and there is no point duplicating this excellent work.

In this compilation, we present the current content (as of 1 September, 1993) of the catalogue of cataclysmic binaries, low-mass X-ray binaries and related objects (Ritter 1984, 1987, 1990; hereafter R84, R87 and R90, respectively), regarding cataclysmic binaries only. Accordingly, this compilation provides the equatorial coordinates, magnitudes, orbital parameters and stellar parameters of the components and other characteristic properties of 231 cataclysmic binaries, together with a comprehensive list of references to the relevant recent literature. In addition, the compilation contains a list of recent references to published finding charts for 218 of the 231 objects and a cross-reference list of aliases of object designations.

The presentation of the material is essentially the same as in R84, R87 and R90, i.e. the compilation consists of (1) a table section, where a few characterizing parameters of each object are tabulated, (2) a reference section, where a selection of recent references to the relevant literature is given, and (3) a list of recent references to published finding charts. Note that earlier references can be found in R84, R87 and R90. In the table section, the objects are listed in sequence of decreasing orbital period, whereas in the reference sections they are listed in lexigraphical order. The sections described so far are preceded by a separate description section in which the quantities listed in the table section and the corresponding abbreviations used in the table and in the table headings are described. The final part of this compilation consists of the 'who's who' section, in which frequently used aliases of the objects' designations are cross-referenced. The organization of this section is basically the same as in R84, R87 and R90, the only difference being that the list of catalogue acronyms has been omitted here. This information can be found in R84, R87 and R90. Supplementary information will be given in the forthcoming sixth edition of the catalogue of cataclysmic binaries, low-mass X-ray binaries and related objects (Ritter 1994).

Wherever applicable, we use as the standard object designation the variable name given in the fourth edition of the general catalogue of variable stars (Kholopov *et al.* 1985a,b, 1987a) and in the name lists of variable stars (up to and including the 71st list, Kholopov *et al.* 1985c, 1987b, 1989; Kazarovets and Samus 1990; Kazarovets, Samus and Goranskij 1993).

A new feature in this compilation is that we now provide limited information about where the values given in the table section are taken from. This is done in the following way: at the end of the reference from which a given quantity, say XYZ, was taken, this quantity is given in parentheses, i.e. (XYZ). The quantities for which this is done are: the periods (ORB.PER., 2. PER., 3. PER., 4. PER.), the mass ratio (M1/M2), the orbital inclination (INCL), and the masses (M1, M2). Whenever available, the cordinates have been taken from Downes and Shara (1993).

References

Downes, R.A. and Shara, M.M.: 1993 *Publ. Astron. Soc. Pacific* **105**, 127

Kazarovets, E.V. and Samus, N.N.: 1990 *Inf. Bull. Variable Stars* no. 3530

Kazarovets, E.V., Samus, N.N. and Goranskij, V.P.: 1993, *Inf. Bull. Variable Stars* no. 3840

Kholopov, P.N., *et al.*: 1985a, *General Catalogue of Variable Stars*, Vol. I, 4th edition (Moscow, Nauka)

Kholopov, P.N., *et al.*: 1985b, *General Catalogue of Variable Stars*, Vol. II, 4th edition (Moscow, Nauka)

Kholopov, P.N., *et al.*: 1985c, *Inf. Bull. Variable Stars* no. 2681

Kholopov, P.N., *et al.*: 1987a, *General Catalogue of Variable Stars*, Vol. III, 4th edition (Moscow, Nauka)

Kholopov, P.N., *et al.*: 1987b, *Inf. Bull. Variable Stars* no. 3058

Kholopov, P.N., *et al.*: 1989, *Inf. Bull. Variable Stars* no. 3323

Ritter, H.: 1984, *Astron. Astrophys. Suppl. Ser.* **57**, 385 (R84)

Ritter, H.: 1987, *Astron. Astrophys. Suppl. Ser.* **70**, 335 (R87)

Ritter, H.: 1990, *Astron. Astrophys. Suppl. Ser.* **85**, 1179 (R90)

Ritter, H.: 1994, in preparation

```
*************************************************************
*                                                           *
*  DESCRIPTION OF TABLE HEADINGS AND OF ABBREVIATIONS       *
*                                                           *
*************************************************************
```

OBJECT NAME: WHEREVER POSSIBLE, THE DESIGNATION OF THE OBJECT
                     GIVEN IN THE GENERAL CATALOGUE OF VARIABLE STARS IS
                     USED HERE.

ALTERN.NAME: IS A FREQUENTLY USED ALTERNATIVE NAME. FURTHER
                     ALTERNATIVE DESIGNATIONS ARE GIVEN IN THE
                     "WHO'S WHO ?" SECTION AT THE END OF THE CATALOGUE

COORDINATES: FIRST ROW   RIGHT ASCENSION (2000) IN HRS MIN SEC
                     SECOND ROW   DECLINATION   (2000) IN DEG '   "

                        THE COORDINATES ARE GIVEN IN THE FOLLOWING FORMAT:
                        RIGHT ASCENSION: HH MM SS.S
                        DECLINATION   : DDD MM SS A
                        IN THE DECLINATION FIELD, "A" IS THE ACCURACY OF
                        OF THE COORDINATES IN SECONDS OF ARC (WRITTEN AS A
                        HEXADECIMAL NUMBER, I.E. 10" = A , 11" = B ,
                        12" = C , 13" = D , 14" = E , 15" = F). IN A CASE
                        WHERE "A" > 15" OR WHERE THE ACCURACY OF THE CO-
                        ORDINATES IS NOT KNOWN, THE "A"-FIELD IS LEFT BLANK.

                        WHEREVER POSSIBLE , THE COORDINATES HAVE BEEN TAKEN
                        FRON THE CATALOGUE OF DOWNES AND SHARA (1993, PUBL.
                        ASTRON.SOC.PACIFIC 105,127). THESE ARE COORDINATES
                        WITH RESPECT TO THE EQUINOX J2000.

TYPE: (FIRST AND SECOND ROW), THE OBJECT TYPE IS COARSELY
                     CHARACTERISED USING THE FOLLOWING ABBREVIATIONS:

                     AC = AM CVN STAR, DOES NOT CONTAIN HYDROGEN, SUBTYPE
                           OF NL
                     AM = POLAR = AM HER SYSTEM, SUBTYPE OF NL, CONTAINS A
                         SYNCHRONOUSLY ROTATING, MAGNETISED WHITE DWARF
                     CP = COHERENT PULSATOR, CONTAINS A COHERENTLY
                         PULSATING WHITE DWARF
                     DD = SYSTEM CONSISTS OF TWO DEGENERATE COMPONENTS
                     DN = DWARF NOVA
                     DQ = DQ HER STAR, CONTAINS A NON-SYNCHRONOUSLY
                         ROTATING MAGNETIZED WHITE DWARF
                     IP = INTERMEDIATE POLAR, SHOWS COHERENT X-RAY PERIOD
                         FROM A NON-SYNCHRONOUSLY SPINNING MAGNETISED
                         WHITE DWARF
                     N = CLASSICAL NOVA OF UNDETERMINED SPEED CLASS
                     NA = FAST NOVA (DECLINE FROM MAX. BY 3 MAG. IN LESS
                         THAN ABOUT 100 DAYS)
                     NB = SLOW NOVA (DECLINE FROM MAX. BY 3 MAG. IN MORE
                         THAN ABOUT 100 DAYS)
                     NC = EXTREMELY SLOW NOVA (TYPICAL TIME SCALE OF THE
                         DECLINE FROM MAXIMUM: DECADES)
                     NL = NOVA-LIKE VARIABLE
                     NR = RECURRENT NOVA
                     SH = NON-SU UMA STAR SHOWING EITHER PERMANENT OR
                         TRANSIENT SUPERHUMPS

```
TYPE:              SU = SU UMA STAR, SUBTYPE OF DN
                   SW = SW SEX STAR, SUBTYPE OF NL
                   UG = DWARF NOVA OF EITHER U GEM OR SS CYG SUBTYPE
                   UX = UX UMA STAR, SUBTYPE OF NL
                   VY = VY SCL STAR (ANTI DAWRF NOVA), SUBTYPE OF NL
                   XS = X-RAY SOURCE
                   ZC = Z CAM STAR, SUBTYPE OF DN
                   *  = OBJECT POSSIBLY RELATED TO THE DQ HER STARS


MAG1, MAG3
MAG2, MAG4:        APPARENT V MAGNITUDE (B MAGNITUDE IF FOLLOWED BY B)
                   WITH THE FOLLOWING MEANING:

                   MAG1 = MAXIMUM BRIGHTNESS OF
                          NOVAE (N,NA,NB,NC,NR) IN MINIMUM
                          DN    (UG,ZC,SU)      IN MINIMUM
                          NL    (UX,AC)         IN NORMAL STATE
                          NL    (AM,VY)         IN HIGH STATE

                   MAG2 = MINIMUM BRIGHTNESS, IN CASE OF ECLIPSES MAGN.
                          AT MIDECLIPSE, OF
                          NOVAE (N,NA,NB,NC,NR) IN MINIMUM
                          DN    (UG,ZC,SU)      IN MINIMUM
                          NL    (UX,AC)         IN NORMAL STATE
                          NL    (AM,VY)         IN HIGH STATE

                   MAG3 = MAXIMUM BRIGHTNESS OF
                          NOVAE (N,NA,NB,NC,NR) IN OUTBURST
                          DN    (UG,ZC)         IN OUTBURST
                          DN    (SU)            IN NORMAL OUTBURST
                          NL    (AM,VY)         IN LOW STATE
                          NL    (DQ,IP)         IN FLARING STATE

                   MAG4 = BRIGHTNESS OF
                          ZC                    IN STANDSTILL
                          SU                    IN SUPEROUTBURST
                        = MINIMUM BRIGHTNESS OF
                          NL    (AM,VY)         IN LOW STATE


T1:                FOR DN (UG, ZC), THE TYPICAL TIME INTERVAL (IN DAYS)
                              BETWEEN TWO SUBSEQUENT OUTBURSTS
                   FOR DN (SU), THE TYPICAL TIME INTERVAL (IN DAYS)
                              BETWEEN TWO SUBSEQUENT NORMAL OUTBURSTS


T2:                FOR DN (SU), THE TYPICAL TIME INTERVAL (IN DAYS)
                              BETWEEN TWO SUBSEQUENT SUPEROUTBURSTS


ORB.PER.:          ORBITAL PERIOD (IN DAYS), IN CASE OF OBJECT TYPE
                      DQ: THE SPECTROSCOPIC PERIOD IS GIVEN HERE IF IT IS
                          DIFFERENT FROM THE PHOTOMETRIC ONE.
                      SU: IF FOLLOWED BY *, THE ORBITAL PERIOD HAS BEEN
                          ESTIMATED FROM THE KNOWN SUPERHUMP PERIOD USING
                          THE EMPIRICAL RELATION GIVEN BY B. STOLZ AND
                          R. SCHOEMBS (1984,ASTRON.ASTROPHYS.132,187)
                      *:  SPECTROSCOPIC PERIOD,
                          PHOTOMETRIC PERIOD IF FOLLOWED BY P.
```

2. PER.:   SECOND PERIOD (IN DAYS), IN CASE OF OBJECT TYPE
     N, NA, NB, NC, NR: ORBITAL PERIOD OF THE PRENOVA,
          UNLESS THE OBJECT IS ALSO OF TYPE DQ OR *. IN
          THESE CASES SEE BELOW.
     DQ: THE PHOTOMETRIC PERIOD IS GIVEN HERE IF IT IS
          DIFFERENT FROM THE SPECTROSCOPIC ONE.
     AM: POLARISATION PERIOD (= SPIN PERIOD OF THE WHITE
          DWARF), IF IT IS DIFFERENT FROM THE PRESUMED
          ORBITAL PERIOD
     SU: SUPERHUMP PERIOD. WHEREVER POSSIBLE, THE SUPER-
          HUMP PERIOD AT THE BEGINNING OF A SUPEROUTBURST
          IS GIVEN.
     SH: PHOTOMETRIC PERIOD, PRESUMABLY SUPERHUMP PERIOD
          OF EITHER PERMANENT OR TRANSIENT SUPERHUMPS
     *:   PHOTOMETRIC PERIOD.

3. PER.
4. PER.:   ADDITIONAL PERIODS IN THE SYSTEM (IN SECONDS),
     IN CASE OF OBJECT TYPE
     CP: 3. PER. = PERIOD OF COHERENT PULSATION,
                    (TRANSIENT IF FOLLOWED BY T).
          4. PER. = SECOND PERIOD OF COHERENT PULSATION,
                    (TRANSIENT IF FOLLOWED BY T).
     DQ: 3. PER. = ROTATION PERIOD OF THE WHITE DWARF.
          4. PER. = OPTICAL PERIOD, (PRESUMABLY DUE TO
                    REPROCESSED X-RAYS).
     IP: 3. PER. = X-RAY PERIOD (PRESUMABLY THE
                    ROTATION PERIOD OF THE WHITE DWARF)
          4. PER. = OPTICAL PERIOD, (PRESUMABLY DUE TO
                    REPROCESSED X-RAYS).

     THE OCCURRENCE OF TRANSIENT QUASI-PERIODIC OSCILL-
     ATIONS (QPO) IN OBJECTS OF TYPE N, DN, NL IS INDI-
     CATED IN THE FIELD "3. PER.".

EB: INDICATES THE OCCURRENCE OF ECLIPSES
     IF BLANK   NO ECLIPSES OBSERVED
     IF   1   1 ECLIPSE PER ORBITAL REVOLUTION OBSERVED
     IF   2   2 ECLIPSES PER ORBITAL REVOLUTION OBSERVED
     IF   D   PERIODIC ECLIPSE-LIKE DIPS OBSERVED

SB: TYPE OF SPECTROSCOPIC BINARY
     IF   1   SINGLE-LINED SPECTROSCOPIC BINARY
     IF   2   DOUBLE-LINED SPECTROSCOPIC BINARY

SPECTR2:   SPECTRAL TYPE OF THE SECONDARY
SPECTR1:   SPECTRAL TYPE OF THE PRIMARY
     THE NUMBER TO THE RIGHT OF THE SLASH INDICATES THE
     LUMINOSITY CLASS , I.E.

                         I = 1
                        II = 2
                       III = 3
                        IV = 4
                         V = 5
                        VI = 6

```
M1/M2:              FIRST ROW    MASS RATIO  M1/M2
                    SECOND ROW   THE CORRESPONDING ERROR


INCL:               FIRST ROW    ORBITAL INCLINATION (IN DEGREES)
                    SECOND ROW   THE CORRESPONDING ERROR


M1:                 FIRST ROW    MASS OF THE PRIMARY (IN SOLAR MASSES)
                    SECOND ROW   THE CORRESPONDING ERROR


M2:                 FIRST ROW    MASS OF THE SECONDARY (IN SOLAR MASSES)
                    SECOND ROW   THE CORRESPONDING ERROR


UNCERTAIN VALUES ARE FOLLOWED BY A COLON.
```

C A T A C L Y S M I C   B I N A R I E S

OBJECT NAME / ALTERN.NAME	COORDINATES (J2000)	TYPE	MAG1 / MAG2	MAG3 / MAG4	T1 / T2	ORB.PER. / 2. PER.	3. PER. / 4. PER.	EB / SB	SPECTR2 / SPECTR1	M1/M2	INCL	M1	M2
V1017 SGR / N SGE 1919	18 32 04.3 / -29 23 13 1	NB DN / XS	13.7 /	7.2 /		5.714 /		/ 1	G5/3 /				
GK PER / N PER 1901	03 31 11.8 / +43 54 17 1	NA DN / DQ XS	10.2 / 14.0	0.2 /		1.996803 /	351.34 /	/ 2	K0/4 /	3.6 / 0.5	<73	0.9 / 0.2	0.25
U SCO / N SCO 1987	16 22 30.7 / -17 52 42 1	NR / *	17.9 / 19.5	8.8 /		1.23452 / 0.05313:		1 / 2					
WY CMA	07 11 43 / -26 58 58	?	14.5 / 15.4			1.14433: /							
V394 CRA / N CRA 1987	18 00 26.0 / -39 00 35 1	NR	18.5B / 20.0B	7.2 /		0.7577 : /							
BV CEN	13 31 19.7 / -54 58 34 1	DN UG / XS	12.6 / 13.3	10.5 /	150 /	0.610116 /		/ 2	G5-8/5 /	0.92 / 0.06	62 / 5	0.83 / 0.10	0.90 / 0.10
V841 OPH / N OPH 1848	16 59 30.3 / -12 53 27 1	NB	13.9B / 14.2B	4.2 /		0.60423: /		1 /					
DI LAC / N LAC 1910	22 35 48.5 / +52 43 00 1	NA	14.3 / 14.6	4.6 /		0.543773 /		/ 1					
V SGE	20 20 14.8 / +21 06 08 1	NL XS	12.2 / 13.9	10.5 / 11.0	550 /	0.514198 /	QPO /	2 / 2	F6-G0/5 / WN5:	0.27	90	0.74	2.8
V442 CEN	11 24 52.0 / -35 54 39 1	DN UG	16.5 /	11.9 /	14-39 /	0.46 : /	QPO /						
QU CAR	11 05 37.4 / -68 38 14 1	NL	11.1 / 11.5			0.454 /		1 /			<60		
DX AND	23 29 46.8 / +43 45 03 1	DN UG	16.4 /	10.9 /		0.44148 /		/ 2	K1/5 /	1.04 / 0.13	45 / 12		<0.8

OBJECT NAME ALTERN.NAME	COORDINATES (J2000)	TYPE	MAG1 MAG2	MAG3 MAG4	T1 T2	ORB.PER. 2. PER.	3. PER. 4. PER.	EB SB	SPECTR2 SPECTR1	M1/M2	INCL	M1	M2
KO VEL 1E 1013-477	10 15 58.4 / -47 58 11 1	NL IP XS	16.7 / 19.0			0.422	4086: / 5330:	1					
1H 0927+501	09 32 15.0 / +49 50 53 1	NL XS	15.			0.4183							
RZ GRU	22 47 12.2 / -42 44 39 1	NL UX	12.3 / 13.4			0.4170 :		1					
AE AQR	20 40 09.7 / -00 52 16 1	NL DQ XS	10.9 / 11.6	9.8		0.411656	33.062 / 33.0767	2	K5/5	1.13 / 0.06	67 / 4	0.78 / 0.06	0.67 / 0.03
SY CNC PG 0853+181	09 01 03.4 / +17 53 55 1	DN ZC	13.5 / 14.5	11.1 / 12.2	22-35	0.380	QPO	1	G8-9/5	0.81 / 0.25	26 / 6	0.89 / 0.28	1.10 / 0.05
RU PEG	22 14 02.6 / +12 42 11 1	DN UG XS	12.7 / 13.1	9.0	75-85	0.3746	QPO	2	K2-3/5	1.29 / 0.20	33 / 5	1.21 / 0.19	0.94 / 0.04
QZ AUR N AUR 1964	05 28 34.1 / +33 18 22 1	NA	18.:	6.:		0.3575		1					
CH UMA PG 1003+678	10 07 00.7 / +67 32 46 1	DN UG	15.9	10.7	204	0.343		2					
MU CEN	12 12 53.9 / -44 28 17 1	DN UG	14.9	11.8		0.342		1		1.22 / 0.23	>45	1.20 / 0.20	0.99 / 0.03
BT MON N MON 1939	06 43 47.2 / -02 01 14 1	NA	15.4 / 18.1	4.52		0.333814 / 0.333801		1 / 1	K5-7				
RX J0515+01	05 15 / +01	XS	15.			0.3333		1					
V363 AUR LANNING 10	05 33 33.4 / +36 59 32 1	NL UX	14.2 / 15.0			0.321242		1 / 2	K0/5	1.12 / 0.04	70 / 2	0.86 / 0.08	0.77 / 0.04
TT CRT	11 34 47.3 / -11 45 31 1	DN	15.9 / 16.3	12.7		0.30428:		2	K5-M0/5	1.25	58:	0.8	

OBJECT NAME ALTERN.NAME	COORDINATES (J2000)	TYPE	MAG1 MAG2	MAG3 MAG4	T1 T2	ORB.PER. 2. PER.	3. PER. 4. PER.	EB SB	SPECTR2 SPECTR1	M1/M2	INCL	M1	M2
AC CNC	08 44 27.5 +12 52 31 1	NL UX	13.8 15.4		204	0.300478		1 2	G8-K2/5	0.81 0.05	72 3	0.82 0.13	1.02 0.14
V838 HER N HER 1991	18 46 31.5 +12 14 01 1	NA	20.6	5.4		0.297635		1					
EM CYG	19 38 40.2 +30 30 27 1	DN ZC XS	13.3 14.4	12.5 12.9	13-46	0.290909	QPO	1 2	K5/5	0.75 0.09	63 10	0.57 0.08	0.76 0.08
Z CAM	08 25 13.4 +73 06 39 1	DN ZC XS	13.6 14.8	10.5 11.7	19-28	0.289840	QPO	2	K7/5	1.41 0.20	57 11	0.99 0.15	0.70 0.03
V426 OPH	18 07 51.8 +05 51 48 1	DN ZC XS	11.5 13.4		17-55	0.2853		2	K2-4/5	1.29 0.10	59 6	0.90 0.19	0.70 0.14
SS CYG	21 42 43.0 +43 35 09 1	DN UG XS	11.4 12.1	8.2	24-63	0.275130	QPO	2	K5/5	1.69 0.06	37 5	1.19 0.02	0.704 0.002
EI UMA PG 0834+488	08 38 22.1 +48 38 01 1	DN UG XS	14.9B			0.2681		1					
U LEO N LEO 1855	10 24 03.3 +14 00 11 1	N?	17.3	10.5		0.2674 :							
AH HER PG 1642+253	16 44 10.1 +25 15 01 1	DN ZC XS	13.9 14.7	11.3 12. :	7-27	0.258116	QPO	2	K2-M0/5	1.25 0.08	46 3	0.95 0.10	0.76 0.08
1H 0253+193	02 56 08.8 +19 26 35	NL IP XS	>23			0.25270	206.298	1			>78		
TW PIC H 0534-581	05 34 50.8 -58 01 42 1	NL IP XS	14.1 15.9			0.2525 :	7560: 7186:	1					
RU LMI CBS 119	10 02 07.6 +33 50 59 1	DN	17.8 19.5	13.8		0.251							
V751 CYG	20 52 12.9 +44 19 25 1	NL VY	13.2 14.5	16. :		0.25 :		1					

OBJECT NAME / ALTERN.NAME	COORDINATES (J2000)	TYPE	MAG1 / MAG2	MAG3 / MAG4	T1 / T2	ORB.PER. / 2. PER.	3. PER. / 4. PER.	EB / SB	SPECTR2 / SPECTR1	M1/M2	INCL	M1	M2
X 0022-7221	00 24 01	NL DQ?	21.			0.25							
CV IN 47 TUC	-72 04	XS				:							
RW SEX	10 19 56.8	NL UX	10.4			0.24507	QPO			1.35	34	0.8:	0.6:
BD -7 3007	-08 41 59 1	XS	10.8					2		0.10	6		
AT CNC	08 28 37.0	DN ZC	15.0B	12.7B	14	0.238691							
TON 23	+25 20 02 1		16.2B										
TX CCL	05 43 20.3	NL IP	15.7			0.2383	1911				25:	1.3:	0.57:
1H 0642-407	-41 01 56 1	XS						1					
RE J0751+14	07 51 17.3	NL IP	14.1			0.2358	833.72						
RX J0751+14	+14 44 23	XS	14.5			:	870						
DO LEO	10 40 51.3	NL	16.0B			0.234515							
PG 1038+155	+15 11 33 1		17.0B					1					
V347 PUP	06 10 33.6	NL DQ?	13.4			0.231936		1		2.2	87	1.2	0.55
LB 1800	-48 44 27 1	XS	15.8					1		0.2	3	0.1	
RW TRI	02 25 36.2	NL UX	12.6			0.231883		1		0.76	82:	0.44	0.58
	+28 05 51 1		15.6					1		0.13		0.08	0.03
V794 AQL	20 17 34.0	NL VY	13.7	20.2B		0.23				1.7	39	0.88	0.53
	-03 39 52 1	XS	16.5			:		1		0.6	17	0.39	0.07
AF CAM	03 32 15.7	DN	17.0	13.4	75	0.23							
	+58 47 22 1		17.3			:							
TV COL	05 29 25.5	NL IP	13.6			0.228600	1911.	1	K1-5/5	0.75	70	0.75	0.56
2A 0526-328	-32 49 05 1	XS	14.1			0.216278	348045.	1		0.15	3	0.15	
BV PUP	07 49 05.3	DN UG	15.6	13.1	19	0.225							
	-23 34 02 1					:							
AY PSC	01 36 55.5	NL UX	15.3			0.217321							
PG 0134+070	+07 16 29 1		17.0					1					

OBJECT NAME ALTERN.NAME	COORDINATES (J2000)	TYPE	MAG1 MAG2	MAG3 MAG4	T1 T2	ORB.PER. 2. PER.	3. PER. 4. PER.	EB SB	SPECTR2 SPECTR1	M1/M2	INCL	M1	M2
V3885 SGR CD -42 14462	19 47 40.6 -42 00 30 1	NL UX XS	9.6 10.3			0.2163	QPO	1		1.0:	<50	0.8: 0.2	0.7: 0.1
AR CNC	09 22 07.6 +31 03 13 1	DN UG?	18.7 >21.2	15.3		0.2146		1	M4-5/5				
CZ ORI	06 16 43.3 +15 24 11 1	DN UG	16.6 17.0	11.8	22-62	0.2146 :							
HL CMA 1E 0643-1648	06 45 17.0 -16 51 35 1	DN ZC XS	13.2 14.5	11.7	17	0.2145		1			45:	1.0:	0.45 0.10
HR DEL N DEL 1967	20 42 20.2 +19 09 40 1	NB XS *	11.9 13.0	3.3		0.214165 0.1775		1		1.2 0.1	40 2	0.67 0.08	0.55 0.03
PW VUL N VUL 1984 I	19 26 05.0 +27 21 58 1	NA		6.4		0.2137 :							
RX AND	01 04 35.6 +41 17 58 1	DN ZC XS	12.6 14.9	10.9 11.8	5-20	0.209893	QPO	1		2.4 0.7	51 9	1.14 0.33	0.48 0.03
V533 HER N HER 1963	18 14 20.3 +41 51 21 1	NA CP	14.3 16.0	3.0		0.2098 :	63.633T						
V825 HER PG 1717+413	17 18 37.1 +41 15 50 1	NL	14.1 14.4			0.206		1					
T AUR N AUR 1891	05 31 59.1 +30 26 45 1	NB	14.9 15.1	4.1		0.204378		1 1			57:	0.68:	0.63:
FO AQR H 2215-086	22 17 55.5 -08 21 05 1	NL IP XS	13.0 14.0			0.202060	1254.45	1 1			70 5		
HX PEG PG 2337+123	23 40 23.8 +12 37 41 1	DN	12.9 16.6			0.2009		2	K/6				
PG 0943+521	09 47 11.9 +51 54 08 1	NL UX	15.2			0.1997 :							

OBJECT NAME ALTERN.NAME	COORDINATES (J2000)	TYPE	MAG1 MAG2	MAG3 MAG4	T1 T2	ORB.PER. 2. PER.	3. PER. 4. PER.	EB SB	SPECTR2 SPECTR1	M1/M2	INCL	M1	M2
EC 1931-5915	19 35 42.8 -59 08 22 1	NL UX	13.4 14.2			0.198096		1 1					
UX UMA	13 36 41.1 +51 54 49 1	NL UX XS	12.7 14.1			0.196671	QPO	1 1	K8-M6/5	0.91 0.21	57 12	0.43 0.10	0.47 0.03
CT SER N SER 1948	15 45 39.0 +14 22 33 1	N	16.6	7.9		0.1950							
IX VEL CPD -48 1577	08 15 19.1 -49 13 21 1	NL UX	9.1 10.0			0.193929	QPO	2		1.54 0.10	60 5	0.82 0.14	0.53 0.09
DQ HER N HER 1934	18 07 30.2 +45 51 32 1	NA DQ	14.2 17.7	1.4		0.193621	71.0745	1 2	M3/5	1.61 0.13	86.5 1.6	0.60 0.07	0.40 0.05
RX J0203+29	02 03 +29	NL AM XS	17.			0.1913							
SS AUR	06 13 22.5 +47 44 25 1	DN UG	14.5 14.8	10.5	40-75	0.1828		2	M1/5	2.8 1.0	38 16	1.08 0.40	0.39 0.02
TW VIR PG 1142-041	11 45 21.2 -04 26 07 1	DN UG XS	15.8 16.3	12.1	15-44	0.18267		1	M2-4/5	2.3 0.6	43 13	0.91 0.25	0.40 0.02
EY CYG	19 54 36.9 +32 21 54 1	DN UG XS	15.5	11.4	96	0.18123:			K0/5				
BD PAV	18 43 12.0 -57 30 45 1	DN UG	15.4 >16.5	12.4		0.17930		1 2		2.27 0.46	>55		
U GEM	07 55 05.4 +22 00 05 1	DN UG XS	14.0 15.2	9.1	118	0.176906	QPO	1 2	M4.5/5	2.17 0.14	69.7 0.7	1.26 0.12	0.57 0.07
PG 1524+622	15 25 32.0 +62 00 59 1	DN:	15.4			0.1766							
WW CET	00 11 24.8 -11 28 43 1	DN ZC XS	15.0 15.7	9.3 13.9	31	0.1765		2		2.1	54 4	0.85 0.11	0.41 0.01

OBJECT NAME ALTERN.NAME	COORDINATES (J2000)	TYPE	MAG1 MAG2	MAG3 MAG4	T1 T2	ORB.PER. 2. PER.	3. PER. 4. PER.	EB SB	SPECTR2 SPECTR1	M1/M2	INCL	M1	M2
CW MON	06 36 54.6 +00 02 15 1	DN UG	16.3	11.9	122	0.1762		1 2	M3/5				
RX J1313-32	13 13 -32	NL AM XS	16.			0.1750 :							
UZ SER	18 11 25.0 -14 55 35 1	DN UG	15.5 16.0	11.9	10-40	0.1730							
PG 1000+667	10 04 34.6 +66 29 14 2	NL:	15.3			0.169							
VY SCL PS 141	23 29 00.5 -29 46 47 1	NL VY	12.9 13.3	18.5		0.1662 :	QPO	1					
CQ DRA BC 4 DRA BC	12 30 07 +69 12 08					0.1656							
DO DRA PG 1140+719	11 43 38.5 +71 41 19 1	DN IP XS	15.6B 16.7B	10.6B		0.165	529 550	2	M3-5/5	2.1 0.3	42 5	0.83 0.18	0.39 0.08
V1193 ORI	05 16 26.7 -00 12 15 1	NL UX	14.1			0.165		1					
V1776 CYG LANNING 90	20 23 30.6 +46 31 29 1	NL UX SW	16.7 17.6		8-38	0.164739		1 1		1.6	75	0.6	0.37
X LEO	09 51 01.6 +11 52 30 1	DN UG	15.8 15.5	12.4		0.1644	QPO	1	M2/5				
AR AND	01 45 03.3 +37 56 33 1	DN UG	16.9 17.6	11.0	25	0.164 :							
UU AQR S 196	22 09 05.8 -03 46 19 1	NL?	13.5 15.5	9.6		0.163579		1 1			77 6	0.88 0.15	0.39 0.09
CN ORI	05 52 07.8 -05 25 02 1	DN UG *	14.2 16.3	11.9 12.8	8-22	0.163199 0.163190	QPO	2	M4/5	1.51 0.09	67 3	0.74 0.10	0.49 0.08

OBJECT NAME ALTERN.NAME	COORDINATES (J2000)	TYPE	MAG1 MAG2	MAG3 MAG4	T1 T2	ORB.PER. 2. PER.	3. PER. 4. PER.	EB SB	SPECTR2 SPECTR1	M1/M2	INCL	M1	M2
KT PER	01 37 08.8 +50 57 19 1	DN ZC	15.4	10.6 12.3	26	0.163							
KR AUR	06 15 44.0 +28 35 08 1	NL VY XS	11.3 14.5	16.9 >18B		0.16280		1		1.7 0.5	38 10	0.59 0.17	0.35 0.02
CM DEL	20 24 57.0 +17 17 54 1	NL UX	13.4 15.3			0.162		1		1.3 0.4	73: 47	0.48 0.15	0.36 0.03
V380 OPH	17 50 13.7 +06 05 28 1	NL?	14.5 >16.1			0.16		1		1.6 0.5	42 13	0.58 0.19	0.36 0.04
EXO 0329-260	03 32 04.6 -25 56 57 1	NL AM XS	17.5			0.1586		1	M4.5/5				
LX SER	15 38 00.2 +18 52 02 1	NL VY SW	14.5 16.5			0.158432	QPO	1 1		1.14 0.25	90	0.41 0.09	0.36 0.02
IP PEG	23 23 08.7 +18 24 59 1	DN UG	14.0 18.5B	12.B	95	0.158206		2 2	M4/5	1.70 0.12	68	1.15 0.10	0.67 0.08
H 0616-818	06 11 44.5 -81 49 25 1	NL IP XS	13.2			0.158 :	936						
BH LYN PG 0818+513	08 22 36.1 +51 05 24 1	NL VY SW	14.5 16.9			0.155875		1 1					
QQ VUL 1E 2003+225	20 05 42.0 +22 39 58 1	NL AM XS	14.5 15.5			0.154520		1	M2-4/5		60 14		
WY SGE N SGE 1783	19 32 43.8 +17 44 55 1	N DN?	19.0B 21.0B	5.4:		0.153634		1 1					
PG 0859+415	09 03 09.0 +41 17 47 1	NL UX SW	14.3			0.152813		1					
AB DRA	19 49 06.8 +77 44 23 1	DN ZC XS	14.5 15.8	12.3	8-22	0.15198		1					

OBJECT NAME ALTERN.NAME	COORDINATES (J2000)	TYPE	MAG1 MAG2	MAG3 MAG4	T1 T2	ORB.PER. 2. PER.	3. PER. 4. PER.	EB SB	SPECTR2 SPECTR1	M1/M2	INCL	M1	M2
S 193	21 51 58.0	NL DQ?	13.0			0.15 :	1140:						
	+14 06 53 1												
AO PSC	22 55 18.0	NL IP	13.3			0.149626	805.20						
H 2252-035	-03 10 41 1	XS	15.0				858.69	1					
V425 CAS	23 03 46.7	NL VY	14.5			0.1496				2.8	25	0.86	0.31
	+53 17 14 1		18					1		1.0	9	0.32	0.02
PX AND	00 30 05.9	NL VY?	15.0			0.146533		1					
PG 0027+260	+26 17 26 1	SW	17.0					1					
RX J2316-05	23 16	NL AM	18.			0.1451 :							
USS 046	-05	XS				0.138 :							
RR PIC	06 35 36.1	NB XS	12.0	1.2:		0.145026	QPO	1			65:	0.95:	0.4:
N PIC 1925	-62 38 23 1		12.5					1					
VZ SCL	23 50 09.2	NL VY	15.6	>18.		0.144622		1		0.7:	90:		0.4:
	-26 22 53 1		18.1					1					
RX J1007-20	10 07	NL AM	18.			0.1444 :							
	-20	XS											
V442 OPH	17 32 15.2	NL VY	14.0	12.6		0.1406				1.1	67:	0.34	0.31
	-16 15 23 1		15.5							0.3	27	0.10	0.02
UU AQL	19 57 18.8	DN UG	16.1	11.0	71	0.14049:							
	-09 19 22 1		16.7					1					
RX J1940-10	19 40 11.4	NL AM	16.			0.14042:							
	-10 25 25 1	XS	17.					1					
BY CAM	05 42 49.0	NL AM	14.6	>17B		0.1403							
H 0538+608	+60 51 31 1	XS				0.138424		1					
V1223 SGR	18 55 02.3	NL IP	12.3	16.:		0.140244	745.8				21	0.5	0.4
	-31 09 49 1	XS	>16.8				794.38	1			6	0.1	

OBJECT NAME ALTERN.NAME	COORDINATES (J2000)	TYPE	MAG1 MAG2 MAG3 MAG4	T1 T2	ORB.PER. 2. PER. 3. PER. 4. PER.	EB SB	SPECTR2 SPECTR1	M1/M2	INCL	M1	M2
V1315 AQL KPD 1911+121	19 13 54.6 +12 18 02 1	NL UX SW	14.4 16.1		0.139690	1 1	1 1	2.9 1.1	82 4	0.73 0.30	0.30 0.01
V1500 CYG N CYG 1975	21 11 36.6 +48 09 02 1	NA NL AM	17.2 18.6 2.2		0.139613 0.137164	1				>0.9	
WX ARI PG 0244+104	02 47 36.3 +10 35 38 1	NL UX SW	15.3		0.13934	1					
BZ CAM 0623+71	06 29 34.1 +71 04 36 1	NL VY *	12.5B 14.0B		0.1390 : 0.15000:	1					
V1668 CYG N CYG 1978	21 42 35.2 +44 01 55 1	NA	19.9B 21.:B 6.7		0.1384	1					
V603 AQL N AQL 1918	18 48 54.5 +00 35 03 1	NA SH XS	11.4 11.9 -1.1		0.138154 0.14649 3682.8	1		2.3 0.9	17 7	0.66 0.27	0.29 0.02
1H 1929+509	19 34 36.6 +51 07 37 1		17.3		0.138 :						
TT ARI	02 06 53.2 +15 17 42 1	NL VY SH?XS	9.5 12.3 14.5 16.3		0.137551 0.13296 QPO	1					
LY HYA 1329-294	13 31 53.9 -29 41 00 1	DN	14.4 18.4		0.13695	1					
DW UMA PG 1030+590	10 33 53.1 +58 46 54 1	NL UX SW	14.9 16.4		0.136607	1 1			80:	0.9:	0.29:
HL AQR PHL 227	22 20 27.0 +02 00 53 1	NL UX	13.5		0.1356 QPO	1					
SW SEX PG 10-2-029	10 15 09.4 -03 08 35 1	NL UX SW	14.8B 16.7B		0.134938	1 1			79 1	0.58 0.20	0.33 0.06
BG CMI 3A 0729+103	07 31 29.0 +09 56 22 1	NL IP XS	14.3 14.7		0.134749 847.03 913.50	1			33 13	0.8 0.2	0.38

OBJECT NAME ALTERN.NAME	COORDINATES (J2000)	TYPE	MAG1 MAG2	MAG3 MAG4	T1 T2	ORB.PER. 2. PER.	3. PER. 4. PER.	EB SB	SPECTR2 SPECTR1	M1/M2	INCL	M1	M2
MV LYR	19 07 16.4 +44 01 07 1	NL VY * XS	12.1 14.1	17.7 18.B		0.1336 0.1379	QPO	1	M5/5				0.17:
AM HER	18 16 13.4 +49 52 03 1	NL AM XS	12.0 13.5	15.0 15.5		0.128927	QPO	1	M4.5/5	2.4: 0.4	60:	0.39:	0.26:
UZ BOO	14 44 01.5 +22 00 56 1	DN	19.B	11.5	360:	0.125 :							
PG 2133+115	21 36 19.2 +11 40 54 1	NL UX	14.7			0.121 :							
TU MEN	04 41 40.9 -76 36 47 1	DN SU	>16	11.6 12.5	37 194	0.1176 0.1262		1		1.7	65 10	0.6	0.35
V2214 OPH N OPH 1988	17 12 02.6 -29 37 33 1	NA	20.5	8.5		0.117515							
DR V211B	20 08 56.5 -65 27 22 1	NL AM	18			0.1111							
V795 HER PG 1711+336	17 12 56.5 +33 31 19 1	NL SH	12.5B 13.2B			0.108265 0.116486		1					
V PER N PER 1887	02 01 53.7 +56 44 04 1	N NL	18.5	9.2		0.10712		1					
V348 PUP 1H 0709-360	07 12 32.9 -36 05 40 1	NL2DQ? XR	15.5 17.0			0.101840		1 1					
RX J1938-46 RE J1938-461	19 38 35.6 -46 12 57 1	NL AM XS	15.2 15.8			0.09723		1					
CC CNC	08 36 19.3 +21 06 1	DN UG	17.4	13.1B		0.0942 :		1					
RX J0531-46 RE J0531-46	05 31 35.8 -46 24 07	NL AM XS	17.0			0.0924 :							

OBJECT NAME / ALTERN.NAME	COORDINATES (J2000)	TYPE	MAG1 / MAG2	MAG3 / MAG4	T1 / T2	ORB.PER. / 2.PER.	3.PER. / 4.PER.	EB / SB	SPECTR2 / SPECTR1	M1/M2	INCL	M1	M2
UZ FOR	03 35 28.7	NL AM	18.2			0.087865		1	M4.5/5	9	86	1.26	0.14
EXO 0333-255	-25 44 23 1	XS	20.5					1		1	3	0.18	
EU CNC	08 51 27.6	NL AM?	20.4			0.0871							
	+11 46 45	XS	21.0										
DM DRA	15 34 12.3	DN UG	20.8	15.5		0.087 :							
	+59 48 31 1												
RX J2137-05	21 07 58.3	NL AM	15.3			0.086820		1	M4.5/5	<5.6		0.95	0.15
RE J2137-05	-05 17 39 1	XS	20.										0.02
YZ CNC	08 10 56.7	DN SU	14.1	11.9	6-16	0.0868	QPO			4.5	38	0.82	0.17
	+28 08 33 1	XS	15.5	10.5	134	0.09204		1			3	0.08	
DV UMA	09 46 36.8	DN SU?	18.6	15.4		0.08597		2	M4.5/5:				
US 943	+44 46 41 1		20.6										
EF PEG	21 15 04.2	DN SU	18.5	10.7		0.0854 *				3.8		0.65:	0.17:
	+14 03 50 1				254	0.0871							
TY PSA	22 49 39.9	DN SU	16.			0.0841	QPO	D					
PS 74	-27 06 55 1		17.			0.08765		1					
KK TEL	20 28 38.5	DN UG	19.3	13.5		0.084							
	-52 18 47 1												
HS VIR	13 43 38.5	DN UG	15.8	13.0		0.0836 :							
PG 1341-079	-08 14 04 1												
1ES 1113+432	11 15 47	NL AM	13.	16.5		0.0805							
	+42 58 50	XS						1					
WW HOR	02 36 11.6	NL AM	17.6			0.080199		1	M6/5:				
EXO 0234-523	-52 19 15 1	XS	21.2					1					
AN UMA	11 04 25.8	NL AM	14.5	16.0		0.079753	QPO						
PG 1101+453	+45 03 14 1	XS		18.9B				1					

OBJECT NAME / ALTERN.NAME	COORDINATES (J2000)	TYPE	MAG1 / MAG2	MAG3 / MAG4	T1 / T2	ORB.PER. / 2. PER.	3. PER. / 4. PER.	EB / SB	SPECTR2 / SPECTR1	M1/M2	INCL	M1	M2
EK UMA / 1E 1048+542	10 51 35.4 / +54 04 31 1	NL AM / XS	18.0 / 20.0			0.07948					56 / 18		
BR LUP	15 35 53 / -40 34 07	DN SU	>17.5	13.1		0.0793 * / 0.0822							
ST LMI / CW 1103+254	11 05 39.8 / +25 06 28 1	NL AM / XS		15.0 / 17.2		0.079089		/ 1	M5-6/5		56 / 4	0.54 / 0.20	
RW UMI / N UMI 1956	16 47 55.6 / +77 01 41 1	NB	18.8 / 21.	6.0		0.079 :							
BL HYI / H 0139-68	01 41 00.4 / -67 53 29 1	NL AM / XS	14.3 / 16.4	16.9 / 17.4		0.078915	QPO	/ 1.	M3-4/5		70 / 10		
MR SER / PG 1550+191	15 52 47.3 / +18 56 27 1	NL AM / XS	14.9 / 15.8	17.		0.078798		/ 1	M5-6/5		35 / 5		
1H 1752+081	18 00 35.6 / +08 10 12 1	NL XS	14.			0.0783		/ 1					
CU VEL	08 58 33.0 / -41 47 52 1	DN SU	15.5	10.7	113 / 386	0.0773 * / 0.0799							
TT BOO	14 57 44.8 / +40 43 40 1	DN SU	<15.6	12.7	45	0.077 :: / 0.067 ::							
SU UMA / PG 0808+627	08 12 28.3 / +62 36 22 1	DN SU / XS	14.2 / 15.0	12.2 / 11.2	5-33 / 160	0.07635 / 0.07904		/ 1					
AW GEM	07 22 40.9 / +28 30 13 1	DN SU	18.8 / 19.4	13.8 / 13.1	98 / 410	0.0762 * / 0.07867							
V503 CYG	20 27 17.5 / +43 41 23 1	DN SU	17.4	13.4	28	0.07599							
CE GRU / GRU V1	21 37 56.5 / -43 42 14 1	NL AM	18.0 / 18.5	20.7B		0.0754							

OBJECT NAME / ALTERN NAME	COORDINATES (J2000)	TYPE	MAG1 / MAG2	MAG3 / MAG4	T1 / T2	ORB.PER. / 2. PER.	3. PER. / 4. PER.	EB / SB	SPECTR2 / SPECTR1	M1/M2	INCL	M1	M2
VZ PYX	08 59 20.0	DN	16.8	12.5		0.075							
1H 0857-242	-24 28 56 1	XS					3108						
WX HYI	02 09 50.1	DN SU	14.7	12.5	14	0.074813				5.5	40	0.9	0.16
	-63 18 41 1	XS	14.8	11.4	140	0.07735		1		1.5	10	0.3	0.05
Z CHA	08 07 28.6	DN SU	15.3	12.4	82	0.074499	QPO	2	M5.5/5	6.7	81.8	0.84	0.125
	-76 32 02 1	XS	17.2	11.9	287	0.07740		2		0.2	0.1	0.09	0.014
VW HYI	04 09 12.1	DN SU	13.4	9.5	27	0.074271	QPO			6	60	0.63	0.11
	-71 17 46 1	XS	13.8	8.5	179	0.07714		1		1	10	0.15	0.02
V1251 CYG	21 40 52.7	DN SU	18.5			0.074 *							
	+48 39 53 1			12.5		0.076							
HT CAS	01 10 13.2	DN SU	16.4	10.8	400	0.073647	QPO	2		6.7	81	0.61	0.09
	+60 04 35 1	XS	18.4			0.076077		1		1.4	1	0.04	0.02
RX J1002-19	10 02	NL AM?	17.			0.0736							
	-19	XS											
AY LYR	18 44 26.8	DN SU	18.4B	13.2	8-43	0.07340*				1.7	44	0.24	0.14
	+37 59 51 1	XS		12.3	205	0.07597				0.4	12	0.06	0.01
VW VUL	20 57 45.1	DN UG?	15.6	13.6	14-24	0.0731		1	M5				
	+25 30 25 1												
T PYX	09 04 41.5	NR	15.3B	6.5		0.073 :							
	-32 22 47 1		15.6B			0.0762		1					
EP DRA	19 07 06.9	NL AM	17.6			0.072656		1			80:		
H 1907+690	+69 08 40 1	XS											
PG 0917+342	09 20 11.3	NL	14.1			0.072 :		1					
CBS 96	+33 56 41 1		15.1										
FO AND	01 15 32.2	DN SU	17.5	13.5		0.07161							
	+37 37 35 1					0.073 :							

OBJECT NAME ALTERN.NAME	COORDINATES (J2000)	TYPE	MAG1 MAG2 MAG3 MAG4	T1 T2	ORB.PER. 2. PER. 3. PER. 4. PER.	EB SB	SPECTR2 SPECTR1	M1/M2	INCL	M1	M2
GD 552	22 50 39.8 +63 28 38 1	DN?	16.5		0.07134			14. 3.	20:	1.4:	0.1:
RZ LEO	11 37 16.1 +01 49 21 1	DN SU?	19 11.5		0.0708	1					
V834 CEN 1E 1405-451	14 09 07.6 -45 17 18 1	NL AM XS	14.2 16.0 17.		0.070498 QPO	1	M6.5/5		50 10	0.66 0.18	
VV PUP	08 15 06.8 -19 03 18 1	NL AM XS	14.5 16.0 17.5 18.0		0.069747 QPO	1	M4/5:	5.5	78 5		
V544 HER	16 38 05.5 +08 37 58 1	DN UG	20.: 14.5		0.069 :						
RX J1957-57	19 57 -57	NL AM? XS	17.		0.0688						
RZ SGE	20 03 18.5 +17 02 51 1	DN SU	16.9 17.4B 12.8 12.2	62-93 266	0.0686 * 0.07035	1					
IR GEM	06 47 34.7 +08 06 22 1	DN SU	16.3 17.0 11.7 11.2	22-48 150	0.0684 0.07076	1					
SS UMI PG 1551+719	15 51 22.4 +71 45 11 1	DN SU	16.9 17.6 12.6	30-48	0.0684 * 0.0701						
TY PSC	01 25 39.4 +32 23 09 1	DN SU	15.3 16.3 12.2 11.7	11-35 370	0.06824 0.070 :						
EX HYA	12 52 24.4 -29 14 56 2	NL IP XS	13.0 14.1 10.0	574	0.068234 4021.62	1 1	M5.5/5:	6.0 1.3	78 1	0.78 0.17	0.13 0.01
BZ UMA PG 0849+580	08 53 44.3 +57 48 40 1	DN	17.8 10.5		0.0679	1	M5.5/5				
RX J0453-42 RE J0453-42	04 53 24.1 -42 14 07	NL AM XS	19.		0.0660						

OBJECT NAME ALTERN.NAME	COORDINATES (J2000)	TYPE	MAG1 MAG2	MAG3 MAG4	T1 T2	ORB.PER. 2. PER.	3. PER. 4. PER.	EB SB	SPECTR2 SPECTR1	M1/M2	INCL	M1	M2
CF GRU 2138-453	21 41 23.0 -45 04 31 1	DN	19.9 20.2			0.065							
FY PER	04 41 56.7 +50 42 36 1	NL?	11.0 14.5			0.064848							
EK TRA	15 14 01.2 -65 05 36 1	DN SU XS	>17	12.1 12.0	231 487	0.0636 * 0.06492							
OY CAR	10 06 22.6 -70 14 06 1	DN SU	15.3 17.3	12.4 11.4	25-50 300	0.063121 0.064631		2 1	M6/5	9.8 0.3	83.3 0.6	0.685 0.011	0.070 0.002
BC UMA	11 52 16.0 +49 14 41 1	DN UG	18.6			0.063							
V436 CEN	11 14 00.1 -37 40 47 1	DN SU XS	15.3 15.5	12.4 11.3	22 335	0.062501 0.063785	QPO	1		4 1	65: 5	0.7: 0.1	0.17:
RX J1844-74 RE J1844-74	18 44 52.2 -74 18 39	NL AM XS	17.6			0.0625							
SX LMI CBS 31	10 54 30.5 +30 06 09 1	NL	16.B			0.0625 :							
RE J1149+28	11 49 55.7 +28 45 08	NL AM XS	16.5B 16.8B			0.0625 :							
V2051 OPH	17 08 19.2 -25 48 35 1	DN UG?	15.0 17.5	13.0		0.062428	QPO	2 1		3.4 0.6	80.5 2.0	0.44 0.05	0.13 0.04
DP LEO 1E 1114+182	11 17 16.1 +17 57 37 1	NL AM XS	17.5B 19.5B	19.5 >22		0.062363		1		6.7	79.6	0.71	0.11
UV PER	02 10 08.2 +57 11 21 1	DN SU	17.5	11.7	360	0.0622 : 0.06641							
CP PUP N PUP 1942	08 11 46.0 -35 21 05 1	NA SH? XS	15.0	0.2		0.06143 0.06834		1					

OBJECT NAME / ALTERN.NAME	COORDINATES (J2000)	TYPE	MAG1 / MAG2	MAG3 / MAG4	T1 / T2	ORB.PER. / 2. PER.	3. PER. / 4. PER.	EB / SB	SPECTR2 / SPECTR1	M1/M2	INCL	M1	M2
V4140 SGR	19 58 49.8 13	DN SU?	17.5	15.5		0.061430		1					
NSV 12615	-38 56 13 1		19.0					1					
AL COM	12 32 25.6	DN	20.8	12.8	325	0.061 :							
	+14 20 58 1												
AQ ERI	05 06 13.1	DN SU	17.7		40:	0.06094							
	-04 08 08 1			12.5		0.06225							
CI UMA	10 18 13.2	DN	18.8	13.8		0.060							
	+71 55 42 1												
GQ MUS	11 52 02.4	NA AM?	17.5	7.2		0.0592		1					
N MUS 1983	-67 12 20 1												
VY AQR	21 12 09.3	DN SU	17.1	8.0B		0.059 :							
	-08 49 37 1		17.5	10.5		0.0644		1					
FS AUR	05 47 48.4	DN UG	16.2	14.4		0.059 :							
	+28 35 10 1												
T LEO	11 38 27.0	DN SU	15.2	11.0	450	0.05882:				1.4	65	0.16	0.11
PG 1135+036	+03 22 07 1		15.7			0.06411:		1		0.3	19	0.04	0.01
V1159 ORI	05 28 59.5	DN	16.0	12.5	5:	0.05872:							
	-03 33 53 2							1					
CY UMA	10 56 57.1	DN SU	17.0	11.9	115:	0.0583 *							
	+49 41 18 1				297:	0.0593							
HV VIR	13 21 03.1	DN SU	19.0	11.5		0.05799							
N VIR 1929	+01 53 30 1					0.05879							
SW UMA	08 36 42.9	DN SU	16.5	10.6		0.056815	954			7.1	45	0.71	0.10
	+53 28 37 1	DQ?XS	17.0	9.	459	0.05833		1		2.0	18	0.22	0.01
WZ SGE	20 07 36.5	DN SU	14.9	7.:		0.056688	27.8682	1		8.7	72	0.45	0.058
	+17 42 15 1	CP XS	15.5		11876	0.05714		1		1.1	2	0.19	0.023

OBJECT NAME ALTERN.NAME	COORDINATES (J2000)	TYPE	MAG1 MAG2	MAG3 MAG4	T1 T2	ORB.PER. 2. PER.	3. PER. 4. PER.	EB SB	SPECTR2 SPECTR1	M1/M2	INCL	M1	M2
EF ERI 2A 0311-227	03 14 13.1 -22 35 42 1	NL AM XS	13.7 15.5	16.5B 17.7B		0.056266	QPO	1			70 5		
HV AND	00 40 55.4 +43 24 58 1	NL	15.3 16.8B			0.05599:							
RX J1307+53 RE J1307+53	13 07 56.4 +53 51 37	NL AM XS	17.			0.05534							
WX CET N CET 1963	01 17 04.2 -17 56 24 1	DN SU	17.5 18.5	9.5	450	0.052 : 0.053 :							
V485 CEN	12 57 23.4 -33 12 08 1	DN UG	18.2	14.	148	0.041 :							
GP COM G61-29	13 05 43.4 +18 01 02 1	NL AC XS	15.7 16.0			0.03231	QPO	1	DBE				
CP ERI	03 10 32.8 -09 45 06 1	NL AC	19.7	16.5		0.01995							
V803 CEN AE 1	13 23 45 -41 44 34 A	NL AC?	13.2 16.8			0.01865:	175		DBP				
CR BOO PG 1346+082	13 48 55.3 +07 57 34 A	NL AC *	13.0 17.5			0.01725:			DBP				
AM CVN HZ 29	12 34 54.4 +37 37 43 1	NL AC SH?XS	14.1 14.2			0.011907 0.012165		1	DBP				
AH ERI	04 22 38.2 -13 21 31 1	DN UG XR	18.4	13.5			2500						

REFERENCES

RX AND   1 SHAFTER,A.W.:1983,PH.D. THESIS,UCLA (M1/M2,INCL,M1,M2)
          2 KAITCHUCK,R.H.:1989,PUBL.ASTRON.SOC.PACIFIC 101,1129
            (ORB.PER.)
          3 SZKODY,P.,PICHE,F.,FEINSWOG,L.:1990,ASTROPHYS.J.SUPPL.
            SER.73,441
          4 WOODS,J.A.,DREW,J.E.,VERBUNT,F.:1990,MONTHLY NOTICES ROY.
            ASTRON.SOC.245,323

AR AND   1 SZKODY,P.,PICHE,F.,FEINSWOG,L.:1990,ASTROPHYS.J.SUPPL.
            SER.73,441 (ORB.PER.)

DX AND   1 DREW,J.E.,HOARE,G.,WOODS,J.A.:1991,MONTHLY NOTICES ROY.
            ASTRON.SOC.250,144
          2 DREW,J.E.,JONES,D.H.P.,WOODS,J.A.:1993,MONTHLY NOTICES
            ROY.ASTRON.SOC.260,803 (M1/M2,INCL,M2)
          3 VRIELMANN,S.,BRUCH,A.:1993,ASTRON.GES.ABSTRACT.SER.9,154
            (ORB.PER.)

FO AND   1 SZKODY,P.,HOWELL,S.B.,MATEO,M.,KREIDL,T.J.N.:1989,PUBL.
            ASTRON.SOC.PACIFIC 101,899 (2. PER.)
          2 THORSTENSEN,J.R.,PATTERSON,J.,THOMAS,G.:1992,IN PREPAR-
            ATION (ORB.PER.)

HV AND   1 ANDRONOV,I.L.,BANNY,M.I.:1985,INF.BULL.VARIABLE STARS
            NO.2763 (ORB.PER.)
          2 SCHWOPE,A.D.,REINSCH,K.:1992,INF.BULL.VARIABLE STARS
            NO.3725

PX AND   1 LI,Y.,JIANG,Z.,CHEN,J.,WEI,M.:1990,INF.BULL.VARIABLE
            STARS NO.3434
          2 LI,Y.,JIANG,Z.,CHEN,J.,WEI,M.:1990,CHIN.ASTRON.ASTROPHYS.
            14,359
          3 THORSTENSEN,J.R.,RINGWALD,F.A.,WADE,R.A.,SCHMIDT,G.D.,
            NORSWORTHY,J.E.:1991,ASTRON.J.102,272 (ORB.PER.)

UU AQR   1 GOLDADER,J.D.,GARNAVICH,P.:1989,INF.BULL.VARIABLE.STARS
            NO.3361 (ORB.PER.)
          2 HAEFNER,R.:1989,INF,BULL.VARIABLE STARS NO.3397
          3 DIAZ,M.P.,STEINER,J.E.:1991,ASTRON.J.102,1417 (INCL,M1,
            M2)

VY AQR   1 AUGUSTEIJN,T.,DELLA VALLE,M.:1990,IAU CIRC.NO.5048
          2 DELLA VALLE,M.,AUGUSTEIJN,T.:1990,THE MESSENGER 61,41
            (ORB.PER.)
          3 PATTERSON,J.,BOND,H.E.,GRAUER,A.D.,SHAFTER,A.W.,
            MATTEI,J.A.:1993,PUBL.ASTRON.SOC.PACIFIC 105,69
            (2. PER.)

AE AQR   1 PATTERSON,J.:1979,ASTROPHYS.J.234,978 (4. PER.)
          2 VAN PARADIJS,J.A.,KRAAKMAN,H.,VAN AMERONGEN,S.:1989,
            ASTRON.ASTROPHYS.SUPPL.SER.79,205 (ERRATUM IN
            ASTRON.ASTROPHYS.SUPPL.SER.90,194)
          3 BRUCH,A.:1991,ASTRON.ASTROPHYS.251,59
          4 DE JAGER,O.C.:1991,ASTROPHYS.J.378,286
          5 ERACLEOUS,M.,PATTERSON,J.,HALPERN,J.:1991,ASTROPHYS.J.
            370,330 (3. PER.)
          6 ROBINSON,E.L.,SHAFTER,A.W.,BALACHANDRAN,S.:1991,
            ASTROPHYS.J.374,298 (M1/M2,INCL,M1,M2)

AE AQR   7 WELSH,W.F.,HORNE,K.,GOMER,R.:1993,ASTROPHYS.J.LETTERS
         410,L39 (ORB.PER.)

FO AQR   1 CHIAPPETTI,L.,BELLONI,T.,BONNET-BIDAUD,J.-M.,
         DEL GRATTA,C.,DE MARTINO,D.,MARASCHI,L.,MOUCHET,M.,
         MUKAI,K.,OSBORNE,J.P.,CORBET,R.H.D.,TANZI,E.G.,
         TREVES,A.:1989,ASTROPHYS.J.342,493 (ORB.PER.)
     2 HELLIER,C.,MASON,K.O.,CROPPER,M.:1989,MONTHLY NOTICES
         ROY.ASTRON.SOC.237,39P (INCL)
     3 OSBORNE,J.P.,MUKAI,K.:1989,MONTHLY NOTICES ROY.ASTRON.
         SOC.238,1233 (4. PER)
     4 STEIMAN-CAMERON,T.Y.,IMAMURA,J.N.,STEIMAN-CAMERON,D.V.:
         1989,ASTROPHYS.J.339,434 (4. PER)
     5 HELLIER,C.,MASON,K.O.,CROPPER,M.:1990,MONTHLY NOTICES
         ROY.ASTRON.SOC.242,250
     6 MARTELL,P.J.,KAITCHUCK,R.H.:1991,ASTROPHYS.J.366,286
     7 MUKAI,K.,HELLIER,C.:1992,ASTROPHYS.J.391,295
     8 NORTON,A.J.,WATSON,M.G.,KING,A.R.,LEHTO,H.J.,
         MC HARDY,I.M.:1992,MONTHLY NOTICES ROY.ASTRON.SOC.
         254,705
     9 KRUSZEWSKI,A.,SEMENIUK,I.:1993,ACTA ASTRON.43,127

HL AQR   1 HAEFNER,R.,SCHOEMBS,R.:1987,MONTHLY NOTICES ROY.ASTRON.
         SOC.224,231 (ORB.PER.)

UU AQL   1 THORSTENSEN,J.R.:1985,PRIVATE COMMUNICATION (ORB.PER.)
     2 SZKODY,P.:1987,ASTROPHYS.J.SUPPL.SER.63,685

V603 AQL   1 SHAFTER,A.W.:1983,PH.D. THESIS,UCLA (M1/M2,INCL,M1,M2)
     2 HAEFNER,R.,METZ,K.:1985,ASTRON.ASTROPHYS.145,311
         (ORB.PER.)
     3 UDALSKI,A.,SCHWARZENBERG-CZERNY,A.:1989,ACTA ASTRON.
         39,125 (3. PER.)
     4 BRUCH,A.:1991,ACTA ASTRON.41,101
     5 PATTERSON,J.,RICHMAN,H.:1991,PUBL.ASTRON.SOC.PACIFIC
         103,735
     6 SCHWARZENBERG-CZERNY,A.,UDALSKI,A.,MONIER,R.:1992,
         ASTROPHYS.J.LETTERS 401,L19
     7 PATTERSON,J.,THOMAS,G.,SKILLMAN,D.R.,DIAZ,M.:1993,
         ASTROPHYS.J.SUPPL.SER.86,235 (2. PER.)

V794 AQL   1 SHAFTER,A.W.:1983,PH.D. THESIS,UCLA (M1/M2,INCL,M1,M2)
     2 SHAFTER,A.W.:1983,INF.BULL.VARIABLE STARS NO.2377
         (ORB.PER.)
     3 MUKAI,K.,MASON,K.O.,HOWELL,S.B.,ALLINGTON-SMITH,J.,
         CALLANAN,P.J.,CHARLES,P.A.,HASSALL,B.J.M.,MACHIN,G.,
         NAYLOR,T.,SMALE,A.P.,VAN PARADIJS,J.:1990,MONTHLY
         NOTICES ROY.ASTRON.SOC.245,385

V1315 AQL   1 SZKODY,P.,PICHE,F.:1990,ASTROPHYS.J.361,235
     2 DHILLON,V.S.,MARSH,T.R.,JONES,D.H.P.:1991,MONTHLY NOTICES
         ROY.ASTRON.SOC.252,342 (ORB.PER.,M1/M2,INCL,M1,M2)
     3 RUTTEN,R.G.M.,VAN PARADIJS,J.,TINBERGEN,J.:1992,ASTRON.
         ASTROPHYS.260,213

TT ARI   1 THORSTENSEN,J.R.,SMAK,J.,HESSMAN,F.V.:1985,PUBL.ASTRON.
         SOC.PACIFIC 97,437 (ORB.PER.)
     2 UDALSKI,A.:1988,ACTA ASTRON.38,315 (2. PER.)
     3 HOLLANDER,A.,VAN PARADIJS,J.:1992,ASTRON.ASTROPHYS.
         265,77

WX ARI   1 BEUERMANN,K.,THORSTENSEN,J.R.,SCHWOPE,A.D.,RINGWALD,F.,
         SAHIN,H.:1992,ASTRON.ASTROPHYS.256,442 (ORB.PER.)

T   AUR   1 BIANCHINI,A.:1980,MONTHLY NOTICES ROY.ASTRON.SOC.192,127
                 (INCL,M1,M2)
               2 BEUERMANN,K.,PAKULL,M.W.:1984,ASTRON.ASTROPHYS.136,250
                 (ORB.PER.)

SS AUR   1 SHAFTER,A.W.:1983,PH.D. THESIS,UCLA (M1/M2,INCL,M1,M2)
               2 SHAFTER,A.W.,HARKNESS,R.P.:1986,ASTRON.J.92,658
                 (ORB.PER.)
               3 FRIEND,M.T.,MARTIN,J.S.,SMITH,R.C.,JONES,D.H.P.:1990,
                 MONTHLY NOTICES ROY.ASTRON.SOC.246,637

FS AUR   1 HOWELL,S.B.,SZKODY,P.:1988,PUBL.ASTRON.SOC.PACIFIC
                 100,224 (ORB.PER.)

KR AUR   1 SHAFTER,A.W.:1983,ASTROPHYS.J.267,222 (ORB.PER.)
               2 SHAFTER,A.W.:1983,PH.D. THESIS,UCLA (M1/M2,INCL,M1,M2)

QZ AUR   1 CAMPBELL,R.D.,SHAFTER,A.W.:1992,IN: PROC. 12TH NORTH
                 AMERICAN WORKSHOP ON CVS AND LMXRBS,A.W.SHAFTER
                 (ED.),MOUNT LAGUNA OBSERVATORY,SDSU,P.4 (ORB.PER.)

V363 AUR   1 SCHLEGEL,E.M.,HONEYCUTT,R.K.,KAITCHUCK,R.H.:1986,
                 ASTROPHYS.J.307,760 (ORB.PER.,M1/M2,INCL,M1,M2)
               2 RUTTEN,R.G.M.,VAN PARADIJS,J.,TINBERGEN,J.:1992,ASTRON.
                 ASTROPHYS.260,213

TT BOO   1 HOWELL,S.B.,SZKODY,P.:1988,PUBL.ASTRON.SOC.PACIFIC
                 100,224 (ORB.PER.,2. PER.)

UZ BOO   1 SZKODY,P.:1987,ASTROPHYS.J.SUPPL.SER.63,685 (ORB.PER.)

CR BOO   1 WOOD,M.A.,WINGET,D.E.,NATHER,R.E.,HESSMAN,F.V.,
                 LIEBERT,J.,KURTZ,D.W.,WESEMAEL,F.,WEGNER,G.:
                 1987,ASTROPHYS.J.313,757 (ORB.PER.)

Z   CAM   1 ROBINSON,E.L.:1973,ASTROPHYS.J.186,347 (ORB.PER.)
               2 SHAFTER,A.W.:1983,PH.D. THESIS,UCLA (M1/M2,INCL,M1,M2)
               3 SZKODY,P.,MATEO,M.:1986,ASTROPHYS.J.301,286

AF CAM   1 SZKODY,P.,HOWELL,S.B.:1989,ASTRON.J.97,1176 (ORB.PER.)
               2 LONG,K.S.,BLAIR,W.D.,DAVIDSEN,A.F.,BOWYERS,C.W.,
                 VAN DYKE DIXON,W.,DURRANCE,S.T.,FELDMAN,P.D.,
                 HENRY,R.C.,KRISS,G.A.,KRUK,J.W.,MOOS,H.W.,VANCURA,O.
                 FERGUSON,H.C.,KIMBLE,R.A.:1991,ASTROPHYS.J.LETTERS
                 381,L25

BY CAM   1 MASON,P.A.,LIEBERT,J.,SCHMIDT,G.D.:1989,ASTROPHYS.J.
                 346,941 (2. PER.)
               2 SZKODY,P.,DOWNES,R.A.,MATEO,M.:1990,PUBL.ASTRON.SOC.
                 PACIFIC 102,1310
               3 ISHIDA,M.,SILBER,A.,BRADT,H.V.,REMILLARD,R.A.,
                 MAKISHIMA,K.,OHASHI,T.:1991,ASTROPHYS.J.367,270
               4 SILBER,A.,BRADT,H.V.,ISHIDA,M.,OHASHI,T.,REMILLARD,R.A.:
                 1992,ASTROPHYS.J.389,704 (ORB.PER.)

BZ CAM   1 LU,W.,HUTCHINGS,J.B.:1985,PUBL.ASTRON.SOC.PACIFIC 97,990
                 (ORB.PER.)
               2 WOODS,J.A.,DREW,J.E.,VERBUNT,F.:1990,MONTHLY NOTICES ROY.
                 ASTRON.SOC.245,323
               3 PAJDOSZ,G.,ZOLA,S.:1992,IN:EVOLUTIONARY PROCCESSES IN
                 INTERACTING BINARY STARS,IAU SYMP.NO.151,Y.KONDO,
                 R.F.SISTERO AND R.S.POLIDAN (EDS.),KLUWER ACADEMIC
                 PUBL.,P.411 (2. PER.)

BZ CAM   4 HOLLIS,J.M.,OLIVERSEN,R.J.,WAGNER,R.M.,FEIBELMAN,W.A.:
         1992,ASTROPHYS.J.393,217

SY CNC   1 SHAFTER,A.W.:1983,PH.D. THESIS,UCLA (ORB.PER.,M1/M2,INCL,
         M1,M2)

YZ CNC   1 PATTERSON,J.:1979,ASTRON.J.84,804 (2. PER.)
       2 SHAFTER,A.W.,HESSMAN,F.V.:1988,ASTRON.J.95,178 (ORB.PER.,
         M1/M2,INCL,M1,M2)

AC CNC   1 OKAZAKI,A.,KITAMURA,M.,YAMASAKI,A.:1982,PUBL.ASTRON.SOC.
         PACIFIC 94,162 (ORB.PER.)
       2 SCHLEGEL,E.M.,KAITCHUCK,R.H.,HONEYCUTT,R.K.:1984,
         ASTROPHYS.J.280,235 (M1/M2,INCL,M1,M2)

AR CNC   1 HOWELL,S.B.,SZKODY,P.,KREIDL,T.J.,MASON,K.O.,
         PUCHNAREWICZ,E.M.:1990,PUBL.ASTRON.SOC.PACIFIC
         102,758 (ORB.PER.)
       2 MUKAI,K.,MASON,K.O.,HOWELL,S.B.,ALLINGTON-SMITH,J.,
         CALLANAN,P.J.,CHARLES,P.A.,HASSALL,B.J.M.,MACHIN,G.,
         NAYLOR,T.,SMALE,A.P.,VAN PARADIJS,J.:1990,MONTHLY
         NOTICES ROY.ASTRON.SOC.245,385
       3 HOWELL,S.B.,BLANTON,S.A.:1993,ASTRON.J.106,311

AT CNC   1 GOETZ,W.:1986,INF.BULL.VARIABLE STARS NO.2918 (ORB.PER.)

CC CNC   1 HOWELL,S.B.,SZKODY,P.,KREIDL,T.J.,MASON,K.O.,
         PUCHNAREWICZ,E.M.:1990,PUBL.ASTRON.SOC.PACIFIC
         102,758
       2 MUNARI,U.,BIANCHINI,A.,CLAUDI,R.:1990,IAU CIRC.NO.5024
         (ORB.PER.)

EU CNC   1 GILLILAND,R.L.,BROWN,T.M.,DUNCAN,T.K.,SUNTZEFF,N.B.,
         LOCKWOOD,G.W.,THOMPSON,D.T.,SCHILD,R.E.,
         JEFFREY,W.A.,PENPRASE,B.E.:1991,ASTRON.J.101,541
         (ORB.PER.)
       2 BELLONI,T.,VERBUNT,F.,SCHMITT,J.H.M.M.:1993,ASTRON.
         ASTROPHYS.269,175

AM CVN   1 SOLHEIM,J.-E.,ROBINSON,E.L.,NATHER,R.E.,KEPLER,S.O.:1984,
         ASTRON.ASTROPHYS.135,1 (2. PER.)
       2 KRUSZEWSKI,A.,SEMENIUK,I.:1992,ACTA ASTRON.42,311
       3 PATTERSON,J.,STERNER,E.,HALPERN,J.P.,RAYMOND,J.C.:1992,
         ASTROPHYS.J.384,234
       4 PATTERSON,J.,HALPERN,J.P.,SHAMBROOK,A.:1993,BULL.AMERICAN
         ASTRON.SOC.25,917 (ORB.PER.)

WY CMA   1 HACKE,G.,RICHERT,M.:1990,VEROEFFENTL.STERNWARTE SONNEBERG
         10,336 (ORB.PER.)

HL CMA   1 HUTCHINGS,J.B.,COWLEY,A.P.,CRAMPTON,D.,WILLIAMS,G.:1981,
         PUBL.ASTRON.SOC.PACIFIC 93,741 (INCL,M1,M2)
       2 WARGAU,W.,BRUCH,A.,DRECHSEL,H.,RAHE,J.:1983,ASTRON.
         ASTROPHYS.125,L1 (ORB.PER.)

BG CMI   1 PENNING,W.R.:1985,ASTROPHYS.J.289,300 (INCL,M1,M2)
       2 CHANMUGAM,G.,FRANK,J.,KING,A.R.,LASOTA,J.-P.:1990,
         ASTROPHYS.J.LETTERS 350,L13
       3 AUGUSTEIJN,T.,VAN PARADIJS,J.,SCHWARZ,H.E.:1991,ASTRON.
         ASTROPHYS.247,64 (ORB.PER.,4. PER.)
       4 SINGH,J.,AGRAWAL,P.C.,APPARAO,K.M.V.,VIVEKANANDA RAO,P.,
         SARMA,M.B.K.:1991,ASTROPHYS.J.380,208

BG CMI   5 NORTON,A.J.,MC HARDY,I.M.,LEHTO,H.J.,WATSON,M.G.:1992,
               MONTHLY NOTICES ROY.ASTRON.SOC.258,697 (3. PER.)
             6 PATTERSON,J.,THOMAS,G.:1993,PUBL.ASTRON.SOC.PACIFIC
               105,59

OY CAR   1 KRZEMINSKI,W.,VOGT,N.:1985,ASTRON.ASTROPHYS.144,124
               (2. PER.)
             2 WOOD,J.H.,HORNE,K.,BERRIMAN,G.,WADE,R.A.:1989,
               ASTROPHYS.J.341,974 (ORB.PER.,M1/M2,INCL,M1,M2)
             3 WOOD,J.H.:1990,MONTHLY NOTICES ROY.ASTRON.SOC.243,219
             4 HESSMAN,F.V.,MANTEL,K.-H.,BARWIG,H.,SCHOEMBS,R.:1992,
               ASTRON.ASTROPHYS.263,147
             5 RUTTEN,R.G.M.,KUULKERS,E.,VOGT,N.,VAN PARADIJS,J.:1992,
               ASTRON.ASTROPHYS.265,159

QU CAR   1 GILLILAND,R.L.,PHILLIPS,M.M.:1982,ASTROPHYS.J.261,617
               (ORB.PER.,INCL)

HT CAS   1 ZHANG,E.-H.,ROBINSON,E.L.,NATHER,R.E.:1986,ASTROPHYS.J.
               305,740 (2. PER.)
             2 MARSH,T.R.:1990,ASTROPHYS.J.357,621
             3 HORNE,K.,WOOD,J.H.,STIENING,R.F.:1991,ASTROPHYS.J.378,271
               (ORB.PER.,M1/M2,INCL,M1,M2)
             4 WOOD,J.H.,HORNE,K.,VENNES,S.:1992,ASTROPHYS.J.385,294

V425 CAS   1 SHAFTER,A.W.:1983,PH.D. THESIS,UCLA (ORB.PER.,M1/M2,INCL,
               M1,M2)
             2 SZKODY,P.,SHAFTER,A.W.:1983,PUBL.ASTRON.SOC.PACIFIC
               95,509

BV CEN   1 GILLILAND,R.L.:1982,ASTROPHYS.J.263,302 (ORB.PER.,M1/M2,
               INCL,M1,M2)

MU CEN   1 FRIEND,M.T.,MARTIN,J.S.,SMITH,R.C.,JONES,D.H.P.:1990,
               MONTHLY NOTICES ROY.ASTRON.SOC.246,654 (ORB.PER.,
               M1/M2,INCL,M1,M2)

V436 CEN   1 VOGT,N.:1981,HABILITATION THESIS,UNIVERSITY BOCHUM
               (ORB.PER.,M1/M2,INCL,M1,M2)
             2 GILLILAND,R.L.:1982,ASTROPHYS.J.254,653 (ORB.PER.)
             3 WARNER,B.:1983,INF.BULL.VARIABLE STARS NO.2397 (2. PER.)

V442 CEN   1 MARINO,B.F.,WALKER,W.S.G.:1984,SOUTHERN STARS 30,389
               (ORB.PER.)

V485 CEN   1 AUGUSTEIJN,T.,VAN KERKWIJK,M.H.,VAN PARADIJS,J.:1993,
               ASTRON.ASTROPHYS.267,L55 (2. PER.)

V803 CEN   1 O'DONOGHUE,D.,WARGAU,W.,WARNER,B.,KILKENNY,D.,
               MARTINEZ,P.,KANAAN,A.,KEPLER,S.O.,HENRY,G.,
               WINGET,D.E.,CLEMENS,J.C.,GRAUER,A.D.:1990,MONTHLY
               NOTICES ROY.ASTRON.SOC.245,140 (ORB.PER.)

V834 CEN   1 PUCHNAREWICZ,E.M.,MASON,K.O.,MURDIN,P.G.,
               WICKRAMASINGHE,D.T.:1990,MONTHLY NOTICES ROY.
               ASTRON.SOC.244,20P
             2 SCHWOPE,A.D.,BEUERMANN,K.:1990,ASTRON.ASTROPHYS.238,173
             3 MIDDLEDITCH,J.,IMAMURA,J.N.,WOLFF,M.T.,STEIMAN-
               CAMERON,T.Y.:1991,ASTROPHYS.J.382,315
             4 SAMBRUNA,R.M.,CHIAPPETTI,L.,TREVES,A.,BONNET-BIDAUD,J.M.,
               BOUCHET,P.,MARASCHI,L.,MOTCH,C.,MOUCHET,M.:1991,
               ASTROPHYS.J.374,744

V834 CEN 5 FERRARIO,L.,WICKRAMASINGHE,D.T.,BAILEY,J.,HOUGH,J.H.,
             TUOHY,I.R.:1992,MONTHLY NOTICES ROY.ASTRON.SOC.
             256,252
          6 LARSSON,S.:1992,ASTRON.ASTROPHYS.265,133
          7 KUBIAK,M.,KRZEMINSKI,W.,UDALSKI,A.,POJMANSKI,G.:1993,
             ACTA ASTRON.43,149
          8 SCHWOPE,A.D.,THOMAS,H.-C.,BEUERMANN,K.,REINSCH,K.:1993,
             ASTRON.ASTROPHYS.267,103 (ORB.PER.,INCL,M1)

WW CET   1 HAWKINS,N.A.,SMITH,R.C.,JONES,D.H.P.:1990,IN:ACCRETION-
             POWERED COMPACT BINARIES,C.W.MAUCHE,CAMBRIDGE
             UNIVERSITY PRESS,CAMBRIDGE,P.113 (ORB.PER.,M1/M2,
             INCL,M1,M2)

WX CET   1 DOWNES,R.A.:1990,ASTRON.J.99,339
          2 HOWELL,S.B.,SZKODY,P.,KREIDL,T.J.,DOBRZYCKA,D.:1991,
             PUBL.ASTRON.SOC.PACIFIC 103,300
          3 O'DONOGHUE,D.,CHEN,A.,MARANG,F.,MITTAZ,J.P.D.,WINKLER,H.,
             WARNER,B.:1991,MONTHLY NOTICES ROY.ASTRON.SOC.
             250,363 (ORB.PER.,2. PER.)

Z   CHA   1 WOOD,J.,HORNE,K.,BERRIMAN,G.,WADE,R.A.,O'DONOGHUE,D.,
             WARNER,B.:1986,MONTHLY NOTICES ROY.ASTRON.SOC.
             219,629 (ORB.PER.)
          2 WADE,R.A.,HORNE,K.:1988,ASTROPHYS.J.324,411 (M1/M2,INCL,
             M1,M2)
          3 WARNER,B.,O'DONOGHUE,D.:1988,MONTHLY NOTICES ROY.ASTRON.
             SOC.233,705 (2. PER.)
          4 O'DONOGHUE,D.:1990,MONTHLY NOTICES ROY.ASTRON.SOC.246,29
          5 WOOD,J.H.:1990,MONTHLY NOTICES ROY.ASTRON.SOC.243,219
          6 KUULKERS,E.,VAN AMERONGEN,S.,VAN PARADIJS,J.,
             ROETTGERING,H.:1991,ASTRON.ASTROPHYS.252,605
          7 HARLAFTIS,E.T.,HASSALL,B.J.M.,NAYLOR,T.,CHARLES,P.A.,
             SONNEBORN,G.:1992,MONTHLY NOTICES ROY.ASTRON.SOC.
             257,607

TV COL   1 SCHRIJVER,J.,BRINKMAN,A.C.,VAN DER WOERD,H.,WATSON,M.G.,
             KING,A.R.,VAN PARADIJS,J.,VAN DER KLIS,M.:1985,
             SPACE SCI.REV.40,121 (3. PER.)
          2 HELLIER,C.,MASON,K.O.,MITTAZ,J.P.D.:1991,MONTHLY NOTICES
             ROY.ASTRON.SOC.248,5P (4. PER.)
          3 HELLIER,C.:1993,MONTHLY NOTICES ROY.ASTRON.SOC.264,132
             (ORB.PER.,2. PER.,INCL,M1,M2)

TX COL   1 BUCKLEY,D.A.H.,TUOHY,I.R.:1989,ASTROPHYS.J.344,376
             (ORB.PER.,3. PER.,INCL,M1,M2)
          2 MOUCHET,M.,BONNET-BIDAUD,J.M.,BUCKLEY,D.A.H.,TUOHY,I.R.:
             1991,ASTRON.ASTROPHYS.250,99

AL COM   1 SZKODY,P.,HOWELL,S.B.,MATEO,M.,KREIDL,T.J.N.:1989,PUBL.
             ASTRON.SOC.PACIFIC 101,899
          2 MUKAI,K.,MASON,K.O.,HOWELL,S.B.,ALLINGTON-SMITH,J.,
             CALLANAN,P.J.,CHARLES,P.A.,HASSALL,B.J.M.,MACHIN,G.,
             NAYLOR,T.,SMALE,A.P.,VAN PARADIJS,J.:1990,MONTHLY
             NOTICES ROY.ASTRON.SOC.245,385
          3 HOWELL,S.B.,SZKODY,P.:1991,INF.BULL.VARIABLE STARS
             NO.3653
          4 ABBOTT,T.M.C.,ROBINSON,E.L.,HILL,G.J.,HASWELL,C.A.:1992,
             ASTROPHYS.J.399,680 (ORB.PER.)

GP COM   1 NATHER,R.E.,ROBINSON,E.L.,STOVER,R.J.:1981,ASTROPHYS.J.
             244,269 (ORB.PER.)

GP COM   2 MARSH,T.R.,HORNE,K.,ROSEN,S.:1991,ASTROPHYS.J.366,535

V394 CRA   1 SCHAEFER,B.:1990.ASTROPHYS.J.LETTERS 355,L39 (ORB.PER.)

TT CRT   1 SZKODY,P.,WILLIAMS,R.E.,MARGON,B.,HOWELL,S.B.,MATEO,M.:
                  1992,ASTROPHYS.J.387,357 (ORB.PER.,M1/M2,INCL,M1)

SS CYG   1 SHAFTER,A.W.:1983,PH.D. THESIS,UCLA (INCL)
              2 BRUCH,A.:1990,ACTA ASTRON.40,369
              3 FRIEND,M.T.,MARTIN,J.S.,SMITH,R.C.,JONES,D.H.P.:1990,
                  MONTHLY NOTICES ROY.ASTRON.SOC.246,654 (ORB.PER.,
                  M1/M2,M1,M2)
              4 JONES,M.H.,WATSON,M.G.:1992,MONTHLY NOTICES ROY.ASTRON.
                  SOC.257,633
              5 VOLOSHINA,I.B.,LYUTYI,V.M.:1993,ASTRON.REP.37,34

EM CYG   1 STOVER,R.J.,ROBINSON,E.L.,NATHER,R.E.:1981,ASTROPHYS.J.
                  248,696 (M1/M2,INCL,M1,M2)
              2 SHAFTER,A.W.:1983,PH.D. THESIS,UCLA (M1/M2,INCL,M1,M2)
              3 BEUERMANN,K.,PAKULL,M.W.:1984,ASTRON.ASTROPHYS.136,250
                  (ORB.PER.)

EY CYG   1 HACKE,G.,ANDRONOV,I.L.:1988,MITT.VERAENDERL.STERNE 11,74
                  (ORB.PER.)
              2 SZKODY,P.,PICHE,F.,FEINSWOG,L.:1990,ASTROPHYS.J.SUPPL.
                  SER.73,441

V503 CYG   1 SZKODY,P.,HOWELL,S.B.,MATEO,M.,KREIDL,T.J.N.:1989,PUBL.
                  ASTRON.SOC.PACIFIC 101,899 (ORB. PER.)
              2 SZKODY,P.,HOWELL,S.B.:1993,ASTROPHYS.J.403,743

V751 CYG   1 BELL,M.,WALKER,M.F.:1980,BULL.AMERICAN ASTRON.SOC.12,63
                  (ORB.PER.)

V1251 CYG   1 KATO,T.:1991,IAU CIRC.NO.5379 (2. PER.)

V1500 CYG   1 KALUZNY,J.,SEMENIUK,I.:1987,ACTA ASTRON.37,349 (ORB.PER.)
              2 HORNE,K.,SCHNEIDER,D.P.:1989,ASTROPHYS.J.343,888 (M1)
              3 SCHMIDT,G.D.,STOCKMAN,H.S.:1991,ASTROPHYS.J.371,749
                  (2. PER.)
              4 PAVLENKO,E.P.,PELT,J.:1992,ASTROPHYSICS 34,77

V1668 CYG   1 KALUZNY,J.:1990,MONTHLY NOTICES ROY.ASTRON.SOC.245,547
                  (ORB.PER.)

V1776 CYG   1 GARNAVICH,P.M.,SZKODY,P.,MATEO,M.,FEINSWOG,L.,BOOTH,J.,
                  GOODRICH,B.,MILLER,H.R.,CARINI,M.T.,WILSON,J.W.:
                  1990,ASTROPHYS.J.365,696 (ORB.PER.,M1/M2,INCL,M1,M2)

CM DEL   1 SHAFTER,A.W.:1983,PH.D. THESIS,UCLA (M1/M2,INCL,M1,M2)
              2 SHAFTER,A.W.:1985,ASTRON.J.90,643 (ORB.PER.)

HR DEL   1 KOHOUTEK,L.,PAULS,R.:1980,ASTRON.ASTROPHYS.92,200
                  (2. PER.)
              2 KUERSTER,M.,BARWIG,H.:1988,ASTRON.ASTROPHYS.199,201
                  (ORB.PER.,M1/M2,INCL,M1,M2)

AB DRA   1 THORSTENSEN,J.R.,FREED,I.W.:1985,ASTRON.J.90,2082
                  (ORB.PER.)
              2 VOLOSHINA,I.B.,SHUGAROV,S.YU.:1989,SOVIET ASTRON.LETTERS
                  15,312

CQ DRA BC   1 REIMERS,D.,GRIFFIN,R.,BROWN,A.:1988,ASTRON.ASTROPHYS.
           193,180 (ORB.PER.)

DM DRA   1 HOWELL,S.B.,SZKODY,P.,KREIDL,T.J.,MASON,K.O.,
           PUCHNAREWICZ,E.M.:1990,PUBL.ASTRON.SOC.PACIFIC
           102,758 (ORB.PER.)

DO DRA   1 FRIEND,M.T.,MARTIN,J.S.,SMITH,R.C.,JONES,D.H.P.:1990,
           MONTHLY NOTICES ROY.ASTRON.SOC.246,637
       2 MUKAI,K.,MASON,K.O.,HOWELL,S.B.,ALLINGTON-SMITH,J.,
           CALLANAN,P.J.,CHARLES,P.A.,HASSALL,B.J.M.,MACHIN,G.,
           NAYLOR,T.,SMALE,A.P.,VAN PARADIJS,J.:1990,MONTHLY
           NOTICES ROY.ASTRON.SOC.245,385
       3 MATEO,M.,SZKODY,P.,GARNAVICH,P.:1991,ASTROPHYS.J.370,370
           (ORB.PER.,M1/M2,INCL,M1,M2)
       4 PATTERSON,J.,SCHWARTZ,D.A.,PYE,J.P.,BLAIR,W.P.,
           WILLIAMS,G.A.,CAILLAULT,J.-P.:1992,ASTROPHYS.J.
           392,233 (3. PER.,4. PER.)

EP DRA   1 REMILLARD,R.A.,STROOZAS,B.A.,TAPIA,S.,SILBER,A.:1991,
           ASTROPHYS.J.379,715 (ORB.PER.,INCL)

AH ERI   1 SZKODY,P.,HOWELL,S.B.,MATEO,M.,KREIDL,T.J.N.:1989,PUBL.
           ASTRON.SOC.PACIFIC 101,899 (3. PER.)

AQ ERI   1 KATO,T.:1991,INF.BULL.VARIABLE STARS NO.3671 (2. PER.)
       2 THORSTENSEN,J.R.,PATTERSON,J.,THOMAS,G.:1992,IN PREPAR-
           ATION (ORB.PER.)

CP ERI   1 HOWELL,S.B.,SZKODY,P.,KREIDL,T.J.,DOBRZYCKA,D.:1991,
           PUBL.ASTRON.SOC.PACIFIC 103,300
       2 ABBOTT,T.M.C.,ROBINSON,E.L.,HILL,G.J.,HASWELL,C.A.:1992,
           ASTROPHYS.J.399,680 (ORB.PER.)

EF ERI   1 CROPPER,M.:1985,MONTHLY NOTICES ROY.ASTRON.SOC.212,709
           (ORB.PER.,INCL)
       2 MUKAI,K.,CHARLES,P.:1985,MONTHLY NOTICES ROY.ASTRON.SOC.
           212,609 (ORB.PER.)
       3 OESTREICHER,R.,SEIFERT,W.,WUNNER,G.,RUDER,H.:1990,
           ASTROPHYS.J.250,324
       4 WICKRAMASINGHE,D.T.,ACHILLEOS,N.,WU,K.,BOYLE,B.J.:1990,
           IAU CIRC.NO.4962
       5 BEUERMANN,K.,THOMAS,H.C.,PIETSCH,W.:1991,ASTRON.
           ASTROPHYS.246,L36

UZ FOR   1 ALLEN,R.G.,BERRIMAN,G.,SMITH,P.S.,SCHMIDT,G.D.:1989,
           ASTROPHYS.J.347,426 (ORB.PER.,M1/M2,INCL,M1,M2)
       2 BAILEY,J.,CROPPER,M.:1991,MONTHLY NOTICES ROY.ASTRON.
           SOC.253,27
       3 RAMSAY,G.,ROSEN,S.R.,MASON,K.O.,CROPPER,M.S.,WATSON,M.G.:
           1993,MONTHLY NOTICES ROY.ASTRON.SOC.262,993

U GEM   1 FRIEND,M.T.,MARTIN,J.S.,SMITH,R.C.,JONES,D.H.P.:1990,
           MONTHLY NOTICES ROY.ASTRON.SOC.246,637 (M1/M2,INCL,
           M1,M2)
       2 MARSH,T.R.,HORNE,K.,SCHLEGEL,E.M.,HONEYCUTT,R.K.,
           KAITCHUCK,R.H.:1990,ASTROPHYS.J.364,637
       3 KIPLINGER,A.L.,SION,E.M.,SZKODY,P.:1991,ASTROPHYS.J.,
           366,569
       4 LONG,K.S.,BLAIR,W.P.,BOWERS,C.W.,DAVIDSEN,A.F.,
           KRISS,G.A.,SION,E.M.,HUBENY,T.1993,ASTROPHYS.J.
           405,327

U   GEM   5 SMAK,J.:1993,ACTA ASTRON.43,121 (ORB.PER.)

AW GEM   1 HOWELL,S.B.,SZKODY,P.:1988,PUBL.ASTRON.SOC.PACIFIC
                   100,224
               2 FUJINO,S.,NAKAI,M.,IIDA,M.,MORIYAMA,M.,MAKIGUCHI,N.,
                   KATO,T.:1989,VARIABLE STAR BULL.JAPAN NO.10,P.37
                   (2. PER.)

IR GEM   1 SZKODY,P.,SHAFTER,A.W.,COWLEY,A.P.:1984,ASTROPHYS.J.
                   282,236 (2. PER.)
               2 FEINSWOG,L.,SZKODY,P.,GARNAVICH,P.:1988,ASTRON.J.96,1702
                   (ORB.PER.)
               3 LAZARO,C.,MARTINEZ-PAIS,I.G.,AREVALO,M.J.,SOLHEIM,J.E.:
                   1991,ASTRON.J.101,196

RZ GRU   1 KELLY,B.D.,KILKENNY,D.,COOKE,J.A.:1981,MONTHLY NOTICES
                   ROY.ASTRON.SOC.196,91P
               2 STICKLAND,D.J.,KELLY,B.D.,COOKE,J.A.,COULSON,I.,
                   ENGELBRECHT,C.,KILKENNY,D.,SPENCER-JONES,J.:1984,
                   MONTHLY NOTICES ROY.ASTRON.SOC.206,819
               3 HAEFNER,R.:1990,PRIVATE COMMUNICATION (ORB.PER)

CE GRU   1 HAWKINS,M.R.S.:1983,NATURE 293,116
               2 TUOHY,I.R.,FERRARIO,L.,WICKRAMASINGHE,D.T.,
                   HAWKINS,M.R.S.:1988,ASTROPHYS.J.LETTERS 328,L59
                   (ORB.PER.)
               3 CROPPER,M.,BAILEY,J.A.,WICKRAMASINGHE,D.T.,FERRARIO,L.:
                   1990,MONTHLY NOTICES ROY.ASTRON.SOC.244,34P
               4 WICKRAMASINGHE,D.T.,FERRARIO,L.,CROPPER,M.,BAILEY,J.:
                   1991,MONTHLY NOTICES ROY.ASTRON.SOC.251,137

CF GRU   1 HAWKINS,M.R.S.,VERON,P.:1987,ASTRON.ASTROPHYS.182,271
               2 HOWELL,S.B.,SZKODY,P.,KREIDL,T.J.,DOBRZYCKA,D.:1991,
                   PUBL.ASTRON.SOC.PACIFIC 103,300 (ORB.PER.)

AH HER   1 HORNE,K.,WADE,R.A.,SZKODY,P.:1986,MONTHLY NOTICES ROY.
                   ASTRON.SOC.219,791 (ORB.PER.,M1/M2,INCL,M1,M2)
               2 BRUCH,A.:1987,ASTRON.ASTROPHYS.172,187

AM HER   1 YOUNG,P.,SCHNEIDER,D.P.,SHECTMAN,S.A.:1981,ASTROPHYS.J.
                   245,1043 (ORB.PER.,M1/M2,INCL,M1,M2)
               2 BAILEY,J.,FERRARIO,L.,WICKRAMASINGHE,D.T.:1991,MONTHLY
                   NOTICES ROY.ASTRON.SOC.251,37P
               3 BHAT,C.L.,KAUL,R.K.,RAWAT,H.S.,SENECHA,V.K.,RANNOT,R.C.,
                   SAPRU,M.L.,TICKOO,A.K.,RAZDAN,H.:1991,ASTROPHYS.J.
                   369,475
               4 BONNET-BIDAUD,J.M.,SOMOVA,T.A.,SOMOV,N.N.:1991,ASTRON.
                   ASTROPHYS.152,L27
               5 WICKRAMASINGHE,D.T.,BAILEY,J.,MEGGITT,S.M.A.,FERRARIO,L.,
                   HOUGH,J.,TUOHY,J.R.:1991,MONTHLY NOTICES ROY.ASTRON.
                   SOC.251,28
               6 SCHAICH,M.,WOLF,D.,OESTREICHER,R.,RUDER,H.:1992,ASTRON.
                   ASTROPHYS.264,529

DQ HER   1 SCHNEIDER,D.P.,GREENSTEIN,J.L.:1979,ASTROPHYS.J.233,935
                   (ORB.PER.)
               2 BALACHANDRAN,S.,ROBINSON,E.L.,KEPLER,S.O.:1983,PUBL.
                   ASTRON.SOC.PACIFIC 95,653 (4. PER.)
               3 HORNE,K.,WELSH,W.F.,WADE,R.A.:1993,ASTROPHYS.J.410,357
                   (M1/M2,INCL,M1,M2)

V533 HER   1 HUTCHINGS,J.B.:1987,PUBL.ASTRON.SOC.PACIFIC 99,57
                   (ORB.PER.)

V544 HER   1 HOWELL,S.B.,SZKODY,P.,KREIDL,T.J.,MASON,K.O.,
               PUCHNAREWICZ,E.M.:1990,PUBL.ASTRON.SOC.PACIFIC
               102,758 (ORB.PER.)

V795 HER   1 SHAFTER,A.W.,ROBINSON,E.L.,CRAMPTON,D.,WARNER,B.,
               PRESTAGE,R.M.:1990,ASTROPHYS.J.354,708 (ORB.PER.,
               2. PER.)
             2 PRINJA,R.K.,ROSEN,S.R.,SUPELLI,K.:1991,MONTHLY NOTICES
               ROY.ASTRON.SOC.248,40
             3 ZHANG,E.,ROBINSON,E.L.,RAMSEYER,T.R.,SHETRONE,M.D.,
               STIENING,R.F.:1991,ASTROPHYS.J.381,534 (2. PER.)
             4 PRINJA,R.K.,DREW,J.E.,ROSEN,S.R.:1992,MONTHLY NOTICES
               ROY.ASTRON.SOC.256,219
             5 PRINJA,R.,ROSEN,S.R.:1993,MONTHLY NOTICES ROY.ASTRON.SOC.
               262,L37

V825 HER   1 FERGUSON,D.H.,GREEN,R.F.,LIEBERT,J.:1984,ASTROPHYS.J.
               287,320
             2 WILSON,J.W.,MILLER,H.R.,AFRICANO,J.L.,GOODRICH,B.D.,
               MAHAFFEY,C.T.,QUIGLEY,R.J.:1986,ASTRON.ASTROPHYS.
               SUPPL.SER.66,323
             3 RINGWALD,F.A.:1991,BULL.AMERICAN ASTRON.SOC.23,1463
               (ORB.PER.)

V838 HER   1 INGRAM,D.,GARNAVICH,P.,GREEN,P.,SZKODY,P.:1992,PUBL.
               ASTRON.SOC.PACIFIC 104,402 (ORB.PER.)
             2 LEIBOWITZ,E.M.,MENDELSON,H.,MASHAL,E.,PRIALNIK,D.,
               SEITTER,W.C.:1992,ASTROPHYS.J.LETTERS 385,L49
             3 LEIBOWITZ,E.M.:1993,ASTROPHYS.J.LETTERS 411,L29

WW HOR   1 BEUERMANN,K.,THOMAS,H.-C.,SCHWOPE,A.,GIOMMI,P.,
               TAGLIAFERRI,G.:1990,ASTRON.ASTROPHYS.238,187
               (ORB.PER.)
             2 BAILEY,J.,WICKRAMASINGHE,D.T.,FERRARIO,L.,HOUGH,J.H.,
               CROPPER,M.:1993,MONTHLY NOTICES ROY.ASTRON.SOC.
               261,L31

EX HYA   1 STERKEN,C.,VOGT,N.,FREETH,R.,KENNEDY,H.D.,MARINO,B.F.,
               PAGE,A.A.,WALKER,W.S.G.:1983,ASTRON.ASTROPHYS.
               118,325 (ORB.PER.)
             2 HELLIER,C.,MASON,K.O.,ROSEN,S.R.,CORDOVA,F.A.:1987,
               MONTHLY NOTICES ROY.ASTRON.SOC.228,463 (M1/M2,INCL,
               M1,M2)
             3 BOND,I.A.,FREETH,R.V.:1988,MONTHLY NOTICES ROY.ASTRON.
               SOC.232,753 (3. PER.)
             4 HILL,K.M.,WATSON,R.D.:1990,ASTROPHYS.SPACE SCI.163,59
             5 REINSCH,K.,BEUERMANN,K.:1990,ASTRON.ASTROPHYS.240,360
             6 ROSEN,S.R.,MASON,K.O.,MUKAI,K.,WILLIAMS,O.R.:1991,MONTHLY
               NOTICES ROY.ASTRON.SOC.249,417
             7 HELLIER,C.,SPROATS,L.N.:1992,INF.BULL.VARIABLE STARS
               NO.3724

LY HYA   1 KUBIAK,M.,KRZEMINSKI,W.:1992,ACTA ASTRON.42,177
             2 HAEFNER,R.,BARWIG,H.,MANTEL,K.-H.:1993,ASTROPHYS.SPACE
               SCI.204,199 (ORB.PER.)

VW HYI   1 SCHOEMBS,R.,VOGT,N.:1981,ASTRON.ASTROHYS.97,185 (M1/M2,
               INCL,M1,M2)
             2 VAN AMERONGEN,S.,DAMEN,E.,GROOT,M.,KRAAKMAN,H.,
               VAN PARADIJS,J.:1987,MONTHLY NOTICES ROY.ASTRON.SOC.
               225,93 (ORB.PER.,? PER.)

VW HYI   3 VAN AMERONGEN,S.,BOVENSCHEN,H.,VAN PARADIJS,J.:1987,
                 MONTHLY NOTICES ROY.ASTRON.SOC.229,245 (2. PER.)
              4 BELLONI,T.,VERBUNT,F.,BEUERMANN,K.,BUNK,W.,IZZO,C.,
                 KLEY,W.,PIETSCH,W.,RITTER,H.,THOMAS,H.C.,VOGES,W.:
                 1991,ASTRON.ASTROPHYS.246,L44
              5 MAUCHE,C.W.,WADE,R.A.,POLIDAN,R.S.,VAN DER WOERD,H.,
                 PAERELS,F.B.S.:1991,ASTROPHYS.J.372,659

WX HYI   1 SCHOEMBS,R.,VOGT,N.:1981,ASTRON.ASTROPHYS.97,185
                 (ORB.PER.,2. PER.,M1/M2,INCL,M1,M2)
              2 KUULKERS,E.,HOLLANDER,A.,OOSTERBROEK,T.,VAN PARADIJS,J.:
                 1991,ASTRON.ASTROPHYS.242,401

BL HYI   1 SCHWOPE,A.D.,BEUERMANN,K.:1989,ASTRON.ASTROPHYS.222,132
                 (ORB.PER.,INCL)

DI LAC   1 KRAFT,R.P.:1964,ASTROPHYS.J.139,457
              2 WEBBINK,R.F.:PRIVATE COMMUNICATION (ORB.PER.)

T LEO   1 SHAFTER,A.W.:1983,PH.D. THESIS,UCLA (M1/M2,INCL,M1,M2)
              2 SHAFTER,A.W.,SZKODY,P.:1984,ASTROPHYS.J.276,305
                 (ORB.PER.)
              3 KATO,T.,FUJINO,S.:1987,VARIABLE STAR BULL.JAPAN NO.3,10
                 (2. PER.)

U LEO   1 DOWNES,R.A.,SZKODY,P.:1989,ASTRON.J.97,1729 (ORB.PER.)

X LEO   1 SHAFTER,A.W.,HARKNESS,R.P.:1986,ASTRON.J.92,658
                 (ORB.PER.)

RZ LEO   1 HOWELL,S.B.,SZKODY,P.:1988,PUBL.ASTRON.SOC.PACIFIC
                 100,224 (ORB.PER.)

DO LEO   1 ABBOTT,T.M.C.,SHAFTER,A.W.,WOOD,J.H.,TOMANEY,A.B.,
                 HASWELL,C.A.:1990,PUBL.ASTRON.SOC.PACIFIC 102,558
                 (ORB.PER.)

DP LEO   1 BIERMANN,P.,SCHMIDT,G.D.,LIEBERT,J.,STOCKMAN,H.S.,
                 TAPIA,S.,KUEHR,H.,STRITTMATTER,P.A.,WEST,S.,
                 LAMB,D.Q.:1985,ASTROPHYS.J.293,303 (ORB.PER.)
              2 CROPPER,M.,MUKAI,K.,MASON,K.O.,SMALE,A.P.,CHARLES,P.A.,
                 MITTAZ,J.P.D.,MACHIN,G.,HASSALL,B.J.M.,
                 CALLANAN,P.J.,NAYLOR,T.,VAN PARADIJS,J.:1990,
                 MONTHLY NOTICES ROY.ASTRON.SOC.245,760
              3 BAILEY,J.,WICKRAMASINGHE,D.T.,FERRARIO,L.,HOUGH,J.H.,
                 CROPPER,M.:1993,MONTHLY NOTICES ROY.ASTRON.SOC.
                 261,L31 (M1/M2,INCL,M1,M2)
              4 CROPPER,M.,WICKRAMASINGHE,D.T.:1993,MONTHLY NOTICES ROY.
                 ASTRON.SOC.260,696

RU LMI   1 WAGNER,R.M.,SION,E.M.,LIEBERT,J.,STARRFIELD,S.G.:1988,
                 ASTROPHYS.J.328,213
              2 HOWELL,S.B.,SZKODY,P.,KREIDL,T.J.,MASON,K.O.,
                 PUCHNAREWICZ,E.M.:1990,PUBL.ASTRON.SOC.PACIFIC
                 102,758 (ORB.PER.)
              3 MUKAI,K.,MASON,K.O.,HOWELL,S.B.,ALLINGTON-SMITH,J.,
                 CALLANAN,P.J.,CHARLES,P.A.,HASSALL,B.J.M.,MACHIN,G.,
                 NAYLOR,T.,SMALE,A.P.,VAN PARADIJS,J.:1990,MONTHLY
                 NOTICES ROY.ASTRON.SOC.245,385

ST LMI   1 CROPPER,M.:1986,MONTHLY NOTICES ROY.ASTRON.SOC.222,853
                 (ORB.PER.,INCL.,M1)

ST LMI   2 PEACOCK,T.,CROPPER,M.,BAILEY,J.,HOUGH,J.H.,
            WICKRAMASINGHE,D.T.:1992,MONTHLY NOTICES ROY.ASTRON.
            SOC.259,583
        3 FERRARIO,L.,BAILEY,J.,WICKRAMASINGHE,D.T.:1993,MONTHLY
            NOTICES ROY.ASTRON.SOC.262,285

SX LMI   1 WAGNER,R.M.,SION,E.M.,LIEBERT,J.,STARRFIELD,S.G.:1988,
            ASTROPHYS.J.328,213 (ORB.PER.)
        2 HOWELL,S.B.,SZKODY,P.:1990,ASTROPHYS.J.356,623

BR LUP   1 O'DONOGHUE,D.:1987,ASTROPHYS.SPACE SCI.136,247 (2. PER.)

BH LYN   1 CHEN,J.-S.,WEI,M.-Z.,LIU,X.-W.:1990,CHIN.ASTRON.
            ASTROPHYS.14,469
        2 WEI,M.-Z.,CHEN,J.-S.,JIANG,Z.-J.:1990,PUBL.ASTRON.SOC.
            PACIFIC 102,698
        3 THORSTENSEN,J.R.,DAVIS,M.K.,RINGWALD,F.A.:1991,ASTRON.J.
            102,683
        4 DHILLON,V.S.,JONES,D.H.P.,MARSH,T.R.,SMITH,R.C.:1992,
            MONTHLY NOTICES ROY.ASTRON.SOC.258,225 (ORB.PER.)

AY LYR   1 UDALSKI,A.,SZYMANSKI,M.:1988,ACTA ASTRON.38,215
            (2. PER.)

MV LYR   1 SCHNEIDER,D.P.,YOUNG,P.,SHECTMAN,S.A.:1981,ASTROPHYS.J.
            245,644 (ORB.PER.,M2)
        2 BORISOV,G.V.:1992,ASTRON.ASTROPHYS.261,154 (2. PER.)
        3 ROSINO,L.,ROMANO,G.,MARZIANI,P.:1993,PUBL.ASTRON.SOC.
            PACIFIC 105,51

TU MEN   1 STOLZ,B.,SCHOEMBS,R.:1984,ASTRON.ASTROPHYS.132,187
            (ORB.PER.,2. PER.,M1/M2,INCL,M1,M2)

BT MON   1 SCHAEFER,B.E.,PATTERSON,J.:1983,ASTROPHYS.J.268,710
            (ORB.PER.,2. PER.)
        2 WILLIAMS,R.E.:1989,ASTRON.J.97,1752

CW MON   1 SZKODY,P.,MATEO,M.:1986,ASTRON.J.92,483 (ORB.PER.)
        2 HOWELL,S.B.,SZKODY,P.:1988,PUBL.ASTRON.SOC.PACIFIC
            100,224

GQ MUS   1 DIAZ,M.P.,STEINER,J.E.:1990,REV.MEX.ASTRON.ASTROF.21,369
            (ORB.PER.)

V380 OPH   1 SHAFTER,A.W.:1983,PH.D. THESIS,UCLA (M1/M2,INCL,M1,M2)
        2 SHAFTER,A.W.:1985,ASTRON.J.90,643 (ORB.PER.)

V426 OPH   1 HESSMAN,F.V.:1988,ASTRON.ASTROPHYS.SUPPL.SER.72,515
            (ORB.PER.,M1/M2,INCL,M1,M2)
        2 SZKODY,P.,KII,T.,OSAKI,Y.:1990,ASTRON.J.100,546

V442 OPH   1 SHAFTER,A.W.:1983,PH.D. THESIS,UCLA (M1/M2,INCL,M1,M2)
        2 SZKODY,P.,SHAFTER,A.W.:1983,PUBL.ASTRON.SOC.PACIFIC
            95,509 (ORB.PER.)

V841 OPH   1 BIANCHINI,A.,FRIEDJUNG,M.,SABBADIN,F.:1989,IN:THE PHYSICS
            OF CLASSICAL NOVAE,IAU COLL.NO.122,A.CASSATELLA,
            R.VIOTTI (EDS.),LECTURE NOTES IN PHYSICS 369,
            SPRINGER VERLAG,BERLIN,P.61 (ORB.PER.)
        2 SHARA,M.M.,POTTER,M.,SHARA,D.J.:1989,PUBL.ASTRON.SOC.
            PACIFIC 101,905

V2051 OPH   1 WATTS,D.J.,BAILEY,J.,HILL,P.W.,GREENHILL,J.G.,
                MC COWAGE,C.,CARTY,T.:1985,ASTRON.ASTROPHYS 154,197
                (M1/M2,INCL,M1,M2)
             2 ECHEVARRIA,J.,ALVAREZ,M.:1993,ASTRON.ASTROPHYS.275,187
                (ORB.PER.)

V2214 OPH   1 BAPTISTA,R.,JABLONSKI,F.J.,CIESLINSKI,D.,STEINER,J.E.:
                1993,ASTROPHYS.J.LETTERS 406,L67 (ORB.PER.)

CN ORI   1 BARRERA,L.H.,VOGT,N.:1989,ASTRON.ASTROPHYS.220,99
                (ORB.PER.,2. PER.)
             2 FRIEND,M.T.,MARTIN,J.S.,SMITH,R.C.,JONES,D.H.P.:1990,
                MONTHLY NOTICES ROY.ASTRON.SOC.246,637 (M1/M2,INCL,
                M1,M2)

CZ ORI   1 SZKODY,P.:1987,ASTROPHYS.J.SUPPL.SER.63,685
             2 BIANCHINI,A.,MARGONI,R.,SPOGLI,C.:1990,IAU CIRC.NO.5031
                (ORB.PER.)
             3 SZKODY,P.,PICHE,F.,FEINSWOG,L.:1990,ASTROPHYS.J.SUPPL.
                SER.73,441

V1159 ORI   1 JABLONSKI,F.J.,CIESLINSKI,D.:1992,ASTRON.ASTROPHYS.
                259,198 (ORB.PER.)

V1193 ORI   1 WARNER,B.,NATHER,R.E.:1988,INF.BULL.VARIABLE STARS
                NO.3140
             2 THORSTENSEN,J.:1990,PRIVATE COMMUNICATION (ORB.PER.)

BD PAV   1 BARWIG,H.,SCHOEMBS,R.:1983,ASTRON.ASTROPHYS.124,287
                (ORB.PER.)
             2 FRIEND,M.T.,MARTIN,J.S.,SMITH,R.C.,JONES,D.H.P.:1990,
                MONTHLY NOTICES ROY.ASTRON.SOC.246,637 (M1/M2,INCL)

RU PEG   1 STOVER,R.J.:1981,ASTROPHYS.J.249,673 (ORB.PER.)
             2 SHAFTER,A.W.:1983,PH.D. THESIS,UCLA (M1/M2,INCL,M1,M2)
             3 FRIEND,M.T.,MARTIN,J.S.,SMITH,R.C.,JONES,D.H.P.:1990,
                MONTHLY NOTICES ROY.ASTRON.SOC.246,654

EF PEG   1 DE YOUNG,J.,SCHMIDT,R.:1991,IAU CIRC.NO.5377 (2. PER.)
             2 HOWELL,S.B.,SCHMIDT,R.,DE YOUNG,J.A.,FRIED,R.,SCHMEER,P.,
                GRITZ,L.:1993,PUBL.ASTRON.SOC.PACIFIC 105,579
                (M1/M2,M1,M2)

HX PEG   1 GREENSTEIN,J.L.,ARP,H.C.,SHECTMAN,S.:1977,PUBL.ASTRON.
                SOC.PACIFIC 89,741
             2 RINGWALD,F.A.:1992,BULL.AMERICAN ASTRON.SOC.24,771
                (ORB.PER.)

IP PEG   1 MARTIN,J.S.,FRIEND,M.T.,SMITH,R.C.,JONES,D.H.P.:1989,
                MONTHLY NOTICES ROY.ASTRON.SOC.240,519 (M1/M2,INCL,
                M1,M2)
             2 WOOD,J.H.,MARSH,T.R.,ROBINSON,E.L.,STIENING,R.F.,
                HORNE,K.,STOVER,R.J.,SCHOEMBS,R.,ALLEN,S.L.,
                BOND,H.E.,JONES,D.H.P.,GRAUER,A.D.,CIARDULLO,R.:
                1989,MONTHLY NOTICES ROY.ASTRON.SOC.239,809
                (ORB.PER.)
             3 WOLF,S.,MANTEL,K.H.,HORNE,K.,BARWIG,H.,SCHOEMBS,R.,
                BAERNBANTNER,O.:1993,ASTRON.ASTROPHYS.273,160

V PER   1 SHAFTER,A.W.,ABBOTT,T.M.C.:1989,ASTROPHYS.J.LETTERS
                339,L75 (ORB.PER.)
             2 WOOD,J.H.,ABBOTT,T.M.C.,SHAFTER,A.W.:1992,ASTROPHYS.J.
                391,729

UV PER   1 KATO,T.:1990,INF.BULL.VARIABLE STARS NO.3522 (ORB.PER.)
                2 UDALSKI,A.,PYCH,W.:1992,ACTA ASTRON.42,285 (2. PER.)

FY PER   1 SAZONOV,A.V.,SUUGAROV,S.YU.:1992,INF.BULL.VARIABLE STARS
                   NO.3744 (ORB.PER.)

GK PER   1 CRAMPTON,D.,COWLEY,A.P.,FISHER,W.A.:1986,ASTROPHYS.J.
                   300,788 (ORB.PER.,M1/M2,M1,M2)
                2 NORTON,A.J.,WATSON,M.G.,KING,A.R.:1988,MONTHLY NOTICES
                   ROY.ASTRON.SOC.231,783 (3. PER.)
                3 PATTERSON,J.:1991,PUBL.ASTRON.SOC.PACIFIC 103,1149
                   (3. PER.)
                4 ISHIDA,M.,SAKAO,T.,MAKISHIMA,K.,OHASHI,T.,WATSON,M.G.,
                   NORTON,A.J.,KAWADA,M.,KOYAMA,K.:1992,MONTHLY NOTICES
                   ROY.ASTRON.SOC.254,647 (3. PER.)

KT PER   1 CLARKE,J.T.,BOWYER,S.:1984,ASTRON.ASTROPHYS.140,345
                2 RATERING,C.,BRUCH,A.,DIAZ,M.:1993,ASTRON.ASTROPHYS.
                   268,694 (ORB.PER.)

RR PIC   1 HAEFNER,R,METZ,K.:1982,ASTRON.ASTROPHYS.109,171 (INCL,
                   M1,M2)
                2 HAEFNER,R.,BETZENBICHLER,W.:1991,INF.BULL.VARIABLE STARS
                   NO.3665 (ORB.PER.)

TW PIC   1 BUCKLEY,D.A.H.,TUOHY,I.R.:1990,ASTROPHYS.J.349,296
                   (3. PER.)
                2 MOUCHET,M.,BONNET-BIDAUD,J.M.,BUCKLEY,D.A.H.,TUOHY,I.R.:
                   1991,ASTRON.ASTROPHYS.250,99
                3 PATTERSON,J.,MOULDEN,M.:1993,PUBL.ASTRON.SOC.PACIFIC
                   105,779 (ORB.PER.,4. PER.)

TY PSC   1 SZKODY,P.,FEINSWOG,L.:1988,ASTROPHYS.J.334,422 (ORB.PER.,
                   2. PER.)
                2 THORSTENSEN,J.R.,PATTERSON,J.,THOMAS,G.:1992,IN PREPAR-
                   ATION (ORB.PER.)

AO PSC   1 VAN AMERONGEN,S.,KRAAKMAN,H.,DAMEN,E.,TJEMKES,S.,
                   VAN PARADIJS,J.:1985,MONTHLY NOTICES ROY.ASTRON.SOC.
                   215,45P (ORB.PER.,3. PER.,4. PER.)
                2 HELLIER,C.,CROPPER,M.,MASON,K.O.:1991,MONTHLY NOTICES
                   ROY.ASTRON.SOC.248,233

AY PSC   1 DIAZ,M.P.,STEINER,J.E.:1990,ASTRON.ASTROPHYS.238,170
                   (ORB.PER.)
                2 HOWELL,S.B.,BLANTON,S.A.:1993,ASTRON.J.106,311
                3 SZKODY,P.,HOWELL,S.B.:1993,ASTROPHYS.J.403,743

TY PSA   1 WARNER,B.,O'DONOGHUE,D.,WARGAU,W.:1989,MONTHLY NOTICES
                   ROY.ASTRON.SOC.238,73 (2. PER.)
                2 O'DONOGHUE,D.,SOLTYNSKI,M.G.:1992,MONTHLY NOTICES ROY.
                   ASTRON.SOC.254,9 (ORB.PER.)

VV PUP   1 SCHNEIDER,D.P.,YOUNG,P.:1980,ASTROPHYS.J.240,871
                   (ORB.PER.)
                2 SZKODY,P.,BAILEY,J.A.,HOUGH,J.H.:1983,MONTHLY NOTICES
                   ROY.ASTRON.SOC.203,749 (M1/M2)
                3 CROPPER,M.S.,WARNER,B.:1986,MONTHLY NOTICES ROY.ASTRON.
                   SOC.220,633 (INCL)
                4 WICKRAMASINGHE,D.T.,FERRARIO,L.,BAILEY,J.:1989,
                   ASTROPHYS.J.LETTERS 342,L35

BV PUP   1 SZKODY,P.,FEINSWOG,L.:1988,ASTROPHYS.J.334,422 (ORB.PER.)

CP PUP   1 O'DONOGHUE,D.,WARNER,B.,WARGAU,W.,GRAUER,A.D.:1989,
                    MONTHLY NOTICES ROY.ASTRON.SOC.240,41 (ORB.PER.)
                2 VOGT,N.,BARRERA,L.H.,BARWIG,H.,MANTEL,K.-H.:1990,IN:
                    ACCRETION-POWERED COMPACT BINARIES,ED. C.W.MAUCHE,
                    CAMBRIDGE UNIVERSITY PRESS,CAMBRIDGE,P.391
                3 DIAZ,M.P.,STEINER,J.E.:1991,PUBL.ASTRON.SOC.PACIFIC
                    103,964
                4 WHITE,J.C.,II,HONEYCUTT,R.K.,HORNE,K.:1993,ASTROPHYS.J.
                    412,278 (2. PER.)

V347 PUP   1 BUCKLEY,D.A.H.,SULLIVAN,D.J.,REMILLARD,R.A.,TUOHY,I.R.,
                    CLARK,M.:1990,ASTROPHYS.J.355,617 (M1/M2,INCL,M1,M2)
                2 BAPTISTA,R.,CIESLINSKY,D.:1991,IAU CIRC.NO.5407
                    (ORB.PER.)

V348 PUP   1 TUOHY,I.R.,REMILLARD,R.A.,BRISSENDEN,R.J.V.,BRADT,H.V.:
                    1990,ASTROPHYS.J.359,204 (ORB.PER.)

T PYX   1 KRAUTTER,J.:1990,PRIVATE COMMUNICATION (2. PER.)
                2 SCHAEFER,B.:1990,ASTROPHYS.J.LETTERS 355,L39
                3 SCHAEFER,B.E.,LANDOLT,A.U.,VOGT,N.,BUCKLEY,D.,WARNER,B.,
                    WALKER,A.R.,BOND,H.E.:1992,ASTROPHYS.J.SUPPL.SER.
                    81,321 (ORB.PER.)

VZ PYX   1 DE MARTINO,D.,GONZALEZ-RIESTRA,E.,RODRIGUEZ,P.,
                    BUCKLEY,D.,DICKSON,J.,REMILLARD,R.E.:1992,IAU CIRC.
                    NO.5481
                2 REMILLARD,R.E.:1992,PRIVATE COMMUNICATION (ORB.PER.,
                    4. PER.)
                3 BARWIG,H.,WIMMER,W.,BUES,I.:1993,IAU CIRC.NO.5689

V SGE   1 HERBIG,G.H.,PRESTON,G.W.,SMAK,J.,PACZYNSKI,B.:1965,
                    ASTROPHYS.J.141,617 (M1/M2,INCL,M1,M2)
                2 KOCH,R.H.,CORCORAN,M.F.,HOLENSTEIN,B.D.,
                    MC CLUSKEY,JR.,G.E.:1986,ASTROPHYS.J.306,618
                    (ORB.PER.)
                3 WILLIAMS,G.A.,KING,A.R.,UOMOTO,A.K.,HILTNER,W.A.:1986,
                    MONTHLY NOTICES ROY.ASTRON.SOC.219,809

RZ SGE   1 BOND,H.E.,KEMPER,E.,MATTEI,J.:1982,ASTROPHYS.J.LETTERS
                    260,L79 (2. PER.)

WY SGE   1 SHARA,M.M.,MOFFAT,A.F.J.,MC GRAW,J.T.,DEARBORN,D.S.,
                    BOND,H.E.,KEMPER,E.,LAMONTAGNE,R.:1984,ASTROPHYS.J.
                    282,763 (ORB.PER.)

WZ SGE   1 ROBINSON,E.L.,NATHER,R.E.,PATTERSON,J.:1978,ASTROPHYS.J.
                    219,168 (ORB.PER.)
                2 PATTERSON,J.:1980,ASTROPHYS.J.241,235 (3. PER.)
                3 PATTERSON,J.,MC GRAW,J.T.,COLEMAN,L.,AFRICANO,J.L.:1981,
                    ASTROPHYS.J.248,1067 (2. PER.)
                4 SION,E.M.,LECKENBY,H.J.,SZKODY,P.:1990,ASTROPHYS.J.
                    LETTERS 364,L41
                5 SMAK,J.:1993,ACTA ASTRON.43,101 (M1/M2,INCL,M1,M2)

V1017 SGR   1 SEKIGUCHI,K.:1992,NATURE 358,563 (ORB.PER.)

V1223 SGR   1 OSBORNE,J.P.,ROSEN,R.,MASON,K.O.,BEUERMANN,K.:1985,SPACE
                    SCI.REV.40,143 (3. PER.)

V1223 SGR   2 PENNING,W.R.:1985,ASTROPHYS.J.289,300 (INCL,M1,M2)
             3 JABLONSKI,F.,STEINER,J.E.:1987,ASTROPHYS.J.323,672
                 (ORB.PER.,4. PER.)
             4 VAN AMERONGEN,S.,AUGUSTEIJN,T.,VAN PARADIJS,J.:1987,
                 MONTHLY NOTICES ROY.ASTRON.SOC.228,377 (4. PER.)

V3885 SGR   1 COWLEY,A.P.,CRAMPTON,D.,HESSER,J.E.:1977,ASTROPHYS.J.
                 214,471 (M1/M2,INCL,M1,M2)
             2 METZ,K.:1989,INF.BULL.VARIABLE STARS NO.3385 (ORB.PER.)

V4140 SGR   1 BAPTISTA,R.,JABLONSKI,F.J.,STEINER,J.E.:1992,ASTRON.J.
                 104,1557 (ORB.PER.)

U   SCO   1 SCHAEFER,B.:1990,ASTROPHYS.J.LETTERS 355,L39
             2 BUDZINOVSKAYA,I.A.,PAVLENKO,E.P.,SHUGAROV,S.YU.:1992,
                 SOVIET ASTRON.LETTERS 18,201 (2. PER.)
             3 JOHNSTON,H.M.,KULKARNI,S.R.:1992,ASTROPHYS.J.396,267
             4 DUERBECK,H.W.,DUEMMLER,R.,SEITTER,W.C.,LEIBOWITZ,E.M.,
                 SHARA,M.M.:1993,PAPER PRESENTED AT THE 2ND HAIFA
                 WORKSHOP ON CVS (ORB.PER.)

VY SCL   1 HUTCHINGS,J.,COWLEY,A.P.:1984,PUBL.ASTRON.SOC.PACIFIC
                 96,559 (ORB.PER.)

VZ SCL   1 WARNER,B.,THACKERAY,A.D.:1975,MONTHLY NOTICES ROY.ASTRON.
                 SOC.172,433 (ORB.PER.,M1/M2,INCL,M2)
             2 WILLIAMS,R.E.:1989,ASTRON.J.97,1752

UZ SER   1 ECHEVARRIA,J.,JONES,D.H.P.,WALLIS,R.E.,MAYO,S.K.,
                 HASSALL,B.J.M.,PRINGLE,J.E.,WHELAN,J.A.J.:1981,
                 MONTHLY NOTICES ROY.ASTRON.SOC.197,565 (ORB.PER.)
             2 ECHEVARRIA,J.:1988,REV.MEXICANA ASTRON.ASTROPHYS.16,37

CT SER   1 RINGWALD,F.A.:1993,PH.D.THESIS,DARTMOUTH COLLEGE,NH
                 (ORB.PER.)

LX SER   1 SHAFTER,A.W.:1983,PH.D. THESIS,UCLA (M1/M2,INCL,M1,M2)
             2 EASON,E.L.E.,WORDEN,S.P.,KLIMKE,A.,AFRICANO,J.L.:1984,
                 PUBL.ASTRON.SOC.PACIFIC 96,372 (ORB.PER)
             3 RUTTEN,R.G.M.,VAN PARADIJS,J.,TINBERGEN,J.:1992,ASTRON.
                 ASTROPHYS.260,213

MR SER   1 LIEBERT,J.,STOCKMAN,H.S.,WILLIAMS,R.E.,TAPIA,S.,
                 GREEN,R.F.,RAUTENKRANZ,D.,FERGUSON,D.H.,SZKODY,P.:
                 1982,ASTROPHYS.J.256,594 (INCL)
             2 ANGELINI,L.,OSBORNE,J.P.,STELLA,L.:1990,MONTHLY NOTICES
                 ROY.ASTRON.SOC.245,652
             3 SCHWOPE,A.D.,,THOMAS,H.-C.,BEUERMANN,K.,NAUNDORF,C.E.:
                 1991,ASTRON.ASTROPHYS.224,373 (ORB.PER.)
             4 WICKRAMASINGHE,D.T.,CROPPER,M.,MASON,K.O.,GARLICK,M.:
                 1991,MONTHLY NOTICES ROY.ASTRON.SOC.250,692

RW SEX   1 BEUERMANN,K.,STASIEWSKI,U.,SCHWOPE,A.D.:1992,ASTRON.
                 ASTROPHYS.256,433 (ORB.PER.,M1/M2,INCL,M1,M2)

SW SEX   1 PENNING,R.W.,FERGUSON,D.H.,MC GRAW,J.T.,LIEBERT,J.,
                 GREEN,R.F.:1984,ASTROPHYS.J.276,233 (ORB.PER.,INCL,
                 M1,M2)
             2 SZKODY,P.,PICHE,F.:1990,ASTROPHYS.J.361,235
             3 RUTTEN,R.G.M.,VAN PARADIJS,J.,TINBERGEN,J.:1992,ASTRON.
                 ASTROPHYS.260,213

KK TEL   1 HOWELL,S.B.,SZKODY,P.,KREIDL,T.J.,DOBRZYCKA,D.:1991,
             PUBL.ASTRON.SOC.PACIFIC 103,300 (ORB.PER.)

RW TRI   1 SHAFTER,A.W.:1983,PH.D. THESIS,UCLA (M1/M2,INCL,M1,M2)
          2 ROBINSON,E.L.,SHETRONE,M.D.,AFRICANO,J.L.:1991,ASTRON.J.
             103,1176 (ORB.PER.)
          3 RUTTEN,R.G.M.,DHILLON,V.S.:1992,ASTRON.ASTROPHYS.253,139
          4 RUTTEN,R.G.M.,VAN PARADIJS,J.,TINBERGEN,J.:1992,ASTRON.
             ASTROPHYS.260,213

EK TRA   1 VOGT,N.,SEMENIUK,I.:1980,ASTRON.ASTROPHYS.89,223
             (2. PER.)

SU UMA   1 THORSTENSEN,J.R.,WADE,R.A.,OKE,J.B.:1986,ASTROPHYS.J.
             309,721 (ORB.PER.)
          2 UDALSKI,A.:1990,ASTRON.J.100,226 (2. PER.)
          3 WOODS,J.A.,DREW,J.E.,VERBUNT,F.:1990,MONTHLY NOTICES ROY.
             ASTRON.SOC.245,323

SW UMA   1 SHAFTER,A.W.:1983,PH.D. THESIS,UCLA (M1/M2,INCL,M1,M2)
          2 SHAFTER,A.W.,SZKODY,P.,THORSTENSEN,J.R.:1986,ASTROPHYS.J.
             308,765 (3. PER.)
          3 ROBINSON,E.L.,SHAFTER,A.W.,HILL,A.J.,WOOD,M.A.,
             MATTEI,J.A.:1987,ASTROPHYS.J.313,772 (2. PER.)
          4 HOWELL,S.B.,SZKODY,P.:1988,PUBL.ASTRON.SOC.PACIFIC
             100,224 (ORB.PER.)
          5 SZKODY,P.,OSBORNE,J.,HASSALL,B.J.M.:1988,ASTROPHYS.J.
             328,243
          6 KATO,T.,HIRATA,R.,MINESHIGE,S.:1992,PUBL.ASTRON.SOC.
             JAPAN 44,L215

UX UMA   1 SHAFTER,A.W.:1983,PH.D. THESIS,UCLA (M1/M2,INCL,M1,M2)
          2 RUBENSTEIN,E.P.,PATTERSON,J.,AFRICANO,J.L.:1991,PUBL.
             ASTRON.SOC.PACIFIC 103,1258 (ORB.PER.)
          3 RUTTEN,R.G.M.,VAN PARADIJS,J.,TINBERGEN,J.:1992,ASTRON.
             ASTROPHYS.260,213
          4 RUTTEN,R.G.M.,DHILLON,V.S.,HORNE,K.,KUULKERS,E.,
             VAN PARADIJS,J.:1993,NATURE 362,518

AN UMA   1 BONNET-BIDAUD,J.M.,MOUCHET,M.,SOMOVA,T.A.,SOMOV,N.N.:
             1992,IAU CIRC.NO.5673 (ORB.PER.)

BC UMA   1 HOWELL,S.B.,SZKODY,P.,KREIDL,T.J.,MASON,K.O.,
             PUCHNAREWICZ,E.M.:1990,PUBL.ASTRON.SOC.PACIFIC
             102,758 (ORB.PER.)
          2 MUKAI,K.,MASON,K.O.,HOWELL,S.B.,ALLINGTON-SMITH,J.,
             CALLANAN,P.J.,CHARLES,P.A.,HASSALL,B.J.M.,MACHIN,G.,
             NAYLOR,T.,SMALE,A.P.,VAN PARADIJS,J.:1990,MONTHLY
             NOTICES ROY.ASTRON.SOC.245,385

BZ UMA   1 CLAUDI,R.,BIANCHINI,A.,MUNARI,U.:1990,IAU CIRC.NO.4975
          2 RINGWALD,F.A.,THORSTENSEN,J.R.:1990,BULL.AMERICAN.ASTRON.
             SOC.22,1291 (ORB.PER.)
          3 RINGWALD,F.A.:1993,PH.D.THESIS,DARTMOUTH COLLEGE,NH
             (ORB.PER.)

CH UMA   1 FRIEND,M.T.,MARTIN,J.S.,SMITH,R.C.,JONES,D.H.P.:1990,
             MONTHLY NOTICES ROY.ASTRON.SOC.246,654 (ORB.PER.,
             M1/M2,INCL,M1,M2)

CI UMA   1 BRUCH,A.:1989,ASTRON.ASTROPHYS.SUPPL.SER.78,145
          2 HOWELL,S.B.,SZKODY,P.,KREIDL,T.J.,MASON,K.O.,
             PUCHNAREWICZ,E.M.:1990,PUBL.ASTRON.SOC.PACIFIC
             102,758 (ORB.PER.)

CI UMA 3 MUKAI,K.,MASON,K.O.,HOWELL,S.B.,ALLINGTON-SMITH,J.,
 CALLANAN,P.J.,CHARLES,P.A.,HASSALL,B.J.M.,MACHIN,G.,
 NAYLOR,T.,SMALE,A.P.,VAN PARADIJS,J.:1990,MONTHLY
 NOTICES ROY.ASTRON.SOC.245,385

CY UMA 1 KATO,T.,FUJINO,S.,IIDA,M.,MAKIGUCHI,N.,KASHIRO,M.:1988,
 VARIABLE STAR BULL.JAPAN NO.5,18 (2. PER.)
 2 WATANABE,M.,HIROSAWA,K.,KATO,T.,NARUMI,H.:1989,VARIABLE
 STAR BULL.JAPAN NO.10,40

DV UMA 1 HOWELL,S.B.,MASON,K.O.,REICHERT,G.A.,WARNOCK,W.,
 KREIDL,T.J.:1988,MONTHLY NOTICES ROY.ASTRON.SOC.
 233,79 (ORB.PER.)
 2 MUKAI,K.,MASON,K.O.,HOWELL,S.B.,ALLINGTON-SMITH,J.,
 CALLANAN,P.J.,CHARLES,P.A.,HASSALL,B.J.M.,MACHIN,G.,
 NAYLOR,T.,SMALE,A.P.,VAN PARADIJS,J.:1990,MONTHLY
 NOTICES ROY.ASTRON.SOC.245,385
 3 HOWELL,S.B.,BLANTON,S.A.:1993,ASTRON.J.106,311
 4 SZKODY,P.,HOWELL,S.B.:1993,ASTROPHYS.J.403,743

DW UMA 1 SHAFTER,A.W.,HESSMAN,F.V.,ZHANG,E.H.:1988,ASTROPHYS.J.
 327,248 (ORB.PER.,INCL,M1,M2)
 2 SZKODY,P.,PICHE,F.:1990,ASTROPHYS.J.361,235
 3 KOPYLOV,I.M.,SOMOV,N.N.,SOMOVA,T.A.:1993,BULL.SPEC.
 ASTROPHYS.OBS.NORTH CAUCASUS 31,12

EI UMA 1 THORSTENSEN,J.R.:1986,ASTRON.J.91,940 (ORB.PER.)

EK UMA 1 MORRIS,S.L.,SCHMIDT,G.D.,LIEBERT,J.,STOCKE,J.,GIOIA,I.,
 MACCACARO,T.:1987,ASTROPHYS.J.314,641 (ORB.PER.,
 INCL)
 2 CROPPER,M.,MASON,K.O.,MUKAI,K.:1990,MONTHLY NOTICES ROY.
 ASTRON.SOC.243,565

RW UMI 1 HOWELL,S.B.,SZKODY,P.,KREIDL,T.J.,DOBRZYCKA,D.:1991,
 PUBL.ASTRON.SOC.PACIFIC 103,300 (ORB.PER.)
 2 CAMPBELL,R.D.,SHAFTER,A.W.:1992,IN: PROC. 12TH NORTH
 AMERICAN WORKSHOP ON CVS AND LMXRBS,A.W.SHAFTER
 (ED.),MOUNT LAGUNA OBSERVATORY,SDSU,P.4

SS UMI 1 CHEN,J.-S.,LIU,X.-W.,WEI,M.-Z.:1990,ASTRON.ASTROPHYS.
 242,397 (2. PER.)
 2 UDALSKI,A.:1990,INF.BULL.VARIABLE STARS NO.3425

CU VEL 1 VOGT,N.:1981,HABILITATION THESIS,UNIVERSITY BOCHUM
 (2. PER.)

IX VEL 1 BEUERMANN,K.,THOMAS,H.-C.:1990,ASTRON.ASTROPHYS.230,326
 (ORB.PER.,M1/M2,INCL,M1,M2)
 2 MAUCHE,C.W.:1991,ASTROPHYS.J.373,624

KO VEL 1 KUBIAK,M.,KRZEMINSKI,W.:1989,PUBL.ASTRON.SOC.PACIFIC
 101,669 (3. PER.,4. PER.)
 2 MUKAI,K.,CORBET,R.H.D.:1991,ASTROPHYS.J.378,701
 3 SAMBRUNA,R.M.,CHIAPPETTI,L.,TREVES,A.,BONNET-BIDAUD,J.M.,
 BOUCHET,P.,MARASCHI,L.,MOTCH,C.,MOUCHET,M.,
 VAN AMERONGEN,S.:1992,ASTROPHYS.J.391,750 (ORB.PER.)

TW VIR 1 SHAFTER,A.W.:1983,INF.BULL.VARIABLE STARS NO.2377
 (ORB.PER.)
 2 SHAFTER,A.W.:1983,PH.D. THESIS,UCLA (M1/M2,INCL,M1,M2)
 3 MANSPERGER,C.S.,KAITCHUCK,R.H.:1990,ASTROPHYS.J.358,260

TW VIR   4 SZKODY,P.,PICHE,F.,FEINSWOG,L.:1990,ASTROPHYS.J.SUPPL.
                  SER.73,441

HS VIR   1 BRUCH,A.:1989,ASTRON.ASTROPHYS.SUPPL.SER.78,145
            2 HOWELL,S.B.,SZKODY,P.,KREIDL,T.J.,MASON,K.O.,
                  PUCHNAREWICZ,E.M.:1990,PUBL.ASTRON.SOC.PACIFIC
                  102,758
            3 RINGWALD,F.A.:1993,PH.D.THESIS,DARTMOUTH COLLEGE,NH
                  (ORB.PER.)

HV VIR   1 INGRAM,D.,SZKODY,P.:1992,INF.BULL.VARIABLE STARS NO.3810
            2 MANTEL,K.H.,BARWIG,H.,RITTER,H.:1992,ASTRON.ASTROPHYS.
                  266,L5
            3 LEIBOWITZ,E.M.,MENDELSON,H.,BRUCH,A.,DUERBECK,H.W.,
                  SEITTER,W.C.,RICHTER,G.A.:1993,ASTROPHYS.J.,IN PRESS
                  (ORB.PER.,2. PER.)

VW VUL   1 SHAFTER,A.W.:1983,PH.D. THESIS,UCLA (M1/M2,INCL,M1,M2)
            2 SHAFTER,A.W.:1985,ASTRON.J.90,643 (ORB.PER.)

PW VUL   1 HACKE,G.:1987,INF.BULL.VARIABLE STARS NO.2979 (ORB.PER.)

QQ VUL   1 NOUSEK,J.A.,TAKALO,L.O.,SCHMIDT,G.D.,TAPIA,S.,HILL,G.J.,
                  BOND,H.E.,GRAUER,A.D.,STERN,R.A.,AGRAWAL,P.C.:1984,
                  ASTROPHYS.J.277,682 (INCL)
            2 ANDRONOV,I.L.,FUHRMANN,B.:1987,INF.BULL.VARIABLE STARS
                  NO.2976 (ORB.PER.)
            3 MUKAI,K.,CHARLES,P.A.,SMALE,A.P.:1988,ASTRON.ASTROPHYS.
                  194,153

DR V211B   1 DRISSEN,L.,SHARA,M.,DOPITA,M.,WICKRAMASINGHE,D.,BELL,J.,
                  BAILEY,J.,HOUGH,J.:1992,IAU CIRC.NO.5609 (ORB.PER.)

GD 552   1 HESSMAN,F.V.,HOPP,U.:1990.ASTRON.ASTROPHYS.228,387
                  (ORB.PER.,M1/M2,INCL,M1,M2)

S 193   1 DOWNES,R.A.,KEYES,C.D.:1988,ASTRON.J.96,777
            2 SZKODY,P.,GARNAVICH,P.,HOWELL,S.,KII,T.:1990,IN:
                  ACCRETION-POWERED COMPACT BINARIES,ED. C.W. MAUCHE,
                  CAMBRIDGE UNIVERSITY PRESS,CAMBRIDGE,P.251
                  (ORB.PER.,3. PER.)

X 0022-7221 1 PARESCE,F.,DE MARCHI,G.,FERRARO,F.R.:1992,NATURE 360,46
                  (ORB.PER.)

RX J0203+29 1 BEUERMANN,K.,THOMAS,H.-C.:1992,IN:PROC. COSPAR SYMP.
                  "RECENT RESULTS ON X-RAY AND EUV ASTRONOMY"
            2 BURWITZ,V.,REINSCH,K.:1993,PRIVATE COMMUNICATION
                  (ORB.PER.)

1H 0253+193 1 KAMATA,Y.,TAWARA,Y.,KOYAMA,K.:1991,ASTROPHYS.J.LETTERS
                  379,L65 (3. PER.)
            2 ZUCKERMAN,B.,BECKLIN,E.E.,MC LEAN,I.S.,PATTERSON,J.:1992,
                  ASTROPHYS.J.400,665 (ORB.PER.,INCL)
            3 KAMATA,Y.,KOYAMA,K.:1993,ASTROPHYS.J.405,307

EXO 0329-260 1 BEUERMANN,K.,THOMAS,H.-C.,GIOMMI,P.,TAGLIAFERRI,G.,
                  SCHWOPE,A.D.:1989,ASTRON.ASTROPHYS.219,L7 (ORB.PER.)

RX J0453-42 1 BEUERMANN,K.,THOMAS,H.-C.:1992,IN:PROC. COSPAR SYMP.
                  "RECENT RESULTS ON X-RAY AND EUV ASTRONOMY"
                  (ORB.PER.)

RX J0515+01   1 BEUERMANN,K.,THOMAS,H.-C.:1992,IN:PROC. COSPAR SYMP.
                  "RECENT RESULTS ON X-RAY AND EUV ASTRONOMY"
                  (ORB.PER.)

RX J0531-46   1 MASON,K.O.,WATSON,M.G.,PONMAN,T.J.,CHARLES,P.A.,
                  DUCK,S.R.,HASSALL,B.J.M.,HOWELL,S.B.,ISHIDA,M.,
                  JONES,D.H.P.,MITTAZ,J.P.D.:1992,MONTHLY NOTICES ROY.
                  ASTRON.SOC.258,749
              2 BEUERMANN,K.:1993,PRIVATE COMMUNICATION (ORB.PER.)

H 0616-818   1 REMILLARD,R.E.:1992,PRIVATE COMMUNICATION (ORB.PER.,
                  4. PER.)

RE J0751+14   1 MASON,K.O.,WATSON,M.G.,PONMAN,T.J.,CHARLES,P.A.,
                  DUCK,S.R.,HASSALL,B.J.M.,HOWELL,S.B.,ISHIDA,M.,
                  JONES,D.H.P.,MITTAZ,J.P.D.:1992,MONTHLY NOTICES ROY.
                  ASTRON.SOC.258,749 (ORB.PER.,3. PER.)
              2 PIIROLA,V.,HAKALA,P.:1993,IAU CIRC.NO.5707 (4. PER.)
              3 PIIROLA,V.,HAKALA,P.,COYNE,G.V.:1993,ASTROPHYS.J.LETTERS
                  410,L107
              4 ROSEN,S.R.,MITTAZ,J.P.D.,HAKALA,P.J.:1993,MONTHLY NOTICES
                  ASTRON.SOC.264,171

PG 0859+415   1 RINGWALD,F.A.:1993,PH.D.THESIS,DARTMOUTH COLLEGE,NH
                  (ORB.PER.)

PG 0917+342   1 HOWELL,S.B.,SZKODY,P.,KREIDL,T.J.,DOBRZYCKA,D.:1991,
                  PUBL.ASTRON.SOC.PACIFIC 103,300
              2 DOBRZYCKA,D.,HOWELL,S.B.:1992,ASTROPHYS.J.388,614
                  (ORB.PER.)

1H 0927+501   1 REMILLARD,R.E.:1992,PRIVATE COMMUNICATION (ORB.PER.)

PG 0943+521   1 RINGWALD,F.A.:1993,PH.D.THESIS,DARTMOUTH COLLEGE,NH
                  (ORB.PER.)

PG 1000+667   1 RINGWALD,F.A.:1993,PH.D.THESIS,DARTMOUTH COLLEGE,NH
                  (ORB.PER.)

RX J1002-19   1 BEUERMANN,K.,THOMAS,H.-C.:1992,IN:PROC. COSPAR SYMP.
                  "RECENT RESULTS ON X-RAY AND EUV ASTRONOMY"
              2 BEUERMANN,K.:1993,PRIVATE COMMUNICATION (ORB.PER.)

RX J1007-20   1 BEUERMANN,K.,THOMAS,H.-C.:1992,IN:PROC. COSPAR SYMP.
                  "RECENT RESULTS ON X-RAY AND EUV ASTRONOMY"
                  (ORB.PER.)

1ES 1113+432 1 REMILLARD,R.A.,SILBER,A.D.,SCHACHTER,J.F.,SLANE,P.:1993,
                  BULL.AMERICAN ASTRON.SOC.25,910 (ORB.PER.)

RE J1149+28   1 MITTAZ,J.P.D.,ROSEN,S.R.,MASON,K.O.,HOWELL,S.B.:1992,
                  MONTHLY NOTICES ROY.ASTRON.SOC.258,277 (ORB.PER.)

RX J1307+53   1 BEUERMANN,K.,THOMAS,H.-C.:1992,IN:PROC. COSPAR SYMP.
                  "RECENT RESULTS ON X-RAY AND EUV ASTRONOMY"
                  (ORB.PER.)

RX J1313-32   1 BEUERMANN,K.,THOMAS,H.-C.:1992,IN:PROC. COSPAR SYMP.
                  "RECENT RESULTS ON X-RAY AND EUV ASTRONOMY"
                  (ORB.PER.)

PG 1524+622 1 RINGWALD,F.A.:1993,PH.D.THESIS,DARTMOUTH COLLEGE,NH
                (ORB.PER.)

1H 1752+081 1 REMILLARD,R.E.:1992,PRIVATE COMMUNICATION (ORB.PER.)

RX J1844-74 1 BEUERMANN,K.,THOMAS,H.-C.:1992,IN:PROC. COSPAR SYMP.
                "RECENT RESULTS ON X-RAY AND EUV ASTRONOMY"
                (ORB.PER.)

1H 1929+509 1 SCHMELZ,J.T.,FEIGELSON,E.D.,SCHWARTZ,D.A.:1986,ASTRON.J.
                92,585
             2 REMILLARD,R.E.:1992,PRIVATE COMMUNICATION (ORB.PER.)

EC 1931-5915 1 BUCKLEY,D.A.H.,O'DONOGHUE,D.,KILKENNY,D.,STOBIE,R.S.,
                REMILLARD,R.A.:1992,MOTHLY NOTICES ROY.ASTRON.SOC.
                258,285 (ORB.PER.)

RX J1938-46 1 BUCKLEY,D.A.H.,O'DONOGHUE,D.,HASSALL,B.J.M.,KELLETT,B.J.,
                MASON,K.O.,SEKIGUCHI,K.,WATSON,M.G.,WHEATLEY,P.J.,
                CHEN,A.:1993,MONTHLY NOTICES ROY.ASTRON.SOC.262,93
                (ORB.PER.)
             2 WARREN,J.K.,VALLERGA,J.V.,MAUCHE,C.W.,MUKAI,K.,
                SIEGMUND,O.H.W.:1993,ASTROPHYS.J.LETTERS 414,L69

RX J1940-10 1 DONE,C.,MADEJSKI,G.M.,MUSHOTZKY,R.F.,TURNER,T.J.,
                KOYAMA,K.,KUNIEDA,H.:1992,ASTROPHYS.J.400.138
                (ORB.PER.)
             2 ROSEN,S.,DONE,C.,WATSON,M.G.,MADEJSKI,G.:1993,IAU CIRC.
                NO.5850 (ORB.PER.)

RX J1957-57 1 BEUERMANN,K.,THOMAS,H.-C.:1992,IN:PROC. COSPAR SYMP.
                "RECENT RESULTS ON X-RAY AND EUV ASTRONOMY"
                (ORB.PER.)

RX J2107-05 1 HAKALA,P.J.,WATSON,M.G.,VILHU,O.,HASSALL,B.J.M.,
                KELLETT,B.J.,MASON,K.O.,PIIROLA,V.:1993,MONTHLY
                NOTICES ROY.ASTRON.SOC.263,61
             2 SCHWOPE,A.D.,THOMAS,H.-C.,BEUERMANN,K.:1993,ASTRON.
                ASTROPHYS.271,L25 (ORB.PER.,M1/M2,M1,M2)

PG 2133+115 1 RINGWALD,F.A.:1993,PH.D.THESIS,DARTMOUTH COLLEGE,NH
                (ORB.PER.)

RX J2316-05 1 BEUERMANN,K.,THOMAS,H.-C.:1992,IN:PROC. COSPAR SYMP.
                "RECENT RESULTS ON X-RAY AND EUV ASTRONOMY"
                (ORB.PER.)
             2 THOMAS,H.-C.:1993,PRIVATE COMMUNICATION (2. PER.)

REFERENCES FOR FINDING CHARTS

IN ORDER TO KEEP THIS LIST AS SHORT AS POSSIBLE, ABBREVIATED REFERENCES
ARE USED FOR FREQUENTLY OCCURING LONG REFERENCES. THE FULL REFERENCES
ARE GIVEN AT THE END OF THIS SECTION.

RX AND	1 DOWNES & SHARA (1993)
AR AND	1 DOWNES & SHARA (1993)
DX AND	1 BRUCH ET AL. (1987) 2 DOWNES & SHARA (1993)
FO AND	1 DOWNES & SHARA (1993)
HV AND	1 DOWNES & SHARA (1993)
PX AND	1 GREEN ET AL. (1982) 2 DOWNES & SHARA (1993) (OBJECT AND 1)
UU AQR	1 DOWNES & SHARA (1993)
VY AQR	1 DOWNES & SHARA (1993)
AE AQR	1 DOWNES & SHARA (1993)
FO AQR	1 DOWNES & SHARA (1993)
HL AQR	1 DOWNES & SHARA (1993)
UU AQL	1 DOWNES & SHARA (1993)
V603 AQL	1 DUERBECK (1987)
V794 AQL	1 MUKAI ET AL. (1990) 2 DOWNES & SHARA (1993)
V1315 AQL	1 DOWNES & SHARA (1993)
TT ARI	1 DOWNES & SHARA (1993)
WX ARI	1 DOWNES & SHARA (1993)
T AUR	1 DUERBECK (1987)
SS AUR	1 DOWNES & SHARA (1993)
FS AUR	1 DOWNES & SHARA (1993)
KR AUR	1 DOWNES & SHARA (1993)
QZ AUR	1 DUERBECK (1987)
V363 AUR	1 DOWNES & SHARA (1993)
TT BOO	1 DOWNES & SHARA (1993)
UZ BOO	1 DOWNES & SHARA (1993)
CR BOO	1 GREEN ET AL. (1986) (PG 1346+082)

```
Z   CAM      1 DOWNES & SHARA (1993)

AF  CAM      1 DOWNES & SHARA (1993)

BY  CAM      1 DOWNES & SHARA (1993)

BZ  CAM      1 DOWNES & SHARA (1993)

SY  CNC      1 DOWNES & SHARA (1993)

YZ  CNC      1 DOWNES & SHARA (1993)

AC  CNC      1 DOWNES & SHARA (1993)

AR  CNC      1 MUKAI ET AL. (1990)
             2 DOWNES & SHARA (1993)

AT  CNC      1 DOWNES & SHARA (1993)

CC  CNC      1 VOGT & BATESON (1982)
             2 DOWNES & SHARA (1993)

EU  CNC      1 GILLILAND,R.L.,BROWN,T.M.,DUNCAN,D.K.,SUNTZEFF,N.B.,
                LOCKWOOD,G.W.,THOMPSON,D.T.,SCHILD,R.E.,
                JEFFREY,W.A.,PENPRASE,B.E.:1991,ASTRON.J.101,541

AM  CVN      1 WILLIAMS (1983)

WY  CMA

HL  CMA      1 CHLEBOWSKI,T.,HALPERN,J.P.,STEINER,J.E.:1981,ASTROPHYS.J.
                LETTERS 247,L35

BG  CMI      1 DOWNES & SHARA (1993)

OY  CAR      1 DOWNES & SHARA (1993)

QU  CAR      1 DOWNES & SHARA (1993)

HT  CAS      1 DOWNES & SHARA (1993)

V425 CAS     1 DOWNES & SHARA (1993)

BV  CEN      1 DOWNES & SHARA (1993)

MU  CEN      1 VOGT & BATESON (1982)
             2 DOWNES & SHARA (1993)

V436 CEN     1 DOWNES & SHARA (1993)

V442 CEN     1 DOWNES & SHARA (1993)

V485 CEN     1 VOGT & BATESON (1982)
             2 DOWNES & SHARA (1993)

V803 CEN     1 ELVIUS,A.:1975,ASTRON.ASTROPHYS.44,117

V834 CEN     1 DOWNES & SHARA (1993)

WW  CET      1 DOWNES & SHARA (1993)

WX  CET      1 DOWNES & SHARA (1993)
```

```
Z   CHA        1 DOWNES & SHARA (1993)

TV COL         1 DOWNES & SHARA (1993)

TX COL         1 DOWNES & SHARA (1993)

AL COM         1 MUKAI ET AL. (1990)
               2 DOWNES & SHARA (1993)

GP COM         1 WILLIAMS (1983)

V394 CRA       1 DUERBECK (1987)

TT CRT         1 DOWNES & SHARA (1993)

SS CYG         1 SHARA ET AL. (1993)
               2 DOWNES & SHARA (1993)

EM CYG         1 DOWNES & SHARA (1993)

EY CYG         1 DOWNES & SHARA (1993)

V503 CYG       1 DOWNES & SHARA (1993)

V751 CYG       1 DOWNES & SHARA (1993)

V1251 CYG      1 DOWNES & SHARA (1993)

V1500 CYG      1 DUERBECK (1987)

V1668 CYG      1 DUERBECK (1987)

V1776 CYG      1 SHAFTER,A.W.,LANNING,H.H.,ULRICH,R.K.:1983,PUBL.ASTRON.
                   SOC.PACIFIC 95,206
               2 GARNAVICH,P.M.,SZKODY,P.,MATEO,M.,FEINSWOG,L.,BOOTH,J.,
                   GOODRICH,B.,MILLER,H.R.,CARINI,M.T.,WILSON,J.W.:
                   1990,ASTROPHYS.J.365,696
               3 DOWNES & SHARA (1993)

CM DEL         1 DOWNES & SHARA (1993)

HR DEL         1 DUERBECK (1987)

AB DRA         1 VOLOSHINA,I.B.,SHUGAROV,S.YU.:1989,SOVIET ASTRON.LETTERS
                   15,312
               2 DOWNES & SHARA (1993)

CQ DRA BC      1 BONNER DURCHMUSTERUNG

DM DRA         1 DOWNES & SHARA (1993)

DO DRA         1 MUKAI ET AL. (1990)
               2 PATTERSON,J.,SCHWARTZ,D.A.,PYE,J.P.,BLAIR,W.P.,
                   WILLIAMS,G.A.,CAILLAULT,J.-P.:1992,ASTROPHYS.J.
                   392,233
               3 DOWNES & SHARA (1993)

EP DRA         1 REMILLARD,R.A.,STROOZAS,B.A.,TAPIA,S.,SILBER,A.:1991,
                   ASTROPHYS.J.379,715
               2 DOWNES & SHARA (1993) (OBJECT DRA 4)
```

```
AH ERI        1 DOWNES & SHARA (1993)

AQ ERI        1 KATO,T.:1991,INF.BULL.VARIABLE STARS NO.3671
              2 DOWNES & SHARA (1993)

CP ERI        1 DOWNES & SHARA (1993)

EF ERI        1 SHARA ET AL. (1993)
              2 DOWNES & SHARA (1993)

UZ FOR        1 DOWNES & SHARA (1993)

U  GEM        1 DOWNES & SHARA (1993)

AW GEM        1 DOWNES & SHARA (1993)

IR GEM        1 DOWNES & SHARA (1993)

RZ GRU        1 KELLY,B.D.,KILKENNY,D.,COOKE,J.A.:1981,MONTHLY NOTICES
                    ROY.ASTRON.SOC.196,91P
              2 DOWNES & SHARA (1993)

CE GRU        1 HAWKINS,M.R.S.:1981,NATURE 293,116 (VARIABLE V1)
              2 DOWNES & SHARA (1993) (OBJECT GRU 2)

CF GRU        1 HAWKINS,M.R.S.,VERON,P.:1987,ASTRON.ASTROPHYS.182,271
              2 DOWNES & SHARA (1993) (OBJECT GRU 3)

AH HER        1 DOWNES & SHARA (1993)

AM HER        1 DOWNES & SHARA (1993)

DQ HER        1 DUERBECK (1987)

V533 HER      1 DUERBECK (1987)

V544 HER      1 VOGT & BATESON (1982)
              2 DOWNES & SHARA (1993)

V795 HER      1 DOWNES & SHARA (1993)

V825 HER      1 GREEN ET AL. (1982)
              2 DOWNES & SHARA (1993)

V838 HER

WW HOR        1 DOWNES & SHARA (1993)

EX HYA        1 DOWNES & SHARA (1993)
              2 SHARA ET AL. (1993)

LY HYA        1 DOWNES & SHARA (1993) (OBJECT HYA 2)

VW HYI        1 SHARA ET AL. (1993)
              2 DOWNES & SHARA (1993)

WX HYI        1 DOWNES & SHARA (1993)

BL HYI        1 SHARA ET AL. (1993)
              2 DOWNES & SHARA (1993)

DI LAC        1 DUERBECK (1987)
```

```
T    LEO       1 DOWNES & SHARA (1993)

U    LEO       1 DUERBECK (1987)
               2 DOWNES,R.A.,SZKODY,P.:1989,ASTRON.J.97,1729

X    LEO       1 DOWNES & SHARA (1993)

RZ   LEO       1 DOWNES & SHARA (1993)

DO   LEO       1 DOWNES & SHARA (1993)

DP   LEO       1 DOWNES & SHARA (1993)

RU   LMI       1 PESCH,P.,SANDULEAK,N.:1986,ASTROPHYS.J.SUPPL.SER.60,543
               2 BRUCH ET AL. (1987)
               3 MUKAI ET AL. (1990)
               4 DOWNES & SHARA (1993)

ST   LMI       1 DOWNES & SHARA (1993)

SX   LMI       1 SANDULEAK,N.,PESCH,P.:1984,ASTROPHYS.J.SUPPL.SER.55,517
               2 DOWNES & SHARA (1993)

BR   LUP       1 DOWNES & SHARA (1993)

BH   LYN       1 DOWNES & SHARA (1993)

AY   LYR       1 DOWNES & SHARA (1993)

MV   LYR       1 ROSINO,L.,ROMANO,G.,MARZIANI,P.:1993,PUBL.ASTRON.SOC.
                    PACIFIC 105,51
               2 DOWNES & SHARA (1993)

TU   MEN       1 DOWNES & SHARA (1993)

BT   MON       1 DUERBECK (1987)

CW   MON       1 DOWNES & SHARA (1993)

GQ   MUS       1 DUERBECK (1987)

V380 OPH       1 DOWNES & SHARA (1993)

V426 OPH       1 DOWNES & SHARA (1993)

V442 OPH       1 DOWNES & SHARA (1993)

V841 OPH       1 DUERBECK (1987)

V2051 OPH      1 DOWNES & SHARA (1993)

V2214 OPH      1 BAPTISTA,R.,JABLONSKI,F.J.,CIESLINSKI,D.,STEINER,J.E.:
                    1993,ASTROPHYS.J.LETTERS 406,L67

CN   ORI       1 DOWNES & SHARA (1993)

CZ   ORI       1 VOGT & BATESON (1982)
               2 WILLIAMS (1983)
               3 DOWNES & SHARA (1993)

V1159 ORI      1 NATSVILSHVILI,R.SH.:1984,INF.BULL.VARIABLE STARS
                    NO.2565
```

V1159 ORI   2 JABLONSKI,F.J.,CIESLINSKI,D.:1992,ASTRON.ASTROPHYS.
                 259,198

V1193 ORI   1 DOWNES & SHARA (1993)

BD PAV   1 DOWNES & SHARA (1993)

RU PEG   1 DOWNES & SHARA (1993)

EF PEG   1 HOWELL,S.B.,SCHMIDT,R.,DE YOUNG,J.A.,FRIED,R.,SCHMEER,P.,
                 GRITZ,L.:1993,PUBL.ASTRON.SOC.PACIFIC 105,579
             2 DOWNES & SHARA (1993)

HX PEG   1 GREEN,R.F.,GREENSTEIN,J.L.,BOKSENBERG,A.:1976,PUBL.
                 ASTRON.SOC.PACIFIC 88,598
             2 DOWNES & SHARA (1993)

IP PEG   1 DOWNES & SHARA (1993)

V PER   1 DUERBECK (1987)

UV PER   1 KATO,T.:1990,INF.BULL.VARIABLE STARS NO.3522
             2 DOWNES & SHARA (1993)

FY PER   1 DOWNES & SHARA (1993)

GK PER   1 DUERBECK (1987)

KT PER   1 WILLIAMS (1983)
             2 BRUCH ET AL. (1987)
             3 DOWNES & SHARA (1993)

RR PIC   1 DUERBECK (1987)

TW PIC   1 DOWNES & SHARA (1993)

TY PSC   1 DOWNES & SHARA (1993)

AO PSC   1 DOWNES & SHARA (1993)

AY PSC   1 DOWNES & SHARA (1993)

TY PSA   1 DOWNES & SHARA (1993)

VV PUP   1 SHARA ET AL. (1993)
             2 DOWNES & SHARA (1993)

BV PUP   1 DOWNES & SHARA (1993)

CP PUP   1 DUERBECK (1987)

V347 PUP   1 BUCKLEY,D.A.H.,SULLIVAN,D.J.,REMILLARD,R.A.,TUOHY,I.R.,
                 CLARK,M.:1990,ASTROPHYS.J.355,617
             2 DOWNES & SHARA (1993) (OBJECT PUP 2)

V348 PUP   1 TUOHY,I.R.,REMILLARD,R.A.,BRISSENDEN,R.J.V.,BRADT,H.V.:
                 1990,ASTROPHYS.J.359,204
             2 DOWNES & SHARA (1993) (OBJECT PUP 1)

T PYX   1 DUERBECK (1987)

VZ PYX   1 DOWNES & SHARA (1993) (OBJECT PYX 1)

```
V    SGE        1 DOWNES & SHARA (1993)

RZ SGE         1 DOWNES & SHARA (1993)

WY SGE         1 DUERBECK (1987)

WZ SGE         1 DOWNES & SHARA (1993)

V1017 SGR      1 VIDAL,N.V.,RODGERS,A.W.:1973,PUBL.ASTRON.SOC.PACIFIC
                    86,26
               2 WILLIAMS (1983)
               3 DUERBECK (1987)

V1223 SGR      1 DOWNES & SHARA (1993)

V3885 SGR      1 METZ,K.:1990,INF.BULL.VARIABLE STARS NO.3413
               2 DOWNES & SHARA (1993)

V4140 SGR      1 DOWNES & SHARA (1993)

U    SCO       1 DUERBECK (1987)
               2 BUDZINOVSKAYA,I.A.,PAVLENKO,E.P.,SHUGAROV,S.YU.:1992,
                    SOVIET ASTRON.LETTERS 18,201

VY SCL         1 DOWNES & SHARA (1993)

VZ SCL         1 DOWNES & SHARA (1993)

UZ SER         1 DOWNES & SHARA (1993)

CT SER         1 WYCKOFF,S.,WEHINGER,P.A.:1978,PUBL.ASTRON.SOC.PACIFIC
                    90,557
               2 DUERBECK (1987)

LX SER         1 DOWNES & SHARA (1993)

MR SER         1 DOWNES & SHARA (1993)

RW SEX         1 DOWNES & SHARA (1993)

SW SEX         1 DOWNES & SHARA (1993)

KK TEL         1 VOGT & BATESON (1982)
               2 DOWNES & SHARA (1993)

RW TRI         1 DOWNES & SHARA (1993)

EK TRA         1 DOWNES & SHARA (1993)

SU UMA         1 DOWNES & SHARA (1993)

SW UMA         1 DOWNES & SHARA (1993)

UX UMA         1 DOWNES & SHARA (1993)

AN UMA         1 SHARA ET AL. (1993)
               2 DOWNES & SHARA (1993)

BC UMA         1 BARBIERI,C.,BARBON,R.,DE BASTIANI,L.,ROMANO,G.,PESCH,P.,
                    SANDULEAK,N.:1985,ASTRON.ASTROPHYS.SUPPL.SER.61,163
                    (NO.43)
```

BC UMA   2 MUKAI ET AL. (1990)
              3 DOWNES & SHARA (1993)

BZ UMA   1 DOWNES & SHARA (1993)

CH UMA   1 DOWNES & SHARA (1993)

CI UMA   1 BRUCH ET AL. (1987)
              2 MUKAI ET AL. (1990)
              3 DOWNES & SHARA (1993)

CY UMA   1 DOWNES & SHARA (1993)

DV UMA   1 MUKAI ET AL. (1990)
              2 DOWNES & SHARA (1993)

DW UMA   1 DOWNES & SHARA (1993)

EI UMA   1 DOWNES & SHARA (1993)

EK UMA   1 SHARA ET AL. (1993)
              2 DOWNES & SHARA (1993)

RW UMI   1 DUERBECK (1987)
              2 KALUZNY,J.,CHLEBOWSKI,T.:1989,ACTA ASTRON.39,35

SS UMI   1 CHEN,J.-S.,LIU,X.-W.,WEI,M.-Z.:1990,ASTRON.ASTROPHYS.
                  242,397
              2 DOWNES & SHARA (1993)

CU VEL   1 DOWNES & SHARA (1993)

IX VEL   1 SHARA ET AL. (1993)
              2 DOWNES & SHARA (1993)

KO VEL   1 DOWNES & SHARA (1993)

TW VIR   1 DOWNES & SHARA (1993)

HS VIR   1 GREEN ET AL. (1982)
              2 DOWNES & SHARA (1993)

HV VIR   1 DUERBECK (1987)

VW VUL   1 DOWNES & SHARA (1993)

PW VUL   1 DUERBECK (1987)

QQ VUL   1 SHARA ET AL. (1993)
              2 DOWNES & SHARA (1993)

DR V211B

GD 552   1 DOWNES & SHARA (1993) (OBJECT CEP 1)

S 193   1 DOWNES & SHARA (1993) (OBJECT PEG 6)

X 0022-7221   1 PARESCE,F.,DE MARCHI,G.,FERRARO,F.R.:1992,NATURE 360,46

RX J0203+29

1H 0253+193

EXO 0329-260 1 DOWNES & SHARA (1993) (OBJECT FOR 1)

RX J0453-42   1 SHARA ET AL. (1993)

RX J0515+01

RX J0531-46   1 SHARA ET AL. (1993)

H 0616-818   1 DOWNES & SHARA (1993) (OBJECT MEN 1)

RE J0751+14   1 ANDRONOV,I.L.:1993,INF.BULL.VARIABLE STARS NO.3828
              2 MASON,K.O.,WATSON,M.G.,PONMAN,T.J.,CHARLES,P.A.,
                  DUCK,S.R.,HASSALL,B.J.M.,HOWELL,S.B.,ISHIDA,M.,
                  JONES,D.H.P.,MITTAZ,J.P.D.:1992,MONTHLY NOTICES ROY.
                  ASTRON.SOC.258,749
              3 SHARA ET AL. (1993)

PG 0859+415   1 DOWNES & SHARA (1993) (OBJECT LYN 1)

PG 0917+342   1 DOWNES & SHARA (1993) (OBJECT CNC 1)

1H 0927+501   1 DOWNES & SHARA (1993) (OBJECT UMA 6)

PG 0943+521   1 DOWNES & SHARA (1993) (OBJECT UMA 1)

PG 1000+667   1 GREEN ET AL. (1986)

RX J1002-19

RX J1007-20

1ES 1113+432

RE J1149+28   1 MITTAZ,J.P.D.,ROSEN,S.R.,MASON,K.O.,HOWELL,S.B.:1992,
                  MONTHLY NOTICES ROY.ASTRON.SOC.258,277
              2 SHARA ET AL. (1993)

RX J1307+53   1 SHARA ET AL. (1993)

RX J1313-32

PG 1524+622   1 DOWNES & SHARA (1993) (OBJECT DRA 3)

1H 1752+081   1 DOWNES & SHARA (1993) (OBJECT OPH 1)

RX J1844-74   1 SHARA ET AL. (1993)

1H 1929-509   1 DOWNES & SHARA (1993) (OBJECT CYG 2)

EC 1931-5915 1 BUCKLEY,D.A.H.,O'DONOGHUE,D.,KILKENNY,D.,STOBIE,R.S.,
                  REMILLARD,R.A.:1992,MONTHLY NOTICES ROY.ASTRON.SOC.
                  258,285

RX J1938-46   1 BUCKLEY,D.A.H.,O'DONOGHUE,D.,HASSALL,B.J.M.,KELLETT,B.J.,
                  MASON,K.O.,SEKIGUCHI,K.,WATSON,M.G.,WHEATLEY,P.J.,
                  CHEN,A.:1993,MONTHLY NOTICES ROY.ASTRON.SOC.262,93
              2 SHARA ET AL. (1993)

RX J1940-10

RX J1957-57

RX J2107-05   1 HAKALA,P.J.,WATSON,M.G.,VILHU,,O.,HASSALL,B.J.M.,
                 KELLETT,B.J.,MASON,K.O.,PIIROLA,V.:1993,MONTHLY
                 NOTICES ROY.ASTRON.SOC.263,61
              2 SCHWOPE,A.D.,THOMAS,H.-C.,BEUERMANN,K.:1993,ASTRON.
                 ASTROPHYS.271,L25
              3 SHARA ET AL. (1993)

PG 2133+115   1 DOWNES & SHARA (1993) (OBJECT PEG 6)

RX J2316-05

ABBREVIATED REFERENCES

BRUCH ET AL. (1987):
         BRUCH,A.,FISCHER,F.-J.,WILMSEN,U.:1987,ASTRON.ASTROPHYS.
            SUPPL.SER.70,481

DOWNES & SHARA (1993):
         DOWNES,R.A.,SHARA,M.M.:1993,PUBL.ASTRON.SOC.PACIFIC 105,
            127

DUERBECK (1987):
         DUERBECK,H.W.:1987,"A REFERENCE CATALOGUE AND ATLAS OF
            GALACTIC NOVAE",D.REIDEL,DORDRECHT; ALSO IN
            SPACE SCI.REV.45,NOS.1-2

GREEN ET AL. (1982):
         GREEN,R.F.,FERGUSON,D.H.,LIEBERT,J.,SCHMIDT,M.:1982,PUBL.
            ASTRON.SOC.PACIFIC 94,560

GREEN ET AL. (1986):
         GREEN,R.F.,SCHMIDT,M.,LIEBERT,J.:1986,ASTROPHYS.J.SUPPL.
            SER.61,305

MUKAI ET AL. (1990):
         MUKAI,K.,MASON,K.O.,HOWELL,S.B.,ALLINGTON-SMITH,J.,
            CALLANAN,P.J.,CHARLES,P.A.,HASSALL,B.J.M.,MACHIN,G.,
            NAYLOR,T.,SMALE,A.P.,VAN PARADIJS,J.:1990,MONTHLY
            NOTICES ROY.ASTRON.SOC.245,385

SHARA ET AL. (1993):
         SHARA,M.M.,SHARA,D.J.,MC LEAN,B.:1993,PUBL.ASTRON.SOC.
            PACIFIC 105,387

WILLIAMS (1983):
         WILLIAMS,G.:1983,ASTROPHYS.J.SUPPL.SER.53,523

VOGT & BATESON (1982):
         VOGT,N.,BATESON,F.M.:1982,ASTRON.ASTROPHYS.SUPPL.SER.
            48,383

```
****************
*              *
*  WHO'S WHO ? *
*              *
****************
```

NOTE THAT X-RAY CATALOGUE DESIGNATIONS INVOLVING THE EQUATORIAL COORDINATES ARE GIVEN IN THE FOLLOWING FORMAT: X HHMMSDDMM (CATALOGUE ACRONYMS), WHERE HHMM IS THE TRUNCATED RIGHT ASCENSION IN HOURS (HH) AND MINUTES (MM), DDMM IS THE TRUNCATED DCLINATION IN DEGREES (DD) AND MINUTES OF ARC (MM), AND S THE SIGN OF THE DECLINATION.

PART I: VARIABLE STAR NAMES IN LEXIGRAPHICAL ORDER

```
RX AND      =  X 0101+4101 (1ES)
PX AND      =  PG 0027+260
UU AQR      =  PB 7078  =  S 196
VY AQR      =  NOVA AQR (1907,1929,1934,1941,1942,1958,1962,1973,1983)
AE AQR      =  X 2037-0102
FO AQR      =  X 2215-0836 (H)
HL AQR      =  PHL 227
V603 AQL    =  NOVA AQL 1918  =  HD 174107  =  KPD 1846+0031
            =  X 1846+0031 (1ES)
V1315 AQL   =  KPD 1911+1212  =  SVS 8130
TT ARI      =  BD +14 341  =  S 14  =  X 0204+1503 (EXO, 1ES, MS)
WX ARI      =  PG 0244+104
T  AUR      =  NOVA AUR 1891  =  AGK3 +30 554  =  BD +30 923A
            =  HD 36294
KR AUR      =  NOVA AUR 1960
QZ AUR      =  NOVA AUR 1964
V363 AUR    =  KPD 0530+3657  =  LANNING 10
CR BOO      =  PG 1346+082
BY CAM      =  X 0538+6050 (H, 1H, 4U)
BZ CAM      =  0623+71
SY CNC      =  BD +18 2101  =  PG 0858+181
AC CNC      =  X 0841+1303 (1H)
AT CNC      =  FBS 388  =  TON 323
EU CNC      =  CV IN M 67
AM CVN      =  CBS 354  =  EG 91  =  HZ 29  =  PG 1232+379
HL CMA      =  X 0643-1648 (CGS, 1E, 1ES)
BG CMI      =  X 0729+1002 (3A)
QU CAR      =  CD -67 1010  =  CPD -67 1645  =  HDE 310376
BV CEN      =  X 1328-5443
V436 CEN    =  X 1111-3724 (1ES)
V803 CEN    =  AE 1
V834 CEN    =  X 1405-4503 (1E, 1ES, H, 1H)
WW CET      =  HV 8002
WX CET      =  NOVA CET 1963
TV COL      =  X 0527-3251 (2A, 3A, CGS, 1ES, 1H)
TX COL      =  X 0541-4103 (1H)
GP COM      =  G 61-29  =  GR 389  =  LTT 18284  =  X 1303+1817 (1ES)
V394 CRA    =  NOVA CRA (1949,1987)
TT CRT      =  FSV 1132-11
SS CYG      =  BD +42 4189A  =  HD 206697  =  KPD 2140+4321
            =  X 2140+4321 (A, 3A, CGS, 1E, H, 1H, RE)
V1500 CYG   =  NOVA CYG 1975
V1668 CYG   =  NOVA CYG 1978
V1776 CYG   =  LANNING 90
HR DEL      =  NOVA DEL 1967
```

```
YY DRA       -> DO DRA
CQ DRA    =  4 DRA BC
DO DRA    =  PG 1140+719 = X 1140+7157 (2A, 3A, 1ES, MS)
             OCCASIONALLY ALSO REFERRED TO AS YY DRA
EP DRA    =  X 1907+6903 (2A, H, 1H, 2H, 4U?)
EF ERI    =  X 0311-2246 (2A, 3A, CGS, 1ES, 1H, RE)
UZ FOR    =  X 0333-2554 (EXO)
U  GEM    =  BD +22 1807  =  HD  64511  =  X 0752+2208 (1E, H)
CE GRU    =  GRU V1
AH HER    =  PG 1642+253
AM HER    =  X 1814+4950 (2A, 3A, CGS, 1E, 1ES, H, 1H, 1M, 4U)
DQ HER    =  NOVA HER 1934
V533 HER  =  NOVA HER 1963
V795 HER  =  PG 1711+336
V825 HER  =  PG 1717+413
V838 HER  =  NOVA HER 1991
WW HOR    =  X 0234-5232 (EXO)
EX HYA    =  X 1249-2858 (2A, 3A, CGS, EXO, 1ES, 1H, 1M?, RE, 4U)
VW HYI    =  X 0409-7125 (RE)
BL HYI    =  X 0139-6808 (3A, 1ES, H, 1H, RE)
DI LAC    =  NOVA LAC 1910  =  HD 214239  =  KPD 2233+5227
T  LEO    =  BD +4 2506A  =  PG 1135+036
U  LEO    =  NOVA LEO 1855  =  BD +14 2239
DO LEO    =  PG 1038+155
DP LEO    =  X 1114+1814 (EXO, 1E, MS)
RU LMI    =  CBS 119  =  TON 1143
ST LMI    =  CW 1103+254
SX LMI    =  CBS 31  =  TON 45
BH LYN    =  PG 0818+513
MV LYR    =  MAC R +43 1  =  X 1905+4356 (H)
BT MON    =  NOVA MON 1939
GQ MUS    =  NOVA MUS 1983
V841 OPH  =  NOVA OPH 1848  =  BD -12 4633
V2214 OPH =  NOVA OPH 1988
V1159 ORI =  NSV 02011
V1193 ORI =  HAMUY'S VARIABLE
BD PAV    =  NOVA PAV 1934
HX PEG    =  PG 2337+123
IP PEG    =  FBS 2320+181
V  PER    =  NOVA PER 1887  =  BD +56 406A  =  HD 12244
GK PER    =  NOVA PER 1901  =  BD +43 740A  =  HD 21629
          =  KPD 0327+4343  =  X 0327+4343 (3A, CGS)
RR PIC    =  NOVA PIC 1925
TW PIC    =  X 0534-5803 (1ES, H)
AO PSC    =  X 2252-0326 (3A, CGS, H, 1H)
AY PSC    =  NSV 00564  =  PG 0134+070  =  PHL 1065
TY PSA    =  PS 74
VV PUP    =  X 0812-1854 (1ES, RE)
CP PUP    =  NOVA PUP 1942
V347 PUP  =  LB 1800  =  X 0609-4844 (4U)
V348 PUP  =  X 0710-3600 (2A, 1H, 1M, 4U)
T  PYX    =  NOVA PYX (1890,1902,1920,1944,1966)
VZ PYX    =  X 0857-2417 (H, 1H)
WY SGE    =  NOVA SGE 1783  =  D'AGELET'S NOVA OF 1783
WZ SGE    =  EG 136  =  NOVA SGE (1913,1946,1978)
          =  X 2005+1733 (1ES)
V1017 SGR =  NOVA SGR 1919  =  HV 3519
V1223 SGR =  X 1851-3113 (3A, CGS, 1E, 1ES, 1H, 4U)
V3885 SGR =  CD -42 14462  =  CPD -42 8912  =  WD 1944-421
V4140 SGR =  NSV 12615
U  SCO    =  BD -17 4554  =  NOVA SCO (1863,1906,1936,1979,1987)
VY SCL    =  GD 1662  =  PHL 538  =  PS 141
VZ SCL    =  TON S 120
CT SER    =  NOVA SER 1948  =  PG 1543+145
```

```
LX SER      =   STEPANIAN'S OBJECT
MR SER      =   PG 1550+191  =  X 1550+1905 (EXO, 1ES)
RW SEX      =   BD -7 3007
SW SEX      =   PG 1012-029
SU UMA      =   PG 0808+627  =  X 0808+6245 (1ES, 1H)
AN UMA      =   PG 1101+453  =  X 1101+4519 (EXO, RE)
BZ UMA      =   PG 0849+580
CH UMA      =   PG 1003+678  =  X 1003+6747 (1E, 1ES)
DV UMA      =   US 943
DW UMA      =   FBS 1031+59  =  PG 1030+590
EI UMA      =   PG 0834+488  =  X 0832+4848 (1H)
EK UMA      =   X 1048+5420 (1E, 1H, MS, RE)
RW UMI      =   NOVA UMI 1956
SS UMI      =   PG 1551+719  =  X 1551+7155 (E)
IX VEL      =   CD -48 3636  =  CPD -48 1577  =   SAO 219684
            =   X 0813-4904 (RE)
KO VEL      =   X 1013-4743 (1E, H)
TW VIR      =   PG 1142-041
HS VIR      =   PG 1341-079
HV VIR      =   NOVA VIR 1929  =  NSV 06201
PW VUL      =   NOVA VUL 1984 I
QQ VUL      =   X 2003+2231 (1E, EXO, H, RE)


PART II: OTHER NAMES IN ALPHABETICAL ORDER
------------------------------------------------


AE 1                      -> V803 CEN
AGK3 +69 515              -> 4 DRA
AGK3 +30 554              -> T  AUR
BD +70 700                -> 4 DRA
BD +56 406A               -> V  PER
BD +43 740A               -> GK PER
BD +42 4189A              -> SS CYG
BD +30 923A               -> T  AUR
BD +22 1807               -> U  GEM
BD +18 2101               -> SY CNC
BD +14 341                -> TT ARI
BD +14 2239               -> U  LEO
BD +4 2506A               -> T  LEO
BD -7 3007                -> RW SEX
BD -12 4633               -> V841 OPH
BD -17 4554               -> U  SCO
CBS 31                    -> SX LMI
CBS 96                    -> 0917+3409
CBS 119                   -> RU LMI
CBS 354                   -> AM CVN
CD -42 14462              -> V3885 SGR
CD -48 3636               -> IX VEL
CD -67 1010               -> QU CAR
CPD -42 8912              -> V3885 SGR
CPD -48 1577              -> IX VEL
CPD -67 1645              -> QU CAR
CV IN M 67                -> EU CNC
D'AGELET'S NOVA OF 1783   -> WY SGE
DR V211B                  =   DRISSEN V211B
EG 91                     -> AM CVN
EG 136                    -> WZ SGE
FBS 388                   -> AT CNC
G 61-29                   -> GP COM
GD 1662                   -> VY SCL
GRU V1                    -> CE GRU
GR 389                    -> GP COM
HAMUY'S VARIABLE          -> V1193 ORI
```

```
HD  12244               -> V   PER
HD  21629               -> GK  PER
HD  36294               -> T   AUR
HD  64511               -> U   GEM
HD  108907              -> 4   DRA
HD  174107              -> V603 AQL
HD  206697              -> SS  CYG
HD  214239              -> DI  LAC
HDE 310376              -> QU  CAR
HV  3519                -> V1017 SGR
HV  8002                -> WW  CET
HZ  29                  -> AM  CVN
LANNING 10              -> V363 AUR
LANNING 90              -> V1776 CYG
LB  1800                -> V347 PUP
LTT 18284               -> GP  COM
MAC R +43 1             -> MV  LYR
NOVA AQR (1907,1929,1934,1941,1942,1958,1962,1973,1983)  -> VY AQR
NOVA AQL 1918           -> V603 AQL
NOVA AUR 1891           -> T   AUR
NOVA AUR 1960           -> KR  AUR
NOVA AUR 1964           -> QZ  AUR
NOVA CET 1963           -> WX  CET
NOVA CRA (1949,1987)    -> V394 CRA
NOVA CYG 1975           -> V1500 CYG
NOVA CYG 1978           -> V1668 CYG
NOVA DEL 1967           -> HR  DEL
NOVA HER 1934           -> DQ  HER
NOVA HER 1963           -> V533 HER
NOVA HER 1991           -> V838 HER
NOVA LAC 1910           -> DI  LAC
NOVA LEO 1855           -> U   LEO
NOVA MON 1939           -> BT  MON
NOVA MUS 1983           -> GQ  MUS
NOVA OPH 1848           -> V841 OPH
NOVA OPH 1988           -> V2214 OPH
NOVA PAV 1934           -> BD  PAV
NOVA PER 1887           -> V   PER
NOVA PER 1901           -> GK  PER
NOVA PIC 1925           -> RR  PIC
NOVA PUP 1942           -> CP  PUP
NOVA PXY (1890,1902,1920,1944,1966)  -> T   PYX
NOVA SGE 1783           -> WY  SGE
NOVA SGE (1913,1946,1978)  -> WZ SGE
NOVA SGR 1919           -> V1017 SGR
NOVA SCO (1863,1906,1936,1979,1987)  -> U   SCO
NOVA SER 1948           -> CT  SER
NOVA UMI 1956           -> RW  UMI
NOVA VIR 1929           -> HV  VIR
NOVA VUL 1984 I         -> PW  VUL
NSV 00564               -> AY  PSC
NSV 02011               -> V1159 ORI
NSV 06201               -> HV  VIR
NSV 12615               -> V4140 SGR
PB  7078                -> UU  AQR
PHL 227                 -> HL  AQR
PHL 538                 -> VY  SCL
PHL 1065                -> AY  PSC
PS  74                  -> TY  PSA
PS  141                 -> VY  SCL
S 14 (S = STEPHENSON)   -> TT  ARI
S 196 (S = STEPHENSON)  -> UU  AQR
SAO 015816              -> 4   DRA
SAO 219684              -> IX  VEL
```

```
STEPANIAN'S OBJECT      -> LX SER
SVS 8130                -> V1315 AQL
TON 45                  -> SX LMI
TON 323                 -> AT CNC
TON 1051                -> 0917+3409
TON 1143                -> RU LMI
TON S 120               -> VZ SCL
US 943                  -> DV UMA
4 DRA                   = AGK +69 515  =  BD +70 700  =  HD 108907
                        = SAO 015816
4 DRA BC                -> CQ DRA
```

PART III: NAMES INVOLVING THE EQUATORIAL COORDINATES

(IN ORDER OF INCREASING B1950 RIGHT ASCENSION)

```
HHMMSDDMM   HHMMSDDMM
  (1950)      (2000)

0027+2600   0030+2617   -> PX AND
0101+4101   0104+4117   -> RX AND
0134+0701   0136+0716   -> AY PSC
0139-6808   0141-6753   -> BL HYI
0204+1503   0206+1517   -> TT ARI
0234-5232   0236-5219   -> WW HOR
0244+1023   0244+1035   -> WX ARI
0312-2246   0314-2235   -> EF ERI
0327+4344   0331+4354   -> GK PER
0333-2554   0335-2544   -> UZ FOR
0409-7125   0409-7117   -> VW HYI
0527-3251   0529-3249   -> TV COL
0530+3657   0533+3659   -> V363 AUR
0534-5803   0534-5801   -> TW PIC
0538+6050   0542+6051   -> BY CAM
0541-4103   0543-4101   -> TX COL
0609-4843   0610-4844   -> V347 PUP
0616-8148   0611-8149   = X 0616-8148 (H, 1H)
0623+7106   0629+7104   -> BZ CAM
0643-1648   0645-1651   -> HL CMA
0710-3600   0712-3605   -> V348 PUP
0728+1002   0731+0956   -> BG CMI
0752+2208   0755+2200   -> U  GEM
0808+6245   0812+6236   -> SU UMA
0812-1854   0815-1903   -> VV PUP
0813-4904   0815-4913   -> IX VEL
0818+5115   0822+5105   -> BH LYN
0834+4848   0838+4838   -> EI UMA
0841+1303   0844+1252   -> AC CNC
0849+5800   0853+5748   -> BZ UMA
0857-2417   0859-2428   -> VZ PYX
0858+1805   0901+1753   -> SY CNC
0917+3409   0920+3356   = CBS 96   =  TON 1051
0928+5004   0932+4950   = X 0928+5004 (H, 1H)
1003+6747   1007+6732   -> CH UMA
1012-0253   1015-0308   -> SW SEX
1013-4743   1015-4758   -> KO VEL
1030+5902   1033+5846   -> DW UMA
1038+1527   1040+1511   -> DO LEO
1048+5420   1051+5404   -> EK UMA
1101+4519   1104+4503   -> AN UMA
1102+2522   1105+2506   -> ST LMI
1111-3724   1114-3740   -> V436 CEN
```

```
1114+1814   1117+1757   -> DP LEO
1132-1128   1134-1145   -> TT CRT
1135+0338   1138+0322   -> T  LEO
1140+7157   1143+7144   -> DO DRA
1142-0409   1145-0426   -> TW VIR
1232+3754   1234+3737   -> AM CVN
1249-2858   1252-2914   -> EX HYA
1303+1817   1305+1801   -> GP COM
1328-5443   1331-5458   -> BV CEN
1341-0759   1343-0814   -> HS VIR
1346+0812   1348+0757   -> CR BOO
1405-4503   1409-4517   -> V834 CEN
1543+1431   1545+1422   -> CT SER
1550+1905   1552+1856   -> MR SER
1551+7155   1550+7146   -> SS UMI
1642+2520   1644+2515   -> AH HER
1711+3334   1712+3331   -> V795 HER
1717+4118   1718+4115   -> V825 HER
1758+0810   1800+0810   =  X 1758+0810 (H, 1H)
1814+4950   1816+4952   -> AM HER
1846+0031   1848+0035   -> V603 AQL
1851-3113   1855-3109   -> V1223 SGR
1905+5356   1907+4401   -> MV LYR
1907+6903   1907+6908   -> EP DRA
1911+1212   1913+1217   -> V1315 AQL
1931-5915   1935-5908   =  EC 19314-5915  =  X 1931-5915 (1H)
1933+5100   1934+5107   =  X 1933+5100 (H, 1H)
1944-4207   1947-4200   -> V3885 SGR
2003+2231   2005+2240   -> QQ VUL
2005+1733   2007+1742   -> WZ SGE
2037-0102   2040-0052   -> AE AQR
2140+4321   2142+4335   -> SS CYG
2215-0836   2217-0821   -> FO AQR
2233+5227   2235+5242   -> DI LAC
2252-0326   2255-0310   -> AO PSC
2320+1808   2323+1825   -> IP PEG
2337+1221   2340+1237   -> HX PEG
```

```
PART IV: REFERENCES TO THE CATALOGUE ACRONYMS
-------------------------------------------------

REFERENCES TO MOST OF THE CATALOGUE ACRONYMS WHICH APPEAR IN THIS
COMPILATION ARE GIVEN IN PREVIOUS EDITIONS OF THE CATALOGUE OF CATA-
CLYSMIC BINARIES, LOW-MASS X-RAY BINARIES AND RELATED OBJECTS, I.E. IN

RITTER,H.:1984,ASTRON.ASTROPHYS.SUPPL.SER.57,385
RITTER,H.:1987,ASTRON.ASTROPHYS.SUPPL.SER.70,335
RITTER,H.:1990,ASTRON.ASTROPHYS.SUPPL.SER.85,1179

REFERENCES TO FURTHER ACRONYMS WILL BE GIVEN IN THE FORTHCOMING 6TH ED-
ITION OF THE CATALOGUE (RITTER 1994, IN PREPARATION)
```

Object index

Sources are listed in order of increasing right ascension and declination. Catalog indications (e.g. 4U) have been dropped for many (but not all) X-ray sources for simplicity. For a listing of special source names, see list starting on p. 640. Primary references are listed in **boldface**.

Ch. 14 is a catalog of X-ray binaries. Ch. 15 is a catalog of cataclysmic binaries with known or suspected orbital periods. Occurrences of source names in those two chapters are not indexed here.

0050−727, *see* SMC X-3
0053−739, *see* SMC X-2
0053+604, 71
0103−762, 114
0114+650, 17
0115−737, *see* SMC X-1
0115+634, 16–19, 25, 32–4, 60, 101, 460
0236+610, 17, 71, 310, 318, 325
0331+530 (=0332+530), 16, 17, 19, 25, 32–4, 101, 127, 165, 293, 460, 474
0332+530, *see* 0331+530
0352+309, *see* X Per
0422+32, 71, 99, 261, 262, 267, 268, 310
0437−47, 460
0521−720, *see* LMC X-2
0527.8−6954, 114
0532−664, *see* LMC X-4
0535−66 (=0538−66), 17, 19, 25, 60, 67, 68, 71, 113, 114, 236, 444, 460
0535+262, 17, 19, 59, 71, 460, 461, 474
0538−641, *see* LMC X-3
0538−66, *see* 0535−66
0540−697, *see* LMC X-1
0543−682, *see* CAL 83
0544−665, 114
0547−711, *see* CAL 87
0614+091, 82
0620−00, 5, 9, 71, 77–9, 82, 91, 93, 98–102, 104–6, 110–12, 127, 129, 131, 133, **137–9**, 151–3, 156, 162, 164, 165, 167, 310, 313, 314, 436, 460, 462
0655+64, *see* PSR 0655+64
0709−36, 436
0745−673, 192
0748−676 (=0748−673), 5–8, **10**, 12, 15, 16, 38, 39, 73, 77, 82, 89, 91, 99, 180, 181, 189, 192, 195–7, 208, 210, 460, 461
0820+02, *see* PSR 0820+02
0834−430, 19
0836−429, 190
0900−403, *see* Vela X-1

0918−549, 82
0921−630, 5, **14**, 73, 74, 76, 82, 86, 116
1048−594, 19, 22, 33
1118−615, 19, 71
1119−603, *see* Cen X-3
1124−68, 5, 9, 71, 91, 98, 100–2, 104, 110, 111, 128, 129, 131, 138, **139–41**, 142, 160–2, 164–7, 259, 261–3, 265, 267–70, 297, 310, 313–15, 437, 460, 462
1145−619, 17, 19, 27, 71
1223−624, *see* GX 301−2
1254−690, 5, 8, **10**, 39, 82, 88, 94, 198, 199
1257+12, *see* PSR 1257+12
1258−613, *see* GX 304−1
1259−63, *see* PSR 1259−63
1323−619, 5, 39, 189
1353−54, 152
1354−645, 76, 100, 129, **152**, 160
1417−624, 19
1455−314, *see* Cen X-4
1509−58, *see* PSR 1509−58
1516−569, *see* Cir X-1
1524−617, *see* TrA X-1
1534+12, *see* PSR 1534+12
1538−522, 17, 19, 32–5, 45, 48, 101, 111, 461
1543−475, 91, 100, 102, 118, 129, **153–4**
1553−542, 16, 17, 19, 101
1556−605, 5, 82
1608−522, 37, 100, 105, 162, 163, 165, 179–82, 184, 186, 189, 190, 192, 193, 195, 196, 199, 206, 210, 275, 277, 278, 296, 299
1617−155, *see* Sco X-1
1624−490, 5, **13–14**, 39, 40
1626−673 (=1627−67), 5, **9**, 19, 27, 31, 45, 74, 82, 95, 96, 101, 178, 295, 460, 514
1630−472, 129, 134, **154–5**, 261, 268
1636−536, 5, 39, 74, 77, 79, 82, 90, 93–5, 162, 165, 177, 179, 180, 184, 185, 188–91, 193–8, 203, 275, 278, 279, 288
1656+354, *see* Her X-1

1657—415, 17, 19
1658—298, 5, 7, **12**, 39, 82, 100, 105, 189, 192
1659—487, *see* GX 339—4
1700—377, 17, 44, 47, 67–9, 71, 72, 101, 109, 111, 294
1702—429, 181, 275, 278
1702—36, *see* GX 349+2
1704+240, 71
1705—440, 190, 193, 275, 277–9
1705—250, 82, 100, 129, **155**, 156
1715—321, 195, 196
1718—19, *see* PSR 1718—19
1722—363, 19
1722—40, 165
1724—307, 188, 195, 196
1728—337, 165, 179, 182, 183, 189, 190, 195, 196, 199, 212, 275, 278
1728—247, *see* GX 1+4
1728—169, *see* GX 9+9
1730—335, *see* Rapid Burster
1735—444, 5, 71, 77, 79, 82, 90, 92, 94, 177, 179, 182, 189–91, 195–7, 207, 210, 275, 278, 279
1740.7—2942, 128, 130, **155–7**
1741—322, 130, **157**
1743—29, 183
1743—28, 190, 208
1744—24A, *see* PSR 1744—24A
1745—248, 190, 208, 210
1746—371, 6, 5, 8, **12**, 39, 188, 195, 196
1747—214, 186
1755—338, 5, 8, **10–11**, 12, 38–40, 82, 88, 89, 130, **157**, 163, 460
1758—258, 130, **157–8**
1813—14, *see* GX 17+2
1820—30, 5, 6, **9**, 15, 16, 26, 27, 120, 162, 179, 181, 188, 189, 192, 195, 196, 202, 207, 275, 278, 279
1820—11, *see* PSR 1820—11
1821—24, *see* PSR 1821—24
1822—371, 5, 6, 8, **11–12**, 13, 15, 40, 41, 49, 71, 73, 76, 82, 88
1822—000, 82
1826—24, 130, 142, **158**, 261, 262
1831—00, 460, 461
1833—076, 19
1836—045, 19
1837+05 (=1836+05), *see* Ser X-1
1843—024, 19
1843+009, 19
1846—031, 130, 261
1850—087, 77–9, 183, 188, 275, 278
1855+09, *see* PSR 1855+09
1900+14, *see* SGR 1900+14
1905+000, 77–9
1907+09, 17, 19, 24, 26, 34, 101
1908+00, *see* Aql X-1
1909+048, *see* SS 433
1913+16, *see* PSR 1913+16
1916—05, 5, 7, 8, **9–10**, 26, 27, 39, 77–9, 118–20, 460
1937+21, *see* PSR 1937+21
1953+29, *see* PSR 1953+29, 460
1956+350, *see* Cyg X-1
1957+20, *see* PSR 1957+20
1957+115 (=1956+11), 5, 82, 90, 130, **158–9**, 163

2000+251, 5, 9, 82, 91, 104, 130, 131, 133, 134, 138, 144, **150–2**, 160–2, 261–3, 310, 313, 314
2023+33, 5, 9, 71, 74, 76, 82, 91, 98, 100–4, 109–11, 120, 127, 130, 131, 138, **142–8**, 163–6, 261–3, 268, 269, 310, 313, 315, 328, 460, 462
2030+375, 16, 17, 19, 20, 23, 27, 29, 30, 45, 101, 293–5
2030+407, *see* Cyg X-3
2127+11A, *see* PSR 2127+11A
2127+11C, *see* PSR 2127+11C
2127+119, 5, 6, **13**, 71, 76, 77, 82, 86, 185, 186, 188, 199, 243, 289
2129+12, 275, 278
2129+47, 5, **11**, 13, 74, 76, 82, 83, 89, 91, 99, 117, 118, 244
2138+568, *see* Cep X-4
2142+380, *see* Cyg X-2
2259+59, 19, 21, 22, 33, 34, 95, 463
2303+46, *see* PSR 2303+46

List of sources by name

1E 1740.7—2942, *see* 1740.7—2942
1E 2259+59, *see* 2259+59
1H 0709—36, *see* 0709—36

A 0535+26, *see* 0535+26
A 0538—66, *see* 0535—66
A 0620—00, *see* 0620—00
A 1118—61, *see* 1118—615
A 1909+04, *see* SS 433
AC 211, *see* 2127+119
AE Aqr, 331, 349, 360, 369, 448
AM Her, 333, 345, 360
Andromeda, *see* M31
Aql X-1, 5, 77, 82, 94, 100, 102, 105, 106, 131, 134, 162, 177, 189, 198, 199, 206, 209, 310, 313, 314
AY Lyr, 362, 363

B 1900+14, *see* SGR 1900+14
B 1916—05, 437
BQ Cam, *see* 0331+530
BV Pup, 355, 356
BW Cir, *see* 1354—645

CAL 83, 5, 12, **14**, 71, 75–8, 82, 86, 90, 114, 116, 117
CAL 87, 5, **12**, 14, 75–8, 82, 86, 88, 114, 117
Cen X-2, 127, 152
Cen X-3, 17, 19, 21, 27, 28, 31, 34, 44, 45, 47, 48, 59, 66, 67, 101, 111, 126, 294, 295, 458, 460, 461, 477
Cen X-4, 5, **12–13**, 71, 77, 82, 91, 98, 100–2, 105, 106, 109, 131, 167, 177, 189, 244, 310, 313, 314, 460
Cep X-4, 19, 33, 34
Cir X-1, 5, **14–15**, 48, 127, 164, 165, 195, 196, 255, 274, 278, 288, **289–91**, 296–9, 310, 317, 318, 327, 460, 462
CN Ori, 364, 365
CP Pup, 381
CRUX, *see* Cen X-2
Cyg X-1, 17, 26, 35, 38, 67, 69, 71, 74, 99, 101, 109,

110, 112, 113, 115, 120, 126, 127, 130, **134–5**, 136, 142, 144, 148, 156, 158, 162, 164–6, 254, 255, 259–64, 266, 268–72, 274, 293, 296, 310, 323–5, 458, 460, 474
Cyg X-2, 4, 5, **14**, 26, 27, 36, 39, 40, 71, 76–8, 82, 83, 86, 90, 91, 97, 106, 116, 160, 162, 275, 280, 283–8, 310, 315, 317, 460, 461
Cyg X-3, 17, **48–9**, 59, 120, 274, 289, **293**, 310, 312, 319, 320, 324, 462, 463, 474, 476

DQ Her, 331, 343, 448

EF Eri, 369, 373, 378
EK TrA, 342
EM Cyg, 360
η Car, 22
EX Hya, 332, 352, 353, 368, 374, 375, 377, 378
EXO 2030+375, *see* 2030+375

FO Aqr, 378

G 109.1−1.0, 21, 33
GK Per, 332, 333, 345, 349, 368, 375, 378, 381
GR 0834−430, *see* 0834−430
GRO J0422+32, *see* 0422+32
GS 2000+25, *see* 2000+25
GS 2023+33, *see* 2023+33
GS/GRS 1124−683, *see* 1124−68
GX 1+4, 19, 21, 27, 76, 460
GX 3+1, 189, 192, 275, 278, 279, 288, 289
GX 5−1, 157, 240, 275, 279, 282–5, 310, 315, 317
GX 9+1, 40, 275, 278, 288, 289
GX 9+9, 5, 82, 90, 275, 278, 288, 289
GX 13+1, 275, 278, 288, 289, 310
GX 17+2, 120, 184, 275, 281, 284, 310, 315–17
GX 301−2, 17, 19, 34, 43, 45–7, 67, 101, 458, 460, 461
GX 304−1, 17, 19
GX 339−4, 5, 9, 82, 90, 99, 129, 131, 134, 142, **148–50**, 156, 158, 160, 162, 166, 255, 260–9, 297
GX 340+0, 275, 310, 315, 317
GX 349+2, 77, 275, 279, 285, 310, 315, 317

HD 153919, *see* 1700−377
HD 226868, *see* Cyg X-1
HD 245770, *see* 0535+262
Hen 715, *see* 1145−619
Her X-1, 5, **14**, 15, 19, 21, 25–8, 30–4, 39, 45–7, 67, 71, 73, 74, 76, 83, 89–91, 93, 95, 96, 101, 111, 112, 118–20, 178, 294, 460, 462, 475, 477, 485, 513
HT Cas, 338, 368–71, 373
HV2554, *see* 0527.8−6954
HZ Her, *see* Her X-1

IC342, 400
IX Vol, 342, 356

J 0422+32, *see* 0422+32

κ Ori, 69
KY TrA, 100

KZ TrA, *see* 1626−673

λ Cep, 72
Large Magellanic Cloud, *see* LMC
LB 1800, 356, 358
LMC, 12, 25, 391, 408, 413, 528
LMC transient, *see* 0535−66
LMC X-1, 17, 71, 109, 113–15, 127, 129, **136–7**, 160, 162, 164–6, 261–3, 266, 267
LMC X-2, 77, 78, 82, 86, 114, 116, 117, 160, 162
LMC X-3, 17, 26, 59, 67, 71, 101, 109, 113, 114, 129, **136**, 160–2, 164, 165, 167, 261–3, 458, 460, 475
LMC X-4, 17–19, 25–7, 59, 67, 68, 71, 101, 111, 114, 120, 460
LS I +61 303, *see* 0236+610

M4, 491
M15, 6, 13, 245, 310, 488
M31, 63, 391, 392, 394–401, 403, 406, 410, 413, 487, 532
M32, 395
M33, 391–3, 400, 402, 403
M33 X-7, 392, 394
M51, 400
M81, 400–3
M82, 400
M83, 400
M101, 400, 402, 403
M106, 400
Magellanic Cloud sources, 60, 61, 78, **113**, 114
Milky Way, 400
MSH15−52, 503

N49, 528, 532
NGC 247, 400
NGC 253, 400, 410
NGC 2403, 400
NGC 3628, 400
NGC 4236, 400
NGC 4258, *see* M106
NGC 4449, 400
NGC 4631, 400
NGC 6342, 246
NGC 6441, 6, 12
NGC 6624, 6, 310
NGC 6712, 310, 491
NGC 6946, 400
NGC 7027, 85
NGC 7078, *see* M15
Nova Cyg 1975, *see* V1500 Cyg
Nova Her 1934, *see* DQ Her
Nova Mon 1975, *see* 0620−00
Nova Mus 1991, *see* 1124−68
Nova Oph 1977, *see* 1705−250
Nova Vul 1988, *see* 2000+251

OY Car, 333, 338–40, 353, 358, 367, 371, 382

PG 0027+260, 344
PSR 0655+64, 460, 461, 463, 474, 512
PSR 0820+02, 460, 461, 512
PSR 1257+12, 245, 460

PSR 1259−63, 444, 460, 461, 473, 475
PSR 1509−58, 503
PSR 1534+12, 108, 111, 460, 461
PSR 1718−19, 246, 248
PSR 1744−24A, 245, 248, 464
PSR 1820−11, 460, 461
PSR 1821−24, 236
PSR 1855+09, 108, 111, 235, 236, 240−2, 460, 461, 512
PSR 1913+16, 108, 111, 233, 245, 460, 461, 463, 475, 476, 511
PSR 1937+21, 233, 236, 242, 460, 484
PSR 1953+29, 233, 236, 240, 242, 461
PSR 1957+20, 233, 236, 243, 245, 460, 461, 464, 485, 486
PSR 2127+11A, 245
PSR 2127+11C, 245
PSR 2303+46, 460, 461, 475

QQ Vul, 370, 371
QV Nor, *see* 1538−522
QZ Vul, *see* 2000+251

Rapid Burster, 120, 175, 176, 192, 199, 201, **211−27**, 289, **291−2**, 296, 299, 430
RE 0751+14, 352, 374, 375, 451
RE J1938−461, 352
RW Sex, 343
RW Tri, 338, 357, 369
RX And, 356, 357, 361, 362

Sco X-1, 1, 4, 5, **13**, 33, 35, 36, 40, 41, 71, 77, 79−83, 85, 90, 92, 96−8, 106, 107, 127, 160, 162, 275, 279, 281, 282, 284, 285, 288, 308, 310, 315, 317, 419, 460, 461
Sct X-1, *see* 1833−076
Ser X-1, 82, 92, 93, 182, 179, 189, 196, 210, 275, 278
SGR 1806−20, 532
SGR 1900+14, 532
Sgr A West, 155
Small Magellanic Cloud, *see* SMC
SMC, 391, 410
SMC X-1, 17−19, 45, 66, 67, 71, 101, 111, 114, 458, 460
SMC X-2, 114
SMC X-3, 114
SN 1987A, 249, 503
SN AD185, 503
SS 433, 17, 25, 26, 48, 59, 67, 115, 116, 120, 310, 320−5, 328, 460, 462, 463, 476
SS Aur, 361, 362
SS Cyg, 345−8, 350, 352−6, 365, 368, 369, 434, 440
SU UMa, 357, 370
SW UMa, 333

SY Cnc, 369

Ter 5, 245
TrA X-1, 100, 129, **152−3**
TT Ari, 333, 356, 369
TU Men, 434, 436
47 Tuc, 487, 491
TV Col, 332, 378

U Gem, 331, 339, 346, 348, 350, 351, 354, 356, 363, 364, 366, 369−73, 378, 382, 434, 435
UX UMa, 357, 333
UY Vol, *see* 0748−676
UZ For, 373

V0331+530, *see* 0331+530
V0332+53, *see* 0331+530
V404 Cyg, *see* 2023+33
V426 Oph, 356, 369
V471 Tau, 379
V603 Aql, 342, 343
V616 Mon, *see* 0620−00
V635 Cas, *see* 0115+634
V662 Cas, *see* 0114+650
V725 Tau, *see* 0535+262
V795 Her, 436
V821 Ara, *see* GX 339−4
V822 Cen, *see* Cen X-4
V1333 Aql, *see* Aql X-1
V1405 Aql, *see* 1916−05
V1408 Aql, *see* 1957+115
V1500 Cyg, 333, 343, 381
V2107 Oph, *see* 1705−250
V2134 Oph, *see* 1658−298
V4134 Sgr, *see* 1755−338
Vela X-1, 17−19, 22, 27, 34, 43−8, 65, 67, 69, 71, 73, 101, 111, 112, 460, 461, 474
Virgo cluster, 412
V Per, 436
VW Hyi, 346−8, 350, 352, 353, 356, **364−6**, 368, 434, 435

W50, 463
Wra 977, 67
WX Hyi, 356, 366

X Leo, 370
X Per, 16, 27, 30, 33, 71, 112

YZ Cancri, 357, 378

Z Cam, 350, 351, 357
Z Cha, 338, 345, 365−8, 370, 371, 422, 431, 434, 435
ZZ Cet, 348

Author index

Chapter numbers given in **boldface** indicate that the chapter was written by the listed author.

Aaronson, M., 403
Ables, J., 308
Abramenko, A., 196
Africano, J., 379
Alexander, S., 33
Alme, M., 485
Alpar, M., 286, 288, 349
Andersen, J., 467
Anderson, B., 504
Anderson, N., 187, 207
Angelini, L., 135, 166, 266, 293, 295, 368, 381
Antia, H., 487
Anzer, U., 447
Aoki, T., 190
Apparao, K., 199
Arefév, V., 144
Arnett, W., 457
Arons, J., 30
Asai, K., 15, 16
Atmanspacher, H., 257
Augusteijn, T., 285
Aurière, M., 13
Ayasli, S., 19, 176, 206

Baan, W., 223–5
Bahcall, J., 5, 126
Bailes, M., 491, 512
Bailyn, C., 5, 9, 105, 141, 335
Bally, J., 156
Balucinska-Church, M., 40
Banit, M., 485
Bardeen, J., 500
Barr, P., 41, 134, 211, 214, 217–19, 222, 225
Barret, D., 127, 153, 298
Basinska, E., 179, 182, 186, 187, 189, 190, 214, 274
Basko, M., 31, 485
Bastian, T., 360
Bateson, F., 346, 356, 368
Bath, G., 132, 199, 337, 363, 428
Baum, S., 320
Baykal, A., 23
Baym, G., 510
Bazzano, A., 155, 156
Beck, R., 401
Becker, R., 17, 45, 346

Becklin, E., 48
Begelman, M., 41, 438, 463
Belian, R., 12, 153, 176
Bell-Burnell, S., 19
Belloni, T., 261–4, 272, 293–5, 347, 353
Benz, A., 360
Berger, M., 293
Bergeron, P., 344
Bernacca, P., 196
Berriman, G., 375
Berry, D., 332
Beuermann, K., 369, 374, 375, 378
Bhat, C., 349
Bhatia, A., 85
Bhattacharya, D., **Ch. 5**, 491, **Ch. 12**, 508, 509
Bianchini, A., 379
Bieging, J., 139
Biehle, G., 478
Bignami, G., 22
Bildsten, L., 177, 201–3, 205
Bisnovatyi–Kogan, G., 26, 30, 441, 514
Blaauw, A., 457, 489
Blandford, R., 475, 502, 503
Blankenship, L., 310
Blitz, L., 404
Blondin, J., 44, 514
Bode, M., 336
Bodenheimer, P., 471
Boersma, J., 472
Boldt, E., 274
Boley, F., 137
Boller, T., 405
Bolton, C., 126, 134, 135
Bonazzola, S., 31
Bond, H., 144, 146
Bonnet-Bidaud, J., 48, 293
Bonsema, P., 484
Bookbinder, J., 360
Bottinelli, L., 401
Bouchet, L., 156
Bowen, I., 83
Bowyer, S., 134, 177
Boynton, P., 14, 23, 25, 26
Bradt, H., 2, 255, 536
Braes, L., 310

Brainerd, J., 31
Brandt, S., 139
Branduardi, G., 36, 44, 274, 280
Branduardi-Raymont, G., 5, 14
Brinkman, A., 41, 262, 268
Brinkmann, W., 294
Brookshaw, L., 485
Brown, R., 3, 23, 319
Bruch, A., 369
Buckley, D., 352
Buie, M., 144, 146
Bulik, T., 35

Callanan, P., 5, 13, 148, 152
Canal, R., 478
Canizares, C., 12, 127, 272, 405, 408, 415
Cannizzo, J., 132, 363
Cannon, R., 478
Canuto, V., 31
Caraveo, P., 22
Carlini, A., 22
Carone, T., 350
Carpenter, G., 25
Carter, D., 144
Casares, J., 5, 103, 120, 144, 146, 147
Casatella, A., 337
Castor, J., 2, 47
Cavaliere, A., 176
Celnikier, L., 214, 222
Chakrabarty, D., 17
Chanmugam, G., 360, 514
Chardin, G., 293
Charles, P., 5, 13, 22, 25, 146, 147, 150–2, 177, 274
Cheng, F., 368
Cherepashchuk, A., 463
Chevalier, C., 5, 11–14, 17, 105, 106, 118, 151, 152, 154
Chiappetti, L., 153, 154
Chitre, S., 199
Chlebowski, T., 41
Chodil, G., 127, 152
Christian, D., 41
Church, M., 40
Ciardullo, R., 396, 397
Ciotti, L., 406, 407
Clark, G., 33–5, 44, 48, 115, 181, 189, 192, 345, 391, 392, 399, 401, 487
Clayton, D., 192
Coe, M., 22, 137
Collmar, W., 281
Collura, A., 396
Cominsky, L., 5, 7, 12, 19, 39, 128, 152, 183, 190
Conner, J., 12, 176
Connors, P., 128
Conti, P., 43
Cook, K., 403
Cook, M., 155, 338, 365
Cooke, B., 127, 152, 155, 285, 310, 317
Cool, A., 168, 335
Cooper, L., 500
Corbet, R., 5, 17–19, 22–4, 278, 447
Cordes, J., 472
Cordova, F., 11, 12, 48, 88, 132, **Ch. 8**, 336, 345–8,

350, 352, 354–6, 358, 360, 369, 370, 373, 374, 377, 378
Coté, J., 244
Courvoisier, T., 5, 10, 39, 198
Cowley, A., 5, 10, 12–14, 17, 26, 79, 105, 109, 136, 148, 157, 285
Crampton, D., 5, 10, 13, 14, 17, 463, 487
Cropper, M., 336, 370, 373, 378, 420, 450
Crosa, L., 14, 26
Czerny, B., 37, 189, 209
Czerny, M., 37, 189, 209

Damen, E., 185, 188, 191, 279
Daugherty, J., 31, 33
Davelaar, J., 21, 127
David, L., 405, 406
Davidsen, A., 96
Davidson, G., 223
Davidson, K., 1, 18, 27, 49
Davies, R., 3, 21, 23
Davis, R., 139
Davison, P., 17, 19
Day, C., 19, 22, 39, 42, 44, 47, 186
De Jager, O., 349
De Kool, M., 139, 481
De Loore, C., 293, 476
Deeter, J., 14–16, 23, 26
Delgado, A., 14, 475
Della Valle, M., 140, 141
Denis, M., 261, 262, 268
D'Ercole, A., 406
Dewey, R., 472
Dickel, J., 315, 327
Dieters, S., 280, 281, 284, 285
Döbereiner, S., 127, 150
Doi, K., 256
Dolan, J., 148
Doll, H., 294
Donahue, M., 415
Dotani, T., 13, 27, 31, 212, 214, 219, 221, 222, 262, 265, 279, 281, 284, 291, 292
Doty, J., 179, 182, 189, 190
Downes, D., 139
Downes, R., 331, 333, 578–80
Doxsey, R., 148, 157, 178
Drew, J., 342, 347, 356–9, 378
Duck, S., 355
Duerbeck, H., 336
Duldig, M., 139
Dulk, F., 360
Duorah, H., 224, 225

Eachus, L., 137
Eardley, D., 38, 224
Ebisawa, K., 136, 137, 148, 159–61, 166, 262, 266, 267
Ebisuzaki, T., 159, 162, 187, 203
Echevarría, J., 359
Edwards, D., 363
Efremov, V., 150
Eggleton, P., 168, 378, 383, 468, 478
Ellwood, J., 412
Elsner, R., 20, 30, 281, 286

Elson, R., 39
Elston, R., 320
Elvis, M., 137, 138
Epstein, A., 17
Eracleous, M., 352
Ergma, E., 485
Evans, B., 336
Evans, W., 12, 176

Fabbiano, G., **Ch. 9**, 391–7, 399–415
Fabian, A., 6, 15, 35, 39, 41, 44, 186, 273, 487
Fabricant, D., 405
Fahlman, G., 19, 21, 44, 463
Faulkner, J., 479, 481
Feast, M., 15
Fedorova, A., 485
Feigelson, E., 404
Ferrario, L., 375
Feynman, R., 515
Fisher, D., 128
Fishman, G., 148
Flannery, B., 467, 476
Flowers, E., 513, 517
Fomalont, E., 308
Forman, W., 14, 17, 25, 391, 394, 396, 400, 405, 406, 408
Fortner, B., 166, 292
Foster, A., 35, 42
Frail, D., 460, 484
Francey, R., 127, 152
Frank, J., 39, 40, 133, 168, 373, 419, 426
Fransson, C., 44
Freese, K., 514
Friedman, H., 274
Friend, D., 2, 47
Friend, M., 359
Frontera, F., 266, 293
Fruchter, A., 485
Fryxell, B., 23, 41, 42, 457
Fu, A., 23
Fujimoto, M., 188, 190, 191, 201, 203, 206, 208
Fuligni, F., 266
Furth, H., 225
Fushiki, I., 201, 204–6, 209, 210

Gabler, R., 382
Gallagher, J., 188
Garcia, M., 11, 118, 310
Garmire, G., 347
Gehrz, R., 360
Geldzahler, B., 48, 308, 310, 320
Gerend, D., 25, 26
Ghosh, P., 19, 20, 22, 286, 287, 363, 445
Giacconi, R., 1, 14, 19, 254, 285, 390
Gies, D., 134, 135
Giles, A., 271
Gilfanov, M., 139
Gilliland, R., 363
Ginzburg, V., 311, 499, 500, 515
Gioia, I., 404, 406, 408
Giommi, P., 6, 12, 39
Glass, I., 15
Goldman, I., 200

Goldreich, P., 496, 518
Goldwurm, A., 139–41
Gonzalez-Riestra, R., 141
Goranskij, V., 579
Gorenstein, P., 405, 412
Gotthelf, E., 144, 146
Gottlieb, E., 1, 5, 13, 90, 285
Gottwald, M., 180, 181, 189, 190, 192, 196, 208, 210
Gould, R., 3
Grabelsky, D., 12, 14
Graham-Smith, F., 327
Gray, A., 156
Grebenev, S., 19, 127, 135, 139, 140, 148, 158, 166, 262, 265–7
Greenstein, J., 479
Gregory, P., 17, 19, 21, 48, 310, 318, 325, 463
Greiner, J., 12, 14, 116, 139, 408, 409
Griffiths, R., 17, 25, 44, 155, 400, 405, 411
Grindlay, J., 5, 9, 11, 12, 37, 119, 176, 177, 189, 196, 209, 217, 274, 288, 310, 335
Gruber, D., 26, 30, 31, 33
Güdel, M., 360
Guilbert, P., 35, 41
Guinan, E., 356
Gunn, J., 496, 503, 510
Gursky, H., 1

Haberl, F., 43, 44, 46–8, 179, 189
Habets, G., 168, 472, 475
Hackwell, J., 196
Hakala, P., 375, 451
Hall, R., 21
Halpern, J., 14, 21, 33, 34, 352
Hamada, T., 31
Hameury, J., 133, 244, 377, 379
Han, X., 146–8, **Ch. 7**, 310, 313, 315, 324
Hanami, H., 224, 225
Hanawa, T., 187, 190, 202–4, 207, 208, 224, 225
Hansen, B., 484
Hansen, C., 176
Harding, A., 31, 33
Harkness, R., 489
Harmon, B., 127, 131, 148, 153, 154, 166
Harpaz, A., 244, 485
Harrison, T., 360
Hashimoto, M., 202
Hasinger, G., 12, 14, 35–7, 166, 261–4, 272, 274, 275, 277, 280, 281, 283–5, 287, 288, 293–5, 316, 408, 409
Hassall, B., 342, 356, 366
Haswell, C., 102, 103, 110, 137, 147
Hatchett, S., 44, 47, 106
Hawkins, F., 17
Hayakawa, S., 199, 223, 224
Haynes, R., 318
Heap, S., 356
Hearn, D., 345
Heckman, T., 405
Heemskerk, M., 26, 128
Heise, J., 345–7, 353, 356, 366, 473
Helfand, D., 12, 14, 391, 399
Hellier, C., 5, 12, 15, 40, 41, 336, 368, 376–8
Herold, H., 33

Hertz, P., 5, 13, 280, 284, 331, 335
Hessman, F., 365
Higdon, J., 531
Hind, J., 331
Hirano, T., 40, 41
Hirose, M., 371, 433
Hirotani, K., 224, 225
Hjellming, R., 146–8, 150, 152, **Ch. 7**, 309, 310, 313, 315, 317, 320–7, 360
Hoare, M., 347
Hodge, P., 394
Hoffman, J., 176, 179, 182, 183, 186, 189, 190, 192, 199, 211, 216
Holberg, J., 350, 355, 356, 366, 368
Holm, A., 357
Holt, S., 7, 11–13, 27, 28, 31–4, 45, 49, 128, 131–4, 154, 155, 157, 274, 425
Honey, W., 148
Horiuchi, R., 223, 225
Horne, K., 176, 336, 338, 344, 345, 368–70, 373, 421
Hoshi, R., 36, 38, 132, 200
Howarth, I., 336, 358
Huang, M., 132
Hubeny, I., 343
Huckle, H., 19
Hurley, K., **Ch. 13**
Hurst, G., 142, 144
Hut, P., 488
Hutchings, J., 14, 17, 109, 112, 136

Iben, I., 478
Ichikawa, S., 364
Icke, V., 352
Iga, S., 148, 260
Ikeuchi, S., 208
Illarionov, A., 23, 225
Illingworth, G., 403
Ilovaisky, S., 5, 6, 13, 14, 17, 106, 148, 149, 151, 152, 166, 260, 261
Imamura, J., 149, 150, 166, 266
Inoue, H., 23, 46, 47, 190, 213, 214, 261, 277, 278, 289, 296, 297, 353
Inoue, I., 181
Isern, J., 478
Ishida, M., 353, 375
Ives, J., 19
Iwasawa, K., 21, 33, 34, 463

Jansen, F., 293
Jensen, K., 359
Johnson, A., 198
Johnson, W., 19
Johnston, H., 102, 137
Johnston, K., 310, 312, 320–3, 325–7
Johnston, M., 17, 25
Johnston, S., 460, 491
Jones, A., 199
Jones, C., 14, 17, 25, 127, 154, 157, 394, 396, 400, 405, 406, 408
Jones, D., 13, 144
Jones, L., 22
Jones, M., 5, 13, 39, 348, 352, 353, 355
Jones, P., 518

Jongert, H., 278
Joss, P., 6, 175, 176, 178, 201, 202, 206, 207, 212, 378, 481, 483
Julian, W., 496

Kadonaga, T., 225
Kahabka, P., 12, 14, 408, 409
Kahn, S., 41, 374
Kallman, T., 41–4, 47, 48, 85, 106, 359
Kaluzienski, L., 2, 5, 12–14, 128, 138, 152, 155, 157
Kaluzny, J., 146, 147
Kaminker, A., 31, 34, 211, 212, 214
Kanno, S., 31
Kaper, L., 72, 73
Kastner, S., 85
Katoh, M., 187
Katz, J., 6, 26
Kawai, N., 26, 155, 211, 212, 214, 218, 219, 224–6
Kazarovets, E., 579
Keel, W., 405
Keliang, H., 286
Kellermann, K., 312
Kelley, R., 16–19, 199
Kembhavi, A., 487
Kemp, J., 135
Kesteven, M., 141, 310
Kholopov, R., 579
Kii, T., 31, 355, 369
Killeen, J., 225
Kim, D., 400, 404–11
Kim, W., 368
King, A., 24, 40, 48, 49, 244, 336, 345, 349, 358, 369, 373–9, **Ch. 10**, 433
Kiplinger, A., 366
Kirshner, R., 403
Kirzhnits, J., 500, 515
Kitamoto, S., 48, 135, 138, 139, 142, 152–4, 160, 166, 261, 262, 268–70, 273, 278, 293, 310
Klebesadel, R., 524
Klein, U., 401
Kleinmann, D., 211
Kleinmann, S., 211
Kley, W., 348, 349
Kluźniak, W., 440
Kochanek, C., 487
Kodonaga, T., 223
Kolb, U., **Ch. 15**
Komberg, B., 514
Kondo, Y., 335, 381
Königl, A., 320
Kouveliotou, C., 135, 166, 262, 266, 267
Koyama, K., 19, 21, 33, 34, 177, 189, 212, 293, 335
Kraft, R., 331, 337, 479
Krause, M., 401
Krezminski, W., 364
Krishnamohan, S., 507
Kriss, G., 25
Krolik, J., 47
Kruskal, M., 225
Kudritzki, R., 382
Kuiper, L., 109
Kulkarni, P., 199
Kulkarni, S., 463, 486, 491

Kundt, W., 128
Kunieda, H., 211, 213–16, 219, 225
Kurfess, J., 19
Kuulkers, E., 275, 279–81, 284, 366, 371

Labay, J., 478
LaDous, C., 346
Lamb, D., 201, 202, 204, 205, 360, 373, 381
Lamb, F., 1, 18–20, 22, 27, 30, 202, 223, 225, 252, 277, 286–8, 349, 445, 475
Lamb, R., 19
Laming, J., 177, 186
Lampton, M., 274
Lang, F., 26
Langer, S., 31
Langmeier, A., 190, 277, 280, 281, 284
Lasota, J., 40, 244, 373–5, 377, 379, 453
Law, W., 475
Lawrence, A., 79, 198, 199
Le Fèvre, O., 13
Lea, S., 23, 30
Leahy, D., 30, 47, 156
Leibowitz, E., 146, 147
Lesh, J.R., 60
Leventhal, M., 156
Levine, A., 18, 27, 157, 178
Lewin, W., 3, 19, 120, **Ch. 3**, 131, 166, **Ch. 4**, 176–9, 181–4, 186–90, 192, 193, 195–8, 207, 208, 210–14, 221, 224, 275, 277, 279–81, 283, 284, 286, 289, 292
Lewis, W., 44, 47, 483
Li, F., 17, 153, 178, 189, 201, 210
Liang, E., 135, 261
Liebert, J., 344
Liedahl, D., 41
Lightman, A., 3, 23, 38, 224
Liller, W., 13, 14, 17, 176, 211
Lin, D., 132, 223, 224, 368
Ling, J., 135
Lingenfelter, R., 156, 531
Livio, M., 23, 199, 335, 336, 339, 364, 365, 371, 378–81, 383, 453, 471
Lloyd, C., 137
Lochner, J., 26, 27, 135, 262, 272
Lodenquai, J., 31
London, R., 41, 42, 159, 162
Long, K., 12, 14, 139, 351, 352, 382, 392, 393, 403, 404, 487
Lubin, L., 211, 214–22, 291, 292
Lubow, S., 2, 40, 433, 453
Lucke, R., 19
Lucy, L., 43
Lund, N., 139
Lyne, A., 327, 504

McClintock, J., 2, 5, 11–13, 19, **Ch. 2**, 84, 105, 133, 134, 137, 141, 148, 157, 167, 177, 178, 196, 273, 536
McCray, R., 31, 33, 44, 47, 85, 106
McDermott, P., 487
Macdonald, A., 405
MacDonald, J., 203
Machin, G., 310

McKee, C., 41
McMillan, S., 487
Maeder, A., 466, 473
Maejima, Y., 148, 257, 262, 268, 269, 271
Magnier, E., 186, 212
Makino, F., 139, 142, 150, 152, 158, 255, 290
Makishima, K., 19, 21, 33, 34, 40, 42, 46, 148, 189, 192, 261, 262, 272, 278–80, 293, 395, 411
Manchester, R., 487
Mandrou, P., 156, 157
Mao, S., 531
Maraschi, L., 16, 176, 458
Margon, B., 11, 26, 146, 310, 320, 350, 463
Marino, B., 368
Markert, T., 148, 152, 260, 392, 415
Marscher, A., 319
Marsden, B., 144
Marsh, T., 339, 341, 379
Marshall, F., 137, 157, 158, 165, 261
Marshall, H., 176, 192, 200, 211, 213, 214, 216–19
Marshall, N., 17, 24
Marti, J., 319, 325
Mason, K., 4, 5, 10–12, 14, 17, 20, 39–41, 44, 48, 88, 198, 199, 214, 274, 288, 336, 345–8, 352, 354, 356, 358, 370, 372–4, 376–8
Masters, A., 373
Mateo, M., 357, 378
Mathews, J., 479
Matilsky, T., 137, 153, 154
Matsuda, T., 23
Matsuoka, M., 12, 47, 95, 189, 198
Matteson, J., 135
Mauche, C., 336, 347, 350, 353, 355, 356, 358, 359
Mayer, M., 10
Mazeh, T., 25
Meegan, C., 529
Meekins, J., 270–2
Meintjes, P., 349
Melia, F., 41, 42, 189, 207
Mendelson, H., 25
Mendez, R., 382
Mereghetti, S., 22
Mersov, G., 26, 30
Mestel, L., 379, 480
Mészáros, P., 1, 31, 33–5, 182
Meurs, E., 488, 489
Meyer, F., 132, 223, 224, 363
Meyer-Hofmeister, E., 132, 363
Meynet, G., 466, 473
Mi, G., 44
Michel, F., 223
Middleditch, J., 5, 9, 96, 283, 284, 369, 370
Migdal, A., 500, 515
Mihara, T., 33, 34, 47
Miley, G., 310
Milgrom, M., 6, 224
Miller, G., 288
Minato, J., 44
Mineshige, S., 132, 133, 144, 150, 363, 368, 371
Mirabel, I., 156–8
Mitsuda, K., 1, 35–8, 159, 160, 163, 274, 277–82, 284, 286, 287
Mittaz, J., 346, 451

Miyaji, S., 203, 208
Miyamoto, S., 135, 148, 166, 257, 261–70, 273
Mobberley, M., 142, 144
Mochnacki, S., 468
Molnar, L., 48, 120, 318, 319
Moneti, A., 15
Morfill, G., 257, 293
Morton, D., 465
Motch, C., 5, 10, 86, 148–50, 166, 262, 266, 267
Mukai, K., 5
Murakami, T., 179, 181, 182, 189, 190, 192, 193, 199, 204, 210
Murdin, P., 15, 17, 126, 134, 137, 153, 289

Nagase, F., **Ch. 1**, 18, 27, 31, 33, 34, 45–8, 178, 293, 295
Nagel, W., 1, 31, 34, 35
Nakamura, H., 48
Nakamura, N., 163, 181, 184, 189
Narayan, R., 441, 491, 506
Nather, E., 369, 371
Naylor, T., 5, 6, 13, 353, 358
Negoro, H., 262, 274
Nelson, J., 96, 346, 347, 352
Neugebauer, G., 35
Nicolson, G., 15
Noguchi, T., 150, 151
Nolan, P., 135, 148, 261–3, 268, 272
Nomoto, K., 335, 381, 489
Norris, J., 284, 286, 287
Norton, A., 336, 349, 374, 375, 377
Nousek, J., 371
Nozakura, T., 208
Nugent, J., 347

Oda, M., 134, 208, 210, 254, 255, 260–2, 268, 271, 272
O'Donoghue, D., 365, 366, 381, 431
Ogawa, M., 138–40, 160, 161
Ogawara, Y., 268, 272
Ögelman, H., 23
Ohashi, T., 45, 47, 179, 190, 203
Okamura, S., 150, 151
Oke, J., 137
Onsager, L., 515
Oosterbroek, T., 181, 185, 290, 291
Orlandini, M., 293
Osaki, Y., 132, 353, 355, 363, 364, 369, 371, 427, 433
Osborne, J., 336, 349
Ostriker, J., 1, 18, 27, 49, 127, 496, 503, 510
Owen, F., 310
Owocki, S., 358

Paciesas, W., 131
Paczyński, B., 12, 116, 136, 187, 207, 378, 379, 383, 441, 466, 468, 469, 471, 479, 481, 531, 532
Padovani, P., 411
Page, C., 264, 269
Pakull, M., 12, 25, 67, 86
Palumbo, G., 400, 404
Panek, R., 358
Parmar, A., **Ch. 1**, 5–7, 10, 12, 14–17, 19, 20, 23–5, 27, 29, 30, 35–40, 45, 154, 155, 158, 189, 261, 268, 295, 370
Parsignault, D., 17, 274, 288
Patterson, J., 331, 333, 336, 348, 349, 352, 369, 370, 379, 481
Paul, J., 156, 157
Pavlenko, E., 151
Payne-Gaposchkin, C., 337, 345
Peacock, A., 35, 40, 412
Pedersen, H., 5, 10, 93, 95, 153, 177, 190, 197, 198
Pederson, H., 152
Pellegrini, S., 407
Penninx, W., 177, 181, 189, 193, 195–7, 207, 275, 280, 281, 284, 285, 310, 316, 317, 325
Penrod, G., 112
Peres, G., 392–4, 396, 400
Pethick, C., 1, 27, 510
Petre, R., 17
Petterson, J., 26, 30, 31, 44
Phinney, E., 460, 484, 486–8
Picklum, R., 176, 201, 202
Pietsch, W., 11, 369, 374, 378
Piirola, V., 375, 451
Pines, D., 1, 27, 510
Pinto, P., 177, 186
Podsiadlowski, P., 244, 485
Polidan, R., 347, 350, 353, 355, 356, 366, 368
Ponman, T., 35, 42, 274, 285, 288, 310, 317, 355
Popham, R., 441
Pounds, K., 44, 48, 49, 152, 352
Pravdo, S., 19, 45
Press, W., 271
Preston, R., 310
Priedhorsky, W., 6, 9, 17, 26, 97, 128, 131–5, 154, 158, 189, 268, 271, 280, 283, 284, 310, 315
Primini, F., 394, 396, 400
Prince, T., 156, 488
Pringle, J., 1, 3, 6, 18, 23, 27, 49, 132, 336–8, 347, 352, 365, 366, 378, 379, 383, 419, 428, 441, 453
Proszynski, M., 187, 207
Pye, J., 17
Pylyser, E., 245, 482

Rallis, A., 392
Ramaty, R., 156
Rappaport, S., 6, 9, 16, 17, 19, 20, 24, 31, 132, 244, 345, 378, 458, 481–5
Ray, A., 487
Raymond, J., 35, 41, 42, 347, 350, 355, 358, 359
Reale, F., 396
Rees, M., 1, 6, 18, 27
Reich, W., 156
Reisenegger, A., 518
Remillard, R., 5, 11, 13, 105, 110, 137, 148, 167
Reynolds, A., 45
Rhoades, C., 128
Richardson, J., 345
Richardson, K., 349
Richter, G., 144
Richter, O., 404
Ricker, G., 19, 178
Ricketts, M., 17, 24, 137, 345

Ritter, H., 6, 331, 336, 344, 475, 481, 578, 579, **Ch. 15**
Robba, N., 293
Roberts, J., 26
Robinson, E., 335, 345
Romaine, S., 400, 403
Romani, R., 168, 514
Rosa, M., 404
Rose, L., 45
Rosen, S., 336, 376, 377, 451
Rosenberg, F., 19
Rosenbluth, M., 225
Rosner, R., 24, 25, 37, 159
Ross, R., 35, 39, 41, 42, 186
Rothschild, R., 26, 30, 31, 155, 271
Rubenstein, E., 379
Ruderman, M., 31, 233, 243, 485, 499, 513, 517–19
Ruffini, R., 128
Rutledge, R., 222, 291
Rutten, R., 367
Ryba, M., 485

Saar, S., 339
Sadeh, D., 199
Saffer, R., 344
Salpeter, E., 201–3, 205, 419
Salter, M., 504
Samimi, J., 127, 148, 289
Samus, N., 579
Sandage, A., 1, 401–3
Sanford, P., 17, 19, 44
Sansom, A., 5, 6, 12, 15, 16, 38
Sarazin, C., 408
Sarna, M., 133
Sato, N., 44–6
Sato, S., 199
Sauls, J., 515, 520
Savonije, G., 2, 42, 352, 474, 481–3
Sawada, K., 23
Scalo, J., 488
Schachter, J., 83, 85
Schaefer, B., 5
Scharlemann, E., 445
Schatzman, E., 478
Schechter, P., 271
Schiffer, F., 358
Schlickeiser, R., 156
Schmidt, B., 403
Schmidt, G., 375, 381
Schmidt, H., 14
Schmidtke, P., 10
Schmitz-Frayasse, M., 157
Schoelkopf, R., 199
Schreier, E., 17, 44, 126
Schreiffer, J., 500
Schulz, N., 36, 37, 275, 288, 328
Schwartz, R., 135
Schwarzenberg-Czerny, A., 26
Schwarzschild, M., 225
Seaquist, E., 310
Serlemitsos, P., 35, 40, 48
Seward, F., 22, 41
Shachter, J., 86

Shafer, R., 15
Shafter, A., 102, 103, 110, 137, 147, 333
Shaham, J., 286, 349, 485
Shakura, N., 2, 36–8, 159, 337
Shapiro, S., 3, 23, 38, 135
Shara, M., 336, 578–80
Share, G., 189
Shaviv, G., 343, 377, 382
Sheffer, E., 26, 30
Shibazaki, N., 36, 274, 279, 280, 286
Shields, G., 41
Shinoda, K., 293, 295
Shklovsky, I., 419, 463
Shlosman, I., 359
Shore, S., 357
Shrader, C., 141, 152
Shu, F., 2, 40
Sienkiewicz, R., 378, 379, 481
Sigurdsson, S., 488
Singh, J., 374
Singh, L., 224, 225
Sion, E., 356, 366, 380
Skinner, G., 19, 25, 155, 157, 158
Slettebak, A., 458
Smak, J., 132, 363–5, 427, 434
Smale, A., 5, 9, 14, 22, 26, 27, 39, 40, 49, 119
Smarr, L., 475
Sofia, S., 112
Soker, N., 23, 35, 41, 42, 471
Solomon, P., 43
Soong, Y., 33, 34, 293, 295
Spencer, R., 310, 320
Spruit, H., 224–6, 379, 445, 480, 481
Srinivasan, G., **Ch. 12**, 507, 516
Stanger, V., 400, 405
Starrfield, S., 188, 337, 381
Steinman-Cameron, T., 150
Stella, L., 5, 6, 9, 12, 16, 17, 19, 24, 25, 27, 30, 35–9, 127, 159, 189, 211, 213–16, 219–23, 277, 279, 281, 284, 286, 291–3
Sterne, T., 112
Stevens, I., 42, 44
Stewart, R., 289, 318, 325, 327
Stiening, R., 338, 369, 370, 373
Stinebring, D., 485, 489
Stockman, H., 375, 381
Stollman, G., 116
Strickman, M., 19
Sugimoto, D., 187, 188, 202, 203, 207
Sunyaev, R., 2, 19, 23, 31, 35–8, 127, 135, 136, 139, 140, 144, 147, 150, 153–9, 225, 262, 337, 485
Supper, R., 400
Sutantyo, W., 462, 487
Sutherland, P., 256, 264, 271, 272, 499
Suzuki, K., 40
Swank, J., 2, 5, 7, 9, 12, 13, 27, 28, 31–5, 39–41, 43, 45, 182, 293–5, 345, 352–5, 374
Syrovatskii, S., 311
Szkody, P., 146, 333, 355, 357, 366, 369, 378
Sztajno, M., 39, 40, 179, 180, 182, 184, 188, 189, 210, 274

Taam, R., 23, 131, 132, **Ch. 4**, 176, 177, 186, 187, 201–3, 206, 207, 223–6, 445, 471, 476, 487, 514
Takagishi, K., 196
Takeshima, T., 293–5
Takizawa, M., 138, 140, 150, 151, 160, 161, 166, 261, 262, 265, 267–70
Tamman, G., 403
Tamura, K., 25, 34
Tan, J., 5, 15, 16, 211, 213–16, 218, 219, 226, 310
Tanaka, Y., 3, **Ch. 3**, 127, 136, 138, 140, 142, 143, 145, 146, 148, 150, 151, 154, 158, 160, 161, 163, 167, 181, 261, 262, 293, 410
Tananbaum, H., 5, 14, 19, 26, 39, 127, 260, 310, 323, 412
Tapia, S., 345
Tavani, M., 42, 485
Tawara, Y., 19, 186, 214, 219, 220, 291, 292
Taylor, A., 17, 35, 310, 318, 325
Taylor, B., 40
Taylor, J., 107, 463, 485, 489
Tennant, A., 15, 127, 289, 290
Terada, K., 166, 261, 262
Terrell, J., 17, 26, 158, 189
Terrell, N., 135, 256, 262, 264, 271, 272
Terzan, A., 13
Thomas, H., 14, 369, 374, 378, 475
Thomas, R., 198
Thorne, K., 478
Thorsett, S., 503
Thorstensen, J., 5, 11, 106, 177, 333, 344
Tillett, J., 203
Titarchuk, L., 35, 38, 154, 159, 485
Tominatsu, A., 224, 225
Toor, A., 289
Treves, A., 22, 136, 262, 264, 458
Trinchieri, G., 391–7, 399, 400, 402–11, 415
Trümper, J., 25, 26, 30, 33, 34, 36, 37, 135, 390, 395
Truran, J., 380
Tsunemi, H., 138, 150, 151, 212
Tucker, W., 405, 408
Tueller, J., 33
Turner, M., 255
Turtle, A., 141, 310
Tylenda, R., 352

Ubertini, P., 135
Udalski, A., 146, 147
Ulmer, M., 5, 11, 17, 198, 218
Unno, W., 257
Urpin, V., 502

Vacca, W., 36, 186, 187, 189
Vader, J., 396
Van Amerongen, S., 132, 366
Van den Heuvel, E., 12, 24, 117, 132, 167, 168, 243, 293, **Ch. 11**, 458, 463, 472, 473, 476, 477, 481, 484, 488, 489, 491, 514
Van der Hucht, K., 475
Van der Klis, M., 15, 16, 35–7, 48, 152, 166, 177, 193, 194, **Ch. 6**, 257–9, 273–5, 277, 279–84, 287–90, 293, 297, 298, 316
Van der Laan, H., 312

Van der Woerd, H., 41, 134, 154, 346–8, 353, 356, 366
Van Kerkwijk, M., 4, 17, 24, 48, 69, 293, 463, 474
Van Paradijs, J., 2, 26, 39, **Ch. 2**, 112, 116, 120, 128, 131, 132, **Ch. 4**, 176, 177, 180, 181, 184–6, 188–91, 193, 195–200, 207, 210, 216, 243, 280, 284, 285, 289, 366, **Ch. 14**
Van Speybroeck, L., 394–6, 403, 404, 406, 487
Van Teeseling, A., 347, 353
Vauclair, G., 380
Vaughan, B., 281, 282
Vedrenne, G., 298
Vennes, S., 338, 368
Ventura, J., 1, 31, 34
Verbunt, F., 2, 9, 39, 132, 168, 178, 342, 347, 353, 356–9, 364–6, 368, 378, 379, 381, **Ch. 11**, 460, 476–8, 480, 481, 484, 487, 491
Vermeulen, R., 320, 322, 463
Vikhlinin, A., 135, 166, 266, 267
Viotti, R., 337
Vitello, P., 359
Voges, W., 5, 14, 33, 34
Vogt, S., 112
Vrtilek, S., 5, 14, 35, 39, 41, 42, 83, 97, 106, 280, 285

Wade, C., 308, 310
Wade, R., 10, 336, 342–5, 347, 350, 353, 365, 382, 481
Wagner, R., 110, 142, 144, 146, 147, 150, 151, 310
Wagoner, R., 440
Waki, I., 177, 185
Walker, G., 44
Walker, M., 182, 224, 369
Walker, W., 368
Wallace, R., 187, 202, 207
Walter, F., 5, 7, 9, 373
Wang, Q., 391
Wang, Y., 1, 23, 30
Ward, M., 17, 25, 336
Warner, B., 335, 336, 345, 369, 371, 431, 448
Warren, J., 352
Warwick, R., 17, 345
Wassermann, I., 201–3, 205
Waters, L., 24
Watson, M., 5, 13, 17, 26, 39, 40, 44, 155, 156, 310, 320, 336, 346, 348, 349, 352, 353, 355, 373–5, 377, 378, 391, 400, 405
Weaver, T., 176, 201, 206–9
Webbink, R., 289, 344, 378, 469, 482, 483, 490
Webster, B., 17, 126, 134
Wehrse, R., 343, 382
Weisberg, J., 107, 463
Weisskopf, M., 262, 268, 271
Welter, G., 1, 30
West, R., 141, 375
Wheaton, W., 33, 34, 135
Wheeler, J., 132, 133, 335, 363, 368, 381, 489
Whelan, J., 15, 137, 310, 478
White, N., **Ch. 1**, 2, 4–7, 9–12, 17, 19, 24, 25, 27, 28, 30–49, 119, 128, 135–7, 155, 157–9, 163, 165, 166, 176, 178, 189, 261, 274, 280, 286, 288, 370, 425, 437

White, R., 408
Whitehurst, R., 132, 369, 431, 433, 436
Wickramasinghe, D., 375
Wijers, R., 328
Williams, R., 343, 381
Willingale, R., 48, 49
Wilson, C., 155
Wilson, J., 485
Wilson, W., 112
Wolfendale, A., 349
Wolszczan, A., 460, 463, 484
Woltjer, L., 499
Woo, J., 48
Wood, J., 333, 338, 340, 345, 356, 359, 368, 370, 373
Wood, K., 5, 7, 12, 39, 157, 284, 285, 524
Woods, J., 342, 357
Woodward, J., 502

Woosley, S., 176, 201, 202, 206–9
Wright, E., 13, 211
Wu, K., 375

Yabe, T., 503
Yahel, R., 31, 34
Yakovlev, D., 502
Yamauichi, S., 212
Yaqoob, T., 158, 163
Yongheng, Z., 286
Yoshida, K., 278, 299, 353

Zamorani, G., 404
Zheleznyakov, V., 309
Zwaan, C., 379, 480
Zylstra, G., 41, 42
Zytkow, A., 478

Subject index

Primary references are set in **boldface**.

AAVSO, 362, 363
Abrikosov fluxoids, 501
accretion, 1, 236, 327, **Ch. 10**
  angular momentum, *see* angular momentum
  centrifugal barrier, 24, 25, 59
  close binaries, Ch. 10
  columns, 377, **451**
  Coulomb interactions, 30, 31
  curtain, 377, 450
  disk, *see* accretion disk
  Eddington limit, 486
  flow, 1, 2, 23, 42, 373, 374, 444
  gas blobs, 448–50, 452
  gas streams, 422, 426
  geometry, 331, 358, 370, 371, 373, **376–7**
  instabilities, 237, 333, 361
  luminosity, 337
  magnetic fields, 335
  mass transfer, *see* mass transfer
  metallicity effects, 392
  models, 27, **444–51**; *see also under* accretion disk
  rate, 25, 38, 59, 339, 348, 427–9
  reversal, 23
  shocks, 31, 452
  simulations, 23
  stellar wind, 2, 23, 24, 30, 59, 419, **441–4**, 446, 470–4
  super-Eddington, 48, 244, 327, 392, 463, 476
  torques, 22, 439, 441, 445, 446, 450
  viscosity, *see* viscosity
  wake, 49
accretion disk, 3, 6, 9, 19, 20, 23, 26, 27, 30, 31, 35, 38, 40, 63, 74, 236, 309, 315, 320, 326, 331–4, 419–23
  absorption lines, 3, 352, 368, 382
  α-disk, *see under* accretion disk, models
  angular momentum, *see under* angular momentum
  boundary layer (BL), 1, 36, 37, 338, 345–8, 350, 352, 382, **438–41**
    luminosity, 338, 347, 353
    structure, 349
    temperature, 347, 348, 359
    UV emission, 366
    X-ray emission, 346, 353

bright spot, 333, 338–40, 365, 366, 368, 369, **370**, 371, 422, 423
bulge, 7, 9, 16, 40
Compton heating/cooling, 35–7, 438
corona, 7, 8, 10–14, 39, 40, **41–2**, 45, 49, 349, 423, 425, **438**, 440
Doppler tomography, 336, **339**, 341, 343, 382
eclipse mapping, 336, **338–9**, 340, 368, 421
eclipses, 334, 337–339, 344, 365
electron scattering, 38
emission lines, 336, 339, 343, 353, 356, 358, 368, 369, 378, 381–3
  formation, 419–20
humps, 333, 338–40, 365, 366, 368–9, **370**, 371, 422, 423, 430–6
imaging, 336, **338–9**, 340, 368, 421
instabilities, 2, 10, 132–4, 334, 348, 349, 361, 363, 364, 369, 419, 426, **427–30**, 436, 445
Keplerian orbits, 236, 237, 240, 339
limit cycles, 427–8
luminosities, 43, 338, 342, 363
magnetic coupling, 236, 237
models, 23, **37–8**, **159–63**, 336, **337–8**, 339, 340, 342, 345, 350, 371, 372, 381, 382
  absorption, 377
  α-disk, 337, 363
  blackbody, 37, 343, 350
  boundary layer (BL), 347
  cold-absorber, 38
  Kurucz stellar atmosphere, 343, 350
  leaky absorber, 374, 375, 450
  occultation, 376–7
  Shakura–Sunyaev, 337, 363
  Shaviv and Wehrse, 343
  thick-disk, 6, 7
  thin-disk, 6, 353, 420
  two-component, 35–7, 39–40, 374, 377
  two-temperature, 353
opacity, 38
partial, 332, 374
precession, 9, 27, 67, 334, 369, 371, 463
radiation pressure, 37, 369, 438
radius determinations, 365–7
resonances, 431–4
shocks, 421, **437**, 452

[accretion disk]
spectra, 3, 38–9, 338, **339–43**
splash, 7, 38
structure, 353, **423–7**
superhumps, 368–9, 430–6
temperature, 337–9, 366
tidal heating, 369
tides, 3, 239, 334, 420, 426, **431**
UV emission, 366
vertical structure, 425
viscosity, *see* viscosity
X-ray screening, 244
accretion-induced collapse, 244, 248, 335, 380, 381, 476, 478, 491
active galactic nuclei, *see under* galaxies
Alfvén radius, *see under* neutron stars
Algols, 437
AM Herculis stars, *see* cataclysmic variables, polars
angular momentum, 18, 23, 24, 30, 43, 48, 419, 420, 431, 435, 438, 443, 448
of accretion disk, 2, 236, 364, 485
capture of, 3, 18, 237, 238, 442
loss of, 16, 238, 239, 440, 441, **479–81**, 482, 485, 486
transport, 437
annihilation lines, 527; *see also* black hole candidates, 511 keV line
ANS, 345
anti-dwarf novae, *see* cataclysmic variables, VY Sculptoris stars
Apollo–Soyuz, 350
Arecibo Observatory, 248
Ariel V, 13, 137, 152, 153, 156, 157, 345, 537
ART-P, 140
ASCA, 10, 200, 391, 410, 411
ASTRO-D, *see* ASCA
atoll sources, 37, 274–7
comparison with Z sources, 288–9
optical vs. X-ray flux, 278
spectral states, 277–8
type I X-ray bursts, 279
X-ray variability, 278–9
AXAF, 391, 411, 412

banana state, *see* atoll sources, spectral states
band-limited noise, *see* black hole candidates; X-ray variability
BATSE, 528–31; *see also* CGRO
BBXRT, 40, 49
BCS theory, 500
Be stars, *see under* companion star
beat frequency models, 350; *see also* Z sources, X-ray variability, horizontal branch quasi-periodic oscillations
β Lyrae systems, 437
BHCs, *see* black hole candidates
black hole candidates, 38, **Ch. 3**, **259–74**, 402, 436, 458
511 keV line, 104, 135, 156
absence of material surface, 163, 166
companion star, *see under* companion star
comparison with AGN, 164

comparison with neutron star, 160–7
descriptions, 134–59
diagnostics, 164
dipping sources, 148
dynamical properties, 102–5
energy spectra, 260–1
in high-mass X-ray binaries, 132
in low-mass X-ray binaries, 131, 460
mass function, 102–4
masses, 107, 128
millisecond bursts, 271
models, 273–4
specific systems, 109–11
optical properties, **104–5**, 148, 150
phenomenology, 127, 128, 273
radial velocity amplitude, 102–5
synchrotron bubble, 147, 152
transient, 102, 128
ultra-soft spectra, 165
variability
energy dependent, 268–70
flickering, 142, 144, 148, 155, 158, 165
'high state' noise, 263–5
'low state' noise, 261–3
ν_{flat}, 261–5, 268–70, 273, 274, 297, 299
timescales, 270–1
power density spectra, 140, 144, 148, 158, 165, 166
quasi-periodic oscillations, 135, 137, 140, 148, 149, 166, 265–70
shot noise, 135, 271
time delays, 135, 166, 268–70
'very high state' noise, 265
X-ray, 261–73
X-ray emission, 135, 157, 159–65
black holes, 1, 2, 23, 419, 423, 425, 440, 465
binary systems, 334, Ch. 11, 475
orbital parameters, 459
orbital periods, 462
birth rate, 490
candidates, *see* black hole candidates
companion star, *see under* companion star
optical and X-ray novae, 461–2
origins, 167, 464
blackbody emission, *see under* accretion disk, models; low-mass X-ray binaries; type I X-ray bursts, spectra
blackbody radius, *see under* Rapid Burster, type I X-ray bursts; type I X-ray bursts; type II X-ray bursts
blue stragglers, 489
bremsstrahlung emission, 35, 309
bulge sources, *see* low-mass X-ray binaries
burst sources, *see* gamma-ray bursts; X-ray burst sources

cataclysmic variables, 4, 10, 12, 27, 74, **Ch. 8**, 396, 419–22, 426, 427, 433, 436, 437, 440, 448, 478, Ch. 15; *see also* accretion disk
accretion rates, 345, 368, 481
accretion states, 332, 339, 346, 356, 378, 444
AM Herculis stars, *see* cataclysmic variables, polars

[cataclysmic variables]
  applications, 334–5
  characteristics, 336–73
  classical novae, 331, 332, **333**, 336, 337, 343, 360, 380, 381, 580
    neon novae, 380
    outbursts, 361, 381
    recurrence times, 333
    subclasses, 580
  companion star, *see under* companion star
  dips, *see under* cataclysmic variables, light curves
  DQ Herculis stars, *see* cataclysmic variables, intermediate polars
  dwarf novae, 331, **332–3**, 340, 344, 356, 358, 380, 381, 421, 423, 427–30, 434, 440, 580
    critical temperature, 363, 368
    declines, 338
    delayed emission, 364
    and disk behavior, 334, 361
    IR emission, 360
    light curves, 360, 361, **362**, 365, 367, 368
    models, 361, 363–4, 367, 368, 383
    oscillations, 348
    outbursts, 337, 342, 345, 347–9, 354–6, **361–9**, 378, 381, 382
    quiescent, 347, 353–6, 366, 370, 378, 379, 382
    recurrence times, 364, 379–81
    spectra, 353
    standstills, 362
    SU UMa subclass, 9, 119, 120, 132, 334, 360, 368, 430, 433, 436, 437, 580, 581
    subclasses, 361–2, 580
    superhumps, 363, **369, 370**, 383
    superoutbursts, 353, 356, 358, **362–3**, 366, 368, 370, 434–6
    U Gem subclass, 581
    UV emission, 366
    X-ray emission, 346, 347, 352, 355, 356
    X-ray pulsations, 345, 349
    Z Cam subclass, 581
  eclipses, 344, 353, 357, 358, 364–7, 369–72, 377, 382, 421
  ejecta, 335
  flickering, 349, **369**, 373, 382
  gamma-ray emission, 349
  intermediate polars, 331–3, 336, 343, 345, 349, 356, **368**, 374–6, **377–8**, 381, 448, 451, 580
    accretion flow, 377, 378
    outbursts, 332, 368
    X-ray pulsations, 346
  IR emission, 359, 375, 382
  lifetimes, 380
  light curves, 353, 354, **371, 372**, 377
    dips, 354, 365, 368, 369, **370–3**
    humps, *see under* accretion disk
  magnetic, 331, 332, 335, 336, 343, 345, 373, 375, 378, 380, 381; *see also* cataclysmic variables, intermediate polars; cataclysmic variables, polars
  magnetic fields, 331, 333, 334, 343, 360, **373–8**, 381, 383
  mass loss, 358, 359, 438

  novalike objects, 331, **333–4**, 356–8, 360, 378, 382, 580
    oscillations, 334, 348
    UX UMa subclass, 581
  optical emission, 355
    flaring, 373, 383
    flickering, 369, 373, 382
    polarization, 375, 381
  orbital parameters, 331, 336, 359
  orbital periods, 242, 331, **376, 378**, 379, 458, 462, 481, 578
  origins, 379, 383
  oscillations, 345, 349, **370**, 382
    long-term modulations, 379
    quasi-periodic, 349, 369, 373, 378
    rapid, 347–9, 360
  polars, 331, 333, 336, 370, 373–5, 377, 381, 420, 436, 450–2, 580
  population, *see under* populations
  radio emission, 360
  recurrent novae, **333**, 580
  S curve, 333, 339, **363**, 377
  spectra, 343, 346, 382
  standard model, 331–4
  stellar winds, 335, 347, **356–9**, 367, 375, 378, 382
  subclasses, 331, 381, 580
  SW Sextantis stars, 334, 343, 344, 382, 581
  UV emission, 336, 346, **350–2**, 355, 356, 358, 361, 375, 382
    spectra, **351**, 356–8
    time delays, 356, 361, 364, 378
  VY Sculptoris stars, 333, 429, 581
  X-ray emission, 336, 345–6, 348, 349, 352–6, 361, 373–5, 382
    flaring, 374
    flickering, 349, **369**
    instabilities, 348
    models, 348, 352, 353, 374, 375
    spectra, 347, 352, 353, **354**, 373–5, 382
    time delays, 356, 361, 364
CGRO, 99, 154, 523, 533, 537
Chandrasekhar limit, 380–1
class I sources, *see* high-mass X-ray binaries
class II sources, *see* low-mass X-ray binaries
color–color diagrams, 315, 316, 328, 408
Columbia (space shuttle), 346
companion star
  B type, 2, 4, 22, 42, 43, 59, 335, 358, 404, 442, 447, 488, 489, 499, 500
  Be type, 16, 18, 22–5, 27, 59, 318, 334, 442–4, 447, 458, 473–5, 488, 489
    optical outbursts, 458
    rotational instabilities, 458
    stellar winds, 473–4
  of black hole, 461
  of cataclysmic variable, 331, 333, **359**, 363, 364, 367, 369, 476
  corotation, 331, 480
  disruption, 487
  F type, 11
  He core, 474, 481
  in high-mass X-ray binaries, 59, 63–4, 70, 458, 460

[companion star]
irradiation of, 64, 133, 233, 234, 243–6, 438, 464, 485
K type, 11, 13
loss, 233, 243
in low-mass X-ray binaries, 238, 239, 243, 476, 477, 478
mass and radius, 469–71, 474, 475, 479, 481, 482
mass transfer, 63, **469**, 470
neutron star, 470
O type, 2, 4, 22, 42, 43, 335, 358, 404, 442, 447, 478, 488, 489, 499, 500
Oe type, 59
Of type, 68
red dwarf, 464
red giant, 477
Roche lobe, *see* Roche lobe
white dwarf, 463, 464, 470, 474, 511, 512
Wolf–Rayet star, 48, 49, 475
Comptonization, *see under* Z sources
corotation, *see under* companion star
critical surface, *see* Roche lobe
CVs, *see* cataclysmic variables
cyclotron lines
absorption, 22, 33, 498, 508
cyclotron scattering resonance features (CSRF), 22, 33, **34**
emission, 31–3, 452, 513

dipping sources, *see under* black hole candidates; low-mass X-ray binaries
DQ Herculis stars, *see* cataclysmic variables, intermediate polars

eclipses, *see under* accretion disk; cataclysmic variables; high-mass X-ray binaries; low-mass X-ray binaries; X-ray binaries
Eddington limit, 60, 78, 234, 235, 237, 399, 425
Einstein X-ray Observatory, 11, 12, 14, 41, 155, 199, 345, 346, 352, 354, 373, 374, 390–6, 401, 403, 405, 408, 410–4, 537
ellipsoidal variations, *see under* high-mass X-ray binaries; low-mass X-ray binaries, light curves
equations of state, *see under* neutron stars
EUVE, 346, 350, 352, 375
evolution
angular momentum, *see* angular momentum
Be X-ray binaries, 475, **489–90**
binary ionization, 242, 247
binary pulsars, 461
binary–single star encounters, **117–20**, 154, 168, 488, 491
cataclysmic variables, 243, 375, **378–1**, 383, 479
circularization, 487
close binaries, **239**, 242
common envelope, *see* evolution, spiral-in
conservative, 469–70, 473
donor expansion, 482
gravitational radiation, 238, 239, 379, 479–81
hibernation, 379
high-mass X-ray binaries, 245, 461, 473–6, 489
intermediate binaries, 239

irradiated companion, 244, 246, 248, 438, 484, **485–6**
kick velocities, 62–3, **475**, 476, 477, 490
low-M phase, 242
low-mass X-ray binaries, **238–40**, 242, **243–5**, 461, 476–7, **479–85**
magnetic braking, 239, 242, 243, 379, 479–81
mass losses, 438, **486**
millisecond pulsars, **243**, 244
non-conservative, 469–73
period gap, 242, 243, 375, 379, 380, **481**
period minimum, 481
physical collision, 247, 248, 487
Roche lobe overflow, *see under* Roche lobe
spiral-in, 118, 470, **471**, 472–4, 476–8, 481, 489, 490
stellar wind, 473, 480
supernova explosion, 471–7, 489, 490
tidal capture, 6, **246**, 247, 335, **487–8**
tidal disruption, 242
triple-star encounters, **117–20**, 154, 168, 488, 491
white dwarfs, 335
wide binaries, **238–9**, 240, 247, 470, 476, 483
Wolf–Rayet stars, 463
X-ray binaries, 460
see also stellar evolution
excretion disk, 245
EXOSAT, 7, 8, 10–13, 16, 26, 40, 41, 45, 48, 154, 177, 181, 184, 185, 188, 190, 198, 216, 217, 219, 220, 223, 255, 274, 277, 291, 300, 346, 347, 353, 354, 356, 361, 366, 370, 372, 374, 377, 537
Explosive Transient Camera, 529, 533
extreme ultra-soft (EUS) sources, 14, 76, 116–17, 408

fast Fourier transform (FFT), *see* X-ray variability
femtolensing, 527
FES, 365
flaring branch, *see* Z sources, spectral states
flip-flops, 148, 265, 443
Fourier transform, *see* X-ray variability

galaxies
active galactic nuclei, 81, 308, 334, 415
'broad line' region, 81
elliptical, 390, 404–9, 412, 415
interstellar medium, 405–7, 409, 415
Local Group, 390, 391, 414
X-ray sources, 391–7
luminosity distributions, 396, **398–404**
metallicity, 399, 401
normal, 410–11
spiral, 405, 406, 408–9, 410, 412, 415
star formation, 404
starburst, 390, 405, 411
surface brightness distribution, 404
X-ray emission, 62, 390, **404–5**
spectra, 405–9
Galaxy (Milky Way), 391, 394, 399, 410, 415
gamma-ray bursts, **Ch. 13**
binary systems, 533
C_{max}/C_{min} distribution, 530–1
counterparts, 523, **528–9**, 533

[gamma-ray bursts]
cyclotron lines, *see* cyclotron lines
dipole moment, 529, 530
distribution, 523, 530
emission regions, 526
energy spectra, **526–8**, 532
absorption features, 527
Landau levels, 527
gravitational lensing, 525
femtolensing, 527
gravitational radiation, 533
magnetic fields, 508
March 5, 1979 burst, 524, 528, 531
models
galactic disk, 531
halo-only, 532
neutron star, 519
two-population, 531
origin
cosmic strings, 525
heliospheric, 532
neutron star, 525, 527, 532
repetition, 531, 532
spectra, *see* gamma-ray bursts, energy spectra
statistical properties, 529–32
quadrupole moment, 529, 530
time histories, **524–7**, 532
periodicities, 525, 526
V/V_{max}, 530
X-ray paucity constraint, 526
X-ray precursors/tails, 527
general relativistic effects, 236, 463
Giant Metrewave Radio Telescope, 248
Ginga, 10, 12, 16, 18, 21, 26, 32, 33, 47, 48, 83, 139,
140, 142, 143, 145, 146, 150–3, 158, 160, 165,
186, 219, 221, 255, 256, 263, 265, 273, 300, 315,
346, 353, 355, 369, 374, 375, 377, 395, 411, 537
globular clusters
binaries, 486–7
formation, 246–7, 335, 488
primordial, 246, 247
disruption of, 248
escape velocity, 248
lower cutoff mass, 248
low-luminosity X-ray sources, 335, 491
low-mass X-ray binaries, 4–6, 234, 245, 247, 484
primordial neutron stars, 248
radio pulsars, 234, **245–8**
ejection of, 245
millisecond, 335
population, 247
recycled, 487
X-ray sources, 12, 13, 395, 396, 464, 477, **486–8**,
491
GRANAT, 135, 139, 157, 265, 537
gravitational lensing, *see under* gamma-ray bursts;
X-ray pulsars, emission mechanisms
gravitational radiation, *see under* evolution
gravity darkening theorem (Von Zeipel), 63–4
gravity waves, *see* evolution, gravitational radiation
Greenbank Telescope, 248
GRO, *see* CGRO
gyro-resonance radiation, 309

Hakucho, 18, 149, 197, 199, 216, 219, 537
hard X-ray transients, *see* X-ray transients
HEAO-1, 6, 11, 13, 33, 45, 155–8, 199, 216, 345,
353, 354, 373, 537
HEAO-2, see *Einstein X-ray Observatory*
helium flash, *see* thermonuclear flash
He-peculiar stars, 335
Hertzsprung–Russell diagram, 466
HEXE, 147
high-mass X-ray binaries, 2–4, 15, **16–18**, 42, 58,
59–73, 234, 328, 334, 392, 404, 458, **473–5**
absorption variability, 4
accretion, *see* accretion
accretion disk, *see* accretion disk
ages, 511
angular momentum, *see* angular momentum
Be-type, 458
birthrate, 488–90
color index, 60, 61
companions, *see under* companion star
cyclotron lines, *see* cyclotron lines
eclipses, 16, 458
ellipsoidal variations, 63–5, 458
evolution, *see under* evolution
lifetimes, 474, 489
luminosity
absolute visual magnitude, 60
distribution of X-ray sources, 60, 62
L_X, 60
optical magnitude, 61
magnetic fields, 510
mass loss rate, 70, 71
mass transfer, *see* mass transfer
neutron stars, *see* neutron stars
optical emission, 59, 460
Hα, 69, 70
light curves, 63, 65–7
long term periodicities, 67
spectra, 68, 69, 72
orbital parameters, 59, 459, 475
orbital periods, **16–17**, 59, 458, **462**, 474
variations, 17
origins, 62–3, 457, 464–73
populations, *see under* populations
Roche lobe, *see* Roche lobe
space velocities, 62–3
spin periods, *see under* neutron stars
spin-up–spin-down, *see* spin-down; spin-up
standard type, 458
transients, *see* X-ray transients
UV emission, 336
observations, 70
P Cygni profiles, 70, 72, 73
Raman scattering, 72
winds, 42–8, 69–71
focusing, 2, 70
terminal velocity, 72
X-ray heating, 63, 64, 69, 70
X-ray absorption, 474
X-ray emission
flaring, 4, 42
outbursts, 4
variability, 458

Hopkins Ultraviolet Telescope, 346, 350–2, 382
horizontal branch, *see* Z sources, spectral states
Hubble Space Telescope, 104, 426
hydrogen flash, *see* thermonuclear flash

IHTFP, 666
instabilities, *see under* accretion; mass transfer
intermediate polars, *see under* cataclysmic variables
IRAS, 360
island state, *see* atoll sources, spectral states
IUE, 72, 83, 97, 104, 107, 141, 315, 336, 345–7, 350, 356, 361, 365, 366

Kvant, 150, 537

Lagrange point, 2, 63, 422, **467**
LAMAR, 412, 413
low-mass X-ray binaries, 2, 3, **4–18**, 22, 27, 39, 45, 48, 58, **73–107**, 234, 237, 328, 334, 346, 361, 390, 395, 396, 404, 406, 425, 430, 433, 440, 458–62
accretion, *see* accretion
accretion disk, *see* accretion disk
accretion rates, 479, 483
ADC sources, 76, 438
ages, 79, 462, 511
angular momentum, *see* angular momentum
atoll sources, *see* atoll sources
birthrate, 234, 241, 490–1
black holes, 460
blackbody, 35, 36
Bowen fluorescence, 74, 83–6
characteristics, 178
CNO abundances, 86
color–color diagram, 36
companions, *see under* companion star
correlated spectral and timing properties, 298–300
dipping sources, 7–10, 13, 14, 38–40, 47, 86, 119, 423, 425, 426
eclipses, 423
X-ray, 6
Eddington limit, *see* Eddington limit
evolution, *see under* evolution
flaring, 35, 36
gamma-ray emission, 486
globular clusters, *see under* globular clusters
high-luminosity systems, *see* Z sources
kinematic properties, 79
lifetimes, 234, 241, 243, 244, 247, 490, 491, 513
light curves, 87–91, 335, 485
dips, 38, 370–3
ellipsoidal variability, 91
low luminosity, absence of, 242, 243
low-luminosity systems, *see* atoll sources
L_X/L_{Edd} (γ), 78, 247
L_X/L_{opt} (ξ), 8, 75, 76
$\dot{M}$, 106
magnetic fields, 240, 510–11
mass function, 108, 109
mass–orbital-period relations, 478
mass transfer, *see under* mass transfer
metallicities, 86

neutron stars, *see* neutron stars
optical emission, 73, 74
absolute visual magnitude, 77, 78
bursts, 74, 92–5
colors, 74, 75
light curves, *see* low-mass X-ray binaries, light curves
luminosities, 74
pulsations, 95, 96
reprocessing, 95
spectra, 74, 79, 80
temperatures, 79
X-ray correlations, 92, 96, 97
orbital parameters, 108, 336, 459, 483
orbital periods, **4–16**, 242, 458, **462**, 479, **480**, 484
minimum, 481
table of, 5
variations, 15–16
orbital timescale $(P_o/\dot{P}_o)$, 15, 17
origins, 246, 335, 381, 457, **476–9**, 490
population, *see under* populations
primary spin rate, 240
pulsations, absence of, 178, 240, 244, 510
quasi-periodic oscillations, **240**, 335, 349
radial velocity determinations, 111–13
Roche lobe, *see* Roche lobe
sky distribution, 179
spin periods, *see under* neutron stars
spin-up, *see* spin-up
transients, *see* X-ray transients
UV emission, 336
lines, 106, 107
observations, 106
spectra, 83
viewing angle, 8
winds, 438
X-ray bursts, *see* X-ray burst sources
X-ray emission, 486
heated gas, 73, 80, 82, 83
reprocessing, 74, 106
spectra, **35–8**, 40, 41
Z sources, *see* Z sources

Magellanic Clouds, *see under* X-ray sources
magnetic braking, *see under* evolution
magnetic field decay, *see under* neutron stars, magnetic field evolution; spin-down
magnetosphere, *see under* neutron stars
March 5, 1979 burst, *see under* gamma-ray bursts
mass determinations, *see under* neutron stars
mass ratio (q), *see* Roche lobe
mass transfer, 469–71
cataclysmic variables, 332
duration, 234, 237, 239
events 337, 367, 436
high-mass X-ray binaries, 59
instabilities 132–4, 334, 363, 364, 369
low-mass X-ray binaries, 239, 243
modes, 240, 333, 475
orbital evolution, 239, 242, 469, 470, 472, 476, 478, 479, 482–4
rates, 59, 239, 242, 244, 379, 436, 479

massive X-ray binaries, *see* high-mass X-ray binaries
Meissner effect, *see* neutron stars, magnetic field evolution, flux explosion
metallicity effects, *see under* accretion
Milky Way, *see* Galaxy (Milky Way)
millisecond pulsars, *see* radio pulsars, millisecond

NASA, 416
neutron stars, 1, 4, 23, 27, 33, 36, 43, 44, 318, 327, 328, 419, 423, 425, 437, 443, 452, 465, 525
  absorption lines, *see under* type I X-ray bursts
  Alfvén radius, 18–21, 237, 244, 444
  angular momentum, *see* angular momentum
  angular momentum equation, 446
  birth rate, 490
  comparison with black hole candidates, 160–7
  core–crust coupling, 515
  crust, 23, 502, 518–19
  cyclotron lines, *see* cyclotron lines
  equations of state, 238
  hot spot, 30
  interior, 500, 510, 513, 515–18
  magnetic field evolution, 241, 246, **Ch. 12**
    accretion, 514, 517
    in binaries, 510–12
    consequences, 518–19
    convective instability, 514
    flux expulsion, 500, 510, 516–18
    Hall drift, 518
    MHD instability, 513–14, 517
    ohmic decay, 496, 503–6, 514
    plate tectonics, 519
    spin-down, *see* spin-down
    timescale, 504–7, **508**, 509, 510, 513, 516
  magnetic fields, 1, 25, 237, 444, 476, 491, Ch. 12
    core, 503
    crustal, 502–3, 514, 517, 518
    estimation, 496–8
    evolution, *see* neutron stars, magnetic field evolution
    fluxoid–vortex interactions, 515–18, **520–1**
    fluxoids, 510, 515, 516, 518, 520
    fossil field, **499–500**, 502, 515, 517
    origins, **499–500**, 502–3, 517
    pressure, 18
    residual, **512–13**, 517
    structure, 500–1
    thermomagnetic instability, 502–3
    transport, 514, 515, 518
  magnetosphere, 19, 20, 236
  mass determinations, 107–8
  mass–radius relation, *see under* X-ray burst sources
  maximum mass, 113, 462
  nuclear processes, *see under* X-ray burst sources
  origins, 248, 464
  precession, 26, 30
  slow rotators, 23
  spin period
    equilibrium, 20, 237, 242
    initial, 23
    minimum, 233, 235, **238**

spin-down, *see* spin-down
spin-up, *see* spin-up
tidal capture, *see under* evolution
X-ray pulsations, *see* X-ray pulsars
normal branch, *see* Z sources, spectral states
novae, *see under* cataclysmic variables

OB stars, *see under* companion star
Onsager–Feynman vortices, 515
optical bursts, *see* gamma-ray bursts, counterparts; X-ray burst sources, optical counterparts
optical novae, *see under* X-ray transients
Orbiting Astronomical Observatory, 537
OSO-7, 260, 537
OSO-8, 8, 40, 45, 176, 353

pivot point, 285, 287
plasma, *see* accretion; accretion disk
polars, *see under* cataclysmic variables
populations, 462
  black-hole binaries, 167–8
  cataclysmic variables, 331, 396
  high-mass X-ray binaries, 61–2, 488–9
  low-mass X-ray binaries, 178, 462, 490
  Population I, 391
  Population II, 39
  radio pulsars, 489
    binary, 491, 492
    millisecond, 241, 491, 513
    steady-state model, 506
    synthesized, 508–10
power density spectra, *see* X-ray variability, Fourier spectra
Pulsar X-1, 147
pulsars, *see* radio pulsars; X-ray pulsars

quasars, 308, 413, 419
quasi-periodic oscillations, *see under* black hole candidates, variability; cataclysmic variables, oscillations; low-mass X-ray binaries; Rapid Burster; type II X-ray bursts; X-ray variability; Z sources, X-ray variability

radial velocity curves, *see under* black hole candidates; low-mass X-ray binaries
radio emission, *see under* cataclysmic variables; type I X-ray bursts; type II X-ray bursts; X-ray binaries
radio flares, *see under* X-ray binaries, radio emission
radio pulsars, Ch. 5, 327, 328, 444
  ages
    kinetic, 496, **504**
    spin-down, 235, 496, 497, 504, 507, 509, 512, 516
  antediluvian, 460
  beaming probability, 506
  binary, 243, 249, 460, 463–4, 488, 510–12
    birth rate, 489
    destruction, 484
    eclipsing, 236, 245, 246
    first-born of, 512, 513
    fraction, 234
    magnetic fields, 511

[radio pulsars, binary]
    orbit diagrams, 461
    orbital periods, **462**, 483
    origins, 476, **483–5**, 487–9
    PSR1913+16 type, 463
    PSR1953+29 type, 463–4, 483
  birth rate, 491, 506
  death, 235, 497, 508, 511
  dispersion measures, 509
  emission properties, 236
    polarization, 495, 497
  glitches, 519
  in globular clusters, *see under* globular clusters
  'Hubble line', 235, 497
  lifetimes, 491, 495
  light cylinder, 244
  low-field, 513
  luminosities, 509
  magnetic field–period diagram, 234, 497, 504, 505
  magnetic fields, 233, 495–7, **503–16**
    geometry, 236, 497
  millisecond, Ch. 5, 328, 383
    ages, 235
    binary, 234
    birth rate, 234, **241**
    birth rate problem, 243, 245, **247**, 512–13
    eclipsing, 464
    globular clusters, *see under* globular clusters, radio pulsars
    lifetimes, 241, 513
    magnetic fields, 234, 237, 240, 244
    origins, 233–5, **236–8**, 240–5, 247, 335, 483, 491
    population, *see under* populations, radio pulsars
    selection effects, 241
    spin-down rate, 234
  models, 236, 496
  planets, 245, 460, 484
  population, *see under* populations
  proper motions, 504
  pulsar current, 506, 507, 512
  pulse profiles, 31, 236
  recycled, 233, 246, 248, 487, 488, 505, 511, 512
    birth rate, 247
    magnetic fields, 496
  spin-up line, *see under* spin-up
  sub-millisecond, 238, 249
  supernova remnant associations, 233, 503
  velocities, 248, 473, 504, **512**
Rapid Burster, 211–27
  accretion, 211
  models, *see under* type II X-ray bursts
  'naked-eye' oscillations, *see* Rapid Burster, quasi-periodic oscillations
  'normal' neutron star behavior, 211, 219
  observations (complete), 211
  persistent emission, 216, 219
  quasi-periodic oscillations, 216, **219–22**
  in quiescence, 217
  type I X-ray bursts, 211
    blackbody radius, 217
    luminosity, 218

  quasi-periodic oscillations, absence of, 219
  spectra, 217
  type II X-ray bursts, *see* type II X-ray bursts
recycled pulsars, *see under* radio pulsars
red noise, *see* X-ray variability, noise
Roche lobe, 2, 12, 42, 44, 48, 63, 74, 328, 431, **467–8**
Roche lobe overflow, 237, 419, 423, 438, 441, 443, 445, 448, 467, 468, **479**, 482
  from He star, 478
  from irradiated companion, 244, 246
  from main-sequence star, 239, 478
  quasi-Roche lobe overflow, 2
  from red giant, 238, 470, 472, 477, 478
  from supergiant, 59, 474
ROSAT, 11, 12, 139, 346, 347, 350, 352, 353, 355, 369, 373–6, 378, 390, 391, 393–6, 401, 408–10, 451, 528, 537

SAS-3, 18, 176, 187, 196, 213, 345, 537
SIGMA, 141, 153, 156
SMM, 528
soft gamma-ray repeaters, 531, 532
soft X-ray transients, *see under* X-ray transients
solar nebula, 334, 336
Space Lab, 537
spin-down, 20–3, 33, 495, 512
  of companion star, 239
  and irradiated companion, 244
  and magnetic field decay, 515–17
  and magnetic field strength, 233, 496, 517, 519
  timescale, 21, 22, 516
  torque, **496**, 497, 516
spin-up, 18–20, 23, 24, 233, 234, **236–8**, 240, 245, 247, 447, 483, 495, 511, 517
  after collision, 248
  and magnetic field strength, 233, 495–6, 498, 511
  spin-up line, 235, 495, 497, 511, 512
  timescale, 20, 446, 447, 495
spiral-in, *see under* evolution
stellar evolution, 464–7
  convection, 465, 466
  end-products, 464–5
  initial mass function, 248, 488
  magnetic field, 499–500
  mass losses, 465, 469
  mass–luminosity relations, 467, 483
  mass–radius relations, 467, 483
  size variations, 465, **466**
stellar wind, *see* angular momentum; X-ray pulsars, wind driven; *see under* accretion; evolution
SU UMa systems, *see under* cataclysmic variables, dwarf novae
superconductivity, 500–1
  Abrikosov fluxoids, 501
  BCS theory, 500
  fluxoid pinning, 520
  Type I, 500
  Type II, 500–1, 510
superfluidity, 515
  Onsager–Feynman vortices, 515
  superfluid drag, 520
  vortex pinning, 520

superhumps, *see under* accretion disk; cataclysmic
    variables, dwarf novae
supernova remnants, 15, 21, 22, 33, 391, 404, 413,
    415
    pulsar associations, 233, 503
supernovae, 335, 380, 443, 462
    in binary systems, 457, **471–3**, 476–8
    Type Ia, 381
    Type Ib, 489
superoutbursts, *see under* cataclysmic variables,
    dwarf novae
symbiotic stars, 333, 335
synchrotron bubble events, 312–15, 319, 320, 324,
    325, 327
synchrotron emission, 308, 309, 311, 318, 319, 328

T Tauri stars, 335
Tenma, 18, 40, 45, 48, 154, 278, 335, 537
thermonuclear flash, 380–1; *see also* cataclysmic
    variables, classical novae; *see also under* type I
    X-ray bursts
Thorne–Zytkow objects, 478
tidal lobe overflow, 443, 444
timescales
    companion star vaporization, 243, 485, 486
    dynamical, 469
    evolutionary, 465, **466**, 470, 477, 481
    gravitational radiation, 242
    magnetic field evolution, *see under* neutron stars
    spin-down, *see under* spin-down
    spin-up, *see under* spin-up
    thermal, 242, **466**, 469, 481
transients, *see* X-ray transients
TTM, 147
two-component models, *see under* accretion disk,
    models
type I sources, *see* high-mass X-ray binaries
type I X-ray bursts, 4, 15, 163, 166, 176, **178–99**
    absorption lines, 177, 185–6, 200
    accretion rate and properties, 192–6
    α-values, 191, 192, 203, 204, 208
    blackbody radius, 193
    comparison with type II X-ray bursts, 211
    double-peaked profiles, 188, 190
    Eddington limit, 179, 182, 184, 187, 188, 193, 207
    energetics, 191–2
    fluence, 179, 181, 190
    intervals, 189–91
      and burst energy, 190
    IR emission, 199
    limit cycle behavior, 207
    maximum luminosity, 181
    optical bursts, 190, **196–8**
    persistent emission, 181, 191–2
    precursors, 187–8, 207
    profiles, 179–81
    radio emission, 198–9
    radius expansion, 175, 177, 179, 180, 183, **186–9**,
      190, 193, 195, 199, 200, 207
      peak flux gap, 188, 203
    rise times, 179, 204
    short time intervals, 208
    spectra, 181–6

    deviations from blackbody, 184–5
    standard candle, 77, 78, 188, 193
    theory, 201–10
      vs. observations, 206–10
    thermal state and burst types, **205–6**, 208
    thermonuclear flash model, 190, 193, 196,
      201–10, 211
    transient sources, 189
    variability, 199
type II sources, *see* low-mass X-ray binaries
type II X-ray bursts, 176
    blackbody radius, 218–19
    burst cycle, 214
    comparison with type I X-ray bursts, 211
    duration, 214
    luminosities , 218–9
    mode I vs. mode II, 214
    models, 219, **222–7**
    observed properties, 213–15
    profiles, 213, 217
    quasi-periodic oscillations, 219–21
    radio/IR emission, 216
    ringing, 214
    spectra, 218–19
    time-invariant profiles, 214–15
    very low fluence, 215

UHURU, 255, 537

Vela 5B, 27
viscosity, 37, 334, 337, 361, 383, 420, 421, 428, 429,
    437, 448
    parameter (α), 337, 339, 363
    viscous heating, 2
VLA, 320, 360, 463
VLBI, 320, 463
von Kármán vortex street, 443
Voyager, 346, 350, 353, 355, 356, 361, 366

white dwarfs, 2, 239, 242, 419, 436, 449, 452
    accreting, *see* cataclysmic variables
    CO, 464, 465
    Hamada–Salpeter, 345
    He, 464, 465
    magnetic fields, 357, 499, 500
    magnetic moments, 420, 444
    mass determinations, 344
    moment of inertia, 441
    ONeMg, 380, 381, 464, 465, 500
    origins, 464
    structure, 335, 348
    temperatures, 380
    UV emission, 366
Wolf–Rayet stars, *see under* companion star;
    evolution

XMM, 391, 411–13
X-ray background, 411
X-ray binaries, **Ch. 1, Ch. 2, 458–63, Ch. 14**
    accretion, *see* accretion
    color–color diagrams, 315
    cyclotron lines, *see* cyclotron lines
    dust scattering, 48

[X-ray binaries]
  eclipses, 3, 10, 11, 12, 13, 15, 47
  Galactic, 391
  galaxies, 406
  gamma-ray emission, 318
  in globular clusters, *see* globular clusters
  IR brightness, 318
  iron lines, 45–8
  lifetimes, 474, 491
  light curves, 462
  L_X/L_{opt} (ξ), 13, 14
  in normal galaxies, **Ch. 9**
  orbit diagrams, 461
  orbital parameters, 443, 459
  orbital periods, 3–18
    determinations, 3
  peculiar systems (1E2259+59, Cir X-1, Cyg X-3,
    Her X-1, SS433), 26, **48–9**, 320, **462–3**, 475–6
    jets, 26, 27, 320, 322, 327, 463
    optical emission, 26
    shells, 463
  pulsing, 19, 445, 446
    pulse periods, table of, 19
  quasi-periodic oscillations, *see under* X-ray
    variability
  radio emission, **Ch. 7**, 462, 463
    flaring, 48, 309, 317–19, 323, 325
    flux density, 311, 313
    jets, 308, 320–5
    models, 309, 323–7
    nebulosities, 308
    periodic, 317–20
    polarization, 309
    transient, 313–15
  soft excess, 47–8
  spin-up, 446
  third periods, 25–7
  transient sources, *see* X-ray transients
X-ray burst sources, 4, 11, 12, **Ch. 4**, 524
  CNO abundance, 177, 191, 202–6
  Eddington limit, *see* Eddington limit
  envelope structure, 202–3, 209–10
  history, 176–7
  $\dot{M}$ and nuclear burning, 203–6
  magnetic fields, 201, 240
  mass–radius relation, 199–200
  nuclear processes, 201–2
  number, 177
  optical counterparts, 25, 177, 178
  origins, 178
  Rapid Burster, *see* Rapid Burster
  recent progress, 175–6
  Roche lobe overflow, *see* Roche lobe overflow
  spectra, *see under* type I X-ray bursts; type II
    X-ray bursts
  standard candle, *see under* type I X-ray bursts
  steady nuclear burning, 190
X-ray bursts, 2, 460
  type I, *see* type I X-ray bursts
  type II, *see* type II X-ray bursts
X-ray novae, *see* X-ray transients, soft
X-ray pulsars, 1, **18–25**, 45, 460, 463, 495, 510
  accretion, *see* accretion

cyclotron lines, *see* cyclotron lines
emission mechanisms, 30–1
  beaming, 30, 31
  gravitational focusing, 31
fast rotators, 20
high-frequency noise, 293
luminosities, 27
magnetic fields, 34, 35, 498
periods, **18**, 19, 516
  evolution, 23, 498
  table of, 19
  variations, 18, 22, 23
pulse frequency vs. ν_{break}, 293–5
pulse profiles, **27–9**, 30
quasi-periodic oscillations, *see under* X-ray
  variability
slow rotators, 20
spectra, **32–5**, 460
transient, 45
wind-driven, 23, 514, 516
X-ray sources
  bulge, 395, 396, 404, 405, 415
  bursters, *see* X-ray burst sources
  class I, *see* high-mass X-ray binaries
  class II, *see* low-mass X-ray binaries
  disk, 395, 401, 404
  extreme ultra-soft, *see* extreme ultra-soft sources
  Galactic, 394
  Galactic Ridge, 335
  globular clusters, *see under* globular clusters
  high-mass X-ray binaries, *see* high-mass X-ray
    binaries
  interstellar medium, 405
  Local Group galaxies, 391–7
  low-mass X-ray binaries, *see* low-mass X-ray
    binaries
  Magellanic Clouds, 12, **113–17**, 391–2, 399, 401,
    410
  soft X-ray transients, *see* X-ray transients, soft
  spiral arm, 395, 401, 404, 415
  super-Eddington, 399
  variability, *see* X-ray variability
  transients, *see* X-ray transients
  type I, *see* high-mass X-ray binaries
  type II, *see* low-mass X-ray binaries
X-ray spectra, *see under* atoll sources; cataclysmic
  variables; galaxies; low-mass X-ray binaries;
  Rapid Burster; type I X-ray bursts; type II
  X-ray bursts; X-ray pulsars; Z sources
X-ray telescopes, ideal, 415, 416
X-ray transients, 244, 308, 312, 313, 315, 328, 436,
  437, 443, 460, 474, 488
  mechanisms, 132–4
  optical novae, 25, 73, 131
  quiescence, 244
  recurrence, 128–31
  soft, 73, 97–9, 334, 430
    dynamical properties, 102–3
    neutron star, 105–6
    optical spectra, 98
X-ray variability, **Ch. 6**, 415–16
  autocorrelation function, 256, 258
  Fourier spectra, 256, **257–9**

[X-ray variability]
  high-frequency noise, 278–81, 284, 290, 293–5
  low-frequency noise, *see under* Z sources, X-ray
      variability
  millisecond bursts, 271, 290
  noise (definitions), 257
  normal branch quasi-periodic oscillations, *see
      under* Z sources, X-ray variability
  overview, 295–300
  power density spectra, 256, **257–9**
  quasi-periodic oscillations, 2, 15, 37, 257, 290,
      295, 316; *see also under* Z sources, X-ray
      variability
  random processes, 256
  root-mean-square (rms) variation, 258–9
  shot noise, 256
  very-low-frequency noise, 278–81
XTE, 300

Z sources, 35–7, 315–17, 328
  comparison with atoll sources, 288–9
  Comptonization, 328
  optical/UV emission, 276, 280, 285–6
  properties, 274–7
  spectral states, 279–80
    and *M*, 280
    rank number, 280
  X-ray variability, 280–8
    flaring branch quasi-periodic oscillations, 283
    horizontal branch quasi-periodic oscillations,
        281–3
    low-frequency noise, 281–3
    models, 286–8
    normal branch quasi-periodic oscillations, 281,
        283
    time lags, 282
Zanstra method, 347